Structures and Infrastructure Systems

Our knowledge to model, design, analyse, maintain, manage, and predict the life-cycle performance of infrastructure systems is continually growing. However, the complexity of these systems continues to increase and an integrated approach is necessary to understand the effect of technological, environmental, economic, social, and political interactions on the life-cycle performance of engineering infrastructure. In order to accomplish this, methods have to be developed to systematically analyse structure and infrastructure systems, and models have to be formulated for evaluating and comparing the risks and benefits associated with various alternatives. Civil engineers must maximize the life-cycle benefits of these systems to serve the needs of our society by selecting the best balance of the safety, economy, resilience, and sustainability requirements despite imperfect information and knowledge. Within the context of this book, the necessary concepts are introduced and illustrated with applications to civil and marine structures. This book is intended for an audience of researchers and practitioners worldwide with a background in civil and marine engineering, as well as people working in infrastructure maintenance, management, cost, and optimization analysis.

All the chapters in this book are authored or co-authored by Dan M. Frangopol. They were originally published as articles in *Structure and Infrastructure Engineering.*

Professor Dan M. Frangopol is the inaugural holder of the Fazlur R. Khan Endowed Chair of Structural Engineering and Architecture at Lehigh University, USA. Before joining Lehigh University in 2006, he was Professor of Civil Engineering at the University of Colorado at Boulder, where he is now Professor Emeritus. He is recognised as a leader in the field of life-cycle engineering of civil and marine structures. Professor Frangopol is the Founding President of the *International Association for Bridge Maintenance and Safety* (IABMAS) and the *International Association for Life-Cycle Civil Engineering* (IALCCE). He has authored/co-authored two books, 42 book chapters, 370 articles in archival journals (including 9 award-winning papers from ASCE, IABSE, and Elsevier), and more than 600 papers in conference proceedings. He is the Founding Editor of *Structure and Infrastructure Engineering* and of the Book Series *Structures and Infrastructures*. Professor Frangopol is the recipient of several medals, awards, and prizes, from ASCE, IABSE, IASSAR and other professional organizations, such as the OPAL Award, the Newmark Medal, the Alfredo Ang Award, the T.Y. Lin Medal, the F.R. Khan Medal, and the Croes Medal (twice), to name a few. He has served as a consultant or advisor to numerous companies. He holds 4 honorary doctorates and 12 honorary professorships from major universities. He is a foreign member of the Academia Europaea and of the Royal Academy of Belgium, an Honorary Member of the Romanian Academy, and a Distinguished Member of the ASCE.

Structures and Infrastructure Systems

Life-Cycle Performance, Management, and Optimization

Dan M. Frangopol

Author & Co-author

Routledge
Taylor & Francis Group

LONDON AND NEW YORK

First published 2018
by Routledge
2 Park Square, Milton Park, Abingdon, Oxon, OX14 4RN, UK

and by Routledge
52 Vanderbilt Avenue, New York, NY 10017

First issued in paperback 2020

Routledge is an imprint of the Taylor & Francis Group, an informa business

British Library Cataloguing in Publication Data
A catalogue record for this book is available from the British Library

ISBN 13: 978-0-367-57165-8 (pbk)
ISBN 13: 978-0-8153-9605-5 (hbk)

Typeset in Times New Roman
by RefineCatch Limited, Bungay, Suffolk

Publisher's Note
The publisher accepts responsibility for any inconsistencies that may have
arisen during the conversion of this book from journal articles to book chapters,
namely the possible inclusion of journal terminology.

Disclaimer
Every effort has been made to contact copyright holders for their permission to
reprint material in this book. The publishers would be grateful to hear from any
copyright holder who is not here acknowledged and will undertake to rectify
any errors or omissions in future editions of this book.

Contents

Citation Information   vii

Notes on Contributors   xi

Preface   1
Dan M. Frangopol

Part I: State-of-the-art   3

1. Life-cycle performance, management, and optimisation of structural systems under uncertainty: accomplishments and challenges   5
Dan M. Frangopol

2. Bridge network performance, maintenance and optimisation under uncertainty: accomplishments and challenges   30
Dan M. Frangopol and Paolo Bocchini

3. Life-cycle of structural systems: recent achievements and future directions   46
Dan M. Frangopol and Mohamed Soliman

4. Bridge life-cycle performance and cost: analysis, prediction, optimisation and decision-making   66
Dan M. Frangopol, You Dong and Samantha Sabatino

Part II: General methodology   85

5. Optimal bridge maintenance planning using improved multi-objective genetic algorithm   87
Hitoshi Furuta, Takahiro Kameda, Koichiro Nakahara, Yuji Takahashi and Dan M. Frangopol

6. Maintenance and management of civil infrastructure based on condition, safety, optimization, and life-cycle cost   96
Dan M. Frangopol and Min Liu

7. Life cycle utility-informed maintenance planning based on lifetime functions: optimum balancing of cost, failure consequences and performance benefit   109
Samantha Sabatino, Dan M. Frangopol and You Dong

8. Efficient multi-objective optimisation of probabilistic service life management   127
Sunyong Kim and Dan M. Frangopol

Part III: Life-cycle performance under corrosion and fatigue   141

9. Probabilistic limit analysis and lifetime prediction of concrete structures   143
Fabio Biondini and Dan M. Frangopol

10. Integration of the effects of airborne chlorides into reliability-based durability design of reinforced concrete structures in a marine environment   157
Mitsuyoshi Akiyama, Dan M. Frangopol and Motoyuki Suzuki

11. Fatigue system reliability analysis of riveted railway bridge connections   167
Boulent M. Imam, Marios K. Chryssanthopoulos and Dan M. Frangopol

12. Fatigue performance assessment and service life prediction of high-speed ship structures based on probabilistic lifetime sea loads 185
Kihyon Kwon, Dan M. Frangopol and Sunyong Kim

13. Experimental investigation of the spatial variability of the steel weight loss and corrosion cracking of reinforced concrete members: novel X-ray and digital image processing techniques 199
Sopokhem Lim, Mitsuyoshi Akiyama, Dan M. Frangopol and Haitao Jiang

14. Reliability-based durability design and service life assessment of reinforced concrete deck slab of jetty structures 216
Mitsuyoshi Akiyama, Dan M. Frangopol and Koshin Takenaka

Part IV: Life-cycle performance under earthquakes 227

15. Life-cycle cost of civil infrastructure with emphasis on balancing structural performance and seismic risk of road network 229
Hitoshi Furuta, Dan M. Frangopol and Koichiro Nakatsu

16. Long-term seismic performance of RC structures in an aggressive environment: emphasis on bridge piers 239
Mitsuyoshi Akiyama and Dan M. Frangopol

17. Performance analysis of Tohoku-Shinkansen viaducts affected by the 2011 Great East Japan earthquake 254
Mitsuyoshi Akiyama, Dan M. Frangopol and Keita Mizuno

18. Probabilistic assessment of an interdependent healthcare–bridge network system under seismic hazard 274
You Dong and Dan M. Frangopol

Part V: Inspection and monitoring 285

19. Application of the statistics of extremes to the reliability assessment and performance prediction of monitored highway bridges 287
Thomas B. Messervey, Dan M. Frangopol and Sara Casciati

20. Probabilistic bicriterion optimum inspection/monitoring planning: applications to naval ships and bridges under fatigue 300
Sunyong Kim and Dan M. Frangopol

21. Integration of structural health monitoring in a system performance based life-cycle bridge management framework 316
Nader M. Okasha and Dan M. Frangopol

22. Critical issues, condition assessment and monitoring of heavy movable structures: emphasis on movable bridges 334
F. Necati Catbas, Mustafa Gul, H. Burak Gokce, Ricardo Zaurin, Dan M. Frangopol and Kirk A. Grimmelsman

Part VI: Redundancy as life-cycle performance indicator 351

23. Time-variant redundancy of structural systems 353
Nader M. Okasha and Dan M. Frangopol

24. Redundancy and robustness of highway bridge superstructures and substructures 376
Michel Ghosn, Fred Moses and Dan M. Frangopol

25. Effects of post-failure material behaviour on redundancy factors for design of structural components in nondeterministic systems 398
Benjin Zhu and Dan M. Frangopol

26. Time-variant redundancy and failure times of deteriorating concrete structures considering multiple limit states 418
Fabio Biondini and Dan M. Frangopol

Index 431

Citation Information

All the chapters in this book are authored or co-authored by Dan M. Frangopol. They were originally published in *Structure and Infrastructure Engineering*. When citing this material, please use the original page numbering for each article, as follows:

Chapter 1
Life-cycle performance, management, and optimisation of structural systems under uncertainty: accomplishments and challenges
Dan M. Frangopol
Structure and Infrastructure Engineering, volume 7, issue 6 (2011), pp. 289–413

Chapter 2
Bridge network performance, maintenance and optimisation under uncertainty: accomplishments and challenges
Dan M. Frangopol and Paolo Bocchini
Structure and Infrastructure Engineering, volume 8, issue 4 (2012), pp. 341–356

Chapter 3
Life-cycle of structural systems: recent achievements and future directions
Dan M. Frangopol and Mohamed Soliman
Structure and Infrastructure Engineering, volume 12, issue 1 (2016), pp. 1–20

Chapter 4
Bridge life-cycle performance and cost: analysis, prediction, optimisation and decision-making
Dan M. Frangopol, You Dong and Samantha Sabatino
Structure and Infrastructure Engineering, volume 13, issue 10 (2017), pp. 1239–1257

Chapter 5
Optimal bridge maintenance planning using improved multi-objective genetic algorithm
Hitoshi Furuta, Takahiro Kameda, Koichiro Nakahara, Yuji Takahashi and
Dan M. Frangopol
Structure and Infrastructure Engineering, volume 2, issue 1 (2006), pp. 33–41

Chapter 6
Maintenance and management of civil infrastructure based on condition, safety, optimization, and life-cycle cost
Dan M. Frangopol and Min Liu
Structure and Infrastructure Engineering, volume 3, issue 1 (2007), pp. 29–41

Chapter 7
Life cycle utility-informed maintenance planning based on lifetime functions: optimum balancing of cost, failure consequences and performance benefit
Samantha Sabatino, Dan M. Frangopol and You Dong
Structure and Infrastructure Engineering, volume 12, issue 7 (2016), pp. 830–847

Chapter 8
Efficient multi-objective optimisation of probabilistic service life management
Sunyong Kim and Dan M. Frangopol
Structure and Infrastructure Engineering, volume 13, issue 1 (2017), pp. 147–159

Chapter 9

Probabilistic limit analysis and lifetime prediction of concrete structures

Fabio Biondini and Dan M. Frangopol

Structure and Infrastructure Engineering, volume 4, issue 5 (2008), pp. 399–412

Chapter 10

Integration of the effects of airborne chlorides into reliability-based durability design of reinforced concrete structures in a marine environment

Mitsuyoshi Akiyama, Dan M. Frangopol and Motoyuki Suzuki

Structure and Infrastructure Engineering, volume 8, issue 2 (2012), pp. 125–134

Chapter 11

Fatigue system reliability analysis of riveted railway bridge connections

Boulent M. Imam, Marios K. Chryssanthopoulos and Dan M. Frangopol

Structure and Infrastructure Engineering, volume 8, issue 10 (2012), pp. 967–984

Chapter 12

Fatigue performance assessment and service life prediction of high-speed ship structures based on probabilistic lifetime sea loads

Kihyon Kwon, Dan M. Frangopol and Sunyong Kim

Structure and Infrastructure Engineering, volume 9, issue 2 (2013), pp. 105–115

Chapter 13

Experimental investigation of the spatial variability of the steel weight loss and corrosion cracking of reinforced concrete members: novel X-ray and digital image processing techniques

Sopokhem Lim, Mitsuyoshi Akiyama, Dan M. Frangopol and Haitao Jiang

Structure and Infrastructure Engineering, volume 13, issue 1 (2017), pp. 118–134

Chapter 14

Reliability-based durability design and service life assessment of reinforced concrete deck slab of jetty structures

Mitsuyoshi Akiyama, Dan M. Frangopol and Koshin Takenaka

Structure and Infrastructure Engineering, volume 13, issue 4 (2017), pp. 468–477

Chapter 15

Life-cycle cost of civil infrastructure with emphasis on balancing structural performance and seismic risk of road network

Hitoshi Furuta, Dan M. Frangopol and Koichiro Nakatsu

Structure and Infrastructure Engineering, volume 7, issue 1–2 (2011), pp. 65–74

Chapter 16

Long-term seismic performance of RC structures in an aggressive environment: emphasis on bridge piers

Mitsuyoshi Akiyama and Dan M. Frangopol

Structure and Infrastructure Engineering, volume 10, issue 7 (2014), pp. 865–879

Chapter 17

Performance analysis of Tohoku-Shinkansen viaducts affected by the 2011 Great East Japan earthquake

Mitsuyoshi Akiyama, Dan M. Frangopol and Keita Mizuno

Structure and Infrastructure Engineering, volume 10, issue 9 (2014), pp. 1228–1247

Chapter 18

Probabilistic assessment of an interdependent healthcare–bridge network system under seismic hazard

You Dong and Dan M. Frangopol

Structure and Infrastructure Engineering, volume 13, issue 1 (2017), pp. 160–170

Chapter 19

Application of the statistics of extremes to the reliability assessment and performance prediction of monitored highway bridges

Thomas B. Messervey, Dan M. Frangopol and Sara Casciati

Structure and Infrastructure Engineering, volume 7, issue 1–2 (2011), pp. 87–99

Chapter 20

Probabilistic bicriterion optimum inspection/monitoring planning: applications to naval ships and bridges under fatigue
Sunyong Kim and Dan M. Frangopol
Structure and Infrastructure Engineering, volume 8, issue 10 (2012), pp. 912–927

Chapter 21

Integration of structural health monitoring in a system performance based life-cycle bridge management framework
Nader M. Okasha and Dan M. Frangopol
Structure and Infrastructure Engineering, volume 8, issue 11 (2012), pp. 999–1016

Chapter 22

Critical issues, condition assessment and monitoring of heavy movable structures: emphasis on movable bridges
F. Necati Catbas, Mustafa Gul, H. Burak Gokce, Ricardo Zaurin, Dan M. Frangopol and
Kirk A. Grimmelsman
Structure and Infrastructure Engineering, volume 10, issue 2 (2011), pp. 261–276

Chapter 23

Time-variant redundancy of structural systems
Nader M. Okasha and Dan M. Frangopol
Structure and Infrastructure Engineering, volume 6, issue 1–2 (2010), pp. 279–301

Chapter 24

Redundancy and robustness of highway bridge superstructures and substructures
Michel Ghosn, Fred Moses and Dan M. Frangopol
Structure and Infrastructure Engineering, volume 6, issue 1–2 (2010), pp. 257–278

Chapter 25

Effects of post-failure material behaviour on redundancy factors for design of structural components in nondeterministic systems
Benjin Zhu and Dan M. Frangopol
Structure and Infrastructure Engineering, volume 11, issue 4 (2015), pp. 466–485

Chapter 26

Time-variant redundancy and failure times of deteriorating concrete structures considering multiple limit states
Fabio Biondini and Dan M. Frangopol
Structure and Infrastructure Engineering, volume 13, issue 1 (2017), pp. 94–106

For any permission-related enquiries please visit:
http://www.tandfonline.com/page/help/permissions

Notes on Contributors

Mitsuyoshi Akiyama is Professor at the Department of Civil and Environmental Engineering, Waseda University, Japan.

Fabio Biondini is Professor of Structural Engineering at the Department of Civil and Environmental Engineering, Politecnico di Milano, Italy.

Paolo Bocchini is the Frank Hook Assistant Professor at the Department of Civil and Environmental Engineering, Lehigh University, USA.

Sara Casciati is Associate Professor at the School of Architecture, University of Catania, Italy.

F. Necati Catbas is Professor at the Department of Civil, Environmental and Construction Engineering, University of Central Florida, USA.

Marios K. Chryssanthopoulos is Professor of Structural Systems at the Department of Civil and Environmental Engineering, University of Surrey, UK.

You Dong is Assistant Professor of Structural Engineering at the Hong Kong Polytechnic University, Hong Kong, China.

Dan M. Frangopol is the inaugural holder of the Fazlur R. Khan Endowed Chair of Structural Engineering and Architecture at the Department of Civil and Environmental Engineering, Lehigh University, USA.

Hitoshi Furuta is Professor at the Department of Informatics, Kansai University, USA.

Michel Ghosn is Professor at the Department of Civil Engineering, City College of New York, USA.

H. Burak Gokce is Chief R&D Engineer, Yapı Merkezi, Turkey.

Kirk A. Grimmelsman is Practice Leader – Evaluation of Performance and Risk, Intelligent Infrastructure Systems, USA.

Mustafa Gul is Associate Professor at the Department of Civil and Environmental Engineering, University of Alberta, Canada.

Boulent M. Imam is Senior Lecturer at the Department of Civil and Environmental Engineering, University of Surrey, UK.

Haitao Jiang is Research Associate at the Department of Civil and Environmental Engineering, Waseda University, Japan.

Takahiro Kameda is Director at Toiie Laboratory, Japan.

Sunyong Kim is Assistant Professor at the Department of Civil and Environmental Engineering, Wonkwang University, South Korea.

Kihyon Kwon is Senior Researcher at the Structural Engineering Research Institute of Korea Institute of Civil Engineering and Building Technology in Goyang, South Korea.

Sopokhem Lim is Assistant Professor at the Department of Civil and Environmental Engineering, Waseda University, Japan.

Min Liu is Assistant Professor at the Department of Civil Engineering, Catholic University of America, USA.

Thomas B. Messervey is CEO at R2M Solution s.r.l., Italy.

Keita Mizuno is a Civil Engineer at Central Nippon Expressway Co. Ltd., Japan.

Fred Moses is a Structural Engineering Consultant in Houston, USA

Koichiro Nakahara is a Civil Engineer at the Civil Engineering Management Division at Kajima Corporation, Japan.

Koichiro Nakatsu is Associate Professor at the Department of Modern Life, Osaka Jonan Women's Junior College, Japan.

Nader M. Okasha is Associate Professor at the University of Hail, Saudi Arabia.

Samantha Sabatino is Assistant Professor at the University of Texas at Arlington, USA.

Mohamed Soliman is Assistant Professor at Oklahoma State University, USA.

Motoyuki Suzuki is Emeritus Professor at the Department of Civil and Environmental Engineering, Tohoku University, Japan.

Yuji Takahashi is a Civil Engineer at the Civil Engineering Management Division at Kajima Corporation, Japan.

Koshin Takenaka is a Civil Engineer at Shimizu Cooperation, Japan.

Ricardo Zaurin is Lecturer at the Department of Civil, Environmental and Construction Engineering, University of Central Florida, USA.

Benjin Zhu is a Structural Engineer at ABS Consulting in Irvine, USA.

Preface

Our knowledge to model, analyse, design, maintain, manage, and predict the life-cycle performance of structures and infrastructure systems is continually growing. However, the complexity of these systems continues to increase and an integrated approach is necessary to understand the effects of technological, economic, environmental, social, and political interactions on the life-cycle performance of engineering infrastructure. In order to accomplish this, methods have to be developed to systematically analyse structure and infrastructure systems, and models have to be formulated for evaluating and comparing the risks and benefits associated with various alternatives. The life-cycle benefits of these systems have to be maximised in order to serve the needs of our society by selecting the best balance of safety, life-cycle cost, risk, and sustainability despite imperfect information and knowledge.

In recognition of the need for such methods and models, this collection of papers from *Structure and Infrastructure Engineering* authored/co-authored by the Editor-in-Chief, Dan M. Frangopol, presents advances in life-cycle performance, management, and optimisation of structures and infrastructure systems. This collection deals with the state-of-the-art, general methodology, life-cycle performance under corrosion and fatigue, life-cycle performance under earthquakes, inspection and monitoring, and redundancy as life-cycle performance indicator.

The 26 chapters of this book provide technical support for moving toward more rational management and optimization approaches to structural engineering considering life-cycle reliability, redundancy, robustness, risk, resilience, cost, and sustainability.

The material is suitable for anyone interested in life-cycle performance, management, and optimization of structures and infrastructure systems, including students, researchers, and practitioners from all areas of engineering and industry.

I am grateful to many colleagues, former post-doctoral researchers, and graduate students who contributed to this volume as chapter co-authors including M. Akiyama, F. Biondini, P. Bocchini, S. Casciati, F. N. Catbas, M. K. Chryssanthopoulos, Y. Dong, H. Furuta, M. Ghosn, H. B. Gokce, K. A. Grimmelsman, M. Gul, B. M. Imam, H. Jiang, T. Kameda, S. Kim, K. Kwon, S. Lim, M. Liu, T. B. Messervey, K. Mizuno, F. Moses, K. Nakahara, K. Nakatsu, N. M. Okasha, S. Sabatino, M. Soliman, M. Suzuki, Y. Takahashi, S. Takenaka, R. Zaurin, and B. Zhu. Finally, my thanks to D. Y. Yang for constructive discussions.

Dan M. Frangopol
Bethlehem, PA, USA
March 2018

Part I
State-of-the-art

Life-cycle performance, management, and optimisation of structural systems under uncertainty: accomplishments and challenges[1]

Dan M. Frangopol

Our knowledge to model, analyse, design, maintain, monitor, manage, predict and optimise the life-cycle performance of structures and infrastructures under uncertainty is continually growing. However, in many countries, including the United States, the civil infrastructure is no longer within desired levels of performance and safety. Decisions regarding civil infrastructure systems should be supported by an integrated reliability-based life-cycle multi-objective optimisation framework by considering, among other factors, the likelihood of successful performance and the total expected cost accrued over the entire life-cycle. The primary objective of this paper is to highlight recent accomplishments in the life-cycle performance assessment, maintenance, monitoring, management and optimisation of structural systems under uncertainty. Challenges are also identified.

1. Introduction

Stable economic growth and social development of most countries are intimately dependent upon the reliable and durable performance of their structures and infrastructures. Natural hazards, ageing, and functionality fluctuations can inflict detrimental effects on the performance of structural systems during their life-cycles. Even the inherently conservative initial design of structural systems may not protect a structure from these threats. Natural phenomena such as earthquakes, hurricanes and floods can create structural disasters. Ageing and/or increased structural performance demand may significantly affect the vulnerability of constructed facilities (Tsompanakis 2010, Esteva et al. 2010, Casciati and Faravelli 2010). Environmental stressors are the primary factors that drive the ageing process. The effect of structural ageing is perhaps most widely apparent in bridge deterioration, exacerbated by increase in traffic over time, but also impacts other civil infrastructure systems such as buildings and nuclear power plants (Ellingwood and Mori 1993, Ellingwood 1998, 2005).

The accurate modelling of structures and the loading conditions to which they are expected to be exposed during their life-cycle as well as their possible deterioration mechanisms are major issues of structural and engineering mechanics, respectively (Schuëller 1998).

Uncertainty in the modelling of structures and randomness in loading phenomena require the use of probabilistic methods in life-cycle analysis. Explicitly distinguishing the two types of uncertainty, namely the aleatory and epistemic, is crucial for the proper handling of a probabilistic analysis approach (Ang and Tang 2007). Whereas randomness (or aleatory uncertainty) cannot be reduced, improvement in knowledge or in the accuracy of predictive models will reduce the epistemic uncertainty (Ang and De Leon 2005).

Ultimately, optimal decisions are to be made that ensure maintaining or improving reliability of structural systems under multiple objectives and various constraints. This can only be achieved through proper integrated risk management planning in a life-cycle comprehensive framework. Figure 1 shows a schematic representation of a life-cycle integrated management framework example. In this framework, tools for structural performance assessment and prediction, structural health monitoring (SHM), integration of new information (from SHM and/or inspection), and optimisation of strategies (inspection, maintenance, monitoring, repair, and replacement) are required. Life-cycle performance assessment is the backbone of the process which requires current evaluation and future prediction. Uncertainty is an integral component in all aspects of this or any life-cycle management (LCM)

---

[1]Based on a keynote paper presented at the 10th International Conference on Structural Safety and Reliability (ICOSSAR 2009), Osaka, Japan, 13–17 September 2009.

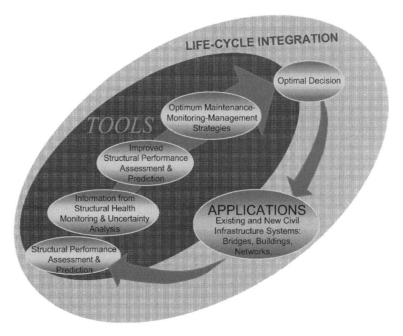

Figure 1. Schematic representation of life-cycle integrated management framework.

framework (Frangopol and Liu 2007a, Frangopol and Okasha 2009).

This paper aims to highlight recent accomplishments in the life-cycle performance assessment, maintenance, monitoring, management, and optimisation of structural systems under uncertainty. Challenges are also identified.

2. System performance assessment and prediction

It is generally recognised that probability-based concepts and methods provide a rational and more scientific basis for treating uncertainties (Ang and Tang 1984, 2007). Uncertainty associated with natural randomness and uncertainties arising from imperfections in modelling and prediction of reality are combined. The results derived from the effects of the combined uncertainties are useful, and indeed have led to the development of more consistent criteria for safety assessment and design of engineered structures and systems (Ang and De Leon 1997, 2005, Ang 2010).

The commonly employed methodology to design and evaluate structural systems based on component analysis reflects the current activities of both structural designers and inspectors who look at structural members to satisfy individual safety checks (Moses 1989, Galambos 1989). This methodology, which is enforced in most structural design and evaluation specifications, leads either (a) to a considerable waste of resources because of over-conservatism in the design and/or evaluation of structural systems which are able to continue to carry loads after the damage or failure

of one or more of their members, or (b) to over-estimation of the actual load carrying capacity and/or underdesign of structural systems which are not able to redistribute loads (Hendawi and Frangopol 1994).

Alternatively, system-based safety measures provide more rational assessment means for structures. The relationship between system safety and member safety depends on the system's configuration and whether the members are modelled in parallel, or in series, or both, the ductility of the members and the degree of mutual dependence among failure modes (Moses 1982, Ghosn et al. 2010). System reliability, redundancy and robustness have been the most widely studied system-based performance measures in structural engineering. Such studies include time-invariant measures (Moses 1982, Frangopol and Curley 1987, Frangopol 1987, Fu 1987, De et al. 1990, Paliou et al. 1990, Frangopol and Nakib 1991, Frangopol et al. 1992, 1998, Ghosn and Moses 1998, Bertero and Bertero 1999, Gharaibeh and Frangopol 2000, Gharaibeh et al. 2000a,b, Liu et al. 2001, Ghosn and Frangopol 2007, Biondini et al. 2008, Ghosn et al. 2010) and time-variant measures (Mori and Ellingwood 1993, Enright and Frangopol 1998a,b, Estes and Frangopol 1999, 2001, 2005, Akgül and Frangopol 2004a,b, Yang et al. 2004, 2006a,b, Okasha and Frangopol 2010a). A novel methodology promoting the system design based on robustness, resistance and sustainability has been proposed by Shinozuka (2008).

The probability of failure of a system is defined as the probability of violating any of the limit state functions that define its failure modes. Limit states of structural systems, as an example, are expressed by equations relating the resistances of the structural components to the load effects acting on these components. Safety margins at a point in time t are expressed as

$$M(t) = R(t) - Q(t) \qquad (1)$$

where $M(t)$ = instantaneous safety margin, $R(t)$ = instantaneous resistance, and $Q(t)$ = instantaneous load effect. By assuming that R and Q are statistically independent random variables, the (instantaneous) probability of failure is (Ellingwood 2005)

$$P_f(t) = P[M(t) < 0] = \int_0^\infty F_R(x,t) f_Q(x,t) dx \qquad (2)$$

where $F_R(x,t)$ is the instantaneous cumulative probability distribution function of the resistance and $f_Q(x, t)$ is the instantaneous probability density function of the load effect. For condition assessment and service life prediction, the probability of satisfactory performance over a service or inter-inspection period is a more relevant metric of performance (Ellingwood 2005). The reliability function provides the probability of survival of a structural component subjected to a sequence of discrete stochastic load events described by a Poisson point process with mean occurrence rate λ during a period of time t_L. This probability can be calculated as (Mori and Ellingwood 1993)

$$L(t_L) = \int_0^\infty \exp\left[-\lambda\left[t_L - \int_0^{t_L} F_s\{r \cdot g(t)\}dt\right]\right] f_{R_0}(r) dr \qquad (3)$$

where R_0 is the initial strength, $f_{R0}(r)$ is the probability density function of R_0, $r \cdot g(t)$ = strength at time t, and $g(t)$ = degradation function. Equation (3) usually must be evaluated numerically by Monte Carlo simulation for realistic deterioration mechanisms (Ciampoli and Ellingwood 2002).

Recent catastrophic structural failures in the United States and throughout the world have reinforced the importance of providing civil structures with system redundancy and robustness to ensure the safety of their users and occupants. Although most engineers agree on the general goals of structural redundancy, issues related to defining these goals and establishing objective measures to quantify redundancy and robustness remain largely unresolved. In this context, system redundancy is defined as the ability of a structural system to redistribute the applied load after reaching the ultimate capacity of its main load-carrying members (Frangopol 1987). Robustness was defined as the

ability of the system to still carry some load after the brittle fracture of one or more critical components (De et al. 1989). It is worth mentioning that research in structural robustness is lagging far behind that in structural redundancy. Even though numerous redundancy measures exist in the literature to date, no standard redundancy measure has been generalised. Most proposed redundancy measures attempt to quantify the amount of reliability available between the first component failure and the system failure. Recent studies conducted by Frangopol and Okasha (2008) showed by example that the two most appropriate time-variant redundancy indices are

$$RI_1(t) = \frac{P_{y(sys)}(t) - P_{f(sys)}(t)}{P_{f(sys)}(t)} \qquad (4)$$

$$RI_2(t) = \beta_{f(sys)}(t) - \beta_{y(sys)}(t) \qquad (5)$$

where $P_{y(sys)}(t)$ = probability of first member failure (e.g. first yield) occurrence in the system at time t; $P_{f(sys)}(t)$ = probability of system failure (e.g. collapse) at time t; $\beta_{y(sys)}(t)$ and $\beta_{f(sys)}(t)$ are reliability indices with respect to first member failure and system failure at time t, respectively. Equations (4) and (5) are based on probabilities of failure or reliability indices calculated using basic reliability theory. When lifetime functions are used, Okasha and Frangopol (2009, 2010d) showed that the most appropriate time-variant redundancy index is

$$RI_3(t) = \frac{An_{wc}(t) - An_s(t)}{An_s(t)} \qquad (6)$$

where $An_s(t)$ and $An_{wc}(t)$ are the unavailability of the system and weakest component, respectively. Unavailability is an indicator of performance similar to the probability of failure.

It is clear from Equations (4), (5) and (6) that the system reliability is the essence in redundancy quantification. In fact, the true redundancy in a system is only obtained by accurate computation of the system reliability. Evaluating the system reliability for complex structures still poses a challenge for researchers today. Regardless of the reliability method implemented, all random variables in the system, their distributions and parameters need to be identified. Depending on the method implemented, different types of limit state functions need to be established.

In the case of component reliability, the resistance formulation is obtained via simple principles of mechanics of materials with sole regard to the component, whereas the load effect requires structural analysis of the entire system. In the case of system reliability, however, two methods are mentioned herein, while others also exist, namely the incremental method (Moses 1982, Rashedi and Moses 1983,1988, Tang and Melchers 1988) and the member replacement method, also known as the failure path approach (Thoft-Christensen and Murotsu 1986, Hendawi and Frangopol 1994). The incremental method enables the finding of an expression for a given failure mode in terms of the component resistances. Detailed structural analyses are required for the system resistance expression. Even though the incremental method is very effective for simple truss and frame structures (Rashedi and Moses 1983, Fu 1987, Fu and Frangopol 1990, Okasha and Frangopol 2010a), and even though Fu (1987) applied it to a deck-girder bridge superstructure with limited number of girders, it is not practical for large and complex systems. The member replacement method follows the failure sequence of a given failure mode and generates a component limit state function for each member after each failure event. These limit state functions are considered in a series-parallel model where each failure sequence represents a branch of parallel components and the failure sequences are connected together in series. Structural analysis is required for finding the load effect expressions for each limit state function. For complex systems such as bridges, this is not an easy task. Estes and Frangopol (1999) and Akgül and Frangopol (2004a,b) used AASHTO specified distribution factors to establish the load effect expressions for various members of bridges due to moving axles of truck loads. Following a failure path requires removal of components from the system and replacing them with their post failure capacity, rendering the AASHTO distribution factors inapplicable. Thus, only the intact component limit state functions are used and a series-parallel system model is assumed based on engineering judgment. The redundancy obtained using such approach is only as good as the assumed series-parallel system model.

Alternatively, and more accurately, incremental nonlinear finite element analyses (INL-FEA) are used to evaluate the system reliability of complex structures. In this approach, the resistance of the entire system can be obtained as the load at which a given failure mode occurs. In fact, serviceability as well as strength failure modes can be considered. The INL-FEA obtained strength has been used in various studies where the system reliability was computed in the reduced random variable space using first-order reliability method (FORM) (Ghosn et al. 2010).

Explicitly considering the functional dependence of the uncertain system resistance on the random variables of the geometric and material properties in INL-FEA is impossible; indeed the INL-FEA obtained resistance is computed through implicit algorithmic form. An ideal solution for this limitation is the direct

estimation of the system reliability using a Monte Carlo Simulation (MCS) of the INL-FEA. Unfortunately, the high reliability associated with practical structural engineering applications demands an enormous number of simulations to accurately compute the system reliability. Besides, the INL-FEA is time demanding itself. Even with current day advances in computational technology, MCS of INL-FEA is too expensive and impractical.

Czarnecki and Nowak (2007) and Strauss et al. (2008a) implemented a hybrid approach to a bridge structure as follows: first the probabilistic distribution of the resistance is generated through Latin hypercube sampling (Olsson et al. 2003, Kyriakidis 2005) of the INL-FEA considering all random variables of the system; second the reliability is calculated using FORM.

The response surface method (RSM) has also been implemented in system reliability of bridge superstructures (Liu et al. 2001), substructures (Ghosn and Moses 1998), and bridge systems (Ghosn et al. 1993, 2010, Moses et al. 1993). The RSM consists of executing INL-FEA analyses at pre-determined values of the random variables and uses a perturbation technique to determine how the response changes as each of the input variables is varied by a known amount. Given the results of the finite-element analysis, an approximate closed-form expression of the response is obtained (Ghosn et al. 2010). The response expression is treated as the system resistance and the system reliability is calculated using FORM.

It is worth mentioning that computation of the reliability function for systems, $L_s(t_L)$, defined as the probability that the system survives through t_L (see Equation (3)) has also been achieved. Mori and Ellingwood (1993) have shown that the reliability of a series system of m components under a Poisson point load process with mean occurrence rate λ can be computed as

$$L_s(t_L) = \underbrace{\int_0^\infty \cdots \int_0^\infty}_{m-fold} L_s(t_L | \mathbf{R}_o = \mathbf{r}) \cdot f_{\mathbf{R}}(\mathbf{r}) d\mathbf{r} \qquad (7)$$

in which

$$L_s(t_L | \mathbf{R}_o = \mathbf{r})$$
$$= \exp\left(-\lambda \cdot \left\{ t_L - \int_0^{t_L} F_S \left[\min_{i=1}^{m} \frac{r_i g_i(t)}{c_i} \right] dt \right\} \right) \qquad (8)$$

where $g_i(t)$ = resistance degradation function for component i, c_i = structural action coefficient for component i, $\mathbf{R}_o = \{R_1, R_2, \ldots, R_m\}$ = initial strength of the structural components assumed deterministic and equal to $\mathbf{r} = \{r_1, r_2, \ldots, r_m\}$, $f_{\mathbf{R}}(\mathbf{r})$ joint probability density function of the initial strength for all the components in the system. They suggested that the m-fold integration in Equation (7) is performed by Monte Carlo simulation (or more efficiently by importance sampling or Latin hypercube sampling) while evaluating Equation (8) numerically in closed form. Enright and Frangopol (1998b) used the failure path method to compute the reliability function of a general (i.e. series-parallel) system and developed the program RELTSYS for this purpose (Enright and Frangopol 2000).

Lifetime functions (Leemis 1995) are a simple alternative to Mori and Ellingwood's (1993, 1994a) time-dependent reliability approach, and have been utilised for the life-cycle performance prediction of bridge structures (Yang et al. 2004, 2006a,b, Okasha and Frangopol 2009, 2010d). Establishing the lifetime function system reliability can be done with various methods such as the minimal path and cut sets (Hoyland and Rausand 1994, Leemis 1995).

Probabilistic descriptors have also been used as life-cycle performance measures (Kong and Frangopol 2003, Neves et al. 2004, Frangopol and Neves 2004, Liu and Frangopol 2005a,b, Neves and Frangopol 2005, Neves et al 2006a,b, Bucher and Frangopol 2006, Frangopol and Liu 2007a, Petcherdchoo et al. 2008). Using these measures, the performance indicator of a structure (or group of structures) is basically a random variable and the variation in the performance over the life-cycle is reflected in the changes of the descriptors of this random variable. A typical model for the life-cycle performance under no maintenance is

$$\theta(t) = \begin{cases} \theta(0) & \text{if } 0 \leq t \leq t_I \\ \theta(0) - \alpha_1(t - t_I)^{\alpha_2}\varepsilon(t) & \text{if } t > t_I \end{cases} \quad (9)$$

where $\theta(t)$ is the life-cycle performance indicator (such as the reliability index), t_I is the time of initiation of deterioration, α_1 and α_2 are the deterioration parameters, $\varepsilon(t)$ is error term often modelled by a lognormal random variable, and t is time (Melchers 2003, Frangopol et al. 2004, van Noortwijk and Frangopol 2004, Ellingwood 2005).

As indicated by Ellingwood (2005), certain deterioration mechanisms, such as general corrosion, are amenable to more precise modelling. Along these lines, Kong et al. (2002) suggested a method for reliability analysis of chloride penetration in saturated concrete. Stewart and Rosowsky (1998), Val et al. (1998), and Vu et al. (1998) investigated the corrosion effects on reliability of concrete bridges, Marsh and Frangopol (2008) developed a reliability model incorporating temporal and spatial variations of probabilistic corrosion rate sensor data, Biondini et al. (2006a,b, 2008)

proposed a structural performance framework for durability analysis of reinforced concrete structures subjected to the diffusive attack from external aggressive agents, and Akiyama et al. (2010) proposed an approach for integration of the effects of airborne chlorides into reliability-based durability design of reinforced concrete structures in a marine environment.

Ship and offshore structures have also had their share of system reliability research. Ship structures are floating box girder beams that are internally stiffened and subdivided, in which the decks and bottom structure are flanges and the side shell and any longitudinal bulkheads are webs (Hughes 1983). Reliability of a ship structure is usually governed by the reliability of the critical section, usually at midspan, with respect to ultimate bending moments (Mansour 1997). Although this is treated as a component reliability problem, the section under analysis is in essence a system of many components, namely the plates and stiffeners. The computation of the flexural resistance of a ship structure is also a complex task. INL-FEA has been reportedly used for this purpose (Chen et al. 1983, Kutt et al. 1985). Alternatively, the Idealised Structural Unit Method is used as an effective tool for nonlinear analysis of large structures (Mansour 1997). The total number of elements and nodal points in this method is much smaller than those associated with the finite element method. Smith (1977) developed a hybrid finite element-incremental curvature method which derives the moment-curvature function for the complete hull. While this method is based on finite element results for each stiffened panel, Gordo et al. (1996) used simple analytical formulas to model this behaviour. Okasha and Frangopol (2010b) have proposed a modification to this method, where instead of incrementation of the curvature, the ultimate moment capacity is found by solving an optimisation problem. The saving in computational time was shown to be significant (Okasha and Frangopol 2010b). A program was developed for this method with capability of Latin hypercube sampling for the generation of the probabilistic distribution of the ultimate strength of ships (Okasha and Frangopol 2010b). Extension of the program for computations of time-variant reliability of ships is underway.

Most research in the reliability of ship structures is based on reliability calculation in the reduced random variable space (Mansour and Hovem 1994, Mansour 1997, Wirsching et al. 1997, Paik et al. 1998, Guedes Soares and Teixeira 2000, Paik and Frieze 2001, Akpan et al 2002). However, Lua and Hess (2006) computed the first-failure based reliability in the full random variable space using a hybrid approach. The first failure flexural strength was computed by means of the incremental curvature method described above. Furthermore, Chen *et al.* (2003) also used the response surface method to calculate the reliability of a ship hull in composite material with respect to the ultimate limit state. The ultimate moment capacity was calculated by a modified incremental method where it was assumed that after the collapse of each stiffened panel no residual strength is present (Chen *et al.* 2003).

Recent developments in reliability-based management of inspection, monitoring, maintenance and repair of offshore structures are described by Moan (2005), with focus on management of hull damage due crack growth and corrosion. Moan (2005) showed how design for robustness, choice of inspection method and scheduling as well as repair strategy, need to be implemented to obtain an acceptable risk for various types of offshore structures.

In any case, an accurate structural performance measure is the means by which the management process of structures and infrastructures is driven. Allocation of funds and decisions on interventions are directly linked to the life-cycle structural performance as indicated by the performance measure with respect to a prescribed performance threshold. As indicated in Figure 2, without monitoring information, inaccurate prediction of future performance might cause tremendous consequences due to failure occurrences if later reaching of the prescribed performance threshold is predicted (see Figure 2(a)), and financial loses due to unnecessary maintenance actions if earlier reaching of the prescribed performance threshold is predicted (see

Figure 2(b)). Preventing such unwanted results and consequences is possible by harnessing the advances in SHM in real-time validation of the predicted results (see Figure 2, where updated prediction based on monitoring information is indicated).

3. Integration of SHM in LCM

3.1. Motivation

The modelling, assessment, and performance prediction of structures and infrastructures over time is by its very nature complex and uncertain. Moreover, because models that treat this issue are very sensitive to changes in their input parameters, SHM provides a powerful and needed mechanism to reduce uncertainty, to calibrate, and to improve structural assessment and performance prediction models (Hess 2007, Frangopol and Messervey 2008, 2009). Until now, visual-based inspections or non-destructive testing have primarily performed this function (Estes *et al.* 2004). Although both provide valuable sources of information, they both also have limitations. Visual inspections are primarily used to estimate structural condition in management programs that employ a condition index as the performance metric for assessment and management decisions. The main disadvantages of such models are that the actual safety level is not explicitly or adequately accounted for and that discrete stochastic transitions between condition states fail to account for previous structural performance and prohibit more accurate continuous modelling approaches (Frangopol and Liu 2007a). In addition,

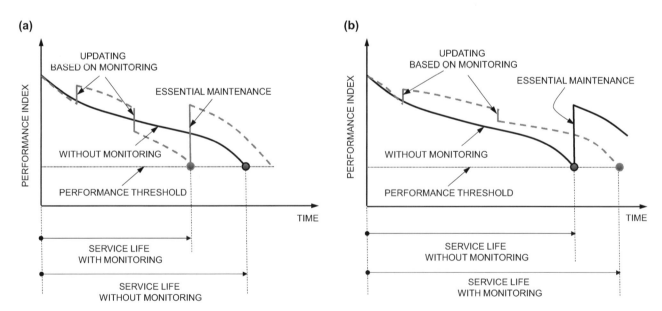

Figure 2.   Performance index profile with and without monitoring: (a) service life with monitoring less than service life without monitoring, and (b) service life with monitoring larger than service life without monitoring.

successful visual inspection depends on considering all possible damage scenarios at all critical locations, not an easily accomplished task even for an experienced inspector (Aktan *et al.* 2001). Human error is also a consideration. One recent study reported that in some cases more than 50% of bridges are being classified incorrectly via visual inspections (Catbas *et al.* 2008). Although certainly not the norm, a separate recent report highlighted the falsification of bridge inspections by contractors to keep up with timelines for reporting purposes (Dedman 2008). Lastly, as the number of deficient bridges increases, it is questionable whether or not existing assessment methods will be adequate (Susoy *et al.* 2007). With respect to non-destructive evaluation (NDE), this mechanism can provide excellent results but has traditionally been scheduled for a particular concern, is conducted at a point in time and is expensive. However, it should be noted that SHM can be considered as simply an improved method to conduct non-destructive evaluation. It provides the ability to conduct distributed (fibre optic), quasi-distributed (multiple sensors), or singular evaluation, at a point in time or over time.

Reductions in size, wireless capabilities, improved energy performance, the development of new sensor types, and reductions in cost are making SHM practical for civil, marine, and aerospace structure applications. Although monitoring devices have existed for some time, they have typically required a controlled environment, hard wired cables, and immense effort to obtain data, making their application in a field environment difficult. Recent improvements in these devices are now making it feasible to obtain site-specific response data cost effectively and offer great potential with respect to the design, assessment, maintenance, and rehabilitation of civil infrastructure.

The assessment of the reliability of an existing bridge based on SHM data has been presented by Frangopol *et al.* (2008a, b) and Liu *et al.* (2009a, b). However, although bridge managers may now have access to a wealth of structural response data, more research is required on how to effectively manage, process, and utilise it. Currently, there is a gap between SHM and bridge inspection and management methods. Whereas SHM has focused primarily on damage detection, bridge managers are focused on answers to serviceability and reliability issues (Glaser *et al.* 2007). These issues require the following questions answered:

(a) Has the load capacity of the structure changed?
(b) What is the probability of failure of individual structural members and the whole structure?
(c) What preventative maintenance needs to be performed?

Therefore, in addition to its focus on damage, the goal of SHM should be to facilitate rational decision-making regarding the safety and reliability of a structure, and proper actions to take when risk concerns are raised.

It is evident that both communities (LCM and SHM) stand to benefit from combining the strengths offered by each approach. LCM approaches offer bridge managers a practical predictive view of cost, safety, and condition, but in many regards lack knowledge of actual structural performance. In contrast, SHM techniques effectively capture structural behaviour and the demands on a structure, but are not as effective in translating this information into actionable data for bridge managers. As such, research into how to best integrate LCM and SHM under uncertainty is timely, appropriate, and needed. The client of such research efforts is the infrastructure manager, who likely has limited funds to maintain multiple structures.

3.2. *Supporting paradigm*

SHM technologies compete with available funds for infrastructure design and management. Consequently, these technologies will not be adopted unless they are (a) required by code, or (b) incentivised, or (c) profitable. Code specifications may not necessarily take the form of mandating SHM itself, but could require that design submissions include a maintenance plan, warranty period, and/or life-cycle cost (LCC) estimate. Incentives could range from additional available funding, favourable tax treatment, or lower insurance premiums (if insurance is or becomes required). Advertising potential or the recognition of a technologically superior product are also possible benefits but are difficult to quantify. Profitability is perhaps the strongest motivator (aside from a direct code mandate) but requires a life-cycle treatment of the problem and a shift away from the lowest cost bid design selection process. The optimal solution of how to best encourage and facilitate SHM/LCM is likely a combination of these possibilities (e.g. available/acceptable design alternatives recognised in the code which include incentives for those who choose to use it and carry the likelihood that over time the system will be profitable through near-optimal management actions).

Although the goal of optimally designed and managed structures that ensure public safety over their useful lifespan is shared by public officials, infrastructure users, and owners, determining how to achieve this goal is more difficult. Differences in methods, assessment metrics, competing interests and competing demands quickly complicate the process. For example, in an environment of limited resources it is likely that researchers would request funding to develop more efficient management techniques whereas infrastructure managers would prefer using this funding to repair or replace existing damaged structures. The trucking industry desires heavier allowable truck weights to improve productivity whereas bridge managers desire lower limits to reduce wear and tear on their structures. Public officials are responsible for public safety but must also accept some level of risk as its elimination is not feasible or affordable. Societal issues in adopting life-cycle concepts within the political system are discussed in Corotis (2009a,b).

Co-ordinated and synchronised actions (adoptions-in-concert) are necessary to facilitate synergistic and efficient solutions. Technologies must be adopted by and worked within the programs utilised for asset management. To support SHM, the asset management programs have to directly account for safety and also to consider costs in a life-cycle context. In turn, asset management must be supported by and exist within regulatory structure that allows for performance-based engineering and one that requires the consideration of total lifetime costs. Instrumental and inherent to the entire hierarchy is that resources are optimised, reliability level is assured, and condition is adequate (Messervey and Frangopol 2008).

Although easily stated, the realisation of such a paradigm is not as easily achieved. Differences in design methods, assessment techniques, management programs, legal systems, and political processes are important when answering the question of how to best integrate health monitoring. Because the creation and implementation of this paradigm will have to span these differences, an introspective analysis within the engineering community is appropriate to agree on several key rules and standards. Particular attention is needed to standardise and include risk in the calculation of LCCs and as a metric to compare alternatives that do and do not employ SHM. Also necessary is a common period of time (warranty period) over which to calculate maintenance costs for newly constructed structures. In doing so, solutions with and without monitoring can be fairly compared. Minimum performance thresholds need to be agreed upon to indicate when corrective actions are required.

3.3. Using monitoring as a catalyst to improve existing design and management methodologies

Although monitoring is associated almost exclusively with structural assessment and most frequently to troubleshoot a particular damage, to not consider how SHM can improve design would be short-sighted.

Instead, it is noted that assessment is a natural and inherent part of the design process and that studies of existing structures have resulted in the codes currently in use. The past several decades have witnessed significant change in the design of civil infrastructure as our understanding of and ability to manage the uncertainties involved with material resistances and load demands have improved. Improvements have been the result of material and loading research, testing, the establishment and study of performance databases, and better computational methods and platforms. Reflecting these advances, design methodologies have shifted from deterministic-based approaches, such as allowable stress design, to the semi-probabilistic approaches found in current codes such as load resistance factored design (LRFD) (AASHTO 2007), the Canadian Highway Bridge Design Code (CHBDC 2006), and the European Highway Agency Eurocodes (EUROCODES 2002). In the future, performance-based design will likely be adopted as progress in material science, design software, construction methods, and SHM are empowering the engineer to better address the uncertainties inherent to the design and operation of civil structures. The evolution of design methods was indicated in Frangopol and Messervey (2008). Despite differences in their treatment of uncertainty, each method (deterministic, semi-probabilistic, and probabilistic) seeks an optimal balance between economical design and safe performance.

The inclusion of monitoring data is best suited for the performance-based approach as monitoring data is uncertain in nature. For LRFD approach, monitoring can provide data to confirm or improve existing load factors, resistance factors, and load combinations for extreme events. In the past, many studies have been undertaken to model the performance of in-service bridges over time (Ghosn and Moses 1998, Enright and Frangopol 1999a,b, Ghosn 2000, Ghosn et al. 2003, Gindy and Nassif 2006). Such studies have sought to better model truck populations, the distribution and frequency of traffic configurations on structures, the load distribution within structural systems, system effects, and the combinations of extreme events. Monitoring presents the opportunity to revisit such studies, improve modelling assumptions and parameters, and recalibrate factors to achieve more accurate design codes.

Although there is much room for improvement in existing codes, the stronger potential of monitoring data is to enabling the adoption and use of performance-based design. By providing the additional information necessary to better quantify the parameters of the random variables associated with performance-based design, such as member resistances and load effects, monitoring may serve as the catalyst for a change in methodology itself. Although advanced reliability-based design methodology has been proposed in several countries and some provisions allow more flexibility for its use, this method to date has not been widely adopted.

Similar to design methodologies, the evolution of civil infrastructure management programs can be related to time, improved knowledge, and the reduction of uncertainty. Currently and almost exclusively, bridge management programs are based on visual inspections. In special cases, non-destructive evaluation (NDE) tests are performed to investigate a specific area or problem of interest. Although relatively few, there is a growing number of monitoring applications. The anticipated evolution of bridge management programs as the capability to obtain data and to treat uncertainty increases was presented in Frangopol and Messervey (2008). At present, the current state-of-the art reflects the predominant use of visual inspection-based condition state models.

In response to the previously discussed limitations associated with condition state models, reliability-based models were developed that specifically assess structural reliability (Casas *et al.* 2002, Watanabe *et al.* 2004, Cruz *et al.* 2006, Koh and Frangopol 2008). Such models seek to take into account all parameters and uncertainties affecting structural performance and are well suited for the computation of structural system reliability. However, due to the amount of input parameters required and the sensitivity of the results to the accuracy of the input, these models have been limited in implementation. In the foreseeable future, it is likely that monitoring technologies will provide the mechanism to make reliability-based models more practical for implementation. A further improvement on reliability-based models is possible by combining the advantages of the condition-state and reliability-based models. As condition-state models do not directly address safety, reliability-based models do not directly address condition where repairs may be required to improve traffic condition despite a high level of structural reliability. This has led to the development of hybrid-type models that account for both condition and reliability (Frangopol 2003, Neves and Frangopol 2005, Bucher and Frangopol 2006, Neves *et al.* 2006a, b). Such models provide a more holistic treatment of the problem but also imply a greater degree of complexity and increased cost.

Although monitoring can benefit any of these bridge management models, it is best suited for the reliability-based or hybrid types. Ideally, a bridge management program allows for individual bridge or bridge network assessment, maintenance, inspection,

and repair planning based on real-time structure-specific data. Such approaches are heavily dependent upon monitoring data. How to best integrate this data into reliability-based LCM programs is being investigated by many researchers worldwide (Budelmann and Hariri 2006, Messervey and Frangopol 2006, Hosser *et al.* 2008). A combined approach to bridge management that incorporates LCM and SHM is attractive as the advantages of each approach offset the other's disadvantages. Although different researchers have developed various models to accomplish this task, all share the general framework presented by Frangopol and Messervey (2008).

3.4. *The top-down approach*

Instead of a bottom-up reaction to specific deficiency, a top-down approach to the development of monitoring systems within a life-cycle context is necessary. Such an approach requires the adoption of methods and metrics suited for probabilistic data and capable of quantifying the benefit of increased levels of safety over time. For an existing structure, this implies a reliability-based LCM approach with the inclusion of risk. For a new structure, this implies performance-based and durability-based design. To ensure the best use of limited resources, common metrics, methodologies, and means of communication must be agreed upon. Despite the pressing need for new innovations, the integration of SHM with LCM will likely be incremental. As such, how these technologies can benefit existing methods while serving as a catalyst for future change is of interest.

The goal is to ensure that monitoring assets are employed effectively, meaning that they are used for the most critical structure, at the appropriate location and at the right time. To do this, the formulation of a monitoring strategy should consider:

(a) historical failures and current assessment of the type of structure of interest,
(b) how the structure fits within a larger network,
(c) the type of measurement desired (global vs. local) and what sensing mechanisms are most appropriate,
(d) what types of uncertainty are present and how they will be modelled, and
(e) how assets will be prioritised at the structure level with respect to member importance, system effects, and time.

Historically, albeit unfortunately, structural failures and collapses have acted as the catalysts that have shaped design codes, construction methods, and management practices. Several studies have been conducted in this area and serve as an excellent resource. Matousek and Schneider (1976) studied about 800 reported failures and errors in the field of structural engineering across several classes of structures. Stewart and Melchers (1997) summarised parts of a number of studies involving structural failures across a variety of structures. More recently, Frangopol and Messervey (2007) observed trends in several bridge failures that occurred in 2006. Perhaps the most thorough recent study is a 2004 analysis of the reasons for reconstruction across 1691 bridges in Japan (Joint Task Committee on Maintenance Engineering 2004). This study reported that serviceability concerns and the upgrade of highway routes accounted for about 49% of reconstruction efforts. Reconstruction resulting from damage and strength concerns accounted for 34% with most problems arising in the superstructure which is subjected to environmental exposure and traffic load effects. Amongst this group of superstructure failures, problems with reinforced concrete accounted for 87% of the failures. From these results, it appears that monitoring strategies for concrete may be of particular interest for bridge managers as slab failure and concrete spalling/cracking accounted for most of the superstructure failures.

Although past failures certainly provide insight, the current condition and classification of existing structures must also be considered when developing a monitoring strategy. In the United States, the National Bridge Inventory (NBI) provides statistics on bridges by bridge type, classification, location, age, and current condition (FHWA 2007). Statistics are also available that detail replacement, rehabilitation, and new construction projects as part of the Highway Bridge Replacement and Rehabilitation Program (HBRRP). Combining this data better enables the design of monitoring approaches for assessing existing structures as well as those newly constructed.

Rarely is the management of a structure considered in isolation. Whenever possible, inspections, assessments, and maintenance actions should be taken in context of where the allocated resources will provide the most benefit. For a transportation network where bridges serve as critical nodes, analysis requires consideration of network connectivity, user satisfaction, and network reliability (Liu and Frangopol 2006a). Monitoring can be allocated to the most important bridge within a network with respect to any of these three metrics or to a bridge with known defects. An appropriate starting point to establish bridge importance is to relate individual bridge reliability to the reliability of the bridge network. The reliability importance factor (RIF) for any bridge is defined as the sensitivity of the bridge network reliability β_{net} to the change in the individual bridge system reliability $\beta_{sys,i}$ as (Liu and Frangopol 2005c)

$$RIF_i = \frac{\partial \beta_{net}}{\partial \beta_{sys,i}} \qquad (10)$$

Using this probabilistic performance indicator, the bridge for which changes in performance have the largest impact on the reliability of the bridge network can be identified for monitoring priority.

Liu and Frangopol (2006b) proposed a comprehensive mathematical model for evaluating the overall performance of a bridge network based on probabilistic analyses of network connectivity, user satisfaction, and structural reliability of the critical bridges in the network. A multi-objective approach based on genetic algorithms can be utilised to optimise bridge network maintenance as presented in Liu and Frangopol (2006a). Stochastic dynamic programming can also be used for bridge network maintenance optimisation (Frangopol and Liu 2007b).

At the structural level, monitoring strategies are broadly categorised in two groups, local and global. Both provide different types of information and in general support different analysis types. Selecting an appropriate strategy might be dictated by the structure, type of analysis, or both (Messervey 2008).

Data collection must be within the broader context and treatment of uncertainty. As noted previously, uncertainty can be partitioned in two broad categories, aleatory and epistemic (Ang and Tang 2007). Both types of uncertainty are of interest as related to infrastructure assessment, LCM and optimisation. In terms of prediction, one must generally focus upon epistemic uncertainty.

Once the type of data desired is determined, the question of how to obtain the data in the most cost-effective manner becomes important. At the structure level, monitoring must be allocated to the most important members, for the critical performance functions, to characterise the most significant random variables, at the appropriate point in time. For structural components (such as bridge decks) and individual members (such as girders), how these elements perform within the context of the larger structure will determine their importance with respect to monitoring. For members arranged in series, where the failure of any member leads to the failure of the system, the weakest member (i.e. member with the highest probability of failure) is the most important. Conversely, if members are arranged in parallel such that the failure of any member does not lead to the failure of the system, then the strongest member (i.e. member with the lowest probability of failure) is the most important. Of course, the degree of correlation between the failure modes must be considered.

Varying rates of reliability deterioration may also affect monitoring priorities or when monitoring is needed. Figure 3 shows the reliability profiles of series, series-parallel, and parallel systems consisting of three members with perfectly correlated failure modes. The monitoring priority has to be given to the member having the highest contribution to the system reliability. For example, as shown in Figure 3(a), monitoring path of the series system model I follows the smallest reliability index (i.e. $\min[\beta_1; \beta_2; \beta_3]$), since the member having the lowest reliability index in a series system has the highest impact on the system reliability. In contrast, monitoring priority of parallel system III has to be given to the member having the highest reliability index (i.e. $\max[\beta_1; \beta_2; \beta_3]$) as shown in Figure 3(c). For the series-parallel system II in Figure 3(b), the monitoring priority has to be determined through comparison between reliability of member 1 and the reliability of the sub-system consisting of members 2 and 3. In this case, the monitoring path will follow the time-dependent function $\min[\beta_1; \max(\beta_2;\beta_3)]$. As shown in Figure 3, the monitoring path changes over time due to the type of system and varying reliability deterioration rates of different members. An example could be exterior steel girder being subject to a higher corrosion rate than interior girders due to a greater exposure to de-icing salts.

Such an analysis could be also utilised to answer the question of when to monitor. If a minimum system reliability threshold β_{min} is established (see Figures 3 and 4), intensive monitoring will be needed at time associated with a critical time on which the monitoring path reaches β_{min}. Given perfect information, Figure 4 compares the monitoring paths for the three systems shown in Figure 3 and indicates the critical monitoring points A, B, and C associated with system models I, II, and III, respectively. In the absence of perfect information, Monte Carlo simulation of the model parameters can be utilised to estimate the earliest possible down-crossing of the minimum system reliability threshold. This would be appropriate for a monitoring system with high operational costs that can be turned on or off, or for a non-permanent monitoring solution that must be scheduled.

3.5. *Monitoring within a life-cycle context*

Structural models and their idealisations, deterioration mechanisms, material resistances, geometries, and loads are uncertain. For these reasons, a life-cycle performance profile can be considered as shown in Figure 5 where uncertainties are associated with initial performance index, deterioration initiation time, deterioration

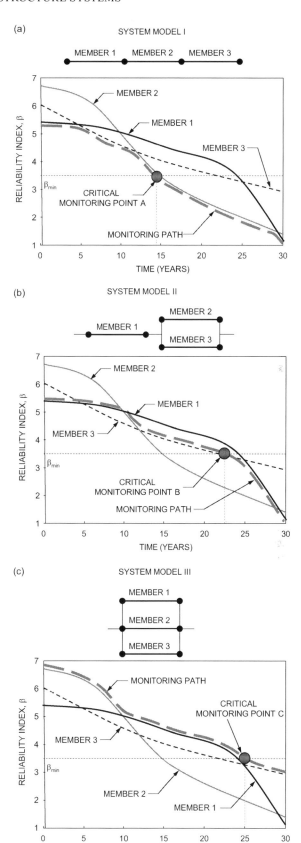

Figure 3. Time-dependent monitoring paths for (a) system model I; (b) system model II; and (c) system model III.

rate, performance indices after preventive and essential maintenances, and service lives with or without maintenance (Frangopol 1998, Frangopol et al. 2001, van Noortwijk and Frangopol 2004, Frangopol et al. 2004, Kong and Frangopol 2005). It is noted that these service lives are uncertain, because the initial performance index, deterioration initiation time, and deterioration rate have uncertainties. In addition, as the model predicts further into the future, the uncertainty associated with prediction of service life increases.

In general, data provided by SHM improves the accuracy of the computed probability of failure of any monitored structural component or system. Figure 6 shows the possible impact of updating using SHM data on load effect and resistance. The impact of updating using SHM on service life prediction was indicated in Figure 2.

The minimum expected LCC with respect to lifetime performance is the most widely used design criterion. The general form of the expected LCC is (Frangopol et al. 1997):

$$C_{ET} = C_T + C_{PM} + C_{INS} + C_{REP} + C_F \qquad (11)$$

where C_{ET} = expected total cost, C_T = initial design/construction cost, C_{PM} = expected cost of routine maintenance, C_{INS} = expect cost of performing inspections, C_{REP} = expected cost of repairs and C_F = expected cost of failure. Inclusion of monitoring into this general form results in (Frangopol and Messervey 2007):

$$C_{ET}^0 = C_T^0 + C_{PM}^0 + C_{INS}^0 + C_{REP}^0 + C_F^0 + C_{MON} \quad (12)$$

where the superscript ^{0} indicates costs in Equation (11) affected by monitoring, and C_{MON} = expected cost of monitoring. The cost of monitoring is expressed as:

$$C_{MON} = M_T + M_{OP} + M_{INS} + M_{REP} \qquad (13)$$

where M_T = expected initial design/construction cost of the monitoring system, M_{OP} = expected operational cost of the monitoring system, M_{INS} = expected inspection cost of the monitoring system, and M_{REP} expected repair cost of the monitoring system. The operational cost of the monitoring system includes the cost of power (battery or electricity) as well as the costs associated with data processing and data management. The benefit of the monitoring system, B_{MON}, is then captured through a comparison of the expected life-

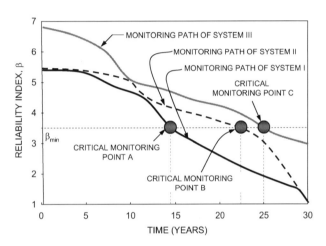

Figure 4. Comparison of monitoring paths and critical monitoring points of system models I, II, and III.

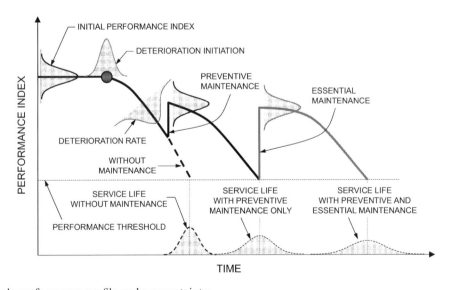

Figure 5. Life-cycle performance profile under uncertainty.

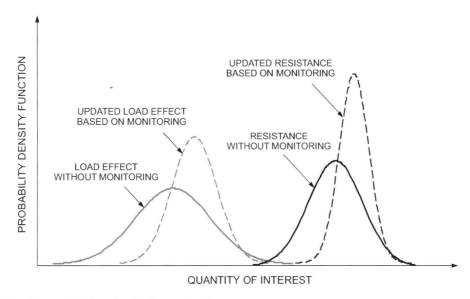

Figure 6.   Possible effect of SHM on load effect and resistance.

cycle total cost with and without monitoring by subtracting Equation (12) from Equation (11):

$$B_{MON} = C_{ET} - C_{ET}^0 \qquad (14)$$

Unless code-driven and using cost as the only criterion, monitoring would only be justified if $B_{MON} > 0$, meaning that monitoring is cost-effective.

If life-cycle costing is incorporated into the consideration of design alternatives, then higher initial costs (corresponding to a lower probability of failure) are paired with lower expected failure costs and lower additional (e.g. maintenance, inspection, and repair) expected costs. Figure 7 schematically shows these relationships with and without monitoring in the near-optimal region. If the monitoring cost is included in maintenance cost, there will be no change in the initial cost, and failure cost is decreased because the accurate prediction of structural performance can lead to less risk. Furthermore, monitoring can allow avoiding unnecessary maintenance, and as a result, if the cost-effective monitoring strategy is applied, there will be a small difference between the maintenance costs with and without monitoring as shown in Figures 7(a) and (b). Therefore, a reduction in the total LCC can be expected (i.e. $COST_B < COST_A$). However, if the monitoring strategy is not cost-effective, it is impossible to obtain the benefit from monitoring as shown in Figure 7(c) (i.e. $COST_C > COST_A$).

Because time-dependent reliability models for civil infrastructure must predict far into the future, SHM has the potential to substantially improve the accuracy of both structural deterioration prediction and load models through parameter characterisation and

parameter updating over time. One area of particular interest is in the characterisation of monitoring-based live loads on bridges. There are significant and compelling advantages to the use of SHM technologies to improve specifically the modelling of live loads on highway bridge structures. These are summarised as follows:

(a) Improvement of existing models.
(b) Efficiency and accuracy.
(c) Bridge specific consideration.
(d) Performance updates over time.
(e) Warning against extreme loads.

First, there is the possibility to improve the accuracy of existing models and code provisions on the basis of more complete and up-to-date data. For example, it is reasonable to assume that the 1975 study of 9250 trucks used for LRFD calibration of the first AASHTO LRFD Bridge Design Specifications (1994) is no longer representative of current truck traffic. Recently, weigh in motion (WIM) studies have been utilised to create much larger databases for truck weights. One such study performed by Gindy and Nassif (2006) examines an 11-year period across 33 WIM sites located in the state of New Jersey and consists of millions of records. Truck volumes, types, and weights, as well as seasonal effects and the implication of short collection periods are addressed. Such studies, based upon monitoring information, can provide the data necessary to re-examine the assumptions utilised in existing codes. Developing a design truck is only part of the problem. Vehicle speeds, vehicle spacing, the consideration of multi-lane

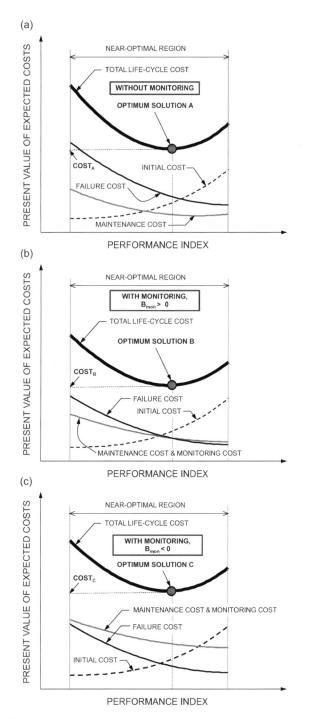

Figure 7. Optimum solution based on LCC minimisation (a) without monitoring, (b) with cost-effective monitoring, $B_{mon} > 0$, and (c) without cost-effective monitoring, $B_{mon} < 0$.

measuring instead of modelling the live load response of a structure, such efforts are bypassed and greater accuracy may be obtained. This is the second main advantage of leveraging SHM. The third is passing from a general to a specific assessment of a structure. Codes and guidelines for both design and assessment must be generalised such that they are applicable across a wide variety of structures although some more recent codes have provisions allowing for a performance based design approach. The collection and use of structure specific data in a probabilistic analysis can consistently account for uncertainties at the local level and develop essentially what could be termed a bridge specific code (Enevoldsen 2008). The last two advantages pertain to the value of increased information over time. Once a monitoring based live load distribution is characterised, it is possible not only to update an existing model but also to track changes as the structure ages. Such changes could either indicate an increase in the live loads placed upon the structure consistent with current trends in traffic or a decrease in the resistance capacity indicating damage. If the monitoring data is continuous, then it also becomes possible to provide a warning to decision makers when target threshold levels are breached.

Two primary challenges are associated with the characterisation of monitoring-based live loads. First, live loads are a function of time. For example, if a structure is monitored for a short period and only light load demands are recorded, one cannot conclude that the safety level of the structure is appropriate. One could conclude that the structure was safe for the loads encountered during the period monitored, but does not adequately convey the safety of the structure over its intended service life. When considering the entire life of a structure, it becomes clear that extreme events such as combinations of overloaded trucks, hurricanes, earthquakes, and other foreseen events must be considered. The second challenge is that the accuracy of any monitoring-based parameter estimates is a function of the amount of data collected. For example, the mean value obtained using one week of monitoring data will vary from that obtained if a second week of data is considered independently or if the two weeks are considered together. Messervey (2008) and Messervey *et al.* (2010) examined how extreme value statistics can be used to address these two challenges.

4. Role of optimisation

As shown in Figure 8, optimisation is the essential tool for providing optimal decision support in the framework of LCM of structural performance under uncertainty. Components of this framework feed their outcomes into this tool, which in turn are

structures, and the frequency of side-by-side occurrences are all aspects that affect the analysis. To this end, many studies have focused on the development of realistic traffic simulations (Zokaie *et al.* 1991, Cohen *et. al* 2003, O'Connor and O'Brien 2005). However, such studies are time consuming. In contrast, by

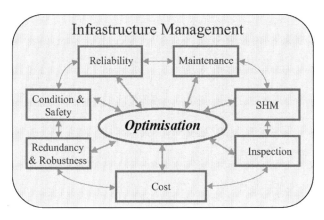

Figure 8. Role of optimisation in infrastructure management.

computationally-intensive processed to find the best solution fulfilling multiple objectives and satisfying prescribed constraints. Furthermore, as shown in Figure 8, optimisation is the core of the management of infrastructure process. All elements of this process interact and sometimes conflict. From all possible solutions, only a multi-criteria optimisation can extract the best solutions that maintain the optimum compromises among these elements.

Traditionally, funds allocated for LCM are concentrated towards the maintenance of structures and infrastructure systems. Indeed, the large number of deficient structures and infrastructures in the face of the scarce funds available for their upkeep is a major challenge for decision makers in charge of properly allocating these funds. For instance, it is estimated that a $17 billion annual investment is needed to substantially improve current bridge condition in the United States; currently only $10.5 billion is spent annually on the construction and maintenance of bridges (ASCE 2009). Accordingly, an integral objective in any efficient problem formulation for the optimisation of the life-cycle of structures is to minimise the total expected maintenance cost. In fact, LCC has been the sole objective in several problem formulations (Frangopol and Moses 1994, Mori and Ellingwood 2004b, Chang and Shinozuka 1995, Ang and De Leon 1997, Estes and Frangopol 1999, Wen and Kang 2001, Yang *et al.* 2006b, Okasha and Frangopol 2009). However, considering multiple objectives in the optimisation provides a set of optimum solutions, which gives a more rational and flexible decision support system. Since cost and performance interact in a conflicting manner, optimisation formulations where both minimising cost and maximising performance are simultaneous objectives have been pursued (Frangopol 1985, Fu and Frangopol 1990, Furuta *et al.* 2006, 2010, Liu and Frangopol 2005a,b, Neves *et al.* 2006a,b, Frangopol and Liu 2007a, Okasha and Frangopol 2009, 2010c).

A maintenance optimisation formulation requires one or more life-cycle performance indicators, such as system reliability (Augusti *et al.* 1998, Estes and Frangopol 1999), system reliability and redundancy (Okasha and Frangopol 2009), lifetime-based reliability (Yang *et al.* 2006b), lifetime-based reliability and redundancy (Okasha and Frangopol 2010c), cost and spacing of corrosion rate sensors (Marsh and Frangopol 2007), and probabilistic condition and safety indices (Liu and Frangopol 2005a,b, Neves *et al.* 2006a,b, Frangopol and Liu 2007a,b). The choice of the performance indicators has a great impact on the computational demand of the optimisation process. Frangopol and Okasha (2009) performed a comparison between the computational time required in a genetic algorithm (GA) optimisation (Goldberg 1989, Deb 2001, Deb *et al.* 2002) process using: (a) instantaneous reliability (see Equation (2)), (b) reliability function (see Equation (3)), (c) probabilistic indicators (see Equation (9)), and (d) lifetime-based reliability. For the investigated example, it was observed that the computational time required using these measures are (a) 19.5 hours, (b) 168.5 hous, (c) 10.81 hours, and (d) 0.08 hours, respectively. On the other hand, it is clear that accuracy of these measures can be sorted in the following ascending order (d), (c), (a) and (b).

Whether an automated optimisation algorithm is required for maintenance optimisation depends on whether preventive maintenance (PM) actions are considered, how essential maintenance (EM) actions are treated, and the number of EM types considered.

Consider the case where only EM types are available. When only one EM type is considered, this maintenance type is applied simply when a performance threshold is reached (Neves *et al.* 2006a). Otherwise, with more than one EM, a choice of which maintenance type to be applied each time the performance threshold is reached has to be made. Two approaches are available for such purpose. In the first approach, the maintenance type to apply once a performance threshold is reached is the one providing the lowest present cost per year of increase of service life. This method was proposed by Estes and Frangopol (1999), used in Yang *et al.* (2006a) and extended for multiple performance measures in Okasha and Frangopol (2009). However, this approach relies on the embedded assumption that the service life extension offered by the maintenance type is the same regardless of when it is applied. The second method compares the total cost of all EM scenarios required to provide a pre-specified service life from start to end. The total cost of all EM scenarios required to provide a pre-specified service life from start to end is compared through an event-tree type of analysis. An algorithm is provided by Okasha and Frangopol

(2010c) for efficiently performing the event-tree analysis for EM optimisation.

Now, consider the case where PM types are available. Such optimisation process requires dealing with two major types of unknowns: the times of application of the PM and the number of applications of PM. By considering the uniform (i.e., equal time intervals) application of PM, the problem is simplified. Neves *et al.* (2006a,b) used two design variables for PM: one for the time of first application of PM and the other for the time interval between subsequent applications. However, non-uniform interval PM strategies have been shown to be more economical than uniform ones (Frangopol *et al.* 1997). To consider non-uniform applications of PM, Liu and Frangopol (2005b) used a real value string with a bit length equal to the designated life-cycle in years and each bit may take an integer value representing one of the identifiers (ids) given for the considered maintenance options. Miyamoto *et al.* (2000) used binary strings for describing the ids of the maintenance options. The result was a binary matrix in which each row represented the maintenance action at a given year and the binary code in that row represented one of the considered maintenance options. Other similar representations were used (Liu *et al.* 1997, Furuta *et al.* 2006, Liu and Frangopol 2005a, Morcous and Lounis 2005). Okasha and Frangopol (2010c) used continuous design variables that represent the times of application of PM, and integer design variables that represent the optimum number applications of each PM type. In fact, the problem is formulated as a nested optimisation problem, where each time the global design variables are provided by the GA, a local nested optimisation takes place to find the optimum set of EM applications using the event-tree algorithm, given the presence of the PM applications and satisfying the given thresholds.

Considering two different performance indicators in addition to the LCC as objective functions provides a rational decision support. By considering the mean condition index, the mean safety index, and LCC as objectives, a decision space of optimum solutions as shown in Figure 9 can be obtained (Neves *et al.* 2006a). The problem formulation providing these solutions is

Find:
The time of the first application of PM and the time interval between the subsequent applications.

To achieve the following three objectives:
Minimise present value of mean cumulative maintenance cost at the end of lifetime.
Minimise maximum mean condition index during lifetime.

Maximise minimum mean safety index during lifetime.

Subject to the following constraints:
Maximum mean condition index $C_{max} \leq 3.0$.
Minimum mean safety index $S_{min} \geq 0.91$

A different decision space of optimum solutions as shown in Figure 10 can be obtained by considering the lifetime unavailability, the lifetime redundancy, and LCC as objectives (Okasha and Frangopol 2010c). The value of the unavailability and redundancy objective functions are the worst value they reach throughout

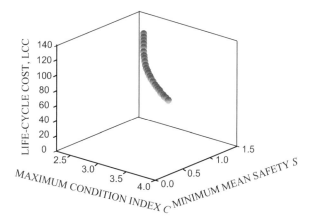

Figure 9. Pareto solutions; three-objective optimization including maximum (worst) mean condition index during lifetime, minimum (worst) mean safety index during lifetime, and present value of mean cumulative cost at the end of lifetime (adapted from Neves *et al.* 2006a).

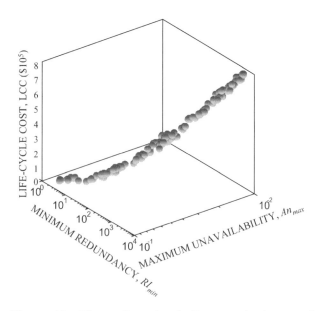

Figure 10. Three dimensional Pareto-optimal set of maintenance solutions (adapted from Okasha and Frangopol 2010c).

the lifetime of a bridge, which is considered as 75 years. The goal of the optimisation is to minimise the maximum (worst) value reached for the unavailability throughout the service life $An_{\max}$, to maximise the minimum (worst) value reached for the redundancy throughout the service life $RI_{\min}$, and to minimise the LCC associated with applying maintenance during 75 years. The formulation is as follows:

Find:
An_{th}, RI_{th}, Td_1, Td_2, ... , Td_{10}, Nd, Tg_1, Tg_2, ... , Tg_{10}, Ng
To achieve the following three objectives:

$$\text{Minimise } An_{\max} \tag{15a}$$

$$\text{Maximise } RI_{\min} \tag{15b}$$

$$\text{Minimise LCC} \tag{15c}$$

Subject to the constraints:

$$10^{-1} \leq An_{th} \leq 10^{-3} \tag{15d}$$

$$0.1 \leq x_r \leq 0.4 \tag{15e}$$

$$2 \leq Tm_i \leq 73 \tag{15f}$$

$$Tm_i - Tm_{i-1} \geq 2 \tag{15g}$$

$$Nd = \{0, 1, 2, \ldots, 10\} \tag{15h}$$

$$Ng = \{0, 1, 2, \ldots, 10\} \tag{15i}$$

where An_{th}, RI_{th} are the unavailability and redundancy thresholds, respectively, Td_1, Td_2, ... , Td_{10} are the times of application (in years) of the 1st, 2nd, ... , and 10th silane treatment on the bridge deck, respectively, Tg_1, Tg_2, ... , Tg_{10} are the times of application (in years) of the 1st, 2nd, ... , and 10th girder re-painting, respectively, Nd is the number of applications of the silane treatment, Ng is the number of applications of the girder re-painting, Tm_i is the time of application (in years) of the ith maintenance action, and $i = 1, 2, \ldots,$ 10, and x_r is a variable related to the redundancy threshold (Okasha and Frangopol 2010c).

A trade-off between performance and cost is evident in the solutions in Figures 9 and 10. In either figure, a wide range of optimum maintenance solutions are presented, from which decision makers can pick the one that satisfies their given budgets and preferences.

In addition to the significant role optimisation plays in the LCM framework, by providing optimum maintenance plans, it can also be exploited for the goal of providing optimum SHM plans. Research towards this goal is recent and much needed. Continuous long-term monitoring of an entire structural system can prevent unexpected failure through accurate assessment of its structural performance. A cost-efficient placement of sensors and effective use of recorded data are required by using probabilistic and statistical methods (Strauss *et al.* 2008b). Kim and Frangopol (2010) proposed the optimal planning of SHM formulated as a bi-objective problem considering both maximisation of availability of monitoring data for prediction of structural performance and

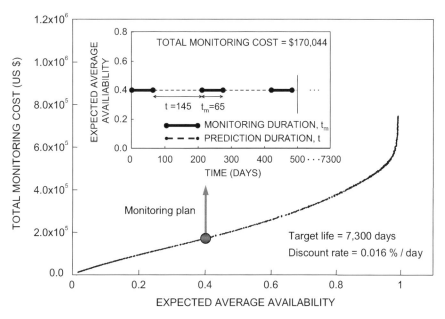

Figure 11.   SHM planning under uncertainty based on conflicting objectives: Objective space with Pareto front and an optimal solution.

minimisation of total monitoring cost. The availability of monitoring data for prediction is defined as the probability that the monitoring data can be usable during non-monitoring period. The formulation of the availability of monitoring data is based on the exceedance probability of the largest difference between the prediction model and the monitoring data. The average availability $\bar{A}$ of the monitoring data for the prediction of structural performance is defined as

$$\bar{A} = [T/t(1 - P_a(t))] + P_a(t) \quad \text{for} \quad 0 \le T \le t \quad (16)$$

where T = time to loose usability of the monitoring data; and $P_a(t)$ = availability of the monitoring data during time period t. Based on Equation (16), the expected average availability can be derived (Ang and Tang 1984). Figure 11 shows 1000 Pareto solutions of the bi-objective problem for a target service life of 20 years. As shown in Figure 11, the required monitoring duration and prediction duration associated with the expected average availability of 0.4 are 65 days and 145 days, respectively.

The monitoring cost (or the expected average availability) can be assigned by structural managers according to the importance and state of deterioration of a structural member. Based on the assigned monitoring cost (or the expected average availability), the optimum monitoring plan can be determined among the Pareto solutions. In Kim and Frangopol (2010), the total monitoring cost of a structural system is allocated to individual components based on their reliability importance factors. As a result, each component of the structural system can have a different monitoring plan. The monitoring planning based on the approach proposed by Kim and Frangopol (2010) can be used as an initial monitoring strategy. It can be updated with new information from monitoring and/or engineering judgment.

Under uncertainty, decision related to management of structures and infrastructures should be made by maximising structural performance (i.e. maximising minimum expected performance index during lifetime and maintaining it above a minimum threshold) and minimising the expected LCC. Therefore, design and maintenance planning can be best formulated as a multi-objective optimisation problem (Frangopol and Liu 2007a). If no SHM is applied during the expected service life of a structure or a network of structures, the group of trade-off solutions of the bi-objective optimisation, with structural performance and LCC as conflicting criteria, is indicated in Figure 12(a). The two groups of optimised trade-off solutions (i.e. without and with SHM) are compared in Figure 12(b). As shown in Figure 12(b), the optimal Pareto front has solutions with SHM (i.e. from A to B, where

(a)

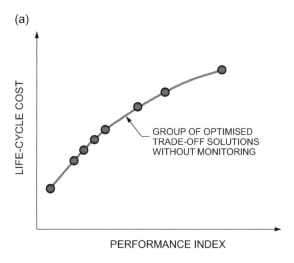

(b)

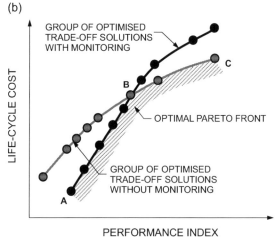

Figure 12. Trade-off solutions between two conflicting objectives: (a) without SHM, and (b) with or without SHM, and optimal Pareto front.

SHM is cost-effective) and without SHM (i.e. from B to C, where the SHM is not cost-effective).

5. Conclusions

In general, there are three main core components of any advanced life-cycle structural management optimisation framework, namely the performance indicators upon which decisions are made, the monitoring and updating tools by which the optimum plan is validated throughout the lifespan of a structure or a network of inter-dependent structures, and the optimisation technique by which the best future plans are determined.

Uncertainties are inherent in all aspects of the assessment process and future prediction of structural performance and, therefore, principles of probability and statistics have to be used. Approaches based on reliability methods have proven their success in quantifying these uncertainties for engineering applications. Hence, reliability-based performance measures have been the primary tool in the proposed, and sometimes implemented, structural management optimisation frameworks. In recent years, the interest in this process has been shifting from the application of component reliability towards the application of system reliability. System reliability provides a rational measure of the overall system safety by accounting for the interaction among the components that form the structural system. Redundancy and robustness are additional system performance measures that may be integrated into the structural management optimisation framework.

SHM can provide powerful means to reduce uncertainties in the prediction models and random variable descriptors of structural performance. In fact, SHM is vital for the proctoring of the structural system performance and the assurance of satisfactory operation. The information obtained from SHM can be used to reduce epistemic uncertainties and, consequently, to better estimate the reliability and lifespan of existing structures, and improve the design of new structures.

Due to the scarcity of financial funds allocated for maintaining and/or improving the reliability of the ever growing stock of deteriorating civil infrastructure systems, especially bridges, optimisation of the resources available is crucial. The significance of using multi-objective optimisation in civil infrastructure maintenance management under uncertainty has to be emphasised when making decisions regarding requirements for continued service of deteriorating structures and networks of structures.

Today, advances in each of the three main core components of structural management optimisation have been accomplished. Yet, challenges remain. Some of these are identified as follows:

(1) Numerous measures of structural system performance, redundancy and robustness have been made available. These measures vary in their ability to capture certain aspects of the structural performance and behaviour, their accuracy in doing so, and their approach of handling uncertainty. Recently developed time-dependent system redundancy measures appears promising for capturing system effects related to the time-dependent ability of a structural system to redistribute loading. Effective and practical methods for capturing system performance including redundancy and robustness in a time-dependent context will continue to present an important challenge.

(2) The discussion in this paper emphasises the fact that an increase in the accuracy of structural performance prediction is met with a disproportionate increase in computational cost. Including the uncertainties in the process makes matters further complex. For instances, harnessing the advances in finite element analysis, and computational and simulation power of today to probabilistically predict the life-cycle performance of a complex and realistic structural system, albeit possible, is still too costly to be practical.

(3) The advances in computation and optimisation technologies, such as those in the field of genetic algorithms, have paved the road for flourishing research towards creating platforms and formulations enabling powerful prioritisation of maintenance activities of deteriorating structures and infrastructures. A constant challenge encountered by such efforts is the limited data available on the probabilistic effects of realistic interventions on the performance indicators. In addition, development of prediction models for the structural performance with higher accuracy will improve the results of any optimisation process. Incorporation of SHM in this process is a field in its infancy.

(4) Improvements in probabilistic and physical models for evaluating and comparing the risks and benefits associated with various alternatives for maintaining or upgrading the reliability of existing structures are needed.

(5) Encouraging education and research in the field of risk communication especially on educating infrastructure stakeholders on their perception and tolerance of risk associated with civil infrastructure is needed (Ellingwood 2005, Frangopol and Ellingwood 2010).

(6) Further and continued promotion of an integrated probability-based approach as the rational basis for understanding the effect of technological, environmental, economical, social and political interactions on the life-cycle performance and cost of engineering infrastructure is required.

Acknowledgements

Preparation of this paper was facilitated by research awards from (a) the US National Science Foundation through grants CMS-0638728 and CMS-0639428, (b) the Commonwealth of Pennsylvania, Department of Community and Economic Development, through the Pennsylvania Infrastructure Technology Alliance (PITA Awards 541535, 541673, 546174, and 541918), (c) the US Federal Highway Administration Cooperative Agreement Award DTFH61-07-H-00040, and (d) the US Office of Naval Research Contract

Number N00014-08-1-0188. This support is gratefully acknowledged. Former and current graduate students including Nader Okasha, Sunyong Kim, and Thomas Messervey, among others, made significant contributions to various aspects of the research. However, the opinions and conclusions presented in this paper are those of the author and do not necessarily reflect the views of the sponsoring organisations.

References

AASHTO (American Association of State Highway and Transportation Officials), 2007. *LRFD bridge design specifications*. 4th edn. Washington, DC.

Akgül, F. and Frangopol, D.M., 2004a. Computational platform for predicting lifetime system reliability profiles for different structure types in a network. *Journal of Computing in Civil Engineering*, 18 (2), 92–104.

Akgül, F. and Frangopol, D.M., 2004b. Bridge rating and reliability correlation: Comprehensive study for different bridge types. *Journal of Structural Engineering*, 130 (7), 1063–1074.

Akiyama, M., Frangopol, D.M., and Suzuki, M., 2010. Integration of the effects of airborne chlorides into reliability-based durability design of reinforced concrete structures in a marine environment. *Structure and Infrastructure Engineering*, DOI: 10.1080/15732470903363313.

Akpan, U.O., Koko, T.S., Ayyub, B. and Dunbar, T.E., 2002. Risk assessment of ageing ship hull structures in the presence of corrosion and fatigue. *Marine Structures*, 15 (3), 211–231.

Aktan, A.E., Chase, S., Inman, D. and Pines, D., 2001. Monitoring and managing the health of infrastructure systems. *In: Proceedings of the 2001 SPIE conference on health monitoring of highway transportation infrastructure*, 6–8 March 2001, Irvine, California, SPIE.

Ang, A.H.-S., 2010. On risk and reliability – Contributions to engineering and future challenges. *In*: H. Furuta, D.M. Frangopol and M. Shinozuka, eds. *Safety, reliability and risk of structures, infrastructures and engineering systems*, 3–4. London: CRC Press/Balkema, Taylor and Francis Group. Full paper on CD-ROM.

Ang, A.H.-S. and De Leon, D., 1997. Determination of optimal target reliabilities for design and upgrading of structures. *Structural Safety*, 19, 91–104.

Ang, A.H.-S. and De Leon, D., 2005. Modelling and analysis of uncertainties for risk-informed decisions in infrastructures engineering. *Structure and Infrastructure Engineering*, 1 (1), 19–21.

Ang, A.H.-S. and Tang, W.H., 1984. *Probability concepts in engineering planning and design. Vol. II*. New York: John Wiley and Sons.

Ang, A.H.-S. and Tang, W.H., 2007. *Probability concepts in engineering: Emphasis on applications to civil and environmental engineering*. 2nd edn. New York: Wiley.

ASCE, 2009. *Report card for America's infrastructure*. Reston, VA: American Society of Civil Engineers.

Augusti, G., Ciampoli, M. and Frangopol, D.M., 1998. Optimal planning of retrofitting interventions on bridges in a highway network. *Engineering Structures*, 20 (11), 933–939.

Bertero, R.D. and Bertero, V.V., 1999. Redundancy in earthquake-resistant design. *Journal of Structural Engineering*, 125 (1), 81–88.

Biondini, F., Frangopol, D.M. and Malerba, P.G., 2006a. Time-variant structural performance of the Certosa cable-stayed bridge. *Structural Engineering International Journal of IABSE*, 16 (3), 235–244.

Biondini, F., Bontempi, F., Frangopol, D.M. and Malerba, P.G., 2006b. Probabilistic service life assessment and maintenance planning of concrete structures. *Journal of Structural Engineering*, 132 (5), 810–825.

Biondini, F., Frangopol, D.M. and Malerba, P.G., 2008. Uncertainty effects on lifetime structural performance of cable–stayed bridges. *Probabilistic Engineering Mechanics*, 23 (4), 509–522.

Biondini, F., Frangopol, D.M. and Restelli, S., 2008. On structural robustness, redundancy and static indeterminacy. *In: Proceedings of the ASCE structures congress*, 24–26 April, Vancouver, Canada. 10 pages on CD-ROM.

Bucher, C. and Frangopol, D.M., 2006. Optimisation of lifetime maintenance strategies for deteriorating structures considering probabilities of violating safety, condition, and cost thresholds. *Probabilistic Engineering Mechanics*, 21 (1), 1–8.

Budelmann, H. and Hariri, K., 2006. A structural monitoring system for RC/PC structures. *In*: H.-N. Cho, D.M. Frangopol and A.H.-S. Ang, eds. *Life-cycle cost and performance of civil infrastructure systems*. London: Taylor and Francis, 3–17.

Casas, J.R., Frangopol, D.M. and Nowak, A.S., eds, 2002. *Bridge maintenance, safety and management*. Book and CD-ROM. Barcelona, Spain: CIMNE.

Casciati, S. and Faravelli, L., 2010. Vulnerability assessment for medieval civic towers. *Structure and Infrastructure Engineering*, 6 (1–2), 193–203.

Catbas, F.N., Susoy, M. and Frangopol, D.M., 2008. Structural health monitoring and reliability estimation: Long span truss bridge application with environmental monitoring data. *Engineering Structures*, 30 (9), 2347–2359.

Chang, S.E. and Shinozuka, M., 1996. Life-cycle cost analysis with natural hazard risk. *Journal of Infrastructure Systems*, 2 (3), 118–126.

CHBDC (Canadian Highway Bridge Design Code), 2006. Ministry of Transportation, Ontario, Canada.

Chen, N.-Z., Sun, H.-H. and Guedes Soares, C., 2003. Reliability analysis of a ship hull in composite material. *Composite Structures*, 62 (1), 59–66.

Chen, Y.-K., Kutt, L.M., Piaszczyk, C.M. and Bieniek, M.P., 1983. Ultimate strength of ship structures. *SNAME Transactions*, 91, 149–168.

Ciampoli, M. and Ellingwood, B.R., 2002. Probabilistic methods for assessing current and future performance of concrete structures in nuclear power plants. *Materials and Structures*, 35, 3–14.

Cohen, H., Fu, G., Dekelbab, W. and Moses, F., 2003. Predicting truck load spectra under weight limit changes and its application to steel bridge fatigue assessment. *Journal of Bridge Engineering*, 8 (5), 312–322.

Corotis, R.B., ed., 2009a. Politics, perception and related social issues of life-cycle cost management for civil infrastructure systems. *Structure and Infrastructure Engineering*, 5 (1) Special issue, 1–65.

Corotis, R.B., 2009b. Societal issues in adopting life–cycle concepts within the political system. *Structure and Infrastructure Engineering*, 5 (1), 59–65.

Cruz, P.J.S., Frangopol, D.M. and Neves, L.C., eds, 2006. *Bridge maintenance, safety, management, life-cycle performance and cost*. Book and CD-ROM. London: Taylor and Francis.

Czarnecki, A.A. and Nowak, A.S., 2007. Reliability-based evaluation of steel girder bridges. *Bridge Engineering, Proceedings of the Institution of Civil Engineers*, 160, (Issue BE1), 9–15.

De, S.R., Karamchandani, A. and Cornell, C.A., 1990. Study of redundancy in near ideal parallel structural systems. *In*: A.H.-S. Ang, M. Shinozuka and G.I. Schüeller, eds. *Structural safety and reliability*. New York: ASCE, 2, 975–982.

Deb, K., 2001. *Multi-objective optimisation using evolutionary algorithms*. UK: John Wiley and Sons.

Deb, K., Pratap, A., Agrawal, S. and Meyarivan, T., 2002. A fast and elitist multiobjective genetic algorithm: NSGA–II. *Transaction on Evolutionary Computation*, 6 (2), 182–197.

Dedman, B., 2008. *Ga. Employee faked bridge inspections* [online]. MSNBC. Available from: http://www.msnbc.msn.com/id/23020686/ [Accessed 24 July 2008].

Ellingwood, B.R., 1998. Issues related to structural ageing in probabilistic risk analysis of nuclear power plants. *Reliability Engineering and System Safety*, 62 (3), 171–183.

Ellingwood, B.R., 2005. Risk-informed condition assessment of civil infrastructure: State of practice and research issues. *Structure and Infrastructure Engineering*, 1 (1), 7–18.

Ellingwood, B.R. and Mori, Y., 1993. Probabilistic methods for condition assessment and life prediction of concrete structures in nuclear power plants. *Nuclear Engineering and Design*, 142, 155–166.

Enevoldsen, I., 2008. Practical implementation of probability based assessment methods for bridges, (keynote paper). *In*: H.-M. Koh and D.M. Frangopol, eds. *Bridge maintenance, safety, management, health monitoring and informatics*. London: CRC Press/Balkema, Taylor and Francis Group, 41–48. Full paper on CD-ROM.

Enright, M.P. and Frangopol, D.M., 1998a. Service-life prediction of deteriorating concrete bridges. *Journal of Structural Engineering*, 124 (3), 309–317.

Enright, M.P. and Frangopol, D.M., 1998b. Failure time prediction of deteriorating fail-safe structures. *Journal of Structural Engineering*, 124 (12), 1448–1457.

Enright, M.P. and Frangopol, D.M., 1999a. Reliability-based condition assessment of deteriorating concrete bridges considering load redistribution. *Structural Safety*, 21, 159–195.

Enright, M.P. and Frangopol, D.M., 1999b. Condition prediction of deteriorating concrete bridges. *Journal of Structural Engineering*, 125 (10), 1118–1125.

Enright, M. and Frangopol, D.M., 2000. RELTSYS: A computer program for life prediction of deteriorating system. *Structural Engineering and Mechanics*, 9 (6), 557–568.

Estes, A.C. and Frangopol, D.M., 1999. Repair optimisation of highway bridges using system reliability approach. *Journal of Structural Engineering*, 125 (7), 766–775.

Estes, A.C. and Frangopol, D.M., 2001. Bridge lifetime system reliability under multiple limit states. *Journal of Bridge Engineering*, 6 (6), 523–528.

Estes, A.C. and Frangopol, D.M., 2005. Chapter 36. Life-cycle evaluation and condition assessment of structures. *In*: W.-F. Chen and E.M. Lui, eds. *Structural engineering handbook*. 2nd edn. Boca Raton, New York: CRC Press, 36.1–36.51.

Estes, A.C., Frangopol, D.M. and Foltz, S.D., 2004. Updating reliability of steel miter gates on locks and dams using visual inspection results. *Engineering Structures*, 26, 319–333.

Esteva, L., Diaz-Lopez, O. and Ismael-Hernandes, E., 2010. Seismic vulnerability functions of multi-storey buildings: estimation and applications. *Structure and Infrastructure Engineering*, 6(1–2), 3–16.

EUROCODES, 2002. *Basis of design and actions on structures. BS EN 1990:2002*. London, UK: British Standards Institution.

FHWA (Federal Highway Administration), 2007. *National bridge inventory* [online]. Available from: http://www.fhwa.dot.gov/bridge/nbi.htm [Accessed May 2007].

Frangopol, D.M., 1985. Multi-criteria reliability-based structural optimization. *Structural Safety*, 3 (1), 23–28.

Frangopol, D.M., ed., 1987. *Effects of damage and redundancy on structural performance*. New York: ASCE.

Frangopol, D.M., 1998. A probabilistic model based on eight random variables for preventive maintenance of bridges. *Progress meeting on optimum maintenance strategies for different bridge types*, October 1998. London, UK: Highways Agency.

Frangopol, D.M., 2003. Preventive maintenance strategies for bridge groups. Analysis, final project report to the Highways Agency, London, UK.

Frangopol, D.M. and Curley, J.P., 1987. Effects of damage and redundancy on structural reliability. *Journal of Structural Engineering*, 113 (7), 1533–1549.

Frangopol, D.M. and Ellingwood, B.R., 2010. Life-cycle performance, safety, reliability and risk of structural systems; A framework for new challenges, Editorial. *Structure*. A joint publication of NCSEA/CASE/SEI, 7 March 2010.

Frangopol, D.M. and Furuta, H., eds, 2001. *Life-cycle cost analysis and design of civil infrastructure systems*. Reston, Virginia: ASCE.

Frangopol, D.M., Gharaibeh, E.S., Hearn, G. and Shing, P.B., 1998. System reliability and redundancy in codified bridge evaluation and design. *In*: N.K. Srivastava, ed. *Structural Engineers World Congress*. Paper Reference T121–2, 18–23 July 1998. San Francisco, California: Elsevier. CD-ROM.

Frangopol, D.M., Kallen, M.-J. and van Noortwijk, J., 2004. Probabilistic models for life–cycle performance of deteriorating structures: review and future directions. *Progress in Structural Engineering and Materials*, 6 (4), 197–212.

Frangopol, D.M., Kong, J.S. and Gharaibeh, E.S., 2001. Reliability-based life-cycle management of highway bridges. *Journal of Computing in Civil Engineering*, 15 (1), 27–34.

Frangopol, D.M., Izuka, M. and Yoshida, K., 1992. Redundancy measures for design and evaluation of structural systems. *Transactions of ASME. Journal of Offshore Mechanics and Arctic Engineering*, 114 (4), 285–290.

Frangopol, D.M., Lin, K.-Y. and Estes, A.C., 1997. Life-cycle cost design of deteriorating structures. *Journal of Structural Engineering*, 123 (10), 1390–1401.

Frangopol, D.M. and Liu, M., 2007a. Maintenance and management of civil infrastructure based on condition, safety, optimisation, and life-cycle cost. *Structure and Infrastructure Engineering*, 3 (1), 29–41.

Frangopol, D.M. and Liu, M., 2007b. Bridge network maintenance optimisation using stochastic dynamic programming. *Journal of Structural Engineering*, 133 (12), 1772–1782.

Frangopol, D.M. and Messervey, T.B., 2007. Integrated life-cycle health monitoring, maintenance, management and cost of civil infrastructure. *In*: F. Lichu, S. Limin, and S. Zhi, eds. *International symposium on integrated life-cycle design and management of infrastructure*, 16–18 May 2007 (Keynote Paper). Shanghai, China: Tongji University Press, 3–20. CD ROM.

Frangopol, D.M. and Messervey, T.B., 2008. Chapter 16. Life-cycle cost and performance prediction: Role of structural health monitoring. *In*: S.-S. Chen and A.H.-S. Ang, eds. *Frontier technologies for infrastructures engineering*. Vol. 4. D.M. Frangopol, book series editor. Boca Raton, London, New York, Leiden: CRC Press/Balkema, 361–381.

Frangopol, D.M. and Messervey, T.B., 2009. Chapter 89. Maintenance principles for civil structures. *In*: C. Boller, F.-K. Chang, and Y. Fujino, eds. *Encyclopedia of structural health monitoring*. Vol. 4. Chichester, UK: John Wiley and Sons, 1533–1562.

Frangopol, D.M. and Moses, F., 1994. Chapter 13, Reliability-based structural optimization. *In*: H. Adeli, ed. *Advances in design optimization*. London, UK: Chapman and Hall, 492–570.

Frangopol, D.M. and Nakib, R., 1991. Redundancy in highway bridges. *Engineering Journal*, 28 (1), 45–50.

Frangopol, D.M. and Neves, L.C., 2004. Chapter 14. Probabilistic maintenance and optimisation strategies for deteriorating civil infrastructures. *In*: B.H.V. Topping and C.A. Mota Soares, eds. *Progress in computational structures technology*. Stirling, Scotland: Saxe-Coburg Publications, 353–377.

Frangopol, D.M. and Okasha, N.M., 2008. Probabilistic measures for time-variant redundancy. *In*: *Proceedings of the inaugural international conference of the engineering mechanics institute (EM08)*. Minneapolis, Minnesota: ASCE. CD-ROM.

Frangopol, D.M. and Okasha, N.M., 2009. Chapter 1. Multi-criteria optimisation of life-cycle maintenance programs using advanced modelling and computational tools. *In*: B.H.V. Topping, L.F. Costa Neves and C. Barros, eds. *Trends in civil and structural computing*. Stirlingshire, Scotland: Saxe-Coburg Publications, 1–26.

Frangopol, D.M., Strauss, A., and Kim, S., 2008a. Bridge reliability assessment based on monitoring. *Journal of Bridge Engineering*, 13 (3), 258–270.

Frangopol, D.M., Strauss, A. and Kim, S., 2008b. Use of monitoring extreme data for the performance prediction of structures: General approach. *Engineering Structures*, 30 (12), 3644–3653.

Fu, G., 1987. Lifetime structural system reliability. Report No. 87–9, Department of Civil Engineering, University of Case Western Reserve University, Cleveland, OH.

Fu, G. and Frangopol, D.M., 1990. Balancing weight, system reliability and redundancy in a multi–objective optimisation framework. *Structural Safety*, 7 (2–4), 165–175.

Furuta, H., Kameda, T., Nakahara, K., Takahashi, Y. and Frangopol, D.M., 2006. Optimal bridge maintenance planning using improved multi-objective genetic algorithm. *Structure and Infrastructure Engineering*, 2 (1), 33–41.

Furuta, H., Frangopol, D.M. and Nakatsu, K., 2010. Life-cycle cost optimisation with emphasis on balancing structural performance and seismic risk of road network. *Structure and Infrastructure Engineering*, DOI: 10.1080/15732471003588346.

Galambos, T.V., 1989. System reliability and structural design. *In*: D.M. Frangopol, ed. *New directions in structural system reliability*. Boulder, CO: University of Colorado Press, 158–166.

Gharaibeh, E.S. and Frangopol, D.M., 2000. Safety assessment of highway bridges based on system reliability and redundancy. *Congress report. 16th congress of IABSE*, 18–21 September 2000, Lucerne, Switzerland, 274–275. 8 pages on CD-ROM.

Gharaibeh, E.S., Frangopol, D.M. and Enright, M.P., 2000a. Redundancy and member importance evaluation of highway bridges. *In*: R.E. Melchers and M.G. Stewart, eds. *Applications of statistics and probability*. Rotterdam: Balkema, 2, 651–658.

Gharaibeh, E.S., Frangopol, D.M., Shing, P.B. and Hearn, G., 2000b. System function, redundancy, and component importance: Feedback for optimal design. *In*: M. Elgaaly, ed. *Advanced technology in structural engineering*. Reston, Virginia: ASCE. 8 pages on CD-ROM.

Ghosn, M. and Frangopol, D.M., 2007. Structural redundancy and robustness measures and their use in assessment and design. *In*: J. Kanda, T. Takada, and H. Furuta, eds. *Applications of statistics and probability in civil engineering*. London: Taylor and Francis, 181–182. 7 pages on CD-ROM.

Ghosn, M. and Moses, F., 1998. *Redundancy in highway bridge superstructures*. Washington, DC: National Academy Press. National Cooperative Highway Research Program, NCHRP Report 406, Transportation Research Board.

Ghosn, M., 2000. Development of truck weight regulations using bridge reliability model. *Journal of Bridge Engineering*, 5 (4), 293–303.

Ghosn, M., Moses, F. and Wang, J., 2003. *Design of highway bridges for extreme events*. Washington, DC: NCHRP TRB Report 489.

Ghosn, M., Moses, F. and Frangopol, D.M., 2010. Redundancy and robustness of highway bridge super-structures and substructures. *Structure and Infrastructure Engineering*, 6 (1/2), 257–278.

Ghosn, M., Moses, F. and Khedekar, N., 1993. Response functions and system reliability of bridges. *In*: P.D. Spanos and Y.T. Wu, eds. *Proceedings of IUTAM symposium*, 7–10 June, San Antonio. *Probabilistic structural mechanics: advances in structural reliability methods*. New York: Springer-Verlag.

Gindy, M. and Nassif, H., 2006. Effect of bridge live load based on 10 years of WIM data. *In*: P.J.S. Cruz, D.M. Frangopol, and L.C. Neves, eds. *Bridge maintenance, safety, management, life-cycle performance and cost*, 16–19 July 2006, Porto, Portugal. London: Taylor and Francis. 9 pages on CD-ROM.

Glaser, S.D., Li, H., Wang, M.L., Ou, J. and Lynch, J., 2007. Sensor technology innovation for the advancement of structural health monitoring: a strategic program of US–China research for the next decade. *Smart Structures and Systems*, 3 (2), 221–244.

Goldberg, D.E., 1989. *Genetic algorithms in search, optimisation and machine learning*. MA: Addison-Wesley.

Gordo, J.M., Guedes Soares, C. and Faulkner, D., 1996. Approximate assessment of the ultimate longitudinal strength of the hull girder. *Journal of Ship Research*, 40 (1), 60–69.

Guedes Soares, C. and Teixeira, A.P., 2000. Structural reliability of two bulk carrier designs. *Marine Structures*, 13 (2), 107–128.

Hendawi, S. and Frangopol, D.M., 1994. System reliability and redundancy in structural design and evaluation. *Structural Safety*, 16 (1+2), 47–71.

Hess, P.E., III, 2007. Structural health monitoring for high-speed naval ships. *In*: F.-K. Chang, ed. *Proceedings of the 6th international workshop on structural health monitoring. Quantification, validation, and implementation*, 11–13 September 2007. Lancaster, PA: DEStech. Stanford University, Stanford, California, 2, 3–15.

Hoyland, A. and Rausand, M., 1994. *System reliability theory: Models and statistical methods*. New York: Wiley-Interscience/John Wiley and Sons.

Hughes, O.F., 1983. *Ship structural design: A rationally-based, computer-aided, optimisation approach*. New York: John Wiley and Sons.

Joint Task Committee on Maintenance Engineering, 2004. *Infrastructure maintenance engineering*. Japan Society of Civil Engineers, University of Tokyo Press (in Japanese).

Kim, S. and Frangopol, D.M., 2010. Optimal planning of structural performance monitoring based on reliability importance assessment. *Probabilistic Engineering Mechanics*, 25 (1), 86–98.

Hosser, D., Klinzmann, C. and Schnetgoke, R., 2008. A framework for reliability-based system assessment based on structural health monitoring. *Structure and Infrastructure Engineering*, 4 (4), 271–285.

Koh, H.-M. and Frangopol, D.M., eds, 2008. *Bridge Maintenance, Safety, Management, Health Monitoring and Informatics*. Set of book and CD-ROM. A Balkema book and CD-ROM, CRC Press. Boca Raton: Taylor and Francis Group.

Kong, J.S., Ababneh, A.N., Frangopol, D.M., and Xi, Y., 2002. Reliability analysis of chloride penetration in saturated concrete. *Probabilistic Engineering Mechanics*, 17 (3), 305–315.

Kong, J.S. and Frangopol, D.M., 2003. Life-cycle reliability-based maintenance cost optimisation of deteriorating structures with emphasis on bridges. *Journal of Structural Engineering*, 129 (6), 818–828.

Kong, J.S. and Frangopol, D.M., 2005. Sensitivity analysis in reliability-based lifetime performance prediction using simulation. *Journal of Materials in Civil Engineering*, 17 (3), 296–306.

Kutt, L.M., Piaszczyk, C.M., Chen, Y.K. and Lin, D., 1985. Evaluation of the longitudinal ultimate strength of various ship hull configurations. *SNAME Transactions*, 93, 33–53.

Kyriakidis, P.C., 2005. Sequential spatial simulation using Latin hypercube sampling. *In*: O. Leuangthong and C.V. Deutsch, eds. *Geostatistics Banff 2004: 7th international geostatistics congress, quantitative geology and geostatistics*. Dordrecht, The Netherlands: Kluwer Academic Publishers, 14 (1), 65–74.

Leemis, L.M., 1995. *Reliability, probabilistic models and statistical methods*. New Jersey: Prentice Hall.

Liu, C., Hammad, A. and Itoh, Y., 1997. Multiobjective optimisation of bridge deck rehabilitation using a genetic algorithm. *Computer-Aided Civil and Infrastructure Engineering*, 12 (6), 431–443.

Liu, D., Ghosn, M., Moses, F. and Neuenhoffer, A., 2001. Redundancy in highway bridge substructures. National Cooperative Highway Research Program. *NCHRP Report 458*. Transportation Research Board, National Academy Press, Washington DC.

Liu, M. and Frangopol, D.M., 2005a. Bridge annual maintenance prioritisation under uncertainty by multi-objective combinatorial optimisation. *Computer-Aided Civil and Infrastructure Engineering*, 20 (5), 343–353.

Liu, M. and Frangopol, D.M., 2005b. Multiobjective maintenance planning optimisation for deteriorating bridges considering condition, safety, and life–cycle cost. *Journal of Structural Engineering*, 131 (5), 833–842.

Liu, M. and Frangopol, D.M., 2005c. Time-dependent bridge network reliability: Novel approach. *Journal of Structural Engineering*, 131 (2), 329–337.

Liu, M. and Frangopol, D.M., 2006a. Optimising bridge network maintenance management under uncertainty with conflicting criteria: Life-cycle maintenance, failure, and user costs. *Journal of Structural Engineering*, 131 (11), 1835–1845.

Liu, M. and Frangopol, D.M., 2006b. Probability-based bridge network performance evaluation. *Journal of Bridge Engineering*, 11 (5), 633–641.

Liu, M., Frangopol, D.M. and Kim, S., 2009a. Bridge system performance assessment from structural health monitoring: A case study. *Journal of Structural Engineering*, 135 (6), 733–742.

Liu, M., Frangopol, D.M. and Kim, S., 2009b. Bridge safety evaluation based on monitored live load effects. *Journal of Bridge Engineering*, 14 (4), 257–269.

Lua, J. and Hess, P., 2006. First failure based reliability assessment and sensitivity analysis of a naval vessel under hogging. *Journal of Ship Research*, 50 (2), 158–170.

Mansour, A.E., 1997. Assessment of reliability of ship structures. *Ship Structure*. Committee Publications, Report No: SSC-398, Washington, DC.

Mansour, A.E. and Hovem, L., 1994. Probability based ship structural analysis. *Journal of Ship Research*, 38 (4), 329–339.

Marsh, P.S. and Frangopol, D.M., 2007. Lifetime multi-objective optimisation of cost and spacing of corrosion rate sensors embedded in a deteriorating reinforced concrete bridge deck. *Journal of Structural Engineering*, 133 (6), 777–787.

Marsh, P.S. and Frangopol, D.M., 2008. Reinforced concrete bridge deck reliability model incorporating temporal and spatial variations of probabilistic corrosion rate sensor data. *Reliability Engineering and System Safety*, 93 (3), 394–409.

Matousek, M. and Schneider, J., 1976. Untersuchungen zur Struktur des Sicherheitsproblems bei Bauwerken (in German). Institut für Baustatik und Konstruktion der ETH Zurich, Bericht No. 59, ETH Zurich.

Melchers, R.E., 2003. Probabilistic model for marine corrosion of steel for structural reliability assessment. *Journal of Structural Engineering*, 129, 1484–1493.

Messervey, T.B., 2008. *Integration of structural health monitoring into the design, assessment, and management of civil infrastructure*. Dissertation (PhD). University of Pavia, Pavia, Italy.

Messervey, T. and Frangopol, D.M., 2006. A framework to incorporate structural health monitoring into reliability-based life-cycle bridge management models. *In*: G. Deodatis and P.D. Spanos, eds. *Proceedings of the 5th computational stochastic mechanics conference*, 21–23 June, Rhodes, Greece. *Computational stochastic mechanics*. Rotterdam: Millpress, 2007, 463–469.

Messervey, T.B. and Frangopol, D.M., 2008. Integration of health monitoring and asset management in a life-cycle perspective. *In*: H.-M. Koh and D.M. Frangopol, eds. *Bridge maintenance, safety, management, health monitoring and informatics*. London: CRC Press/Balkema, Taylor and Francis, 391. Full paper on CD–ROM. Taylor and Francis Group plc, London, 1836–1844.

Messervey, T.B., Frangopol, D.M. and Casciati, S., 2010. Application of statistics of extremes to the reliability assessment of monitored highway bridges. *Structure and Infrastructure Engineering*, DOI: 10.1080/157324 71003588619.

Miyamoto, A., Kawamura, K. and Nakamura, H., 2000. Bridge management system and maintenance optimisation for existing bridges. *Computer-Aided Civil and Infrastructure Engineering*, 15 (1), 45–55.

Moan, T., 2005. Reliability-based management of inspection, maintenance and repair of offshore structures. *Structure and Infrastructure Engineering*, 1 (1), 33–62.

Morcous, G. and Lounis, Z., 2005. Maintenance optimisation of infrastructure networks using genetic algorithms. *Automation in Construction*, 14, 129–142.

Mori, Y. and Ellingwood, B.R., 1993. Reliability-based service-life assessment of ageing concrete structures. *Journal of Structural Engineering*, 119 (5), 1600–1621.

Mori, Y. and Ellingwood, B.R., 1994a. Maintaining reliability of concrete structures. I: Role of inspection/repair. *Journal of Structural Engineering*, 120, 824–845.

Mori, Y. and Ellingwood, B.R., 1994b. Maintaining reliability of concrete structures II: Optimum inspection / repair strategy. *Journal of Structural Engineering*, 120, 846–862.

Moses, F., 1982. System reliability developments in structural engineering. *Structural Safety*, 1 (1), 3–13.

Moses, F., 1989. New directions and research needs in system reliability research. *In*: D.M. Frangopol, ed. *New directions in structural system reliability*. Boulder, CO: University of Colorado Press, 6–16.

Moses, F., Khedekar, N. and Ghosn, M., 1993. System reliability of redundant structures using response functions. *In*: *Proceedings of the international conference on structural safety and reliability (ICOSSAR'93)*. Innsbruck, 9–13 August 1994, Rotterdam: A.A. Balkema, 1301–1307.

Neves, L.C., Frangopol, D.M. and Cruz, P.J.S., 2004. Cost of life extension of deteriorating structures under reliability-based maintenance. *Computers and Structures*, 82 (13/14), 1077–1089.

Neves, L.C. and Frangopol, D.M., 2005. Condition, safety and cost profiles for deteriorating structures with emphasis on bridges. *Reliability Engineering and System Safety*, 89 (2), 185–198.

Neves, L.C., Frangopol, D.M. and Cruz, P.J., 2006a. Probabilistic lifetime-oriented multi-objective optimisation of bridge maintenance: Single maintenance type. *Journal of Structural Engineering*, 132 (6), 991–1005.

Neves, L.C., Frangopol, D.M. and Petcherdchoo, A., 2006b. Probabilistic lifetime-oriented multi-objective optimisation of bridge maintenance: Combination of maintenance types. *Journal of Structural Engineering*, 132 (11), 1821–1834.

O'Connor, A. and O'Brien, E.J., 2005. Traffic load modelling and factors influencing the accuracy of predicted extremes. *Canadian Journal of Civil Engineering*, 32, 270–278.

Okasha, N.M. and Frangopol, D.M., 2009. Lifetime-oriented multi-objective optimisation of structural maintenance considering system reliability, redundancy and life-cycle cost using GA. *Structural Safety*, 31 (6), 460–474.

Okasha, N.M. and Frangopol, D.M., 2010a. Time-variant redundancy of structural systems. *Structure and Infrastructure Engineering*, 6 (1–2), 279–301.

Okasha, N.M. and Frangopol, D.M., 2010b. Efficient method based on optimisation and simulation for the probabilistic strength computation of the ship hull. *Journal of Ship Research*, 54 (4)

Okasha, N.M. and Frangopol, D.M., 2010c. Novel approach for multi-criteria optimization of life-cycle preventive and essential maintenance of deteriorating structures. *Journal of Structural Engineering* (in press).

Okasha, N.M. and Frangopol, D.M., 2010d. Redundancy of structural systems with and without maintenance: An approach based on lifetime functions. *Reliability Engineering and System Safety*, 95 (5), 520–533.

Olsson, A., Sandberg, G. and Dahlblom, O., 2003. On Latin hypercube sampling for structural reliability analysis. *Structural Safety*, 25 (1), 47–68.

Paik, J.K. and Frieze, P.A., 2001. Ship structural safety and reliability. *Progress in Structural Engineering and Materials*, 3 (2), 198–210.

Paik, J.K., Thayamballi, A.K., Kim, S.K. and Yang, S.H., 1998. Ship hull ultimate strength reliability considering corrosion. *Journal of Ship Research*, 42 (2), 154–165.

Paliou, C., Shinozuka, M. and Chen, Y.-N., 1990. Reliability and redundancy of offshore structures. *Journal of Engineering Mechanics*, 116 (2), 359–378.

Petcherdchoo, A., Neves, L.C. and Frangopol, D.M., 2008. Optimising lifetime condition and reliability of deteriorating structures with emphasis on bridges. *Journal of Structural Engineering*, 134 (4), 544–552.

Rashedi, M.R. and Moses, F., 1983. Studies on reliability of structural systems. Report R83–3. Department of Civil Engineering, Case Western Reserve University, Cleveland, OH.

Rashedi, M.R. and Moses, F., 1988. Identification of failure modes in system reliability. *Journal of Structural Engineering*, 114 (2), 292–313.

Schuëller, G.I., 1998. Structural reliability–Recent advances. *In*: N. Shiraishi, M. Shinozuka, and Y.K. Wen, eds. *Proceedings of the 7th international conference on structural safety and reliability (ICOSSAR'97)* November 1998, Kyoto, Japan. The Netherlands: A.A. Balkema Publications, 3–35.

Shinozuka, M., 2008. Chapter 12. Resilience and sustainability of infrastructure systems. *In*: S.-S. Chen and A.H.-S. Ang, eds. *Frontier technologies for infrastructures engineering*. Vol. 4. D.M. Frangopol, Book Series Editor. Boca Raton, London, New York, Leiden: CRC Press/Balkema, 245–270.

Smith, C., 1977. Influence of local compressive failure on ultimate longitudinal strength of a ship's hull. *Proceedings of the PRADS: International symposium on practical design in shipbuilding*, Tokyo, Japan. 153–158.

Stewart, M.G. and Melchers, R.E., 1997. *Probabilistic risk assessment of engineering systems*. London: Chapman and Hall.

Stewart, M.G. and Rosowsky, D.V., 1998. Time-dependent reliability of deteriorating reinforced concrete bridge decks. *Structural Safety*, 20, 91–109.

Strauss, A., Bergmeister, K., Hoffmann, S., Pukl, R. and Novák, D., 2008a. Advanced life-cycle analysis of existing concrete bridges. *Journal of Materials in Civil Engineering*, 20 (1), 9–19.

Strauss, A., Frangopol, D.M. and Kim, S., 2008b. *Statistical, probabilistic and decision analysis aspects related to the efficient use of structural monitoring systems*. Berlin: Beton- und Stahlbetonbau, Ernst and Sohn, 103. Special Edition, 23–28.

Susoy, M., Catbas, N. and Frangopol, D.M., 2007. SHM development using system reliability. *In*: J. Kanda, T. Takada, and H. Furuta, eds. *Proceedings of the 10th international conference on applications of statistics and probability in civil engineering (ICASP 10)*, 31 July–3 August, Tokyo, Japan. Applications of statistics and probability in civil engineering. London: Taylor and Francis, 79–80 and CD-ROM.

Tang, K. and Melchers, R.E., 1988. Incremental formulation for structural reliability analysis. *Civil Engineering Systems*, 5, 153–158.

Thoft-Christensen, P. and Murotsu, Y., 1986. *Application of structural systems reliability theory*. Berlin: Springer.

Tsompanakis, Y., ed. 2010. Vulnerability assessment of structures and infrastructures, *Structure and Infrastructure Engineering*, 6 (1–2), Special Issue, 1–301.

Val, V.D., Stewart, M.G. and Melchers, R.E., 1998. Effect of reinforcement corrosion on reliability of highway bridges. *Engineering Structures*, 20, 1010–1019.

van Noortwijk, J.M. and Frangopol, D.M., 2004. Two probabilistic life–cycle maintenance models for deteriorating civil infrastructures. *Probabilistic Engineering Mechanics*, 19 (4), 345–359.

Vu, K. and Stewart, M.G., 2000. Structural reliability of concrete bridges including improved chloride–induced corrosion models. *Structural Safety*, 22, 313–333.

Watanabe, E., Frangopol, D.M., and Utsunomiya, T., eds., 2004. *Bridge maintenance, safety, management and cost*. Lisse, The Netherlands; A.A. Balkema, Swets and Zeitlinger BV. Book and CD-ROM.

Wen, Y.K. and Kang, Y.J., 2001. Minimum building life-cycle cost design criteria. I. Methodology and II. Applications. *Journal of Structural Engineering*, 127 (3), 330–346.

Wirsching, P.H., Ferensic, J. and Thayamballi, A.K., 1997. Reliability with respect to ultimate strength of a corroded ship hull. *Marine Structures*, 10 (7), 501–518.

Yang, S.-I., Frangopol, D.M. and Neves, L.C., 2004. Service life prediction of structural systems using lifetime functions with emphasis on bridges. *Reliability Engineering and System Safety*, 86 (1), 39–51.

Yang, S.-I., Frangopol, D.M. and Neves, L.C., 2006a. Optimum maintenance strategy for deteriorating structures based on lifetime functions. *Engineering Structures*, 28 (2), 196–206.

Yang, S.-I., Frangopol, D.M., Kawakami, Y. and Neves, L.C., 2006b. The use of lifetime functions in the optimisation of interventions on existing bridges considering maintenance and failure costs. *Reliability Engineering and System Safety*, 91 (6), 698–705.

Zokaie, T., Imbsen, R.A. and Osterkamp, T.A., 1991. Distribution of wheel loads on highway bridges. *Transportation Research Record 1290*, 119–126, Washington, DC.

Bridge network performance, maintenance and optimisation under uncertainty: accomplishments and challenges

Dan M. Frangopol and Paolo Bocchini

This article presents a critical review of the state-of-the-art in the field of bridge network performance analysis, reliability assessment, maintenance management and optimisation. Previous accomplishments and results are summarised while stressing the aspects of the analysis of a transportation network, which are more challenging for a structural engineer. For instance, the bridge network is described as a spatially distributed system at a scale that is much larger than any individual structure. This requires a different perspective in the modelling of natural extreme events and involves models for the interaction between the individual components (i.e. bridges) and the overall network. Then, the time domain is investigated, and the problem of structural deterioration and its effects on the network performance are addressed. The most important differences between a transportation network and other lifelines are considered next, and techniques for the transportation network analysis and performance assessment are summarised. Finally, a set of selected problems and applications is considered and additional challenges are identified and suggested for future developments.

1. Introduction

According to the American Society of Civil Engineers (ASCE 2009), 'More than 26%, or one in four, of the nation's bridges are either structurally deficient or functionally obsolete', and out of a total investment needs for bridges and roads in the next 5 years of $930 billion, the shortfall will be $549.5 billion, that is almost 60%. It is widely acknowledged that the social and economic growth of any nation strongly depends on the extension, the efficiency and the reliability of its infrastructure systems. Among these, the transportation network plays a role of utmost importance for everyday life, as well as for the emergency response and recovery activities after the occurrence of an extreme event. Therefore, it is required that the lifelines are designed to maintain their serviceability even in the case of so-called 'low-probability high-consequence' events, such as hurricanes, floods and earthquakes, to ensure prompt interventions and a fast restoration of the normal activities (Duwadi 2010).

In every network, there are specific components that are crucial. For instance, this article places emphasis on transportation networks, and for this kind of infrastructure, the crucial components are often represented by bridges. In fact, even in transportation networks for which most of the unserviceable joints can be easily bypassed, bridges are usually bottlenecks. Moreover, bridges are the most fragile component in the case of major natural hazards. In fact, a bridge network is defined as a transportation network in which bridges are the only components that can suffer structural damage. Therefore, it is very important that bridges, as crucial components, preserve their serviceability even in the case of extreme events and all along their lives. Otherwise, in the case of collapse or major damage, the entire network system would suffer severe traffic disruption, retard the post-event activities and negatively impact both the short- and long-term socio-economic development of the region.

In the last decades, considerable research has been conducted towards a better understanding of the above-mentioned aspects and, consequently, towards a better structural design practice that also accounts for these issues. However, a truly holistic approach is still missing. For an overall analysis of networks, different perspectives have to be merged in a unified framework. The aspects that should be accounted for include the following: (a) the interaction of the individual components (i.e. bridges) in a spatially distributed system; (b) the variability in time of the characteristics of the infrastructure; (c) the multi-hazard nature of the analysis; (d) the network effects and (e) the economic constraints all along the life-cycle of the infrastructure.

The essence of the network (as opposed to individual components or groups of independent components) consists in the correlations, interactions and mutual dependencies of its components. When dealing with a transportation network, the various components are spread over very large areas. Therefore, the infrastructure manager is required to model the relationship among the various components at a scale (the spatially distributed system) that usually overwhelms the dimensions of a single structure.

The variability in the time domain must also be considered. Although the effects of long-term structural deterioration on the performance of structures and infrastructures have already been recognised (Chang and Shinozuka 1996, PONTIS 2005, Pandey *et al.* 2009), the implementation of methodologies taking into account time-dependent effects is less common. In most existing procedures and guidelines for design and performance analysis (e.g. seismic) of civil structures, the structural capacities are assumed to be time invariant when a post-event reactive retrofit is carried out to partially or fully repair damaged structures. Obviously, this assumption does not reflect the realistic time-dependent structural capacity evolution due to progressive deterioration. Therefore, these procedures predict incorrect capacities and damage levels that are likely to be on the unconservative side. Safety and economic consequences due to these failures, especially when combined at the network level, are in general much more costly than the correction of structural deficiency due to deterioration. Therefore, it is advantageous and economical to proactively apply maintenance interventions for enhancing the structural performance. In order to efficiently plan the activities, an accurate predictive tool that also considers the network performance and the relative importance of the individual components should be adopted.

When dealing with low-probability high-consequence events, it is always desirable to choose a multi-hazard approach. In fact, the same structures and infrastructure systems can be subject to several types of extreme events. Even if these events can induce different mechanical stresses, their effects from many points of view are very similar. For instance, they can lead to the unavailability of the infrastructure and determine the same socio-economic impact, they can be included in the same probabilistic analysis framework, and their effects can be mitigated with similar proactive interventions. Moreover, they are often interdependent, for example fires that follow earthquakes, tsunamis that follow earthquakes and floods

that follow hurricanes. Therefore, it is necessary to develop a general approach that accounts for all of them.

As mentioned previously, there are critical components in the networks, but a study that aims at providing a useful decision tool for structural designers, urban planners and risk managers should not disregard the network system and the network effects. In fact, the condition of every component influences the flows on the entire network and, in turn, modifies the demands applied to the other components. For instance, for bridges that have already suffered an extreme event, this can lead to congestion-based disruptions. A recent study (Hernández-Fajardo and Dueñas-Osorio 2009) estimated that the fragility of components of coupled lifeline systems can be almost doubled if the network dependencies are considered in the computation. Moreover, this is another possible source of delays in the post-event interventions.

Finally, every decision has to account for a limited budget. For instance, according to ASCE (2009), in the next 5 years, the investment shortfall for civil infrastructure will be $1.176 trillion. In order to improve the performance of the network, the decision makers can act in different ways. For instance, they can improve the reliability and the robustness of critical components (Ghosn *et al.* 2010); they can change the topology of the network, making it more redundant and they can monitor more efficiently the structural health of some components to be able to better plan the proactive maintenance. All these tasks should be performed, but when the budget is limited, it is necessary to choose only the more effective ones and prioritise the interventions (Frangopol and Bocchini 2011a). These decisions cannot be taken arbitrarily; they have to be supported by a numerical tool that solves this multi-criteria optimisation problem under uncertainty with limited financial resources. In most cases, such a tool provides a series of Pareto optimal solutions and the final decision is taken considering also socio-economic aspects and previous experience.

In short, scientific research has provided significant advances in many areas associated with infrastructural networks and their constitutive elements, including structural health monitoring (Kim and Frangopol 2010), life-cycle analysis (Frangopol 2011), maintenance (Neves *et al.* 2006a,b), reliability analysis (Liu and Frangopol 2005c), vulnerability assessment (Murray *et al.* 2008, Matisziw and Murray 2009) and optimisation (Bocchini and Frangopol 2011b). In the following sections, some of these contributions are summarised and compared, emphasising accomplishments and challenges. Moreover, to the authors' knowledge, there are no investigations on the possible interactions of the results that all the above-mentioned analyses provide. Therefore, this article is also an occasion to promote an integrated framework that considers all these aspects in a fully probabilistic perspective. This would be a tool that can dramatically improve the allocation of the available resources and it will result in increased serviceability, safety and security of infrastructure networks, better response and lower economic impact for every possible hazard.

2. Spatially distributed systems

The analysis of spatially distributed systems brings the focus of the structural engineer to a new scale. In fact, 'large-scale buildings', such as skyscrapers and sport arenas, have dimensions in the order of hundreds of metres (Fairweather et al. 2004); similarly, 'long-span bridges' are up to a few kilometres long structures, even including the abutments (Chen and Duan 2000). On the contrary, when dealing with an infrastructure network or a lifeline, the engineer is handling an entity spread over several kilometres, sometimes even hundreds of kilometres. Some of the analyses on the individual network components (e.g. individual bridges) are not affected by this change in scale, but they only require the repeated use of traditional methodologies. However, many other analyses have to be performed with a completely different perspective.

A first example of this issue can be the effect of extreme events, such as earthquakes. Several studies have focused on the spatial variation of seismic ground motion and its effects on the demand for large structures (Zhang and Deodatis 1996, Shinozuka et al. 1999, Jankowski and Wilde 2000, Zerva and Zervas 2002, Kim and Feng 2003, Zerva and Beck 2003, Jankowski 2006, Zerva 2009), while others have studied the spatial variation of different external loads (e.g. wind velocity: Davenport 1968, Simiu and Scanlan 1996, Gioffré et al. 2000; rainfall: Korving et al. 2009). However, when dealing with a whole network, even these approaches are not sufficient. When an earthquake occurs, the bridges of a network experience different ground motions (Chang et al. 2000). Their structural responses are different because they have different structural characteristics and they are subject to different external solicitations. However, these bridges cannot be analysed only individually, because the traffic disruption caused by simultaneous damage on several bridges is not just the superposition of the detours and delays caused by the unserviceability of each bridge. Therefore, problems such as the risk assessment, the social cost caused by traffic disruption, the optimal economic resources allocation, both for preventive (i.e. pre-event) and post-event intervention require to model the structural demand and the structural response at the network scale.

At this large scale, the interactions between the individual components and the correlation of their in/out of service states should also be considered. Most of the probabilistic network analysis techniques proposed in the literature assume total independence between the serviceability of the various bridges (Augusti et al. 1998, Akgül and Frangopol 2004c, Liu and Frangopol 2005c, 2006a).This assumption is certainly not realistic, since the bridges of a network are very likely to have similar traffic loads, to be designed with the same codes, to have similar environmental conditions and to have experienced the same extreme events. Moreover, Bocchini and Frangopol (2010, 2011a) have shown that the degree of correlation among the bridge serviceability states, all across the network, has a strong impact on the network performance. Therefore, every analysis at the network scale should account not only for the interactions between the various components but also for the correlation of the structural conditions.

When structural engineers start to work at this new scale, one of the first problems they have to face is the need of a strong loss of details in the model. In fact, computational efficiency as well as epistemic uncertainties and insufficient data often require a drastic reduction in the level of detail of the models. This can be felt as an unacceptable loss of accuracy in the results, but this perception is usually wrong. In fact, a similar process of synthesis and abstraction is followed very often in the everyday practice. When an engineer analyses a steel connection with a finite element model, very small elements are employed, the geometry of the plates is described with utmost precision and several layers of elements are connected to discretise the thickness of each plate. However, when the same engineer models the structure to which the connection belongs, much larger elements are preferable and the joint itself might even disappear. This is universally accepted as a part of the art of approximation: deep attention should be paid to the critical components, but to capture the overall behaviour of the system, a coarser description of its functional parts is required. Similarly, when dealing with a network, the individual structures, such as entire bridges, become just components of a more complex system. Therefore, only a synthetic description of their characteristics and of their structural response should be considered.

A practical application of this process is the use of analytical models for the time-dependent reliability of the individual bridges. It is well known that the structural characteristics deteriorate over time and, therefore, the structural reliability tends to decrease along the life-cycle. Specific analyses can be performed to assess this decay in the reliability performance indicators for different types of bridges. Akgül and

Frangopol (2004a,b, 2005a,b) have proposed such analyses for steel girder bridges, prestressed concrete bridges, and reinforced concrete bridges. However, only the most important bridges of a network are usually thoroughly modelled and sometime monitored, so that their predicted reliability profiles can be considered realistic. For all the other bridges, this information is, in general, unavailable and a less-detailed description is required. In these cases, the above-mentioned life-cycle reliability models (Frangopol et al. 2001a,b, Frangopol 2003, van Noortwijk and Frangopol 2004, Kong and Frangopol 2004b) can be used. This kind of model, that includes uncertainties, can be assessed knowing some basic characteristics of the individual bridge. Therefore, the reliability profile can be assessed without the need (and the cost) of thorough studies on every bridge.

Another issue related to the analysis under poor information and loss of details in the network model arises when dealing with maintenance management. In fact, most of the studies on bridge maintenance tend to classify the interventions in Preventive Maintenance (PM), also called 'time-based' or 'proactive', and Essential Maintenance (EM), also called 'performance-based' or 'threshold-based'. PM consists of those interventions that are scheduled at predefined time intervals in order to always keep the bridge in a good service level. This kind of intervention is also characterised by the lowest impact on the bridge safety and the lowest cost. EM, instead, is applied when an indicator of the bridge performance crosses a pre-defined (acceptable) threshold. The most used indicator is certainly the bridge reliability index, even if several other indicators can be considered, such as redundancy and availability (Okasha and Frangopol 2010). In order to analytically define the EM application, a bridge limit state has to be chosen, for instance it can be the excessive deformation of the main girders or the collapse. Event E_1 is defined as the bridge that reaches the investigated limit state. Given a specific event E_1, it is possible to define the time to failure associated with that limit state TF_{E_1} as the time between a reference instant $t = 0$ and the moment at which E_1 occurs (i.e. the limit state is reached). Then, by definition, the reliability at time t is the probability that E_1 does not occur in the interval $[0, t]$:

$$\mathrm{REL}_{E_1}(t) = P(t < \mathrm{TF}_{E_1}) \qquad (1)$$

where $P(\cdot)$ denotes the probability of occurrence of the event in brackets. Under the usual Gaussian assumption, the reliability index β_{E_1} is computed as:

$$\beta_{E_1}(t) = \Phi^{-1}[\mathrm{REL}_{E_1}(t)] \qquad (2)$$

where Φ^{-1} is the inverse standard Gaussian cumulative distribution function. A lower threshold $\bar{\beta}$ for the reliability index allows the definition of a second event E_2 as $\beta_{E_1}(t) \le \bar{\beta}$. EM, by definition, is applied whenever event E_2 occurs. However, when an entire transportation network is considered, the previously mentioned issue of incomplete information arises, and, for most of the bridges, thorough studies on the time-dependent reliability profile β_{E_1} could be unavailable. Therefore, if the moment in which β_{E_1} down crosses the threshold $\bar{\beta}$ is unknown there will certainly be no EM interventions applied at that unknown instant. Therefore, in these cases, it appears much more realistic to consider a third type of maintenance, called Required Maintenance (RM). This kind of intervention is applied only when the structure manifests an imminent state of distress or when the distress has occurred (Bocchini and Frangopol 2011b). Using the classical distinction between 'preventive' and 'corrective' interventions (Kallen and van Noortwijk 2006), both PM and EM fall in the first category, while RM represents corrective maintenance, also including the limit case in which the failure has not occurred yet, but it is so imminent that the restoration is exactly as if the failure had occurred. Therefore, using again the previously defined events, for a specific limit state, RM is applied when event E_1 is actually imminent, or when it has just occurred. With this definition, restoration can be seen as a special case of RM, where event E_1 is assumed to be the bridge collapse. However, the definition of RM is more general than restoration, since it includes minor interventions if E_1 is a different limit state (e.g. excessive corrosion, excessive deformation and serviceability limit). Since the occurrence of event E_2 (that triggers EM) is based on the definition of event E_1 (that triggers RM), it is evident that these two types of maintenance are strongly interconnected. However, they are not the same. Actually, on one hand, E_2 can occur even if E_1 never happened. In fact, the threshold $\bar{\beta}$ is usually assumed high, therefore when $\beta_{E_1}(t)$ down crosses it (event E_2 occurs), the reliability $\mathrm{REL}_{F_1}(t) = \Phi[\beta_{E_1}(t)]$ is still very close to one (i.e. event E_1 is still very unlikely to occur). On the other hand, E_1 can happen at any time, because the bridge reliability is never equal to unity, which means that there is always a chance of failure. When the focus of a study is on an individual bridge, it makes perfect sense to consider EM. In fact, if a single bridge is studied, $\beta_{E_1}(t)$ profiles will likely be available, and not only the distress caused by the occurrence of event E_1 should be avoided but even the probability of having a low reliability can be avoided. However, in the case of a bridge network, when the detailed information on many individual bridge performance indexes is unavailable, it is not possible to know when the $\beta_{E_i}(t)$ of individual bridges cross an acceptable threshold. Therefore, it is not realistic to apply EM at that instant. If EM is included in a model and then actually not applied, the final result would be an unconservative maintenance schedule.

While the large scale of bridge networks yields the above-mentioned issues, it can also be seen as an opportunity to develop very attractive unified models. For instance, Figure 1 shows a schematic representation of a possible approach in which the hazard induced by several natural extreme events is modelled as a random field and the time-space superposition is used as basis for a multi-hazard approach.

In conclusion, the network analyses require a solid background in the most traditional techniques for solution of structural problems, but also a novel set of tools and expertise that provide a better overall description of the system and a deeper insight on the various interactions and dynamics that happen at the network scale.

3. From individual structures to the network

Every time an analysis involves different scales, it is necessary to have models for the interactions between the various scales. In the case of bridge networks, the most important of these models is the one that links the structural condition and the damage level of an individual bridge (structure scale) to its traffic flow capacity (relevant at the network scale).

Very different models have been introduced in the literature. For instance, Shinozuka *et al.* (2006) and Zhou *et al.* (2010) have proposed the use of a 'bottleneck assumption', according to which the practical flow capacity of a highway segment depends only on the residual practical flow capacity of its most damaged bridge. Given the damage state of the bridge in the worse condition, they provide the percentage flow capacity of the highway segment. Three possible

sets of residual capacity are presented to perform a sensitivity analysis. For instance, the median set assumes that the residual flow capacity is 100% in the cases of no damage or minor damage; 50% in the case of moderate damage; and 25% in the case of major damage or collapse (considering local detour routes). As expected, the results of the study (that consists in an exhaustive benefit/cost analysis of seismic retrofit) show a strong dependence on the chosen set of residual capacity values. This further enlightens the importance of an accurate and a well-calibrated model for the interaction between the bridge scale and the network scale.

Song and his co-workers have developed the 'matrix-based system reliability method' and have proficiently applied it to the solution of several problems involving different networks and lifelines (Kang *et al.* 2008, Kim *et al.* 2009b, Lee *et al.* 2009, Song and Kang 2009). These authors have also used the bottleneck assumption. In the studies in which only two bridge states are considered (failure/non-failure), the collapse of a bridge implies the unserviceability of the entire highway segment to which it belongs. In the studies in which multiple damage states are considered, arbitrary percentage of the residual flow capacity are assumed for every damage level (i.e. 100% for no damage, 70% for moderate damage, 30% for heavy damage and 0% for collapse).

A similar approach, based again on the bottleneck assumption has been presented by Bocchini and Frangopol (2011a). In this case, the individual bridge damage state was described with a real-valued variable in the interval [0, 4], where 0 and 4 represent the 'no damage' and 'collapse' states, respectively. Therefore, the model for the interaction between the two scales had to be described by a continuous function:

$$f_{ij}^c = f_{ij}^{c0} \cdot \exp[-\alpha^{\mathrm{RC}} \cdot (\max l)^{\beta^{\mathrm{RC}}}] \qquad (3)$$

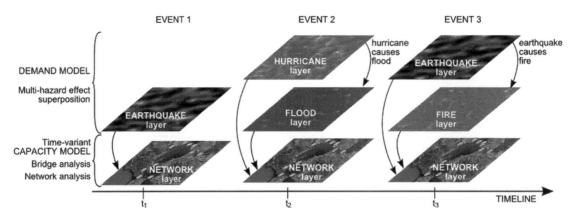

Figure 1.   Space and time interconnection among the various extreme events.

where f_{ij}^c is the residual practical flow capacity of highway segment ij; f_{ij}^{c0} is the original practical flow capacity of highway segment ij; max l is the maximum damage level among the bridges on highway segment ij; α^{RC} and β^{RC} are parameters, assumed equal to 0.03 and 4, respectively. Also in this case, the values assumed for the parameters affect the final results, so they need to be thoroughly calibrated.

In a series of studies on resilience of networks (see Section 5), Bruneau and Reinhorn (2007) have proposed another continuous relationship between the damage state of a network component and its serviceability. Even though they have studied different kinds of networks, and the relationship between damage level and serviceability (better called 'functionality' in this case) is implicitly included in a sophisticated probabilistic framework, the continuous functions that describe the recovery in 'quality of infrastructure' during the repair activities suggest an underlying assumption conceptually similar to the one in Equation (3).

For rural highway systems, with very limited detour routes, the bottleneck assumption appears reasonable. However, when more detours are available, such as for urban highway systems, it is too conservative. In fact, if a 10-km highway segment with five bridges and secondary roads is considered as an example, the model has to yield very different results when considering the collapse of an individual bridge or the collapse of all five bridges. The bottleneck assumption treats these two cases in the same way. On the contrary, if just one bridge has collapsed, the vehicles can bypass that single bridge and use the rest of the highway segment with only a limited delay. Instead, if all the bridges are collapsed, the entire highway segment is practically unserviceable. These considerations led Bocchini and Frangopol (2011c) to propose a new model for the bridge scale – network scale interaction. Only two possible states for every bridge are considered (in and out of service) and the residual practical capacity of a highway segment is given by:

$$f_{ij}^c = f_{ij}^{c0} \cdot \max\left[0; (1 - \mathrm{NUB}_{ij} \cdot \alpha^{UB})\right] \qquad (4)$$

where NUB_{ij} is the number of unserviceable bridges on highway segment ij and α^{UB} is a parameter that was proposed to be set equal to 0.1. This means that every unserviceable bridge causes a reduction in the practical traffic flow capacity of 10%. Also in this case, the value of the parameter α^{UB} should be carefully calibrated for the specific highway system, depending on the length of the segments and the possible alternative routes. For rural highways, for instance, it

should be increased to 0.2, or even up to 0.5. It should be noted that when only two bridge states are considered (i.e. in/out of service), this model can be seen as a generalisation of the bottleneck assumption. In fact, the bottleneck assumption can be obtained as a special case, by setting $\alpha^{UB} = 1$. An even more sophisticate model has been proposed by Bocchini and Frangopol (2011d). In this case, the delay caused by each bridge out of service is explicitly computed as a function of the associated detour length, the flow capacity of secondary roads, the bridge damage condition and the presence of ongoing restoration interventions on the bridge. Data required to calibrate such model for bridges in the United States are provided by the National Bridge Inventory (FHWA 2009) and by the database distributed with the software HAZUS-MH MR4 (DHS 2009).

All the models that have been presented are characterised by advantages and drawbacks. Each of them is suitable for certain networks and inappropriate for others. Depending on the specific network and the type of analysis that is performed, the choice of the bridge scale–network scale interaction model and the careful calibration of its parameters are critical to obtain accurate results. These choices can benefit very much from empirical studies, data collection and traffic flow monitoring. The acquisition of this kind of data can be very challenging, especially after an extreme event, but novel techniques, such as the use of mobile phones geolocation (Herrera *et al.* 2010), appear very promising.

4. Time-dependent problems

In Section 2, some of the issues related to the large scale of bridge networks in the space domain have been discussed. Similarly, an aging infrastructure poses several problems due to the variability in time of its characteristics, during its usually long life-cycle. Therefore, a bridge network should always be analysed as a spatially distributed system with time-dependent properties.

This topic has already been widely investigated for individual bridges. For instance, Mori and Ellingwood (1993) pioneered the study of service-life assessment of deteriorating concrete structures. Czarnecki and Nowak (2008) developed life-cycle profiles of bridge reliability based on the model of the deterioration process and on a probabilistic load description. Frangopol *et al.* (1997) and later Estes and Frangopol (2001) have proposed a life-cycle analysis of deteriorating structures based on system reliability and aimed at assessing the risk of failure (for various limit states), ranking the components depending on their importance and minimising the total maintenance and inspection cost. Frangopol *et al.* (2004) presented a comprehensive state-of-the-art review of probabilistic models for the time-dependent degradation of bridge performance. Akgül and Frangopol (2004a,b, 2005a,b) have analysed the time-dependent reliability of several structural types of bridges. Kong and Frangopol (2003, 2004a, 2005) worked on the management optimisation of aging structures. Orcesi and Frangopol (2010, 2011) and Okasha and Frangopol (2011) studied the integration of monitoring into life-cycle performance optimisation techniques. Frangopol and Messervey (2007a,b, 2008, 2009) and Messervey *et al.* (2011) focused on the use of structural health monitoring to reduce the uncertainties, update the data, and thus optimise the management of bridges. Indeed, the Structural Engineering Institute of the American Society of Civil Engineers recently founded the Technical Council on Life-Cycle Performance, Safety, Reliability and Risk of Structural Systems to study all these topics (Frangopol and Ellingwood 2010).

Fewer studies have addressed the effects of the individual component deterioration or damage induced by extreme events on the network reliability. Akgül and Frangopol (2003, 2004c,d) have studied the effects of the time-dependent reliability of individual bridges on the network connectivity and the relative importance of bridges. Shinozuka *et al.* (2006) and Zhou *et al.* (2010) analysed the effectiveness of retrofit interventions with a cost-benefit analysis. Lee *et al.* (2009) investigated the issue of the network flow capacity estimation considering both extreme events and deterioration. Kim *et al.* (2009a) proposed optimal maintenance management techniques for railroad bridge networks. Liu and Frangopol (2005c) and Bocchini and Frangopol (2011c) have proposed techniques for the computation of the time-dependent network reliability (in terms of the ability to maintain a good level of service) as a function of the time-dependent reliability of the individual bridges.

Further studies in this field are certainly desirable, since the first investigations have proved that the impact of the degradation process on the network reliability can be very significant.

5. Network analysis and performance indicators

Among all lifelines, transportation networks have a unique characteristic: all the nodes can be source and destination of flows. This property, for instance, does not apply to power/water distribution systems, where the flow origins (power plants/reservoirs) and the destinations (final users) are fixed. For this reason, transportation networks require numerical analysis tools that are totally different from those used for other lifelines. Moreover, these tools cannot be based only on physical models, but have to take into account also sociological aspects (e.g. drivers tend to adapt their routes and sometimes even their destinations depending on the traffic conditions).

A consequence of the fact that all the nodes can be source and destination is the impossibility (in general) to use the traditional series-parallel system reliability models. Figure 2 shows a simple bridge network, where the numbers in square boxes represent four cities, the lines represent highway segments and the letters represent two bridges. For a driver that goes from 1 to 4, the two bridges A and B are in parallel. However, for a driver that goes from 2 to 3, the two bridges are in series configuration. This trivial example shows that the reliability of a transportation network cannot be computed as the reliability of a system built by components in series-parallel configuration, whose individual reliabilities are known.

In most of the studies on the structural reliability of bridge networks, this problem is not addressed. On the contrary, several approaches have selected a single node as origin and another one as destination of the traffic flows or, at least, they have fixed a very limited number of origins and destinations (Liu and Frangopol 2005c, 2006a, Kang *et al.* 2008, Golroo *et al.* 2010). These approaches are computationally very efficient and accurate. They can give reliable information on the important ranking of the bridges for a few specific routes and useful indications for the maintenance planning, if the analysed routes are far more important than any other origin–destination pair in the network. However, they cannot describe the real dynamics of the traffic flows and, therefore, are not suitable for a truly complete model of the bridge network.

More appropriate analysis techniques have been developed by transportation engineers. One of the most comprehensive, advanced and popular methodologies is the one presented by Evans (1976). This methodology has the advantage to be able to solve the combined traffic assignment and distribution problem without the need of an entire enumeration of the possible paths, which is computationally impractical for large networks (Rubin 1974). The transportation network has to be described according to graph theory

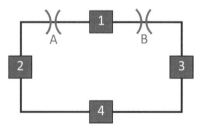

Figure 2. Simple bridge network layout.

(Gibbons 1985), while the travels of the network users are distributed among the various cities by means of a 'gravitational model' (Levinson and Kumar 1994). This computational procedure imposes the 'user equilibrium' (Frank and Wolfe 1956, Lee and Machemehl 2005) by satisfying the following two principles (Wardrop 1952): (i) travel times on all routes actually used are equal, and less than those which would be experienced by a single vehicle on any unused route and (ii) the total network travel time is minimum. Moreover, this technique is particularly suitable for the analysis of bridge networks that also take into account extreme events. In fact, it does not assume that the travel origin and destination are fixed a priori, instead they are adaptive to the conditions of the network (i.e. the in/out of service state of the bridges and/or their damage level). Therefore, combining this network analysis technique with one of the methodologies for the components–network interaction presented in Section 3, it is possible to assess the network performance after the occurrence of extreme events, bridge failures and all along the life-cycle (Bocchini and Frangopol 2010, 2011a,c).

The most common transportation network performance indicator is the total travel time (TTT), defined as the time spent to reach the destination by all the users that depart in a fixed time window, such as one hour (Bocchini and Frangopol 2010):

$$TTT_1 = \sum_{i \in I} \sum_{j \in J} \int_0^{f_{ij}} \tau_{ij}(f) df \qquad (5)$$

or sometimes computed as (Scott et al. 2006):

$$TTT_2 = \sum_{i \in I} \sum_{j \in J} f_{ij} \tau_{ij} \qquad (6)$$

In both Equations (5) and (6), i and j are nodes of the network; I is the entire set of network nodes; J is the subset of nodes that can be reached from node i using a single highway segment ij; f_{ij} is the traffic flow on segment ij as computed by the network analysis and τ_{ij} is the time required to cover segment ij, as computed by the network analysis. Similarly, the total travel distance (TTD) can be computed as:

$$TTD = \sum_{i \in I} \sum_{j \in J} f_{ij} \lambda_{ij} \qquad (7)$$

where λ_{ij} is the length of segment ij. Since the time required to cover a segment and its length are linked by

a strongly non-linear function of the traffic flow, TTT and TTD are not proportional to each other. Other performance indicators, mostly based on the total travel time, have been reviewed by Lomax et al. (2003).

Another important indicator of the network performance is the measure of the possibility to reach every node, from every other node. Several analytical measures of this ability have been proposed in the literature. For instance, Liu and Frangopol (2006a) computed 'connectivity' as the probability to be able to reach a specific destination from a fixed origin. Ng and Efstathiou (2006) defined the 'network disconnectedness' as:

$$NetDis = \frac{\varepsilon}{l_{max}} \qquad (8)$$

where ε is the number of unreachable pairs of nodes and l_{max} is the maximum possible number of links. Bocchini and Frangopol (2011a) have introduced the 'Fully Connected Ratio', computed through Monte Carlo Simulation as:

$$FCR = \frac{\text{Number of samples where all the nodes are reachable}}{\text{Total number of samples}} \times 100 \qquad (9)$$

Scott et al. (2006) presented a critical review of the literature about the use of the ratio

$$\frac{V}{C} = \frac{\text{traffic volume on a high way segment}}{\text{high way segment capacity}} \qquad (10)$$

as an index of the important ranking for highway network segments.

In the same study, the network robustness index (NRI) is proposed as a better indicator, since it takes into account the benefits for the overall network due to the improvement of an individual segment capacity:

$$NRI_a = \sum_a f_a \tau_a \delta_a \quad TTT_2 \qquad (11)$$

where index a runs over the highway segments; f_a is the traffic flow on segment a as computed by the network analysis; τ_a is the time required to cover segment a, as computed by the network analysis; δ_a is equal to 1 if segment a is not the one considered (removed) and 0 otherwise; and TTT_2 is given by Equation (6), with all the bridges intact. NRI_a is computed for every highway segment and it quantifies the performance loss due to the removal of a highway segment (caused for instance, by the failure of its bridges). The segments with the highest NRI_a are those that should be maintained first.

Finally, among the new network performance indicators that are gaining popularity, one of the most interesting is resilience (Rose 2004, Rose and Liao 2005). Resilience is the ability of a network, a group of structures or an individual structure to return in a good state of service after a disruptive event. More details are given in Section 6.3.

6. Selected problems and applications

6.1. *Time-dependent reliability*

As already mentioned, bridges deteriorate over time and this, in turn, determines a reduction of their structural reliability. At the bridge network scale, this means that it is more probable to have bridges out of service that cause a larger number of detours. Golroo *et al.* (2010) and Bocchini and Frangopol (2011a) independently proposed two very similar indicators for the bridge network reliability. They can be defined as the probability that the ratio between total travel time in the degraded network over the total travel time when the bridges are all intact and in service is lower than a specified threshold:

$$R_{network}(t, \bar{\theta}) = P\left[\frac{TTT_{degraded}(t)}{TTT_{intact}} \leq \bar{\theta}\right]$$
$$= P[TTT_{degraded}(t) < \bar{\theta} \cdot TTT_{intact}] \quad (12)$$

where $R_{network}$ is the network reliability; t is time; $\bar{\theta}$ is a predefined acceptable threshold; $P[\cdot]$ indicates the probability of occurrence of the event in brackets; $TTT_{degraded}$ is the total travel time associated with the network in the deteriorated condition, as computed by Equations (5) or (6) and TTT_{intact} is the total travel time when the bridges are all intact and in service, as computed by Equations (5) or (6).

Bocchini and Frangopol (2011c) applied the proposed indicator to a bridge network located in Colorado and obtained the time-dependent reliability profile for a 75-year life-cycle. A qualitative network reliability curve, under the assumption that no maintenance interventions are applied, is shown in Figure 3. Actually, these profiles are also useful to the purpose of optimal maintenance planning at the network level. Besides all the requirements on the individual bridges, one of the goals of the network management should be to guarantee a constantly high level of service. The profile in Figure 3, for instance, shows that for the first half of the life-cycle, the network reliability is very high, but after that time, the probability to have an unacceptably low service level increases dramatically. Therefore, bridge maintenance interventions have to be applied, so that the network reliability during the last years is improved. A better insight on the problem

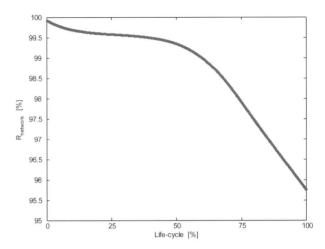

Figure 3. Time–variant bridge network reliability.

of maintenance optimisation is given in the next Section.

6.2. *Maintenance optimisation*

In the literature, it is possible to find many studies that have dealt with the optimal maintenance planning for individual bridges – see, for instance, the review by Frangopol (2011), among others. However, the maintenance management is usually planned by institutions and agencies that are in charge for entire transportation networks or, at least, for several bridges. For this reason, many studies have been focusing on maintenance at the bridge network level (Moghtaderi-Zadeh and Der Kiureghian 1983, Augusti *et al.* 1998, Liu and Frangopol 2005a, 2006b) and the interest is strongly increasing in the last years (Kim *et al.* 2009a, Gao *et al.* 2010, Golroo *et al.* 2010, Peeta *et al.* 2010, Bocchini and Frangopol 2011b, Bocchini *et al.* 2011).

As for the maintenance of individual bridges, the problem is usually formulated as a multi-objective optimisation with conflicting criteria, such as the minimisation of the life-cycle maintenance cost, the optimisation of the network reliability or other performance indicators and the minimisation of the user cost. Genetic algorithms (Goldberg 1989, Tomassini 1995) are the most popular computational tool for this kind of problem. In particular, multi-objective genetic algorithms (Deb 2001, Deb *et al.* 2002) automatically provide a Pareto front of optimal solutions, among which decision makers can choose the one that best fits the needs of the network and the economic constraints. For instance, Figure 4 shows a Pareto front obtained with two conflicting criteria: (i) the maximisation of a network performance index and (ii) the minimisation of the total present maintenance cost (see, for instance, Bocchini and Frangopol 2011b). Each dot represents an optimal solution, which means a list of years of PM

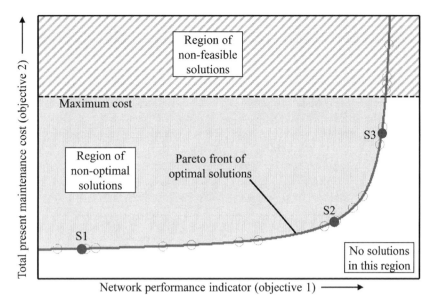

Figure 4. Pareto front of optimal solutions for the network level maintenance.

application for each bridge of the network. The grey area is the domain of solutions that are compliant with the constraints but are non-optimal. Three dots are highlighted and labelled S1, S2 and S3. The Pareto front presents a rapid increase after the region where solution S2 is located. This means that all the optimal solutions on the right hand side of the group to which S2 belongs, such as S3, determine a small increase in the network performance associated, with a large increase in the expected total cost. On the contrary, the optimal solutions on the left hand side, such as S1, yield a limited reduction in the total cost associated with a large loss in the network performance. Therefore, in this specific example, the overall most convenient solutions are those belonging to the cluster of S2.

This kind of study is a very valuable tool for decision makers. It provides an overall view of the effects of the maintenance interventions and drives the choice of the optimal one. However, it should always be combined with other studies on the optimal maintenance of individual bridges (Liu and Frangopol 2005d,e, Bucher and Frangopol 2006, Furuta *et al.* 2006, 2011, Neves *et al.* 2006a,b, Yang *et al.* 2006a,b, Frangopol and Liu 2007, Petcherdchoo *et al.* 2008, Orcesi *et al.* 2010) and on studies on the bridge importance ranking (see, for instance, Akgül and Frangopol 2003, 2004e, Liu and Frangopol 2005b).

6.3. Post-event recovery and resilience

A particular case of intervention is the restoration of the network after an extreme event that has damaged the bridges. In this case, the most interesting network

performance indicator is resilience. Several definitions of resilience in civil engineering can be found in the literature, but the most widely accepted is the one provided by Bruneau *et al.* (2003): 'resilience is defined as the ability of social units (e.g. organisations, communities) to mitigate hazards, contain the effects of disasters when they occur, and carry out recovery activities in ways that minimise social disruption and mitigate the effects of future earthquakes'. In the same article and in subsequent ones (Cimellaro *et al.* 2006, 2010, Bruneau and Reinhorn 2007), an exhaustive description of the properties of resilience and an analytical formulation are provided,

$$r_1 = \int_{t_0}^{t_0 + t_h} Q(t)\mathrm{d}t \qquad (13)$$

where r_1 is the resilience associated to a certain extreme event, t_0 is the time of occurrence of the extreme event, t_h is the investigated time horizon and Q is the functionality. Alternatively, the value of resilience can be normalised with respect to the investigated time horizon in order to have a non-dimensional index:

$$r_2 = \frac{1}{t_h} \cdot \int_{t_0}^{t_0 + t_h} Q(t)\mathrm{d}t \qquad (14)$$

Several studies have used resilience as restoration performance index or as prioritisation criterion for the restoration of bridge networks (Bocchini and Frangopol 2011d, Frangopol and Bocchini 2011b) and other lifelines (Çağnan *et al.* 2006, Xu *et al.* 2007) subject to extreme events.

The properties of resilience and the technical, organisational, social and economic factors by which it is affected are thoroughly defined qualitatively in the mentioned references. However, to the best of the authors' knowledge, the quantification of these properties is strongly dependent on the application and a complete framework for the resilience analysis of bridge networks is still missing.

6.4. *Networks of networks*

Another interesting problem is the one of the interaction between several networks of the same type or of different types.

Networks of the same type (e.g. a network of transportation networks) should be considered a numerical convenient solution that allows to reduce the computational cost analysing the sub-networks separately and considering only minor interactions at the macroscale.

Different types of lifelines can be strongly interdependent too. For this reason, several articles in this field have been published in the last years. Dueñas-Osorio et al. (2007a,b) proposed a combined lifelines analysis with interactions based on geographical proximity; Peeta and Zhang (2009) proposed a computational framework featuring a multilayer superposition of the various infrastructure networks; Alexoudi et al. (2009) applied fuzzy logic to the assessment of importance ranking of interdependent lifelines; Kim et al. (2009b) used the already mentioned matrix-based system reliability method for the seismic analysis of several interacting infrastructures and Hernández-Fajardo and Dueñas-Osorio (2009) performed the seismic analysis of two coupled lifelines.

The general conclusion of these studies is that only the simultaneous consideration of all the critical infrastructure systems in an area can provide a correct assessment of the consequences of an extreme event, of its impact on the recovery activities and on the possibility to promptly restore the normal socio-economic assett.

7. Conclusions and future challenges

Lifelines and bridge networks are a topic that raises increasing interest in the scientific literature. Several outstanding accomplishments and contributions have provided the solid basis for the study of these complex systems, but a lot more work is certainly required and several challenges should still be faced.

In some cases, the techniques used for individual network components can also be exploited for network analyses. However, usually the network introduces interdependencies among the components that cannot be disregarded. Therefore, the network analysis cannot consist entirely of a repeated application of analyses on the individual components. A clear example has been given in Section 6.2; the bridge maintenance schedule that yields the optimal network performances is not obtained as the collection of the optimal schedules for individual bridges.

When all the components of a networked system have to be analysed together, computational issues arise very frequently and often they become the main challenge in the solution of the problem. In particular, when uncertainties are involved, or when a numerical optimisation is performed, the solution has to be achieved by using very efficient techniques. For this reason, some topics such as reliability analysis and optimal maintenance planning are still objects of numerous studies and they are likely to be investigated even more in the future years, focusing on the numerical aspects.

Another topic that has not been mentioned yet, even though it has an important role, is structural health monitoring. Structural testing data are always related to an individual network component (e.g. a bridge). However, data analysis and post-processing can take advantage of the information obtained for other components of the same network, that usually share similar conditions (e.g. correlated traffic loads, comparable environmental conditions and same history of natural extreme events). Moreover, in addition to the maintenance planning, the monitoring management can also be optimised at the network level, with a consequent reduction in the costs and/or an improvement in the overall network serviceability. This goal is certainly another future challenge for researchers in this field.

Most of the concepts presented in this review deal with bridge networks, but they are general enough to be applicable also to other critical components of the transportation infrastructure. In fact, the general framework of 'transportation asset management' (FHWA 2007) includes bridges, culverts, tunnels and the entire pavement of the transportation network. The extension and validation of the presented approaches in the cases of mixed networks (meaning networks with different types of components) is another interesting topic for further research.

However, the most challenging problem that has to be faced when addressing transportation network analysis is probably the need for a strongly multi-disciplinary approach. Transportation engineering, bridge engineering, life-cycle cost analysis, structural deterioration modelling, maintenance management, multi-hazard analysis, probabilistic analysis, socio-economic surveying, multi-objective optimisation, structural health monitoring, operational research and decision making are just a few of the skill sets that are usually required. For all these topics, analytical and computational proficiency, as well as experimental facilities are required. Therefore, it appears evident that these studies can only be performed by strong research groups that involve people with different backgrounds. Moreover, the scientific community plays a role of utmost importance, providing knowledge and advancements in the fields that cannot be covered by an individual research group. The future of scientific research on lifelines and networks strongly relies on the ability of researchers around the world to operate as a strongly interconnected, redundant and efficient network.

Acknowledgements

This article is dedicated to the memory of Dr. Jan M. van Noortwijk and to the legacy of his outstanding scholarly contributions. The support from (a) the National Science Foundation through grant CMS-0639428; (b) the Commonwealth of Pennsylvania, Department of Community and Economic Development, through the Pennsylvania Infrastructure Technology Alliance (PITA); (c) the U.S. Federal Highway Administration Cooperative Agreement Award DTFH61–07-H-00040; and (d) the U.S. Office of Naval Research Contract Number N-00014–08–0188 is gratefully acknowledged. The opinions and conclusions presented in this article are those of the authors and do not necessarily reflect the views of the sponsoring organisations.

References

Akgül, F. and Frangopol, D.M., 2003. Rating and reliability of existing bridges in a network. *Journal of Bridge Engineering*, 8 (6), 383–393.

Akgül, F. and Frangopol, D.M., 2004a. Lifetime performance analysis of existing steel girder bridge superstructures. *Journal of Structural Engineering*, 130 (12), 1875–1888.

Akgül, F. and Frangopol, D.M., 2004b. Lifetime performance analysis of existing prestressed concrete bridge superstructures. *Journal of Structural Engineering*, 130 (12), 1889–1903.

Akgül, F. and Frangopol, D.M., 2004c. Computational platform for predicting lifetime system reliability profiles for different structure types in a network. *Journal of Computing in Civil Engineering*, 18 (2), 92–104.

Akgül, F. and Frangopol, D.M., 2004d. Time-dependent interaction between load rating and reliability of deteriorating bridges. *Engineering Structures*, 26 (12), 1751–1765.

Akgül, F. and Frangopol, D.M., 2004e. Bridge rating and reliability correlation: comprehensive study for different bridge types. *Journal of Structural Engineering*, 130 (7), 1063–1074.

Akgül, F. and Frangopol, D.M., 2005a. Lifetime performance analysis of existing reinforced concrete bridges. I. Theory. *Journal of Infrastructure Systems*, 11 (2), 122–128.

Akgül, F. and Frangopol, D.M., 2005b. Lifetime performance analysis of existing reinforced concrete bridges. II. Application. *Journal of Infrastructure Systems*, 11 (2), 129–141.

Alexoudi, M.N., Kakderi, K.G., and Pitilakis, K.D., 2009. Seismic risk and hierarchy importance of interdependent lifeline systems using fuzzy reasoning. *In*: H. Furuta, D.M. Frangopol, and M. Shinozuka, eds. *Safety, reliability and risk of structures, infrastructures and engineering systems*. Osaka, Japan: CRC Press, 2881–2888.

American Society of Civil Engineers (ASCE), 2009. *Report card for America's infrastructure*. American Society of Civil Engineers.

Augusti, G., Ciampoli, M., and Frangopol, D.M., 1998. Optimal planning of retrofitting interventions on bridges in a highway network. *Engineering Structures*, 20 (11), 933–939.

Bocchini, P., Frangopol, D.M., and Deodatis, G., 2011. Computationally efficient simulation techniques for bridge network maintenance multi-objective optimization under uncertainty. *In*: G. Deodatis and P.D. Spanos, eds. *Computational Stochastic Mechanics*. Singapore: Research Publishing Services, 93–101.

Bocchini, P. and Frangopol, D.M., 2010. On the applicability of random field theory to transportation analysis. *In*: *Proceedings of the 5th international conference on bridge maintenance, safety, and management*, IABMAS2010, 11–15 July 2010, Philadelphia, USA. *In*: D.M. Frangopol, R. Sause, and C.S. Kusko, eds. *Bridge maintenance, safety, management, health monitoring and optimization*. London: CRC Press/Balkema, Taylor and Francis Group, 3025–3032; full paper on CD-ROM, Taylor and Francis Group plc.

Bocchini, P. and Frangopol, D.M., 2011a. A stochastic computational framework for the joint transportation network fragility analysis and traffic flow distribution under extreme events. *Probabilistic Engineering Mechanics*, 26 (2), 182–193.

Bocchini, P. and Frangopol, D.M., 2011b. A probabilistic computational framework for bridge network optimal maintenance scheduling. *Reliability Engineering and System Safety*, 96 (2), 332–349.

Bocchini, P. and Frangopol, D.M., 2011c. Generalized bridge network performance analysis with correlation and time-variant reliability. *Structural Safety*, doi: 10.1016/j.strusafe.2011.02.002.

Bocchini, P. and Frangopol, D.M., 2011d. Optimal resilience- and cost-based post-disaster intervention prioritization for bridges along a highway segment. *Journal of Bridge Engineering, ASCE*. Available online. doi: 10.1061/(ASCE)BE.1943-5592.0000201.

Bruneau, M., *et al.*, 2003. A framework to quantitatively assess and enhance the seismic resilience of communities. *Earthquake Spectra*, 19 (4), 733–752.

Bruneau, M. and Reinhorn, A., 2007. Exploring the concept of seismic resilience for acute care facilities. *Earthquake Spectra*, 23 (1), 41–62.

Bucher, C. and Frangopol, D.M., 2006. Optimization of lifetime maintenance strategies for deteriorating structures considering probabilities of violating safety, condition, and cost thresholds. *Probabilistic Engineering Mechanics*, 21 (1), 1–8.

Çağnan, Z., Davidson, R.A., and Guikema, S.D., 2006. Post-earthquake restoration planning for Los Angeles electric power. *Earthquake Spectra*, 22 (3), 589–608.

Chang, S.E. and Shinozuka, M., 1996. Life cycle cost analysis with natural hazard risk. *Journal of Infrastructure Systems*, 2 (3), 118–126.

Chang, S.E., Shinozuka, M., and Moore, J.E., II, 2000. Probabilistic earthquake scenarios: extending risk analysis methodologies to spatially distributed systems. *Earthquake Spectra*, 16 (3), 557–572.

Chen, W.-F. and Duan, L., 2000. *Bridge engineering handbook*. Boca Raton: CRC Press.

Cimellaro, G.P., Reinhorn, A.M., and Bruneau, M., 2006. Quantification of seismic resilience. *In: Proceedings of the 8th national conference of earthquake engineering*, 18–22 April 2006, San Francisco, California.

Cimellaro, G.P., Reinhorn, A.M., and Bruneau, M., 2010. Seismic resilience of a hospital system. *Structure and Infrastructure Engineering*, 6 (1), 127–144.

Czarnecki, A.A. and Nowak, A.S., 2008. Time-variant reliability profiles for steel girder bridges. *Structural Safety*, 30 (1), 49–64.

Davenport, A.G., 1968. The dependence of wind load upon meteorological parameters. *In: Proceedings of the International Research Seminar on Wind Effects on Buildings and Structures*, Toronto, Canada: University of Toronto Press, 19–82.

Deb, K., 2001. *Multi-objective optimization using evolutionary algorithms*. Chichester, UK: John Wiley and Sons.

Deb, K., *et al.*, 2002. A fast and elitist multiobjective genetic algorithm: NSGA-II. *IEEE. Transactions on Evolutionary Computation*, 6 (2), 182–197.

Department of Homeland Security (DHS), 2009. *HAZUS-MH MR4 earthquake model user manual*. Washington, DC: Department of Homeland Security, Federal Emergency Management Agency, Mitigation Division.

Dueñas-Osorio, L., Craig, J.I., and Goodno, B.J., 2007a. Seismic response of critical interdependent networks. *Earthquake Engineering and Structural Dynamics*, 36, 285–306.

Dueñas-Osorio, L., *et al.*, 2007b. Interdependent response of networked systems. *Journal of Infrastructure Systems*, 13 (3), 185–194.

Duwadi, S.R., 2010. Recognizing and reducing vulnerabilities of transportation infrastructure. *In*: D.M. Frangopol, R. Sause, and C.S. Kusko, eds. *Bridge maintenance, safety, management and life-cycle optimization*. UK: CRC Press, Taylor and Francis, 2620–2624.

Estes, A.C. and Frangopol, D.M., 2001. Bridge lifetime system reliability under multiple limit states. *Journal of Bridge Engineering*, 6 (6), 523–528.

Evans, S.P., 1976. Derivation and analysis of some models for combining trip distribution and assignment. *Transportation Research*, 10 (1), 37–57.

Fairweather, V., Tomasetti, R., and Thornton, C., 2004. *Expressing structure: the technology of large-scale buildings*. Birkhäuser Basel, 184.

FHWA, 2007. *Asset management overview*. Washington, DC: U.S. Department of Transportation, Federal Highway Administration.

FHWA, 2009. *National bridge inventory*. Washington, DC: U.S. Department of Transportation, Federal Highway Administration. Available from: http://www.fhwa.dot.gov/bridge/nbi.htm [Accessed 2 April 2010].

Frangopol, D.M., 2003. *Preventive maintenance strategies for bridge groups*. Technical report, Highways Agency, London.

Frangopol, D.M., 2011. Life-cycle performance, management, and optimization of structural systems under uncertainty: accomplishments and challenges. *Structure and Infrastructure Engineering*, 7 (6), 389–413.

Frangopol, D.M. and Bocchini, P., 2011a. Integrated maintenance-monitoring-management framework for optimal decision making in bridge life-cycle performance: emphasis on SHM and bridge networks. *Proceedings of the 2011 NSF Engineering Research and Innovation Conference*, 4–7 January 2011, Atlanta, GA. *In: Engineering for Sustainability and Prosperity* Disk 1 (Data DVD): Papers, 9.

Frangopol, D.M. and Bocchini, P., 2011b. Resilience as optimization criterion for the rehabilitation of bridges belonging to a transportation network subject to earthquake. *In: SEI-ASCE 2011 Structures Congress*. 14–16 April 2011, Las Vegas, NV.

Frangopol, D.M. and Ellingwood, B.R., 2010. Life-cycle performance, safety, reliability and risk of structural systems. *Structure Magazine*, March, 7.

Frangopol, D.M., *et al.*, 2001a. Reliability based evaluation of rehabilitation rates of bridge groups. *In: Proceedings of the international conference on safety, risk and reliability trends in engineering*, IABSE, Malta, March 21–23; Safety, Risk and Reliability – Trends in Engineering, Conference Report, IABSE-CIB-ECCS-fib-RILEM, 267-272, also on CD-ROM, 1001-1006.

Frangopol, D.M., Kong, J.S., and Gharaibeh, E.S., 2001b. Reliability-based life-cycle management of highway bridges. *Journal of Computing in Civil Engineering*, 15 (1), 27–34.

Frangopol, D.M., Kallen, M.-J., and van Noortwijk, J., 2004. Probabilistic models for life-cycle performance of deteriorating structures: review and future directions. *Progress in Structural Engineering and Materials*, 6 (4), 197–212.

Frangopol, D.M., Lin, K.-Y., and Estes, A.C., 1997. Life-cycle cost design of deteriorating structures. *Journal of Structural Engineering*, 123 (10), 1390–1401.

Frangopol, D.M. and Liu, M., 2007. Maintenance and management of civil infrastructure based on condition, safety, optimization, and life-cycle cost. *Structure and Infrastructure Engineering*, 3 (1), 29–41.

Frangopol, D.M. and Messervey, T.B., 2007a. Integrated lifecycle health monitoring, maintenance, management and cost of civil infrastructure. *In: Proceedings of the international symposium on integrated life-cycle design and management of infrastructures*, (keynote paper). Shanghai, China: Tongji University.

Frangopol, D.M. and Messervey, T.B., 2007b. Risk assessment for bridge decision making. *Proceedings of the fourth civil engineering conference in the Asian Region*. CECAR 4, 25–28 June 2007, Taipei, Taiwan. *In*: ASCE Tutorial & Workshop on Quantitative Risk Assessment, Taipei, Taiwan, 37–42.

Frangopol, D.M. and Messervey, T.B., 2008. Use of structural health monitoring for improved civil infrastructure management under uncertainty. *In: Proceedings of WG7.5 reliability and optimization of structural systems* (keynote paper), 6–9 August 2008, Toluca, Mexico.

Frangopol, D.M. and Messervey, T.B., 2009. Life-cycle cost and performance prediction: Role of structural health monitoring. *In*: S.-S. Chen and A.H.-S. Ang, eds. *Frontier technologies for infrastructures engineering: structures and infrastructures book series*, chapter 16, vol. Boca Raton, London, New York, Leiden: CRC Press/Balkema, 361–381.

Frank, M. and Wolfe, P., 1956. An algorithm for quadratic programming. *Naval Research Logistics Quarterly*, 3 (1–2), 95–110.

Furuta, H., *et al.*, 2006. Optimal bridge maintenance planning using improved multi-objective genetic algorithm. *Structure and Infrastructure Engineering*, 2 (1), 33–41.

Furuta, H., Frangopol, D.M., and Nakatsu, K., 2011. Life-cycle cost of civil infrastructure with emphasis on balancing structural performance and seismic risk of road network. *Structure and Infrastructure Engineering*, 7 (1–2), 65–74.

Gao, L., Xie, C., and Zhang, Z., 2010. Network-level multi-objective optimal maintenance and rehabilitation scheduling. *In: Proceedings of the 89th annual meeting of the Transportation Research Board of the national academies*, 10–14 January 2010, Washington, DC, USA.

Ghosn, M., Moses, F., and Frangopol, D.M., 2010. Redundancy and robustness of highway bridge super-structures and substructures. *Structure and Infrastructure Engineering*, 6 (1–2), 257–278.

Gibbons, A.M., 1985. *Algorithmic graph theory*. Cambridge: Cambridge University Press.

Gioffré, M., Gusella, V., and Grigoriu, M., 2000. Simulation of non-Gaussian field applied to wind pressure fluctuations. *Probabilistic Engineering Mechanics*, 15, 339–345.

Goldberg, D.E., 1989. *Genetic algorithms in search, optimization, and machine learning*. Reading, MA: Addison-Wesley Professional.

Golroo, A., Mohaymany, A.S., and Mesbah, M., 2010. Reliability based investment prioritization in transportation networks. *In: Proceedings of the 89th annual meeting of the Transportation Research Board of the national academies*, 10–14 January 2010, Washington, DC, USA.

Hernández-Fajardo, I. and Dueñas-Osorio, L., 2009. Time sequential evolution of interdependent lifeline systems. *In*: H. Furuta, D.M. Frangopol, and M. Shinozuka, eds. *Safety, reliability and risk of structures, infrastructures and engineering systems*. Osaka, Japan: CRC Press, 2864–2871.

Herrera, J.C., *et al.*, 2010. Evaluation of traffic data obtained via GPS-enabled mobile phones: The Mobile Century field experiment. *Transportation Research Part C: Emerging Technologies*, 18 (4), 568–583.

Jankowski, R., 2006. Numerical simulations of space-time conditional random fields of ground motions. *Computational Science – ICCS 2006*, 56–59.

Jankowski, R. and Wilde, K., 2000. A simple method of conditional random field simulation of ground motions for long structures. *Engineering Structures*, 22 (5), 552–561.

Kallen, M.-J. and van Noortwijk, J., 2006. Optimal periodic inspection of a deterioration process with sequential condition states. *International Journal of Pressure Vessels and Piping*, 83, 249–255.

Kang, W.-H., Song, J., and Gardoni, P., 2008. Matrix-based system reliability method and application to bridge networks. *Reliability Engineering and System Safety*, 93 (11), 1584–1593.

Kim, L., Cho, H.-N., and Cho, C., 2009a. Life cycle performance-based optimal allocation methodology for railroad bridge networks. *In*: H. Furuta, D.M. Frangopol, and M. Shinozuka, eds. *Safety, reliability and risk of structures, infrastructures and engineering systems*. Osaka, Japan: CRC Press, 2033–2040.

Kim, S. and Frangopol, D.M., 2010. Optimal planning of structural performance monitoring based on reliability importance assessment. *Probabilistic Engineering Mechanics*, 25 (1), 86–98.

Kim, S.-H. and Feng, M.Q., 2003. Fragility analysis of bridges under ground motion with spatial variation. *International Journal of Non-Linear Mechanics*, 38 (5), 705–721.

Kim, Y., *et al.*, 2009b. Seismic risk assessment of complex interacting infrastructures using matrix-based system reliability method. *In*: H. Furuta, D.M. Frangopol, and M. Shinozuka, eds. *Safety, reliability and risk of structures, infrastructures and engineering systems*. Osaka, Japan: CRC Press, 2889–2893.

Kong, J.S. and Frangopol, D.M., 2003. Evaluation of expected life-cycle maintenance cost of deteriorating structures. *Journal of Structural Engineering*, 129 (5), 682–691.

Kong, J.S. and Frangopol, D.M., 2004a. Cost-reliability interaction in life-cycle cost optimization of deteriorating structures. *Journal of Structural Engineering*, 130 (11), 1704–1712.

Kong, J.S. and Frangopol, D.M., 2004b. Prediction of reliability and cost profiles of deteriorating structures under time- and performance-controlled maintenance. *Journal of Structural Engineering*, 130 (12), 1865–1874.

Kong, J.S. and Frangopol, D.M., 2005. Probabilistic optimization of aging structures considering maintenance and failure costs. *Journal of Structural Engineering*, 131 (4), 600–616.

Korving, H., *et al.*, 2009. Risk-based design of sewer system rehabilitation. *Structure and Infrastructure Engineering*, 5 (3), 215–227.

Lee, C. and Machemehl, R.B., 2005. *Combined traffic signal control and traffic assignment: algorithms, implementation and numerical results*. Southwest Region University Transportation Center, Center for Transportation Research University of Texas at Austin, Research Report No. SWUTC/05/472840-00074-1.

Lee, Y.-J., Song, J., and Gardoni, P., 2009. Post-hazard flow capacity of bridge transportation network considering structural deterioration. *In*: H. Furuta, D.M. Frangopol, and M. Shinozuka, eds. *Safety, reliability and risk of structures, infrastructures and engineering systems*. Osaka, Japan: CRC Press, 2894–2901.

Levinson, D.M. and Kumar, A., 1994. Multimodal trip distribution: structure and application. *Transportation Research Record*, 1466, 124–131.

Liu, M. and Frangopol, D.M., 2005a. Balancing connectivity of deteriorating bridge networks and long-term maintenance cost through optimization. *Journal of Bridge Engineering*, 10 (4), 468–481.

Liu, M. and Frangopol, D.M., 2005b. Bridge annual maintenance prioritization under uncertainty by multi-objective combinatorial optimization. *Computer Aided Civil and Infrastructure Engineering*, 20 (5), 343–353.

Liu, M. and Frangopol, D.M., 2005c. Time-dependent bridge network reliability: novel approach. *Journal of Structural Engineering*, 131 (2), 329–337.

Liu, M. and Frangopol, D.M., 2005d. Multiobjective maintenance planning optimization for deteriorating bridges considering condition, safety and life-cycle cost. *Journal of Structural Engineering*, 131 (5), 833–842.

Liu, M. and Frangopol, D.M., 2005e. Maintenance planning of deteriorating bridges by using multiobjective optimization. *Transportation Research Record: Journal of the Transportation Research Board*, 11 (S), 491–500.

Liu, M. and Frangopol, D.M., 2006a. Probability-based bridge network performance evaluation. *Journal of Bridge Engineering*, 11 (5), 633–641.

Liu, M. and Frangopol, D.M., 2006b. Optimizing bridge network maintenance management under uncertainty with conflicting criteria: life-cycle maintenance, failure, and user costs. *Journal of Structural Engineering*, 132 (11), 1835–1845.

Lomax, T., *et al.*, 2003. *Selecting travel reliability measures.* Texas Transportation Institute Monograph, College Station, Texas.

Matisziw, T.C. and Murray, A.T., 2009. Modeling s-t path availability to support disaster vulnerability assessment of network infrastructure. *Computers and Operations Research*, 36 (1), 16–26.

Messervey, T.B., Frangopol, D.M., and Casciati, S., 2011. Application of the statistics of extremes to the reliability assessment and performance prediction of monitored highway bridges. *Structure and Infrastructure Engineering*, 7 (1–2), 87–99.

Moghtaderi-Zadeh, M. and Der Kiureghian, A., 1983. Reliability upgrading of lifeline networks for post-earthquake serviceability. *Earthquake Engineering & Structural Dynamics*, 11 (4), 557–566.

Mori, Y. and Ellingwood, B.R., 1993. Reliability-based service-life assessment of aging concrete structures. *Journal of Structural Engineering*, 119 (5), 1600–1621.

Murray, A.T., Matisziw, T.C., and Grubesic, T.H., 2008. A methodological overview of network vulnerability analysis. *Growth and Change*, 39 (4), 573–592.

Neves, L.A.C., Frangopol, D.M., and Cruz, P.J., 2006a. Probabilistic lifetime-oriented multiobjective optimization of bridge maintenance: single maintenance type. *Journal of Structural Engineering*, 132 (6), 991–1005.

Neves, L.A.C., Frangopol, D.M., and Petcherdchoo, A., 2006b. Probabilistic lifetime-oriented multi-objective optimization of bridge maintenance: combination of maintenance types. *Journal of Structural Engineering*, 132 (11), 1821–1834.

Ng, A.K.S. and Efstathiou, J., 2006. Structural robustness of complex networks. *In*: *Proceedings of the international workshop and conference on network science –NetSci2006*, Bloomington, IN.

Okasha, N.M. and Frangopol, D.M., 2010. Novel approach for multi-criteria optimization of Life-Cycle preventive and essential maintenance of deteriorating structures. *Journal of Structural Engineering*, 136 (8), 1009–1022.

Okasha, N.M. and Frangopol, D.M., 2011. Integration of structural health monitoring in a system performance based life-cycle bridge management framework. *Structure and Infrastructure Engineering*, doi: 10.1080/15732479.2010.485726.

Orcesi, A.D. and Frangopol, D.M., 2010. Optimization of bridge management under budget constraints: Role of structural health monitoring. *Transportation Research Record: Journal of the Transportation Research Board*, 1–3 (2202), 148–158.

Orcesi, A.D. and Frangopol, D.M., 2011. Optimization of bridge maintenance strategies based on structural health monitoring information. *Structural Safety*, 33 (1), 26–41.

Orcesi, A.D., Frangopol, D.M., and Kim, S., 2010. Optimization of bridge maintenance strategies based on multiple limit states and monitoring. *Engineering Structures*, 32 (3), 627–640.

Pandey, M.D., Yuan, X.-X., and van Noortwijk, J.M., 2009. The influence of temporal uncertainty of deterioration on life-cycle management of structures. *Structure and Infrastructure Engineering*, 5 (2), 145–156.

Peeta, S. and Zhang, P., 2009. Modeling infrastructure interdependencies: theory and practice. *In*: H. Furuta, D.M. Frangopol, and M. Shinozuka, eds. *Safety, reliability and risk of structures, infrastructures and engineering systems*. Osaka, Japan: CRC Press, 2872–2880.

Peeta, S., *et al.*, 2010. Pre-disaster investment decisions for strengthening a highway network. *Computers and Operations Research*, 37 (10), 1708–1719.

Petcherdchoo, A., Neves, L.A.C., and Frangopol, D.M., 2008. Optimizing lifetime condition and reliability of deteriorating structures with emphasis on bridges. *Journal of Structural Engineering*, 134 (4), 544–552.

PONTIS, 2005. *User's manual, Release 4.4.* Cambridge, MA: Cambridge Systematics, Inc.

Rose, A., 2004. Defining and measuring economic resilience to disasters. *Disaster Prevention and Management*, 13 (4), 307–314.

Rose, A. and Liao, S., 2005. Modelling regional economic resilience to disasters: acomputable general equilibrium analysis of water service disruptions. *Journal of Regional Science*, 45 (1), 75–112.

Rubin, F., 1974. A search procedure for Hamilton paths and circuits. *Journal of the Association for Computing Machinery*, 21 (4), 576–580.

Scott, D.M., *et al.*, 2006. Network Robustness Index: a new method for identifying critical links and evaluating the performance of transportation networks. *Journal of Transport Geography*, 14 (3), 215–227.

Shinozuka, M., *et al.*, 1999. Modeling, synthetics and engineering applications of strong earthquake wave motion. *Soil Dynamics and Earthquake Engineering*, 18 (3), 209–228.

Shinozuka, M., *et al.*, 2006. Cost-effectiveness of seismic bridge retrofit. *In*: P.J.S. Cruz, D.M. Frangopol, and L.A.C. Neves, eds. *Bridge maintenance, safety, management, life-cycle performance and cost*. London: Taylor & Francis.

Simiu, E. and Scanlan, R., 1996. *Wind effects on structures.* 3rd ed. New York: John Wiley and Sons.

Song, J. and Kang, W.-H., 2009. System reliability and sensitivity under statistical dependence by matrix-based system reliability method. *Structural Safety*, 31 (2), 148–156.

Tomassini, M., 1995. A survey of genetic algorithms. *Annual Reviews of Computational Physics*, 3 (1), 87–118.

van Noortwijk, J.M. and Frangopol, D.M., 2004. Two probabilistic life-cycle maintenance models for deteriorating civil infrastructures, *Probabilistic Engineering Mechanics*, 19 (4), 345–359.

Wardrop, J.G., 1952. Some theoretical aspects of road traffic research. *ICE Proceedings, Engineering Divisions*, 1, 325–362.

Xu, N., *et al.*, 2007. Optimizing scheduling of post-earthquake electric power restoration tasks. *Earthquake Engineering & Structural Dynamics*, 36 (2), 265–284.

Yang, S.I., *et al.*, 2006a. The use of lifetime functions in the optimization of interventions on existing bridges considering maintenance and failure costs. *Reliability Engineering and System Safety*, 91 (6), 698–705.

Yang, S.I., Frangopol, D.M., and Neves, L.C., 2006b. Optimum maintenance strategy for deteriorating bridge structures based on lifetime functions. *Engineering Structures*, 28 (2), 196–206.

Zerva, A., 2009. *Spatial Variation of Seismic Ground Motions*. Boca Raton, USA: CRC Press, Taylor & Francis Group.

Zerva, A. and Beck, J.L., 2003. Identification of parametric ground motion random fields from spatially recorded seismic data. *Earthquake Engineering and Structural Dynamics*, 32, 771–791.

Zerva, A. and Zervas, V., 2002. Spatial variation of seismic ground motions: an overview. *Applied Mechanics Reviews*, 55 (3), 271–297.

Zhang, R. and Deodatis, G., 1996. Seismic ground motion synthetics of the 1989 Loma Prieta earthquake. *Earthquake Engineering and Structural Dynamics*, 25, 465–481.

Zhou, Y., Banerjee, S., and Shinozuka, M., 2010. Socio-economic effect of seismic retrofit of bridges for highway transportation networks: a pilot study. *Structure and Infrastructure Engineering*, 6 (1), 145–157.

Life-cycle of structural systems: recent achievements and future directions[†]

Dan M. Frangopol and Mohamed Soliman

Structural systems are under deterioration due to ageing, mechanical stressors, and harsh environment, among other threats. Corrosion and fatigue can cause gradual structural deterioration. Moreover, natural and man-made hazards may lead to a sudden drop in the structural performance. Inspection and maintenance actions are performed to monitor the structural safety and maintain the performance over certain thresholds. However, these actions must be effectively planned throughout the life-cycle of a system to ensure the optimum budget allocation and maximum possible service life without adverse effects on the structural system safety. Life-cycle engineering provides rational means to optimise life-cycle aspects, starting from the initial design and construction to dismantling and replacing the system at the end of its service life. This paper presents a brief overview of the recent research achievements in the field of life-cycle engineering for civil and marine structural systems and indicates future directions in this research field. Several aspects of life-cycle engineering are presented, including the performance prediction under uncertainty and optimisation of life-cycle cost and intervention activities, as well as the role of structural health monitoring and non-destructive testing techniques in supporting the life-cycle management decisions. Risk, resilience, sustainability, and their integration into the life-cycle management are also discussed.

1. Introduction

Structures and infrastructure systems play a significant role in improving the economic, social, and environmental welfare of nations. These systems are subjected to deterioration due to ageing effects (e.g. corrosion), natural hazards (e.g. seismic events and hurricanes) and man-made extreme events (e.g. collisions and terrorist attacks). A sudden failure or loss of functionality of these systems may have severe economic, social, and environmental impacts (Chang, McDaniels, Mikawoz, & Peterson, 2007; Decò & Frangopol, 2011, 2013; Ellingwood, 2006; Rinaldi, 2004; Zhu & Frangopol, 2013a). Recent studies suggest that the consequences of failure are significantly more than just the cost of rebuilding or replacing the dysfunctional component, especially if the social and environmental impacts are included (Bocchini, 2013; Dong, Frangopol, & Saydam, 2013, 2014a; Padgett & Tapia, 2013; Zhou, Banerjee, & Shinozuka, 2010). Moreover, it has been shown that failures due to extreme events may have significant long-term consequences (Bocchini, 2013). Therefore, in order to minimise the number of failures and their consequences, infrastructure managers have to implement various strategies to maintain adequate long-

term performance and functionality while considering financial constraints. These activities include periodic inspections, maintenance and retrofit actions, in addition to structural health monitoring (SHM) which can provide an accurate indication on the actual structural responses and aid in the performance prediction and the evaluation of future maintenance needs (Frangopol, Strauss, & Kim, 2008a; Strauss, Frangopol, & Kim, 2008).

Although these actions aid in maintaining the system performance within acceptable limits, they may impose a major financial burden. Accordingly, these actions should be rationally scheduled along the life-cycle of the system within an integrated framework capable of simultaneously considering various economic and safety requirements. Uncertainties associated with the performance prediction, damage initiation and propagation, damage detection capabilities, and the effect of maintenance and retrofit on the structural performance should be included for the proper life-cycle management (Ang & De Leon, 2005; Ellingwood & Kinali, 2009; Frangopol & Bocchini, 2012; Frangopol et al., 2012; Kim & Frangopol, 2011a; Li & Ellingwood, 2006; Liu & Frangopol, 2005a; Okasha & Frangopol, 2010a).

[†]Based on the Fazlur R. Khan plenary lecture and the associated paper presented at the Fourth International Symposium on Life-Cycle Civil Engineering (IALCCE2014), Tokyo, Japan, 16–19 November 2014.

Within the last two decades, several studies introduced techniques which can assist the infrastructure management of multiple types of structural systems, including bridges (Biondini, Bontempi, Frangopol, & Malerba, 2006; Biondini, Camnasio, & Palermo, 2014; Biondini & Frangopol, 2008; Enright & Frangopol, 1999; Estes & Frangopol, 2001; Frangopol & Estes, 1997; Frangopol & Okasha, 2009; Kong & Frangopol, 2003b, 2004a, 2004b; Neves, Frangopol, & Cruz, 2006; Okasha & Frangopol, 2010a), bridge networks (Bocchini & Frangopol, 2013; Chang, Peng, Ouyang, Elnashai, & Spencer, 2012; Liu & Frangopol, 2005b; Rokneddin, Ghosh, Dueñas-Osorio, & Padgett, 2013; Saydam, Bocchini, & Frangopol, 2013; Shinozuka, Murachi, Dong, Zhou, & Orlikowski, 2003), and lifelines such as water networks and power grids (Jayaram & Baker, 2010; Lundie, Peters, & Beavis, 2004; Turconi, Simonsen, Byriel, & Astrup, 2014). These studies considered various deteriorating mechanisms, such as fatigue (Chung, Manuel, & Frank, 2006; Garbatov & Guedes Soares, 2001; Kim & Frangopol, 2011a; Kwon & Frangopol, 2010), corrosion (Akiyama, Frangopol, & Suzuki, 2012; Biondini et al., 2014; Biondini & Frangopol, 2008; Kendall, Keoleian, & Helfand, 2008; Kim & Frangopol, 2011b), seismic effects (Chang et al., 2012; Decò & Frangopol, 2013; Dong et al., 2014a;Frangopol & Akiyama, 2011; Furuta, Frangopol, & Nakatsu, 2011; Ghosh & Padgett, 2010; Padgett, Dennemann, & Ghosh, 2010; Rokneddin et al., 2013), scour (Stein, Young, Trent, & Pearson, 1999), tsunami (Akiyama, Frangopol, Arai, & Koshimura, 2013) and a combination of hazards (Akiyama, Frangopol, & Matsuzaki, 2011; Decò & Frangopol, 2011; Dong et al., 2013; Frangopol & Akiyama, 2011; Zhu & Frangopol, 2013a). Several of these studies investigated the life-cycle performance of deteriorating systems, while others focused on evaluating the life-cycle cost considering maintenance and repair actions, in addition to scheduling these actions to yield optimum life-cycle decisions.

Because most infrastructure management decisions are made under strict budgetary constraints, optimisation is an essential tool for the proper life-cycle management. By employing optimisation techniques, trade-offs between conflicting life-cycle management criteria such as minimising the life-cycle cost and maximising the expected service life can be identified. Indeed, this process can be computationally demanding especially for large-scale infrastructure systems such as networks of damaged structures. However, the recent increase in computational capabilities have made it possible to conduct complex, large-scale simulations, and paved the road for sophisticated probabilistic techniques to be applied to infrastructure management problems.

A comprehensive life-cycle management framework should be composed of integrated modules responsible for performing various management computational tasks. These tasks include performance prediction under multiple hazards, optimisation of interventions, and reliability- and cost-informed decision-making, among others. An example of such a framework is presented in Frangopol (2011) and Frangopol et al. (2012). This framework, shown in Figure 1, has been applied to various types of structural systems such as bridges, bridge networks, and naval vessels. The development of this framework required a parallel development of an integrated computational platform, shown schematically in Figure 2 for civil structural systems, which combines different modules of the life-cycle framework to establish the optimum life-cycle decisions. As shown in the figure, the platform consists of a central user interface [e.g. MATLAB (MathWorks, 2014a) or VisualDOC (VisualDOC, 2012)] responsible for the data flow to/from separate computational modules which perform different tasks of the life-cycle analysis such as the structural analysis [e.g. ABAQUS (ABAQUS, 2009)], reliability analysis [e.g. RELSYS (Estes & Frangopol, 1998)] and structural optimisation [e.g. MATLAB (MathWorks, 2014b)], among others. The numbers shown in Figure 2 indicate the input and output data processed by each of the computational modules.

This paper, which is based on the Fazlur R. Khan plenary lecture presented by the first author at the Fourth International Symposium on Life-Cycle Civil Engineering (IALCCE2014), Tokyo, Japan, 16–19 November 2014, presents a brief overview of the recent research achievements in the field of life-cycle engineering for civil and marine infrastructure systems with emphasis on bridges, bridge networks and naval vessels. Different aspects of the life-cycle management are presented. These aspects include performance prediction under uncertainty, optimisation of life-cycle cost and intervention activities, as well as the role of SHM and non-destructive testing techniques in supporting life-cycle management decisions. Risk, resilience, sustainability aspects and their integration into the life-cycle management are also discussed.

2. Performance of structures and systems

The performance of a structural/infrastructure system deteriorates due to the effect of mechanical stressors, harsh environment, and extreme events. As shown in Figure 3, performance deteriorates gradually due to effects such as corrosion, whereas seismic events, hurricanes and/or other extreme events may cause a sudden drop in the performance. Significant research work has been done in past decades to improve the accuracy of models which can predict structural performance under time-dependent deterioration. Specifically, corrosion and fatigue have received significant attention. Corrosion of steel reinforcement is one of the main factors that causes the deterioration of reinforced concrete (RC) structures. Its effect is accelerated when a member is subjected to de-icing salt spray. Corrosion can

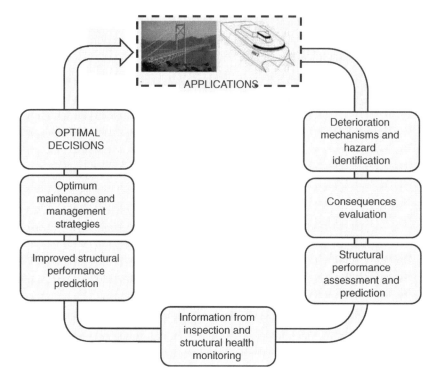

Figure 1. General life-cycle management procedure.

damage an RC member in various ways such as cracking, spalling, and loss of steel section. Given the randomness associated with various aspects of corrosion initiation and crack initiation and propagation, structural performance assessment under corrosion deterioration should be per-

formed on a stochastic basis. Papakonstantinou and Shinozuka (2013) presented a stochastic approach, capable of considering the spatial variability in the model parameters, for predicting the corrosion damage and its effect on the structure. Bastidas-Arteaga, Bressolette, Chateauneauf, and

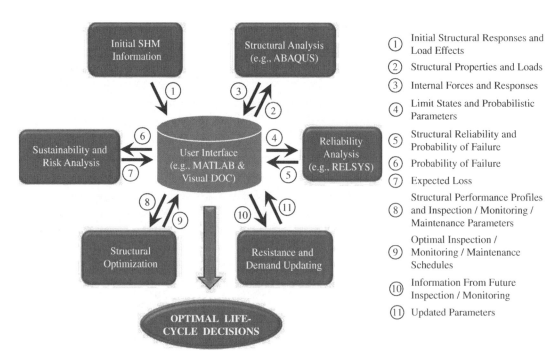

Figure 2. Developed computational framework for life-cycle management under uncertainty.

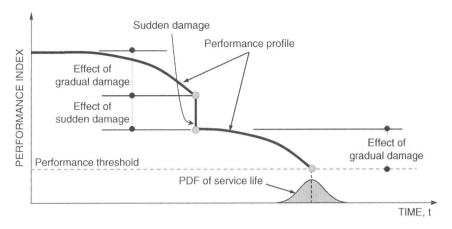

Figure 3. Effect of gradual deterioration and sudden damage on the structural performance.

Sánchez-Silva (2009) studied the reliability of RC structures under corrosion-fatigue deterioration process. Their study showed that the combined action of corrosion and fatigue can significantly reduce the service life of RC structures. A review of the available corrosion initiation and propagation models can be found in Otieno, Beushausen, and Alexander (2011) and Papakonstantinou and Shinozuka (2013).

Fatigue is another deterioration mechanism which can significantly affect the performance of bridges and naval vessels. For a component subjected to elastic stress fluctuations, fatigue damage may accumulate at regions of stress concentration where the local stress exceeds the yield limit of the material. Stress concentrations can occur at the component due to the presence of initial flaws in the material, welding process or fabrication. Initiation and propagation of cracks in the plastic localised region occur due to the cumulative damage acting over a certain number of stress fluctuations. These cracks can eventually cause the fracture of the component. The fatigue deterioration can be reduced by adopting improved structural details, avoiding stress concentrations and decreasing the number of welded attachments, among others. Currently, design specifications (see American Association of State Highway and Transportation Officials, 2014) provide guidelines for maximising the fatigue life and offer means for selecting details associated with higher fatigue resistance. For studying the crack condition at a given critical location, the fracture mechanics approach can be used to predict the crack growth (see British Standards Institute, 2005). It should be noted that the research in the field of time-dependent performance prediction of deteriorating structures is massive. Accordingly, a separate review paper would be required to cover each deterioration mechanism. Therefore, the comprehensive treatment of these aspects is out of the scope of this paper.

Recent research has shown that climate change can increase the risk of deterioration of infrastructure due to the temperature increase and the escalation in likelihood of carbonation-induced corrosion (Stewart, Wang, &

Nguyen, 2011). Evaluating the effect of climate change on corrosion initiation and propagation, as well as the structural reliability, is currently an active research area with studies focusing on evaluating the increase in the probability of corrosion initiation due to climate changes (Bastidas-Arteaga, Chateauneauf, Sanchez-Silva, Bressolette, & Schoefs, 2010), evaluating the reliability of deteriorating structures under the effect of climate change (El Hassan, Bressolette, Chateauneauf, & El Tawil, 2010; Stewart et al., 2011), and developing adaptation strategies to mitigate the risk of additional deterioration due to climate change (Bjarnadottir, Li, & Stewart, 2011; Stewart, Wang, & Nguyen, 2012).

For deteriorating structures, if no maintenance is performed, the performance will continue to deteriorate after an extreme event until the safety threshold is reached. At this point, a major maintenance action, denoted as essential maintenance (EM), should be performed to restore the structural performance to a level close to its initial condition. In some cases, this major maintenance may be very expensive to perform and, depending on the extent of maintenance (i.e. level of restoration), it may be cheaper to replace the entire structure. As a result, infrastructure managers may choose to perform several preventive maintenance (PM) actions along the life-cycle of the structure in order to extend its safe service life. Preventive maintenance can be performed to improve structural performance by reducing the damage level, or to stop the damage propagation for a period of time. The effects of various maintenance types on the performance of a deteriorating structure are schematically shown in Figure 4. Whether to perform only preventive, essential or both maintenance types is a difficult question that can be answered through the life-cycle management techniques. A major part of this difficulty is attributed to the presence of uncertainties associated with the performance prediction models, damage initiation and propagation, occurrence of extreme events and their effect on the structure, and the effect of maintenance actions on the performance.

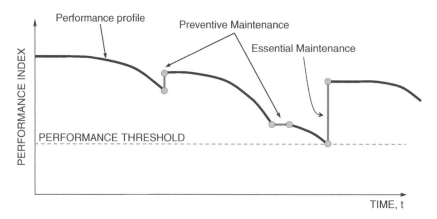

Figure 4. Effect of different maintenance types on the structural performance.

Due to these uncertainties, it is essential to perform the life-cycle management on a probabilistic platform.

Several performance indicators can be used for life-cycle management under uncertainty. Reliability, risk, robustness and redundancy have been widely used as performance indicators within the last decades. Many of these indicators are based on identifying the probability of failure of a system. Among these are the reliability index and risk. The probability of failure of a system is defined as the probability of violating any of the limit states that define its failure modes. A limit-state consisting of the capacity and demand terms representing structural resistance, $R(t)$, and load effects, $S(t)$, respectively, is defined as

$$g(t) = R(t) - S(t) = 0, \qquad (1)$$

where $R(t)$ and $S(t)$ are expressed in terms of the governing random variables (e.g. yield stress, modulus of elasticity, live-load effects and parameters of the deterioration model) at time t. Under the assumption that $R(t)$ and $S(t)$ are statistically independent random variables, the instantaneous structural probability of failure is (Melchers, 1999)

$$P_f(t) = P(g(t) < 0) = \int_0^\infty F_R(x, t) f_S(x, t) \mathrm{d}x, \qquad (2)$$

where $F_R(x, t)$ is the instantaneous cumulative distribution function (CDF) of the resistance and $f_S(x, t)$ is the instantaneous probability density function (PDF) of the load effects. The corresponding reliability index $\beta(t)$ can be computed as

$$\beta(t) = \Phi^{-1}(1 - P_f(t)), \qquad (3)$$

in which $\Phi(\cdot)$ is the standard normal CDF. In general, due to various ageing phenomena, the structural resistance decreases with time. On the other hand, load effects may increase with time. As a result, the reliability index

generally decreases with time, in the manner shown schematically in Figure 3.

Based on the probability of failure and the reliability index previously discussed, several useful probabilistic measures such as risk, redundancy, robustness, and vulnerability can be quantified (Saydam & Frangopol, 2011). Vulnerability, which measures the structural tolerance to damage, can be defined as the ratio of the failure probability of the damaged system to that of the intact system. Redundancy is another performance indicator which measures the reserve capacity of the structure and gives an indication on the presence of alternative load paths within the structure. A comprehensive review of such indicators is given in Frangopol (2011) and Saydam and Frangopol (2011).

Another class of performance indicators is based on the lifetime functions (Leemis, 1995). Lifetime functions offer multiple indicators for evaluating the structural performance of components and systems. These indicators have been successfully used for the life-cycle management of bridges under different deteriorating mechanisms (Barone & Frangopol, 2013a, 2013b, 2014; Barone, Frangopol, & Soliman, 2014; Okasha & Frangopol, 2010b). Multiple lifetime functions can be defined, based on the PDF of the time-to-failure, including the cumulative probability of failure, the survivor function, the hazard function, and the cumulative hazard function. Each of these functions represents a distinctive feature that can be implemented within the general life-cycle management framework.

The random time-to-failure T of a component is defined as the time elapsing from placing the component into operation until it fails for the first time (Rausand & Høyland, 2004). The PDF of the time-to-failure $f(t)$ can be found through performing statistical analysis considering the damage propagation model and it is the first step to calculate the rest of the lifetime reliability measures. The cumulative probability of failure $F(t)$ represents the probability that the component is not functioning at any time t and is expressed as

$$F(t) = P(T \leq t) = \int_0^t f(x)\mathrm{d}x. \qquad (4)$$

The survivor function $S(t)$, on the other hand, represents the probability that the component will be functioning at any time t, and it is calculated as the complement of the cumulative probability of failure

$$S(t) = 1 - F(t) = P(T > t) = \int_t^\infty f(x)\mathrm{d}x. \qquad (5)$$

The survivor function provides a measure of the reliability of the component because it constitutes the definition that the structure is functioning at time t. Figure 5 represents schematically the relation between $f(t)$, $F(t)$ and $S(t)$. As shown, at a certain time t_i, the cumulative probability of failure $F(t_i)$ is represented by the area A_1, whereas the survivor function value $S(t_i)$ is represented by the area $A_2 = 1 - A_1$.

2.1 Risk

Risk is defined as the combination of chances and consequences of events generated by hazards in a given context. Risk, in contrast to other performance indicators such as the reliability index, provides an indication on the potential loss associated with failures (Modarres, 2006). These losses can be due to natural hazards, man-made extreme events, or due to normal ageing and environmental attacks. In general, risk-based decision-making and management should cover three main fronts: methodologies to mitigate risk before the hazard occurrence (e.g. maintenance and rehabilitation), effective consequences reduction processes during the hazard (e.g. improving the

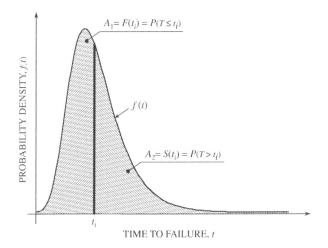

Figure 5. Schematic showing the relationship between the PDF of time to failure, cumulative probability of failure, and survivor function.

emergency response protocols) and optimisation to ensure the best possible resilience after the occurrence of the extreme event (Joint Committee on Structural Safety, 2008).

Therefore, risk represents an important tool for the life-cycle performance evaluation of structures and infrastructure systems. With the ability to consider the consequences of failure, a feature especially beneficial with respect to hazard analyses, risk has been widely used for assessing the safety of structures and systems under various types of hazards such as seismic events (Chang & Shinozuka, 1996; Decò & Frangopol, 2011; Karmakar, Ray-Chaudhuri, & Shinozuka, 2014; Shiraki et al., 2007) and scour effects (Stein et al., 1999). In addition, it has been used for the life-cycle evaluation and optimisation of bridge maintenance and retrofit under seismic events (Decò & Frangopol, 2013; Padgett et al., 2010; Zhu & Frangopol, 2013a), decision-making (Ang & De Leon, 2005), and condition assessment of civil infrastructure (Ellingwood, 2005).

In addition, risk-based assessment approaches are capable of being performed on a multi-hazard basis. For a structure with multiple failure modes, the probability of system failure under a given hazard is

$$P(F_t|H) = P\big([\text{any } g_i(t) < 0]|H\big), \quad i = 1, 2, \ldots, n, \quad (6)$$

in which $P[F_t|H]$ is the probability of failure under a given hazard H at time t and is calculated as the probability of violating any of the limit states, where $g_i(t) < 0$ is the failure event associated with the ith limit state at time t. Therefore, the probability of structural failure $P_f(t)$ is

$$P_f(t) = P(F_t|H) \cdot P(H)$$

where $P(H)$ is the probability of occurrence of the hazard. Accordingly, the risk of failure under a given hazard is expressed as (Ellingwood, 2006; Zhu & Frangopol, 2013a)

$$RI(t) = P_f(t) \cdot C_f(t), \qquad (8)$$

where $C_f(t)$ represents the time-variant consequences associated with the failure of the structure. For the proper sustainability and risk analyses, these consequences should include various direct and indirect losses associated with failure of the investigated system.

For assessing the life-cycle performance of single bridges, Decò and Frangopol (2011) proposed a risk-based approach that considers bridges under the effects of traffic, corrosion, scour and seismic events. The output of such approach is the life-cycle risk profile for an ageing bridge. Zhu and Frangopol (2013a) also used risk as a performance indicator to find the optimum EM and PM times and types for bridges subjected to ageing effects, in addition to seismic hazards. In Saydam et al. (2013), a methodology for quantifying lifetime risk of bridge superstructures is presented. The risk was quantified in terms of the expected direct and indirect losses. A scenario-based approach, which uses the Pontis element condition rating system, was used for identifying the expected losses. The deterioration process of bridge components was modelled as a Markov process. In addition, a reliability-based approach was used to compute the probabilities of component and system failure, given the condition states.

Risk has also been used for assessing the performance of spatially distributed bridges, bridge networks (Furuta et al., 2011; Shiraki et al., 2007), and lifelines (Jayaram & Baker, 2010). Saydam et al. (2013) presented a methodology for assessing the time-dependent expected losses of deteriorated highway bridge networks. A five-state Markov model was used to predict the time-dependent performance of bridges. The direct consequences were assessed on the basis of scenarios characterised by individual bridge failures and maintenance shutdowns. The indirect consequences were quantified by solving the traffic assignment problem. The time-variant direct, indirect, and total expected losses were computed for a given time horizon. In Decò and Frangopol (2013), bridge vulnerability was evaluated with respect to seismic and abnormal traffic hazards. The effects induced by seismic hazard were investigated by means of fragility analysis. Random earthquakes were generated using Latin hypercube sampling (Iman, 2008), and probabilities of exceeding specific structural damage states were computed for each specific seismic scenario. Traffic hazard was assessed considering Weibull distributed time-to-failure of the bridge superstructure. Consequence analysis included different levels of bridge serviceability (fully serviceable, partly serviceable, closed and collapsed), and monetary values were used to evaluate direct and indirect consequences.

Furuta et al. (2011) performed life-cycle cost analysis which considers the socio-economic effects due to the collapse of structures and used a stochastic model to evaluate the structural response under earthquake effects. Shiraki et al. (2007) presented a methodology to establish transportation network system risk curves that provide the plots of the annual probability of exceedance for different levels of network delays. Monte Carlo simulation was used with bridge fragility analyses to evaluate the bridge damage index after a seismic event. Frangopol and Bocchini (2012) presented a comprehensive review of the methodologies for the performance analyses, management, and optimisation of bridge networks.

2.1.1 Sustainability

The American Society of Civil Engineers defines sustainability as 'a set of environmental, economic and social conditions in which all of society has the capacity and opportunity to maintain and improve its quality of life indefinitely without degrading the quantity, quality or availability of natural, economic, and social resources' (American Society of Civil Engineers, 2014a). As evidenced by this definition, sustainability is an integral property that is quantified in terms of social, environmental, and economic metrics. Although sustainability assessment approaches for buildings have been established since the early 1990s, this field is not well developed for the case of infrastructure systems (Bocchini, 2013).

For infrastructure systems, research has been performed in an attempt to formulate quantitative measures for the three sustainability metrics, especially for bridges and bridge networks. In order to use a unified metric for sustainability evaluation, social and environmental metrics of the sustainability are often expressed in their associated monetary value. This can be done by evaluating the consequences associated with these two metrics (Bocchini, Frangopol, Ummenhofer, & Zinke, 2014; Dong et al., 2013). In this manner, sustainability evaluation can be seen as a more detailed risk analysis which considers a wider range of consequences. Extreme events, such as earthquakes, can have severe consequences which may reduce infrastructure functionality and affect emergency responses and recovery operations. These consequences may be much higher than the repair or rebuilding costs of the damaged infrastructure system. Following a strong earthquake and its aftershocks, infrastructure systems, such as bridge networks, can be damaged causing human injuries, deaths, property loss and traffic detours.

In this regard, several studies aimed to identify the life-cycle cost and environmental and social impacts of bridge failures and retrofit actions. Padgett and Tapia (2013) proposed a risk-based method for life-cycle environmental sustainability analysis. Their study was an attempt to quantify environmental indicators of sustainability for structures subjected to multiple hazards; the proposed model was applied to a case study to quantify sustainability metrics for an existing bridge. In addition, the impact of seismic retrofit on lifetime sustainability metrics was investigated. In Dong et al. (2013), the time-variant sustainability of bridges associated with multiple hazards, in addition to the effects of structural deterioration was presented. The approach accounted for the effects of flood-induced scour on the seismic fragility of bridges. Sustainability evaluation included the expected downtime, number of fatalities, expected energy waste and carbon dioxide emissions, and the expected economic loss. The effects of corrosion on the reinforcement bars and the subsequent concrete cover spalling were considered. Nonlinear finite element analyses were used to obtain the seismic fragility curves at different points in time.

Dong et al. (2014a) extended their approach to investigate the time-variant seismic sustainability and risk assessment of highway bridge networks. The bridge network was considered to be composed of bridges and links which may be damaged during a seismic event. The methodology considers the occurrence probability of a set of seismic scenarios reflecting the seismic activity of the region. The sustainability and risk depend on the damage states of the links and the bridges within the network following an earthquake scenario. The time variation of the sustainability metrics and risk due to deterioration was identified.

A challenge associated with the sustainability assessment is related to computing the monetary values associated with environmental and social impacts. In general, models for evaluating the life-cycle cost considering the environmental impact of pollutants associated with different construction types can be found in the literature. For instance, Kendall et al. (2008) evaluated the life-cycle cost of different bridge deck alternatives considering agency, user and environmental costs. Soliman and Frangopol (2015) computed the life-cycle cost associated with steel bridge material alternatives and Churchill and Panesar (2013) computed the life-cycle cost of noise barriers constructed by using photocatalytic cement and compared this cost to that arising from traditional concrete. However, in some cases, it may be difficult to specify several of the cost parameters associated with the sustainability assessment. For instance, the cost of carbon dioxide carries a large variability and it may be difficult to quantify (Pindyck, 2013). However, for structural engineering applications, an estimate can usually be done (see for example Churchill & Panesar, 2013; Kendall et al., 2008). Furthermore, if the assessment aims to find the best design among certain available alternatives, a comparative study will be performed, and in this case, the cost of carbon dioxide will not have a significant effect on the final decision.

2.1.2 Resilience

The previously discussed approaches for risk management focused on the infrastructure performance assessment and management before the occurrence of a disruptive event. These approaches provide the expected loss associated with the failure and planning for maintenance/retrofit activities that can mitigate the effects of extreme events. However, after the occurrence of a disruptive event, another indicator should be defined to model the infrastructure performance restoration in order to allow for the optimisation of restoration and repair activities for damaged components. Resilience, as a performance indicator, has been introduced to assist in performing such a task. Resilience in civil engineering can be defined as (Bruneau et al., 2003) 'the ability of social units (e.g. organisations, communities) to mitigate hazards, contain the effects of disasters when they occur, and carry out

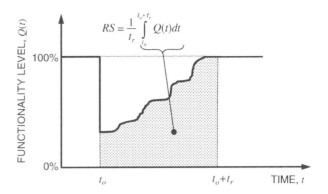

Figure 6. Graphical representation of the resilience.

recovery activities in ways that minimise social disruption and mitigate the effects of future earthquakes'. A normalised measure to quantify the resilience of a system or a network is to find the integration over time of the functionality (Bocchini & Frangopol, 2012; Cimellaro, Reinhorn, & Bruneau, 2010a; Frangopol & Bocchini, 2011):

$$RS = \frac{1}{t_r} \int_{t_o}^{t_o+t_r} Q(t) \mathrm{d}t, \qquad (9)$$

in which Q is the functionality, t_o is the occurrence time of the extreme event and t_r is the investigated time horizon. This definition of resilience can be represented graphically as indicated in Figure 6. Several studies used this definition of resilience to support decision-making and infrastructure management after the occurrence of an extreme event. Applications covered bridge networks (Bocchini & Frangopol, 2012), healthcare facilities (Cimellaro, Reinhorn, & Bruneau, 2010b; Jacques et al., 2014), and lifelines (Çagnan, Davidson, & Guikema, 2006; Chang & Shinozuka, 2004; Liu, Giovinazzi, MacGeorge, & Beukman, 2013; Ouyang, Dueñas-Osorio, & Min, 2012).

The methodologies mentioned previously can predict the performance of infrastructure systems under the effects of sudden hazards and gradual deterioration; however, a generalised integrated multi-hazard performance evaluation approach is still missing.

3. Life-cycle optimisation and decision-making

Optimisation represents the core of the life-cycle management process. This process is responsible for presenting the output of the life-cycle management plans to support the decision-making process. Optimal management actions that fulfil the management objectives under the given budgetary, time, safety, and functionality constraints are established in this process. The complexity of the optimisation depends heavily on the scale of the problem (i.e. component-level analysis vs. system level) where the optimisation becomes significantly more computationally demanding when analysing networks of damaged structures.

Studies on the life-cycle optimisation can consider different deterioration phases along the service life of a system. Several studies aimed to perform the inspection, maintenance and retrofit scheduling under gradual deterioration phenomena (Garbatov & Guedes Soares, 2001; Kim & Frangopol, 2011a, 2011b; Kong & Frangopol, 2003a, 2003b; Kong & Frangopol, 2005; Kwon & Frangopol, 2011, 2012; Soliman & Frangopol, 2014) while others considered also the effects of sudden hazards (Padgett et al., 2010; Rokneddin et al., 2013; Venkittaraman & Banerjee, 2013; Zhu & Frangopol, 2013a).

3.1 Inspection, maintenance and retrofit optimisation

Establishing the best inspection, maintenance and retrofit schedules requires a robust optimisation process which integrates the proper damage occurrence and propagation models with the previous knowledge about the safety and financial constraints. For instance, for the given system performance profiles presented in Figure 7, two possible life-cycle maintenance plans can be identified. Maintenance Plan 2 considers performing only one EM during the service life, while Maintenance Plan 1 proposes performing two PM actions through the service life. Although both plans provide the same service life, the life-cycle cost of these options may be significantly different. This is especially true if all the aspects of life-cycle maintenance cost (e.g. traffic delays and environmental impact) are considered. As can be expected, for a real structure, a very large number of potential maintenance options are available, and evaluating all these options individually may be time-consuming. Therefore, a robust optimisation technique is essential for obtaining the optimum life-cycle maintenance strategy. Similarly, an efficient optimisation

procedure is required to find the optimum inspection, monitoring, repair, and retrofit schedules. Single and multi-objective optimisation procedures can be constructed to fulfil the management objectives. For a single structure, these objectives include maximising the service life (Kim, Frangopol, & Soliman, 2013), minimising the damage detection delay (Kim & Frangopol, 2011b, 2011c), minimising the total life-cycle cost (Kim et al., 2013; Okasha & Frangopol, 2010a; Zhu & Frangopol, 2013a) and maximising the probability of damage detection (PoD) (Chung et al., 2006; Soliman, Frangopol, & Kim, 2013). For networks of damaged structures, these objectives include minimising the life-cycle cost and maximising the network reliability, which are often defined in terms of the connectivity (Bocchini & Frangopol, 2011, 2013; Liu & Frangopol, 2005b). Various probabilistic performance indicators can be used for this process such as the reliability index and risk, among others. Moreover, the probabilistic damage level (e.g. crack size or corrosion depth) can be used as the performance indicator.

For structures subjected to fatigue effects, Kwon and Frangopol (2011) proposed a reliability-based approach for scheduling the life-cycle inspection and repair activities. Their approach used a probabilistic crack growth model integrated into a reliability-based approach. The main goal of their study was to extend the life of critical details by applying the appropriate and timely inspection and maintenance actions such that they meet the target service life or a target reliability level. The outputs of the management process are the optimum inspection and repair times when adopting different inspection qualities and maintenance types. The appropriate maintenance actions were selected using the PoD functions associated with each inspection method and a crack size based threshold. Kim and Frangopol (2011c) proposed an approach for obtaining the optimum inspection schedule of fatigue critical steel details based

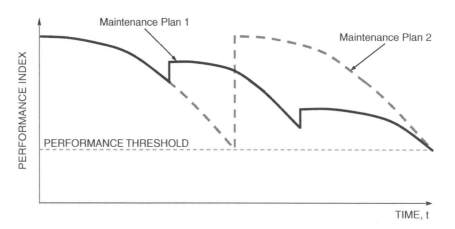

Figure 7.   Two life-cycle maintenance plans.

on minimising the damage detection delay. This delay is measured as the time elapsed between the damage occurrence and detection and it can be critical for structural deterioration phenomena characterised by high propagation rates (e.g. fatigue in aluminium vessels).

Probabilistic event tree analysis was used in their study to formulate the damage detection delay with different inspection scenarios. The model represents all the possible inspection events and their particular outcomes (i.e. detecting or not detecting the damage during an inspection). The formulation of the damage detection delay can be integrated in a single objective optimisation process to find the optimum inspection times which minimises the damage detection delay in structural details subjected to fatigue. This optimisation problem is formulated as (Kim & Frangopol, 2011c)

$$\text{Find } \mathbf{t_{ins}} = \left\{ t_{\text{ins},1}, t_{\text{ins},2}, \ldots, t_{\text{ins},n} \right\}, \quad (10\text{a})$$

$$\text{to minimize } E(t_{\text{del}}), \quad (10\text{b})$$

$$\text{such that } t_{\text{ins},i} - t_{\text{ins},i-1} \geq 1 \text{ year}, \quad (10\text{c})$$

$$\text{given } n, f_T(t), \text{ parameters of PoD function.} \quad (10\text{d})$$

where $\mathbf{t_{ins}}$ is a vector consisting of n design variables representing inspection times $t_{\text{ins},1}, t_{\text{ins},2}, \ldots, t_{\text{ins},n}$; $t_{\text{ins},i}$ is the ith inspection time (years); $f_T(t)$ is the PDF of the initial service life (i.e. time-to-failure) and $E(t_{\text{del}})$ is the expected damage detection delay.

In this procedure, the only objective was to minimise the damage detection delay for a given number of inspections with no consideration of the inspection cost. This cost can be very high in some cases as inspections may induce traffic delays that normally lead to high users' costs. Moreover, some inspections may require placing the structure out of service for a certain amount of time. As a result, finding the optimum balance between the cost and safety can be a challenging process that can be solved by multi-objective optimisation procedures. A typical solution of a bi-objective optimisation problem comes in the form of the Pareto-optimal solution set (i.e. optimum trade-offs among all objectives) presented in Figure 8. A solution is Pareto-optimal if there does not exist another solution that improves at least one objective without worsening another one (Arora, 2011). Genetic algorithms (GAs) have been successfully used to solve such complex optimisation problems mainly due to (a) their dependency only on the objective function and not on its derivatives, which are, in many cases, difficult or impossible to be evaluated; (b) the ease of handling discrete variables; (c) the search from a population of points rather than from a single point; and (d) the ease of implementation in a parallel computing environment, which significantly reduces the computational effort.

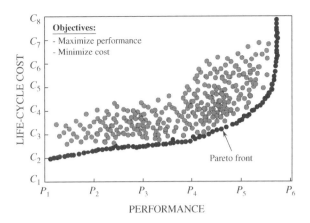

Figure 8. Pareto-optimal solution set of bi-objective optimisation for maximising the performance and minimising the life-cycle cost.

Kim and Frangopol (2012) proposed a bi-objective optimisation process to find a well-balanced inspection plan that can simultaneously minimise the damage detection delay and inspection cost. The formulation is capable of finding the Pareto-optimal solution set when the same inspection method is applied during the life cycle and when different inspection methods are applied. The optimiser in this application can be programmed to find the optimum number of inspections, inspection times, and the quality of each inspection. In addition, the approach can simultaneously consider the scheduling for inspection and monitoring interventions. The outcomes of this optimisation process are the inspection times, quality of inspection, monitoring times and monitoring durations for a given number of inspections and/or monitoring actions. Kim et al. (2013) proposed another approach for the inspection/maintenance planning that can be applied to various types of structures and deteriorating mechanisms. The model used the physical damage level and PoD functions to evaluate the damage detection at each inspection. It was assumed in their study that maintenance is performed as soon as the damage is detected. The objectives of this management procedure were maximising the expected service life simultaneously with minimising the expected total life-cycle cost.

Considering multiple types of maintenance actions within the optimisation scheme, in which the optimiser should select the times and types of maintenance actions, significantly increases the complexity of the optimisation process. Multiple approaches for maintenance optimisation have been proposed over the last decade. Some of them only optimise the EM actions, others optimise PM actions, while few of them optimise both. Regardless of the performance index used in the optimisation process, whenever more than one maintenance option is available, a method has to be defined to select the best maintenance options and their application times. Regarding EM optimisation, Estes and Frangopol (1999) and Okasha and Frangopol (2009) identified the EM type to be applied as the one that provides the lowest present cost per year of increase in the service life. Okasha and Frangopol (2010a) proposed an algorithm based on event tree analysis which calculates and compares the cost of all available EM scenarios required to reach the pre-specified service life of the structure.

For PM scheduling, multiple approaches have been proposed. Neves et al. (2006) considered uniform time intervals between the applications of PM actions. Two design variables were used: one for the first application time and the second for the time interval between subsequent applications. Okasha and Frangopol (2010a) used continuous design variables for the time of application of PM and integer variables for the optimum number of applications of each maintenance type. The performance of the structure was modelled by using lifetime functions (Leemis, 1995). Specifically, the unavailability and redundancy were used as performance indicators, based on which optimum maintenance strategies were sought.

Models based on Markov decision process have also been introduced to solve inspection and maintenance optimisation problems. Such models can incorporate the outcomes of inspections and SHM actions into the life-cycle cost and performance optimisation process. Uncertainties associated with the state of the system at different decision stages can be accounted for in a partially observable Markov decision process which does not assume that the state of a system is precisely known. Examples of such approaches implementing stationary Markov models can be found in Faddoul, Raphael, and Chateauneuf (2011) and Corotis, Ellis, and Jiang (2005) and non-stationary Markov models in Papakonstantinou and Shinozuka (2014).

Several other applications of multi-objective optimisation in the life-cycle management can be found in the literature to obtain the optimal trade-offs between conflicting life-cycle management aspects. These aspects included the inspection cost and the PoD of inspection plans (Soliman et al., 2013), transportation network performance and maintenance cost (Bocchini & Frangopol, 2011), expected loss of transportation network failures (as a measure of sustainability), and retrofit cost (Dong, Frangopol, & Saydam, 2014b). Other studies for maintenance and retrofit optimisation include Brown et al. (2013), Faturechi and Miller-Hooks (2013), Frangopol and Okasha (2009), Furuta, Kameda, Fukuda, & Frangopol (2004), Liu and Frangopol (2005a, 2005b, 2005c, 2006) and Neves et al. (2006).

3.2 Post-hazard functionality restoration

In the post-hazard functionality line of research, optimisation procedures are aimed to provide decision-makers with the optimum solutions to restore the system functionality after a disruptive event. Within the last decade, several studies have been performed to study such problems (Bocchini, 2013; Bocchini & Frangopol, 2012; Chang et al., 2012; Mackie, Wong, & Stojadinovic, 2010; Padgett & DesRoches, 2007; Venkittaraman & Banerjee, 2013). The quantitative resilience index given by Equation (9) represents one of the best measures to address this functionality restoration problem. Thus, it has been adopted in many studies to develop a network performance indicator. Because bridges are critical components within transportation networks, maintaining a minimum acceptable level of functionality during an extreme event plays a key role in ensuring that disaster response teams will reach their destination.

In this context, Bocchini and Frangopol (2012) presented an approach for the intervention prioritisation for bridges along a highway segment subjected to extreme events. Based on the information about the damage levels after the event and the bridge characteristics, the proposed computational approach can establish the optimal intervention schedule. In this model, the design variables are the delay between the extreme event and the start of the restoration and the speed of the restoration. The outcomes of the process are the starting time of intervention and the progress pace of the restoration. The objectives of the optimisation were to maximise the resilience of the highway stretch and minimise the total cost of interventions. Because the two objectives were conflicting, GAs were used to find the Pareto-optimal solution set. This approach was focused on restoring the network functionality to its pre-event level and is more suitable for the long-term recovery planning. Bocchini (2013) generalised this approach by making it applicable to different phases of the recovery.

The approaches presented previously have been successful in addressing the infrastructure management problem with a wide range of applications. Future research efforts should focus on formulating the necessary tools needed to put these approaches in the hands of decision-makers and disaster managers.

3.3 Utility-based life-cycle decision-making

Although the optimisation processes discussed above provide decision-making support for infrastructure managers, the desirability of a given alternative may depend on several attributes such as monetary value, time factors, environmental measures, and social impact (Ang & Tang, 1984). In these cases, quantifying the monetary value associated with these alternatives may not be possible and an alternative methodology for supporting the decision-making process is required. The concept of utility can be used to transfer various attribute values into a uniform scale always ranging from 0 (worst case) to 1 (best case). In such cases, utility can provide a true measure of decision-makers' desirability of a given alternative. In addition, this model enables the integration of the infrastructure manager's risk attitude into the decision-making process.

Several forms of the utility function can be found in the literature. Among those are the exponential, logarithmic and quadratic utility functions. The exponential utility function, for instance, can be defined as (Ang & Tang, 1984)

$$u(x) = \frac{1}{1 - \exp(-\gamma)}\left[1 - \exp\left(-\gamma \frac{x_{max} - x}{x_{max}}\right)\right], \quad (11)$$

where x is the risk under investigation, γ indicates the risk attitude of the decision-maker (e.g. $\gamma > 0$ indicates risk-aversion) and x_{max} denotes the maximum value which is utilised to normalise the utility function so that it always takes values between 0 and 1. Accordingly, the risk and sustainability can be evaluated in terms of utility while incorporating the attitude of the decision-maker.

Because the sustainability assessment of infrastructure systems requires the analysis of various aspects related to the social, economic, and environmental metrics, the utility theory lends itself to such analysis. Once the utility function associated with each of the sustainability attributes is appropriately established, a multi-attribute utility that effectively represents all aspects of sustainability can be obtained. By using an additive formulation for the multi-attribute utility function, utility values associated with each attribute are multiplied by weighting factors and summed over all attributes involved (Stewart, 1996). Hence, the multi-attribute utility associated with a structural system can be computed as (Jiménez, Ríos-Insua, & Mateos, 2003)

$$u_S = w_{Ec}u_{Ec}(Ec) + w_S u_S(S) + w_{En}u_{En}(En), \quad (12)$$

where w_{Ec}, w_S and w_{En} are the weighting factors corresponding to each sustainability metric; u_{Ec}, u_S and u_{En} are the utility functions for the economic, social, and environmental attributes, respectively; and Ec, S and En are the values of the three sustainability metrics.

In order to integrate utility into the life-cycle management process in which the interventions (e.g. inspection, maintenance and retrofit actions) are optimised along the life cycle of a system, the change in the utility (i.e. benefit) arising from these interventions has to be quantified. For instance, the benefit of a seismic retrofit action implemented to a bridge may be evaluated considering a certain time interval, earthquake model and the seismic performance of the bridge. Next, the utility associated with the sustainability metrics with and without retrofit actions can be evaluated using the aforementioned multi-attribute utility model.

In order to directly compare the cost and benefit associated with an intervention alternative and determine the overall effectiveness of a particular life-cycle management plan, Dong, Frangopol, and Sabatino (2015) adopted a cost–benefit indicator defined in terms of the benefit utility associated with an intervention alternative and the cost utility associated with same alternative. Their model was used to find the optimum retrofit plan for bridge networks under seismic hazards. An optimisation process was formulated and solved to find the optimum type of retrofit action performed on each bridge within a network such that the utility associated with the retrofit cost for the entire bridge network is maximised and the utility associated with the retrofit benefit considering a specific time interval is maximised. Their study concluded that the risk attitude of the decision-maker has a significant impact on the optimal solutions resulting from the proposed utility-based life-cycle management approach.

4. Role of SHM and inspection information in the performance updating and damage detection

Information from SHM and non-destructive inspection provides a clear indication on the current condition state of the structure under investigation (Okasha, Frangopol, & Orcesi, 2012; Soliman & Frangopol, 2014; Zhu & Frangopol, 2013b). Moreover, it can provide an early warning regarding structural damage before significant maintenance or repairs are required. Although one of the main objectives of SHM is to identify the actual structural responses under normal operational loads, other objectives also include detecting damage in various structural components, predicting the structural performance, and assisting in the optimisation of management interventions, among others.

Information obtained from SHM and non-destructive testing provides useful data that can be used to update performance prediction models. This updated performance prediction will result in an updated intervention schedule by adopting the process schematically shown in Figure 9. The updating process is essential for the effective implementation of life-cycle management because inspection and SHM information carry unavoidable uncertainties. Thus, maintenance and repair decisions should be made based on both the inspection/SHM data and the results of prediction models (Zheng & Ellingwood, 1998). Within the last two decades, multiple approaches have been proposed for updating performance prediction models based on inspection outcomes and/or SHM data. Moan and Song (2000) proposed an approach to update the reliability of a system of fatigue critical details based on the results of the inspection of several components in the system. Other approaches in this field use the Bayesian updating of model parameters. In this process, information

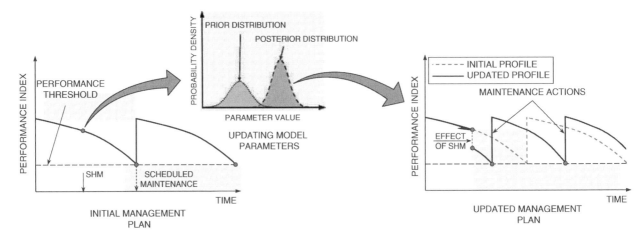

Figure 9. Performance updating based on SHM or inspection information.

from inspection or SHM is used to represent the likelihood function which can be combined with the prior information on model parameters to find their posterior distributions. The posterior distributions can be found as (Ang & Tang, 2007)

$$P(\boldsymbol{\theta}|\mathbf{d}) = \frac{P(\boldsymbol{\theta}) \cdot P(\mathbf{d}|\boldsymbol{\theta})}{\int P(\boldsymbol{\theta}) \cdot P(\mathbf{d}|\boldsymbol{\theta}) \mathrm{d}\boldsymbol{\theta}}, \qquad (13)$$

where $P(\boldsymbol{\theta}|\mathbf{d})$ is the posterior distribution of model parameters $\boldsymbol{\theta}$ given the additional information $\mathbf{d}$; $P(\boldsymbol{\theta})$ represents the prior distribution of model parameters; $P(\mathbf{d}|\boldsymbol{\theta})$ is the likelihood function of obtaining information $\mathbf{d}$ conditioned by $\boldsymbol{\theta}$; and $\mathbf{d}$ and $\boldsymbol{\theta}$ are the vectors of observed data and model parameters, respectively.

By knowing the prior distributions and the likelihood function, the posterior distributions can be established using sampling approaches based on Markov chain Monte Carlo simulation such as the Metropolis algorithm (Metropolis, Rosenbluth, Rosenbluth, Teller, & Teller, 1953) or the slice sampling algorithm (Neal, 2003). Although this updating process may be computationally expensive, it has been successfully included within the proposed life-cycle management framework.

Soliman and Frangopol (2014) proposed an approach for obtaining the updated optimal inspection times based on information collected during previous inspections performed within the initially designed inspection plan. They considered the following likelihood function (Perrin, Sudret, & Pendola, 2007):

$$P(\mathbf{d}|\boldsymbol{\theta}) = \prod_{i=1}^{n} \left[\frac{1}{\sqrt{2\pi} \cdot \sigma_e} \cdot \exp\left(-\frac{1}{2} \left(\frac{d_i - a_{p,i}}{\sigma_e} \right)^2 \right) \right], \qquad (14)$$

where d_i and $a_{p,i}$ are the observed and predicted data, respectively, at the ith inspection; and σ_e represents a single error term combining the measurement and modelling errors which is assumed to follow a normal distribution with zero mean and standard deviation σ_e (i.e. $N(0, \sigma_e)$). Based on studying the sample space of inspection outcomes, subsequent inspection times can be found.

Okasha and Frangopol (2012) proposed another approach for integrating the information obtained from SHM in the life-cycle bridge management framework. In their approach, bridge performance is predicted using extreme value statistics. Advanced modelling tools and techniques were used for lifetime reliability computations, including incremental nonlinear finite element analyses, quadratic response surface modelling using design of experiments concepts and Latin hypercube sampling, among other techniques. The new information is considered in the form of discrete sample values such as the maximum daily bending moment affecting a certain cross section over a given period of time. Accordingly, the likelihood function was defined as (Ang & Tang, 2007)

$$P(\mathbf{d}|\theta) = \prod_{i=1}^{N} f_X(x_i|\theta), \qquad (15)$$

in which $f_X(x_i|\theta)$ is the PDF of X evaluated at the SHM data value x_i. This likelihood function was used to update the extreme value distributions of the bending moments of the bridge girders, and accordingly, the time-dependent probability of failure profile was updated. A similar approach was used in Okasha, Frangopol, and Decò (2010) to evaluate the reliability of ship structures with the integration of SHM information. Okasha et al. (2012) proposed another approach for the automated finite element updating using strain data recorded by SHM crawl tests. The automated finite element model updating technique was used for updating the resistance parameters of the structure. The results from the crawl tests were used to update the finite element model and, in turn, to update the lifetime reliability.

The inclusion of monitoring results into reliability evaluation and life-cycle assessment of highway bridges has major advantages. However, monitoring systems can, in some cases, add economic burden to the life-cycle management process. The cost of monitoring generally consists of (a) preparation cost, (b) maintenance cost of monitoring system (sensors, wiring and data acquisition systems), (c) analysis and report preparation costs and (d) continuous review of data for long-term monitoring programmes (Frangopol et al., 2008a). Each of the discussed items contains personnel and material costs. Among these items, the continuous review of monitoring data contributes as the maximum portion of the monitoring cost for monitoring actions lasting up to 3 years (Frangopol et al., 2008a). Therefore, the proper life-cycle management technique should seek the optimal balance between the safety requirements and the financial constraints.

Another benefit of integrating SHM into the life-cycle assessment is that it can enable automated damage detection. Using the SHM information, damage can be detected at early stages before causing any significant reliability loss to the monitored structure. Multiple approaches have been proposed for such task and many of them rely on vibration-based methods. In these methods, the change in the measured dynamic response is used as an indication of damage occurrence in the monitored structure (Sohn, Farrar, Hunter, & Worden, 2001). Statistical pattern recognition approaches are mostly used for such purposes, in which signatures obtained from the recorded signals are used to extract features which change with the damage occurrence (Farrar, Duffey, Doebling, & Nix, 1999; Nair, Kiremidjian, & Law, 2006). Several damage detection approaches based on time-series methods have been proposed within the last decades (Da Silva & Dias Junior, 2007; Farrar et al., 1999; Gul & Catbas, 2009, 2011; Nair et al., 2006; Sohn, Czarnecki, & Farrar, 2000; Sohn & Farrar, 2001). Most of these models construct autoregressive models by using the data acquired from each sensor.

Other models based on vector autoregressive modelling (Mattson & Pandit, 2006) have been proposed. These models describe a particular signal in terms of its own past values in addition to the past vales of other sensors (Okasha, Frangopol, Saydam, & Salvino, 2011). The main benefit of these methods is that they do not require the modelling of the entire structure, a task that can be difficult and time-consuming especially for large-scale structures (Zheng & Mita, 2007). In addition to these methods, the change point detection method has been recently applied for the damage detection in civil structural systems (Nigro, Pakzad, & Dorvash, 2014; Noh, Rajagopal, & Kiremidjian, 2013). A change point detection method sequentially receives features extracted from structural vibration data and conducts a hypothesis test (Tartakovsky & Veeravalli, 2005).

Novel monitoring systems used in structural engineering contain sensors providing a very large amount of data. The proper handling of the continuously provided monitoring data is one of the main difficulties in SHM (Frangopol, Strauss, & Kim, 2008b; Strauss et al., 2008; Zheng & Mita, 2007; Liu, Frangopol, & Kwon, 2010). Methodologies for processing the large amount of data are still an active research topic.

5. Discussion

Within the past two decades, major developments in the field of infrastructure management of deteriorating structures have been achieved. The parallel advancements in the computational capabilities enabled complex system- and network-level simulations and optimisation problems to be solved. This led to more robust life-cycle management approaches that can accurately predict the structural performance under uncertainty and identify the optimal life-cycle interventions schedule. These approaches have been successfully performed at the component, system, and network levels. This paper provided a brief overview of the recent achievements in the field of the life-cycle management of deteriorating infrastructure systems. Probabilistic performance prediction and evaluation approaches have been discussed. In addition, the optimisation of the life-cycle interventions under the effects of gradual and sudden deterioration has been presented. SHM information and its integration for the efficient life-cycle management have also been presented. Recent achievements associated with the performance evaluation and life-cycle optimisation of structural systems under uncertainty have been addressed.

As evidenced by the previous sections of the paper, the performance prediction process can be considered as the foundation of life-cycle management of infrastructure. This prediction process depends, to a great extent, on the accuracy of the performance prediction models and the descriptors of their probabilistic parameters. However, in some cases, the accurate information regarding some model parameters does not exist. Therefore, future efforts to quantify these parameters are crucial. In addition to experimental investigations, long-term SHM can aid in improving the performance assessment process of civil and marine structures. Although approaches for performance assessment based on SHM information have been well developed, damage detection based on SHM is still an active line of research. Methodologies for processing the large amount of data for damage diagnosis and prognosis in existing structures are still required. This is especially true for the case of naval structures where more research is still needed to enhance the decision-making process based on SHM information.

It can be seen that most of the studies dealing with maintenance and rehabilitation optimisation impose assumptions with respect to the degree of performance improvements and damage-level reduction due to maintenance application, for example, assuming that performance is restored to the initial level after maintenance actions. However, information regarding the effects of maintenance on the probabilistic performance indicators is generally hard to obtain. More research is needed to capture the true effects of maintenance on the overall system performance. Once obtained, this information can be easily utilised in the presented life-cycle management approaches.

Among the future challenges, the most insurmountable is perhaps related to efforts needed to reduce the gap between the theory and practice in the life-cycle analysis and management field. A large number of efficient and effective life-cycle management techniques exist; however, the real-world application of these methodologies does not exist. The calibration and verification of such methods are still necessary. However, with no real-world use of such methods, this calibration process becomes more difficult. Although funding decisions are still made on a short-term basis, it has been recently recognised that life-cycle analysis can significantly reduce long-term cost and increase the sustainability and resilience of our infrastructure (American Society of Civil Engineers, 2014b). Accordingly, future research efforts should focus on formulating the necessary tools to put these approaches in the hands of decision-makers and infrastructure managers. Focus should also be placed on developing codes and guidelines for life-cycle assessment and management for various types of infrastructure systems.

Disclosure statement

No potential conflict of interest was reported by the authors.

Funding

The support by grants from (a) the National Science Foundation (NSF) Award [CMS-0639428], (b) the Commonwealth of Pennsylvania, Department of Community and Economic Development, through the Pennsylvania Infrastructure Technology Alliance (PITA), (c) the US Federal Highway Administration (FHWA) Cooperative Agreement Award [DTFH61-07-H-00040], (d) the US Office of Naval Research (ONR) Awards [N00014-08-1-0188, N00014-12-1-0023] and (e) the National Aeronautics and Space Administration (NASA) Award [NNX10AJ20G] is gratefully acknowledged. The opinions presented in this paper are those of the authors and do not necessarily reflect the views of the sponsoring organisations.

References

ABAQUS (2009). *ABAQUS/standard version 6.9 user's manuals*. Pawtucket, RI: Hibbit, Karlsson, and Soreson.

Akiyama, M., Frangopol, D.M., Arai, M., & Koshimura, S. (2013). Reliability of bridges under tsunami hazard: Emphasis on the 2011 Great East Japan earthquake. *Earthquake Spectra, 29*, S295–S314.

Akiyama, M., Frangopol, D.M., & Matsuzaki, H. (2011). Life-cycle reliability of RC bridge piers under seismic and airborne chloride hazards. *Earthquake Engineering and Structural Dynamics, 40*, 1671–1687.

Akiyama, M., Frangopol, D.M., & Suzuki, M. (2012). Integration of the effects of airborne chlorides into reliability-based durability design of reinforced concrete structures in a marine environment. *Structure and Infrastructure Engineering, 8*, 125–134.

American Association of State Highway and Transportation Officials (2014). *AASHTO LRFD bridge design specifications* (7th ed.). Washington, DC: American Association of State Highway and Transportation Officials.

American Society of Civil Engineers (2014a). *ASCE and sustainability*. Retrieved from http://www.asce.org/Sustainability/ASCE-and-Sustainability/ASCE--Sustainability

American Society of Civil Engineers (2014b). *Maximizing the value of investments using life cycle cost analysis* (ASCE Life Cycle Cost Analysis Report). Retrieved from http://www.asce.org/Infrastructure/Life-Cycle-Cost-Analysis-Report/

Ang, A.H-S., & De Leon, D. (2005). Modeling and analysis of uncertainties for risk-informed decisions in infrastructures engineering. *Structure and Infrastructure Engineering, 1*, 19–31.

Ang, A.H.-S., & Tang, W.H. (1984). *Probability concepts in engineering planning and design: Decision, risk and reliability* (Vol. II). New York, NY: Wiley.

Ang, A.H.-S., & Tang, W.H. (2007). *Probability concepts in engineering: Emphasis on applications to civil and environmental engineering* (2nd ed.). Hoboken, NJ: John Wiley & Sons.

Arora, J. (2011). *Introduction to optimum design* (3rd ed.). New York, NY: Elsevier Academic Press.

Barone, G., & Frangopol, D.M. (2013a). Hazard-based optimum lifetime inspection and repair planning for deteriorating structures. *Journal of Structural Engineering, 139*, 04013017.

Barone, G., & Frangopol, D.M. (2013b). Reliability, risk and lifetime distributions as performance indicators for life-cycle maintenance of deteriorating structures. *Reliability Engineering and System Safety, 123*, 21–37.

Barone, G., & Frangopol, D.M. (2014). Life-cycle maintenance of deteriorating structures by multi-objective optimization involving reliability, risk, availability, hazard and cost. *Structural Safety, 48*, 40–50.

Barone, G., Frangopol, D.M., & Soliman, M. (2014). Optimization of life-cycle maintenance of deteriorating bridges considering expected annual system failure rate and expected cumulative cost. *Journal of Structural Engineering, 140*, 04013043.

Bastidas-Arteaga, E., Bressolette, P.H., Chateauneauf, A., & Sánchez-Silva, M. (2009). Probabilistic lifetime assessment of RC structures under coupled corrosion-fatigue deterioration processes. *Structural Safety, 31*, 84–96.

Bastidas-Arteaga, E., Chateauneauf, A., Sanchez-Silva, M., Bressolette, P.H., & Schoefs, F. (2010). Influence of weather and global warming in chloride ingress into concrete: A stochastic approach. *Structural Safety*, *32*, 238–249.

Biondini, F., Bontempi, F., Frangopol, D.M., & Malerba, P.G. (2006). Probabilistic service life assessment and maintenance planning of concrete structures. *Journal of Structural Engineering*, *132*, 810–825.

Biondini, F., Camnasio, E., & Palermo, A. (2014). Lifetime seismic performance of concrete bridges exposed to corrosion. *Structure and Infrastructure Engineering*, *10*, 880–900.

Biondini, F., & Frangopol, D.M. (2008). Probabilistic limit analysis and lifetime prediction of concrete structures. *Structure and Infrastructure Engineering*, *4*, 399–412.

Bjarnadottir, S., Li, Y., & Stewart, M.G. (2011). A probabilistic-based framework for impact and adaptation assessment of climate change on hurricane damage risk and costs. *Structural Safety*, *33*, 173–185.

Bocchini, P. (2013). Computational procedure for the assisted multi-phase resilience-oriented disaster management of transportation systems. In G. Deodatis, B.R. Ellingwood, & Frangopol (Eds.), *Safety, reliability, risk, and life-cycle performance of structures and infrastructures* (pp. 581–588). London: CRC Press, Taylor and Francis Group.

Bocchini, P., & Frangopol, D.M. (2011). A probabilistic computational framework for bridge network optimal maintenance scheduling. *Reliability Engineering and System Safety*, *96*, 332–349.

Bocchini, P., & Frangopol, D.M. (2012). Optimal resilience- and cost-based postdisaster intervention prioritization for bridges along a highway segment. *Journal of Bridge Engineering*, *17*, 117–129.

Bocchini, P., & Frangopol, D.M. (2013). Connectivity-based optimal scheduling for maintenance of bridge networks. *Journal of Engineering Mechanics*, *139*, 170–769.

Bocchini, P., Frangopol, D.M., Ummenhofer, T., & Zinke, T. (2014). Resilience and sustainability of the civil infrastructure: Towards a unified approach. *Journal of Infrastructure Systems*, *20*, 0414004.

British Standards Institute (2005). *Guide to methods for assessing the acceptability of flaws in metallic structures* (BS7910). London: British Standards Institute.

Brown, N.J.K., Gearhart, J.L., Jones, D.A., Nozick, L.K., Romero, N., & Xu, N. (2013). Multi-objective optimization for bridge retrofit to address earthquake hazards. In *Proceedings of the 2013 Winter Simulation Conference* (pp. 2475–2486). Catonsville, MD: INFORMS Simulation Society.

Bruneau, M., Chang, S.E., Eguchi, R.T., Lee, G.C., O'Rourke, T. D., Reinhorn, A.M., Shinozuka, M., Tierney, K., Wallace, W.A., & von Winterfeldt, D. (2003). A framework to quantitatively assess and enhance the seismic resilience of communities. *Earthquake Spectra*, *19*, 733–752.

Chang, L., Peng, F., Ouyang, Y., Elnashai, A.S., & Spencer, B.F. Jr (2012). Bridge seismic retrofit program planning to maximize postearthquake transportation network capacity. *Journal of Infrastructure Systems*, *18*, 75–88.

Chang, S.E., McDaniels, T.L., Mikawoz, J., & Peterson, K. (2007). Infrastructure failure interdependencies in extreme events: Power outage consequences in the 1998 ice storm. *Natural Hazards*, *41*, 337–358.

Chang, S.E., & Shinozuka, M. (1996). Life-cycle cost analysis with natural hazard risk. *Journal of Infrastructure Systems*, *23*, 118–126.

Chang, S.E., & Shinozuka, M. (2004). Measuring improvements in the disaster resilience of communities. *Earthquake Spectra*, *20*, 739–755.

Chung, H., Manuel, L., & Frank, K. (2006). Optimal inspection scheduling of steel bridges using nondestructive testing techniques. *Journal of Bridge Engineering*, *113*, 305–319.

Churchill, C.J., & Panesar, D.K. (2013). Life-cycle cost analysis of highway noise barriers designed with photocatalytic cement. *Structure and Infrastructure Engineering*, *9*, 983–998.

Cimellaro, G.P., Reinhorn, A.M., & Bruneau, M. (2010a). Framework for analytical quantification of disaster resilience. *Engineering Structures*, *32*, 3639–3649.

Cimellaro, G.P., Reinhorn, A.M., & Bruneau, M. (2010b). Seismic resilience of a hospital system. *Structure and Infrastructure Engineering*, *6*, 127–144.

Corotis, R.B., Ellis, J.H., & Jiang, M. (2005). Modeling of risk-based inspection, maintenance and life-cycle cost with partially observable Markov decision process. *Structure and Infrastructure Engineering*, *1*, 75–84.

Çagnan, Z., Davidson, R.A., & Guikema, S.D. (2006). Post-earthquake restoration planning for Los Angeles electric power. *Earthquake Spectra*, *22*, 589–608.

Da Silva, S., & Dias Junior, M. (2007). Statistical damage detection in a stationary rotor systems through time series analysis. *Latin American Applied Research*, *37*, 243–246.

Decò, A., & Frangopol, D.M. (2011). Risk assessment of highway bridges under multiple hazards. *Journal of Risk Research*, *14*, 1057–1089.

Decò, A., & Frangopol, D.M. (2013). Life-cycle risk assessment of spatially distributed aging bridges under seismic and traffic hazards. *Earthquake Spectra*, *29*, 127–153.

Dong, Y., Frangopol, D.M., & Sabatino, S. (2015). Optimizing bridge network retrofit planning based on cost-benefit evaluation and multi-attribute utility associated with sustainability. *Earthquake Spectra* (in press).

Dong, Y., Frangopol, D.M., & Saydam, D. (2013). Time-variant sustainability assessment of seismically vulnerable bridges subjected to multiple hazards. *Earthquake Engineering and Structural Dynamics*, *42*, 1451–1467.

Dong, Y., Frangopol, D.M., & Saydam, D. (2014a). Sustainability of highway bridge networks under seismic hazard. *Journal of Earthquake Engineering*, *18*, 41–66.

Dong, Y., Frangopol, D.M., & Saydam, D. (2014b). Pre-earthquake multi-objective probabilistic retrofit optimization of bridge networks based on sustainability. *Journal of Bridge Engineering*, *19*, 04014004.

El Hassan, J., Bressolette, P., Chateauneuf, A., & El Tawil, K. (2010). Reliability-based assessment of the effect of climate conditions on the corrosion of RC structures subjected to chloride ingress. *Engineering Structures*, *32*, 3279–3287.

Ellingwood, B.R. (2005). Risk-informed condition assessment of civil infrastructure: State of practice and research issues. *Structure and Infrastructure Engineering*, *1*, 7–18.

Ellingwood, B.R. (2006). Mitigating risk from abnormal loads and progressive collapse. *Journal of Performance of Constructed Facilities*, *20*, 315–323.

Ellingwood, B.R., & Kinali, K. (2009). Quantifying and communicating uncertainty in seismic risk assessment. *Structural Safety*, *31*, 179–187.

Enright, M.P., & Frangopol, D.M. (1999). Maintenance planning for deteriorating concrete bridges. *Journal of Structural Engineering*, *125*, 1407–1414.

Estes, A.C., & Frangopol, D.M. (1998). RELSYS: A computer program for structural system reliability analysis. *Structural Engineering and Mechanics, 6*, 901–919.

Estes, A.C., & Frangopol, D.M. (1999). Repair optimization of highway bridges using system reliability approach. *Journal of Structural Engineering, 125*, 766–775.

Estes, A.C., & Frangopol, D.M. (2001). Minimum expected cost-oriented optimal maintenance planning for deteriorating structures: Application to concrete bridge decks. *Reliability Engineering and System Safety, 73*, 281–291.

Faddoul, R., Raphael, W., & Chateauneuf, A. (2011). A generalised partially observable Markov decision process updated by decision trees for maintenance optimisation. *Structure and Infrastructure Engineering, 7*, 783–796.

Farrar, C.R., Duffey, T.A., Doebling, S.W., & Nix, D.A. (1999). A statistical pattern recognition paradigm for vibration-based structural health monitoring. In *Proceedings of the 2nd international workshop on structural health monitoring* (pp. 764–773). Stanford, CA: Structures And Composites Laboratory, Stanford University.

Faturechi, R., & Miller-Hooks, E. (2013). A mathematical framework for quantifying and optimizing protective actions for civil infrastructure systems. *Computer-Aided Civil and Infrastructure Engineering.* doi:10.1111/mice.12027

Frangopol, D.M. (2011). Life-cycle performance, management, and optimization of structural systems under uncertainty: Accomplishments and challenges. *Structure and Infrastructure Engineering, 7*, 389–413.

Frangopol, D.M., & Akiyama, M. (2011). Lifetime seismic reliability analysis of corroded reinforced concrete bridge piers, Chapter 23. In M. Papadrakakis, M. Fragiadakis, & N.D. Lagaros (Eds.), *Computational methods in earthquake engineering.* In E. Onate (Series Ed.), Computational methods in applied sciences, (Vol. 21, pp. 527–537). Dordrecht: Springer Science+Business Media B.V.

Frangopol, D.M., & Bocchini, P. (2011). Resilience as optimization criterion for the bridge rehabilitation of a transportation network subject to earthquake. In D. Ames, T.L. Droessler, & M. Hoit (Eds.), *Proceedings of the 2011 ASCE structures congress* (pp. 2044–2055). Reston, VA: ASCE.

Frangopol, D.M., & Bocchini, P. (2012). Bridge network performance, maintenance and optimisation under uncertainty: Accomplishments and challenges. *Structure and Infrastructure Engineering, 8*, 341–356.

Frangopol, D.M., Bocchini, P., Deco, A., Kim, S., Kwon, K., Okasha, N.M., & Saydam, D. (2012). Integrated life-cycle framework for maintenance, monitoring, and reliability of naval ship structures. *Naval Engineers Journal, 124*, 89–99.

Frangopol, D.M., & Estes, A.C. (1997). Lifetime bridge maintenance strategies based on system reliability. *Structural Engineering International, 7*, 193–198.

Frangopol, D.M., & Okasha, N.M. (2009). Multi-criteria optimisation of life-cycle maintenance programs using advanced modelling and computational tools. In *Trends in civil and structural computing* (pp. 1–26). Stirlingshire: Saxe-Coburg Publications.

Frangopol, D.M., Strauss, A., & Kim, S. (2008a). Bridge reliability assessment based on monitoring. *Journal of Bridge Engineering, 13*, 258–270.

Frangopol, D.M., Strauss, A., & Kim, S. (2008b). Use of monitoring extreme data for the performance prediction of structures: General approach. *Engineering Structures, 30*, 3644–3653.

Furuta, H., Frangopol, D.M., & Nakatsu, K. (2011). Life-cycle cost of civil infrastructure with emphasis on balancing structural performance and seismic risk of road network. *Structure and Infrastructure Engineering, 7*, 65–74.

Furuta, H., Kameda, T., Fukuda, Y., & Frangopol, D.M. (2004). Life-cycle cost analysis for infrastructure systems: Life cycle cost vs. safety level vs. service life. *Life-Cycle Performance of Deteriorating Structures: Assessment, Design and Management, ASCE*, 19–25.

Garbatov, Y., & Guedes Soares, C. (2001). Cost and reliability based strategies for fatigue maintenance planning of floating structures. *Reliability Engineering and System Safety, 73*, 293–301.

Ghosh, J., & Padgett, J.E. (2010). Aging considerations in the development of time-dependent seismic fragility curves. *Journal of Structural Engineering, 136*, 1497–1511.

Gul, M., & Catbas, F.N. (2009). Statistical pattern recognition for structural health monitoring using time series modeling: Theory and experimental verifications. *Mechanical Systems and Signal Processing, 23*, 2192–2204.

Gul, M., & Catbas, F.N. (2011). Structural health monitoring and damage assessment using a novel time series analysis methodology with sensor clustering. *Journal of Sound and Vibration, 330*, 1196–1210.

Iman, R.L. (2008). Latin hypercube sampling. In E.L. Melnick & B.S. Everitt (Eds.), *Encyclopedia of quantitative risk analysis and assessment* (pp. 1–5). Hoboken, NJ: John Wiley & Sons, Ltd.

Jacques, C.C., McIntosh, J., Giovinazzi, S., Kirsch, T.D., Wilson, T., & Mitrani-Reiser, J. (2014). Resilience of the Canterbury hospital system to the 2011 Christchurch earthquake. *Earthquake Spectra.* doi:10.1193/032013EQS074M

Jayaram, N., & Baker, J.W. (2010). Efficient sampling and data reduction techniques for probabilistic seismic lifeline risk assessment. *Earthquake Engineering and Structural Dynamics, 39*, 1109–1131.

Jiménez, A., Ríos-Insua, S., & Mateos, A. (2003). A decision support system for multiattribute utility evaluation based on imprecise assignments. *Decision Support Systems, 36*, 65–79.

Joint Committee on Structural Safety (2008). Risk assessment in engineering: Principles, system representation and risk criteria. In M.H. Faber (Ed.), *Risk assessment in engineering* (pp. 5–13). Denmark: Joint Committee on Structural Safety.

Karmakar, D., Ray-Chaudhuri, S., & Shinozuka, M. (2014). Finite element model development, validation and probabilistic seismic performance evaluation of Vincent Thomas suspension bridge. *Structure and Infrastructure Engineering.* doi:10.1080/15732479.2013.863360

Kendall, A., Keoleian, G., & Helfand, G. (2008). Integrated life-cycle assessment and life-cycle cost analysis model for concrete bridge deck applications. *Journal of Infrastructure Systems, 14*, 214–222.

Kim, S., & Frangopol, D.M. (2011a). Cost-based optimum scheduling of inspection and monitoring for fatigue sensitive structures under uncertainty. *Journal of Structural Engineering, 137*, 1319–1331.

Kim, S., & Frangopol, D.M. (2011b). Inspection and monitoring planning for RC structures based on minimization of expected damage detection delay. *Probabilistic Engineering Mechanics, 26*, 308–320.

Kim, S., & Frangopol, D.M. (2011c). Optimum inspection planning for minimizing fatigue damage detection delay of ship hull structures. *International Journal of Fatigue, 33*, 448–459.

Kim, S., & Frangopol, D.M. (2012). Probabilistic bicriterion optimum inspection/monitoring planning: Application to naval ships and bridges under fatigue. *Structure and Infrastructure Engineering, 8*, 912–927.

Kim, S., Frangopol, D.M., & Soliman, M. (2013). Generalized probabilistic framework for optimum inspection and maintenance planning. *Journal of Structural Engineering, 139*, 435–447.

Kong, J.S., & Frangopol, D.M. (2003a). Life-cycle reliability-based maintenance cost optimization of deteriorating structures with emphasis on bridges. *Journal of Structural Engineering, 129*, 818–828.

Kong, J.S., & Frangopol, D.M. (2003b). Evaluation of expected life-cycle maintenance cost of deteriorating structures. *Journal of Structural Engineering, 129*, 682–691.

Kong, J.S., & Frangopol, D.M. (2004a). Prediction of reliability and cost profiles of deteriorating structures under time- and performance-controlled maintenance. *Journal of Structural Engineering, 130*, 1865–1874.

Kong, J.S., & Frangopol, D.M. (2004b). Cost-reliability interaction in life-cycle cost optimization of deteriorating structures. *Journal of Structural Engineering, 130*, 1704–1712.

Kong, J.S., & Frangopol, D.M. (2005). Probabilistic optimization of aging structures considering maintenance and failure costs. *Journal of Structural Engineering, 131*, 600–616.

Kwon, K., & Frangopol, D.M. (2010). Bridge fatigue reliability assessment using probability density functions of equivalent stress range based on field monitoring data. *International Journal of Fatigue, 32*, 1221–1232.

Kwon, K., & Frangopol, D.M. (2011). Bridge fatigue assessment and management using reliability-based crack growth and probability of detection models. *Probabilistic Engineering Mechanics, 26*, 471–480.

Kwon, K., & Frangopol, D.M. (2012). Fatigue life assessment and lifetime management of aluminum ships using life-cycle optimization. *Journal of Ship Research, 56*, 91–105.

Leemis, L.M. (1995). *Reliability, probabilistic models and statistical methods*. Upper Saddle River, NJ: Prentice Hall.

Li, Y., & Ellingwood, B.R. (2006). Hurricane damage to residential construction in the US: Importance of uncertainty modeling in risk assessment. *Engineering Structures, 28*, 1009–1018.

Liu, M., & Frangopol, D.M. (2005a). Bridge annual maintenance prioritisation under uncertainty by multiobjective combinatorial optimisation. *Computer-Aided Civil and Infrastructure Engineering, 20*, 343–352.

Liu, M., & Frangopol, D.M. (2005b). Balancing connectivity reliability of deteriorating bridge networks and long-term maintenance cost through optimization. *Journal of Bridge Engineering, 10*, 468–481.

Liu, M., & Frangopol, D.M. (2005c). Multiobjective maintenance planning optimisation for deteriorating bridges considering condition, safety, and life-cycle cost. *Journal of Structural Engineering, 131*, 833–842.

Liu, M., & Frangopol, D.M. (2006). Optimising bridge network maintenance management under uncertainty with conflicting criteria: Life-cycle maintenance, failure and user costs. *Journal of Structural Engineering, 131*, 1835–1845.

Liu, M., Frangopol, D.M., & Kwon, K. (2010). Fatigue reliability assessment of retrofitted steel bridges integrating monitoring data. *Structural Safety, 32*, 77–89.

Liu, M., Giovinazzi, S., MacGeorge, R., & Beukman, P. (2013). Wastewater network restoration following the Canterbury,

NZ earthquake sequence: Turning post-earthquake recovery into resilience enhancement. *International Efforts in Lifeline Earthquake Engineering.* doi:10.1061/9780784413234.021

Lundie, S., Peters, G.M., & Beavis, P.C. (2004). Life cycle assessment for sustainable metropolitan water systems planning. *Environmental Science and Technology, 38*, 3465–3473.

Mackie, K.R., Wong, J., & Stojadinovic, B. (2010). Post-earthquake bridge repair cost and repair time estimation methodology. *Earthquake Engineering and Structural Dynamics, 39*, 281–301.

MathWorks (2014a). *MATLAB programming fundamentals.* Natick, MA: The MathWorks.

MathWorks (2014b). *Global optimization toolbox user's guide.* Natick, MA: The MathWorks.

Mattson, S.G., & Pandit, S.M. (2006). Statistical moments of autoregressive model residuals for damage localisation. *Mechanical Systems and Signal Processing, 20*, 627–645.

Melchers, R.E. (1999). *Structural reliability analysis and prediction* (2nd ed.). Chichester: John Wiley & Sons Ltd.

Metropolis, N., Rosenbluth, A., Rosenbluth, M., Teller, A., & Teller, M. (1953). Equation of state calculations by fast computing machines. *The Journal of Chemical Physics, 21*, 1087–1092.

Moan, T., & Song, R. (2000). Implications of inspection updating on system fatigue reliability of offshore structures. *Journal of Offshore Mechanics and Arctic Engineering, 122*, 173–180.

Modarres, M. (2006). *Risk analysis in engineering: Techniques, tools, and trends.* London: CRC Press, Taylor and Francis Group.

Nair, K.K., Kiremidjian, A.S., & Law, K.H. (2006). Time series-based damage detection and localization algorithm with application to the ASCE benchmark structure. *Journal of Sound and Vibration, 291*, 349–368.

Neal, R.M. (2003). Slice sampling. *The Annuals of Statistics, 31*, 705–767.

Neves, L.C., Frangopol, D.M., & Cruz, P.J. (2006). Probabilistic lifetime-oriented multi-objective optimisation of bridge maintenance: Single maintenance type. *Journal of Structural Engineering, 132*, 991–1005.

Nigro, M., Pakzad, S., & Dorvash, S. (2014). Localized structural damage detection: A change point analysis. *Computer-Aided Civil and Infrastructure Engineering.* doi:10.1111/mice. 12059

Noh, H., Rajagopal, R., & Kiremidjian, A.S. (2013). Sequential structural damage diagnosis algorithm using a change point detection method. *Journal of Sound and Vibration, 332*, 6419–6433.

Okasha, N.M., & Frangopol, D.M. (2009). Lifetime oriented multi-objective optimisation of structural maintenance considering system reliability, redundancy and life-cycle cost using GA. *Structural Safety, 31*, 460–474.

Okasha, N.M., & Frangopol, D.M. (2010a). Novel approach for multi-criteria optimization of life-cycle preventive and essential maintenance of deteriorating structures. *Journal of Structural Engineering, 136*, 1009–1022.

Okasha, N.M., & Frangopol, D.M. (2010b). Redundancy of structural systems with and without maintenance: An approach based on lifetime functions. *Reliability Engineering and System Safety, 95*, 520–533.

Okasha, N.M., & Frangopol, D.M. (2012). Integration of structural health monitoring in a system performance based life-cycle bridge management framework. *Structure and Infrastructure Engineering, 8*, 999–1016.

Okasha, N.M., Frangopol, D.M., & Decò, A. (2010). Integration of structural health monitoring in life-cycle performance assessment of ship structures under uncertainty. *Marine Structures*, 23, 303–321.

Okasha, N.M., Frangopol, D.M., & Orcesi, A.D. (2012). Automated finite element updating using strain data for the lifetime reliability assessment of bridges. *Reliability Engineering and System Safety*, 99, 139–150.

Okasha, N.M., Frangopol, D.M., Saydam, D., & Salvino, L.W. (2011). Reliability analysis and damage detection in high-speed naval craft based on structural health monitoring data. *Structural Health Monitoring*, 10, 361–379.

Otieno, M.B., Beushausen, H.D., & Alexander, M.G. (2011). Modelling corrosion propagation in reinforced concrete structures — A critical review. *Cement and Concrete Composites*, 33, 240–245.

Ouyang, M., Dueñas-Osorio, L., & Min, X. (2012). A three-stage resilience analysis framework for urban infrastructure systems. *Structural Safety*, 36, 23–31.

Padgett, J.E., Dennemann, K., & Ghosh, J. (2010). Risk-based seismic life-cycle cost-benefit LCC-B analysis for bridge retrofit assessment. *Structural Safety*, 32, 165–173.

Padgett, J.E., & DesRoches, R. (2007). Bridge functionality relationships for improved seismic risk assessment of transportation networks. *Earthquake Spectra*, 23, 115–130.

Padgett, J.E., & Tapia, C. (2013). Sustainability of natural hazard risk mitigation: Life cycle analysis of environmental indicators for bridge infrastructure. *Journal of Infrastructure Systems*, 19, 395–408.

Papakonstantinou, K.G., & Shinozuka, M. (2013). Probabilistic model for steel corrosion in reinforced concrete structures of large dimensions considering crack effects. *Engineering Structures*, 57, 306–326.

Papakonstantinou, K.G., & Shinozuka, M. (2014). Optimum inspection and maintenance policies for corroded structures using partially observable Markov decision processes and stochastic, physically based models. *Probabilistic Engineering Mechanics*, 37, 93–108.

Perrin, F., Sudret, B., & Pendola, M. (2007). Bayesian updating of mechanical models-application in fracture mechanics. In *Proceedings of the 18 Ème Congrès Frances De Mècanique* (pp. 27–32). Courbevoie, France: AFM, Maison de la Mécanique.

Pindyck, R.S. (2013). Pricing carbon when we don't know the right price. *Regulation* (Cato Institute) 36, 43–46.

Rausand, M., & Høyland, A. (2004). *System reliability theory: Models, statistical methods, and applications*. New York, NY: John Wiley & Sons.

Rinaldi, S.M. (2004). Modeling and simulating critical infrastructures and their interdependencies. In *Proceedings of the 37th annual Hawaii international conference on system sciences*. IEEE. doi:10.1109/HICSS.2004.1265180

Rokneddin, K., Ghosh, J., Dueñas-Osorio, L., & Padgett, J.E. (2013). Bridge retrofit prioritisation for ageing transportation networks subject to seismic hazards. *Structure and Infrastructure Engineering*, 9, 1050–1066.

Saydam, D., Bocchini, P., & Frangopol, D.M. (2013). Time-dependent risk associated with deterioration of highway bridge networks. *Engineering Structures*, 54, 221–233.

Saydam, D., & Frangopol, D.M. (2011). Time-dependent performance indicators of damaged bridge superstructures. *Engineering Structures*, 33, 2458–2471.

Shinozuka, M., Murachi, Y., Dong, X., Zhou, Y., & Orlikowski, M.J. (2003). Effect of seismic retrofit of bridges on transportation networks. *Earthquake Engineering and Engineering Vibration*, 2, 169–179.

Shiraki, N., Shinozuka, M., Moore, J.E., Chang, S.E., Kameda, H., & Tanaka, S. (2007). System risk curves: Probabilistic performance scenarios for highway networks subject to earthquake damage. *Journal of Infrastructure Systems*, 13, 43–54.

Sohn, H., Czarnecki, J.A., & Farrar, C.R. (2000). Structural health monitoring using statistical process control. *Journal of Structural Engineering*, 126, 1356–1363.

Sohn, H., & Farrar, C.R. (2001). Damage diagnosis using time series analysis of vibration signals. *Smart Materials and Structures*, 10, 446–451.

Sohn, H., Farrar, C.R., Hunter, N.F., & Worden, K. (2001). Structural health monitoring using statistical pattern recognition techniques. *Journal of Dynamic Systems, Measurement, and Control*, 123, 706–711.

Soliman, M., & Frangopol, D.M. (2014). Life-cycle management of fatigue sensitive structures integrating inspection information. *Journal of Infrastructure Systems*, 20, 04014001.

Soliman, M., & Frangopol, D.M. (2015). Life-cycle cost evaluation of conventional and corrosion-resistant steel for bridges. *Journal of Bridge Engineering*, 20, 0614005.

Soliman, M., Frangopol, D.M., & Kim, S. (2013). Probabilistic optimum inspection planning of steel bridges with multiple fatigue sensitive details. *Engineering Structures*, 49, 996–1006.

Stein, S., Young, G., Trent, R., & Pearson, D. (1999). Prioritizing scour vulnerable bridges using risk. *Journal of Infrastructure Systems*, 5, 95–101.

Stewart, M.G., Wang, X., & Nguyen, M.N. (2011). Climate change impact and risks of concrete infrastructure deterioration. *Engineering Structures*, 33, 1326–1337.

Stewart, M.G., Wang, X., & Nguyen, M.N. (2012). Climate change adaptation for corrosion control of concrete infrastructure. *Structural Safety*, 35, 29–39.

Stewart, T.J. (1996). Robustness of additive value function methods in MCDM. *Journal of Multi-Criteria Decision Analysis*, 5, 301–309.

Strauss, A., Frangopol, D.M., & Kim, S. (2008). Use of monitoring extreme data for the performance prediction of structures: Bayesian updating. *Engineering Structures*, 30, 3654–3666.

Tartakovsky, A., & Veeravalli, V. (2005). General asymptotic Bayesian theory of quickest change detection. *Theory of Probability and Its Applications*, 49, 458–497.

Turconi, R., Simonsen, C.G., Byriel, I.P., & Astrup, T. (2014). Life cycle assessment of the Danish electricity distribution network. *The International Journal of Life Cycle Assessment*, 19, 100–108.

Venkittaraman, A., & Banerjee, S. (2013). Enhancing resilience of highway bridges through seismic retrofit. *Earthquake Engineering and Structural Dynamics*. doi:10.1002/eqe.2392

VisualDOC (2012). *VisualDOC user's manual, version 7.1*. Colorado Springs, CO: Vanderplaats Research and Development.

Zheng, H., & Mita, A. (2007). Two-stage damage diagnosis based on the distance between ARMA models and pre-whitening filters. *Smart Materials and Structures*, 16, 1829–1836.

Zheng, R., & Ellingwood, B.R. (1998). Role of non-destructive evaluation in time-dependent reliability analysis. *Structural Safety*, 20, 325–339.

Zhou, Y., Banerjee, S., & Shinozuka, M. (2010). Socio-economic effect of seismic retrofit of bridges for highway transportation networks: A pilot study. *Structure and Infrastructure Engineering*, *6*, 145–157.

Zhu, B., & Frangopol, D.M. (2013a). Risk-based approach for optimum maintenance of bridges under traffic and earth-quake loads. *Journal of Structural Engineering*, *139*, 422–434.

Zhu, B., & Frangopol, D.M. (2013b). Incorporation of structural health monitoring data on load effects in the reliability and redundancy assessment of ship cross-sections using Bayesian updating. *Structural Health Monitoring*, *12*, 377–392.

Bridge life-cycle performance and cost: analysis, prediction, optimisation and decision-making

Dan M. Frangopol , You Dong and Samantha Sabatino

ABSTRACT

The development of a generalised framework for assessing bridge life-cycle performance and cost, with emphasis on analysis, prediction, optimisation and decision-making under uncertainty, is briefly addressed. The central issue underlying the importance of the life-cycle approach to bridge engineering is the need for a rational basis for making informed decisions regarding design, construction, inspection, monitoring, maintenance, repair, rehabilitation, replacement and management of bridges under uncertainty which is carried out by using multi-objective optimisation procedures that balance conflicting criteria such as performance and cost. A number of significant developments are summarised, including time-variant reliability, risk, resilience, and sustainability of bridges, bridge transportation networks and interdependent infrastructure systems. Furthermore, the effects of climate change on the probabilistic life-cycle performance assessment of highway bridges are addressed. Moreover, integration of SHM and updating in bridge management and probabilistic life-cycle optimisation considering multi-attribute utility and risk attitudes are presented.

1. Introduction

The condition of civil infrastructure systems around the world is degrading due to a variety of deteriorating mechanisms, including ageing, environmental stressors, man-made hazards (e.g. blasts and fires) and natural hazards (e.g. earthquakes and hurricanes), among others. Consequently, improving the overall condition and safety of deteriorating infrastructure systems is a key concern worldwide. For example, the American Society of Civil Engineers reported, within the 2013 Report Card for America's Infrastructure, that the average age of the United States' 607,380 bridges was 42 years (ASCE, 2013). Additionally, nearly a quarter of these highway bridges were classified as either structurally deficient or functionally obsolete (FHWA, 2013). Therefore, it is crucial to implement rational management strategies that maintain performance of highway bridges within acceptable levels through their life-cycle. Life-cycle management is widely recognised as an effective tool for maximising the cost-effectiveness of implementing intervention actions that improve condition and safety, and extend the service life of deteriorating infrastructure systems.

In order to predict performance of structural systems during their life-cycle under uncertainty, deterioration mechanisms for the investigated systems (e.g. corrosion and fatigue) must be carefully considered. Aggressive environmental conditions and natural ageing processes facilitate a gradual reduction in the performance (e.g. system reliability) of existing structures.

Alternatively, there are extreme events that cause an abrupt reduction in the functionality of structures such as blasts, fires, earthquakes, hurricanes and terrorist attacks. During their life-cycle, bridges can be subjected to multiple hazards. Thus, it is necessary to consider the performance of bridges under multiple hazards in the risk assessment and mitigation procedure, all in a life-cycle context. Life-cycle assessment of deteriorating highway bridges includes aleatory and epistemic uncertainties associated with natural randomness and inaccuracies in the prediction or estimation of reality, respectively (Ang & Tang, 2007). These are present within modelling the structural resistance (e.g. material properties and geometrical characteristics), the occurrence and magnitude of hazards that may impact the structure (e.g. corrosion, fatigue, earthquakes, floods and hurricanes), operating conditions, and loading cases, among others; uncertainties are also associated with the interventions performed during the service life of structures (e.g. inspection, maintenance, monitoring, repair and replacement) and their costs. Due to these uncertainties, it is imperative for structural engineers to accurately model and assess the structural performance and total cost within a probabilistic life-cycle context. Furthermore, the effects of maintenance, repair and rehabilitation on structural life-cycle performance must be well understood. The influence of maintenance and repairs on structural performance can be incorporated in a generalised framework for multi-criteria optimisation of the life-cycle management of infrastructure systems (Ellingwood

Based on the T.Y. Lin plenary lecture and the associated paper presented at the 8th International Conference on Bridge Maintenance, Safety and Management (IABMAS2016), Iguassu Falls, Paraná, Brazil, 26–30 June, 2016.

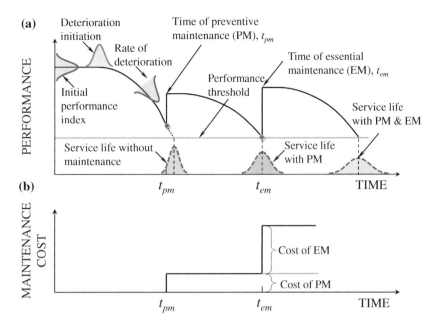

Figure 1. Effects of gradual deterioration and preventive and essential maintenance on (a) structural performance and (b) maintenance cost.

& Frangopol, 2016; Frangopol, 2011; Frangopol & Liu, 2007; Frangopol & Soliman, 2016). Within the last two decades, several studies introduced probabilistic techniques which can assist the bridge management process (Biondini, Camnasio, & Palermo, 2014; Biondini & Frangopol, 2016; Biondini, Frangopol, & Malerba, 2008; Enright & Frangopol, 1999a, 1999b; Estes & Frangopol, 2001; Frangopol, Kallen, & van Noortwijk, 2004; Frangopol & Kim, 2011; Frangopol, Kong, & Gharaibeh, 2001; Frangopol, Lin, & Estes, 1997; Frangopol & Liu, 2007; Frangopol & Okasha, 2009; Frangopol & Soliman, 2016; Ghosn et al., 2016a, 2016b; Kim & Frangopol, 2017; Koh & Frangopol, 2008; Kong, Ababneh, Frangopol, & Xi, 2002; Kong & Frangopol, 2003; Lim, Akiyama, & Frangopol, 2016; Miyamoto, Kawamura, & Nakamura, 2000; Morcous & Lounis, 2005; Neves, Frangopol, & Cruz, 2006; Okasha & Frangopol, 2010a, 2010b; Sánchez-Silva, Frangopol, Padgett, & Soliman, 2016; Stewart & Rosowsky, 1998; Thanapol, Akiyama, & Frangopol, 2016, among others).

The effects of maintenance on the probabilistic performance profile (such as reliability index) and cost are depicted in Figure 1. Within this figure, the probabilistic aspect of performance prediction is illustrated by the probability density functions (PDFs) of the initial performance index, deterioration initiation, rate of deterioration, and service life (a) without maintenance, (b) with preventive maintenance (PM) only and (c) with both preventive and essential maintenance (EM). In general, preventive maintenance is applied to slightly improve the performance or delay the deteriorating process of a bridge in order to keep the bridge above the required level of structural performance. Preventive maintenance actions for a deteriorating bridge includes replacing small parts, patching concrete, repairing cracks, changing lubricants, and cleaning and painting exposed parts, among others. On the other hand, essential maintenance is typically a performance-based intervention. As depicted in Figure 1(a), essential maintenance is applied when the bridge performance level reaches a predefined threshold. Essential maintenance actions lead to much higher levels of bridge performance than preventive maintenance actions,

but they typically cost more. Strengthening and replacement of bridge components are examples of essential maintenance actions. Furthermore, the effects of maintenance on the total cost of bridge management must be considered. Figure 1(b) shows the cumulative maintenance cost as a function of time for preventive and essential maintenance interventions.

Performance of bridge systems may be represented by a variety of indicators. Approaches for the life-cycle management of bridges involving reliability performance indicators consider uncertainties associated with loads, resistance and modelling, but are not able to account for the consequences incurred from bridge failure. Risk-based indicators provide the means to combine the probability of structural failure with the consequences associated with this event (Ang & De Leon, 2005; Ellingwood, 1998, 2005, 2006; Lounis & McAllister, 2016; Saydam & Frangopol, 2014; Saydam, Frangopol, & Dong, 2013; Zhu & Frangopol, 2013b). Several risk approaches within a generalised life-cycle framework are presented in this paper. Furthermore, methodologies considering sustainability as a performance indicator are discussed. The incorporation of sustainability in the life-cycle performance assessment and management procedures allows for the effective integration of economic, social and environmental aspects. A sustainability performance metric may be established considering multi-attribute utility theory, which facilitates the combination of several risks while incorporating the risk attitude of the decision-maker (Jiménez, Ríos-Insua, & Mateos, 2003). This particular sustainability performance indicator has been applied to the life-cycle management of bridges (Sabatino, Frangopol, & Dong, 2015a, 2015b) and bridge networks (Dong, Frangopol, & Sabatino, 2015). Additionally, risk and sustainability concepts may be successfully integrated within optimal bridge management planning. A general process outlining the use of reliability, risk, multi-attribute utility and sustainability concepts within a robust decision-making process regarding bridge management is shown in Figure 2. The goals of implementing optimal bridge management plans are to improve the performance and functionality of bridges, mitigate detrimental consequences and

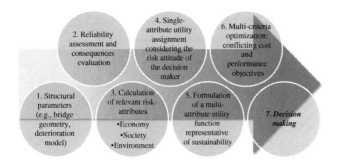

Figure 2. General procedure for sustainability-informed decision-making.

minimise costs. These ultimate aims are satisfied with a comprehensive life-cycle framework, like the one shown in Figure 2.

Resilience is another structural performance indicator that accounts for structural performance, along with recovery patterns under hazard effects (Bruneau et al., 2003). Presidential Policy Directive (PPD, 2013) defined resilience as 'the ability to prepare for and adapt to changing conditions and withstand and recover from disruptions.' Considering the effects of uncertainties, it is crucial for the quantification of resilience at the holistic level to be processed through a probabilistic framework. Several deterministic and few probabilistic studies have been reported to analyse the resilience of individual bridges and bridge networks (Biondini, Camnasio, & Titi, 2015; Bocchini & Frangopol, 2012; Decò, Bocchini, & Frangopol, 2013; Decò & Frangopol, 2013; Dong & Frangopol, 2015, 2016b; Titi & Biondini, 2013).

The main effects of climate change on the performance prediction of bridges are investigated and summarised herein. Measurements taken over the last decades indicate that the sea level, global temperature and ocean temperature are all rising at elevated rates (Allison et al., 2009; Church & White, 2006, 2011; Levitus et al., 2009; Peterson & Baringer, 2009). Additionally, a significant increase in carbon dioxide (CO_2) concentration in the atmosphere has been observed (IPCC, 2007, 2014). Since these trends will continue within the near future, it is crucial to determine the effects of climate change on the performance and life-cycle assessment of deteriorating infrastructure systems.

This paper presents an overview of life-cycle management concepts for bridge systems under uncertainty and the application of such concepts in bridge sustainability considering the risk attitude of the decision-maker. Risk- and sustainability-informed management of bridges under the effects of both gradual and sudden deteriorations is investigated. Quantifying the life-cycle performance, risk, and sustainability of bridges at the component and network levels is also addressed. Additionally, the effect of climate change on probabilistic performance is examined herein. Moreover, bridge management planning and optimisation under a constrictive budget and performance constraints are presented through a probabilistic management framework. This framework can serve as a useful tool in risk mitigation and, in general, decision-making associated with bridges. The approach presented can provide optimal intervention strategies to the decision-maker that will allow for risk- and sustainability-informed decisions regarding maintenance of individual bridges, bridge networks and interdependent infrastructure systems during their lifetime.

2. Performance evaluation and prediction

Performance of bridge structures can be quantified at the cross-section, component, whole structure (system), group of structures (network) and network of networks levels. In most of the current bridge design and assessment codes, performance requirements are based on component strength. Typically, performance assessment activities associated with bridge components rely on visual inspections results. For bridges, visual inspection results are usually employed to establish a condition rating index to indicate the bridge's remaining load-carrying capacity. The bridges in the United States are rated using two different methods based on visual inspection. The first method uses the National Bridge Inventory (NBI) condition rating system (FHWA, 2013). According to the NBI condition rating system, the condition evaluation corresponds to the physical state of the deck, superstructure and substructure components of a bridge. The second method uses the element-level condition rating method to represent the conditions of bridge components. Generally, bridge management systems characterize the performance of structural elements with discrete condition states which incorporate predefined degrees of damage (Hawk & Small, 1998; Thompson, Small, Johnson, & Marshall, 1998). Based on the identified condition states, maintenance interventions may be prioritized among all inspected structural components.

The Pontis (Thompson et al., 1998) and another bridge management system BRIDGIT (Hawk & Small, 1998) consider discrete condition states and Markovian deterioration modelling. Research efforts have integrated these discrete condition states within the life-cycle management and intervention optimisation associated with deteriorating infrastructure systems. Most of these approaches incorporate Markov chain models to depict the structural deterioration process. The main element of a Markov chain model is the transition matrix that specifies the probability that the state of a component changes to another state within a specified period of time. Note that the condition index is a subjective measure which may not realistically reflect the true load-carrying capacity of structural members (Liu & Frangopol, 2006b; Saydam et al., 2013).

Although such an approach may ensure an adequate level of safety of components, it does not provide information about the interaction between the components and overall performance of the whole structure (Saydam & Frangopol, 2011). Accordingly, other performance indicators capable of properly modelling the structural performance, while considering various uncertainties associated with resistance and load effects, have been developed and adopted in the life-cycle management of deteriorating infrastructure systems. Structural reliability theory offers a rational framework for quantification of system performance by including both aleatory and epistemic uncertainties, and correlations among random variables.

2.1. Reliability

Structural reliability can be defined as the probability that a component or a system will adequately perform its specified purpose for a prescribed period of time under particular conditions (Leemis, 1995; Paliou, Shinozuka, & Chen, 1990). Component, as well as system reliability can be computed for the investigated infrastructure considering that failure of a single component or a combination of individual components may initiate the failure of the system. For instance, if R and S represent the resistance and the load effect, respectively, the PDFs f_R and f_S, characterising these respective random variables may be established. The probability that S will not exceed R, $P(R > S)$, represents the reliability. As a general case, the time-variant probability of failure $p_F(t)$ can be expressed in terms of joint PDF of the random variables $R(t)$ and $S(t)$, $f_{R,S}(t)$, as:

$$p_F(t) = \int_0^\infty \left(\int_0^s f_{R,S}(t) \mathrm{d}r \right) \mathrm{d}s \qquad (1)$$

Furthermore, the reliability index can be expressed as:

$$\beta(t) = \Phi^{-1}(1 - p_F(t)) \qquad (2)$$

where $\Phi^{-1}(\cdot)$ is the inverse of the standard normal cumulative distribution function (CDF). In addition to evaluating the probability of structural failure at a given point in time, it is also possible to consider various functionality aspects that affect infrastructure systems such as serviceability limit states.

In general, bridge performance can be evaluated by modelling the bridge system as a series, parallel, or series-parallel combination of bridge components (Hendawi & Frangopol, 1994). It is possible to evaluate the reliability of entire bridge structural system by making appropriate assumptions (e.g. series, parallel or series-parallel assumptions) (Ditlevsen & Bjerager, 1986; Rashedi & Moses, 1988; Thoft-Christensen & Murotsu, 1986) regarding the interaction among individual components. Another approach for reliability assessment of bridges makes use of finite element (FE) analysis, if the overall non-linear system behaviour is of interest. A proper statistical distribution for the output of FE analysis (e.g. stress, displacement, bending moment) can be obtained by repeating the analysis for a large number of samples of the random variables associated with the structure. However, for complex structures, the time required to repeat FE analysis many times may be impractical. In such cases, response surface methods (RSMs) can be used to approximate the relation between the desired output of FE analysis and random variables by performing analyses for a significantly less number of samples. The RSM has also been implemented in system reliability of bridge superstructures (Liu, Ghosn, Moses, & Neuenhoffer, 2001), substructures (Ghosn & Moses, 1998), and bridge systems (Okasha & Frangopol, 2010a; Yang, Frangopol, & Neves, 2004). Additionally, Enright and Frangopol (1999a, 1999b) used the failure path method to compute the reliability function of a general (i.e. series-parallel) system and developed the computer program RELTSYS for this purpose (Enright & Frangopol, 2000). Lifetime functions (Leemis, 1995) are adopted for the time-dependent reliability approach, and have been utilised for the life-cycle performance prediction of bridge structures (Barone & Frangopol, 2013a, 2013b, 2014a, 2014b). Establishing the lifetime function system reliability may be carried out utilising various methods such as the minimal path and cut sets approaches (Rausand & Hoyland, 2004; Leemis, 1995).

2.2. Life-cycle cost

One of the most important measures in the evaluation of bridge performance is life-cycle cost. The proper allocation of resources can be achieved by minimising the total cost while keeping structural safety at a desired level. The expected total cost during the lifetime of a bridge structure can be expressed as (Frangopol et al., 1997):

$$C_{ET} = C_T + C_{PM} + C_{INS} + C_{REP} + C_F \qquad (3)$$

where C_T is the initial cost, C_{PM} is the expected cost of routine maintenance cost, C_{INS} is the expected cost of inspections, C_{REP} is the expected cost of repair and C_F is expected failure cost. Assuming the occurrence of the hazard (e.g. earthquake, flood) as a Poisson process, the total life-cycle failure loss of a bridge during the time interval $[0, t_{int}]$ can be computed (Dong & Frangopol, 2016b):

$$C_F(t_{int}) = \sum_{k=1}^{N(t_{int})} l(t_k) \cdot e^{-\gamma t_k} \qquad (4)$$

where t_{int} is investigated time interval; $N(t_{int})$ is the number of hazard events that occur during the time interval; $l(t_k)$ is the expected annual hazard loss at time t_k given the occurrence of the hazard; and γ is the monetary discount rate. Based on Yeo and Cornell (2005), given the Poisson model with mean rate equal to λ_f, the time t_k follows a uniform distribution over the interval $[0, t_{int}]$. Given $N(t_{int}) = \lambda_f \times t_{int}$, the total expected failure loss under hazard effects can be computed (Ross, 2000; Yeo & Cornell, 2005):

$$E[C_F(t_{int})] = \frac{\lambda_f \cdot E(l)}{\gamma} \cdot (1 - e^{-\gamma \cdot t_{int}}) \qquad (5)$$

where $E(l)$ is the expected value of annual loss l of bridge given a hazard event. The expected total loss under different hazard scenarios in a life-cycle context is shown in Figure 3. As indicated, various hazard scenarios may dominate the expected total loss at different time intervals during a structure's life-cycle. Numerous research efforts have focused on balancing cost and performance to determine optimum planning for life-cycle management of civil infrastructure systems (Ang & De Leon, 2005; Chang & Shinozuka, 1996; Estes & Frangopol, 2005; Estes, Frangopol, & Foltz, 2004; Frangopol & Furuta, 2001; Frangopol et al., 1997, 2001; Okasha & Frangopol, 2010a).

2.3. Risk

Risk is quantified by combining the probability of occurrence and the consequences of events generated by hazards. In general, the instantaneous total risk R of a structural system can be formulated as (CIB, 2001):

$$\begin{aligned} R = \iint \cdots \int \kappa\left(x_1, x_2, \ldots, x_m\right) \\ f_X\left(x_1, x_2, \ldots, x_m\right) \cdot \mathrm{d}x_1 \cdot \mathrm{d}x_2 \ldots \mathrm{d}x_m \end{aligned} \qquad (6)$$

where $\kappa(x)$ denotes the consequences associated with events resulting from hazards and $f_X(x)$ is the joint PDF of the random variables involved. The m-fold integral in Equation (6) is difficult to assess and often cannot be solved. Therefore, assumptions are established in order to obtain a simpler expression for total risk. A simplistic approach for calculating instantaneous total risk R is (Ellingwood, 2005):

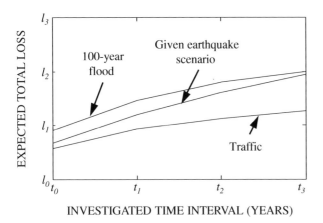

Figure 3. Expected total hazard loss under different hazard scenarios in a life-cycle context.

$$RISK(t) = \sum_{i=1}^{n} C_m(t) \cdot P_{F|H_i} \cdot P(H_i) \qquad (7)$$

where C_m represents the consequences of failure, $P(H_i)$ describes the probability of occurrence of a hazard, $P_{F|H_i}(t)$ is the conditional failure probability given the occurrence of a hazard and n is the total number of independent hazards considered within the analysis.

Several research efforts have been conducted on the risk assessment of bridge structures. Cesare, Santamarina, Turkstra, and Vanmarcke (1993) calculated the total risk associated with a bridge using the reliability and consequences of closure of the bridge. Stein, Young, Trent, and Pearson (1999) used risk concepts for prioritising scour-vulnerable bridges. Adey, Hajdin, and Brühwiler (2003) focused on the risk assessment of bridges affected by multiple hazards. Lounis (2004) presented a multi-criteria approach regarding bridge structural assessment with emphasis on risk. Similarly, Stein and Sedmera (2006) proposed a risk-based approach for bridges performance evaluation in the absence of information on bridge foundations. Ang (2011) focused on life-cycle considerations in risk-informed decision-making for the design of civil infrastructure. Decò and Frangopol (2011) developed a rational framework for the quantitative risk assessment of highway bridges under multiple hazards. Saydam et al. (2013) presented an illustrative example for the time-variant expected losses associated with the flexural failure of girders; a risk-based robustness index was calculated for an existing bridge.

Furthermore, risk analysis was utilised to assess the performance of networks of infrastructure systems (Bocchini & Frangopol, 2012, 2013; Dong, Frangopol, & Saydam, 2014a; Frangopol & Bocchini, 2012). For example, the time-dependent expected losses of deteriorated highway bridge networks were investigated by Saydam et al. (2013). Additionally, Decò and Frangopol (2011, 2013) and Dong et al. (2014a), Dong, Frangopol, and Saydam (2014b) proposed a computational framework for the quantitative assessment of life-cycle risk of multiple bridges within a transportation network including the effects of seismic and abnormal traffic hazards. Overall, risk, as a performance indicator, can offer valuable information regarding the performance of individual structures or spatially distributed systems, such as buildings, bridges and bridge networks.

2.4. Sustainability

Within the field of life-cycle engineering, two definitions of sustainability are usually referred to when developing appropriate sustainability metrics. The first defines it as: 'meeting the needs of the present without comprising the ability of future generations to meet their own needs' (Adams, 2006). The second definition complements the first one by emphasising that economic, environmental and social objectives must be simultaneously satisfied within a sustainable design or plan (Elkington, 2004). It is important to quantify the performance of bridges and networks of structural systems whose functionality is vital for economic and social purposes. Generally, sustainability should be quantified in terms of economic, social, and environmental metrics as indicated in Figure 4.

Recent research efforts have considered a wide variety of risks in order to effectively quantify sustainability. For instance, Dong, Frangopol, and Saydam (2013) presented a framework for assessing the time-variant sustainability of bridges associated with multiple hazards considering the effects of structural deterioration. Their approach was illustrated on a reinforced concrete (RC) bridge and the consequences considered within the risk assessment were the expected downtime and number of fatalities, expected energy waste and carbon dioxide emissions, and the expected loss. Overall, the inclusions of societal and environmental impacts along with economic consequences effectively encompass the concept of sustainability within the risk analysis framework. Combining the economic, societal and environmental risk metrics allows engineers and decision-makers to make informed decisions based on sustainability by providing them with a complete picture of system performance (Dong & Frangopol, 2016c; Lundie, Peters, & Beavis, 2004; Shinozuka, 2008).

Generally, a structure is more sustainable if its life-cycle cost (i.e. design, construction, inspection, maintenance, repair, failure and replacement costs) is low and energy waste, carbon dioxide emissions and user delays arising from its maintenance and repair are low. The social metrics can include downtime and fatalities. The downtime due to detour associated with bridge failure can be computed as (Stein et al., 1999):

$$DT = d \cdot ADT \cdot \frac{D}{S} \qquad (8)$$

where d is the duration of the detour (days), ADT is the average daily traffic to follow detour (number of vehicles), D is the detour length (km) and S is the detour speed (km/h). Here, the downtime can be referred to as the social metric of sustainability. The environmental metric includes the energy consumption, global warming potential and air pollutant emission, among others. Commonly considered environmental metrics including energy waste and carbon dioxide emissions are emphasised herein. The environmental metric associated with traffic detour is expressed as (Kendall, Keoleian, & Helfand, 2008):

$$EN_{DT} = ADT \cdot D \cdot d \cdot \left[Enp_{car} \cdot \left(1 - \frac{T}{100}\right) + Enp_{truck} \cdot \frac{T}{100} \right] \qquad (9)$$

where Enp_{car} and Enp_{truck} are environmental metric per unit distance for cars and trucks (e.g. carbon dioxide kg/km) and T is daily truck traffic ratio (i.e. percentage of average daily total traffic). The environmental metric associated with the repair action is computed as:

$$EN_{RE} = \left(Enp_{steel} \cdot V_{steel} + Enp_{conc} \cdot V_{conc} \right) \cdot RCR \qquad (10)$$

where Enp_{steel} and Enp_{conc} are environmental metric per unit volume for steel and concrete, respectively (e.g. carbon dioxide emissions kg/m^3), V_{steel} and V_{conc} are the volume of steel and concrete, respectively (m^3), and RCR is the repair cost ratio associated with a certain damage state. The fatalities associated with the failure of a highway bridge can also be computed considering its damage states. The time-variant sustainability of a bridge under a given hazard in terms of economic, social and environmental metrics is qualitatively shown in Figure 5 (Dong et al., 2014a). As indicated in Figure 5, the horizontal axis represents time and vertical axis is the expected loss associated with economic, social, and environmental metrics. Given no repair and/ or retrofit actions, these values increase significantly with time. In this figure, the social and environmental metrics are measured in monetary units and compared with the economic loss.

2.5. Utility

Utility theory is utilised in order to depict the relative desirability of maintenance strategies to the decision-maker. In general, utility is defined as a measure of value to the decision-maker.

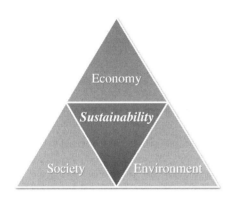

Figure 4. Metrics of sustainability.

Utility theory provides a framework that can measure, combine, and consistently compare these relative values (Ang & Tang, 1984). Multi-attribute utility theory may be used to transfer the marginal utility of each attribute involved in the performance assessment (e.g. economic, social and environmental risks) into one utility value that effectively combines the effects of all risks investigated as shown in Figure 6 (Dong, Frangopol, & Sabatino, 2016; Dong et al., 2015; Sabatino et al., 2015a). Next, all possible solution alternatives are identified and the uncertainties associated with the investigated decision-making problem are accounted for using a probabilistic approach. Since technical and economic uncertainties are both expected and unavoidable in the life-cycle assessment of bridges, decisions regarding life-cycle management must consider all relevant uncertainties associated with failure and its corresponding consequences. In this process, it is usually assumed that there is a single decision-maker who possesses a predetermined risk attitude with respect to a specific system.

Utility theory is employed herein in order to effectively capture the sustainability performance of highway bridges and bridge networks and impact of the decision-maker's risk attitude. Once the utility function associated with each attribute of sustainability is appropriately established, a multi-attribute utility that effectively represents all aspects of sustainability can be obtained by combining the utility functions associated with each attribute (Sabatino et al., 2015b). Within the additive formulation for the multi-attribute utility function, utility values associated with each attribute are multiplied by weighting factors and summed over all attributes involved. The multi-attribute utility associated with a structural system can be computed as (Jiménez et al., 2003):

$$u_S = k_{Eco} u_{Eco} + k_{Soc} u_{Soc} + k_{Env} u_{Env} \qquad (11)$$

where k_{Eco}, k_{Soc} and k_{Env} are the weighting factors corresponding to each sustainability metric and u_{Eco}, u_{Soc} and u_{Env} are the marginal utilities for the economic, social and environmental attributes, respectively. Overall, the proposed global strategy may be adopted for a variety of applications, including but not limited to bridges, buildings, and infrastructure networks.

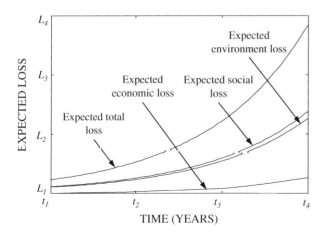

Figure 5. Time-variant annual sustainability metrics of a highway bridge in terms of monetary loss.

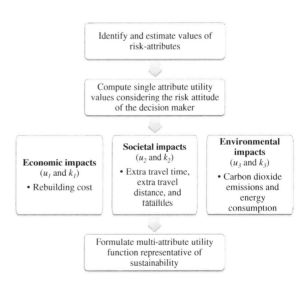

Figure 6. Multi-attribute utility assessment and optimization.

3. Consideration of hazards within a life-cycle context

3.1. Live load and corrosion

The structural performance associated with a specific bridge limit state varies with respect to time due to the increasing live load effects (e.g. by the growing demand of increasing traffic volume) and the progressive deterioration of the mechanical properties (e.g. due to corrosion). The investigated flexural and shear failure modes are those related to the bridge superstructure members (e.g. deck and girders). The deterioration of the flexural and shear capacities over time is induced by corrosion. Several researchers have studied probabilistic models for predicting the time-dependent deterioration of structural members due to corrosion (Akiyama, Frangopol, & Suzuki, 2012; Budelmann & Hariri, 2006; Budelmann, Holst, & Wichmann, 2014; Cavaco, Casas, Neves, & Huespe, 2013; Marsh & Frangopol, 2008; Stewart, 2012; Val & Chernin, 2012; Val, Stewart, & Melchers, 1998; Vu & Stewart, 2000). AASHTO (2015) specifications are adopted for the estimation of the load effects and capacities at each critical section. Additionally, the increase in over time of the live load moments is predicted considering traffic data, such as the average daily truck traffic, and by applying the statistics of extremes (Akgül & Frangopol, 2004a, 2004b; Cohen, Fu, Dekelbab, & Moses, 2003; O'Connor & O'Brien, 2005).

System reliability and redundancy have been extensively studied. Such studies include time-invariant measures (Bertero & Bertero, 1999; Biondini et al., 2008; Frangopol, 1985, 1997; Frangopol & Curley, 1987; Frangopol & Estes, 1997; Frangopol & Nakib, 1991; Frangopol et al., 2001; Ghosn & Frangopol, 2007; Ghosn & Moses, 1998; Ghosn, Moses, & Wang, 2003; Imai & Frangopol, 2002; Liu et al., 2001; Mori & Ellingwood, 1993; Moses, 1982; Paliou et al., 1990) and time-variant measures (Akgül & Frangopol, 2003, 2004a, 2004b; Ellingwood & Mori, 1993; Enright & Frangopol, 1999a, 1999b; Estes & Frangopol, 1999, 2001, 2005; Okasha & Frangopol, 2010a; Yang, Frangopol, Kawakami, & Neves, 2006b; Yang, Frangopol, & Neves, 2004, 2006a).

3.2. Fatigue and fracture

Application of loads on structural components may produce fracture and cause failure if the load is cyclically applied a large number of times. Fatigue failure is due to the progressive propagation of flaws in structural materials under cyclic loading. Fatigue failure is particularly common at the tip of cracks where the stress concentrations are high. These stress concentrations may occur in the component due to discontinuities in the material itself and are not serious when a ductile material like steel is subjected to a static load, as the stresses redistribute themselves to other adjacent elements within the structure.

Fatigue failure involves four stages (Sumi, 1998): (1) crack initiation at points of stress concentration, (2) crack growth, (3) crack propagation and (4) rupture. Generally, fatigue failures are classified into two categories: low-cycle and high-cycle failures, depending upon the number of cycles. Low-cycle fatigue failure occurs under high stress/strain ranges. On the other hand, high-cycle fatigue failure requires very large number of cycles. The most common form of fatigue damage is evaluated using the S-N curve, where the total cyclic stress (S) is plotted against the number of cycles to failure (N) in a logarithmic scale as shown in codes and standards. To carry out fatigue life predictions, a linear fatigue damage model is used in conjunction with relevant S-N curves.

Kwon and Frangopol (2010) investigated the bridge fatigue reliability assessment using PDFs of equivalent stress range based on field monitoring data. Newhook and Edalatmanesh (2013) integrated reliability and structural health monitoring in the fatigue assessment of concrete bridge decks. Stamatopoulos (2013) proposed a general approach to consider the fatigue assessment and strengthening measures of a steel railway bridge. Maekawa and Fujiyama (2013) investigated crack – water interaction and fatigue life assessment of RC bridge decks. Nagy, De Backer, and Van Bogaert (2013) presented an approach to improve the fatigue life of orthotropic bridge decks based on fracture mechanics. Pipinato (2014) investigated the high-cycle fatigue behaviour of riveted connections for railway metal bridges. Furthermore, the fatigue damage deterioration has been investigated by Garbatov and Guedes Soares (2001), Bastidas-Arteaga, Bressolette, Chateauneuf, and Sánchez-Silva (2009), Kim and Frangopol (2011a), and Kwon and Frangopol (2011).

3.3. Extreme events

The United Nations Office for Disaster Risk Reduction (UNISDR) reported in 2011 that natural disasters (e.g. earthquakes, floods and tsunamis) resulted in $366 billion direct economic losses and 29,782 fatalities worldwide (Ferris & Petz, 2011). These staggering statistics highlight the need for effective hazard recovery strategies associated with urban structural systems. Within the last few decades, the occurrence of disruptive, low-probability, high-consequences extreme events across the globe has shifted the focus of scientific communities and decision-makers to develop approaches which can improve the resilience of infrastructure to disasters. In general, earthquake resilience in civil engineering can be defined as (Bruneau et al., 2003) 'the ability of social units (e.g. organizations and communities) to mitigate hazards, contain the effects of disasters when they occur, and carry out recovery activities in ways that minimise social disruption and mitigate the effects of future earthquakes.' The most widely adopted approach to quantify the resilience of an individual structure, a group of structures, or a network of interrelated structures is to compute the resilience as (Bocchini, Frangopol, Ummenhofer, & Zinke, 2014; Cimellaro, Reinhorn, & Bruneau, 2010; Frangopol & Bocchini, 2011):

$$\text{RE} = \frac{1}{t_r} \int_{t_0}^{t_0+t_r} Q(t)\mathrm{d}t \qquad (12)$$

in which $Q(t)$ is the functionality, t_0 is the occurrence time of the extreme event and t_r is the investigated time horizon. The resilience, as computed by Equation (12), is illustrated graphically as shown in Figure 7 for multiple extreme events during the life-cycle of a system (Dong, Frangopol, & Sabatino, 2014c). Regarding the seismic performance analysis, the first step in seismic vulnerability assessment is to identify the

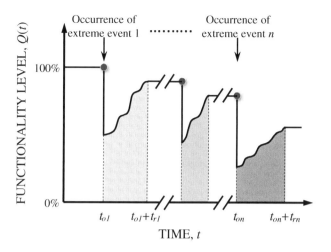

Figure 7. Life-cycle resilience.

seismic intensity associated with the location of the structural system under investigation (Dong & Frangopol, 2015). A number of seismic scenarios should be generated within the region of interest. The generated scenarios should be able to approximate the actual seismic activity of the geographical area. Subsequently, an attenuation equation is used to predict the ground-motion intensity at a certain location (Campbell & Bozorgnia, 2008).

Fragility curves are commonly used to predict structural performance under seismic hazard. Due to time effects, the fragility curves should be evaluated throughout the lifetime of a structure. The time-variant fragility curves can be computed as:

$$P_{S \geq DS_i \mid IM}(t) = \Phi\left(\frac{\ln(IM) - \ln(m_i(t))}{\beta_i(t)} \right) \quad (13)$$

where $\Phi(.)$ is the standard normal cumulative distribution function, IM is the seismic intensity measure (e.g. peak ground acceleration), $\beta_i(t)$ is the standard deviation in the damage state i of the structural fragility at time t and m_i is the median value of ground motion intensity associated with damage state i.

For a given ground motion intensity, the probability of a bridge being in a damage state i is given by the difference between the probabilities of exceedance of damage states i and $i + 1$, where damage state $i + 1$ is more severe than damage state i. These conditional probabilities can be mapped to the bridge damage index (BDI) value (Shiraki et al., 2007). BDI can be evaluated by mapping the bridge damage states given the ground acceleration based on realisation of a value between 0 and 1. A BDI of 1.0 indicates collapse and 0 corresponds to no damage following an earthquake. The expected BDI can be obtained by multiplying the probability of being in each damage state with the corresponding damage factor. Accordingly, the time-variant expected BDI of a bridge with four damage states DS_i for a certain ground motion intensity IM is:

$$\begin{aligned} BDI(t) &= BDI_1 \cdot P_{DS_1 \mid IM}(t) + BDI_2 \cdot P_{DS_2 \mid IM}(t) \\ &+ BDI_3 \cdot P_{DS_3 \mid IM}(t) + BDI_4 \cdot P_{DS_4 \mid IM}(t) \end{aligned} \quad (14)$$

where BDI_i is the bridge damage index for the respective damage state i.

A transportation network is defined in terms of nodes and links. A link is considered to be a single element connecting the nodes of a network. Bridges are typically the most vulnerable structures in a network and should be specially considered (Liu & Frangopol, 2005a). Following an earthquake, the damaged bridges can be open, closed or partially open within a bridge network. Consequently, traffic flow in the links can be different and speed limits might be reduced for various damage conditions of the link. As there may be several bridges located on the link, the damage state of each bridge can affect the functionality of the investigated link. The performance of a link after an earthquake can be expressed in terms of link damage index (LDI) which depends on the BDIs of the bridges on the link. Due to the fact that the seismic vulnerability of a bridge deteriorates with time, LDI should also be updated during the investigated time horizon of the transportation networks. The time-variant LDI can be expressed as (Chang, Shinozuka, & Moore, 2000):

$$LDI(t) = \sqrt{\sum_{j=1}^{n} (BDI_j(t))^2} \quad (15)$$

where n is the number of the bridges located in the link, and BDI_j is the expected damage index for bridge j. The level of link traffic flow capacity and flow speed for a damaged link depends on LDI. The intact, slight, moderate and major damage states are associated with LDI ≤ 0.5, $0.5 < $ LDI ≤ 1.0, $1.0 < $ LDI ≤ 1.5, and LDI > 1.5, respectively (Chang et al., 2000). The increase in the damage state of the link will reduce the link traffic capacity and speed limit.

Strong earthquakes can destroy infrastructure systems and cause injuries and/or fatalities. Therefore, it is important to investigate the seismic performance of interdependent healthcare – bridge network systems to guarantee immediate medical treatment after earthquakes. The assessment of healthcare – bridge network system performance depends on the seismic vulnerability of a hospital and bridges located in a surrounding bridge network, in addition to the ground motion intensity. After a destructive earthquake, the functionality of a highway network can be affected significantly; this, in turn, may hinder emergency management. Additional travel time would result due to the damaged bridges and links; consequently, injured persons may not receive treatment in time. Thus, it is important to account for the effects of damage condition associated with highway bridge networks on the healthcare system performance. Myrtle, Masri, Nigbor, and Caffrey (2005) carried out a series of surveys on performance of hospitals during several earthquakes to identify the important components; Yavari, Chang, and Elwood (2010) investigated performance levels for interacting components (i.e. structural, nonstructural, lifeline and personnel) using data from past earthquakes; Achour, Miyajima, Kitaura, and Price (2011) investigated the physical damage of structural and non-structural components of a hospital under seismic hazard; and Cimellaro, Reinhorn, and Bruneau (2011) introduced a model to describe the hospital performance under earthquake considering waiting time. Dong and Frangopol (2017) investigated the functionality of healthcare system considering the damage conditions associated with bridge networks and the correlation effects. The process used to compute the performance of interdependent healthcare-bridge network is shown in Figure 8.

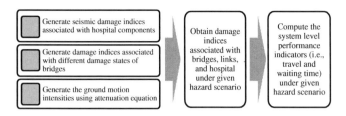

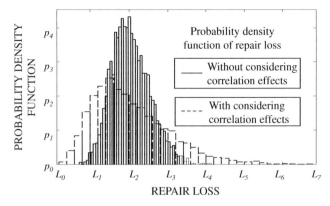

Figure 8. Computational procedure for system level performance indicators of interdependent healthcare-bridge networks.

Figure 9. Probabilistic repair loss with and without considering the correlation effects.

The damage to electric power, telecommunications, transportation, and water systems due to hazard effects can cause enormous social disruption. Therefore, it is of vital importance to investigate the performance of these interdependent networks subjected to hazard effects considering interdependencies in a large scale. Modeling the interaction between component and system is also important for assessing the risk and resilience of infrastructure systems. In order to understand the behaviour of these essential networks (e.g. power, communication, transportation, and water systems), their properties in terms of global connectivity, local clustering and overall shape should be evaluated considering the failure modes associated with both individual components and the interdependent systems. Then, methods and metrics to assess the performance of infrastructure networks, the evolution of their performance over time, and the interdependencies among different networks should be developed. This will contribute to the improvement in the performance-based design and management methods of interdependent infrastructure systems at the community level considering the interdependency among these infrastructure systems (Dueñas-Osorio, Craig, Goodno, & Goodno, 2007; Franchin, 2014; Ghosn et al., 2016a).

The consequences associated with the structural damage/failure under natural hazards (e.g. seismic events) include both direct and indirect consequences (Ellingwood, 2006), and can be expressed in terms of economic, social and environmental metrics. Earthquakes can disrupt traffic flow and affect emergency responses and recovery operations which may yield much higher consequences than the repair or rebuilding costs of a damaged infrastructure system. For the proper sustainability and risk analyses, the consequences associated with structural failures should include the economic, social and environmental metrics, including rebuilding, running, time loss and environmental costs, among others. The uncertainty in the parameters associated with the consequence evaluation should be incorporated within the assessment process. Accordingly, uncertainties associated with hazard scenarios and consequence evaluation are considered within the seismic loss assessment process. Additionally, the correlation among these random variables (e.g. ground motion intensities at different locations of bridges subjected to the same earthquake) is also accounted for within the evaluation process. Monte Carlo simulation is adopted to generate these correlated random variables. Subsequently, the probability density functions of the repair loss with and without considering the correlation effects are qualitatively shown in Figure 9 (Dong et al., 2014a). As indicated, correlation has a large effect on the dispersion of the repair loss.

Bridges also suffer exposure of their pier foundations under scour, which significantly reduces the foundation bearing capacity and can cause structural damage or even collapse during floods (Dong & Frangopol, 2016b). Scour is one of the main bridge failure causes in the United States accounting for about 58% of all failures (Briaud, Chen, Li, Nurtjahyo, & Wang, 2004). It is of vital importance to evaluate the performance of bridges under flood. Generally, there are three types of scour: long-term aggradation and degradation, contraction scour and local scour (Lagasse et al., 2009).

As bridges are exposed to flood-induced scour, bearing capacities of their foundations can be reduced significantly causing bridge damage or even collapse. Extensive research has been conducted on the prediction of local scour depth and a number of predictive methods have been proposed (Briaud et al., 1999, 2004; Melville, 1997; Richardson & Davis, 2001). Given the flood intensity and occurrence probability, the bridge vulnerability under flood can be analysed considering both vertical and lateral failure modes (Dong & Frangopol, 2016b). The load capacity of a bridge pile is directly related to the interaction between the piles and the surrounding soil. A lack of lateral confinement could result in lateral failure of the pile under flow-induced load and the axial load arising from the weight of the superstructure (Zhang, Silva, & Grismala, 2005). Vertical failure refers to the bridge failure in the vertical direction, which can be caused by inadequate soil support or pile instability.

During their life-cycle, bridges can be subjected to multiple hazards. Thus, it is necessary to consider the performance of bridges under multiple hazards in the risk assessment and mitigation procedure, all in a life-cycle context. For example, the flood-induced scour can reduce lateral support of a bridge at foundation and has a major effect on the seismic bridge vulnerability as indicated in Figure 10. The local scour induces the erosion of the soil around the pier and reduces the capacity of the foundation. Although the joint probability of occurrence of multiple hazards is small, past experience shows that simultaneous occurrences of extreme events happen. Due to the effects of global warming and climate change, the frequency, intensity and magnitude of the hazards are increasing. Hence, it is required to consider the effects of flood-induced scour in the seismic loss assessment, especially for bridges located in seismically flood-prone zones.

Additionally, for bridges that span rivers, traffic loading and scour are the two primary causes of failure and lane closure (Zhu & Frangopol, 2016a, 2016b). Therefore, these two hazards need to be considered in the risk assessment process. An

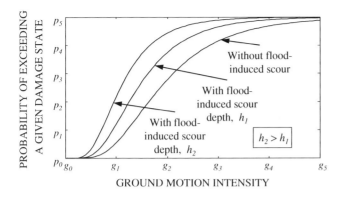

Figure 10. Qualitative fragility curves with and without considering flood-induced scour.

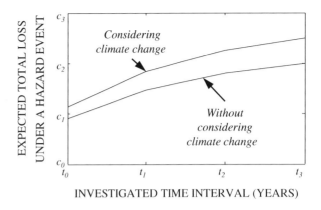

Figure 11. Expected total loss in a life-cycle context under given hazard scenarios with and without considering climate change.

efficient approach for assessing time-variant risks associated with the closure of bridge lanes due to traffic loading and scour is needed. The effects of hazards on bridges have been investigated including scour (Stein et al., 1999; Zhu & Frangopol, 2016a, 2016b), airborne chlorides (Akiyama et al., 2012; Titi, Biondini, & Frangopol, 2014, 2015), tsunami (Akiyama, Frangopol, Arai, & Koshimura, 2013) and a combination of hazards (Akiyama, Frangopol, & Matsuzaki, 2011; Decò & Frangopol, 2011; Dong et al., 2013; and Zhu & Frangopol, 2013b, 2016a, 2016b).

3.4. Climate change

According to the Intergovernmental Panel on Climate Change, the 'scientific evidence for warming of the climate system is unequivocal (NASA, 2015)'. Measurements taken over the last decades indicate that the sea level, global temperature and ocean temperature are all rising at elevated rates (Allison et al., 2009; Church & White, 2006, 2011; Levitus et al., 2009; Peterson & Baringer, 2009). Additionally, the sea ice in the arctic region is rapidly melting and glaciers are retreating almost everywhere around the world (Kwok & Rothrock, 2009; Polyak et al., 2010). Moreover, a significant increase in carbon dioxide (CO_2) concentration in the atmosphere has been observed (IPCC, 2007). Since these trends are projected to continue within the near future, it is crucial to determine the effects of climate change on the performance and life-cycle assessment of deteriorating infrastructure systems. The United States Global Change Research Program (USGCRP, 2008) reported that the average precipitation has increased 5% during a 50-year interval; consequently, the frequency of hazards (e.g. flood) has increased as well as they have become more intense. In general, climate change and increase in hazard intensity contribute to an increase in the probability of bridge failure due to hazard effects. The effects of climate change on the loss of bridges under hazard effects in a life-cycle context are qualitatively shown in Figure 11. Understanding how climate change affects the life-cycle performance of bridges can lead to improved preparedness prior to extreme disasters.

Although scientists agree that the climate is, in general, changing, there is a significant uncertainty associated with identifying the location, timing, and magnitude of changes over the lifetime of bridges and other infrastructure systems. In order to account for the uncertainties associated with the performance assessment

of highway bridges considering climate change, it is crucial to utilise risk methodologies to incorporate detrimental consequences of structural failure and identify the critical infrastructure that is most threatened by change climate in a given region (Committee on Adaptation to a Changing Climate, 2015).

One of the greatest concerns regarding climate change of highway bridges is the rising global temperature. If bridges are subjected to more days with sustained air temperature above 32 °C, the integrity of the pavement may suffer and deterioration in roadway and bridge expansion joints may occur (Schwartz et al., 2014). Furthermore, the construction productivity and costs of management activities, such as repair and rehabilitation interventions, maybe adversely affected by forcing shortened workdays or overnight work periods (TRB, 2014). In conjunction to rising temperature, the effect of increased levels of atmospheric CO_2 on highway bridges is significant. Stewart, Wang, and Nguyen (2011) illustrated that the increase in air temperature and CO_2 levels associated with climate change will increase the likelihood and rate of carbonation-induced corrosion. They also presented an approach that predicts the probability of corrosion initiation and damage for concrete infrastructure subjected to carbonation and chloride-induced corrosion resulting from elevated CO_2 levels and temperatures. The effects of increases in the rate and occurrence of carbonation-induced corrosion on the performance of concrete bridges are significant and cannot be ignored. Carbonation-induced damage risks may increase by more than 16%, which indicates that one in six structures may be subjected to additional corrosion damage by 2100 (Stewart, Wang, & Nguyen, 2012).

In addition to rising temperatures and CO_2 in the atmosphere, climate predictions indicate that the frequency of heavy precipitation events may increase over time. For highway bridges, an increased amount of precipitation may cause increases in soil erosion rates and soil moisture levels, causing road washouts and damage to foundations of roads, bridges and other transportation infrastructure systems (TRB, 2008). Overall, bridge failure due to scour during a heavy precipitation event is an extremely significant concern.

Moreover, bridges located in coastal regions are the most vulnerable to adverse climate change effects. Rising sea levels, combined with potentially more intense storm events and regional subsidence pose great threats to coastal deteriorating infrastructure systems (Schwartz et al., 2014; TRB, 2014). Storm surge paired with increased wave action can lead to bridge scour and increased erosion of roads, supporting structures and foundations (TRB, 2008). The rising sea levels can facilitate saltwater intrusion that accelerates corrosion and ultimately causes a reduction in predicted service-life, an increase in maintenance costs and an increase in probability of structural failure during extreme events (TRB, 2014). Due to its significant effects on the global temperature, atmospheric CO_2 measurements and sea levels, climate change must be considered within the life-cycle assessment and management of deteriorating civil infrastructure, within a probabilistic context.

4. Integration of SHM and updating in bridge management

Structural health monitoring (SHM), inspection and updating provide a powerful method to reduce uncertainty, calibrate and improve structural assessment and performance prediction models (Bucher & Frangopol, 2006; Catbas, Susoy, & Frangopol, 2008; Frangopol & Kim, 2014b; Frangopol & Messervey, 2007, 2008; Frangopol, Strauss, & Kim, 2008; Gul & Catbas, 2011; Klinzmann, Schnetgöke, & Hosser, 2006; Onoufriou & Frangopol, 2002). Life-cycle management approaches offer bridge managers a practical predictive view of cost, safety and condition, but in many regards lack knowledge of actual structural performance. In contrast, SHM techniques effectively capture structural behavior and the demands on a structure, but are not as effective in translating this information into actionable data for bridge managers. Consequently, it is of vital importance to incorporate SHM and updating in the life-cycle management framework.

Monitoring can provide data to confirm or improve existing load factors, resistance factors and load combinations for extreme events. In the past, many studies have been undertaken to model the performance of in-service bridges over time (Enright & Frangopol, 1999a, 1999b; Frangopol et al., 2008; Ghosn, 2000; Ghosn & Moses, 1998; Ghosn et al., 2003; Glaser, Li, Wang, Ou, & Lynch, 2007; Liu, Hammad, & Itoh, 1997). Bush, Omenzetter, Henning, and McCarten (2013) presented an innovative approach to bridge management that provides guidance on the type of data to collect, the accuracy and precision required in the data collection process, the frequency of inspections and the recommended SHM techniques to be used. Similarly, Sousa, Sousa, Neves, Bento, and Figueiras (2013) discussed the application of a SHM system to an RC bridge. The extraction of useful information from SHM data from highway bridges was reviewed by Westgate, Koo, Brownjohn, and List (2013). A novel SHM data processing technique, denoted as singular spectrum analysis, was utilised by Chao and Loh (2013) and applied to a bridge foundation to determine scour and pier settlement. Additionally, Huston, Cui, Burns, and Hurley (2011) studied the non-destructive evaluation of a bridge by comparing five different methods: (a) visual inspection and photographic recording, (b) half-cell electrochemical potential, (c) impulse type multipoint scanning ground penetrating radar, (d) chain drag and (e) impact echo. Ko, Pan, and Chiou (2013) aimed to enhance facility management efficiency using radio frequency identification technology.

Information associated with inspection events can be used to update deterioration models of a structural system to reduce uncertainty. The structural details associated with a given system are correlated due to common parameters associated with materials, design, fabrication, loading and operational conditions. Based on these correlations, the inspection information of one particular component can be used to update deterioration performance of others uninspected components. Probabilistic models have been used to evaluate and update the fatigue reliability using inspection information and are emphasised herein. In order to make the inspection plans more efficient and economic, optimisation can be adopted to determine the optimal number of the details that need to be inspected within a probabilistic manner. Moan and Song (2000) investigated reliability-based fatigue damage assessment and updating details in parallel/series systems. Chen, Sun, and Guedes Soares (2003) proposed a methodology for inspection planning on the basis on Palmgren–Miner's rule. Huang and Garbatov (2013) computed the reliability index of a complex welded structure as a series model under multiple cracks. Maljaars and Vrouwenvelder (2014) presented a reliability-based updating considering multiple critical locations of a bridge.

Based on the bridge component/system reliability and risk, the inspection planning and repair priority among the investigated sensitive systems can be identified. In turn, the inspection results can be used to update risk and the timing for the following inspection and management plans based on risk. In this paper, the updating associated with fatigue sensitive details is illustrated. Generally, if no fatigue crack is detected, the updating can be performed within the original fatigue limit state. If repair actions are conducted, the physical changes need to be considered in the estimation of limit state function. The updated probability of failure of the ith component under fatigue damage given inspection event j can be formulated as follows (Moan & Song, 2000):

$$
\begin{aligned}
P_{i,\,up} &= P[M_i(t) \le 0 \,|\, IE_j(t_{IE})] \\
&= \frac{P[(M_i(t) \le 0) \cap IE_j(t_{IE})]}{P[IE_j(t_{IE})]}
\end{aligned}
\tag{16}
$$

where M_i is limit state function associated with detail i, IE_j is the inspection event j, and t_{IE} is the inspection time.

The results of inspection are utilised for reliability and risk ranking updating associated with inspected and uninspected fatigue sensitive details. It is assumed that once a fatigue crack is detected, the repair action is applied to the cracked detail. The repair action considered herein is replacing the detected cracked detail with a new one. At $t = t_{ins}$ years, if a crack is detected using magnetic particle inspection (MPI), the updated reliability index of the inspected detail is qualitatively shown in Figure 12(a). As indicated, given that a crack is detected, the reliability of the inspected detail decreases significantly without repair and, in turn, the risk associated with the detected cracked detail would increase considerably, as shown in Figure 12(b) (Dong & Frangopol, 2016a).

5. Probabilistic life-cycle optimisation

Life-cycle optimisation is an essential task within the life-cycle management (LCM) framework (Ang & De Leon, 2005; Chang & Shinozuka, 1996; Estes & Frangopol, 1999; Frangopol, 1998, 1999; Frangopol & Soliman, 2016; Kong & Frangopol, 2005; Okasha & Frangopol, 2009; Sabatino, Frangopol, & Dong, 2015a; Soliman & Frangopol, 2014; Soliman, Frangopol, & Kim, 2013;

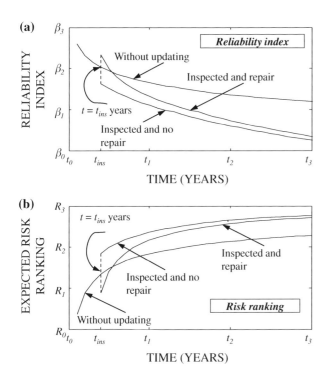

Figure 12. (a) Updated reliability and (b) risk ranking of an inspected detail under fatigue damage with crack detected at year t_{ins}.

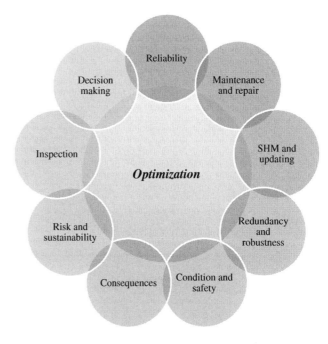

Figure 13. Integrated life-cycle management framework.

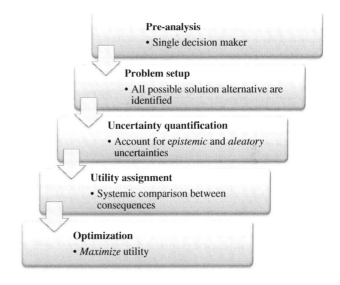

Figure 14. Utility-based decision-making procedure.

Wen & Kang, 2001; Yang et al., 2006b). This process is performed using a probabilistic platform considering various uncertainties associated with LCM as shown in Figure 13. A maintenance optimisation formulation requires one or more life-cycle performance indicators (Frangopol & Saydam, 2014), such as system reliability, redundancy and cost indicators (Augusti, Ciampoli, & Frangopol, 1998; Estes & Frangopol, 1999; Frangopol & Kim, 2014a; Marsh & Frangopol, 2007; Morcous, Lounis, & Cho, 2010; Okasha & Frangopol, 2009, 2010a; Yang et al., 2006b), condition indicators (Frangopol & Liu, 2007; Liu & Frangopol, 2005b, 2006a, 2006b; Neves & Frangopol, 2005; Neves, Frangopol, & Cruz, 2004), probabilistic damage detection delay indicators (Kim & Frangopol, 2011a, 2011b; Soliman et al., 2013), and risk and sustainability-informed performance indicators (Dong et al., 2014b; Sabatino et al., 2015a, 2015b; Zhu & Frangopol, 2013a). Powerful optimisation algorithms are also needed (e.g. Deb, 2001; Deb, Pratap, Agarwal, & Meyarivan, 2002; Frangopol & Soliman, 2013, 2015; Goldberg, 1989).

Planning retrofit actions on bridge networks under tight budget constraints were investigated by Dong et al. (2014b). They presented a probabilistic methodology to establish optimum pre-earthquake retrofit plans for bridge networks based on sustainability. A multi-criteria optimisation problem was formulated to find the optimum timing of retrofit actions for bridges within a network. The role of optimisation is to identify the most effective retrofit strategy in terms of which bridges to be retrofitted and the optimal times for retrofit actions. Utility-informed decision-making is necessary for optimum allocation of limited resources. In general, as shown in Figure 14, utility-informed decision-making may be divided into five separate stages (Keeney & Raiffa, 1993): the pre-analysis, problem set-up, uncertainty quantification, utility assignment and optimisation. The application of utility-informed decision-making in the optimal lifetime

intervention on bridges is a topic of paramount importance and is experiencing growing interest within the field of life-cycle infrastructure engineering. This methodology can be used in assisting decision-making regarding the maintenance/retrofit activities to improve the performance of highway bridge network.

Sabatino et al. (2015a) presented a framework for life-cycle maintenance optimisation of highway bridges that utilises multi-attribute utility theory to quantify the sustainability performance metrics. The ultimate aim of implementing maintenance throughout the lifetime of a bridge is to mitigate the detrimental impacts of structural failure to the economy, society and the environment. Optimum maintenance plans were obtained by carrying out a multi-criteria optimisation procedure where the utility associated with total maintenance cost and utility corresponding to sustainability performance were considered as conflicting objectives. An existing highway bridge was utilised to

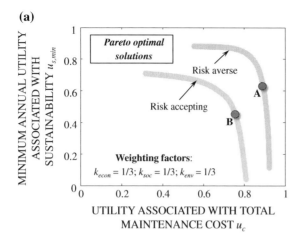

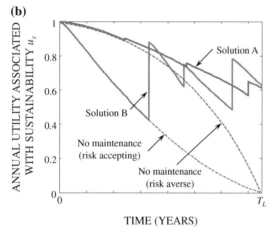

Figure 15. (a) Effect of risk attitude on the Pareto optimal fronts for lifetime maintenance and (b) time-variant utility associated with sustainability for representative solutions A and B.

illustrate the capabilities of the proposed decision support system for maintenance planning. The optimisation was performed by simultaneously maximising the utility associated with total maintenance cost and the annual minimum utility corresponding to the sustainability over the lifetime of the bridge.

The main results of the optimisation procedure are the types maintenance actions performed on the bridge components and their respective times of application. The Pareto optimal solutions obtained considering three maintenance actions with a risk accepting and risk-averse decision-maker are shown in Figure 15. A solution is Pareto-optimal if there does not exist another solution that improves at least one objective without worsening another one. The weighting factors k_{econ}, k_{soc}, and k_{env} are all assumed to be the same (i.e. 1/3), representing equal contribution of detrimental economic, societal and environmental impacts. The Pareto-optimal representative solutions A and B, denoting typical optimum maintenance plans resulting from a risk-averse and risk accepting decision-maker, respectively, are shown in Figure 15(a). The time-variant multi-attribute utilities associated with sustainability corresponding to representative solutions A and B are shown in Figure 15(b).

6. Conclusions

This paper, which is based on the T. Y. Lin plenary lecture presented by the first author at the Eighth International Conference

on Bridge Maintenance, Safety and Management (IABMAS2016), Iguassu Falls, Paraná, Brazil, June 26–30, 2016, presents a brief overview of the integration of risk, sustainability and resilience measures into the LCM of deteriorating infrastructure systems with emphasis on bridges, bridge networks and interdependent infrastructure systems considering climate change effects. The framework covers predicting the time-variant structural performance and the future interventions scheduling, including inspections, monitoring, maintenance and/or repairs actions, such that an optimal management solution which satisfies the goals and constraints is achieved. Moreover, this generalised approach integrates risk and life-cycle loss assessment with multi-objective optimisation techniques to determine optimum bridge and bridge network management plans to assist the decision-maker. Various aspects of the LCM framework are briefly explained with special attention given to the performance assessment and the life-cycle optimiSation processes. The performance assessment of interdependent infrastructure systems under hazard effects is also incorporated within the LCM framework. By considering the probability of occurrence of hazard and structural deteriorations, the performance of interdependent systems in a life-cycle context could be investigated. Overall, the performance assessment of an interdependent healthcare – bridge network system under hazard effects provides system-level probabilistic measures that can aid the emergency management process.

Additionally, the effects of natural hazards and climate change on infrastructure assessment and management were discussed in this paper. Structural deterioration has a significant effect on system performance under hazards. The difference between the life-cycle hazard loss with and without considering ageing effects increases with the investigated time interval. Moreover, because of ageing effects, the resilience of damaged bridges under hazards decreases significantly with time. Given various hazard occurrence models and monetary discount rates, the total hazard loss associated with these hazards during the investigated time interval differs, and the contribution of the hazards changes. Therefore, different hazards could dominate the total life-cycle loss. Consequently, specific risk mitigation strategies associated with specific hazards could be determined. Given more information regarding climate change, this approach could be easily updated by incorporating additional data. It is also noted that irrespective of the amount of data associated with climate change, improved judgmental assessments are necessary to reduce the epistemic uncertainties associated with imperfect models of this change.

Furthermore, this paper presents available methodologies for quantifying the economic, social and environmental metrics to evaluate the sustainability of bridges, bridge networks, and interdependent infrastructure systems. In general, a utility-based performance metric can provide an in-depth understanding of the current and future sustainability associated with infrastructure systems. The presented framework supports the sustainable development of infrastructure systems and provides the optimal intervention strategies to the decision-maker that will ultimately allow for risk-informed decision-making regarding life-cycle management of highway bridges, bridge networks, and interdependent infrastructure systems. Overall, the key objectives of a probabilistic framework are to determine the likelihood of successful performance, find the total expected cost accrued over the entire life-cycle, and make optimal risk-, resilience-, and sustainability-informed decisions related to design, construction, inspection, maintenance, monitoring, repair and replacement of civil infrastructure systems under multiple objectives and constraints.

Acknowledgements

National Science Foundation, Pennsylvania Infrastructure Technology Alliance, U.S. Federal Highway Administration, and U.S. Office of Naval Research are gratefully acknowledged. The opinions and conclusions presented in this paper are those of the authors and do not necessarily reflect the views of the sponsoring organizations.

Disclosure statement

No potential conflict of interest was reported by the authors.

Funding

This work was supported by the National Science Foundation [grant number CMS-0639428] and [grant number CMMI-1537926]; the Commonwealth of Pennsylvania, Department of Community and Economic Development, through the Pennsylvania Infrastructure Technology Alliance (PITA); the U.S. Federal Highway Administration Cooperative Agreement [grant number DTFH61-07-H-00040] and [grant number DTFH61-11-H-00027]; the U.S. Office of Naval Research [grant number N00014-08-1-0188], [grant number N00014-12-1-0023], and [grant number N00014-16-1-2299] (Structural Reliability Program).

References

AASHTO. (2015). *LRFD bridge design specifications*. (7th ed.). Washington, DC: American Association of State Highway & Transportation Officials.

Achour, N., Miyajima, M., Kitaura, M., & Price, A. (2011). Earthquake-induced structural and nonstructural damage in hospital. *Earthquake Spectra, 37*, 617–634.

Adams, W.M. (2006). The future of sustainability: Re-thinking environment and development in the twenty-first century. *Report of the IUCN Renowned Thinkers Meeting*, 29–31 January 2006.

Adey, B., Hajdin, R., & Brühwiler, E. (2003). Supply and Demand system approach to development of bridge management strategies. *Journal of Infrastructure Systems, 9*(3), 117–131.

Akgül, F., & Frangopol, D.M. (2003). Rating and reliability of existing bridges in a network. *Journal of Bridge Engineering, 8*, 383–393.

Akgül, F., & Frangopol, D.M. (2004a). Computational platform for predicting lifetime system reliability profiles for different structure types in a network. *Journal of Computing in Civil Engineering, 18*, 92–104.

Akgül, F., & Frangopol, D.M. (2004b). Bridge rating and reliability correlation: Comprehensive study for different bridge types. *Journal of Structural Engineering, 130*, 1063–1074.

Akiyama, M., Frangopol, D.M., Arai, M., & Koshimura, S. (2013). Reliability of bridges under tsunami hazards: Emphasis on the 2011 Tohoku-Oki earthquake. *Earthquake Spectra, 29*, S295–S314.

Akiyama, M., Frangopol, D.M., & Matsuzaki, H. (2011). Life-cycle reliability of RC bridge piers under seismic and airborne chloride hazards. *Earthquake Engineering & Structural Dynamics, 40*, 1671–1687.

Akiyama, M., Frangopol, D.M., & Suzuki, M. (2012). Integration of the effects of airborne chlorides into reliability-based durability design of reinforced concrete structures in a marine environment. *Structure and Infrastructure Engineering, 8*, 125–134.

Allison, I., Bindoff, N.L., Bindschadler, R.A., Cox, P.M., de Noblet, N., England, M.H., Francis, J.E., Gruber, N., Haywood, A.M., Karoly, D.J., Kaser, G., Le Quéré, C., Lenton, T.M., Mann, M.E., McNeil, B.I., Pitman, A.J., Rahmstorf, S., Rignot, E., Schellnhuber, H.J., Schneider, S.H., Sherwood, S.C., Somerville, R.C.J., Steffen, K., Steig, E.J., Visbeck, M., & Weaver, A.J. (2009). *The copenhagen diagnosis: Updating the world on the latest climate science*. Sydney: UNSW Climate Change Research Center.

Ang, A.H.-S. (2011). Life-cycle considerations in risk informed decisions for design of civil infrastructures. *Structure and Infrastructure Engineering, 7*, 3–9.

Ang, A.H.-S., & De Leon, D. (2005). Modeling and analysis of uncertainties for risk-informed decisions in infrastructures engineering. *Structure and Infrastructure Engineering, 1*, 19–31.

Ang, A.H.-S., & Tang, W.H. (1984). Probability concepts in engineering planning and design – Volume II – Decision, risk, and reliability. New York, NY: John Wiley & Sons.

Ang, A.H.-S., & Tang, W.H. (2007). *Probability concepts in engineering*. (2nd ed.). New York, NY: Wiley.

ASCE. (2013). *Report card for America's infrastructure*. Reston, VA: American Society of Civil Engineers.

Augusti, G., Ciampoli, M., & Frangopol, D.M. (1998). Optimal planning of retrofitting interventions on bridges in a highway network. *Engineering Structures, 20*, 933–939.

Barone, G., & Frangopol, D.M. (2013a). Hazard-based optimum lifetime inspection and repair planning for deteriorating structures. *Journal of Structural Engineering, 139*, 04013017.

Barone, G., & Frangopol, D.M. (2013b). Reliability, risk and lifetime distributions as performance indicators for life-cycle maintenance of deteriorating structures. *Reliability Engineering and System Safety, 123*, 21–37.

Barone, G., & Frangopol, D.M. (2014a). Life-cycle maintenance of deteriorating structures by multi-objective optimization involving reliability, risk, availability, hazard and cost. *Structural Safety, 48*, 40–50.

Barone, G., & Frangopol, D.M. (2014b). Reliability, risk and lifetime distributions as performance indicators for life-cycle maintenance of deteriorating structures. *Reliability Engineering & System Safety, 123*, 21–37.

Bastidas-Arteaga, E., Bressolette, P.H., Chateauneuf, A., & Sánchez-Silva, M. (2009). Probabilistic lifetime assessment of RC structures under coupled corrosion-fatigue deterioration processes. *Structural Safety, 31*, 84–96.

Bertero, R., & Bertero, V. (1999). Redundancy in earthquake-resistant design. *Journal of Structural Engineering, 125*, 81–88.

Biondini, F., Camnasio, E., & Palermo, A. (2014). Lifetime seismic performance of concrete bridges exposed to corrosion. *Structure and Infrastructure Engineering, 10*, 880–900.

Biondini, F., Camnasio, E., & Titi, A. (2015). Seismic resilience of concrete structures under corrosion. *Earthquake Engineering and Structural Dynamics, 44*, 2445–2466.

Biondini, F., Frangopol, D.M., & Malerba, P.G. (2008). Uncertainty effects on lifetime structural performance of cable stayed bridges. *Probabilistic Engineering Mechanics, 23*, 509–522.

Biondini, F., & Frangopol, D. (2016). Life-cycle performance of deteriorating structural systems under uncertainty: Review. *Journal of Structural Engineering, 142*(9), F4016001, 1–17.

Bocchini, P., & Frangopol, D.M. (2012). Optimal resilience- and cost-based postdisaster intervention prioritization for bridges along a highway segment. *Journal of Bridge Engineering, 17*, 117–129.

Bocchini, P., & Frangopol, D.M. (2013). Connectivity-based optimal scheduling for maintenance of bridge networks. *Journal of Engineering Mechanics, 139*, 170–769.

Bocchini, P., Frangopol, D.M., Ummenhofer, T., & Zinke, T. (2014). Resilience and sustainability of the civil infrastructure: Towards a unified approach. *Journal of Infrastructure Systems, 20*, 0414004.

Briaud, J.L., Ting, F.C.K., Chen, H.C., Gudavalli, R., Perugu, S., & Wei, G. (1999). SRICOS: Prediction of scour rate in cohesive soils at bridge piers. *Journal of Geotechnical and Geoenvironmental Engineering, 125*, 237–246.

Briaud, J.L., Chen, H.C., Li, Y., Nurtjahyo, P., & Wang, J. (2004). *Pier and contraction scour in cohesive soils* (NCHRP Report 516). Washington, DC: National Cooperative Highway Research Program.

Bruneau, M., Chang, S.E., Eguchi, R.T., Lee, G.C., O'Rourke, T.D., Reinhorn, A.M., Shinozuka, M., Tierney, K., Wallace, W.A., & Winterfeldt, D.V. (2003). A framework to quantitatively assess and enhance the seismic resilience of communities. *Earthquake Spectra, 19*, 733–752.

Bucher, C., & Frangopol, D.M. (2006). Optimization of lifetime maintenance strategies for deteriorating structures considering probabilities of violating safety, condition, and cost thresholds. *Probabilistic Engineering Mechanics, 21*(1), 1–8.

Budelmann, H., & Hariri, K. (2006). A structural monitoring system for RC/PC structures. In H.-N. Cho, D.M. Frangopol, & A.H.-S. Ang (Eds.), *Life-cycle cost and performance of civil infrastructure systems* (pp. 3–17). London: Taylor & Francis.

Budelmann, H., Holst, A., & Wichmann, H. (2014). Non-destructive measurement toolkit for corrosion monitoring and fracture detection of bridge tendons. *Structure and Infrastructure Engineering, 10*, 492–507.

Bush, S., Omenzetter, P., Henning, T., & McCarten, P. (2013). A risk and criticality-based approach to bridge performance data collection and monitoring. *Structure and Infrastructure Engineering, 9*, 329–339.

Campbell, K.W., & Bozorgnia, Y. (2008). NGA ground motion model for the geometric mean horizontal component of PGA, PGV, PGD and 5% damped linear elastic response spectra for periods ranging from 0.01 to 10 s. *Earthquake Spectra, 24*, 139–171.

Catbas, F.N., Susoy, M., & Frangopol, D.M. (2008). Structural health monitoring and reliability estimation: Long span truss bridge application with environmental monitoring data. *Engineering Structures, 30*, 2347–2359.

Cavaco, E.S., Casas, J.R., Neves, L.A.C., & Huespe, A.E. (2013). Robustness of corroded reinforced concrete structures – A structural performance approach. *Structure and Infrastructure Engineering, 9*, 42–58.

Cesare, M., Santamarina, J.C., Turkstra, C.J., & Vanmarcke, E. (1993). Risk-based Bridge management. *Journal of Transportation Engineering, 119*, 742–750.

Chang, S.E., & Shinozuka, M. (1996). Life-cycle cost analysis with natural hazard risk. *Journal of Infrastructure Systems, 2*, 118–126.

Chang, S.E., Shinozuka, M., & Moore, J.E. (2000). Probabilistic earthquake scenarios: extending risk analysis methodologies to spatially distributed systems. *Earthquake Spectra, 16*, 557–572.

Chao, S., & Loh, C. (2013). Application of singular spectrum analysis to structural monitoring and damage diagnosis of bridges. *Structure and Infrastructure Engineering, 10*, 708–727.

Chen, N.-Z., Sun, H.-H., & Guedes Soares, C. (2003). Reliability analysis of a ship hull in composite material. *Composite Structures, 62*, 59–66.

Church, J.A., & White, N.J. (2006). A 20th century acceleration in global sea-level rise. *Geophysics Research Letters, 33*, L01602.

Church, J.A., & White, N.J. (2011). Sea-level rise from the late 19th to the early 21st century. *Surveys in Geophysics, 32*, 585–602.

CIB. (2001). *Risk assessment and risk communication in civil engineering. TG 32* (Report 259). Rotterdam: Council for Research and Innovation in Building and Construction.

Cimellaro, G.P., Reinhorn, A.M., & Bruneau, M. (2010). Framework for analytical quantification of disaster resilience. *Engineering Structures, 32*, 3639–3649.

Cimellaro, G.P., Reinhorn, A.M., & Bruneau, M. (2011). Performance-based metamodel for healthcare facilities. *Earthquake Engineering and Structural Dynamics, 40*, 1197–1217.

Cohen, H., Fu, G., Dekelbab, W., & Moses, F. (2003). Predicting truck load spectra under weight limit changes and its application to steel bridge fatigue assessment. *Journal of Bridge Engineering, 8*, 312–322.

Committee on Adaptation to a Changing Climate. (2015). In J.R. Olsen (Ed.). *Adapting infrastructure and civil engineering practice to a changing climate.* Reston, VA: American Society of Civil Engineers. doi: 10.1061/9780784479193

Deb, K. (2001). *Multi-objective optimization using evolutionary algorithms.* Chichester: John Wiley and Sons.

Deb, K., Pratap, A., Agarwal, S., & Meyarivan, T. (2002). A fast and elitist multiobjective genetic algorithm: NSGA-II. *IEEE Transactions on Evolutionary Computation, 6*, 182–197.

Decò, A., & Frangopol, D.M. (2011). Risk assessment of highway bridges under multiple hazards. *Journal of Risk Research, 14*, 1057–1089.

Decò, A., & Frangopol, D.M. (2013). Life-cycle risk assessment of spatially distributed aging bridges under seismic and traffic hazards. *Earthquake Spectra, 29*, 127–153.

Decò, A., Bocchini, P., & Frangopol, D.M. (2013). A probabilistic approach for the prediction of seismic resilience of bridges. *Earthquake Engineering and Structural Dynamics, 42*, 1469–1487.

Ditlevsen, O., & Bjerager, P. (1986). Methods of structural systems reliability. *Structural Safety, 3*, 195–229.

Dong, Y., & Frangopol, D.M. (2015). Risk and resilience assessment of bridges under mainshock and aftershocks incorporating uncertainties. *Engineering Structures, 83*, 198–208.

Dong, Y., & Frangopol, D.M. (2016a). Incorporation of risk and updating in inspection of fatigue-sensitive details of ship structures. *International Journal of Fatigue, 82*, 676–688.

Dong, Y., & Frangopol, D.M. (2016b). Probabilistic time-dependent multihazard life-cycle assessment and resilience of bridges considering climate change. *Journal of Performance of Constructed Facilities, 30*, 04016034, 1–12.

Dong, Y., & Frangopol, D.M. (2016c). Performance-based seismic assessment of conventional and base-isolated steel buildings including environmental impact and resilience. *Earthquake Engineering and Structural Dynamics, 45*, 739–756.

Dong, Y., & Frangopol, D.M. (2017). Probabilistic assessment of an interdependent healthcare–bridge network system under seismic hazard. *Structure and Infrastructure Engineering, 13*, 160–170.

Dong, Y., Frangopol, D.M., & Sabatino, S. (2015). Optimizing bridge network retrofit planning based on cost-benefit evaluation and multi-attribute utility associated with sustainability. *Earthquake Spectra, 31*, 2255–2280.

Dong, Y., Frangopol, D.M., & Saydam, D. (2013). Time-variant sustainability assessment of seismically vulnerable bridges subjected to multiple hazards. *Earthquake Engineering and Structural Dynamics, 42*, 1451–1467.

Dong, Y., Frangopol, D.M., & Saydam, D. (2014a). Sustainability of highway bridge networks under seismic hazard. *Journal of Earthquake Engineering, 18*, 41–66.

Dong, Y., Frangopol, D.M., & Saydam, D. (2014b). Pre-earthquake multi-objective probabilistic retrofit optimization of bridge networks based on sustainability. *Journal of Bridge Engineering, 19*, 04014004.

Dong, Y., Frangopol, D.M., & Sabatino, S. (2014c). Risk-informed decision making for disaster recovery incorporating sustainability and resilience. *Proceedings of the 3rd International Conference on Urban Disaster Reduction (3ICUDR).* Boulder, Colorado: EERI. September 28–October 1, 2014, 4 pages.

Dong, Y., Frangopol, D.M., & Sabatino, S. (2016). A decision support system for mission-based ship routing considering multiple performance criteria. *Reliability Engineering & System Safety, 150*, 190–201.

Dueñas-Osorio, L., Craig, James I., Goodno, Barry J., & Goodno, B. (2007). Seismic response of critical interdependent networks. *Earthquake Engineering & Structural Dynamics, 36*, 285–306.

Elkington, J. (2004). *Enter the triple bottom line.* In A. Henriques & J. Richardson (Eds.), *The triple bottom line: does it all add up* (pp. 1–16) Sterling, VA: Earthscan.

Ellingwood, B.R. (1998). Issues related to structural aging in probabilistic risk assessment of nuclear power plants. *Reliability Engineering and System Safety, 62*, 171–183.

Ellingwood, B.R. (2005). Risk-informed condition assessment of civil infrastructure: State of practice and research issues. *Structure and Infrastructure Engineering, 1*, 7–18.

Ellingwood, B.R. (2006). Mitigating risk from abnormal loads and progressive collapse. *Journal of Performance of Constructed Facilities, 20*, 315–323.

Ellingwood, B.R., & Frangopol, D.M. (2016). Introduction to the state of the art collection: Risk-based lifecycle performance of structural systems. *Journal of Structural Engineering, 142*(9), F2016001, 1.

Ellingwood, B.R., & Mori, Y. (1993). Probabilistic methods for condition assessment and life prediction of concrete structures in nuclear power plants. *Nuclear Engineering and Design, 142*, 155–166.

Enright, M., & Frangopol, D.M. (2000). RELTSYS: A computer program for life prediction of deteriorating systems. *Structural Engineering and Mechanics, 9*, 557–568.

Enright, M.P., & Frangopol, D.M. (1999a). Reliability-based condition assessment of deteriorating concrete bridges considering load redistribution. *Structural Safety, 21*, 159–195.

Enright, M.P., & Frangopol, D.M. (1999b). Condition prediction of deteriorating concrete bridges using bayesian updating. *Journal of Structural Engineering, 125*, 1118–1125.

Estes, A.C., & Frangopol, D.M. (1999). Repair optimization of highway bridges using system reliability approach. *Journal of Structural Engineering, 125*, 766–775.

Estes, A.C., & Frangopol, D.M. (2001). Bridge lifetime system reliability under multiple limit states. *Journal of Bridge Engineering, 6*, 523–528.

Estes, A.C., & Frangopol, D.M. (2005). Life-cycle evaluation and condition assessment of structures. In W-F. Chen and E. M. Lui (Eds.), Chapter 36 *Structural Engineering Handbook* (2nd ed., pp. pp. 36–51). CRC Press.

Estes, A.C., Frangopol, D.M., & Foltz, S.D. (2004). Updating reliability of steel miter gates on locks and dams using visual inspection results. *Engineering Structures, 26*, 319–333.

Ferris, E., & Petz, D. (2011). *The year that shook the rich: A review of natural disaster in 2011*. Project on Internal Displacement: The Brookings Institution-London School of Economics.

FHWA. (2013). Deficient bridges by state and highway system 2013. Washington, DC: Federal Highway Administration. Retrieved from http://www.fhwa.dot.gov/bridge/nbi/no10/defbr13.cfm

Franchin, P. (2014). A computational framework for systemic seismic risk analysis of civil infrastructural systems. *Geotechnical, Geological and Earthquake Engineering, 31*, 23–56.

Frangopol, D.M. (1985). Sensitivity of reliability-based optimum design. *Journal of Structural Engineering, 111*, 1703–1721.

Frangopol, D.M. (1997). Application of life-cycle reliability based criteria to bridge assessment and design. In P.C. Das (Ed.), *Safety of bridges* (pp. 151–157). London: Thomas Telford.

Frangopol, D.M. (1998). *A probabilistic model based on eight random variables for preventive maintenance of bridges*. Presented at the progress meeting on optimum maintenance strategies for different bridge types. London: Highways Agency.

Frangopol, D.M. (1999). Life-cycle cost analysis for bridges. In D.M. Frangopol (Ed.), Chapter 9 *Bridge Safety and Reliability* (pp. 210–236). Reston: ASCE.

Frangopol, D.M. (2011). Life-cycle performance, management, and optimization of structural systems under uncertainty: accomplishments and challenges. *Structure and Infrastructure Engineering, 7*, 389–413.

Frangopol, D.M., & Bocchini, P. (2011, April 14–16). *Resilience as optimization criterion for the rehabilitation of bridges belonging to a transportation network subject to earthquake*. Proceedings of SEI-ASCE 2011 Structures Congress, Las Vegas, NV.

Frangopol, D.M., & Bocchini, P. (2012). Bridge network performance, maintenance and optimisation under uncertainty: accomplishments and challenges. *Structure and Infrastructure Engineering, 8*, 341–356.

Frangopol, D.M., & Curley, J.P. (1987). Effects of damage and redundancy on structural reliability. *Journal of Structural Engineering, 113*, 1533–1549.

Frangopol, D.M., & Estes, A.C. (1997). Lifetime bridge maintenance strategies based on system reliability. *Structural Engineering International, 7*, 193–198.

Frangopol, D.M., & Furuta, H. (Eds.). (2001). *Life-cycle cost analysis and design of civil infrastructure systems*. Reston: ASCE.

Frangopol, D.M., Kallen, M.-J., & van Noortwijk, J. (2004). Probabilistic models for life-cycle performance of deteriorating structures: Review and future directions. *Progress in Structural Engineering and Materials, 6*, 197–212.

Frangopol, D.M., Kim, S. (2011). Service life, reliability and maintenance of civil structures. In L.S. Lee & V. Karbari (Eds.), Chapter 5 *Service Life Estimation and Extension of Civil Engineering Structures* (pp. 145–178). Cambridge: Woodhead Publishing Ltd.

Frangopol, D.M., Kim, S. (2014a). Life-cycle analysis and optimization. In W-F. Chen & L. Duan, (Eds.), Chapter 18 *Bridge engineering handbook – Second edition, Vol. 5 Construction and Maintenance* (pp. 537–566). Boca Raton: CRC Press.

Frangopol, D.M., Kim, S. (2014b). Bridge health monitoring. In W-F. Chen & L. Duan, (Eds.), Chapter 10 in *Bridge engineering handbook – Second edition, Vol. 5 Construction and maintenance* (pp. 247–268). Boca Raton: CRC Press.

Frangopol, D.M., Kong, J.S., & Gharaibeh, E.S. (2001). Reliability-based life-cycle management of highway bridges. *Journal of Computing in Civil Engineering, 15*, 27–34.

Frangopol, D.M., Lin, K.-Y., & Estes, A.C. (1997). Life-cycle cost design of deteriorating structures. *Journal of Structural Engineering, 123*, 1390–1401.

Frangopol, D.M., & Liu, M. (2007). Maintenance and management of civil infrastructure based on condition, safety, optimization, and life-cycle cost. *Structure and Infrastructure Engineering, 3*, 29–41.

Frangopol, D.M., & Messervey, T.B. (2007). Lifetime oriented assessment and design optimization concepts under uncertainty: Role of structural health monitoring. In F. Stangenberg, O.T. Bruhns, D. Hartmann, & G. Meschke (Eds.), *Lifetime oriented design concepts* (pp. 133–145). Freiburg: Aedificatio Publishers.

Frangopol, D.M., & Messervey, T.B. (2008, October 24–25). Life-cycle cost and performance prediction: Role of structural health monitoring. *Proceedings of the International Workshop on Frontier Technologies for Infrastructures Engineering*. In S-S. Chen & A.H-S. Ang (Eds.) Taipei: Taiwan Building Technology Center, National TaiwanUniversity of Science and Technology, pp. 323–342.

Frangopol, D.M., & Nakib, R. (1991). Redundancy in highway bridges. *Engineering Journal, 28*, 45–50.

Frangopol, D.M., & Okasha, N.M. (2009). Lifetime-oriented multiobjective optimization of structural maintenance considering system reliability, redundancy and life-cycle cost using GA. *Structural Safety, 31*, 460–474.

Frangopol, D.M., & Saydam, D. (2014). Structural performance indicators for bridges. In W.-F. Chen & L. Duan (eds.), Chapter 9 *Bridge engineering handbook – Second edition, Vol. 1 Fundamentals* (pp. 185-205). Boca Raton: CRC Press.

Frangopol, D.M., & Soliman, M. (2013). Application of genetic algorithms to the life-cycle management optimization of civil and marine infrastructure systems. In Y. Tsompanakis, P. Iványi, & B.H.V. Topping (Eds.), Chapter 6 *Civil and structural engineering computational methods* (pp. 117–128). Stirlingshire: Saxe-Coburg Publications.

Frangopol, D.M., & Soliman, M. (2015). Application of soft computing techniques in life-cycle optimization of civil and marine structures. In J. Kruis, Y. Tsompanakis, & B.H.V. Topping (Eds.), Chapter 2 *Computational techniques for civil and structural engineering* (pp. 43–58). Stirlingshire: Saxe-Coburg Publications.

Frangopol, D.M., & Soliman, M. (2016). Life-cycle of structural systems: Recent achievements and future directions. *Structure and Infrastructure Engineering, 12*(1), 1–20.

Frangopol, D.M., Strauss, A., & Kim, S. (2008). Bridge reliability assessment based on monitoring. *Journal of Bridge Engineering, 13*, 258–270.

Garbatov, Y., & Guedes Soares, C.G. (2001). Cost and reliability based strategies for fatigue maintenance planning of floating structures. *Reliability Engineering & System Safety, 73*, 293–301.

Ghosn, M. (2000). Development of truck weight regulations using bridge reliability model. *Journal of Bridge Engineering, 5*, 293–303.

Ghosn, M., Dueñas-Osorio, L., Frangopol, D. M., McAllister, T., Bocchini, P., Manuel, L., Biondini, F., Hernandez, S., and Tsiatas, G. (2016a). Performance indicators for structural systems and infrastructure networks. *Journal of Structural Engineering, F4016003, 142*, 1–18.

Ghosn, M., Frangopol, D. M., McAllister, T., Shah, M., Diniz, S., Ellingwood, B., Manuel, L., Biondini, F., Catbas, N., Strauss, A., & Zhao, X. (2016b). Reliability-based performance indicators for structural members. *Journal of Structural Engineering, 142*(9), f4016002, 1–13.

Ghosn, M., & Frangopol, D.M. (2007). Redundancy of structures: A retrospective. In D.M. Frangopol, M. Kawatani, & C.-W. Kim (Eds.), *Reliability and optimization of structural systems: Assessment, design, and life-cycle performance* (pp. 91–100). London: Taylor & Francis Group. ISBN: 978-0-415-40655-0.

Ghosn, M., & Moses, F. (1998). *Redundancy in highway bridge superstructures* (National Academy Press, National Cooperative Highway Research Program, NCHRP Report 406), Transportation Research Board.

Ghosn, M., Moses, F., & Wang, J. (2003). *Design of highway bridges for extreme events* (nchrp trb report 489), Washington, DC.

Glaser, S.D., Li, H., Wang, M.L., Ou, J., & Lynch, J. (2007). Sensor technology innovation for the advancement of structural health monitoring: A strategic program of US-China research for the next decade. *Smart Structures and Systems, 3*, 221–244.

Goldberg, D.E. (1989). *Genetic algorithms in search, optimization and machine learning*. MA: Addison-Wesley.

Gul, M., & Catbas, F.N. (2011). Structural health monitoring and damage assessment using a novel time series analysis methodology with sensor clustering. *Journal of Sound and Vibration, 330*, 1196–1210.

Hawk, H., & Small, E.P. (1998). The BRIDGIT bridge management system. *Structural Engineering International, 8*, 309–314.

Hendawi, S., & Frangopol, D.M. (1994). System reliability and redundancy in structural design and evaluation. *Structural Safety, 16*, 47–71.

Huang, W., & Garbatov, Y. (2013). Fatigue reliability assessment of a complex welded structure subjected to multiple cracks. *Engineering Structures, 56*, 868–879.

Huston, D., Cui, J., Burns, D., & Hurley, D. (2011). Concrete bridge deck condition assessment with automated multisensor techniques. *Structure and Infrastructure Engineering, 7*, 613–623.

Imai, K., & Frangopol, D.M. (2002). System reliability of suspension bridges. *Structural Safety, 24*, 219–259, Elsevier.

IPCC. (2007). *Fourth assessment report of the intergovernmental panel in climate change*. Intergovernmental Panel on Climate Change, UK: Cambridge University Press.

IPCC. (2014). *Climate Change 2014*, Synthesis Report, Contribution of Working Groups I, II and III to the Fifth Assessment Report of the IPCC, Intergovernmental Panel on Climate Change, Geneva, Switzerland.

Jiménez, A., Ríos-Insua, S., & Mateos, A. (2003). A decision support system for multiattribute utility evaluation based on imprecise assignments. *Decision Support Systems, 36*, 65–79.

Keeney, R.L., & Raiffa, H. (1993). *Decisions with multiple objectives*. Cambridge: Cambridge University Press.

Kendall, A., Keoleian, G.A., & Helfand, G.E. (2008). Integrated life-cycle assessment and life-cycle cost analysis model for concrete bridge deck applications. *Journal of Infrastructure Systems, 14*, 214–222.

Kim, S., & Frangopol, D.M. (2011a). Cost-based optimum scheduling of inspection and monitoring for fatigue-sensitive structures under uncertainty. *Journal of Structural Engineering, 137*, 1319–1331.

Kim, S., & Frangopol, D.M. (2011b). Inspection and monitoring planning for RC structures based on minimization of expected damage detection delay. *Probabilistic Engineering Mechanics, 26*, 308–320.

Kim, S., & Frangopol, D.M. (2017). Efficient multi-objective optimisation of probabilistic service life management. *Structure and Infrastructure Engineering, 13*, 147–159.

Klinzmann, C., Schnetgöke, R., & Hosser, D. (2006). *A framework for reliability-based system assessment based on structural health monitoring*. Proceedings of the 3rd European Workshop on Structural Health Monitoring, Granada, Spain.

Ko, C., Pan, N., & Chiou, C. (2013). Web-based radio frequency identification facility management systems. *Structure and Infrastructure Engineering, 9*, 465–480.

Koh, H.-M., & Frangopol, D.M. (Eds.) (2008). *Bridge maintenance, safety, management, health monitoring and informatics, Set of Book and CD-ROM*. Boca Raton, London, New York, Leiden: A Balkema Book and CD-ROM, CRC Press, Taylor & Francis Group.

Kong, J.S., Ababneh, A.N., Frangopol, D.M., & Xi, Y. (2002). Reliability analysis of chloride penetration in saturated concrete. *Probabilistic Engineering Mechanics, 17*, 305–315.

Kong, J.S., & Frangopol, D.M. (2003). Life-cycle reliability-based maintenance cost optimization of deteriorating structures with emphasis on bridges. *Journal of Structural Engineering, 129*, 818–828.

Kong, J.S., & Frangopol, D.M. (2005). Sensitivity analysis in reliability-based lifetime performance prediction using simulation. *Journal of Materials in Civil Engineering, 17*, 296–306.

Kwok, R., & Rothrock, D.A. (2009). Decline in Arctic sea ice thickness from submarine and ICESAT records: 1958-2008. *Geophysical Research Letters, 36*, L15501.

Kwon, K., & Frangopol, D.M. (2010). Bridge fatigue reliability assessment using probability density functions of equivalent stress range based on field monitoring data. *International Journal of Fatigue, 32*, 1221–1232.

Kwon, K., & Frangopol, D.M. (2011). Bridge fatigue assessment and management using reliability-based crack growth and probability of detection models. *Probabilistic Engineering Mechanics, 26*, 471–480.

Lagasse, P.F., Clopper, P.E., Pagán-Ortiz, J.E., Zevenbergen, L.W., Arneson, L.A., Schall, J.D., & Girar, L.G. (2009). Bridge scour and stream instability countermeasures: Experience, selection, and design guidance – third edition. Hydraul. Des. Ser. No. 23, *FHWA Pub. No. FHWA-NHI-09-111, FHWANHI-09-112*, Federal Highway Administration, Washington, DC.

Leemis, L.M. (1995). *Reliability, probabilistic models and statistical methods*. Upper Saddle River, NJ: Prentice Hall.

Levitus, S., Antonov, J.I., Boyer, T.P., Locarnini, R.A., Garcia, H.E., & Mishonov, A.V. (2009). Global ocean heat content 1955–2008 in light of recently revealed instrumentation problems. *Geophysics Research Letters, 36*, L07608.

Lim, S., Akiyama, M., & Frangopol, D.M. (2016). Assessment of the structural performance of corrosion-affected RC members based on experimental study and probabilistic modeling. *Engineering Structures, 127*, 189–205.

Liu, C., Hammad, A., & Itoh, Y. (1997). Multiobjective optimization of bridge deck rehabilitation using a genetic algorithm. *Computer-Aided Civil and Infrastructure Engineering, 12*, 431–443.

Liu, D., Ghosn, M., Moses, F., & Neuenhoffer, A. (2001). *Redundancy in highway bridge substructures* (National Cooperative Highway Research Program, NCHRP Report 458), Washington, DC: Transportation Research Board, National Academy Press.

Liu, M., & Frangopol, D.M. (2005a). Time-dependent bridge network reliability: Novel approach. *Journal of Structural Engineering, 131*, 329–337.

Liu, M., & Frangopol, D.M. (2005b). Bridge annual maintenance prioritization under uncertainty by multiobjective combinatorial optimization. *Computer Aided Civil and Infrastructure Engineering, 20*, 343–353.

Liu, M., & Frangopol, D.M. (2006a). Optimizing bridge network maintenance management under uncertainty with conflicting criteria: Life-cycle maintenance, failure, and user costs. *Journal of Structural Engineering, 132*, 1835–1845.

Liu, M., & Frangopol, D.M. (2006b). Probability-based bridge network performance evaluation. *Journal of Bridge Engineering, 11*, 633–641.

Lounis, Z. (2004). Risk-based maintenance optimization of bridge structures. Retrieved from http://irc.nrc-cnrc.gc.ca/fulltext/nrcc47063/nrcc47063.pdf

Lounis, Z., & McAllister, T. (2016). Risk-based decision making for sustainable and resilient infrastructure systems. *Journal of Structural Engineering, 142*(9), F4016005, 1–14.

Lundie, S., Peters, G.M., & Beavis, P.C. (2004). Life cycle assessment for sustainable metropolitan water systems planning. *Environmental Science and Technology, 38*, 3465–3473.

Maekawa, K., & Fujiyama, C. (2013). Crack water interaction and fatigue life assessment of rc bridge decks. Poromechanics V. *Proceedings of the Fifth Biot Conference on Poromechanics*, pp. 2280–2289.

Maljaars, J., & Vrouwenvelder, A.C.W.M. (2014). Probabilistic fatigue life updating accounting for inspections of multiple critical locations. *International Journal of Fatigue, 68*, 24–37.

Marsh, P.S., & Frangopol, D.M. (2007). Lifetime multiobjective optimization of cost and spacing of corrosion rate sensors embedded in a deteriorating reinforced concrete bridge deck. *Journal of Structural Engineering, 133*, 777–787.

Marsh, P.S., & Frangopol, D.M. (2008). Reinforced concrete bridge deck reliability model incorporating temporal and spatial variations of probabilistic corrosion rate sensor data. *Reliability Engineering and System Safety, 93*, 394–409.

Melville, B.W. (1997). Pier and abutment scour: Integrated approach. *Journal of Hydraulic Engineering, 123*, 125–136.

Miyamoto, A., Kawamura, K., & Nakamura, H. (2000). Bridge management system and maintenance optimization for existing bridges. *Computer-Aided Civil and Infrastructure Engineering, 15*, 45–55.

Moan, T., & Song, R. (2000). Implications of inspection updating on system fatigue reliability of offshore structures. *Journal of Offshore Mechanics and Arctic Engineering, 122*, 173–180.

Morcous, G., & Lounis, Z. (2005). Maintenance optimization of infrastructure networks using genetic algorithms. *Automation in Construction, 14*, 129–142.

Morcous, G., Lounis, Z., & Cho, Y. (2010). An integrated system for bridge management using probabilistic and mechanistic deterioration models: Application to bridge decks. *KSCE Journal of Civil Engineering, 14*, 527–537.

Mori, Y., & Ellingwood, B.R. (1993). Reliability-based service-life assessment of aging concrete structures. *Journal of Structural Engineering, 119*, 1600–1621.

Moses, F. (1982). System reliability developments in structural engineering. *Structural Safety, 1*, 3–13.

Myrtle, R.C., Masri, S.E., Nigbor, R.L., & Caffrey, J.P. (2005). Classification and prioritization of essential systems in hospitals under extreme events. *Earthquake Spectra, 21*, 779–802.

Nagy, W., De Backer, H., & Van Bogaert, P. (2013). Fracture mechanics as an improvement of fatigue life assessment orthotropic bridge decks. Research and application: Structural engineering, mechanics and computation. *Proceeding of 5th International Conference on Structure Engineering*, Mechanics and Computation, SEMC 2013, pp. 579–584.

NASA. (2015). *Climate change: How do we know?* National Aeronautics and Space Administration. Retrieved October 9, 2015, from http://climate.nasa.gov/evidence/

Neves, L.C., & Frangopol, D.M. (2005). Condition, safety and cost profiles for deteriorating structures with emphasis on bridges. *Reliability Engineering & System Safety, 89*, 185–198.

Neves, L.C., Frangopol, D.M., & Cruz, P.J.S. (2004). Cost of life extension of deteriorating structures under reliability-based maintenance. *Computers & Structures, 82*, 1077–1089.

Neves, L.C., Frangopol, D.M., & Cruz, P.J. (2006). Probabilistic lifetime-oriented multiobjective optimization of bridge maintenance: Single maintenance type. *Journal of Structural Engineering, 132*, 991–1005.

Newhook, J.P., & Edalatmanesh, R. (2013). Integrating reliability and structural health monitoring in the fatigue assessment of concrete bridge decks. *Structure and Infrastructure Engineering, 9*, 619–633.

O'Connor, A., & O'Brien, E.J. (2005). Traffic load modelling and factors influencing the accuracy of predicted extremes. *Canadian Journal of Civil Engineering, 32*, 270–278, NRC Research Press.

Okasha, N.M., & Frangopol, D.M. (2009). Lifetime-oriented multi-objective optimization of structural maintenance considering system reliability, redundancy and life-cycle cost using GA. *Structural Safety, 31*, 460–474.

Okasha, N.M., & Frangopol, D.M. (2010a). Time-variant redundancy of structural systems. *Structure and Infrastructure Engineering, 6*, 279–301.

Okasha, N.M., & Frangopol, D.M. (2010b). Novel Approach for multicriteria optimization of life-cycle preventive and essential maintenance of deteriorating structures. *Journal of Structural Engineering, 136*, 1009–1022.

Onoufriou, T., & Frangopol, D.M. (2002). Reliability-based inspection optimization of complex structures: a brief retrospective. *Computers & Structures, 80*, 1133–1144.

Paliou, C., Shinozuka, M., & Chen, Y.-N. (1990). Reliability and redundancy of offshore structures. *Journal of Engineering Mechanics, 116*, 359–378.

Peterson, T.C., & Baringer, M.O. (Eds). (2009). State of the climate in 2008. *Special Supplement to the Bulletin of the American Meteorological Society*, 90, S17–S18.

Pipinato, A. (2014, July 7–11). Orthotropic steel deck design to extend the lifetime of plate and box girder bridge and viaducts. IABMAS 2014 – Bridge Maintenance, Safety, Management, Resilience and Sustainability – *Proceedings of the Seventh International Conference on Bridge Maintenance, Safety and Management*. Shangai, China.

Polyak, L., Alley, R.B., Andrews, J.T., Brigham-Grette, J., Cronin, T.M., Darby, D.A., Dyke, A.S., Fitzpatrick, J.J., Funder, S., Holland, M., Jennings, A.E., Miller, G.H., O'Regan, M., Savelle, J., Serreze, M., St. John, K., White, J.W.C., & Wolff, E. (2010). History of sea ice in the arctic. *Quaternary Science Reviews, 29*, 1757–1778.

PPD. (2013). *Critical infrastructure security and resilience.* Presidential Policy Directive PPD 21. Retrieved February 12, from http://www.whitehouse.gov/the-pressoffice/2013/02/12/presidential-policy-directive-critical-infrastructure-security-and-resil

Rashedi, R., & Moses, F. (1988). Identification of failure modes in system reliability. *Journal of Structural Engineering, 114*, 292–313.

Rausand, M., & Hoyland, A. (2004). *System reliability theory: Models and statistical methods.* (2nd ed.). Hoboken, NJ: Wiley Series in Probability and Statistics.

Richardson, E.V., & Davis, S.R. (2001). Evaluating scour at bridges, 4th ed. *FHWA NHI 01- 001 (HEC 18)*, Washington, DC: Federal Highway Administration.

Ross, S.M. (2000). *Introduction to probability models.* (7th ed.). Academic Press.

Sabatino, S., Frangopol, D.M., & Dong, Y. (2015a). Sustainability-informed maintenance optimization of highway bridges considering multi-attribute utility and risk attitude. *Engineering Structures, 102*, 310–321.

Sabatino, S., Frangopol, D.M., & Dong, Y. (2015b). Life-cycle utility-informed maintenance planning based on lifetime functions: Optimum balancing of cost, failure consequences, and performance benefit. *Structure and Infrastructure Engineering, 12*, 830–847.

Sánchez-Silva, M., Frangopol, D., Padgett, J., & Soliman, M. (2016). Maintenance and operation of infrastructure systems: review. *Journal of Structural Engineering, 142*(9), F4016004, 1–16.

Saydam, D., & Frangopol, D.M. (2011). Time-dependent performance indicators of damaged bridge superstructures. *Engineering Structures, 33*, 2458–2471.

Saydam, D., & Frangopol, D.M. (2014). Risk-based maintenance optimization of deteriorating bridges. *Journal of Structural Engineering, 141*(4), 04014120, 1–10.

Saydam, D., Frangopol, D.M., & Dong, Y. (2013). Assessment of risk using bridge element condition ratings. *Journal of Infrastructure Systems, 19*, 252–265.

Schwartz, H.G., Meyer, M., Burbank, C.J., Kuby, M., Oster, Posey, C.J., Russo, E.J. & Rypinski, A. (2014). Ch. 5: Transportation. Climate Change Impacts in the United States: The Third National Climate Assessment. *Global Change Research Program*, 130–149.

Shinozuka, M. (2008, October 24–25). Resilience and sustainability of infrastructure systems. In S-S. Chen & A.H-S. Ang (Eds.), *Proceedings of the International Workshop on Frontier Technologies for Infrastructures Engineering*, Taipei: Taiwan Building Technology Center, National Taiwan University of Science and Technology, pp. 225–244.

Shiraki, N., Shinozuka, M., Moore, J.E., Chang, S.E., Kameda, H., & Tanaka, S. (2007). System risk curves: Probabilistic performance scenarios for highway networks subject to earthquake damage. *Journal of Infrastructure Systems, 13*, 43–54.

Soliman, S.M., & Frangopol, D.M. (2014). Life-cycle management of fatigue-sensitive structures integrating inspection information. *Journal of Infrastructure Systems, 20*, 04014001.

Soliman, M., Frangopol, D.M., & Kim, S. (2013). Probabilistic optimum inspection planning of steel bridges with multiple fatigue sensitive details. *Engineering Structures, 49*, 996–1006.

Sousa, H., Sousa, C., Neves, A.S., Bento, J., & Figueiras, J. (2013). Long-term monitoring and assessment of a precast continuous viaduct. *Structure and Infrastructure Engineering, 9*, 777–793.

Stamatopoulos, G.N. (2013). Fatigue assessment and strengthening measures to upgrade a steel railway bridge. *Journal of Constructional Steel Research, 80*, 346–354.

Stein, S., & Sedmera, K. (2006). Risk-based management guidelines for scour at bridges with unknown foundations. *Final Report for NCHRP Project*, 24–25.

Stein, S.M., Young, G.K., Trent, R.E., & Pearson, D.R. (1999). Prioritizing scour vulnerable bridges using risk. *Journal of Infrastructure Systems, 5*, 95–101.

Stewart, M.G. (2012). Spatial and time-dependent reliability modelling of corrosion damage, safety and maintenance for reinforced concrete structures. *Structure and Infrastructure Engineering, 8*, 607–619.

Stewart, M.G., & Rosowsky, D.V. (1998). Time-dependent reliability of deteriorating reinforced concrete bridge decks. *Structural Safety, 20*, 91–109.

Stewart, M.G., Wang, X., & Nguyen, M.N. (2011). Climate change impact and risks of concrete infrastructure deterioration. *Engineering Structures, 33*, 1326–1337.

Stewart, M.G., Wang, X., & Nguyen, M.N. (2012). Climate change adaptation for corrosion control of concrete infrastructure. *Structural Safety, 35*, 29–39.

Sumi, Y. (1998). Fatigue crack propagation and computational remaining life assessment of ship structures. *Journal of Marine Science and Technology, 3*, 102–112.

Thanapol, Y., Akiyama, M., & Frangopol, D.M. (2016). Updating the seismic reliability of existing RC structures in a marine environment by incorporating the spatial steel corrosion distribution: Application to bridge piers. *Journal of Bridge Engineering, 21*(7), 04016031, 1–17.

Thoft-Christensen, P., & Murotsu, Y. (1986). *Application of structural systems reliability theory*. Berlin: Springer.

Thompson, P.D., Small, E.P., Johnson, M., & Marshall, A.R. (1998). The Pontis bridge management system. *Structural Engineering International, 8*, 303–308.

Titi, A., & Biondini, F. (2013, June 16–20). Resilience of concrete frame structures under corrosion. In G. Deodatis, B.R. Ellingwood, & D.M. Frangopol (Eds.), *11th International Conference on Structural, Safety & Reliability (ICOSSAR 2013)*, New York, NY, *Safety, Reliability, Risk and Life-Cycle Performance of Structures and Infrastructures*, Balkema, London: CRC Press, Taylor & Francis Group.

Titi, A., Biondini, F., & Frangopol, D.M. (2014, July 7–11). Lifetime resilience of aging concrete bridges under corrosion. In A. Chen, D.M. Frangopol, & X. Ruan (Eds.), *Proceedings of the Seventh International Conference on Bridge Maintenance, Safety, and Management*, IABMAS2014, Shanghai, China, in *Bridge Maintenance, Safety, Management and Life Extension* (pp. 1691–1698). Balkema, London: CRC Press, Taylor & Francis Group plc, and full paper on DVD, 426.

Titi, A., Biondini, F., & Frangopol, D.M. (2015, April 23–25). Seismic resilience of deteriorating concrete structures. In N. Ingraffea & N. Libby (Eds.), *Proceedings of the 2015 Structures Congress* (pp. 1649–1660), Portland, OR.

TRB. (2008). Potential impacts of climate change on U.S. transportation, Transportation Research Board Special Report 290, Transportation Research Board, Washington, DC.

TRB. (2014). Strategic issues facing transportation, Volume 2: Climate change, extreme weather events and the highway system: Practitioner's guide and research report, National Cooperative Highway Research Program, Transportation Research Board, Washington, DC.

USGCRP. (2008). Global climate change impacts in the 634 United States. U.S. Global Change Research Program. Retrieved from http://www.globalchange.gov/publications/reports/scientific-assessments/us-635 impacts/full-report/national-climate-change

Val, D.V., & Chernin, L. (2012). Cover cracking in reinforced concrete elements due to corrosion. *Structure and Infrastructure Engineering, 8*, 569–581.

Val, D.V., Stewart, M.G., & Melchers, R.E. (1998). Effect of reinforcement corrosion on reliability of highway bridges. *Engineering Structures, 20*, 1010–1019.

Vu, K., & Stewart, M.G. (2000). Structural reliability of concrete bridges including improved chloride-induced corrosion models. *Structural Safety, 22*, 313–333.

Wen, Y.K., & Kang, Y.J. (2001). Minimum building life-cycle cost design criteria. I: Methodology. *Journal of Structural Engineering, 127*, 330–337.

Westgate, R., Koo, K., Brownjohn, J., & List, D. (2013). Suspension bridge response due to extreme vehicle loads. *Structure and Infrastructure Engineering, 10*, 821–838.

Yang, S.-I., Frangopol, D.M., & Neves, L.C. (2004). Service life prediction of structural systems using lifetime functions with emphasis on bridges. *Reliability Engineering & System Safety, 86*, 39–51.

Yang, S.-I., Frangopol, D.M., & Neves, L.C. (2006a). Optimum maintenance strategy for deteriorating bridge structures based on lifetime functions. *Engineering Structures, 28*, 196–206.

Yang, S.-I., Frangopol, D.M., Kawakami, Y., & Neves, L.C. (2006b). The use of lifetime functions in the optimization of interventions on existing bridges considering maintenance and failure costs. *Reliability Engineering & System Safety, 91*, 698–705.

Yavari, S., Chang, S., & Elwood, K.J. (2010). Modeling post-earthquake functionality of regional health care facilities. *Earthquake Spectra, 26*, 869–892.

Yeo, G.L., & Cornell, C.A. (2005). Stochastic characterization and decision bases under time-dependent aftershock risk in performance-based earthquake engineering. *PEER Report*, University of California, Berkeley, CA.

Zhang, L., Silva, F., & Grismala, R. (2005). Ultimate lateral resistance to piles in cohesionless soils. *Journal of Geotechnical and Geoenvironmental Engineering, 131*, 78–83.

Zhu, B., & Frangopol, D.M. (2013a). Reliability, redundancy and risk as performance indicators of structural systems during their life-cycle. *Engineering Structures, 41*, 34–49.

Zhu, B., & Frangopol, D.M. (2013b). Risk-based approach for optimum maintenance of bridges under traffic and earthquake loads. *Journal of Structural Engineering, 139*, 422–434.

Zhu, B., & Frangopol, D.M. (2016a). Time-variant risk assessment of bridges with partially and full closed lanes due to traffic loading and scour. *Journal of Bridge Engineering, 21*(6), 04016021, 1–15.

Zhu, B., & Frangopol, D.M. (2016b). Time-dependent risk assessment of bridges based on cumulative-time failure probability. *Journal of Bridge Engineering, 06016009*, 1–7. doi:10.1061/(ASCE)BE.1943-5592.0000977

Part II
General methodology

Optimal bridge maintenance planning using improved multi-objective genetic algorithm

HITOSHI FURUTA, TAKAHIRO KAMEDA, KOICHIRO NAKAHARA, YUJI TAKAHASHI
and DAN M. FRANGOPOL

In order to establish a rational bridge management program, it is necessary to develop a cost-effective decision-support system for the maintenance of bridges. In this paper, an attempt is made to develop a bridge management system that can provide practical maintenance plans by using an improved multi-objective genetic algorithm. A group of bridges is analyzed to demonstrate the applicability and efficiency of the proposed method.

1. Introduction

Recently, bridge maintenance work in Japan has become very important and expensive. This is mainly due to the fact that the number of existing bridges requiring repair or replacement has increased drastically in the last decade. In order to establish a rational bridge maintenance program, it is necessary to develop a cost-effective decision-support system that can provide practical and economical life-cycle maintenance plans (Frangopol and Furuta 2001). In this paper, an attempt is made to develop such a system by using a multi-objective genetic algorithm (MOGA).

Although low-cost maintenance plans are desirable for bridge owners, it is necessary to consider various constraints when choosing an optimal maintenance program. For example, the minimization of cumulative maintenance cost requires the prescription of the target safety level and the expected service life. The predetermination of these requirements may be detrimental. Namely, using the same amount of cumulative maintenance cost, it may be possible to find solutions that can largely extend the target expected service life if the safety level can be slightly decreased. Alternatively, using the same amount of cumulative maintenance cost, it may be possible to find solutions that can substantially increase the safety level if the target expected service life is slightly reduced.

In this paper, by introducing the concept of multi-objective bridge maintenance planning optimization, it is intended to find several near-optimal maintenance plans. Although single-objective optimization can provide various solutions by changing the constraints, it requires enormous computation time. When selecting a practical maintenance plan, it is desirable to compare feasible optional solutions obtained under various conditions. This process has to be imposed in order to meet the accountability requirement through the disclosure of information. Thus, in this study an attempt is made to develop a decision support system that can provide several alternative maintenance plans by applying a MOGA. Since the bridge maintenance planning is a complex combinatorial problem, it is quite difficult to obtain Pareto solutions by simple genetic algorithms. Introducing the technique of non-dominated sorting into a genetic algorithm (GA) operation, a new MOGA is developed in order to improve the convergency and reduce the computation time. Several numerical examples are presented to demonstrate the applicability and efficiency of the proposed method.

2. Concrete bridge model

A group of 10 concrete highway bridges are considered in this study. The locations of all these bridges along the coast of Japan are indicated in figure 1. Maintenance management planning for 10 consecutive piers and floor slabs (composite structure of steel girders and reinforced concrete (RC) slabs) is considered here (Ito *et al.* 2002). Each bridge has the same structure and is composed of six main structural components: upper part of pier, lower part of pier, shoe, girder, bearing section of floor slab, and central section of floor slab (figure 2).

Environmental conditions can significantly affect the degree of deterioration of the structures and may vary from location to location according to geographical characteristics such as wind direction, amount of splash, etc. To take the environmental conditions into account, the deterioration type and year from completion of each bridge are summarized in table 1.

In this study, environmental corrosion due to neutralization of concrete, chloride attack, frost damage, chemical corrosion, or alkali–aggregate reaction are considered as major deteriorations. The structural performance of each bridge component i is evaluated by the associated safety level (also called durability level) P_i, which is defined as the ratio of the current safety level to the initial safety level.

Deterioration of a bridge due to corrosion depends on the concrete cover of its components and environmental conditions, among other factors. For each component, the major degradation mechanism and its rate of deterioration are assume to correspond to associated environmental conditions. Figures 3, 4, and 5 show the decreasing patterns of safety levels for RC slabs, shoes and girders, and piers, respectively. Average values are employed here as representative values for each level of chloride attack because the deteriorating rates can vary even in the same environment.

The decrease of RC slab performance is assumed to depend on corrosion. Hence, the safety level depends on the remaining cross section of reinforcement bars.

For shoe and girder performance, the major deterioration mechanism is considered to be fatigue due to repeated loadings. The decrease in performance occurs as the rubber bearing of the shoe or the paint coating of the girder deteriorates.

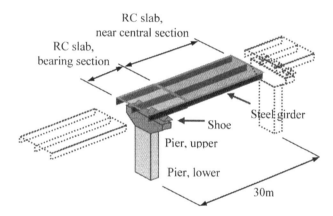

Figure 2. Main components of a bridge.

Table 1. Years from completion and type of deterioration caused by environmental conditions.

Bridge number	Years from completion	Deterioration type
B01	2	Neutralization of concrete
B02	2	Neutralization of concrete
B03	0	Chloride attack (slight)
B04	0	Chloride attack (medium)
B05	0	Chloride attack (severe)
B06	0	Chloride attack (medium)
B07	0	Chloride attack (severe)
B08	1	Chloride attack (medium)
B09	1	Chloride attack (slight)
B10	1	Chloride attack (slight)

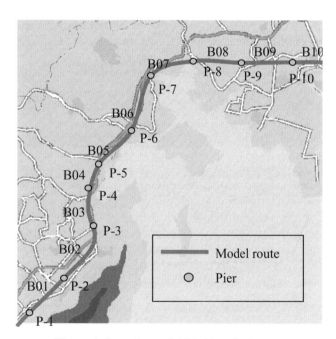

Figure 1. Locations of 10 bridges in Japan.

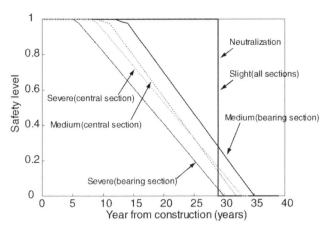

Figure 3. Typical performance of reinforced concrete slabs.

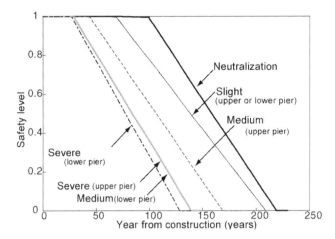

Figure 5. Typical performance of piers.

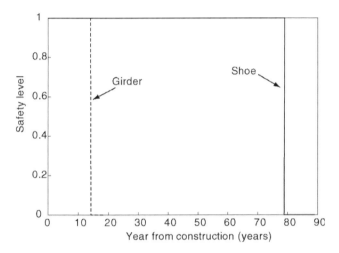

Figure 4. Typical performances of shoes and girders.

For piers, the major mechanism for deterioration is assumed to be only corrosion. Thus the reduced performance of a pier is expressed by the remaining section of reinforcement bars. The development of reinforcement corrosion is determined in accordance with Standard Specification for Design and Construction of Concrete in Japan (Japan Society of Civil Engineers 2001).

3. Maintenance strategies and life-cycle cost

In order to prevent deterioration in structural performance, several options such as repair, restoring, and reconstruction are considered. Their applicability and effects on each component are shown in table 2. Since the effects may differ even under the same conditions, average results are

adopted here. Maintenance methods applicable to RC slab may vary according to the environmental conditions and are determined considering several assumptions (Furuta et al. 2004a, b).

Life-cycle cost (LCC) is defined as the total maintenance cost for the entire bridge group during its life. This is obtained by the summation of the annual maintenance costs through the service life of all the bridges. The future costs are discounted to their present values. However, the discount rate is assumed to be zero in this study. Figure 6 shows a breakdown of the annual maintenance cost.

Other costs, such as indirect construction costs, general costs, and administrative costs, etc., are calculated in accordance with the Cost Estimation Standards for Civil Construction (MLIT 2001).

The direct construction costs consist of material and labor costs and the cost of scaffold. The breakdown of the material and labor costs and the cost of the scaffold is shown in table 3. The construction costs are based upon the market prices.

For calculating the construction costs, the following assumptions are taken into account.

(a) The cost of scaffold can be reduced by sharing. For example, scaffold can be shared for repairing the bearing and the bearing section of RC slab, consequently reducing the scaffolding cost.

(b) Indirect construction costs, such as general administrative costs, can be saved by implementing several repairs in the same year. The ratio of indirect to maintenance costs decreases as the direct costs increase, as shown in figure 7. The value of LCC is reduced when multiple components are repaired simultaneously.

Table 2. Effects of repair, restoring, and reconstruction.

Structural component	Maintenance type	Average effect
Pier or slab	Surface painting	Delays P_i decrease for 7 years
	Surface covering	Delays P_i decrease for 10 years
	Section restoring	Restores P_i to 1.0, and then allows it to deteriorate with the same slope as the initial deterioration curve
	Desalting (Re-alkalization)	P_i deteriorates with the same slope as the initial deterioration curve
	Cathodic protection	Delays P_i decrease for 40 years
	Section restoring with surface covering	Restores P_i to 1.0, delays P_i decrease for 10 years, and then P_i deteriorates with the same slope as the initial deterioration curve
Girder	Painting	Maintains initial performance until the end of the specified lifetime
Shoe	Replacement of bearing	Maintains initial performance until the end of the specified lifetime
Slab	Recasting	Maintains initial performance until the end of the specified lifetime
All	Reconstruction	Restore P_i to 1.0, delays P_i decrease for 10 years, and then P_i deteriorates with the same slope as the initial deterioration curve

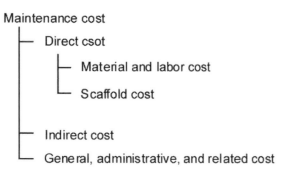

Figure 6. Breakdown of maintenance cost.

4. Multi-objective problem and MOGA

4.1 *Multi-objective problem*

A multiple-objective optimization problem has two or more objective functions that cannot be integrated into a single objective function (Iri and Konno 1982). In general, the objective functions cannot be simultaneously minimized (or maximized). It is the essence of the problem that trade-off

relations exist among the objective functions. The concept of "Pareto optimum" becomes important in order to balance the trade-off relations. The Pareto optimum solution is a solution that cannot improve an objective function without sacrificing other functions (figures 8 and 9). A dominated, also called non-dominant, solution is indicated in figure 8.

4.2 *Multi-Objective Genetic Algorithm (MOGA)*

Genetic algorithm (GA) is an evolutionary computing technique, in which solution candidates are mapped into GA space by encoding. The following steps are employed to obtain the optimal solutions (Goldberg 1989): (a) initialization, (b) crossover, (c) mutation, (d) natural selection, and (e) reproduction. Individuals, which are solution candidates, are initially generated at random. Then, steps (b), (c), (d), and (e) are repeatedly implemented until the termination condition is fulfilled. Each individual has a fitness value to the environment. The environment corresponds to the problem space and the fitness value corresponds to the evaluation value of objective function. Each individual has two aspects: gene type (GTYPE) expressing the chromosome or DNA, and phenomenon type (PTYPE) expressing the solution. GA operations are applied to GTYPE and generate new children from parents (individuals) by effective searches in the problem space, and extend the search space by mutation to enhance the possibility of individuals other than the neighbor of the solution. GA operations that generate useful children from their parents are performed by crossover operation of chromosome or genes (GTYPE) without using special knowledge and intelligence. This characteristic is considered to be one of the reasons for successful application of GA (Furuta and Sugimoto 1997).

In this study, NSGA2 (non-dominated sorting GA-2) (Kitano 1995) is applied as MOGA. In NSGA2, Pareto solutions are obtained from current candidates that are defined as Front *1*. The next Pareto solutions are formed from the candidates excluding Front *1* solutions. This procedure is repeated until all candidates are classified. NSGA2 selects solutions successively from Front *n*. When *n* exceeds the prescribed number, the final solutions are selected from Front *n*. Figure 10 shows the concept of non-dominated sorting.

5. Application of MOGA to maintenance planning

It is desirable to determine an appropriate life-cycle maintenance plan by comparing several solutions for various conditions (Furuta *et al.* 2004a, b). A new decision support system is developed here from the viewpoint of multi-objective optimization, in order to provide various solutions needed for the decision-making.

Table 3. Material, labor, and scaffold costs.

(a) Material and labor costs

Maintenance action	Upper pier (¥/m²)	Lower pier (¥/m²)	Shoe (¥/Part)	Girder (¥/m²)	Slab (central section) (¥/m²)	Slab (bearing section) (¥/m²)
Surface painting	780,000	1,920,000	–	–	1,640,000	3,280,000
Surface covering	2,730,000	6,720,000	–	–	4,100,000	8,200,000
Section restoring	20,670,000	50,880,000	–	–	22,140,000	44,280,000
Desalting (Re-alkalization)	3,510,000	8,640,000	–	–	7,380,000	14,760,000
Cathodic protection	3,900,000	9,600,000	–	–	8,200,000	16,400,000
Section restoring with surface covering	22,620,000	55,680,000	–	–	26,240,000	52,480,000
Reconstruction	–	–	4,200,000	5,400,000	12,300,000	24,600,000

(b) Scaffold cost

Upper pier	Lower pier	Shoe	Girder	Slab (central section)	Slab (bearing section)
¥360,000	¥190,000	¥360,000	¥4,830,000	¥690,000	¥510,000

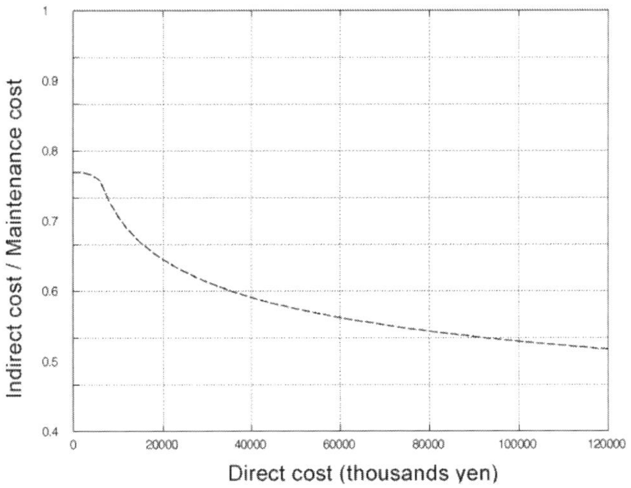

Figure 7. Effect of direct cost on the ratio of indirect/maintenance costs.

In this study, LCC, safety level and service life are used as objective functions. LCC is minimized, safety level is maximized, and service life is maximized. There are trade-off relations among the three objective functions. For example, LCC increases when service life is extended, and safety level and service life decrease due to the reduction of LCC. Then, multi-objective optimization can provide a set of Pareto solutions that can not improve an objective function without making other objective functions worse (Liu and Frangopol 2004a, b).

In the proposed system, DNA structure is constituted as shown in figure 11, in which the DNA of each individual consists of three parts, such as repair method, interval of repair, and shared service life (figure 12). In this figure,

service life is calculated as the sum of repairing years and their interval years. In figure 12, service life is obtained as 67 years which is expressed as the sum of 30 years and 37 years. The repair part and the interval part have the same length. The gene of the repair part has the ID number of repair method. The interval part has enough length to consider service life. In this system, ID 1 means surface painting, ID 2 surface coating, ID 3 section restoring, ID 4 desalting (re-alkalization) or cathodic protection, and ID 5 section restoring with surface covering. The DNA of service life part has a binary expression with six bits and its value is changed to a decimal number.

In crossover, the system generates new candidates by using the procedure shown in figure 13. For mutation, the system shown in figure 14 is used. Then, objective functions are defined as follows:

$$\text{Objective function 1}: C_{total} = \Sigma LCC_i \rightarrow min \tag{1}$$
$$\text{where } LCC_i = \text{LCC for bridge } i$$

$$\text{Objective function 2}: Y_{total} = \Sigma Y_i \rightarrow max$$
$$\text{Constraints}: Y_i > Y_{required} \tag{2}$$
$$\text{where } Y_i = \text{service life of bridge } i,$$
$$Y_{required} = \text{required service life}$$

$$\text{Objective function 3}: P_{total} = \Sigma P_i \rightarrow max$$
$$\text{Constraints}: P_i > P_{target} \tag{3}$$
$$\text{where } P_{target} = \text{target safety level}$$

The above objective functions have trade-off relations to each other. Namely, the maximization of safety level or maximization of service life cannot be realized without increasing LCC. On the other hand, the minimization of

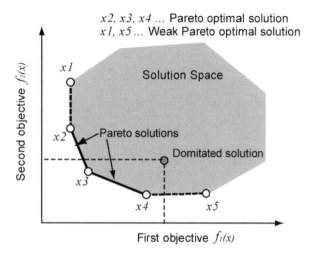

Figure 8. Pareto solutions.

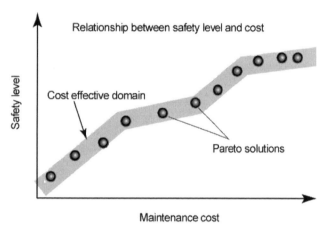

Figure 9. Cost-effective domain including Pareto solutions.

LCC is possible only if the service life and/or the safety level decreases.

6. Numerical example

In the implementation of MOGA, the GA parameters considered are as follows: number of individuals = 2000, crossover rate = 0.60, mutation rate = 0.05, and number of generations = 5000.

Figures 15 to 18 present the results obtained by MOGA. Each figure shows the comparison of the results of the 1st generation (iteration number) and the 5000th generation.

In figure 15, the solutions at the 1st generation spread over the design space. This means that the initial solutions can be generated uniformly. After the 5000th generation, the solutions tend to converge to a surface, which finally forms the Pareto set as the envelope of all solutions. The number of solutions at the 5000th generation is much larger than that at the 1st generation. This indicates that MOGA could obtain various optimal solutions with different LCC values, safety levels, and service lives. From figure 15, it is seen that MOGA can find good solutions, all of which evolve for all the objective functions, and the final solutions are sparse and have discontinuity. In other words, the surfaces associated with the trade-off relations are not smooth. This implies that an appropriate long-term maintenance plan cannot be created by the repetition of the short-term plans.

In figure 16, the vertical axis represents safety level, whereas the horizontal axis represents LCC. Although at the 1st generation, the solutions may have a rather linear relation between safety level and LCC, the relation shows non-linearity through the convergence process. This implies that the safety level may be significantly increased if the LCC can be slightly increased, when the service life is fixed.

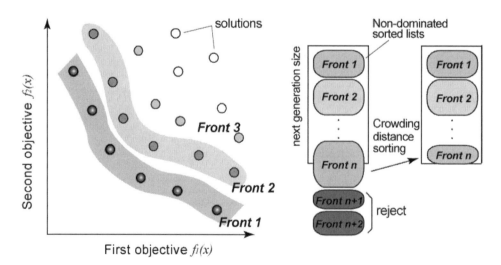

Figure 10. Non-dominated sorting (NSGA2).

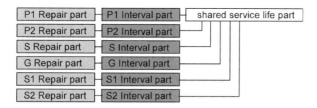

P1 : upper pier, P2 : lower pier, S : shoe, G : girder, S1 : slab (central section), S2 : slab (bearing section)

Figure 11. Structure of DNA.

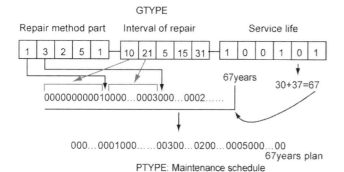

Figure 12. Coding rule.

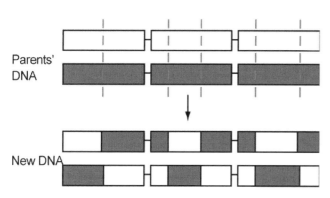

Figure 13. Crossover.

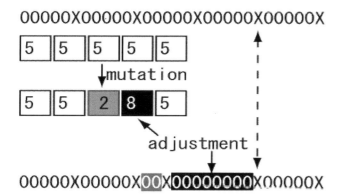

Figure 14. Mutation.

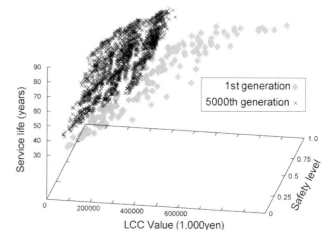

Figure 15. Pareto solutions obtained by multi-objective genetic algorithm.

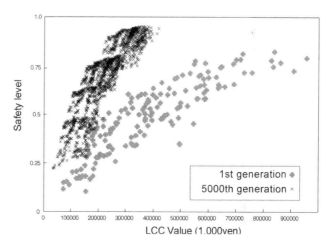

Figure 16. Relation between life-cycle cost and safety level.

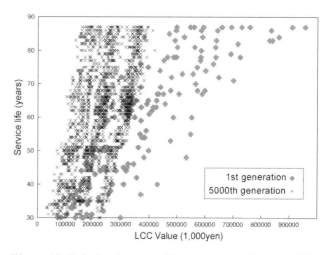

Figure 17. Relation between life-cycle cost and service life.

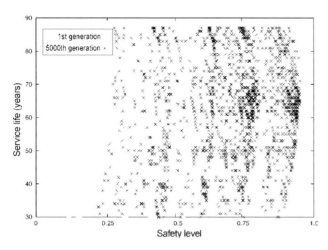

Figure 18. Relation between safety level and service life.

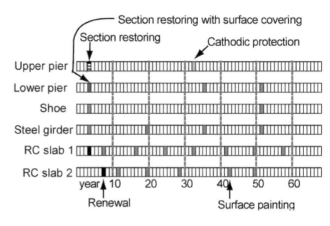

LCC value : 330,464,000 yen
Safety level : 0.923
Service life : 68 years

Figure 20. Representative maintenance plan 2 (Plan 2).

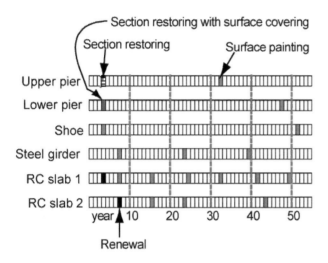

LCC value : 287,588,000 yen
Safety level : 0.858
Service life : 55 years

Figure 19. Representative maintenance plan 1 (Plan 1).

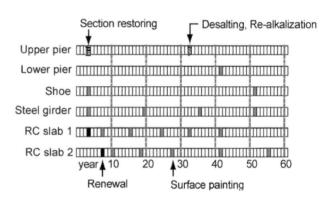

LCC value : 248,607,000 yen
Safety level : 0.782
Service life : 61 years

Figure 21. Representative maintenance plan 3 (Plan 3).

Figure 17 presents the relation between service life and service life. Since LCC and service life have a rather perfect positive linear correlation, it can be said that the service life can be extended if LCC can be increased. On the other hand, there is no distinct relation between safety level and service life, as shown in figure 18. It should be noted that the safety level may not be raised even if the service life is shortened, under a constant LCC. Namely, the relation between safety level and service life is so unclear that the extension of service life should be done with careful examination.

Figures 19 to 22 show four representative long-term maintenance plans extracted from the solutions presented in figure 15. The maintenance plan for only a bridge with a severe environmental condition is presented. In these figures, all LCC values are presented in thousands of yen. Comparing the four maintenance plans, Plan 2 has the largest LCC value, Plan 1 is the next largest, Plan 3 is the third, and Plan 4 is associated with the least cost. As expected, the safety levels correspond to the order of LCC values. However, the service life shows the reverse order between Plan 1 and Plan 3. Namely, Plan 3 has a longer service life than Plan 1, though the LCC value of Plan 3 is smaller than that of Plan 1. This is due to the fact that the safety level of Plan 3 is smaller than that of Plan 1. Thus, it is proved that the proposed method can provide many useful solutions with different characteristics for determining an appropriate maintenance plan available for practical use. From these figures, it is clear that the LCC can be reduced by adopting simultaneous repair works. For

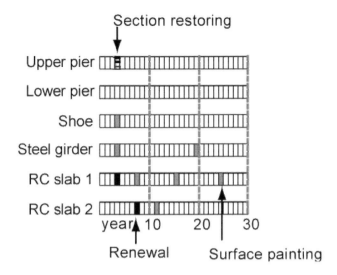

LCC value : 151,313,000 yen
Safety level : 0.720
Service life : 31 years

Figure 22. Representative maintenance plan 4 (Plan 4).

instance, Plan 3 presented in figure 21 implements various works in the same year. Finally, it is confirmed that the proposed method using NSGA2 can provide many near-optimal maintenance plans with various reasonable LCC values, safety levels and service lives. Note that it is quite difficult to obtain such near-optimal solutions by the current MOGA.

7. Conclusions

In this paper, an attempt was made to formulate the optimal maintenance planning as a multi-objective optimization. By considering LCC, safety level, and service life as objective functions, it is possible to obtain the relationships among these three performance indicators and provide bridge maintenance management engineers with various maintenance plans with appropriate allocations of resources. Based on the results presented in this paper, the following conclusions may be drawn:

1. Since the optimal maintenance problem is a very complex combinatorial problem, it is difficult to obtain reasonable solutions by the current optimization techniques.
2. Although a GA is applicable to solve multi-objective problems, it is difficult to apply it to large and very complex bridge network maintenance problems. By introducing the technique of non-dominated sorting GA-(NSGA2), it is possible to obtain efficient near-optimal solutions for the maintenance planning of a group of bridge structures.

3. The Pareto solutions obtained by the proposed method show discontinuity. This means that the surfaces constructed by the trade-off relationships are not smooth, so that an appropriate long-term maintenance plan can not be created by the simple repetition of short term plans
4. In the examples presented, the relation between safety level and LCC is non-linear. The increase of LCC hardly contributes to the improvement of safety level. On the other hand, LCC and service life are almost linearly related. Therefore, service life can be extended if LCC can be increased. However, there is no distinct relation between safety level and service life.
5. LCC can be reduced by adopting simultaneous repair works. The proposed method using NSGA can provide many near-optimal maintenance plans with various reasonable LCC values, safety levels and service lives.

References

Frangopol, D.M. and Furuta, H. (eds), *Life-Cycle Cost Analysis and Design of Civil Infrastructure Systems*, 2001 (ASCE, Reston, VA).

Furuta, H. and Sugimoto, H., *Applications of Genetic Algorithm to Structural Engineering*, 1997 (Morikita Publishing: Tokyo) (in Japanese).

Furuta, H., Kameda, T., Fukuda, Y., and Frangopol, D.M., Life-cycle cost analysis for infrastructure systems: life cycle cost vs. safety level vs. service life, in Proceedings of Joint International Workshops LCC03/IABMAS and fip/JCSS, EPFL, Lausanne, 24–26 March 2003 (keynote lecture); in *Life-Cycle Performance of Deteriorating Structures: Assessment, Design and Management*, edited by Frangopol, D.M., Brühwiler, E., Faber, M.H. and Adey, B., 2004a, pp. 19–25 (ASCE, Reston, VA).

Furuta, H., Kameda, T., and Frangopol, D.M., Balance of structural performance measures, in *Proceedings of the Structures Congress, Nashville, Tennessee, ASCE*, 2004b, CD-ROM (ASCE: Reston, VA).

Goldberg, D.E., *Genetic Algorithms in Search, Optimization and Machine Learning*, 1989 (Addison Wesley: New York).

Iri, M. and Konno, H. (eds.), *Multi-Objective Optimization*, 1982 (Sangyo-tosho: Tokyo) (in Japanese).

Ito, H., Takahashi Y., Furuta, H., and Kameda, T., An optimal maintenance planning for many concrete bridges based on life-cycle cost, in *Proceedings of IABMAS*, Barcelona, Spain, 2002, CD-ROM (HMNE: Barcelona).

Japan Society of Civil Engineers, Standard Specification for Design and Construction of Concrete in Japan (Maintenance) 2001 (Japan Society of Civil Engineers: Tokyo) (in Japanese).

Kitano, H. (ed.), Genetic Algorithm 3, 1995 (Sangyo-tosho: Tokyo) (in Japanese).

Liu, M. and Frangopol, D.M., Probabilistic maintenance prioritization for deteriorating bridges using a multiobjective genetic algorithm, in *Proceedings of the Ninth ASCE Joint Specialty Conference on Probabilistic Mechanics and Structural Reliability*, hosted by Sandia National Laboratories, 2004a, 6 pages on CD-ROM (Omnipress: Albuquerque, New Mexico).

Liu, M. and Frangopol, D.M., Optimal bridge maintenance planning based on probabilistic performance prediction. *Eng. Struct.*, 2004b, **26**, 991–1002.

MLIT, *Cost Estimation Standards for Civil Constructions*, 2001 (Ministry of Land, Infrastructure and Transportation: Japan) (in Japanese).

Maintenance and management of civil infrastructure based on condition, safety, optimization, and life-cycle cost*

DAN M. FRANGOPOL and MIN LIU

Cost-competent maintenance and management of civil infrastructure requires balanced consideration of both the structure performance and the total cost accrued over the entire life-cycle. Most existing maintenance and management systems are developed on the basis of life-cycle cost minimization only. The single maintenance and management solution thus obtained, however, does not necessarily result in satisfactory long-term structure performance. Another concern is that the structure performance is usually described by the visual inspection-based structure condition states. The actual structure safety level, however, has not been explicitly or adequately considered in determining maintenance management decisions. This paper reviews the recent development of life-cycle maintenance and management planning for deteriorating civil infrastructure with emphasis on bridges using optimization techniques and considering simultaneously multiple and often competing criteria in terms of condition, safety and life-cycle cost. This multiple-objective approach leads to a large pool of alternative maintenance and management solutions that helps active decision-making by choosing a compromise solution of preferably balancing structure performance and life-cycle cost.

1. Introduction

Satisfactory lifetime performance of civil infrastructure is of critical importance to sustained economic growth and social development of a modern society. In particular, the highway transportation system is considered to be one of the society's critical foundations. Among the many elements of this complex system, bridges are especially important and the most vulnerable because of their distinct function of joining highways as the crucial nodes. The bridge infrastructure throughout the United States, especially in the dense urban regions of the country, has been constantly exposed to aggressive environments. It faces continuously increasing traffic volumes and heavier truck-loads, which degrade the long-term performance of highway structures at an alarming rate. In the United States, 27.1% of the 590,750 existing bridges nationwide are labeled structurally deficient or functionally obsolete; these bridges must be replaced or upgraded immediately with an estimated annual investment of US$ 9.4 billion required for the next 20 years in order to eliminate all bridge deficiencies (ASCE 2005).

A deteriorating public works infrastructure leads to increased costs for business and ultimately for consumers.

*Based on a keynote paper presented at the International Workshop on Integrated Life-Cycle Management of Infrastructures, Hong Kong, December 9–11, 2004.

Catastrophic failure of civil infrastructure can cause widespread social and economic consequences. Nowadays, the homeland security has become a heightened concern (FHWA 2003). Well-maintained civil infrastructure can substantially increase a country's competitiveness in a global economy and enhance resilience to adverse circumstances such as natural hazards (e.g. earthquakes, hurricanes and floods) and manmade disasters (e.g. vehicular collision and explosive blasts due to terrorists' attacks). In addition to developing more advanced maintenance technologies, structure managers urgently need methodologies to cost-effectively allocate limited budgets for maintaining and managing aging and deteriorating civil infrastructure over the specified time period, with the goal of optimally balancing lifetime structure performance and whole-life maintenance cost.

Researchers and engineers around the world have been developing and implementing different maintenance management programs to achieve desirable solutions to maintain satisfactory infrastructure performance from the long-term economical point of view. In particular, most existing methodologies for the maintenance and management of the highway transportation infrastructure are based on least life-cycle cost while enforcing relevant performance thresholds. There exist practical difficulties with these treatments. For example, if the available budgets are more than the computed minimum life-cycle cost, the infrastructure performance can be maintained at a higher level than what was previously prescribed for deriving the minimum life-cycle cost solution. On the other hand, if the available financial resources are not enough to meet the computed minimum life-cycle cost, bridge managers have to accept a solution that can improve the structure performance to the highest possible level under the budget constraints. Indeed an overall satisfactory maintenance planning solution should therefore be determined by optimally balancing competing objectives of, for example, improving structure performance and reducing the long-term cumulative cost.

Another problem regarding the existing maintenance management systems is that visual inspection-based infrastructure condition states are mostly used to describe the infrastructure performance. The actual infrastructure safety level, however, has not been explicitly or adequately accounted for. Maintenance management decisions made on the basis of infrastructure condition states alone are not necessarily cost-effective and, in some cases, may even cause serious safety consequences. For example, it is possible that a reinforced concrete structure with satisfactory visual condition states could actually be structurally unsafe when invisible flaws such as corrosion of the embedded reinforcement exist. Failure to identify these facts definitely raises severe safety concerns. It is also possible that a structure may be structurally sound despite the poor visual condition status. Under this situation, only minor or moderate repair of these visual defects may be necessary and thus costly upgrading or total replacement can be avoided. Savings of maintenance funds can then be enormous.

In response to the above concerns, all merit metrics relevant to civil infrastructure performance need to be considered *simultaneously* to determine the maintenance management solutions. The life-cycle cost is one such merit measure. Structure performance in terms of condition and safety states should also be treated as merit measures. Traditional methodologies treat the prescribed levels of these performance measures as constraints that valid maintenance management solutions should comply with. However, the advanced methodologies based on a multiobjective optimization formulation treat all these performance measures as additional merit objective functions that are not restrictive in nature and the actual performance levels are at the discretion of civil infrastructure managers. As a result, the present performance-based maintenance management methodologies lead to a group of non-dominated solutions, each of which represents a unique optimized tradeoff between life-cycle cost minimization and infrastructure performance maximization. That is, there does not exist another solution that can improve one objective of the optimized solution being considered without sacrificing at least one of the other objectives. This large set of alternative solutions helps civil infrastructure managers' active decision-making by selecting a maintenance solution with the most desirable balance between the conflicting life-cycle cost and infrastructure performance objectives. It is clear that this will have significant impacts on developing the next-generation of civil infrastructure maintenance management systems, which should integrate the concept of structure safety and reliability as well as consider other relevant conflicting objectives in order to meet maintenance managers' specified requirements and budget constraints.

Research on the use of multiobjective optimization techniques in maintenance management of civil infrastructure has appeared in the recent literature. Multiple and conflicting performance indicators such as condition, safety and durability along with life-cycle cost are simultaneously considered as separate criteria (Liu *et al.* 1997, Miyamoto *et al.* 2000, Furuta *et al.* 2004). Interestingly all these research efforts use genetic algorithms (GAs) as numerical optimization tools. This is because the practical maintenance management problems can be best posed as combinatorial optimization problems. For example, maintenance managers often require a list of prioritized maintenance interventions for civil infrastructure on an annual basis (Das 1999). Due to their inherent features (as will be discussed in a later section), GAs are very effective for solving these kinds of problems.

In order to make rational decisions on preserving deteriorating civil infrastructure, maintenance engineers need to appropriately address sources of uncertainty associated with the deterioration processes of civil infrastructure under no maintenance and under different maintenance interventions. Sources of uncertainty include imperfect description of the mechanical loadings and environmental stressors as well as the inexact prediction of the deteriorating structure performance. There are two general types of uncertainty (Wen *et al.* 2003): aleatory and epistemic. The aleatory uncertainty is caused by the inherent variation of the deterioration due to the combined effects of complex traffic loadings and environmental stresses as well as structural aging. The epistemic uncertainty stems from the randomness caused by subjective assumption in evaluating demand and load-carrying capacity of structures or insufficient knowledge in understanding, for example, the deterioration mechanisms. This type of uncertainty can be lessened through, for example, Bayesian updating, when additional information is gathered (Enright and Frangopol 1999, Estes *et al.* 2004, Rafiq *et al.* 2004). The ensuing maintenance interventions over the specified time period of deteriorating structures add additional uncertainty in the prediction of time-dependent structure performance.

In this paper, the significance of using multiobjective optimization techniques in civil infrastructure maintenance management is emphasized. In particular, the recent development of life-cycle maintenance and management for deteriorating highway bridge infrastructure is discussed, in which the combinatorial optimization is used and multiple and competing criteria in terms of condition, safety and life-cycle cost are considered simultaneously. Uncertainties associated with the deterioration process under no maintenance and under maintenance interventions are treated by Monte Carlo simulation, which produces sample mean predictions of structure performances as well as of life-cycle maintenance cost. As numerical examples, GA-based automated maintenance management procedures are applied to prioritize maintenance resources for deteriorating structures over designated time periods. The first example deals with project-level maintenance management of preserving a population of similar highway reinforced concrete crossheads subject to various degrees of deterioration. The second example discusses the network-level bridge maintenance management, which considers individual bridges acting as the nodes of a highway transportation network based on network analysis under uncertainty. Finally, the research needs of integrating structural health monitoring with maintenance management systems are briefly discussed.

2. Performance and deterioration of civil infrastructure

2.1 *Performance indicators*

Appropriate performance indicators are needed to describe the time-dependent civil infrastructure performance, based on which maintenance management decisions can be made. Some of the commonly used performance indicators are discussed as follows.

2.1.1 Condition index. Visual inspection-based condition rating index is traditionally used to measure the bridge's remaining load-carrying capacity (Pontis 2001). The current bridge management systems (BMSs) characterize structural elements by discrete condition states noting deterioration (Thompson *et al.* 1998). Typically maintenance interventions should be prioritized to civil infrastructure with unacceptable and poor condition rating levels. For reinforced concrete elements under corrosion attack in the United Kingdom, Denton (2002) classifies the visual inspection-based condition states into four discrete levels, denoted as 0, 1, 2 and 3, that represent no chloride contamination, onset of corrosion, onset of cracking and loose concrete/significant delamination, respectively. A value larger than 3 indicates an unacceptable condition state. As a subjective measure, however, the condition index may not faithfully reflect the true load-carrying capacity of structural members.

2.1.2 Safety index. According to bridge specifications in the United Kingdom, the safety index is defined as the ratio of available to required live load capacity (DB12/01 2001). The structure performance is considered unacceptable if the value of safety index drops below 0.91.

2.1.3 Reliability index. The reliability index is used to quantify the structure safety level and can thus be used as a performance indicator (Melchers 1999, Nowak and Collins 2000). This is a more objective means to assess the performance of an existing structure, provided the probabilistic nature of structural load and capability can be properly modeled. For example, Estes and Frangopol (1999) evaluated the system reliability of a bridge structure that comprises both super- and sub-structure and is modeled as a series-parallel system. Using the reliability concept, Ghosn and Johnson (2000) assess the safety level of bridges subject to flood scour and seismic loads. Akgül and Frangopol (2003) conducted reliability analysis of bridge components using performance limit state functions defined in terms of standard code formulations in AASHTO (1996) specifications.

2.2 *Deterioration mechanism and modeling*

Accurate modeling of the structure deterioration process is the most critical component in maintenance management of civil infrastructure, because allocation of maintenance needs must be determined by predicting performance deterioration of civil infrastructure under no maintenance and under different maintenance scenarios. In doing so, it is essential to understand the mechanism of structure deterioration.

2.2.1 Deterioration mechanism. The civil infrastructure deterioration is a very complex process and is affected by many factors, which concern physical, chemical and electrical phenomena, among others. For highway infrastructure such as bridges and roads, aggressive environmental conditions, including alkali-silica reaction, chloride contamination, and sulfate attack contribute to progressive performance deterioration over time. The ever-increasing traffic load effects further aggravate the structure performance. It should be noted that the identified mechanisms cannot explain the entire performance deterioration. Uncertainties are unavoidable in the understanding and proper mathematical treatment of the deterioration process.

2.2.2 Deterioration modeling. The deterioration process of civil infrastructure has been widely depicted using Markov chain models. This mathematical model is built under the premise that current states are only dependent upon a finite number of previous states. If the current state is only a function of its immediately past state, it is called a one-step Markov chain model. The core element of a Markov chain model is a transition matrix that dictates the probability that one state component changes to another state. In most practical cases for maintenance management of civil infrastructure, a stationary Markov chain model is used in which the transition matrix remains the same throughout the specified time period.

The time-dependent performance deterioration of civil infrastructure under no maintenance and under maintenance interventions can also be predicted by continuous computational models. For example, Frangopol *et al.* (2001) proposed a model to describe the deterioration process under no maintenance by a function characterized by an initial performance level, time to damage initiation, and a deterioration curve governed by appropriate functions and in the simplest form, a linear function with a constant deterioration rate. Effects of a generic maintenance action include immediate performance enhancement, suppression of the deterioration process for a specified time interval, reduction in the deterioration rate, and another specified duration of maintenance effect beyond which the deterioration rate resumes to the original one. Uncertainties associated with the deterioration process are considered in terms of probabilistic distribution of the controlling parameters of this computational model. Monte Carlo simulation is then used to account for these uncertainties by obtaining the statistical time-varying performance profiles of deteriorating structures under various single-cyclic and combined maintenance actions.

The performance deterioration may also be predicted if reasonable mechanistic models are available. For example, the deterioration of the reinforcement in reinforced concrete structures is usually caused by de-icing chemicals related corrosion, whose initiation in concrete can be reasonably predicted by Fick's law of diffusion (Thoft-Christensen 1998). Effects of corrosion are considered as a gradual reduction of the reinforcement area. The time-dependent structural capacity can then be predicted. If the information on time-variant loads is available, the structural reliability profiles can be obtained as a result. For example, Enright and Frangopol (1999), Val *et al.* (2000) and Akgül and Frangopol (2003) predict the time-dependent reliability deterioration of highway bridges. Due to the existence of uncertainty in initiation time and rate of corrosion as well as other relevant parameters, Monte Carlo simulations are needed to account for such uncertainty effects on the prediction of deteriorating performance.

3. Numerical optimization for maintenance management

Maintenance management can be readily formulated as a combinatorial optimization problem because there always exist objectives to be optimized. The objectives can be life-cycle cost minimization under structure performance constraints, structure performance maximization under budget constraints, and simultaneous consideration of reducing life-cycle cost and increasing structure performance. Clearly the last criterion is the generalization of the first two objectives. As discussed previously, this paper is focused on the multiobjective optimization formulation. Therefore, effective numerical optimization tools for this formulation are needed.

3.1 *Traditional algorithms*

Most traditional optimization algorithms including mathematical programming are problem-dependent. They usually make use of gradient information to guide the search process and often assume continuous-valued design variables. This may lead to significant difficulties when design variables can only take discrete values such as in the present maintenance management problems. More importantly, most traditional methods by themselves can only handle single-objective based optimization problems, for which a single final optimal solution is sought. In order to solve optimization problems with multiple and usually conflicting objective functions, one has to convert, with non-trivial inconvenience, the original multiobjective problem into a series of equivalent single-objective optimization problems, which in total produce a set of solutions that displays optimal tradeoff among all competing objectives. There are two basic approaches commonly used for this treatment. One approach is the ε-constraint method, which keeps only one objective at a time and converts the other objectives into constraints. The difficulty associated with this method lies in the determination of constraint values, especially the upper and lower limits. The other approach is the weighted sum method, which forms a single composite objective function as a weighted sum of the original multiple objectives, using a specified set of weight coefficients. Because traditional optimization methods usually search for the optimized solution on a point-by-point basis, multiple algorithm runs are needed, each at best generating one of those optimized tradeoff solutions.

3.2 *Heuristic algorithms*

Genetic algorithms (GAs) as one type of heuristic algorithms are stochastic search and optimization tools that follow the survival-of-the-fittest principle from the biological sciences (Goldberg 1989). Since they formally appeared in the 1960s, GAs have been successfully applied to a wide range of problems because of their ease of implementation and robust performance. The population-based GAs are general-purpose numerical tools. Gradients are no longer needed and discrete-valued design variables can be handled without any difficulty. This is particularly useful for the present maintenance planning problems where different maintenance actions have to be scheduled at discrete years. More importantly, GAs can handle multiple conflicting objectives directly and simultaneously. Because GAs maintain many solutions at the same time, a distribution of optimized tradeoff solutions can be obtained with a single algorithm run.

A successful multiobjective optimization algorithm should obtain a non-dominated set of solutions close to the global Pareto optimal front and have this solution set as diverse as possible in order to best possibly prevent solution clustering. Unlike single-objective problems where the objective function itself may be used as the fitness measure to rank the merit of individual solutions, multiobjective GAs need a single fitness measure that reflects the overall merit of multiple competing objectives (Deb 2001). Many of the existing multiobjective GAs adopt the non-dominated sorting technique (Goldberg 1989) to rank all solutions in a population based on the concept of domination.

4. Life-cycle cost oriented design and maintenance management of civil infrastructure

4.1 *Overview*

Consideration of lifetime expenses during the initial design stage of civil infrastructure has received fruitful research attention in recent years. A life-cycle cost analysis consists of calculation of not only the initial construction cost, but also costs due to operation, inspection, maintenance, repair, and damage/failure consequences during a specified lifetime. Future costs are usually discounted to the present value for meaningful comparison. The goal is to produce a cost-effective design solution that balances the initial cost and lifetime cost in the preferred manner. Chang and Shinozuka (1996) reviewed the life-cycle cost analysis for civil infrastructure with emphasis on natural hazard risk mitigation. Hassanain and Loov (2003) discussed cost optimization of concrete bridge components and systems and review developments in life-cycle cost analysis and design of concrete bridges.

The minimum expected life-cycle cost has been the most widely used criterion in design optimization of new structural systems considering lifetime performance. The general form of the expected life-cycle cost can be calculated as (Frangopol *et al.* 1997):

$$C_{ET} = C_T + C_{PM} + C_{INS} + C_{REP} + C_F \qquad (1)$$

where C_{ET} = expected total cost, C_T = initial design/construction cost, C_{PM} = expected cost of routine maintenance, C_{INS} = expected cost of performing inspections, C_{REP} = expected cost of repairs and C_F = expected cost of failure. Application of expected life-cycle cost minimization has been made for the design of both bridges and buildings. Note that for maintenance planning of existing structures over a prescribed time period, the expected total intervention cost, which comprises all but the initial cost, needs to be used instead.

4.2 *Application to design of bridges and buildings*

Frangopol *et al.* (1997) proposed a conceptual framework for reliability-based life-cycle cost optimal design of deteriorating reinforced concrete bridges considering inspection and repair planning, based on the following formulation:

$$\text{Minimize } C_{ET} \text{ subject to } P_{f,life} \leqslant P_{f,life}^* \qquad (2)$$

where $P_{f,life}$ and $P_{f,life}^*$ are the computed and acceptable lifetime failure probabilities, respectively. The reliability-based lifetime approach considers the quality of various inspection techniques, all repair possibilities based on an event tree, effects of aging and corrosion deterioration, damage intensity, effects of repair on structural reliability, and the time value of money. It is pointed out that the quantification of uncertainties in the input data, including the rate of corrosion, quality of inspection methods and failure cost, is necessary for obtaining a reliable structural design integrated with economic and safety considerations. Bayesian updating of damage detection probabilities, use of improved time-varying bridge reliability models, and selection of realistic target reliability levels are all needed.

The studies by Frangopol *et al.* (1997, 2000, 2001) are concerned with designing or maintaining individual bridges. It is known that bridges represent the most critical components of highway networks. Not only under disastrous natural hazards such as flood and earthquakes but also under daily traffic conditions, malfunctioning of bridges could likely disrupt the normal transportation. This causes significant problems to the welfare of the entire society. From this point of view, it is more important to maintain the normal functionality as well as extend life expectancy of the bridge network as a whole instead of dealing with each bridge separately. Therefore, life-cycle cost oriented design and maintenance optimization of all bridges within a network poses a natural extension of the previous studies and deserves future research (Augusti *et al.* 1994, 1998, Akgül and Frangopol 2003). When analyzing the bridge network, user costs in terms of, for example, travel delay and additional vehicle fuel consumption as well as accidents that are caused by reduced serviceability of a bridge network due to maintenance and/or inadequate capacity, need also to be addressed (Chang and Shinozuka 1996).

Recent earthquakes reveal that economic losses caused by less drastic structural damages of building structures as well as ensuing functional disruptions could be comparable to the structure's initial cost. A life-cycle cost analysis provides a convenient means for decision makers to compare seismic design alternatives using economic terms. Both initial construction expenses and lifetime cost due to seismic damage/failure consequences as well as associated repair and retrofitting efforts need to be properly addressed in life-cycle cost analysis. A cost-effective building design should appropriately balance initial and future costs. The minimum expected life-cycle cost has been the most widely used criterion for the seismic design of steel (Kang and Wen 2000) and reinforced concrete buildings (Ang and Lee 2001). A group of conventional trial structural designs with different seismic capacities are investigated and no automated numerical optimization is performed. Cheng and Ang (1999) suggested a reliability-based optimization procedure using GA for cost-effective design and upgrading of seismic-resistant reinforced concrete building structures; the target reliability and the minimum expected life-cycle cost are selected as the objective functions.

Liu *et al.* (2004) proposed a multiobjective optimization approach for solving life-cycle cost oriented seismic design of steel moment frame structures. Initial cost and lifetime seismic damage cost are considered as separate objective functions subject to simultaneous minimization. The lifetime seismic damage cost is computed as (Kang and Wen 2000):

$$C_{seismic} = (C_1 P_1 + C_2 P_2 + \cdots C_j P_j + \cdots + C_N P_N)\frac{v}{\lambda}(1 - e^{-\lambda T})$$

$$(3)$$

where N = number of seismic damage states considered; C_j = cost function of the j-th seismic damage state, which consists of direct structural/nonstructural damage and repair cost, cost due to loss of contents, relocation cost, direct/indirect economic loss, and human injury and fatality cost; P_j = j-th damage state probability given seismic occurrence; v = annual occurrence rate of major seismic events modeled by a Poisson process; λ = annual monetary discount rate; T = service life. This formula implicitly assumes that a damaged structure is immediately retrofitted to its original intact condition after each damage-inducing seismic event. Seven damage states are defined in terms of maximum interstory drift ratio as the seismic performance index. Effects of randomness and uncertainty in estimating seismic demand and capacity as well as in describing seismic hazards are considered in accordance with SAC/FEMA guidelines (BSSC 2000).

4.3 *Application to bridge maintenance management*

The life-cycle cost analysis provides a rational approach to cost-effective prioritization under budget constraints for maintaining satisfactory lifetime performance of deteriorating highway bridges. In the United States, bridge management systems (BMSs), such as Pontis (Thompson *et al.* 1998) and BRIDGIT (Hawk and Small 1998), are used by state departments of transportation. As stated previously, these BMSs are built upon visual inspection based discrete condition states and Markovian deterioration modeling using stationary transition probabilities; lifetime reliability concepts are usually not directly incorporated. These inherent deficiencies cause limitations of these existing BMSs (Frangopol *et al.* 2001).

As an advance over these BMSs, some researchers proposed reliability-based life-cycle maintenance management using optimization (e.g. Frangopol and Das 1999, Thoft-Christensen 1999). Frangopol (1999) presented a systematic framework for cost-effective life-cycle bridge management that integrates both bridge lifetime reliability and life-cycle cost analysis. A review on these subjects can be found in the literature (Frangopol *et al.* 2001). The recently published NCHRP Report 483 'Bridge life-cycle cost analysis' of NCHRP Project 12-43 (Hawk 2003) provides guidelines for transportation agencies to conduct life-cycle cost analysis of major bridges. Representative studies done by the first author and his former graduate students include: (a) Frangopol *et al.* (2000) present realistic examples of optimal bridge maintenance planning based on minimum expected life-cycle cost criterion; and (b) Kong and Frangopol (2003) use a modified event tree analysis to compute probabilities of maintenance actions and the expected life-cycle cost of deteriorating highway bridges.

5. Maintenance management at the project level

5.1 Overview

Tremendous efforts have been devoted by civil engineering researchers and practitioners to develop methodologies and techniques for long-term maintenance management of deteriorating civil infrastructures, in particular bridges. Most of the previous research on bridge maintenance management can be categorized as project-level based because only individual deteriorating bridges or a group of similar bridges are dealt with. Many of the existing maintenance planning optimization procedures rely on deterministic evaluation of bridge performance only. Uncertainties, however, inevitably exist in describing the time-dependent deterioration process under no maintenance as well as under different maintenance interventions. A life-cycle cost analysis would be questionable if constraints are imposed on relevant bridge performance for which the associated sources of uncertainty are not taken into account. In order to carry out a realistic life-cycle cost analysis, uncertainties have to be considered in predicting lifetime bridge performance. As previously indicated, both aleatory and epistemic uncertainties have to be considered in predicting lifetime bridge performance.

5.2 Numerical example

5.2.1 Problem statement. In the first numerical example, a GA-based procedure is used to prioritize maintenance needs for deteriorating reinforced concrete highway crossheads through simultaneous optimization of both structure performance (in terms of condition and safety states as defined in sections 2.1.1 and 2.1.2) and life-cycle maintenance cost with a prescribed discount rate. The maintenance management problem is thus posed as a combinatorial multiobjective optimization problem in that, for any year over the specified time period, at most one of the five different maintenance strategies (as defined in the next subsection) may be carried out. The optimization problem is conceptually stated as:

Goal Obtain a set of tradeoff lists of maintenance actions applied over the specified time period that:

Objectives

(a) minimize the largest (i.e. worst) lifetime condition index value;
(b) maximize the smallest (i.e. worst) lifetime safety index value; and
(c) minimize the present value of life-cycle maintenance cost.

Subject to

(i) condition index value $\leqslant 3.0$; and
(ii) safety index value $\geqslant 0.91$.

Design variables of the present optimization problem are discrete years and types of maintenance applications. A generic maintenance solution is then a list of maintenance interventions prioritized on an annual basis. Modeling of the structure deterioration processes under no maintenance and under different maintenances follows Frangopol *et al.* (2001); uncertainties associated with this modeling are accounted for by means of Monte Carlo simulation.

5.2.2 Maintenance strategies. The condition and safety deterioration processes of a large population of reinforced concrete crossheads under no maintenance are characterized with the data provided elsewhere (Denton 2002, Liu and Frangopol 2005b); information of the five maintenance strategies used in this numerical example, i.e. replacement of expansion joints, silane, cathodic protection, minor concrete repair and 'do nothing but rebuild' is also given. All random variables are assumed to be triangularly distributed and, consequently, defined in terms of minimum, mode and maximum values, respectively. Replacement of expansion joints is statistically the least costly maintenance strategy. It does not improve performance or delay deterioration but alleviates the severity of deterioration of both condition and safety performance. The silane treatment reduces chloride penetration but does not correct existing defects or replace deteriorated structural components. Silane does not improve performance or delay deterioration either. Statistically, silane reduces the deterioration of the condition more efficiently than replacing the expansion joints while having the same effects on safety deterioration. However, silane usually costs more than replacing the expansion joints. Cathodic protection replaces anodes and thus suppresses the corrosion of the reinforcing bars almost completely. It postpones the deterioration of both condition and safety for a relatively long time period (12.5 years) upon application. The minor concrete repair strategy is applied to replace all cover concrete with visual defects but not corroded reinforcing bars. It significantly improves the condition performance while the deterioration in safety is delayed until the condition index exceeds 1.0. Finally, the 'do nothing but rebuild' strategy increases both condition and safety levels to those values typical of a new structural component. Deterioration processes in condition and safety are both delayed; the safety delay is ended when the condition index exceeds 1.0.

5.2.3 Numerical results. The GA-based maintenance prioritization procedure is carried out for deteriorating reinforced concrete crossheads. The initial GA population consists of 1,000 randomly generated trial solutions and each of the subsequent generations contains 200 offspring solutions plus the non-dominated solutions from the previous (i.e. parent) generation. The non-dominated sorting technique (Goldberg 1989) combined with the crowding distance niching (Deb 2001) is applied to determine solutions' fitness values. The constrained binary tournament selection scheme (Deb 2001) is adopted to reproduce fitter solutions as well as to account for possible violation of constraints imposed on condition and safety indices. The uniform crossover and mutation probabilities are 50% and 5%, respectively. Monte Carlo simulation with a sample size of 1,000 is used to consider the effects of uncertainty on prediction of both structure performance and life-cycle maintenance cost. All three objective functions are evaluated in terms of sample mean values. The service life is considered to be 50 years and the discount rate is 6%. Details can be found elsewhere (Liu and Frangopol 2005b).

A total of 129 different optimized maintenance planning solutions are generated at the 30th generation. All of these solutions are feasible in that the minimum performance levels are strictly satisfied, which are enforced as constraints. These solutions lead to, in a Pareto optimal sense, different levels of performance enhancement and maintenance needs. Based on this large set of solutions, civil infrastructure managers can actively select a solution that balances condition, safety and life-cycle maintenance cost objectives in a preferred way. Three representative maintenance solutions are illustrated in figure 1. Solution I uses three applications of replacing expansion joints at years 2, 21 and 32, and three silane treatments at years 27, 37 and 45. It requires the lowest life-cycle maintenance cost of 1016. The condition index objective value is 2.97, which makes the condition constraint almost active (i.e. equal 3.0)

at year 50; its safety index objective value of 1.34 is the lowest. As a comparison, solution II uses three applications of expansion joint replacement in addition to two silane and two cathodic protection actions. It improves the condition and safety objective values to 2.25 and 1.51, respectively. However, the associated life-cycle maintenance cost is increased to 2207. Solution III further enhances the condition and safety values to 2.00 and 1.61, respectively. As a tradeoff, the life-cycle cost drastically increases to 5577.

6. Maintenance management at the network level

6.1 *Overview*

The civil infrastructure maintenance management methodologies discussed so far are of project-level types, that is, they deal with individual structures or a group of similar structures. Maintenance planning for deteriorating civil highway infrastructure from a transportation network perspective provides more rational solutions because the ultimate objective of maintenance management is to improve the performance of the entire transportation network instead of merely that of individual structures in the network. In this section, the maintenance management of a group of bridges that form nodes of a transportation network is discussed. Performance evaluation of degrading highway networks in the transportation engineering community has been focused on road performance in terms of travel time and capacity reliabilities (Bell and Iida 1997). Recently, Shinozuka *et al.* (2003) evaluated the performance of highway bridge networks under earthquakes and demonstrated the importance of seismic retrofit in improving network performance in terms of drivers' travel delay reduction. Maintenance planning optimization for

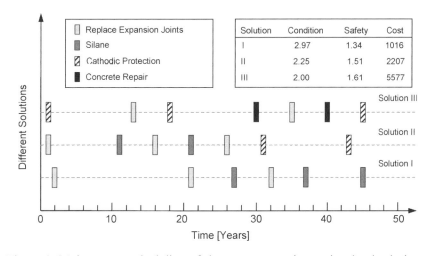

Figure 1. Maintenance scheduling of three representative project-level solutions.

deteriorating bridge networks, which is important for daily operation other than under disastrous events such as earthquakes and floods, has not received adequate attention. Augusti *et al.* (1994, 1998) investigated retrofitting efforts allocation for seismic protection of highway bridge networks. Adey *et al.* (2003) developed a network bridge management system using a supply-and-demand system approach. In general, there exist two types of measures for the overall performance of a bridge network: the network reliability and the long-term costs consisting of both agency cost and user cost.

6.1.1 Network reliability. The network reliability measures the level of satisfactory network performance. Most studies on assessment of reliability for transportation highway infrastructure have focused on maintenance management of degradable road networks (Nicholson and Du 1997), for which a travel path consists of multiple links (i.e. roadways between any two nodes) with binary states (either operating or failed). There are three variations of reliability measures with ascending levels of sophistication: the connectivity reliability, the travel time reliability and the capacity reliability (Bell and Iida 1997, Chen and Recker 2000). The connectivity reliability is associated with the probability that nodes in a highway network are connected. In particular, Iida and Wakabayashi (1989) define the terminal connectivity that refers to the existence of at least one operational path that connects the origin and destination (OD) nodes of interest. The travel time reliability indicates the probability that a successful trip between a specified OD pair can be made within given a time interval and level-of-service (Asakura and Kashiwadani 1991). Based on this reliability measure, the appropriate level of service that should be maintained in the presence of network deterioration can be determined (Chen and Recker 2000). The third alternative is the capacity reliability, which reflects the possibility of the network to accommodate given traffic demands at a specified service level (Chen *et al.* 2002). In this formulation, link capacities may be treated as random variables to consider the time-dependent probabilistic capacity degradation due to physical (e.g. deterioration) and operational (e.g. lane blockage caused by traffic accidents) factors. Inherent in the last two reliability measures are the determination of risk-taking route choice models for simulating travelers' behavior in the presence of both perception error and network uncertainty (Chen and Recker 2000).

6.1.2 Cumulative long-term cost. For maintenance management of deteriorating highway networks, it is of significant importance to use monetary terms as a measure of the overall network performance. There are two basic types of costs: the agent cost and the user cost. The agent cost is composed of direct material and labor expenses needed to perform routine and preventive maintenance, rehabilitation and replacement of existing transportation facilities. The user costs are caused by a loss of adequate service due to, for example, congestion and detour. In some situations the user cost may be a dominating factor in evaluating the overall life-cycle costs for a transportation network. A multiobjective formulation in which the agent cost and user cost are subject to balanced minimization provides a useful means to generate maintenance solutions for satisfactory long-term network performance and expenditures (Liu and Frangopol 2005c). The uncertainty associated with capacity degradation and demand variation should be integrated in the analysis in order to obtain a probabilistic cost measure.

6.2 *Numerical example*

6.2.1 Problem statement. In this section, the network-level maintenance management is illustrated using a real bridge network that is located near the Denver metropolitan area, Colorado (Akgül and Frangopol 2003). A highway transportation network consists of individual bridges and roads that link pairs of bridges to form the network. In the present example, it is assumed that bridges are the only vulnerable elements that are subject to failure in the network, that is, roads always operate at full capacities. As shown in figure 2, this network consists of 13 bridges of different types, including prestressed girder bridges, steel rolled I-beam bridges and combined welded steel plate/reinforced concrete girder bridges. For brevity, two bridge groups are identified: group 1 (G1) includes six bridges (i.e. E-16-Q, E-16-LA, E-16-NM, E-16-LY, E-16-FL and E-16-FK, denoted, in short, as Q, LA, NM, LY, FL and FK, respectively) in series and group 2 (G2) includes four bridges (i.e. E-17-HR, E-17-HS, E-17-MW and D-16-DM, denoted, in short, as HR, HS, MW and DM, respectively) in series. The network performance is evaluated in terms of the terminal reliability for connectivity between the cities of Denver and Lafayette. Flexure failure of bridge slabs is considered as the only failure mode in this example. Based on an event tree analysis that enumerates all combinations of operating and failed bridges, the total probability that the bridge network is disconnected between Denver and Lafayette is obtained by adding occurrence probabilities associated with all disconnected branches. Detailed formulations can be found elsewhere (Liu and Frangopol 2005a). The optimization problem is conceptually stated as:

Goal Prioritize maintenance needs to bridges that are most important to the network performance and over the specified time period that:

Objectives

(i) maximize the overall bridge network performance, which is measured by the lowest value of the lifetime

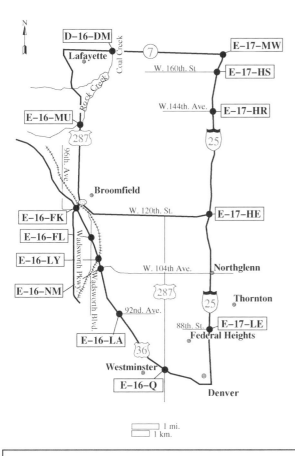

Group 1 includes Q, LA, NM, LY, FL, and FK.
Group 2 includes HR, HS, MW, and DM

Figure 2. A 13-bridge transportation network (adapted from Akgül and Frangopol 2003).

reliability index of connectivity between the origin and destination locations; and

(ii) minimize the present value of total life-cycle maintenance cost.

6.2.2 Maintenance strategies. Four different maintenance strategies are considered in this numerical example to improve network performance: resin injection, increasing the slab thickness, attaching steel plate and replacement. Instead of carrying out detailed reliability analysis of bridges upon application of different maintenance, empirical data based on those in the literature (Furuta *et al.* 2004, Kong and Frangopol 2004) are used to capture the effects of each strategy for maintenance planning purposes. Resin injection is the cheapest maintenance strategy among the four alternatives, with a unit cost of US$200/m^2. It injects epoxy resin into voids and seals cracks in concrete, which repairs the aging deck slabs by reducing the corrosion of reinforcement due to exposure to the open air. The other

three maintenance strategies instantly improve the structure reliability level by various amounts upon application. Increasing the slab thickness and attaching steel plate increase the structure reliability index values by a maximum of 0.7 and 2.0, respectively. As a tradeoff, the unit costs associated with these two maintenance strategies are US$300/m^2 and US$600/m^2, respectively. The most efficient yet the most expensive maintenance strategy is the replacement of the entire aging bridge slabs. It is assumed that this strategy restores the structural system to the initial reliability level with a unit cost of US$900/m^2.

6.2.3 Numerical results. The GA-based maintenance prioritization procedure is described elsewhere (Liu and Frangopol 2005a). The service life is considered to be 30 years and the discount rate is 6%. Figure 3 shows the evolution of the non-dominated solution set for every ten generations up to the 40th generation. There is a steady improvement of the non-dominated tradeoff solutions. In particular, the optimized solutions at the 40th generation represent a wide spread between the conflicting network connectivity and the total maintenance cost objectives. Equivalently, the tradeoff between the maximum network disconnectivity probability and the present value of total maintenance cost is also shown (figure 3b). The resulting optimized maintenance prioritization solutions lead to different improvement of the network performance that requires different monetary budgets. This set of optimized alternative maintenance solutions enables bridge managers to choose one that preferably compromises the competing network performance and maintenance cost. Two representative optimized maintenance solutions associated with different present values of life-cycle maintenance cost and lowest values of reliability index of network connectivity during the 30-year life-cycle are presented in figure 4.

7. Maintenance management systems integrated with health monitoring for civil infrastructure

One very interesting, yet challenging, research topic is to develop a systematic framework for the next-generation of the intelligent structure management system (SMS) through innovative integration of recent advancement in structural health monitoring (SHM) of civil infrastructure such as highway bridges and other essential constructed facilities. Using advanced sensing/information technology and structural modeling/identification schemes, SHM detects, locates and quantifies structural damages caused by catastrophic natural or manmade events as well as by long-term deterioration due to aggressive environmental stressors and the increasing traffic loading. These results assist structure managers in assessing the health states of existing civil infrastructure and eventually determining immediate or future applications of possible maintenance

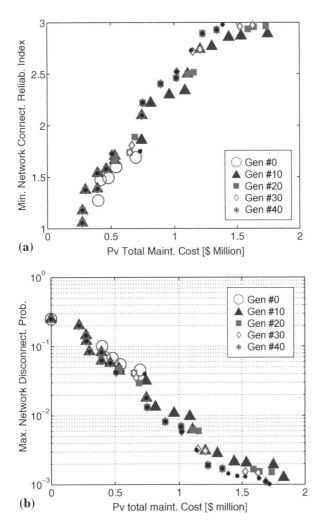

(a)

(b)

Figure 3. Generation-wise evolution of optimized tradeoff maintenance solutions between the present value (PV) of total (i.e. life-cycle) maintenance cost and network performance in terms of (a) the minimum network connectivity reliability index and equivalently (b) the maximum network disconnectivity probability.

and rehabilitation interventions for safety consideration and lifespan extension. Functionalities of SMS and SHM are naturally interwoven and should be treated as such. Most of the existing research and practice activities in SMS and SHM, however, are carried out in a disjointed manner. Therefore, a unified framework is necessary to bridge the gap between these two research areas, and this is crucial to best ensure a sustainable civil infrastructure.

Successful completion of this mission will be a crucial step forward to revolutionize the traditional visual-inspection based SMS methodologies by providing structure managers with an efficient tool to make timely and intelligent decisions on monitoring, evaluation, and maintenance of deteriorating civil infrastructure. This is

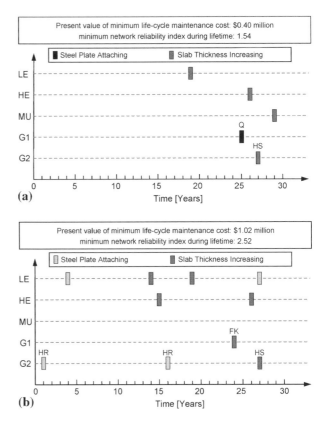

(a)

(b)

Figure 4. Maintenance scheduling of two representative network-level solutions.

achieved by exploring the interaction between SHM and SMS strategies in terms of whole-life costing and structural safety/health/reliability. Impacts of this research are manifold:

(a) The SMS becomes smart by planning maintenance activities based on monitored immediate structure health states as well as updated prediction of future structure health states. The scarce financial resources can then be more rationally prioritized to maximize emergency response capabilities in case of natural/manmade disasters as well as to improve the utility of deteriorating civil infrastructure;

(b) The integrated framework has the potential to identify the SHM strategies that are most cost-efficient to detect deterioration due to different mechanisms (e.g. crack, corrosion and fatigue). Also, SHM strategies of different costs and efficacies, for example, could be appropriately combined in order to provide the most cost-effective health assessment tool;

(c) Novel materials (Wu 2005) and maintenance/rehabilitation approaches could be attempted to accommodate the sophistication level of SHM strategies under consideration; and

(d) The general framework can be extended to intelligently monitor and manage multiple structures within a highway network as well as a broad spectrum of constructed civil facilities such as buildings, highway pavements, railroads, and tunnels.

8. Conclusions

This paper reviews recent developments in civil infrastructure maintenance and management with an emphasis on simultaneous consideration of multiple objectives concerning structure performance and life-cycle cost. Sources of uncertainty associated with the deterioration process are considered in the probabilistic prediction of civil infrastructure performance under no maintenance and under different maintenance interventions. Genetic algorithms are selected as the effective optimization tool because of their significant convenience for the posed combinatorial multi-objective optimization problems. Two numerical examples are provided for illustration purposes. In each example, a set of alternative solutions is produced that exhibits the optimized tradeoff among all competing objectives. Therefore, structure managers' preference on the balance between the lifetime structure performance and life-cycle cost can be integrated into the decision-making process. The interaction among structure condition, safety and life-cycle cost is addressed; the importance of incorporating network effects into civil infrastructure management is emphasized. Finally, research needs of integrating maintenance management and health monitoring for improving long-term decision-making are briefly discussed.

Acknowledgements

The support of the UK Highways Agency, the US National Science Foundation through grants CMS-9912525 and CMS-0217290, and the Colorado Department of Transportation is gratefully acknowledged. The opinions and conclusions presented in this paper are those of the writers and do not necessarily reflect the views of the sponsoring organizations.

References

AASHTO, *Standard Specifications for Highway Bridges* (16th edn), 1996 (American Association of State Highway and Transportation Officials: Washington, DC).

Adey, B., Hajdin, R. and Brühwiler, E., Supply and demand system approach to development of bridge management strategies. *J. Infrastr. Syst.*, ASCE, 2003, **9**, 117–131.

Asakura, Y. and Kashiwadani M., Road network reliability caused by daily fluctuation of traffic flow, *Proceedings of the 19th PTRC Summer Annual Meeting*, Brighton, UK, 1991.

Akgül, F. and Frangopol, D.M., Rating and reliability of existing bridges in a network. *J. Bridge Engng*, ASCE, 2003, **8**, 383–393.

Ang, A.H.-S. and Lee, J.-C., Cost optimal design of R/C buildings. *Reliab. Engng Syst. Safety*, 2001, **73**, 233–238.

ASCE, *Report card for America's Infrastructure*, American Society of Civil Engineers, Reston, VA. Available online at www.asce.org/reportcard/2005/index.cfm (accessed April 2005).

Augusti, G., Borri, A. and Ciampoli, M., Optimal allocation of resources in reduction of the seismic risks of highway networks. *Engng Struct.*, 1994, **16**, 485–497.

Augusti, G., Ciampoli, M. and Frangopol, D.M., Optimal planning of retrofitting interventions on bridges in a highway network. *Engng Struct*, 1998, **20**, 933–939.

Bell, M.G.H. and Iida, Y., *Transportation Network Analysis*, pp. 226, 1997 (John Wiley & Sons: Chichester, UK).

BSSC, *NEHRP Recommended Seismic Design Criteria for New Steel Moment-frame Buildings*, FEMA-350, Building Seismic Safety Council, Federal Emergency Management Agency, Washington, DC, 2000.

Chang, S.E. and Shinozuka, M., Life-cycle cost analysis with natural hazard risk. *J. Infrastr. Syst.*, ASCE, 1996, **2**, 118–126.

Cheng, F.Y. and Ang, A.H.-S., Cost-effectiveness optimization for aseismic design criteria of RC buildings. In *Case Studies in Optimal Design and Maintenance Planning of Civil Infrastructure Systems*, edited by D.M. Frangopol, pp. 13–25, 1999 (ASCE: Reston, VA).

Chen, A. and Recker, W.W., Considering risk taking behavior in travel time reliability, *Report No. UCI-ITS-WP-00-24*, Institute of Transportation Studies, University of California, Irvine, CA, 2000.

Chen, A., Yang, H., Lo, H.K. and Tang, W.H., Capacity reliability of a road network: an assessment methodology and numerical results. *Transport. Res.*, Part B 2002, **36**, 225–252.

Das, P.C., Prioritization of bridge maintenance needs. In *Case Studies in Optimal Design and Maintenance Planning of Civil Infrastructure Systems*, edited by D.M. Frangopol, pp. 26–44, 1999 (ASCE: Reston, VA).

DB12/01, *The Assessment of Highway Bridge Structures*, 2001 (Highways Agency Standard for Bridge Assessment: London, UK).

Deb, K., *Multi-objective Optimization Using Evolutionary Algorithms*, pp. 518, 2001 (John Wiley & Sons: Chichester, UK).

Denton, S., *Data Estimates for Different Maintenance Options for Reinforced Concrete Cross Heads*, (Personal communication) 2002 (Parsons Brinckerhoff Ltd: Bristol, UK).

Enright, M.P. and Frangopol, D.M., Condition prediction of deteriorating concrete bridges using Bayesian updating. *J. Struct. Engng*, ASCE, 1999, **125**, 1118–1125.

Estes, A.C. and Frangopol, D.M., Repair optimization of highway bridges using system reliability approach. *J. Struct. Engng*, ASCE, 1999, **125**, 766–775.

Estes, A.C., Frangopol, D.M. and Foltz, S.D., Updating reliability of steel miter gates on locks and dams using visual inspection results. *Engng Struct.*, 2004, **26**, 319–333.

FHWA, *Recommendations for bridge and tunnel security*, The Blue Ribbon Panel on Bridge and Tunnel Security, Federal Highway Administration, Gaithersburg, MD. Available online at www.fhwa.dot.gov/bridge/security/brp.pdf, 2003 (accessed April 2005).

Frangopol, D.M., Life-cycle cost analysis for bridges. In *Bridge Safety and Reliability*, edited by D.M. Frangopol, pp. 210–236, 1999 (ASCE: Reston, VA).

Frangopol, D.M. and Das, P.C., Management of bridge stocks based on future reliability and maintenance costs. In *Current and Future Trends in Bridge Design, Construction, and Maintenance*, edited by P.C. Das, D.M. Frangopol and A.S. Nowak, pp. 45–58, 1999 (Thomas Telford: London, UK).

Frangopol, D.M., Lin, K.Y. and Estes, A.C., Life-cycle cost design of deteriorating structures. *J. Struct. Engng*, ASCE, 1997, **123**, 1390–1401.

Frangopol, D.M., Gharaibeh, E.S., Kong, J.S. and Miyake, M., Optimal network-level bridge maintenance planning based on minimum expected cost. *J. Transport. Res. Board*, 2000, **2**, 26–33.

Frangopol, D.M., Kong, J.S. and Gharaibeh, E.S., Reliability-based life-cycle management of highway bridges. *J. Comput. Civil Engng*, ASCE, 2001, **15**, 27–34.

Furuta, H., Kameda, T., Fukuda, Y. and Frangopol, D.M., Life-cycle cost analysis for infrastructure systems: life cycle cost vs. safety level vs. service life. In *Life-Cycle Performance of Deteriorating Structures: Assessment, Design and Management*, edited by D.M. Frangopol, E. Brühwiler, M.H. Faber and B. Adey, pp. 19–25, 2004 (ASCE: Reston, VA).

Goldberg D.E., *Genetic Algorithms in Search, Optimization and Machine Learning*, pp. 432, 1989 (Addison-Wesley: Reading, MA).

Ghosn, M. and Johnson, P., Reliability analysis of bridges under the combined effect of scour and earthquakes, *8th ASCE Specialty Conference on Probabilistic Mechanics and Structural Reliability*, Notre Dame, IN, 2000, 6 pages on CD-ROM.

Hassanain, M.A. and Loov, R.E., Cost optimization of concrete bridge infrastructure. *Can. J. Civil Engng*, 2003, **30**, 841–849.

Hawk, H., Bridge life-cycle cost analysis, *Report 483*, 2003 (National Cooperative Highway Research Program, Transportation Research Board: Washington, DC).

Hawk, H. and Small, E.P., The BRIDGIT bridge management system. *Struct. Engng Intl*, IABSE, 1998, **8**, 309–314.

Iida, Y. and Wakabayashi, H., An approximation method of terminal reliability of a road network using partial minimal path and cut set, *Proceedings of the 5th World Conference on Transport Research*, 1989, pp. 367–380, Yokohama, Japan.

Kang, Y.-J. and Wen, Y.K., Minimum life-cycle cost structural design against natural hazards, *Structural Research Series No. 629*, 2000 (Department of Civil and Environmental Engineering, University of Illinois at Urbana-Champaign, Urbana, IL).

Kong, J.S. and Frangopol, D.M., Evaluation of expected life-cycle maintenance cost of deteriorating structures. *J. Struct. Engng*, ASCE, 2003, **129**, 682–691.

Kong, J.S. and Frangopol, D.M., Cost-reliability interaction in life-cycle cost optimization of deteriorating structures. *J. Struct. Engng*, ASCE, 2004, **30**, 1704–1712.

Liu, M. and Frangopol, D.M., Balancing the connectivity reliability of deteriorating bridge networks and long-term maintenance cost using optimization. *J. Bridge Engng*, ASCE, 2005a, **10**, 468–481.

Liu, M. and Frangopol, D.M., Bridge annual maintenance prioritization under uncertainty by multiobjective combinatorial optimization. *Comput.-Aided Civil Infrast. Engng*, 2005b, **20**, 343–353.

Liu, M. and Frangopol, D.M., Optimizing bridge network maintenance management under uncertainty with conflicting criteria: life-cycle maintenance, failure, and user costs. *J. Struct. Engng*, ASCE, 2005c, submitted.

Liu, C., Hammad, A. and Itoh, Y., Multiobjective optimization of bridge deck rehabilitation using a genetic algorithm. *Comput.-Aided Civil Infrastr. Engng*, 1997, **12**, 431–443.

Liu, M., Wen, Y.K. and Burns, S.A., Life cycle cost oriented seismic design optimization of steel moment frame structures with risk-taking preference. *Engng Struct.*, 2004, **26**, 1407–1421.

Melchers, R.E., *Structural Reliability, Analysis and Prediction*, 2nd ed., pp. 456, 1999 (Wiley & Sons: New York).

Miyamoto, A., Kawamura, K. and Nakamura, H., Bridge management system and maintenance optimization for existing bridges. *Comput.-Aided Civil Infrastr. Engng*, 2000, **15**, 45–55.

Nicholson, A.J. and Du, Z.P., Degradable transportation systems: an integrated equilibrium model. *Transport. Res. B*, 1997, **31**, 209–23.

Nowak, A.S. and Collins, K.R., *Reliability of Structures*, pp. 360, 2000 (McGraw-Hill: New York).

Pontis, *User's Manual, Release 4.0*, 2001 (Cambridge Systematics, Inc.: Cambridge, MA).

Rafiq, M.I., Chryssanthopoulos, M.K. and Onoufriou T., Performance updating of concrete bridges using proactive health monitoring methods. *Reliab. Engng & Syst. Safety*, 2004, **86**, 247–256.

Shinozuka, M., Murachi, Y., Dong, X., Zhou., Y. and Orlikowski, M.J., Seismic performance of highway transportation network, *Proceedings of China–US Workshop on Protection of Urbana Infrastructure and Public Buildings against Earthquakes and Manmade Disasters*, Beijing China, 2003.

Thoft-Christensen, P., Assessment of the reliability profiles for concrete bridges. *Engng Struct.*, 1998, **20**, 1004–1009.

Thoft-Christensen, P., Future trends in reliability-based bridge management, *Proceedings of XXI World Road Congress*, Kuala Lumpur, Malaysia, 1–9 October, 1999, 15 pages in CD-ROM (IAPCR/PIARC: La Defense, France).

Thompson, P.D., Small, E.P., Johnson, M. and Marshall, A.R., The Pontis bridge management system. *Struct. Engng Intl*, IABSE, 1998, **8**, 303–308.

Val, D., Stewart, M.G. and Melchers, R.E., Life-cycle performance of reinforced concrete bridges: probabilistic approach. *J Comput. Aided Civil Infrastr. Enging*, 2000, **15**, 14–25.

Wen, Y.K., Ellingwood, B.R., Veneziano, D. and Bracci, J., *Uncertainty Modeling in Earthquake Engineering*, pp. 113, 2003 (Mid-America Earthquake Engineering Center: Urbana, IL).

Wu, H.F., *Composites in Civil Applications*, National Institute of Standards and Technology, Gaithersburg, MD. Available online at www.atp.nist.gov/atp/focus/99wp-ci.htm (accessed April 2005).

Life cycle utility-informed maintenance planning based on lifetime functions: optimum balancing of cost, failure consequences and performance benefit

Samantha Sabatino, Dan M. Frangopol and You Dong

ABSTRACT

Decision-making regarding the optimum maintenance of civil infrastructure systems under uncertainty is a topic of paramount importance. This topic is experiencing growing interest within the field of life cycle structural engineering. Embedded within the decision-making process and optimum management of engineering systems is the structural performance evaluation, which is facilitated through a comprehensive life cycle risk assessment. Lifetime functions including survivor, availability, and hazard at component and system levels are utilised herein to model, using closed-form analytical expressions, the time-variant effect of intervention actions on the performance of civil infrastructure systems. The presented decision support framework based on lifetime functions has the ability to quantify maintenance cost, failure consequences and performance benefit in terms of utility by considering correlation effects. This framework effectively employs tri-objective optimisation procedures in order to determine optimum maintenance strategies under uncertainty. It provides optimum lifetime intervention plans allowing for utility-informed decision-making regarding maintenance of civil infrastructure systems. The effects of the risk attitude, correlation among components and the number of maintenance interventions on the optimum maintenance strategies are investigated. The capabilities of the proposed decision support framework are illustrated on five configurations of a four-component system and an existing highway bridge.

1. Introduction

In 2013, the American Society of Civil Engineers reported, within the Report Card for America's Infrastructure, that the average age of the US' 607,380 bridges was 42 years (American Society of Civil Engineers [ASCE], 2013). Additionally, nearly a quarter of these highway bridges were classified as either structural deficient or functionally obsolete (Federal Highway Administration [FHWA], 2013). These staggering statistics highlight the dire need to implement rational mitigation strategies that maintain structural performance within acceptable levels through the life cycle of deteriorating civil infrastructure. The consequences associated with structural failure can be widespread and significant. In order to avoid the detrimental effects of structural failure, *lifetime functions*, paired with risk and sustainability indicators, are utilised within an efficient life cycle maintenance optimisation procedure to find intervention strategies that balance maintenance cost, failure consequences and performance benefit.

Decision-making regarding the optimum maintenance of civil and marine infrastructures is a topic of paramount importance and is experiencing growing interest within the field of life cycle structural engineering. In general, decision-making may be divided into five separate stages: the pre-analysis, problem set-up, uncertainty quantification, utility assignment and optimisation

(Keeney & Raiffa, 1993). Since technical and economic uncertainties are both expected and unavoidable in the life cycle assessment of civil structures, decisions regarding infrastructure must consider all relevant uncertainties associated with the probability of structural failure and its corresponding consequences (Ang 2011; Ang & De Leon, 2005; Ellingwood, 2005; Frangopol & Soliman, in press). For deteriorating highway bridges, uncertainties are present within modelling the structural resistance (e.g. material properties and element dimensions), the occurrence and magnitude of hazards that may impact the structure (e.g. corrosion, fatigue, earthquakes, floods and hurricanes), operating conditions and loading cases (Stewart, 2001).

The decision-maker may assign utility values to the attributes associated with each alternative considering his/her risk attitude. The attributes associated with each alternative investigated herein include cost (i.e. cost of essential maintenance), failure consequences (i.e. direct and indirect risks) and performance benefit (i.e. improvement in lifetime functions). In this paper, these attributes are identified as cost, risk and benefit, respectively. In order to effectively combine certain attributes for further use in a tri-objective optimisation procedure, a multi-attribute utility function is developed that considers the weighted relative utility value corresponding to each attribute involved. In the last step of the decision-making process, optimisation is performed in order to find the alternative that maximises utility. The proposed framework incorporates three conflicting and interrelated objectives into a maintenance optimisation procedure that has the capability of determining the best intervention schedules (i.e. maintenance plans detailing the components to be maintained and timing of interventions) that simultaneously balance cost, risk and benefit.

Within the proposed decision support system for life cycle maintenance planning of civil structures, several separate detrimental impacts are considered in quantifying the consequences of structural failure to the economy, society and surrounding environment in terms of risk. In its most basic form, risk is calculated as the probability of occurrence of a specific event multiplied by the consequence associated with this event. Previous research efforts have included risk analyses in both qualitative (Ellingwood, 2001; Hessami, 1999) and quantitative manners (Barone & Frangopol, 2014a; Barone, Frangopol, & Soliman, 2014; Decò & Frangopol, 2011, 2012, 2013; Pedersen, 2002) considering a wide variety of hazards and many different structures (Arunraj & Maiti, 2007).

In this paper, risk and performance benefit indicators that incorporate different attributes are investigated. The utility values associated with risk and benefit are computed considering the difference between the risk and benefit attributes after and before maintenance is applied. A similar approach for the risk assessment was employed by Sabatino, Frangopol, and Dong (in press), in which a decision-making support tool was developed for the maintenance of existing highway bridges considering multi-attribute utility theory within an optimisation process that balances the utility values associated with maintenance cost and sustainability. However, lifetime functions including survivor, availability, and hazard at component and system levels and correlation effects were not considered. Multi-attribute utility theory was also employed by Dong, Frangopol, and Sabatino (in press) to determine optimum retrofit actions for a network of bridges subjected to seismic hazard. Although previous studies employed multi-attribute utility to represent the risks that plague structures, the work herein focuses on establishing a comprehensive decision support framework that (a) utilises computationally efficient lifetime functions, including correlation effects, (b) incorporates a large variety of criteria within a tri-objective optimisation procedure that balances maintenance cost, risks and benefit, and (c) may be applied to a wide array of engineering systems.

Lifetime distributions are used to represent the performance of a structural system over its life cycle through functions that probabilistically characterise the system's time-to-failure, which is regarded as a continuous, non-negative random variable (Leemis, 1995). These functions have been utilised to model the time-variant effect of intervention actions on the performance of structural systems in several studies including van Noortwijk and Klatter (2004), Yang, Frangopol, and Neves (2004), Yang, Frangopol, and Neves (2006), Yang, Frangopol, Kawakami and Neves (2006), Okasha and Frangopol (2009, 2010b), Orcesi and Frangopol (2011), and Barone and Frangopol (2013, 2014a, 2014b). Moreover, research regarding lifetime functions has emphasised their power as an effective tool in quantifying the lifetime reliability of highway bridges (Yang et al. 2004). Barone and Frangopol (2014a, 2014b) used system hazard and availability to quantify structural performance and determine optimum

maintenance schedules for an existing highway bridge. However, lifetime functions were not used to calculate risk. These functions are employed herein to determine the benefit provided by optimised lifetime maintenance plans based upon minimum lifetime system availability and maximum lifetime system hazard. Furthermore, these functions and correlation effects are directly incorporated within risk calculations herein in order to compute both direct and indirect risks.

As indicated previously, the proposed decision support framework has the ability to quantify cost, risk and benefit in terms of utility and effectively employs tri-objective optimisation procedures in order to determine optimum maintenance strategies. The effects of the risk attitude, correlation and preferences of the decision-maker, in addition to the number of maintenance interventions on the optimum maintenance strategies, are investigated. A genetic algorithm (GA)-based optimisation procedure is employed to find optimum maintenance schedules. The capabilities of the presented decision support framework are illustrated on five representative 4-component series–parallel systems and an existing highway bridge located in Colorado.

2. Lifetime functions

The main advantage of employing lifetime distributions within reliability calculations is the extreme mathematical flexibility that is provided by their closed-formulation expression of the distribution of time-to-failure. Due to their mathematical versatility, computations involving lifetime distributions are efficient; thus, these distributions are particularly suitable for problems involving optimisation (Okasha & Frangopol, 2010a). Generally, lifetime distributions are used to represent time-variant structural performance through continuous functions that are established in closed form by considering the time-to-failure of components as a continuous random variable (Okasha & Frangopol, 2010b). More specifically, the time-to-failure random variable is defined as the time elapsing from the instance a component is put into operation until it fails for the first time (Hoyland & Rausand, 1994). The type of distribution used to represent the probability density function (PDF) associated with the time-to-failure of a particular component is determined based upon its failure characteristics and historical material behaviour. Commonly employed distributions used to represent the PDF of time-to-failure are the Weibull and exponential distributions (Jiang & Murthy, 1995; Lai & Murthy, 2003; Okasha & Frangopol 2010b; van Noortwijk and Klatter, 2004). With knowledge of the PDFs describing the times-to-failure of the components comprising an engineering system, the distribution of the system's time-to-failure can be calculated considering the type and configuration (e.g. series, parallel or series–parallel) of the system.

2.1. Component analysis

The most commonly used lifetime functions include the PDF of the time-to-failure and survivor, availability and hazard functions. The Weibull PDF of component time-to-failure is (Leemis, 1995):

$$f(t) = k \cdot \lambda \cdot (\lambda \cdot t)^{k-1} \exp\left[-(\lambda \cdot t)^k\right] \quad \text{for } t > 0 \quad (1)$$

where k and λ are the shape and scale parameters of the Weibull distribution, respectively. In general, the shape and scale parameters used in the Weibull PDF of the time-to-failure associated with the components of a structural system are determined based upon failure data related to specific materials and deterioration mechanisms. Next, the probability that the time-to-failure T_F of the investigated component is less than a given time instant t, denoted as the cumulative-time failure probability $F(t)$, is:

$$F(t) = P[t > T_{\mathrm{F}}] = \int_0^t f(x)\mathrm{d}x, \qquad (2)$$

where T_{F} = time-to-failure of the investigated component or system. The complement of $F(t)$ is the survivor function $S(t)$, which expresses the probability of a component or system surviving (i.e. not failed) before the time instant t. $S(t)$ has been utilised to assess bridge lifetime performance and to facilitate the implementation of maintenance strategies to an existing structure (Yang, Frangopol, & Neves, 2004; Yang, Frangopol, & Neves, 2006; Yang, Kawakami & Neves, 2006). The survivor function, in its most general form, is (Leemis, 1995):

$$S(t) = 1 - F(t) = P[t \le T_{\mathrm{F}}] = \int_t^\infty f(x)\mathrm{d}x \qquad (3)$$

If $f(t)$ follows a Weibull distribution, the associated survivor function is:

$$S(t) = 1 - F(t) = \exp\left[-(\lambda \cdot t)^k\right]. \qquad (4)$$

The survivor function may be used as a basis to calculate other lifetime functions. For example, the availability of a component $A(t)$, which is defined as the probability that the component is functioning at a given time instant, coincides with $S(t)$ when no maintenance is considered (Ang & Tang, 1984; Leemis, 1995). The availability function has been employed in assessing the effects of implementing intervention strategies to existing civil infrastructure (Barone and Frangopol 2014a, 2014b; Biswas, Sarkar, & Sarkar, 2003).

In addition to availability, the hazard function $h(t)$ is also investigated as a suitable performance indicator in the context of lifetime functions. This function provides information regarding how fast a component becomes non-functional over time. Overall, the hazard metric is representative of the occurrence of failure as a function of time for any component. There are several other common terms utilised to describe the hazard function including hazard rate, failure rate, rate function and intensity function. The hazard function has been successfully applied to several areas of science and engineering, including biological sciences (Heidenreich, Luebeck, & Moolgavkar, 1997; Horová, Pospíšil, & Zelinka, 2009; Tanner & Wong, 1984), performance assessment of engineering systems (Amari, Pham, & Misra, 2012; Thies, Smith, & Johanning, 2012; Zhang & Li, 2010), fatigue analysis (Lawson & Chen, 1995) and optimum maintenance planning of complex systems (Barone & Frangopol, 2014a, 2014b; Cui, Kuo, Loh, & Xie, 2004).

The hazard function $h(t)$ for a particular component is calculated by considering the probability of failure between t and

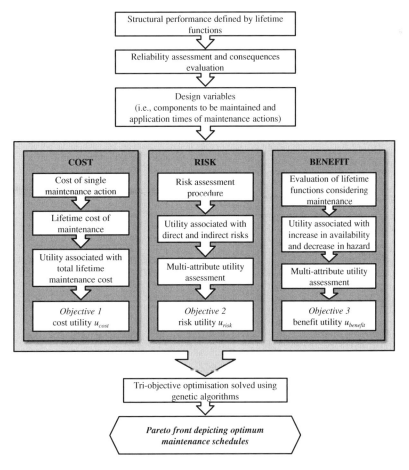

Figure 1. Flowchart outlining the computations involved in the decision support tool.

$t + \Delta t$ conditioned on the event that the component is functioning at time t (Leemis, 1995):

$$h(t) = \lim_{\Delta t \to 0} \frac{P\left[t \le T_F \le t + \Delta t | t \le T_F\right]}{\Delta t} = -\frac{S'(t)}{S(t)} = \frac{f(t)}{S(t)} = \frac{f(t)}{1 - F(t)}.$$

(5)

The units of $h(t)$ are typically given in failures per time unit. Therefore, in order to determine the number of expected failures within a certain interval, $h(t)$ is multiplied by that time interval. The shape of the hazard function is indicative of how a component ages. In general, when the hazard function is large, the component is subject to a significant risk of failure, and when the hazard function is small, the component has less chance of experiencing failure. Considering the assumption that the PDF of time-to-failure $f(t)$ follows the Weibull distribution, the expression for $h(t)$ is:

$$h(t) = k \cdot \lambda^k \cdot t^{k-1}.$$

(6)

2.2. System analysis

Thus far, the mathematical expressions for the survivor $S(t)$, availability $A(t)$ and hazard $h(t)$ functions for the component level have been presented. In order to formulate system-level expressions, the configuration of the system must be considered. For systems comprised of components with known lifetime distributions, the closed-form system survivor $S_{sys}(t)$, availability $A_{sys}(t)$ and hazard $h_{sys}(t)$ functions may be obtained for both statistically independent and perfectly correlated components

based upon cut set techniques (Leemis 1995) considering the system configuration (e.g. series, parallel or series–parallel).

2.3. Effects of essential maintenance

The main advantage of employing lifetime functions to facilitate efficient maintenance planning lies in the fact that it is possible to perform direct computations in analytical form. In general, the type of maintenance applied to structural components greatly depends upon the deterioration mechanisms and their evolution in time. Accordingly, different maintenance types may be applied during the lifetime of a structural component. Assuming that only essential maintenance (i.e. full replacement of component(s)) is implemented and that interventions are performed N times during the lifetime of a deteriorating system, the resulting change to component lifetime functions may be evaluated. $S_i(t) = 1 - F_i(t)$, $A_i(t)$ and $h_i(t)$ represent the ith component's lifetime functions with no maintenance, while $S_{i,m}(t) = 1 - F_{i,m}(t)$, $A_{i,m}(t)$ and $h_{i,m}(t)$ denote the same component's lifetime functions considering maintenance. Assuming that an essential maintenance is implemented at times $T_1, \ldots, T_N$ on the ith component and $T_0 = 0$ (i.e. the initial observation time), the ith component's resulting survivor function adjusted for maintenance $S_{i,m}(t)$, is:

$$S_{i,m}(t) = \begin{cases} S_i(t) & \text{for } t < T_1 \\ S_i(t - T_N) \prod_{j=1}^{N} S_i(T_j - T_{j-1}) & \text{for } T_N \le t \le T_{N+1}, j \ge 1 \end{cases},$$

(7)

where $S_i(t) = 1 - F_i(t)$ = survivor function associated with component i considering no maintenance. Each time essential maintenance is applied to component i, the magnitude of the slope corresponding to $S_{i,m}(t) = 1 - F_{i,m}(t)$ decreases. Although the survivor function is a continuous, monotonically decreasing function of time, the availability function abruptly increases in value whenever maintenance is performed. More specifically, it is assumed that component availability is restored to its original value (i.e. $A_{i,m}(t) = 1$) when essential maintenance (i.e. full replacement) is performed on a particular component. The availability function corresponding to a maintained component i, $A_{i,m}(t)$, is obtained from the unmaintained component survivor function as (Okasha & Frangopol, 2010a):

$$A_{i,m}(t) = \begin{cases} S_i(t) & \text{for } t < T_1 \\ S_i(t - T_N) & \text{for } T_N \le t \le T_{N+1} \end{cases}.$$

(8)

In a similar way, component hazard is restored to its initial value (i.e. $h_{i,m}(t) = 0$) after each replacement. The hazard function for the maintained component i is:

$$h_{i,m}(t) = h_i(t - T_N) \quad \text{for } T_N \le t \le T_{N+1}$$

(9)

where $h_i(t)$ = hazard function associated with component i without maintenance.

After component lifetime functions are established for a particular maintenance plan, the lifetime functions at the system level can provide information regarding the structural system performance considering maintenance. Calculations

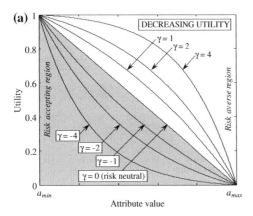

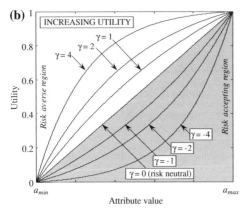

Figure 2. Exponential utility functions that are monotonically (a) decreasing and (b) increasing as the attribute value increases.

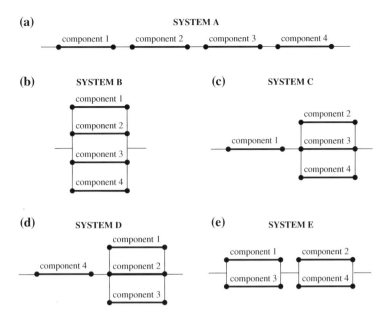

Figure 3. Five configurations of a four-component system.

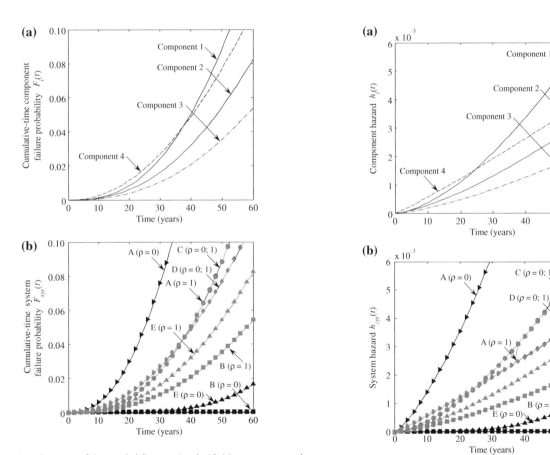

Figure 4. Cumulative-time failure probability associated with (a) components and (b) systems in Figure 3 considering independence ($\rho = 0$) and full correlation ($\rho = 1$) among components.

Figure 5. Hazard associated with (a) components and (b) systems in Figure 3 considering independence ($\rho = 0$) and full correlation ($\rho = 1$) among components.

of system survivor $S_{sys,m}(t)$, availability $A_{sys,m}(t)$ and hazard $h_{sys,m}(t)$ are highly dependent upon system configuration and the correlation among components (i.e. statistically independent $\rho = 0$ or perfectly correlated $\rho = 1$, where $\rho =$ correlation coefficient).

3. Attributes evaluation

This section highlights the three separate objectives utilised within the optimisation procedure embedded within the proposed decision support framework for lifetime maintenance planning. Maintenance cost, failure consequences quantified

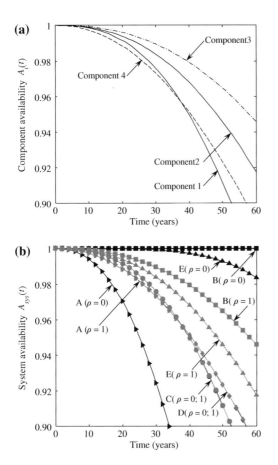

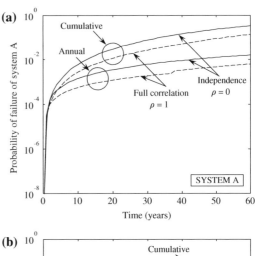

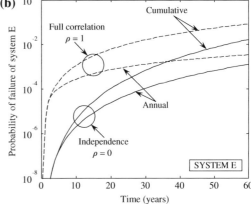

Figure 6. Availability associated with (a) components and (b) systems in Figure 3 considering independence ($\rho = 0$) and full correlation ($\rho = 1$) among components.

Figure 7. Cumulative and annual probability of system failure considering extreme correlation cases among components of (a) system A and (b) system E.

via risk metrics and performance benefit measured by the improvement in minimum annual system availability and maximum annual system hazard are regarded as the three objectives investigated herein.

3.1. Cost

The cost of a lifetime maintenance plan is computed in terms of USD in the year the structure was built. The total cost of a lifetime essential maintenance strategy is determined as (Frangopol, 1999):

$$C_{\text{maint}} = \sum_{j=1}^{N} \frac{C_{\text{EM},j}(t)}{(1 + r_m)^t}, \qquad (10)$$

where $C_{\text{EM},j}$ = cost of maintenance action j applied at year t (USD), r_m = annual discount rate of money and N = total number of essential maintenance actions considered throughout the lifetime of a structure.

3.2. Consequences

Risk, which combines the probability of occurrence of a specific event with the consequence associated with this event, is a crucial performance indicator for civil infrastructure. A simple formulation of risk is (Ang & De Leon, 2005):

$$R = p \cdot \chi, \qquad (11)$$

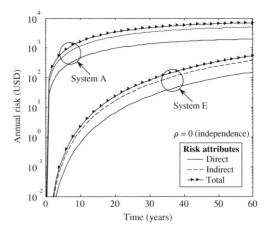

Figure 8. Time-variant profiles of annual direct, indirect and total risk attributes for systems A and E, considering statistical independence among components ($\rho = 0$).

where p = probability of occurrence of an adverse event and χ = consequences of the event. The effects of direct (i.e. economic impacts) and indirect (i.e. social and environmental losses) consequences may be incorporated into the calculation of the total risk. Overall, the evaluation of a wide variety of consequences associated with structural failure plays a fundamental role in the decision-making process regarding infrastructure management planning. Time-variant direct and indirect risk attributes are calculated considering the following expressions:

Table 1. Minimum and maximum annual values of attributes involved in the risk and benefit assessment of the five four-component systems shown in Figure 3.

System configuration (see Figure 3)	Attribute	Independence ($\rho = 0$)		Full correlation ($\rho = 1$)	
		Minimum a_{min}	Maximum a_{max}	Minimum a_{min}	Maximum a_{max}
System A	Direct risk (USD)	0	1974	0	784
	Indirect risk (USD)	0	4935	0	1959
	System hazard h_{sys}	0	.0162	0	.0064
	System availability A_{sys}	.6656	1	.8621	1
System B	Direct risk (USD)	0	1	0	271
	Indirect risk (USD)	0	3	0	678
	System hazard h_{sys}	0	1×10^{-5}	0	.0022
	System availability A_{sys}	.999	1	.9459	1
System C	Direct risk (USD)	0	790	0	783
	Indirect risk (USD)	0	1976	0	1959
	System hazard h_{sys}	0	.0065	0	.0064
	System availability A_{sys}	.8617	1	.8621	1
System D	Direct risk (USD)	0	508	0	501
	Indirect risk (USD)	0	1270	0	1252
	System hazard h_{sys}	0	.0042	0	.0041
	System availability A_{sys}	.8890	1	.8896	1
System E	Direct risk (USD)	0	152	0	420
	Indirect risk (USD)	0	381	0	1050
	System hazard h_{sys}	0	.0012	0	.0034
	System availability A_{sys}	.9385	1	.9175	1

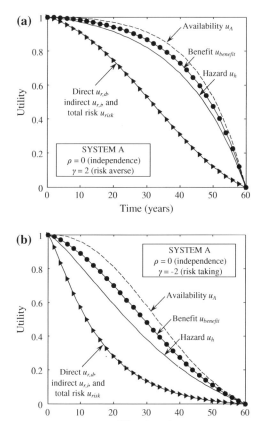

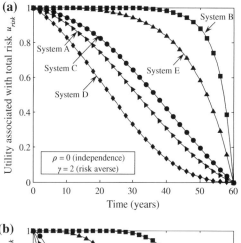

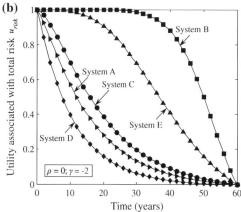

Figure 9. Time-variant profiles of availability, hazard, risk and benefit utilities considering system A, independence among components ($\rho = 0$), and (a) risk-averse ($\gamma = 2$) or (b) risk-taking attitude ($\gamma = -2$).

Figure 10. Time-variant profiles of total risk utilities considering all systems in Figure 3, independence among components ($\rho = 0$), and (a) risk-averse ($\gamma = 2$) or (b) risk-taking attitude ($\gamma = -2$).

$$RA_{direct}(t) = p_{f.sys}(t) \cdot C_{direct}(t) \qquad (12)$$

$$RA_{indirect}(t) = p_{f.sys}(t) \cdot C_{indirect}(t), \qquad (13)$$

where $P_{f.sys}(t)$ = probability of system failure during year t, $C_{direct}(t)$ = direct consequences associated with system failure and $C_{indirect}(t)$ = indirect consequences. The total risk associated with a particular system is calculated considering the sum of its direct and indirect risks. In this paper, the annual probability of system failure $P_{f.sys}(t)$ is calculated using the system hazard function $h_{sys}(t)$ and the cumulative-time failure probability is calculated using the complement of the survivor system function $F_{sys}(t) = 1 - S_{sys}(t)$. As indicated in the illustrative highway bridge example presented herein, these two ways of calculating the system probability of failure (i.e. annual and cumulative) yield distinctly different optimum maintenance plans considering all other parameters remain the same.

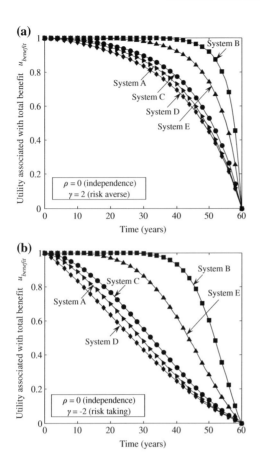

Figure 11. Time-variant profiles of benefit utilities considering all systems in Figure 3, independence among components ($\rho = 0$), and (a) risk-averse ($\gamma = 2$) or (b) risk-taking attitude ($\gamma = -2$).

4. Utility assessment

Utility functions that depict the relative value of each attribute to the decision-maker considering his/her particular risk attitude play vital roles within the proposed decision-making support system. Maintenance strategies associated with relatively high utility values are typically the most desirable solutions. This section provides the process for formulating single- and multi attribute utility functions that effectively depict the decision-maker's value of lifetime maintenance schedules in terms of cost, risk and benefit. Overall, the procedure adopted herein for computing attributes and their corresponding utility values, in relation to decision-making, is shown in Figure 1.

4.1. Single attribute utility assignment

Two types of attributes are considered: (a) one that possesses decreased desirability when its attribute value is increased and (b) another that exhibits increased desirability as the attribute value increases. The investigated attributes that fall into the first category are maintenance cost, direct and indirect risks and system hazard. The system availability is the only attribute herein that causes an increase in desirability as the attribute value is increased. Figure 2a depicts a qualitative representation of typical exponential utility functions that are monotonically decreasing for the attributes that experience a decrease in utility (i.e. desirability) as the attribute value increases. Conversely, Figure 2(b) shows representative exponential utility functions that are monotonically increasing for the system availability; as the system availability increases, the utility also increases. The governing equation for the utility functions shown in Figure 2(a) is (Ang & Tang, 1984):

$$u_{a,dec} = \frac{1}{1 - \exp[-\gamma]}\left(1 - \exp\left[-\gamma\frac{a_{max} - a}{a_{max} - a_{min}}\right]\right), \quad (14)$$

where γ = risk attribute of the decision-maker (i.e. $\gamma > 0$ indicates risk aversion and $\gamma < 0$ denotes risk acceptance), a = expected value of the attribute value under investigation, a_{min} = minimum value of the attribute and a_{max} = maximum value of the attribute. The minimum and maximum values of the investigated attribute are utilised to normalise the utility so that it always takes values between 0 and 1. Considering the same solution alternative, a risk-averse attitude will yield a utility higher than that of a risk-accepting attitude. The monotonically increasing utility functions shown in Figure 2(b) are governed by the following equation:

$$u_{a,inc} = \frac{1}{1 - \exp[-\gamma]}\left(1 - \exp\left[-\gamma\frac{a - a_{min}}{a_{max} - a_{min}}\right]\right). \quad (15)$$

For a single attribute, $u_a = 1$ and $u_a = 0$ correspond to the most and least desirable value that the investigated attribute may take, respectively. In general, a risk-averse attitude produces a concave utility function, while a risk-accepting attitude is represented by a convex utility function.

4.2. Multi-attribute utility

Once the utility function associated with each attribute is appropriately established, multi-attribute utility theory may be employed to combine them into single utility values that

Although specific risks related to the economic, social and environmental impacts are considered herein, the decision-maker may include his/her desired number of risk attributes within the related lifetime maintenance optimisation procedure. For example, if the main concern of the decision-maker is related to the effects of structural failure on environmental and social consequences, then he/she may decide to employ only the indirect consequences within the developed risk calculation approach.

3.3. Benefit

The performance benefit utilised within the proposed tri-objective maintenance optimisation framework is calculated based on the improvement in lifetime function values when essential maintenance strategies are implemented. Two indicators are utilised as benefit attributes herein: (a) an increase in the minimum lifetime availability and (b) a decrease in the maximum lifetime hazard when comparing the no maintenance case with the optimum maintenance case. Therefore, when considering just the increase in minimum lifetime availability as the sole benefit, the difference between minimum lifetime availability offered by the cases with and without maintenance serves as the index employed to measure the performance benefit. A similar procedure may be carried out to determine the performance benefit in terms of system hazard. Although the benefits included herein are related to availability and hazard, other benefits of implementing lifetime maintenance strategies may be included in the proposed decision-making framework.

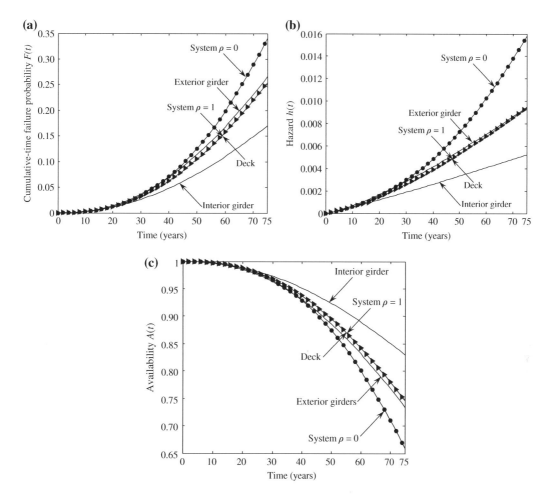

Figure 12. (a) Cumulative-time failure probability, (b) hazard and (c) availability associated with both the components and the system (i.e. superstructure of the bridge E-17-HS) considering no maintenance and extreme correlation cases.

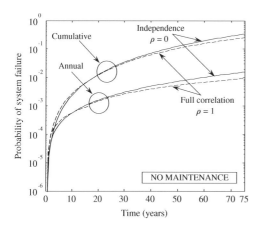

Figure 13. Cumulative and annual probability of system (i.e. superstructure of bridge E-17-HS) failure considering extreme correlation cases among components.

effectively represent each of the three separate utility objectives employed in the optimisation procedure presented herein (i.e. cost u_{cost}, risk u_{risk} and benefit $u_{benefit}$). Although there are various types of multi-attribute utility functions, the additive formulation is employed herein. This formulation is obtained by multiplying marginal utility values associated with each attribute by weighting factors and summing over all attributes investigated

(Stewart, 1996). The utility associated with indirect risk $u_{r,i}$ is computed considering equal contributions of social and environmental consequences. Subsequently, the utility associated with the total risk is quantified as:

$$u_{r,tot} = \frac{1}{2}\left(u_{r,d} + u_{r,i}\right), \quad (16)$$

where $u_{r,d}$ = utility associated with direct risk and $u_{r,i}$ = utility associated with indirect risk.

Similarly, the first step in formulating the multi-attribute utility function representative of the performance benefit considering maintenance is calculating the utility of the minimum lifetime system availability and maximum lifetime system hazard. If the decision-maker would like the multi-criterion optimisation procedure to include the effects of both system availability and hazard, then the following formulation of the benefit utility under maintenance is used:

$$u_{b,m} = \frac{1}{2}\left(u_A + u_h\right), \quad (17)$$

where $u_{b,m}$ = utility associated with the performance benefit considering maintenance, u_A = utility associated with minimum lifetime system availability and u_h = utility associated with maximum lifetime system hazard. Additionally, if the decision-maker desires only the effect of system availability or hazard to be

Table 2. Parameters used in the evaluation of the risk attributes associated with the E-17-HS bridge.

Parameter	Mean value	Reference
Rebuilding cost parameter C_1	1292 USD/m²	Dong, Frangopol, and Saydam (2014)
Width of the bridge W	10.4 m	Akgül (2002)
Length of the bridge L	64.5 m	Akgül (2002)
Occupancy rate for non-truck vehicles O^r	1.56	Stein, Young, Trent, and Pearson (1999); Barone and Frangopol (2014a, 2014b)
Percentage of average daily traffic that is trucks TT^p	4%	Barone and Frangopol (2014a, 2014b)
Detour length L^d	10 km	Barone and Frangopol (2014a, 2014b)
Average daily traffic ADT	400 vehicles	Barone and Frangopol (2014a, 2014b)
Duration of detour D^d	365 days	Barone and Frangopol (2014a, 2014b)
Average detour speed S^d	64 km/h	Barone and Frangopol (2014a, 2014b)
Safe following distance f^d	55 m	Colorado State Patrol (2011)
Carbon dioxide emissions associated with cars CPD^C	.22 kg/km	Dong, Frangopol, and Saydam (2014)
Carbon dioxide emissions associated with trucks CPD^T	.56 kg/km	Dong, Frangopol, and Saydam (2014)
Energy consumption associated with each vehicle EPD	3.80 MJ/km	Dong, Frangopol, and Saydam (2014)
Carbon dioxide emissions associated with rebuilding CD_{REB}	159 kg/m²	Dequidt (2012)
Energy consumption associated with rebuilding EC_{REB}	2.05 GJ/m²	Dequidt (2012)

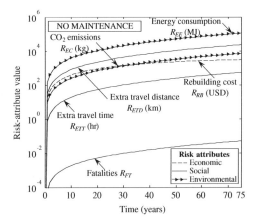

Figure 14. Time-variant profile of each risk attribute considering no maintenance, statistical independence among components ($\rho = 0$) and the annual probability of system failure.

included within the performance benefit utility, then $u_{b,m} = u_A$ or $u_{b,m} = u_h$, respectively.

The utility associated with cost u_{cost} is calculated using the exponential form of the single-attribute utility function. Cost attributes can easily be incorporated within the decision-making process if reliable data regarding these cost metrics are available.

4.3. Utility associated with performance benefit

The total utility associated with the performance benefit, $u_{benefit}$, is:

$$u_{benefit} = u_{b,m} - u_{b,0} \qquad (18)$$

where $u_{b,0}$ = performance utility without maintenance.

5. Tri-objective optimisation framework for lifetime maintenance planning

Overall, the three utility functions integrated within the presented multi-criterion optimisation framework represent the relative value of solution alternatives to the decision-maker, considering his/her risk attitude. The three objectives that are all simultaneously maximised are: (a) utility corresponding to lifetime maintenance investment cost u_{cost}, (b) utility associated with risk u_{risk} and (c) utility associated with performance benefit $u_{benefit}$. The general methodology embedded within the proposed optimisation procedure is shown in Figure 1. Within

the proposed framework, three separate modules compute the objective values utilised within the multi-criterion optimisation process, whose results come in the form of Pareto optimum solutions outlining bridge maintenance planning. Although the numerical examples presented herein address the life cycle essential maintenance planning problem, the proposed multi-criterion decision support system has the capability to optimise non-essential maintenance interventions (e.g. preventive measures which may delay deterioration or temporarily reduce the rate of deterioration) if information regarding the effects of these actions becomes available. A set of Pareto optimum solutions is obtained utilising GAs within an adequate number of generations (Dong, Frangopol, and Saydam, 2015; Frangopol, 2011; Okasha & Frangopol, 2009).

The tri-objective optimisation problem is formulated as follows
Given:
- Lifetime functions representing time-variant structural performance of components comprising an engineering system (information associated with Equations (1)–(6))
- Monetary cost associated with specific essential maintenance actions (input to Equation (10))
- Risk associated with structural failure (Equations (11)–(13))
- Risk attitude of the decision-maker (γ in Equations (14) and (15))
- Desired attributes to be included (e.g. direct, indirect or total risk within the formulation of u_{risk})
- Lifetime under investigation (T_L)
- Total number of maintenance actions (N)

Find:
- Components to be maintained
- Time of application of maintenance actions

So that:
- Utility associated with the total maintenance cost u_{cost} is maximised
- Minimum utility associated with risk u_{risk} over the system's lifetime is maximised
- Minimum utility associated with performance benefit $u_{benefit}$ over the system's lifetime is maximised

Subjected to:
- Maximum allowable total maintenance cost
- Constraints on the allowable minimum and maximum values of each attribute (a_{min} and a_{max}, respectively, in Equations (14) and (15))

Table 3. Minimum and maximum annual values of attributes involved in the risk and benefit assessment of the E-17-HS bridge.

Attribute type	Attribute	Independence ($\rho = 0$)		Full correlation ($\rho = 1$)	
		Minimum a_{min}	Maximum a_{max}	Minimum a_{min}	Maximum a_{max}
Risk (economic)	Rebuilding cost R_{RB} (USD)	0	3105	0	1848
Risk (social)	Extra travel time R_{ETT} (h)	0	555	0	329
	Extra travel distance R_{ETD} (km)	0	23,094	0	13,711
	Fatalities R_{FT}	0	.0528	0	.0314
Risk (environmental)	CO_2 emissions R_{EC} (kg)	0	7165	0	4939
	Energy consumption R_{EE} (MJ)	0	109,510	0	65,015
Benefit	System hazard h_{sys}	0	.0158	0	.0094
Benefit	System availability A_{sys}	.6599	1	.7457	1

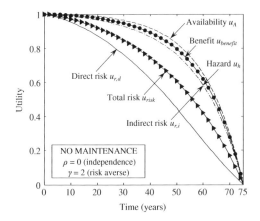

Figure 15. The time-variant profile of risk and benefit utilities considering independence among components ($\rho = 0$) and a risk-averse attitude ($\gamma = 2$).

- Constraints on the application times of maintenance actions

6. Illustrative example

In order to illustrate the capabilities of the proposed decision support system, the presented methodology is demonstrated on five systems consisting of four components, as shown in Figure 3. For these five configurations, the utility assignment is carried out and the evaluation of the no maintenance case is conducted in order to determine the effect of system configuration on the utility profiles. The shape k_i and scale λ_i parameters corresponding to the Weibull distribution are used to define the PDF of time-to-failure of the four components investigated in this example, as follows: component 1, $k_1 = 2.6$ and $\lambda_1 = 8 \times 10^{-3}$; component 2, $k_2 = 2.4$ and $\lambda_2 = 6 \times 10^{-3}$; component 3, $k_3 = 2.4$ and $\lambda_3 = 5 \times 10^{-3}$; and component 4, $k_4 = 2.1$ and $\lambda_4 = 6 \times 10^{-3}$. For the systems in Figure 3, the closed-form system survivor $S_{sys}(t)$, availability $A_{sys}(t)$ and hazard $h_{sys}(t)$ functions are obtained for both statistically independent (i.e. $\rho = 0$) and perfectly correlated (i.e. $\rho = 1$) components. The cumulative-time failure probability, $F(t)$, hazard $h(t)$ and availability $A(t)$ of each of the four components in Figure 3 are shown in Figures 4(a), 5(a) and 6(a), respectively. Considering the five system configurations shown in Figure 3, the system cumulative-time failure probability $F_{sys}(t) = 1 - S_{sys}(t)$, availability $A_{sys}(t)$ and hazard $h_{sys}(t)$ for the extreme correlation cases are shown in Figures 4(b), 5(b) and 6(b), respectively. Since the calculation of the probability of failure may use either $F_{sys}(t)$ or $h_{sys}(t)$, Figure 7 depicts the effects of correlation among components on both the cumulative-time system failure probability and annual system failure probability of systems A and E. The most conservative cases are associated with the cumulative-time

failure probability for the independence and full correlation assumptions, for systems A and E, respectively.

Next, the consequences of structural failure are examined and the corresponding risk attributes are evaluated for systems A to E. The economic impact (i.e. direct loss) is measured in terms of the risk associated with the rebuilding cost C_{direct} during a certain year. Similarly, the indirect losses (i.e. consequences to society and environment) are measured in terms of the risk associated with the monetary indirect consequences $C_{indirect}$. For illustrative purposes, in this example, C_{direct} and $C_{indirect}$ are assumed as 400,000 USD and 1,000,000 USD, respectively. Considering these assumptions, direct, indirect and total risk profiles for systems A and E are shown in Figure 8.

The last step involves formulating the time-variant utility values associated with risk and benefit. Once the time-variant risk and benefit attributes are determined, they may be transferred to utility considering the exponential formulations in Equations (14) and (15). For example, the utility corresponding to each attribute is calculated considering the range of risk attribute values shown in Table 1. In this table, a_{min} and a_{max} are the expected attribute values at $t = 0$ and $t = T_L$ (i.e. 60 years). u_{risk} and $u_{benefit}$ are computed using Equations (16)–(18). Figure 9 depicts utility profiles for system A with risk-averse and risk-taking attitudes. It is evident from this figure that in this case risk-averse and risk-taking attitudes yield convex and concave utility functions, respectively. Furthermore, time-variant total risk and benefit utilities of the five systems A to E are shown in Figures 10 and 11, respectively. The effect of the system configuration is quite significant. For systems with a high level of redundancy (i.e. systems B and E), both total risk and benefit utilities tend to remain at high levels for a longer period of time than those associated with the other systems.

7. Case study

The case study presented herein applies the developed framework to the E-17-HS bridge, an existing reinforced concrete highway bridge located in Adams County, Colorado. The bridge deck is supported by four reinforced concrete T-girders, as detailed in Akgül (2002). The superstructure of the bridge (i.e. the system) is modelled with a series–parallel model that defines system failure as either failure of the deck or any two adjacent girders.

It is assumed that the PDFs of time-to-failure of the three main components of the bridge superstructure, the deck (D), exterior girders (G_E) and interior girders (G_I), follow the Weibull distribution with the following shape k_i and scale λ_i parameters: (a) deck, $k_1 = 2.4$ and $\lambda_1 = 8 \times 10^{-3}$; (b) exterior gird-

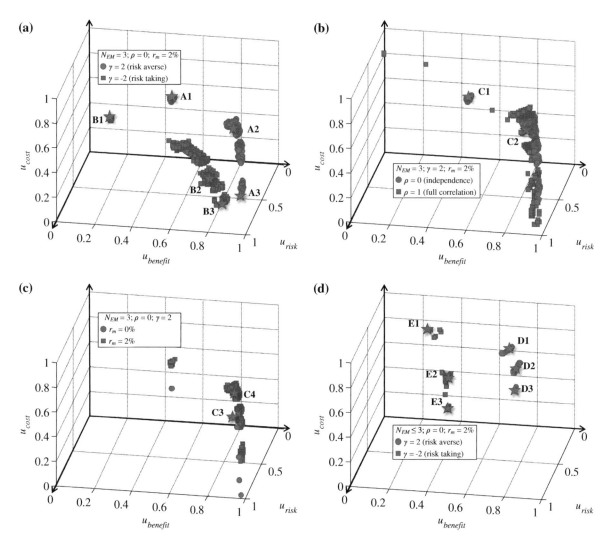

Figure 16. Pareto optimum solutions considering the effects of (a) risk attitude γ, (b) correlation among components ρ, (c) discount rate of money r_m and (d) number of maintenance actions N_{EM}.

Table 4. Maintenance plans corresponding to the six optimum solutions on the Pareto fronts shown in Figure 16(a).

Solution (see Figure 16(a))	Maintenance actions	Time of application (years)
A1	$[G_I, G_E, G_I]$	[24, 47, 70]
A2	$[D, G_I, D]$	[29, 46, 64]
A3	$[S, S' D]$	[16, 36, 57]
B1	$[G_I, G_I', G_I]$	[23, 46, 70]
B2	$[D, G, D]$	[23, 41, 52]
B3	$[D, S, S]$	[18, 33, 54]

Note: D = replace the deck; G_I = replace all interior girders; G_E = replace all exterior girders; G = replace all girders; S = replace the entire superstructure.

ers, $k_2 = 2.3$ and $\lambda_2 = 8 \times 10^{-3}$; and (c) interior girders, $k_3 = 2.1$ and $\lambda_3 = 6 \times 10^{-3}$ (Barone & Frangopol, 2014a). The functions $F(t)$, $h(t)$ and $A(t)$ associated with no maintenance considering extreme correlations and a lifetime $T_L = 75$ years are reported in Figure 12. Figure 13 depicts the annual and cumulative system probability of failure profiles under extreme correlation conditions. The five possible maintenance options and their associated costs considered herein are (Barone & Frangopol, 2014a, 2014b): replacing the deck (D), 100,000 USD; replacing the two exterior

(G_E) or the two interior (G_I) girders, 80,000 USD; replacing all girders (G), 140,000 USD; and replacing the entire superstructure (S), 200,000 USD. An annual discount rate of money $r_m = .02$ is assumed unless otherwise noted.

The next step includes evaluating the detrimental consequences associated with structural failure of the system. Based on Dong, Frangopol and Sabatino (in press), the expected value of all the risk attributes are calculated in terms of their respective units. Namely, the economic impact (i.e. direct loss) is measured in terms of the risk associated with the rebuilding cost R_{RB} during a certain year, while the social consequences of bridge failure include the extra travel time R_{ETT} and distance R_{ETD} experienced by vehicle operators, in addition to any fatalities that may occur R_{FT}. The third type of risk examined encompasses the detrimental effects of structural failure on the environment. More specifically, the environmental metric accounts for two impacts: (a) carbon dioxide emissions R_{EC} and (b) energy consumption associated with detour and bridge repair R_{EE}. In general, indirect risks integrate the effects of both social and environmental consequences of structural failure.

Within this paper, the investigated risk attribute values may be computed in two different ways: one in which the annual prob-

Table 5. Optimum utility values corresponding to the six optimum solutions within the Pareto fronts in Figure 16(a).

Solution (see Figure 16(a))	Cost utility u_{cost}	Optimum utility						
		Direct risk $u_{r,d}$	Indirect risk $u_{r,i}$	Total risk u_{risk}	Hazard u_h	Availability u_A	Benefit $u_{benefit}$
A1	.8232	.5291	.6061	.5676	.6061	.4581	.5347
A2	.7648	.7756	.8729	.8508	.8729	.9471	.9010
A3	.2725	.9088	.9424	.9377	.9424	.9829	.9626
B1	.3865	.1313	.1609	.1515	.1609	.1023	.1316
B2	.1886	.4219	.6222	.5842	.6222	.8153	.7393
B3	.0439	.5269	.7302	.6730	.7302	.9067	.8184

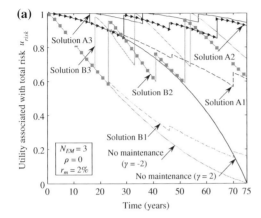

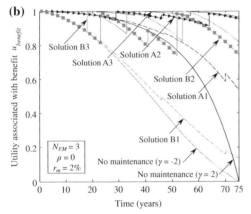

Figure 17. (a) Risk and (b) benefit utility profiles associated with the six optimum solutions shown in Figure 16(a).

ability of failure (i.e. $P_{f,sys}$ in Equations (12) and (13)) is derived directly from the system hazard and one that utilises the system's cumulative distribution function as a measure of system failure (i.e. $F_{sys}(t) = 1 - S_{sys}(t)$). Using the parameters in Table 2, the time-variant consequences considering the no maintenance case are determined. The time-variant profile of the expected value corresponding to each risk attribute considering statistical independence among components ($\rho = 0$) is shown in Figure 14.

Within the proposed tri-objective optimisation procedure, for each alternative, it is necessary to evaluate the utilities associated with total maintenance cost u_{cost}, risk u_{risk} and benefit $u_{benefit}$. The value associated with u_{cost} is obtained from Equation (14) with minimum cost $a_{min} = 0$ and maximum budget $a_{max} = 400,000$ USD. Once the time-variant risk and benefit attributes are determined, they may be transferred to utility considering the exponential formulations in Equations (14) and (15). The utility value corresponding to each attribute is calculated considering the range of risk attribute values shown in Table 3, in which the

annual risk formulation is based on the system hazard. Within this case study, these minimum and maximum attribute values are obtained considering the no maintenance case; a_{min} and a_{max} are the expected attribute values at $t = 0$ and $t = T_L$ (i.e. 75 years) assuming no maintenance. In other cases, the decision-maker may directly assign values to a_{min} and a_{max} in order to reflect personal risk tolerances. Furthermore, u_{risk} and $u_{benefit}$ are computed considering Equations (16)–(18). Figure 15 depicts the time-variant utility profiles corresponding to the risk and benefit considering no maintenance. In this case, all utilities decrease continuously over the lifetime of the structure. The application of essential maintenance interventions improves the performance of the structure and reduces risk; lifetime maintenance planning can effectively mitigate a variety of risks, while simultaneously ensuring that performance is within acceptable levels.

There are several inputs that influence the final results of the proposed decision support tool. For this case study, the optimisation is performed by simultaneously maximising the utility associated with total maintenance cost u_{cost}, the minimum utility corresponding to the risk u_{risk} and the minimum utility associated with the benefit $u_{benefit}$ over the lifetime of the bridge (i.e. $T_L = 75$ years). The main outputs of this optimisation procedure are the bridge components to be maintained and their respective times of application. The following constraints are also considered herein: (a) total maintenance cost should not exceed 400,000 USD, (b) constraints on the allowable minimum a_{min} and maximum a_{max} values of each risk and benefit attribute are defined in Table 3, (c) essential maintenance may not be performed before $t = 5$ years or after $t = 70$ years and (d) consecutive maintenance actions must be performed at least 3 years apart. The tri-objective maintenance planning problem is solved using a GA-based optimisation approach. MATLAB's Global Optimization Toolbox (MathWorks, 2013) is utilised in order to determine optimum lifetime maintenance strategies for the highway bridge investigated.

The first set of optimum solutions presented herein employs an annual risk formulation that incorporates the system hazard function. In this particular example, both the hazard and availability improvements experienced from maintenance are used to establish the utility associated with benefit. Three-dimensional Pareto fronts obtained considering different risk attitudes γ, correlations among components ρ, number of maintenance actions N_{EM} and discount rates of money r_m are shown in Figure 16.

Considering the two Pareto fronts depicted in Figure 16(a) (i.e. $N_{EM} = 3$, $\rho = 0$, $r_m = 2\%$ and variable risk attitude γ), the maintenance strategies corresponding to solutions A1, A2, A3, B1, B2 and B3 are detailed in Table 4. The optimum solutions associated with a risk-averse (i.e. $\gamma = 2$) and risk-taking ($\gamma = -2$) decision-maker are A1, A2, A3 and B1, B2, B3, respectively.

Table 6. Attribute values corresponding to the six optimum solutions within the Pareto fronts in Figure 16(a).

| | Maximum values | | | | | | Minimum value |
| | Economic attribute | Social attributes | | | Environmental attributes | | System hazard | System availability |
Solution	R_{RB} (USD)	R_{ETD} (km)	R_{ETT} (h)	R_{FT}	R_{EC} (kg)	R_{EE} (MJ)	h_{sys}	A_{sys}
A1	2155	14,521	349	.0332	4505	68,858	.0099	.7456
A2	1381	6865	165	.0157	2130	32,554	.0047	.9505
A3	712	3621	87	.0083	1123	17,170	.0025	.9823
B1	2159	14,928	359	.0342	4632	70,788	.0102	.7454
B2	1076	4567	110	.0105	1417	21,656	.0031	.9704
B3	817	3067	74	.0070	952	14,544	.0021	.9857

Note: R_{RB}, R_{ETD}, R_{ETT}, R_{FT}, R_{EC} and R_{EE} are defined in Table 3.

Table 7. Maintenance plans corresponding to the four optimum solutions within the Pareto fronts in Figure 16(b) and (c).

| Solution (see Figure 16(b) and (c)) | Correlation coefficient ρ | Discount rate r_m (%) | Maintenance actions | Time of application (years) | Cost utility u_{cost} | Minimum utility | |
						Total risk u_{risk}	Benefit $u_{benefit}$
C1	0	2	$[G_I, G_E, G_I]$	[24, 48, 70]	.8231	.5676	.5353
C2	1	2	$[D, G_I, D]$	[22, 31, 53]	.7388	.8684	.9078
C3	0	0	$[D, G_E, G_I]$	[37, 53, 59]	.5822	.8818	.9103
C4	0	2	$[D, G_I, D]$	[28, 42, 62]	.7611	.8565	.9157

Table 8. Maintenance plans corresponding to the six optimum solutions within the two Pareto fronts shown in Figure 16(d).

| Solution (see Figure 16(d)) | Maintenance actions | Time of application (years) | Cost utility u_{cost} | Optimum utility | |
				Total risk u_{risk}	Benefit $u_{benefit}$
D1	[D]	[50]	.9680	.6056	.7574
D2	[D, D]	[40, 68]	.8870	.7367	.8162
D3	[D, D, D]	[35, 58, 70]	.7472	.7942	.8210
E1	[D]	[56]	.8039	.1374	.2449
E2	[D, D]	[42, 68]	.5148	.2906	.3817
E3	[D, D, D]	[27, 53, 70]	.2913	.3326	.3866

Table 9. Maintenance plans corresponding to the two optimum solutions within the two Pareto fronts shown in Figure 22.

| Solution (see Figure 22) | Maintenance actions | Time of application (years) | Cost utility u_{cost} | Minimum utility | |
				Total risk u_{risk}	Benefit $u_{benefit}$
F1	$[D, G_I, D]$	[29, 46, 64]	.7648	.8508	.9010
F2	$[G_I, G_E, G_I]$	[23, 46, 70]	.8232	.3688	.5338

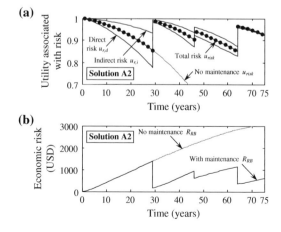

Figure 18. Time-variant (a) risk utilities and (b) economic risk values associated with the optimum solution A2 in Figure 16(a).

They represent maintenance strategies that correspond to different values of utility associated with cost, risk and benefit. The maintenance plan for solution A1 entails replacing the interior girders at years 24 and 70, and replacing the exterior girders at year 47. The maintenance plan associated with solution A3 includes replacing the entire superstructure at years 16 and 57, and the deck at year 36.

Furthermore, the optimum values of the utilities associated with the six representative solutions in Figure 16(a) are indicated in Table 5. The profiles of the utility associated with total risk u_{risk} and performance benefit $u_{benefit}$ for the representative solutions A1, A2, A3 and B1, B2, B3 are shown in Figure 17(a) and (b), respectively. In addition to the specific optimum maintenance plans and utility values, the maximum and minimum lifetime risk and benefit attributes corresponding to each Pareto alternative are also examined; Table 6 summarises the maximum and minimum attribute values corresponding to the representative

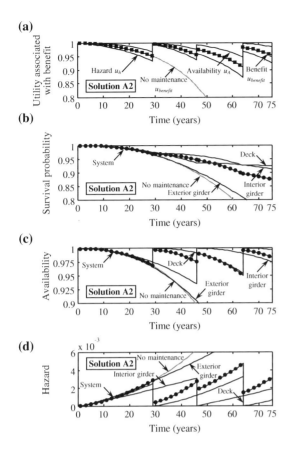

(a)

(b)

(c)

(d)

Figure 19. Time-variant (a) utilities associated with benefit, (b) survival probability, (c) availability and (d) hazard associated with the optimum solution A2 in Figure 16(a).

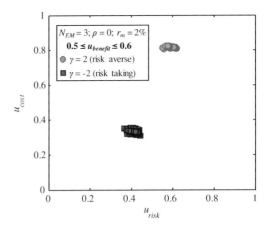

Figure 20. Pareto optimum solutions associated with benefit utilities $u_{benefit}$ within the range .5–.6 associated with Figure 16(a).

optimum maintenance strategies highlighted in Figure 16(a). In order to illustrate time effects associated with the optimum solution A2 for the cases with and without maintenance, the risk utility and economic risk profiles are shown in Figure 18. Furthermore, for the same solution, the performance benefit utility, survival probability, availability and hazard associated with bridge components and the bridge system are indicated in Figure 19.

The Pareto fronts contained within Figure 16(a) are further detailed in Figure 20 for the case $.5 \leq u_{benefit} \leq .6$. As shown, for Pareto solutions exhibiting a benefit utility between .5 and .6, the risk utility for a risk-averse (i.e. $\gamma = 2$) and risk-accepting (i.e. $\gamma = -2$) decision-maker always falls in the range of $.55 \leq u_{risk} \leq .62$ and $.37 \leq u_{risk} \leq .45$, respectively. Similarly, the cost utilities associated with the specific solutions outlined in Figure 20 fall in the range $.81 \leq u_{cost} \leq .82$ and $.31 \leq u_{cost} \leq .36$ for a risk averse and risk accepting attitude, respectively.

In addition to the effect of the decision-maker's risk attitude on the Pareto solutions, the influence of the assumed correlation among components ρ, the discount rate of money r_m and the number of essential maintenance actions N_{EM} are investigated. Figure 16(b) depicts Pareto fronts for a risk-averse decision-maker considering two extreme cases of correlation among the components of the system. The Pareto fronts associated with statistical independence and perfect correlation are examined, and two representative solutions C1 and C2 from these fronts are highlighted in Figure 16(b). Similarly, the influence of the discount rate of money on Pareto solutions is indicated in Figure 16(c).

The comparison of the optimum solutions C3 and C4 in Figure 16(c) with solutions C1 and C2 is given in Table 7. In this table, the cost utilities, corresponding minimum lifetime risk and benefit utilities and maintenance schedules associated with solutions C1, C2, C3 and C4 are summarised.

Next, the effect of the number of essential maintenance actions on the Pareto solutions is examined. Figure 16(d) shows Pareto solutions considering a variable number of essential maintenance actions (i.e. $N_{EM} = 1$, 2 or 3), a discount rate of 2% and a variable risk attitude. The maintenance plans corresponding to the three representative solutions on each of the Pareto fronts in Figure 16(d), solutions D1, D2, D3, E1, E2, and E3, in addition to their associated cost, risk and benefit utilities, are detailed in Table 8. In general, maintenance plans that consider only one intervention have the lowest cost (i.e. high cost utility) but, as a limit, can only achieve certain levels of risk and benefit utilities. Intervention strategies that contain two or more essential maintenance actions can achieve higher levels of utilities associated with risk and benefit but possess higher maintenance costs (i.e. lower cost utility) when compared to the plans containing only one maintenance action. This trend can also be observed in Figure 21, which contains the time-variant risk and benefit utilities corresponding to solutions D1, D2 and D3. Solution D3 dictates maintenance that frequently restores the performance benefit and risk utilities and allows them to remain relatively large throughout the lifetime, while solution D1 only contains one intervention that dramatically increases the risk and performance benefit utilities but is unable to sustain high levels throughout the remaining life.

The final part of this case study includes the comparison of the Pareto fronts obtained by carrying out the lifetime maintenance optimisation considering annual or cumulative-time system failure probability, calculated with $h_{sys}(t)$ and $F_{sys}(t)$, respectively. Figure 22 presents the output of these two optimisations assuming a risk-averse decision-maker (i.e. $\gamma = 2$), a 2% discount rate of money and three essential maintenance actions. The maintenance plans and utility values associated with two representative solutions associated with annual and cumulative-time system failure probability, F1 and F2, respectively, are reported in Table 9. In general, the optimum maintenance plans considering cumulative-time system failure probability exhibit smaller utilities than those associated with annual system failure probability.

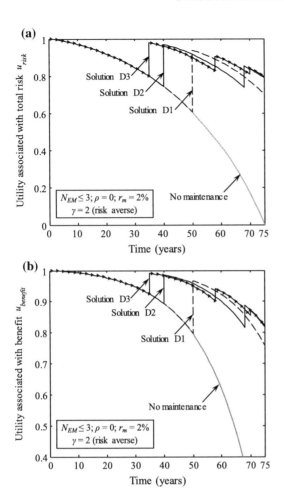

Figure 21. Time-variant (a) risk and (b) benefit utilities corresponding to optimum solutions D1, D2 and D3 in Figure 16(d).

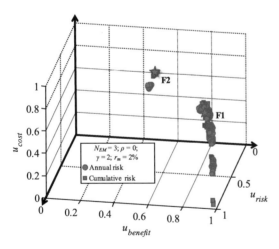

Figure 22. Pareto optimum solutions associated with both annual failure probability and cumulative-time failure probability considering a risk-averse attitude ($\gamma = 2$), independence among components ($\rho = 0$), discount rate of money of 2% ($r_m = .02$) and three maintenance actions ($N_{EM} = 3$).

8. Conclusions

This paper presents a decision support framework that has the ability to quantify cost, risk and benefit in terms of utility and effectively employs tri-objective optimisation procedures in order to determine the best maintenance strategies for struc-

tures with deteriorating components characterised by lifetime functions including survivor, availability, and hazard at component and system levels and correlation effects. The effects of the risk attitude and preferences of the decision-maker, number of maintenance interventions, discount rate of money, correlation among components and computational type of system failure probability (i.e. annual or cumulative-time) on the optimum maintenance strategies are investigated. The flexibility of the multi-attribute utility evaluation process is demonstrated by examining five systems with different configurations comprised of four components. Additionally, the capabilities of the presented optimisation and decision support framework are illustrated on an existing highway bridge.

Overall, a comprehensive approach for the multi-objective life cycle maintenance optimisation of deteriorating structural systems based on multi-attribute utility theory considering lifetime functions, correlation effects and the decision-maker's risk attitude is developed.

The following conclusions are drawn:

(1) Optimum essential maintenance strategies are determined using a multi-criterion optimisation approach based on lifetime functions that balances three objectives: the utilities associated with maintenance cost, risk and performance benefit.

(2) Employing lifetime functions within risk and life cycle optimisation under uncertainty assessment provides mathematical flexibility due to their closed-form expression of the distribution of time-to-failure.

(3) The way system failure probability is calculated, annual or cumulative-time, influences the final Pareto solutions.

(4) System modelling greatly influences the optimum maintenance plans. Depending upon how system failure is modelled, the optimum time-variant utilities can vary significantly.

(5) Further research is needed on life cycle utility-informed intervention planning in connection with bridge networks (Akgül & Frangopol, 2003; Frangopol & Liu, 2007) incorporating various types of maintenance actions (Frangopol & Estes, 1997; Frangopol & Das, 1999; Neves, Frangopol, & Cruz, 2006; Neves, Frangopol, & Petcherdchoo, 2006).

Acknowledgements

The work was supported by the National Science Foundation [grant number CMS-0639428]; the Commonwealth of Pennsylvania, Department of Community and Economic Development, through the Pennsylvania Infrastructure Technology Alliance (PITA), and the U.S. Federal Highway Administration Cooperative Agreement Award [DTFH61-07-H-00040]. The opinions and conclusions presented in this paper are those of the authors and do not necessarily reflect the views of the sponsoring organisations.

Disclosure statement

No potential conflict of interest was reported by the authors.

References

Akgül, F. (2002). *Lifetime system reliability prediction for multiple structure types in a bridge network* (PhD thesis). Department of Civil, Environmental, and Architectural Engineering, University of Colorado, Boulder, CO.

Akgül, F., & Frangopol, D.M. (2003). Rating and reliability of existing bridges in a network. *Journal of Bridge Engineering, 8*, 383–393.

Amari, S.V., Pham, H., & Misra, R.B. (2012). Reliability characteristics of k-out-of-n warm standby systems. *IEEE Transactions on Reliability, 61*, 1007–1018.

Ang, A.H.S. (2011). Life-cycle considerations in risk-informed decisions for design of civil infrastructures. *Structure and Infrastructure Engineering, 7*, 3–9.

Ang, A.H.S., & De Leon D (2005). Modeling and analysis of uncertainties for risk-informed decisions in infrastructures engineering. *Structure and Infrastructure Engineering, 1*, 19–31.

Ang, A.H.-S., & Tang, W.H. (1984). *Probability concepts in engineering planning and design volume II – Decision, risk and reliability*. New York, NY: Wiley.

Arunraj, N.S., & Maiti, J. (2007). Risk-based maintenance – Techniques and applications. *Journal of Hazardous Materials, 142*, 653–661.

ASCE. (2013). *2013 Report card for America's infrastructure*. Reston, VA: Author. http://www.infrastructurereportcard.org/a/documents/2013-Report-Card.pdf

Barone, G., & Frangopol, D.M. (2013). Hazard-based optimum lifetime inspection and repair planning for deteriorating structures. *Journal of Structural Engineering, 139*(12), 1–12.

Barone, G., & Frangopol, D.M. (2014a). Life-cycle maintenance of deteriorating structures by multi-objective optimization involving reliability, risk, availability, hazard and cost. *Structural Safety, 48*, 40–50.

Barone, G., & Frangopol, D.M. (2014b). Reliability, risk and lifetime distributions as performance indicators for life-cycle maintenance of deteriorating structures. *Reliability Engineering & System Safety, 123*, 21–37.

Barone, G., Frangopol, D.M., & Soliman, M. (2014). Optimization of life-cycle maintenance of deteriorating bridges with respect to expected annual system failure rate and expected cumulative cost. *Journal of Structural Engineering, 140*(2), 1–13.

Biswas, A., Sarkar, J., & Sarkar, S. (2003). Availability of a periodically inspected system, maintained under an imperfect-repair policy. *IEEE Transactions on Reliability, 52*, 311–318.

Colorado State Patrol. (2011). *Colorado driver handbook*. Denver, CO.

Cui, L., Kuo, W., Loh, H.T., & Xie, M. (2004). Optimal allocation of minimal & perfect repairs under resource constraints. *IEEE Transactions on Reliability, 53*, 193–199.

Decò, A., & Frangopol, D.M. (2011). Risk assessment of highway bridges under multiple hazards. *Journal of Risk Research, 14*, 1057–1089.

Decò, A., & Frangopol, D.M. (2012). Lifetime risk assessment of bridges affected by multiple hazards. *The sixth international Conference on Bridge Maintenance, Safety and Management* (pp. 2922–2929). Stresa, Lake Maggiore.

Decò, A., & Frangopol, D.M. (2013). Life-cycle risk assessment of spatially distributed aging bridges under seismic and traffic hazards. *Earthquake Spectra, 29*, 127–153.

Dequidt, T. (2012). *Life cycle assessment of a Norwegian bridge* (MS thesis). Department of Civil and Transport Engineering, Norwegian University of Science and Technology, Trondheim.

Dong, Y., Frangopol, D.M., & Saydam, D. (2014). Sustainability of highway bridge networks under seismic hazard. *Journal of Earthquake Engineering, 18*, 41–66.

Dong, Y., Frangopol, D.M., & Saydam, D. (2015). Pre-earthquake probabilistic retrofit optimization of bridge networks based on sustainability. *Journal of Bridge Engineering, 19*(6), 1–10.

Dong, Y., Frangopol, D. M., and Sabatino, S. (in press). Optimizing bridge network retrofit planning based on cost-benefit evaluation and multi-attribute utility associated with sustainability. *Earthquake Spectra*. doi:10.1193/012214EQS015M

Ellingwood, B.R. (2001). Acceptable risk bases for design of structures. *Progress in Structural Engineering and Materials, 3*, 170–179.

Ellingwood, B.R. (2005). Risk-informed condition assessment of civil infrastructure: State of practice and research issues. *Structure and Infrastructure Engineering, 1*, 7–18.

FHWA. (2013). *Deficient bridges by state and highway system 2013*. Washington, DC: Author. http://www.fhwa.dot.gov/bridge/nbi/no10/defbr13.cfm

Frangopol, D.M. (1999). Life-cycle cost analysis for bridges. Chapter 9. In D.M. Frangopol (Ed.), *Bridge safety and reliability* (pp. 210–236). Reston, VA: ASCE.

Frangopol, D.M. (2011). Life-cycle performance, management, and optimisation of structural systems under uncertainty: Accomplishments and challenges. *Structure and Infrastructure Engineering, 7*, 389–413.

Frangopol, D.M., & Das, P.C. (1999). Management of bridge stocks based on future reliability and maintenance costs. In: P. C. Das, D. M. Frangopol & A. S. Nowak (Eds.), *Current and future trends in bridge design, construction, and maintenance* (pp. 45–58). London: The Institution of Civil Engineers, Thomas Telford.

Frangopol, D.M., & Estes, A.C. (1997). Lifetime bridge maintenance strategies based on system reliability. *Structural Engineering International, 7*, 193–198.

Frangopol, D.M., & Liu, M. (2007). Maintenance and management of civil infrastructure based on condition, safety, optimization, and life-cycle cost. *Structure and Infrastructure Engineering, 3*, 29–41.

Frangopol, D.M., & Soliman, M. (in press). Life-cycle of structural systems: Recent achievements and future directions. *Structure and Infrastructure Engineering*. doi:10.1080/15732479.2014.999794

Heidenreich, W.F., Luebeck, E.G., & Moolgavkar, S.H. (1997). Some properties of the hazard function of the two-mutation clonal expansion model. *Risk Analysis, 17*, 391–399.

Hessami, A.G. (1999). Risk management: A systems paradigm. *Systems Engineering, 2*, 156–167.

Horová, I., Pospíšil, Z., & Zelinka, J. (2009). Hazard function for cancer patients and cancer cell dynamics. *Journal of Theoretical Biology, 258*, 437–443.

Hoyland, A., & Rausand, M. (1994). *System reliability theory: Models and statistical methods*. New York, NY: Wiley.

Jiang, R., & Murthy, D.N.P. (1995). Reliability modeling involving two Weibull distributions. *Reliability Engineering & System Safety, 47*, 187–198.

Keeney, R.L., & Raiffa, H. (1993). *Decisions with multiple objectives: Preferences and value tradeoffs*. Cambridge: Cambridge University Press.

Lai, C.D., & Murthy, D.N.P. (2003). A modified Weibull distribution. *IEEE Transactions on Reliability, 52*, 33–37.

Lawson, L., & Chen, E.Y. (1995). Microcracks: The hazard function and reliability inspection. *Journal of Testing and Evaluation, 23*, 315–318.

Leemis, L.M. (1995). *Reliability probabilistic models and statistical methods*. Englewood Cliffs, NJ: Prentice Hall.

MathWorks. (2013). *Statistics ToolboxTM 7 User's Guide*. Natick, MA: The Math Works.

Neves, L.A.C., Frangopol, D.M., & Cruz, P.J.S. (2006). Probabilistic lifetime-oriented multiobjective optimization of bridge maintenance: Single maintenance type. *Journal of Structural Engineering, 132*, 991–1005.

Neves, L.A.C., Frangopol, D.M., & Petcherdchoo, A. (2006). Probabilistic lifetime-oriented multiobjective optimization of bridge maintenance: Combination of maintenance types. *Journal of Structural Engineering, 132*, 1821–1834.

Okasha, N.M., & Frangopol, D.M. (2009). Lifetime-oriented multi-objective optimization of structural maintenance considering system reliability, redundancy and life-cycle cost using GA. *Structural Safety, 31*, 460–474.

Okasha, N.M., & Frangopol, D.M. (2010a). Novel approach for multicriteria optimization of life-cycle preventive and essential maintenance of deteriorating structures. *Journal of Structural Engineering, 136*, 1009–1022.

Okasha, N.M., & Frangopol, D.M. (2010b). Redundancy of structural systems with and without maintenance: An approach based on lifetime functions. *Reliability Engineering & System Safety, 95*, 520–533.

Orcesi, A., & Frangopol, D.M. (2011). Use of lifetime functions in the optimization of nondestructive inspection strategies for bridges. *Journal of Structural Engineering, 137*, 531–539.

Pedersen, P.T. (2002). Collision risk for fixed offshore structures close to high-density shipping lanes. *Proceedings of the Institution of Mechanical Engineers, Part M: Journal of Engineering for the Maritime Environment, 216*, 29–44.

Sabatino, S., Frangopol, D.M., & Dong, Y. (in press). Sustainability-informed maintenance optimization of highway bridges considering multi-attribute utility and risk attitude. *Engineering Structures*. doi: 10.1016/j.engstruct.2015.07.030

Stein, S.M., Young, G.K., Trent, R.E., & Pearson, D.R. (1999). Prioritizing scour vulnerable bridges using risk. *Journal of Infrastructure Systems, 5*, 95–101.

Stewart, T.J. (1996). Robustness of additive value function methods in MCDM. *Journal of Multi-Criteria Decision Analysis, 5*, 301–309.

Stewart, M.G. (2001). Reliability-based assessment of ageing bridges using risk ranking and life cycle cost decision analyses. *Reliability Engineering and System Safety, 74*, 263–273.

Tanner, M.A., & Wong, W.H. (1984). Data-based nonparametric estimation of the hazard function with applications to model diagnostics and exploratory analysis. *Journal of the American Statistical Association, 79*, 174–182.

Thies, P.R., Smith, G.H., & Johanning, L. (2012). Addressing failure rate uncertainties of marine energy converters. *Renewable Energy, 44*, 359–367.

van Noortwijk, J.M., & Klatter, H.E. (2004). The use of lifetime distributions in bridge maintenance and replacement modelling. *Computers & Structures, 82*, 1091–1099.

Yang, S., Frangopol, D.M., & Neves, L.A.C. (2004). Service life prediction of structural systems using lifetime functions with emphasis on bridges. *Reliability Engineering & System Safety, 86*, 39–51.

Yang, S., Frangopol, D.M., & Neves, L.A.C. (2006). Optimum maintenance strategy for deteriorating bridge structures based on lifetime functions. *Engineering Structures, 28*, 196–206.

Yang, S., Frangopol, D.M., Kawakami, Y., & Neves, L.A.C. (2006). The use of lifetime functions in the optimization of interventions on existing bridges considering maintenance and failure costs. *Reliability Engineering & System Safety, 91*, 698–705.

Zhang, Z., & Li, X. (2010). Some new results on stochastic orders and aging properties of coherent systems. *IEEE Transactions on Reliability, 59*, 718–724.

Efficient multi-objective optimisation of probabilistic service life management

Sunyong Kim and Dan M. Frangopol

ABSTRACT

The inspection and maintenance plans to ensure the structural safety and extend the service life of deteriorating structures can be established effectively through an optimisation process. When several objectives are required for inspection and maintenance strategies, a multi-objective optimisation process needs to be used in order to consider all objectives simultaneously and to rationally select a well-balanced solution. However, as the number of objectives increases, additional computational efforts are required to obtain the Pareto solutions, for decision-making to select well-balanced solutions, and for visualisation of the solutions. This paper presents a novel approach to multi-objective optimisation process of probabilistic service life management with four objectives: minimising the damage detection delay, minimising the probability of failure, maximising the extended service life and minimising the expected total life-cycle cost. With these four objectives, the single, bi-, tri- and quad-objective optimisation processes are investigated using the weighted sum method and genetic algorithms. The objective reduction approach with the Pareto optimal solutions is applied to estimate the degree of conflict among the objectives, and to identify the redundant objectives and minimum essential objective set. As a result, the efficiency in decision-making and visualisation for service life management can be improved by removing the redundant objectives.

Introduction

The structural performance of a deteriorating structure can be improved by the application of appropriate inspection and maintenance actions (IAEA, 2002; Liang, Lin, & Liang, 2002; NCHRP, 2006; Stewart & Rosowsky, 1998). Efficient and effective inspection and maintenance planning can be made based on an optimisation process (Onoufriou & Frangopol, 2002; Kong & Frangopol, 2004; Luki & Cremona, 2001; Miyamoto, Kawamura, & Nakamura, 2000). In the last few decades, significant efforts have been made in the field of the optimum inspection and maintenance planning of deteriorating structures (Frangopol, 2011; Frangopol & Soliman, 2016). The formulation of optimisation for this planning requires damage occurrence and propagation prediction, and the estimation of the effects of inspection and maintenance on the service life of a deteriorating structure under uncertainty (Ellingwood & Mori, 1997; Madsen, Torhaug, & Cramer, 1991; Soliman & Frangopol, 2014). The objectives of minimising the total life-cycle cost, maximising the service life, maximising the structural performance indicators such as reliability index, availability and redundancy during the service life, and minimising the damage detection delay have been applied to determine the optimum inspection and maintenance planning (Frangopol, Lin, & Estes, 1997; Furuta, Kameda, Nakahara, Takahashi, & Frangopol, 2006; Kim & Frangopol, 2011a, 2011c, Kim, Frangopol, & Zhu, 2011; Liu, Li, Huang, & Yuanhui, 2009; Okasha & Frangopol, 2010).

Inspection and maintenance planning are functions of the optimisation problem. For example, the inspection and maintenance planning associated with the minimisation of the total life-cycle cost will not be the same as the plan associated with the minimisation of the probability of failure during the service life or the maximisation of the expected extended service life of a structure. Practically, structural managers will need to make a decision on which objectives need to be used, or how all the possible objectives will be considered in a rational way because practical structural management might require only one plan for inspection and maintenance of a single structural system (Brockhoff & Zitzler, 2009). The multi-objective optimisation provides multiple trade-off solutions for the decision-makers, and one of the obtained solutions can be selected using higher level information to be considered (Deb, 2001; Fonseca & Fleming, 1998). Most existing studies on optimum inspection and maintenance planning have been based only on a few (i.e. one, two or three) objectives (Frangopol & Kim, 2014; Frangopol & Liu, 2007). The development of new concepts and approaches for optimum inspection and maintenance planning can increase the number of objectives to be optimised simultaneously. In general, optimisation with a large number of objectives involves difficulties with respect to computation, decision-making and visualisation (Saxena, Duro, Tiwari, Deb, & Zhang, 2013; Verel, Liefooghe, Jourdan, & Dhaenens, 2011). Recently, objective reduction approaches to address these difficulties have been developed

Table 1. Objectives, design variables, required estimations and given conditions in optimum service life management.

Objective	Design variables	Required estimations	Given conditions
Minimisation of expected damage detection delay $E(t_{delay})$	Optimum inspection times	E① and E②	C① and C②
Minimisation of probability of failure P_{fail}	Optimum inspection times	E① and E②	C① and C②
Maximisation of expected extended service life $E(t_{life})$	Optimum inspection and maintenance times	E①, E② and E③	C①, C② and C③
Minimisation of expected total life-cycle cost C_{lcc}	Optimum inspection and maintenance times	E①, E②, E③ and E④	C①, C② and C③

Notes: E① = Damage occurrence/propagation under uncertainty.
E② = Relation between degree of damage and probability of damage detection of inspection method.
E③ = Effect of available maintenance on service life extension.
E④ = Life-cycle cost over time.
C① = Number and quality of inspections.
C② = Available inspection types.
C③ = Available maintenance types and damage criteria to determine the types of maintenance.

by Deb and Saxena (2006), Brockhoff and Zitzler (2009), and Singh, Isaacs, and Ray (2011), among others.

This paper deals with the efficient multi-objective optimisation of probabilistic service life management associated with four objectives: minimising the damage detection delay, minimising the probability of failure, maximising the extended service life and minimising the expected total life-cycle cost. With these objectives, the single, bi-, tri- and quad-objective optimisation processes are investigated based on both the preference-based and Pareto front-based approaches represented by the weighted sum method and genetic algorithms (GA), respectively. In the Pareto front-based approach, the objective reduction approach is applied to estimate the degree of conflict among the objectives, and to identify both the redundant objectives and the minimum essential objective set. As a result, the efficiency in decision-making and visualisation for service life management is significantly improved by removing the redundant objectives. In this paper, the multi-objective optimisation of service life management is applied to an existing reinforced concrete bridge under corrosion.

Objectives for optimum service life management

The optimum service life management of a deteriorating structure is based on objectives related to structural performance, total life-cycle cost and service life (Fu & Frangopol, 1990; Frangopol & Kim, 2014; Frangopol & Liu, 2007). In this study, minimisation of the expected damage detection delay, minimisation of the probability of failure, maximisation of extended service life and minimisation of expected total life-cycle cost are used as the objectives for optimum service life management. Table 1 shows the required estimations and conditions for optimisation formulations with these four objectives. As indicated, if only the optimum inspection plan needs to be found, and the probabilistic models are available (which represent damage occurrence and propagation over time and quantify the relation between the degree of damage and probability of damage detection of an inspection method), then the optimisation can be performed with the objective of minimising the expected damage detection delay or minimising the probability of failure. Furthermore, in the case where optimum inspection and maintenance planning needs to be addressed, considering the effect of maintenance on service life or life-cycle cost of a deteriorating structure, managers can use the objective of maximising the expected service life or minimising the expected total life-cycle cost. Detailed formulations of these four objectives are summarised in this section.

Expected damage detection delay

The damage detection delay, defined as the time-lapse from damage occurrence to the time for the damage to be detected, can be used to establish the optimum inspection plan. Maintenance actions generally follow inspection, and the type of maintenance/repair depends on the inspection results. Therefore, a reduced damage detection delay can lead to effective and timely maintenance actions. If the damage occurrence time is known, and inspection methods are perfect, there will be no damage detection delay. However, due to the uncertainties associated with damage occurrence/propagation and inspection methods, damage detection delay cannot be avoided. Considering these uncertainties, the expected damage detection delay was formulated and applied to deteriorating structures subjected to corrosion and fatigue (Kim & Frangopol, 2011a, 2011b).

When N_{insp} inspections are applied to detect damage, the expected damage detection delay $E(t_{delay})$ is formulated as:

$$E(t_{delay}) = \sum_{i=1}^{N_{insp}+1} \left(\int_{t_{insp,i-1}}^{t_{insp,i}} \left[t_{delay,i} \cdot f_T(t) \right] dt \right) \qquad (1)$$

where $t_{delay,i}$ = damage detection delay for the damage to occur in the time interval $t_{insp,i-1} \leq t < t_{insp,i}$; $t_{insp,i}$ = ith inspection time; and $f_T(t)$ = probability density function (PDF) of the damage occurrence time t. The damage detection delay $t_{delay,i}$ for a given damage occurrence time of $t_{insp,i-1} \leq t < t_{insp,i}$ is:

$$t_{delay,i} = \sum_{k=i}^{N_{insp}+1} \left(\prod_{j=1}^{k} \left(1 - P_{insp,j-1} \right) \right) \cdot P_{insp,k} \cdot \left(t_{insp,k} - t \right) \quad (2)$$

where $P_{insp,i}$ = probability of damage detection of the ith inspection, $P_{insp,0} = 0$ for $j = 1$, and $P_{insp,Ninsp+1} = 1.0$ for $k = N_{insp+1}$. This is the conditional probability that damage is detected at the jth inspection when the damage intensity at time t is $\delta(t)$.

The probability of damage detection P_{insp} can be expressed based on the normal cumulative distribution function (CDF) as (Frangopol et al., 1997):

$$P_{insp} = \Phi \left(\frac{\delta(t) - \delta_{.5}}{\sigma_\delta} \right) \qquad (3)$$

where $\Phi(\cdot)$ is the standard normal CDF, and σ_δ is the standard deviation (SD) of the damage intensity δ. The damage intensity δ ranges from 0 (i.e. no damage) to 1.0 (i.e. full

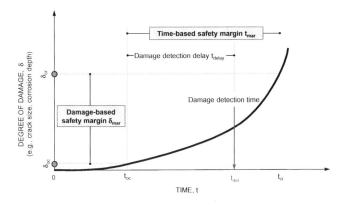

Figure 1. Damage-based and time-based safety margins.

damage), and can be estimated in terms of crack size or corrosion depth. σ_δ is assumed to be $.1\delta_{.5}$ herein, where $\delta_{.5}$ is the damage intensity at which the inspection method has a probability of detection of .5. The quality of an inspection method is represented by $\delta_{.5}$. The times $t_{\text{insp}, 0}$ and $t_{\text{insp}, N\text{insp} + 1}$ in Equation (1) are defined as:

$$t_{\text{insp}, 0} = F_T^{-1}(\Phi(-u)) \qquad (4a)$$

$$t_{\text{insp}, N_{\text{insp}+1}} = F_T^{-1}(\Phi(u)) \qquad (4b)$$

where $F_T^{-1}(\cdot)$ = inverse CDF of the damage occurrence time t, and $u > 0$, and $t_{\text{insp}, 0}$ and $t_{\text{insp}, N\text{insp} + 1}$ serve as the lower and upper bounds of damage occurrence time t, respectively. These bounds can be determined, assuming that the probabilities of damage occurrence and damage detection are both very small before the lower bound $t_{\text{insp}, 0}$, and that these probabilities are very close to one near the upper bound $t_{\text{insp}, N\text{insp} + 1}$. For example, if the damage occurrence time t is represented by the normal distribution with the mean of 5 years and SD of 1 year, and u in Equations (4a) and (4b) is 3.0, $t_{\text{insp}, 0}$ and $t_{\text{insp}, N\text{insp} + 1}$ are 2 and 8 years, respectively (i.e. $F_T^{-1}(\Phi(-3)) = 2$, $F_T^{-1}(\Phi(3)) = 8$). The probabilities that the damage will occur before 2 and 8 years are .0013 and .9987, respectively (i.e. $F_T(2) = .0013$, $F_T(8) = .9987$). In this study, u is assumed to be 3.0 (Kim & Frangopol, 2011a). The expected damage detection delay $E(t_{\text{delay}})$ can be reduced by increasing the number of inspections and/or inspection quality. Detailed application and results regarding minimising the expected damage detection delay for the optimum inspection planning can be found in Kim and Frangopol (2011a, 2011b).

Time-based safety margin and probability of failure

During the evolution of the damage process, the safety margin of deteriorated structures can be expressed by both damage intensity and time. The safety margin based on the damage intensity δ_{mar} is expressed as:

$$\delta_{\text{mar}} = \delta_{cr} - \delta_{oc} \qquad (5)$$

where δ_{cr} = critical damage intensity that can lead to structural failure; and δ_{oc} = damage intensity when the damage can be detected. A time-based safety margin t_{mar} is defined as:

$$t_{\text{mar}} = t_{cr} - t_{oc} \qquad (6)$$

where t_{cr} = time when the damage intensity reaches the critical damage (i.e. damage threshold) δ_{cr}; and t_{oc} = damage occurrence time.

As shown in Figure 1, the damage intensities δ_{cr} and δ_{oc} in Equation (5) are associated with t_{cr} and t_{oc}, respectively. The damage has to be detected before reaching δ_{cr}. This means that the damage detection delay $t_{\text{delay}} = t_{\text{det}} - t_{oc}$ should be less than t_{mar} of Equation (6), where t_{det} is the time for the damage to be detected. Accordingly, assuming that the appropriate and immediate maintenances are applied after damage is detected, the time-based state function g can be expressed as:

$$g = t_{cr} - t_{\text{det}} = t_{\text{mar}} - t_{\text{delay}} \qquad (7)$$

The relations among the critical time t_{cr}, damage detection time t_{det}, time-based safety margin t_{mar} and damage detection delay t_{delay} are illustrated in Figure 1. Considering the uncertainties associated with damage detection and damage initiation/propagation, t_{cr}, t_{det} t_{mar} and t_{delay} in Equation (7) can be treated as random variables. The time-based probability of failure is defined as:

$$P_{\text{fail}} = P(t_{cr} - t_{\text{det}} \leq 0) = P(t_{\text{mar}} - t_{\text{delay}} \leq 0) \qquad (8)$$

The inspection plan can be established based on the optimisation problem with the objective of minimising the time-based probability of failure P_{fail}. Increasing the number of inspections and/or inspection quality can lead to a reduction of P_{fail}.

Extended service life under uncertainty

The service life of a deteriorating structure is limited; it can be extended through appropriate and timely maintenance actions. The extended service life with inspection and maintenance is formulated using the decision tree model shown in Figure 2. Considering the damage occurrence time, inspection time, damage detection and maintenance types, the extended service life after the ith scheduled inspection $t_{\text{life}, i}$ is formulated as:

$$t_{\text{life}, i} = (1 - P_{\text{insp}}) \cdot t_{\text{life}, i-1} \qquad (9a)$$
$$+ P_{\text{insp}} \cdot (t_{\text{life}, i-1} + t_{ex}) \quad \text{for } t_{\text{insp}, i} \leq t_{\text{life}, i-1}$$

$$t_{\text{life}, i} = t_{\text{life}, i-1} \quad \text{for } t_{\text{insp}, i} > t_{\text{life}, i-1} \qquad (9b)$$

where t_{ex} = service life extension after damage detection. If the damage is detected by the scheduled inspection at time $t_{\text{insp}, i}$, the damage intensity is measured by the in-depth inspection following the scheduled inspection, and a maintenance type is determined according to the damage intensity δ. Furthermore, if the damage is not detected or the inspection application time is later than the service life, the service life will not be extended (i.e. $t_{ex} = 0$) as shown in Equation (9b) and Figure 2. When two maintenance types are available and the service life extensions associated with these two maintenance types are $t_{ex, I}$ and $t_{ex, II}$, respectively, the service life extension t_{ex} is estimated as:

$$t_{ex} = 0 \quad \text{for } \delta(t_{\text{insp}, i}) \leq \delta_I \qquad (10a)$$

$$t_{ex} = t_{ex, I} \text{ for } \delta_I < \delta(t_{\text{insp},i}) \leq \delta_{II} \qquad (10b)$$

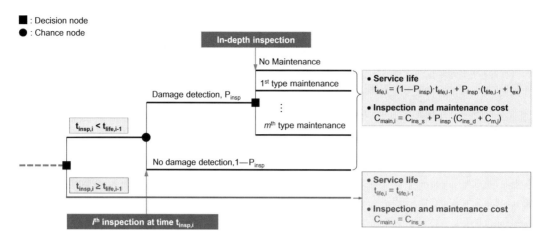

Figure 2. Decision tree for extended service life and the associated cost formulation.

$$t_{ex} = t_{ex,\text{II}} \quad \text{for } \delta(t_{\text{insp},i}) > \delta_{\text{II}} \qquad (10c)$$

where δ_{I} and δ_{II} are the damage thresholds. In the case of $\delta(t_{\text{insp},i}) \leq \delta_{\text{I}}$, no maintenance is applied, and t_{ex} becomes zero.

Life-cycle cost analysis

The life-cycle cost analysis has been applied in various fields of engineering for effective and efficient inspection and maintenance management considering structural performance and service life under uncertainty (Frangopol, Lin et al., 1997; Frangopol, Saydam, & Kim, 2012). Decision-makers can allocate the optimum financial resources and establish the optimum inspection and maintenance actions while minimising the life-cycle cost during the service life of a deteriorating structure (Kong & Frangopol, 2003). The life-cycle cost analysis requires evaluating and integrating the initial construction cost C_{ini}, total inspection and maintenance cost C_{main} and the expected failure cost C_{fail} of deteriorating structures. The expected total life-cycle cost C_{lcc} during the target lifetime t_{life}^{*} can be expressed as (Chang & Shinozuka, 1996; Enright & Frangopol, 1999; Frangopol et al., 1997; Thoft-Christensen & Sørensen, 1987; Wallbank, Tailor, & Vassie, 1999):

$$C_{\text{lcc}} = C_{\text{ini}} + C_{\text{main}} + C_{\text{fail}} \qquad (11)$$

Using the decision tree shown in Figure 2, the inspection and maintenance cost $C_{\text{main},i}$ associated with ith inspection time $t_{\text{insp},i}$ is:

$$C_{\text{main},i} = C_{\text{ins\_s}} + P_{\text{insp}}(C_{\text{ins\_d}} + C_{m,i}) \quad \text{for } t_{\text{insp},i} \leq t_{\text{life},i-1} \quad (12a)$$

$$C_{\text{main},i} = C_{\text{ins\_s}} \quad \text{for } t_{\text{insp},i} > t_{\text{life},i-1} \qquad (12b)$$

where $C_{\text{ins\_s}}$ = cost of the scheduled inspection; $C_{\text{ins\_d}}$ = cost of the in-depth inspection; and $C_{m,i}$ = maintenance cost. As mentioned previously, a maintenance type is determined according to the damage intensity δ measured by the in-depth inspection following the scheduled inspection, and the maintenance cost $C_{m,i}$ in Equation (12a) is expressed in a similar way when the service life extension t_{ex} is estimated as indicated in Equations (10a)–(10c).

When the number of scheduled inspections N_{insp} is available, the total inspection and maintenance cost C_{main} in Equation (11) is:

$$C_{\text{main}} = \sum_{i=1}^{N_{\text{insp}}} C_{\text{main},i} \qquad (13)$$

Furthermore, the expected failure cost C_{fail} in Equation (11) is computed as:

$$C_{\text{fail}} = C_{f} P_{\text{fail}} \qquad (14)$$

where C_{f} is the expected monetary loss associated with structural failure. P_{fail} is the probability that the extended service life $t_{\text{life},i}$ after the ith inspection is less than the target lifetime t_{life}^{*}, computed as:

$$P_{\text{fail}} = P(t_{\text{life},i} \leq t_{\text{life}}^{*}) \qquad (15)$$

Efficient multi-objective optimisation

Depending on the objective of the optimisation process, the inspection and maintenance planning can be varied. Only one inspection and maintenance strategy is able to be applied for a single structural system. For this reason, the structural managers should: (a) select the most appropriate objective among the possible objectives to establish the best inspection and maintenance plan through the single objective optimisation process, or (b) consider all the possible objectives simultaneously using the multi-objective optimisation approach so that the structural managers select the most well-balanced solution among the multiple trade-off solutions. In general, the methods used to solve the multi-objective optimisation problem are categorised into preference-based and Pareto front-based approaches, which are represented by the weighted sum method and evolution algorithm, respectively (Saxena et al., 2013; Verel et al., 2011).

Preference-based approach

The most common preference-based approach is the weight sum method, in which a relative preference vector is used to convert multiple objectives into a single objective. The weight sum method with M objective functions to be minimised can be formulated as (Arora, 2012):

$$\text{Minimise } f_w(\mathbf{x}) = \sum_{i=1}^{M} w_i f_i^{\text{norm}}(\mathbf{x}) \qquad (16)$$

where $\mathbf{x}$ = vector of design variables; f_i^{norm} = ith normalised objective function; and w_i = weight factor reflecting the relative preference among the objective set $\Phi_M = \{f_1, f_2, ..., f_M\}$ such that $\sum_{i=1}^{M} w_i = 1$ and $w_i \geq 0$. In order to estimate in a rational way the weight factor, fuzzy preference approach can be used. More detailed information on this approach can be found in Parmee, Cvetkovic, Watson, and Bonham (2000). The normalised objective function $f_i^{\text{norm}}(\mathbf{x})$ is:

$$f_i^{\text{norm}}(\mathbf{x}) = \frac{f_i(\mathbf{x}) - f_i^{\text{min}}}{f_i^{\text{max}} - f_i^{\text{min}}} \qquad (17)$$

where $f_i(\mathbf{x})$ is the ith objective function; and f_i^{max} and f_i^{min} are the maximum and minimum of all the $f_i(\mathbf{x})$ values, respectively. It is known that the weight sum method is easy to use even though the number of objectives is large. However, the complete Pareto optimal set cannot be obtained, even though the continuous weight factor w_i is applied (Arora, 2012; Deb, 2001).

Pareto front-based approach

In a Pareto front-based approach, a number of Pareto optimal solutions (i.e. Pareto front) are obtained (Jin & Sendhoff, 2008). Multi-objective genetic algorithms (MOGA) has been known as the most common and attractive Pareto front-based approach (Singh et al., 2011). The main reasons for using MOGA are: (a) it is not necessary to consider the convexity, concavity and/or continuity of the objective function and (b) the representative Pareto front can be obtained after a sufficient number of generations (Arora, 2012). More detailed information on MOGA is summarised in Deb (2001).

Recent studies on MOGA (Deb & Saxena, 2006; Purshouse & Fleming, 2007; Teytaud, 2007) show that increasing the number of objectives leads to high computational cost, low search efficiency and difficulty in visualisation and decision-making. With an increase in the number of objectives, the dimensionality of the Pareto front increases, the required number of points needed to represent the Pareto optimal front increases and the Pareto front searching ability of MOGA deteriorates. Even though the data used to represent the Pareto front are computed, visualising more than a three-dimensional (3D) Pareto front will be difficult for proper decision-making.

Objective reduction approach

In order to address the aforementioned difficulties, several objective reduction approaches have been developed recently. The objective reduction approach is the process to identify the essential and redundant objectives among the original objective set (Brockhoff & Zitzler, 2009; Deb & Saxena, 2006; Singh et al., 2011). The Pareto front is only affected by the essential objectives. The redundant objectives can be removed without changing the Pareto front obtained with the original objectives. In general, the objective reduction approach is used during the Pareto front search process (Deb & Saxena, 2006; Singh et al., 2011) or during the decision-making process as a posteriori (Brockhoff & Zitzler,

2009; Saxena et al., 2013). Two recent approaches for objective reduction are summarised in this section.

Correlation-based objective reduction approach

The multi-objective optimisation provides the Pareto front, only when the objectives of multi-criteria decision analysis conflict with each other (Deb, 2001). According to Carlsson and Fullér (1995), the relations between the two objectives of minimising both f_1 and f_2 are defined as:

(i) f_1 is in conflict with f_2 on $\mathbf{X}$ if $f_1(x_1) \geq f_1(x_2)$ implies $f_2(x_1) \leq f_2(x_2)$ for all $x_1, x_2 \in \mathbf{X}$.

(ii) f_1 supports f_2 on $\mathbf{X}$ if $f_1(x_1) \geq f_1(x_2)$ implies $f_2(x_1) \geq f_2(x_2)$ for all $x_1, x_2 \in \mathbf{X}$.

(iii) f_1 and f_2 are independent on $\mathbf{X}$ otherwise,

where $\mathbf{X}$ is a feasible solution set.

The correlation coefficient can be used to quantify the conflict among the objectives (Jaimes, Aguirre, Tanaka, & Coello, 2010). If the correlation coefficient between the objective functions f_1 and f_2 is negative, f_1 decreases while f_2 increases and vice versa. This means that f_1 and f_2 conflict with each other. On the other hand, a positive correlation coefficient indicates that f_1 and f_2 support each other because f_1 and f_2 decrease or increase simultaneously. The objective space of f_1 and f_2, and the possible Pareto front associated with the above cases of (i), (ii) and (iii) are illustrated in Figure 3. When f_1 supports f_2 with the correlation coefficient of 1.0 (see Figures 3(b)), a single optimum solution is available instead of the Pareto front. Therefore, in this case, f_1 or f_2 is a redundant objective, because the optimum solution is not changed, even though one of these objectives is removed. Based on this concept, the correlation-based objective reduction approaches, such as the principal component analysis (Deb & Saxena, 2006; Saxena et al., 2013) and objective partitioning method (Jaimes et al., 2010), have been developed and applied to search the Pareto front using MOGA.

Dominance relation-based objective reduction approach

The objective reduction approach developed by Brockhoff and Zitzler (2006, 2009) is based on the dominance relation among the objective values. In this approach, a solution $\mathbf{x}_1 \in \mathbf{X}$ is said to dominate another solution $\mathbf{x}_2 \in \mathbf{X}$ if $\mathbf{x}_1 \prec_{\Phi^*} \mathbf{x}_2$ and $\mathbf{x}_2 \not\prec_{\Phi^*} \mathbf{x}_1$ for a particular objective function set $\Phi^* \subseteq \Phi_M := \{f_1, f_2, ..., f_M^{\Phi^*}\}$, in which f_i is the ith objective function among M objective functions, $\mathbf{X}$ is the design space and $\subseteq$ denotes the subset. All the objective functions of Φ_M have to be minimised. The notation $\mathbf{x}_1 \prec_{\Phi^*} \mathbf{x}_2$ is defined as:

$$\mathbf{x}_1 \prec_{\Phi^*} \mathbf{x}_2 \colon \Leftrightarrow \forall f_i \in \Phi^* \colon f_i(\mathbf{x}_1) \leq f_i(\mathbf{x}_2) \qquad (18)$$

If no solution can be found that dominates $\mathbf{x}^* \in \mathbf{X}$, the solution $\mathbf{x}^*$ is called the Pareto optimal. An objective of minimising f_i is considered to be redundant and non-conflicting with the other objectives in a given objective set Φ_M, when the objective function f_i can be removed without changing the dominance relation among the objective values. Since the dominance relation is represented by the Pareto front, it can be said that the Pareto front remains unchanged, when the redundant objectives are removed (Brockhoff & Zitzler, 2006, 2009).

(a)

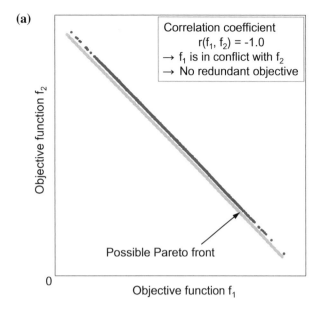

(b)

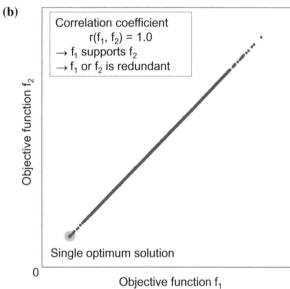

(c)

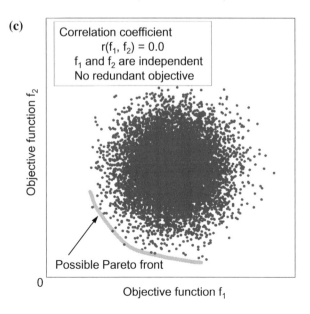

Figure 3. Objective space and Pareto front for minimising f_1 and f_2: (a) conflict; (b) support; (c) independence.

The dominance relation-based objective reduction approach is useful for investigating the effect of adding or removing a particular objective set $\Phi^* \subseteq \Phi_M$ on the Pareto front, to estimate the degree of conflict between Φ^* and Φ_M, and to find the redundant objectives and minimum essential objective set. The degree of conflict between Φ^* and Φ_M is expressed as the extent Δ to which the Pareto front changes when Φ^* is removed. Δ needs to be normalised with objectives that have various units and orders of magnitudes. The normalised degree of conflict is denoted as Δ_{norm}. The value of Δ_{norm} ranges from 0 (i.e. non-conflict) to 1.0 (i.e. full-conflict). In this paper, the algorithms associated with the dominance relation-based reduction approach, which are provided by Brockhoff and Zitzler (2006, 2009), are applied for an efficient decision-making process with the Pareto optimal solution set obtained from MOGA.

Application to an existing highway bridge

The approach for efficient service life management using multi-objective optimisation is applied to the I-39 Northbound Bridge over the Wisconsin River. More information on this bridge is given in Mahmoud, Connor, and Bowman (2005) and Kim et al. (2011). This application focuses on the corrosion of the top transverse reinforcement bars of the interface between spans, where the maximum negative moment can occur as indicated in Kim et al. (2011).

Initial service life estimation

The corrosion initiation and propagation under uncertainty are estimated based on both Fick's second law (Crank, 1975) and the pitting corrosion model (Val & Melchers, 1997). For a given corrosion rate r_{corr} (mm/year), the maximum pit depth $PT(t)$ at time t is expressed as (Val & Melchers, 1997):

$$PT(t) = r_{corr}R(t - t_{oc}) \quad \text{for } t > t_{oc} \tag{19}$$

where R = ratio of maximum pit depth to average pit depth; t_{oc} = corrosion initiation time (mm/year). Moreover, the damage intensity $\delta(t)$ based on the pitting corrosion model is (Kim et al., 2011):

$$\delta(t) = \frac{PT(t)}{d_0} \tag{20}$$

where d_0 (mm) = initial diameter of a reinforcement bar. In this paper, the initial service life is considered to be the time when the maximum pit depth PT reaches the allowable maximum pit depth PT_{allow} = 4.43 mm. The random variables to estimate the initial service life are provided in Table 2. Using Equation (19) and Monte Carlo simulation with 10^6 samples, the initial service life $t_{life, 0}$ when the PT reaches PT_{allow} is computed with a mean and SD of 22.14 and 4.91 years, respectively (denoted as LN (22.14; 4.91 years), where LN denotes lognormal distribution.

Formulation of objective functions

In this study, four objectives are used: (a) minimising the expected damage detection delay $f_1 = E(t_{delay})$; (b) minimising the time-based probability of failure $f_2 = P_{fail}$; (c) maximising the

Table 2. Variables for estimating initial service life.

Random variables	Notation and units	Mean	Coefficient of variation	Type of distribution
Corrosion initiation time	t_{oc} (years)	8.59	.27	Lognormal
Rate of corrosion	r_{corr} (mm/year)	.06	.3	Lognormal
Coefficient representing ratio between maximum and average corrosion penetrations	R	6	.1	Normal
Initial diameter of reinforcement	d_0	19.05	.02	Lognormal
Allowable maximum pit depth	PT_{allow} (mm)	4.43	–	Deterministic

Notes: Based on information provided in González, Andrade, Alonso, and Feliu (1995), Torres-Acosta and Martinez-Madrid (2003), and Kim, Frangopol, and Zhu (2011).

expected extended service life $f_3 = E(t_{life})$; and (d) minimising the expected total life-cycle cost $f_4 = C_{lcc}$. In this application, corrosion initiation is considered as the damage occurrence time. As defined in Equation (1), $E(t_{delay})$ can be formulated using the PDF $f_T(t)$ of corrosion initiation time in Table 2, probability of corrosion damage detection P_{insp} (see Equation (3)) and damage intensity (see Equation (20)). The objective function P_{fail} is defined in Equation (8), where the time-based safety margin t_{mar} is the difference between the time associated with the critical degree of corrosion damage t_{cr} and corrosion initiation time t_{oc} as shown in Figure 1. The initial service life $t_{life,0} = t_{cr}$ is considered lognormal with a mean and SD of 22.14 and 4.91 years, respectively. The descriptors of random variable t_{oc} are provided in Table 2.

The formulation of $E(t_{life})$ is based on Equations (9a) and (9b) considering the uncertainty associated with the corrosion damage occurrence and propagation. In this application, it is assumed that when the damage intensity δ is between damage intensity bounds $\delta_I = .03$ and $\delta_{II} = .05$, corrosion protection is used, and the associated service life extension t_{ex} is treated as a lognormal random variable: LN (6; .5 years). If the damage intensity δ is larger than $\delta_{II} = .05$, the deck repair is applied, and the original structural performance is expected (i.e. $t_{ex} = t_{life,0}$) (Kim, Frangopol, & Soliman, 2013).

Furthermore, in order to formulate the objective function of the expected total life-cycle cost C_{lcc} (see Equation (11)), it is assumed that the initial cost C_{ini}, in-depth inspection C_{ins} and expected monetary loss C_f are 1000, 20 and 10,000, respectively. The scheduled inspection cost C_{ins_s} in Equations (12a) and (12b) was estimated by Mori and Ellingwood (1994):

$$C_{ins\_s} = C_{ins}(1 - .7\delta_{.5})^{10} \qquad (21)$$

where $C_{ins} = 30$. As mentioned previously, the damage intensity $\delta_{.5}$ at which the given inspection method has 50% probability of damage detection represents the quality of inspection in this paper. The maintenance costs $C_{m,i}$ for corrosion protection and deck repair are assumed 60 and 300, respectively (Kim et al., 2013).

Correlation among objective functions

As mentioned previously, the coefficient of correlation can be used to quantify the conflict among objective functions. The coefficient of correlation between the objective functions f_i and f_j $r(f_i, f_j)$ is defined as:

$$r(f_i, f_j) = \frac{E\left[\{f_i(\mathbf{x}) - E(f_i(\mathbf{x}))\} \cdot \{f_j(\mathbf{x}) - E(f_j(\mathbf{x}))\}\right]}{\sigma_i \cdot \sigma_j} \qquad (22)$$

where $E(\cdot)$ denotes the expected value, $\mathbf{x}$ is the variable in the design space, and σ_i and σ_j are the SDs of $f_i(\mathbf{x})$ and $f_j(\mathbf{x})$, respectively. When $\delta_{.5}$, representing the quality of inspections, is .03, and the design space is associated with the time interval between the inspection times (i.e. 1 year $\leq t_{insp,i} - t_{insp,i-1} \leq 20$ years), the correlation coefficients among the four objective functions of $f_1 = E(t_{delay})$, $f_2 = P_{fail}$, $f_3 = E(t_{life})$ and $f_4 = C_{lcc}$ are computed as listed in Table 3. Considering the two objective functions of f_1 and f_2, the correlation coefficients for one, two or three inspections, $N_{insp} = 1, 2$ and 3, are .939, .926 and .929, respectively.

Single-objective optimisation

The single-objective optimisation problem with f_1, f_2, f_3 or f_4 is formulated as:

$$\text{Find } t_{insp} = \{t_{insp,1}, t_{insp,2}, \dots, t_{insp,N_{insp}}\} \qquad (23a)$$

$$\text{for minimising } f_1 = E(t_{delay}) \qquad (23b)$$

$$\text{or for minimising } f_2 = P_{fail} \qquad (23c)$$

$$\text{or for maximising } f_3 = E(t_{life}) \qquad (23d)$$

$$\text{or for minimising } f_4 = C_{lcc} \qquad (23e)$$

$$\text{such that } 1 \text{ year} \leq t_{insp,i} - t_{insp,i-1} \leq 20 \text{ years} \qquad (23f)$$

$$\text{given } N_{insp} = 2 \text{ and } \delta_{.5} = .03 \qquad (23g)$$

The design variables of this problem are the inspection times t_{insp} (years). The time interval between inspections has to be between 1 year and 20 years (see Equation (23f)). The number of inspections $N_{insp} = 2$ and $\delta_{.5} = .03$ representing the quality of inspections are given as indicated in Equation (23g). The optimisation problem was solved using the global optimisation toolbox (i.e. GA) of MATLAB® version R2015b (MathWorks, 2015) with 200 generations.

Figure 4 depicts the optimum inspection times with objectives of Equations (23b)–(23e). If two inspections with $\delta_{.5} = .03$ are applied to minimise $E(t_{delay})$, the inspection must be performed at 10.13 and 13.12 years, and the corresponding $E(t_{delay})$ will be 3.65 years. Two inspections applied at 17.02 and 34.04 years

Table 3. Coefficients of correlation among objective functions.

		f_1	f_2	f_3	f_4
Number of inspections	f_1	1.0	.939	−.338	.931
$N_{insp} = 1$	f_2	.939	1.0	−.469	.954
	f_3	−.338	−.469	1.0	−.650
	f_4	.931	.954	−.650	1.0
Number of inspections	f_1	1.0	.926	−.074	.300
$N_{insp} = 2$	f_2	.926	1.0	−.172	.421
	f_3	−.074	−.172	1.0	−.794
	f_4	.300	.421	−.794	1.0
Number of inspections	f_1	1.0	.929	−.190	−.055
$N_{insp} = 3$	f_2	.929	1.0	−.323	.061
	f_3	−.190	−.323	1.0	−.786
	f_4	.055	.061	−.786	1.0

Notes: f_1 = expected damage detection delay $E(t_{delay})$.
f_2 = probability of failure P_{fail}.
f_3 = expected extended service life $E(t_{life})$.
f_4 = expected total life-cycle cost C_{lcc}.

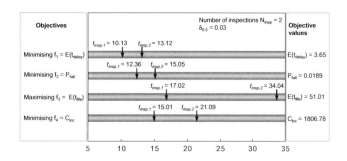

Figure 4. Inspection times based on single-objective optimisations.

result in the maximum expected extended service life $E(t_{life})$ of 51.01 years.

Bi-objective optimisation

The inspection planning can be formulated as the bi-objective optimisation consisting of three combinations of two objectives among the four objective functions f_1, f_2, f_3 and f_4 (i.e. $\{f_1, f_2\}$, $\{f_1, f_3\}$ and $\{f_1, f_4\}$). Such bi-objective optimisation for optimum inspection planning is:

$$\text{Find } t_{insp} = \{t_{insp, 1}, t_{insp, 2}, \ldots, t_{insp, N_{insp}}\} \quad (24a)$$

$$\text{for } \Phi_2 := \{f_1, f_2\} = \text{minimising both } E(t_{delay}) \text{ and } P_{fail} \quad (24b)$$

$$\text{or for } \Phi_2 := \{f_1, f_3\} = \text{minimising } E(t_{delay}) \text{ and maximising } E(t_{life}) \quad (24c)$$

$$\text{or for } \Phi_2 := \{f_1, f_4\} = \text{minimising both } E(t_{delay}) \text{ and } C_{lcc} \quad (24d)$$

The design variables are the inspection times t_{insp}. The constraints and given conditions of this problem are the same as those in Equations (23f) and (23g).

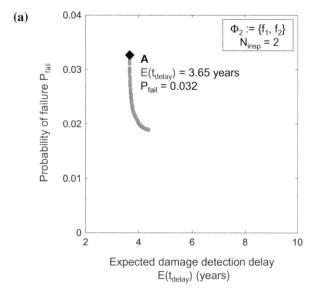

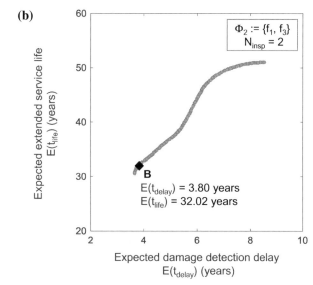

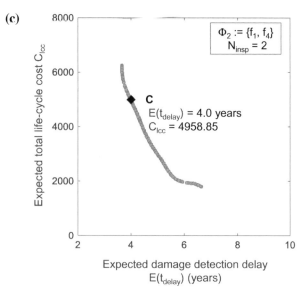

Figure 5. Pareto solution sets of bi-objective optimisation: (a) $\Phi_2 := \{f_1, f_2\}$; (b) $\Phi_2 := \{f_1, f_3\}$; (c) $\Phi_2 := \{f_1, f_4\}$.

Table 4. Design variable and objective function values associated with Pareto optimum solutions in Figures 5–8.

Pareto opti-mum solution	Objective function values				Design variables			Given condi-tions	
	Expected damage detection delay $E(t_{delay})$	Probability of failure P_{fail}	Expected extended service life (years) $E(t_{life})$	Expected total life-cy-cle cost C_{lcc}	Optimum inspection times (years)			Number of inspections N_{insp}	Number of objectives to be consid-ered
					$t_{insp,\,1}$	$t_{insp,\,2}$	$t_{insp,\,3}$		
A	3.65	.032	–	–	10.13	13.12	–	2	2
B	3.80	–	32.02	–	10.26	14.30	–	2	2
C	4.00	–	–	4958.85	11.46	14.93	–	2	2
D	3.65	.032	30.55	–	10.13	13.12	–	2	3
E	4.82	.056	30.17	1497.66	11.84	–	–	1	4
F	3.65	.032	30.55	6251.06	10.13	13.12	–	2	4
G	3.13	.023	30.29	10,624.2	9.32	11.36	13.91	3	4

In this application, MOGA provided in MATLAB® version R2015b (MathWorks, 2015) is used to solve the multi-objective optimisation. After 300 generations with 100 populations, the Pareto optimal set are found as shown in Figure 5. Table 4 provides the values of objective functions and design variables of three representative solutions (A, B and C) in Figure 5. If the bi-objective optimisation with minimising both $E(t_{delay})$ and P_{fail} is considered for inspection planning, the Pareto optimal solutions are shown in Figure 5(a). Solution A leads to $E(t_{delay})$ = 3.65 years and P_{fail} = .032. The inspection plan for solution A requires two inspections with $\delta_{.5}$ = .03 at 10.13 and 13.12 years.

As indicated in Table 3, the correlation coefficient between these two objectives is .926. Consequently, these two objectives support each other, but are not perfectly correlated. Therefore, the Pareto set exists as shown in Figure 5(a), and f_1 or f_2 cannot be redundant. Furthermore, when the bi-objective optimisation is formulated with both f_1 and f_3, the Pareto optimal solutions in Figure 5(b) can be obtained. For Pareto point B in Figure 5(b), the associated mean values $E(t_{delay})$ and $E(t_{life})$ are 3.80 and 32.02 years, respectively. The Pareto optimal solutions including solution C for the bi-objective optimisation formulated with Equation (24d) can be found in Figure 5(c).

Tri-objective optimisation

By considering three objectives f_1, f_2 and f_3, the tri-objective optimisation is formulated as:

$$\text{Find } t_{insp} = \{t_{insp,\,1}, t_{insp,\,2}, \ldots, t_{insp,\,N_{insp}}\} \tag{25a}$$

for $\Phi_3:\ = \{f_1, f_2, f_3\}$

$\quad$ = minimising both $E(t_{delay})$ and P_{fail}, and maximising $E(t_{life})$ (25b)

By considering N_{insp} = 2 and $\delta_{.5}$ = .03 and the constraint associated with Equation (23f), the Pareto optimal set of 300 populations is obtained through the MOGA with 1000 generations as shown in Figure 6. Figure 6(a) illustrates this Pareto solution set in the 3D Cartesian coordinate system. Also, this solution set is visualised in the parallel coordinate system as shown in Figure 6(b). This parallel coordinate plot consists of three vertical axes indicating the values of $E(t_{delay})$, P_{fail} and $E(t_{life})$ of Figure 6(a).

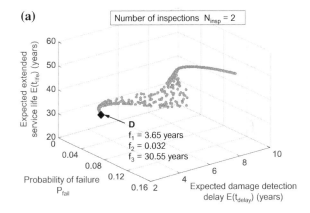

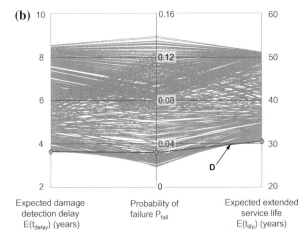

Figure 6. Pareto solution sets of tri-objective optimisation with $\Phi_3:\ = \{f_1, f_2, f_3\}$ for N_{insp} = 2: (a) 3D Cartesian coordinate system; (b) parallel coordinate system.

Each point in Figure 6(a) is represented by a polyline connecting the corresponding value on the vertical axis in Figure 6(b).

For example, if solution D in Figure 6(a) is selected for optimum inspection planning, the values of $E(t_{delay})$, P_{fail} and $E(t_{life})$ are 3.65 years, .032 and 30.55 years, respectively, and the required inspection times can be found in Table 4. In Figure 6(b), the solution D is visualised as the polyline connecting these three values (i.e. $E(t_{delay})$ = 3.65 years, P_{fail} = .032 and $E(t_{life})$ = 30.55 years) on vertical axes. In the parallel coordinate plot, the number of objectives able to be visualised is not limited (Liebscher, Witowski, & Goel, 2009).

Table 5. Optimum solutions obtained using the weighted sum method for $N_{insp} = 2$.

Normalised weighting factor				Design variables (years)		Objective values			
w_1	w_2	w_3	w_4	$t_{insp,1}$	$t_{insp,2}$	$E(t_{delay})$ (years)	C_{fail}	$E(t_{life})$ (years)	C_{lcc}
1	0	0	0	10.134	13.119	3.654	.032	30.550	6251.056
.4	.2	.2	.2	12.732	16.010	4.685	.021	33.824	3693.087
0	1	0	0	12.358	15.051	4.391	.019	33.031	4262.287
.2	.4	.2	.2	13.164	16.117	4.967	.022	34.173	3298.787
0	0	1	0	17.020	34.039	8.612	.144	51.012	3025.736
.2	.2	.4	.2	14.850	29.700	6.925	.080	49.224	2523.269
0	0	0	1	15.009	21.093	6.643	.060	36.669	1806.777
.2	.2	.2	.4	14.134	25.334	6.314	.074	43.135	2070.887

(a)

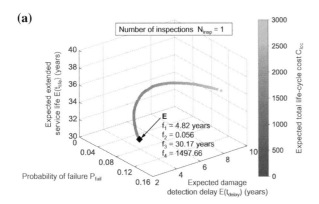

(b)

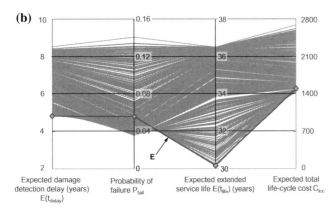

Figure 7. Pareto solution sets of quad-objective optimisation with $\Phi_4 := \{f_1, f_2, f_3, f_4\}$ for $N_{insp} = 1$: (a) 3D Cartesian coordinate system; (b) parallel coordinate system.

(a)

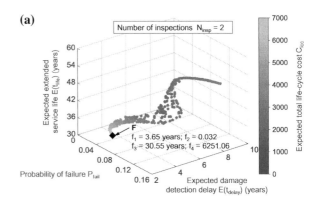

(b)

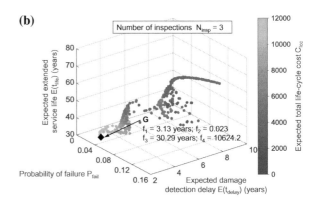

Figure 8. Pareto solution sets of quad-objective optimisation with $\Phi_4 := \{f_1, f_2, f_3, f_4\}$ in 3D Cartesian coordinate system: (a) $N_{insp} = 2$; (b) $N_{insp} = 3$.

Quad-objective optimisation

The four objectives of f_1, f_2, f_3 and f_4 are considered simultaneously as a quad-objective optimisation. This quad-objective optimisation is formulated as:

$$\text{Find } t_{insp} = \{t_{insp,1}, t_{insp,2}, \cdots, t_{insp,N_{insp}}\} \quad (26a)$$

$$
\begin{aligned}
\text{for } \Phi_4 := \{f_1, f_2, f_3, f_4\} \\
= \text{ minimising both } E(t_{delay}) \text{ and } P_{fail}, \text{ maximising} \\
E(t_{life}), \text{ and minimising } C_{lcc}
\end{aligned}
\quad (26b)
$$

$$\text{given } N_{insp} = 1, 2 \text{ or } 3 \text{ and } \delta_{.5} = .03 \quad (26c)$$

The constraint of this problem is provided by Equation (23f). The number of inspections and value of $\delta_{.5}$ are given in Equation (26c). In order to solve the quad-objective optimisation problem, both the weight sum method and MOGA are used.

Using the weight sum method with various normalised weighting factors, eight Pareto optimal solutions for the number of inspections $N_{insp} = 2$ are obtained as indicated in Table 5. When the normalised factor w_1 is equal to 1, only the objective function f_1 is taken into account for the quad-objective optimisation problem, and the associated application times of the two inspections are 10.13 and 13.12 years (see Table 5 and Figure 4).

MOGA, with the number of generations fixed at 1000 with 300 populations, is applied in order to find the Pareto optimal solution set. Because the dimensionality of the obtained Pareto solutions is four, the solutions are visualised using both the 3D Cartesian coordinate and parallel coordinate systems as shown in Figures 7 and 8. In the 3D Cartesian coordinate system, the x, y and z axes correspond to $E(t_{delay})$, P_{fail} and $E(t_{life})$, respectively,

Table 6. Normalised degree of conflict between the quad-objective set $\Phi_4 = \{f_1, f_2, f_3, f_4\}$ and the reduced objective set Φ_R.

Number of inspections $N_{insp} = 1$		Number of inspections $N_{insp} = 2$		Number of inspections $N_{insp} = 3$	
Reduced objective set Φ_R	Normalised degree of conflict Δ_{norm}	Reduced objective set Φ_R	Normalised degree of conflict Δ_{norm}	Reduced objective set Φ_R	Normalised degree of conflict Δ_{norm}
$\{f_1, f_2\}$	1.0	$\{f_1, f_2\}$	1.0	$\{f_1, f_2\}$	1.0
$\{f_1, f_3\}$	.0	$\{f_1, f_3\}$	.57	$\{f_1, f_3\}$	.64
$\{f_1, f_4\}$	1.0	$\{f_1, f_4\}$	.78	$\{f_1, f_4\}$	.64
$\{f_2, f_3\}$	.51	$\{f_2, f_3\}$	.51	$\{f_2, f_3\}$	.51
$\{f_2, f_4\}$	1.0	$\{f_2, f_4\}$	.78	$\{f_2, f_4\}$	.57
$\{f_3, f_4\}$	.21	$\{f_3, f_4\}$	1.0	$\{f_3, f_4\}$	1.0
$\{f_1, f_2, f_3\}$	.0	$\{f_1, f_2, f_3\}$	.19	$\{f_1, f_2, f_3\}$	.50
$\{f_1, f_2, f_4\}$	1.0	$\{f_1, f_2, f_4\}$	.78	$\{f_1, f_2, f_4\}$	.57
$\{f_1, f_3, f_4\}$	.0	$\{f_1, f_3, f_4\}$	.53	$\{f_1, f_3, f_4\}$	.64
$\{f_2, f_3, f_4\}$	.21	$\{f_2, f_3, f_4\}$	.50	$\{f_2, f_3, f_4\}$	.49

(a)

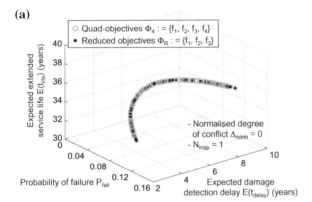

(b)

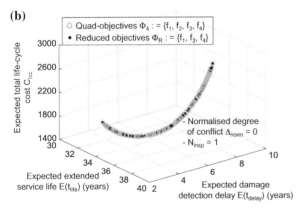

(c)

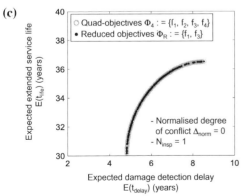

Figure 9. Pareto solution sets of the quad-objective and reduced objective optimisation problems with the normalised degree of conflict $\Delta_{norm} = 0$: (a) $\Phi_R := \{f_1, f_2, f_3\}$; (b) $\Phi_R := \{f_1, f_3, f_4\}$; (c) $\Phi_R := \{f_1, f_3\}$.

and the value of C_{lcc} is also provided. Table 4 provides the values of design variables and objective functions associated with three representative solutions for $N_{insp} = 1, 2$ and 3 (i.e. E in Figure 7(a),

F in Figure 8(a), G in Figure 8(b)). It should be noted that the inspection application times of the three representative solutions A, D and F are identical, as indicated in Table 4, and these solutions result in the minimum $E(t_{delay})$ of Figure 4 for $N_{insp} = 2$ and $\delta_{.5} = .03$.

Table 6 provides the normalised degrees of conflict among the four objectives Δ_{norm}, which are estimated through the dominance relation-based approach. As shown, the normalised degree of conflict between quad-objective set $\Phi_4 := \{f_1, f_2, f_3, f_4\}$ and the reduced objective set $\Phi_R := \{f_1, f_2, f_3\}$ is zero (i.e. $\Delta_{norm} = 0$) for $N_{insp} = 1$. Figure 9(a) illustrates a comparison between the Pareto fronts of $\Phi_4 := \{f_1, f_2, f_3, f_4\}$ and $\Phi_R := \{f_1, f_2, f_3\}$. From this result, it can be seen that there will be no change in the Pareto front when the objective f_4 is removed from the quad-objective set Φ_4. The normalised degree of conflict Δ_{norm} between $\Phi_4 := \{f_1, f_2, f_3, f_4\}$ and $\Phi_R := \{f_1, f_3, f_4\}$ is zero, and the Pareto fronts associated with $\Phi_4 := \{f_1, f_2, f_3, f_4\}$ and $\Phi_R := \{f_1, f_3, f_4\}$ are compared in Figure 9(b). This means that f_2 has no effect on the Pareto front. Furthermore, Table 6 and Figure 9(c) show that, considering the relation between $\Phi_4 := \{f_1, f_2, f_3, f_4\}$ and $\Phi_R := \{f_1, f_3\}$, the normalised degree of conflict Δ_{norm} is zero, and there will be no change if the objectives f_2 and f_4 are omitted. It can be concluded that the objectives f_2 and f_4 are redundant, the minimum essential objective set $\Phi_R := \{f_1, f_3\}$ for $N_{insp} = 1$, and in the decision-making process and visualisation for service life management, only the minimum essential objectives of f_1 and f_3 can therefore be considered instead of the four objectives of f_1, f_2, f_3 and f_4.

For $N_{insp} = 2$ and 3, any normalised degree of conflict Δ_{norm} is not zero, as shown in Table 6. However, if the allowable normalised degree of conflict $\Delta_{norm, all}$ for $N_{insp} = 2$ is .2, the minimum essential objective set can be $\Phi_R := \{f_1, f_2, f_3\}$, and the Pareto solutions computed from the tri-objective optimisation problem associated with $\Phi_R := \{f_1, f_2, f_3\}$ can be used instead of the solutions from the quad-objective optimisation.

Conclusions

In this paper, an efficient multi-objective probabilistic optimisation of service life management of deteriorating structures is presented. Through the objective reduction approach, the degree of conflict among objectives is estimated, and the minimum essential objectives are identified to improve efficiency in decision-making and visualisation. Multi-objective optimisation of service life management is investigated using four objectives: minimising the expected damage detection delay, minimising the probability of failure, maximising the extended service life and minimising the expected total life-cycle cost. The optimum inspection plans resulting from single and multi-objective optimisation processes with these four objectives are computed using the weighted sum method and GAs, respectively.

Based on the presented study, the following conclusions are drawn:

(1) The correlation among objective functions can be used as an indicator to show how the objectives are interrelated. It should be noted that the correlation coefficient does not provide sufficient information to identify the redundant objectives or the minimum essential objectives. From the correlation results provided in this paper, it can be seen that the expected damage detection delay and the probability of failure are strongly interrelated. This is because the probability of failure is based on the state function, which is expressed by the difference between the time-based safety margin and damage detection delay, and the increase in damage detection delay leads to the increase in the probability of failure. For this reason, even though the damage propagation is slow, the expected damage detection delay and probability of failure can be strongly interrelated.

(2) In the case of a single inspection, minimising the probability of failure and minimising the expected total life-cycle cost are redundant objectives. Therefore, decision-makers can use the Pareto solution set associated with minimising the expected damage detection delay and maximising the extended service life. Therefore, the required dimensions of the Pareto front decreases from four to two.

(3) As the number of objectives for service life management of deteriorating structures increases, more rational and well-balanced solutions can be expected, but more effort is required for computation to obtain the Pareto solutions, for decision-making to select well-balanced solutions, and for the visualisation of the solutions. The objective reduction approach used in this paper focuses on improving the efficiency in decision-making and visualisation. Further research is needed to integrate the algorithms for improving both the ability to search the Pareto front and the efficiency in decision-making and visualisation.

(4) This paper is limited to the four objectives associated with damage detection, probability of failure, service life and life-cycle cost. However, additional objectives for optimum inspection and maintenance planning can be considered using other structural performance indicators for deteriorating structures such as redundancy, robustness, risk, resilience and sustainability.

Acknowledgements

The opinions presented in this paper are those of the authors and do not necessarily reflect the views of the sponsoring organisations.

Disclosure statement

No potential conflict of interest was reported by the authors.

Funding

This work was supported by the National Science Foundation (NSF) [Award CMMI-1537926]; the Commonwealth of Pennsylvania, Department of Community and Economic Development, through the Pennsylvania Infrastructure Technology Alliance (PITA); the U.S. Federal Highway Administration (FHWA) [Cooperative Agreement Award DTFH61-07-H-00040]; the U.S. Office of Naval Research (ONR) [Awards N00014-08-1-0188, N00014-12-1-0023, and N00014-16-1-2299]; the National Aeronautics and Space Administration (NASA) [Award NNX10AJ20G]; and the Wonkwang University.

References

Arora, J. S. (2012). *Introduction to optimum design* (3rd ed.). London: Elsevier.

Brockhoff, D., & Zitzler, E. (2006). *Dimensionality reduction in multiobjective optimization with (partial) dominance structure preservation: Generalized minimum objective subset problems* (TIK Report 247). Zurich: ETH Zurich.

Brockhoff, D., & Zitzler, E. (2009). Objective reduction in evolutionary multiobjective optimization: Theory and applications. *Evolutionary Computation, 17*, 135–166.

Carlsson, C., & Fullér, R. (1995). Multiple criteria decision making: The case for interdependence. *Computers & Operations Research, 22*, 251–260.

Chang, S. E., & Shinozuka, M. (1996). Life-cycle cost analysis with natural hazard risk. *Journal of Infrastructure Systems, 2*, 118–126.

Crank, J. (1975). *The mathematics of diffusion* (2nd ed.). Oxford: Oxford University Press.

Deb, K. (2001). *Multi-objective optimization using evolutionary algorithms.* New York, NY: Wiley.

Deb, K., & Saxena, D. (2006, July 16–21). Searching for Pareto-optimal solutions through dimensionality reduction for certain large-dimensional multi-objective optimization problems. *Proceedings of the IEEE congress on evolutionary computation (CEC2006)*, Vancouver, Canada.

Ellingwood, B. R., & Mori, Y. (1997). Reliability-based service life assessment of concrete structures in nuclear power plants: optimum inspection and repair. *Nuclear Engineering and Design, 175*, 247–258.

Enright, M. P., & Frangopol, D. M. (1999). Maintenance planning for deteriorating concrete bridges. *Journal of Structural Engineering, 125*, 1407–1414.

Fonseca, C. M., & Fleming, P. J. (1998). Multiobjective optimization and multiple constraint handling with evolutionary algorithms. I. A unified formulation, *IEEE Transactions on Systems, Man, and Cybernetics – Part A: Systems and Humans, 28*, 26–37.

Frangopol, D. M. (2011). Life-cycle performance, management, and optimization of structural systems under uncertainty: Accomplishments and challenges. *Structure and Infrastructure Engineering, 7*, 389–413.

Frangopol, D. M., & Kim, S. (2014). Life-cycle performance analysis and optimization, chapter 18. In Wai-Fah Chen & Lian Duan (Eds.), *Bridge engineering handbook*(2nd Ed., Vol. 5, pp. 537–566). London: CRC Press-Taylor & Francis Group.

Frangopol, D. M., Lin, K. Y., & Estes, A. C. (1997). Life-cycle cost design of deteriorating structures. *Journal of Structural Engineering, 123*, 1390–1401.

Frangopol, D. M., & Liu, M. (2007). Maintenance and management of civil infrastructure based on condition, safety, optimization, and life-cycle cost. *Structure and Infrastructure Engineering, 3*, 29–41.

Frangopol, D. M., Saydam, D., & Kim, S. (2012). Maintenance, management, life-cycle design and performance of structures and infrastructures: A brief review. *Structure and Infrastructure Engineering, 8*(1), 1–25.

Frangopol, D. M., & Soliman, M. (2016). Life-cycle of structural systems: Recent achievements and future directions. *Structure and Infrastructure Engineering, 12*(1), 1–20.

Fu, G., & Frangopol, D. M. (1990). Reliability-based vector optimization of structural systems. *Journal of Structural Engineering, 116,* 2143–2161.

Furuta, H., Kameda, T., Nakahara, K., Takahashi, Y., & Frangopol, D. M. (2006). Optimal bridge maintenance planning using improved multi-objective genetic algorithm. *Structure and Infrastructure Engineering, 2,* 33–41.

González, J. A., Andrade, C., Alonso, C., & Feliu, S. (1995). Comparison of rates of general corrosion and maximum pitting penetration on concrete embedded steel reinforcement. *Cement and Concrete Research, 25,* 257–264.

IAEA. (2002). *Safety and effective nuclear power plant life cycle management towards decommissioning* (IAEA-TECDOC-1305). Vienna: International Atomic Energy Agency.

Jaimes, A., Aguirre, H., Tanaka, K., & Coello, C. A. C. (2010). Objective space partitioning using conflict information for many-objective optimization. In Robert Schaefer, Carlos Cotta, Joanna Kołodziej, & Günter Rudolph (Eds.), *Parallel problem solving from nature, PPSN XI* (pp. 657–666). Berlin: Springer.

Jin, Y., & Sendhoff, B. (2008). Pareto-based multiobjective machine learning: An overview and case studies. *Systems, Man, and Cybernetics, Part C: Applications and Reviews, 38,* 397–415.

Kim, S., & Frangopol, D. M. (2011a). Optimum inspection planning for minimizing fatigue damage detection delay of ship hull structures. *International Journal of Fatigue, 33,* 448–459.

Kim, S., & Frangopol, D. M. (2011b). Inspection and monitoring planning for RC structures based on minimization of expected damage detection delay. *Probabilistic Engineering Mechanics, 26,* 308–320.

Kim, S., & Frangopol, D. M. (2011c). Cost-based optimum scheduling of inspection and monitoring for fatigue-sensitive structures under uncertainty. *Journal of Structural Engineering, 137,* 1319–1331.

Kim, S., Frangopol, D. M., & Soliman, M. (2013). Generalized probabilistic framework for optimum inspection and maintenance planning. *Journal of Structural Engineering, 139,* 435–447.

Kim, S., Frangopol, D. M., & Zhu, B. (2011). Probabilistic optimum inspection/repair planning to extend lifetime of deteriorating RC structures. *Journal of Performance of Constructed Facilities, 25,* 534–544.

Kong, J., & Frangopol, D. M. (2003). Evaluation of expected life-cycle maintenance cost of deteriorating structures. *Journal of Structural Engineering, 129,* 682–691.

Kong, J., & Frangopol, D. M. (2004). Cost–reliability interaction in life-cycle cost optimization of deteriorating structures. *Journal of Structural Engineering, 130,* 1704–1712.

Liang, M.-T., Lin, L.-H., & Liang, C.-H. (2002). Service life prediction of existing reinforced concrete bridges exposed to chloride environment. *Journal of Infrastructure Systems, 8,* 76–85.

Liebscher, M., Witowski, K., & Goel, T. (2009, June 1–5). Decision making in multiobjective optimization for industrial applications – Data mining and visualization of Pareto data. *Proceedings of the 8th world congress on structural and multidisciplinary optimization,* Lisbon, Portugal.

Liu, Y., Li, Y., Huang, H.-Z., & Yuanhui, K. (2009). An optimal sequential preventive maintenance policy under stochastic maintenance quality. *Structure and Infrastructure Engineering, 7,* 315–322.

Luki, M., & Cremona, C. (2001). Probabilistic optimization of welded joints maintenance versus fatigue and fracture. *Reliability Engineering & System Safety, 72,* 253–264.

Madsen, H. O., Torhaug, R., & Cramer E. H. (1991). Probability-based cost benefit analysis of fatigue design, inspection and maintenance.

Proceedings of the marine structural inspection, maintenance and monitoring symposium (pp. 1–12). Arlington, VA: SSC/SNAME.

Mahmoud, H. N., Connor, R. J., & Bowman, C. A. (2005). *Results of the fatigue evaluation and field monitoring of the I-39 Northbound Bridge over the Wisconsin River* (ATLSS Report No. 05-04). Bethlehem, PA: Lehigh University.

MathWorks. (2015). *Optimization Toolbox^{TM} user's guide.* Natick, MA: Author.

Miyamoto, A., Kawamura, K., & Nakamura, H. (2000). Bridge management system and maintenance optimization for existing bridges. *Computer-Aided Civil and Infrastructure Engineering, 15,* 45–55.

Mori, Y., & Ellingwood, B. R. (1994). Maintaining reliability of concrete structures. II: Optimum inspection/repair. *Journal of Structural Engineering, 120,* 846–862.

NCHRP. (2006). *Manual on service life of corrosion-damaged reinforced concrete bridge superstructure elements* (NCHRP-Report 558). Washington, DC: Transportation Research Board, National Cooperative Highway Research Program.

Okasha, N. M., & Frangopol, D. M. (2010). Redundancy of structural systems with and without maintenance: An approach based on lifetime functions. *Reliability Engineering & System Safety, 95,* 520–533.

Onoufriou, T. & Frangopol, D. M. (2002). Reliability-based inspection optimization of complex structures: A brief retrospective. *Computers & Structures, 80,* 1133–1144.

Parmee, I. C., Cvetkovic, D., Watson, A. H., & Bonham, C. R. (2000). Multiobjective satisfaction within an interactive evolutionary design environment. *Evolutionary Computation, 8,* 197–222.

Purshouse, R. C., & Fleming, P. J. (2007). On the evolutionary optimization of many conflicting objectives. *Evolutionary Computation, 11,* 770–784.

Saxena, D. K., Duro, J. A., Tiwari, A., Deb, K., & Zhang, Q. (2013). Objective reduction in many-objective optimization: linear and nonlinear algorithms. *Evolutionary Computation, 17,* 77–99.

Singh, H. K., Isaacs, A., & Ray, T. (2011). A Pareto corner search evolutionary algorithm and dimensionality reduction in many-objective optimization problems. *Evolutionary Computation, 15,* 539–556.

Soliman, M., & Frangopol, D. M. (2014). Life-cycle management of fatigue-sensitive structures integrating inspection information. *Journal of Infrastructure Systems, 20,* 04014001.

Stewart, M. G., & Rosowsky, D. V. (1998). Time-dependent reliability of deteriorating reinforced concrete bridge decks. *Structural Safety, 20,* 91–109.

Teytaud, O. (2007). On the hardness of offline multi-objective optimization. *Evolutionary Computation, 15,* 475–491.

Thoft-Christensen, P., & Sørensen, J. D. (1987). Optimal strategy for inspection and repair of structural systems. *Civil Engineering and Environmental Systems, 4,* 94–100.

Torres-Acosta, A. A., & Martinez-Madrid, M. (2003). Residual life of corroding reinforced concrete structures in marine environment. *Journal of Materials in Civil Engineering, 15,* 344–353.

Val, D. V., & Melchers, R. E. (1997). Reliability of deteriorating RC slab bridges. *Journal of Structural Engineering, 123,* 1638–1644.

Verel, S., Liefooghe, A., Jourdan, L., & Dhaenens, C. (2011, January 17–21). Analyzing the effect of objective correlation on the efficient set of MNK-landscapes. *Proceedings of the 5th conference on learning and intelligent optimization (LION 5),* Rome, Italy.

Wallbank, E. J., Tailor, P., & Vassie, P. R. (1999). Strategic planning of future maintenance needs. In P. C. Das (Ed.), *Management of highway structures* (pp. 163–172). London: Thomas Telford.

Part III

Life-cycle performance under corrosion and fatigue

Probabilistic limit analysis and lifetime prediction of concrete structures

FABIO BIONDINI and DAN M. FRANGOPOL

This paper presents a general approach to the probabilistic prediction of the lifetime of reinforced concrete frames with respect to structural collapse. The structural system is considered to be exposed to an aggressive environment and the effects of the structural damaging process are described by the corresponding evolution in time of the axial force – bending moment resistance domains. The collapse load is computed by means of limit analysis. Monte Carlo simulations are used to account for the randomness of the main structural parameters. In this way, both the time-variant probability of failure, as well as the expected structural lifetime associated with a prescribed reliability level, are evaluated. An application to the probabilistic time-variant limit analysis and lifetime prediction of a reinforced concrete arch bridge is presented.

1. Introduction

The safety level associated with the ultimate limit state of structural collapse can be adequately evaluated only after suitable structural models, which are able to describe the fundamental behaviour of the structural system, have been selected. In many cases, this aim can be achieved by assuming perfectly plastic behaviour and neglecting second order effects, which make the general theory of limit analysis applicable to steel structures. In spite of such idealizations, this theory can also be successfully applied to concrete structures, at least for the prediction of the collapse loads, if the concrete tensile strength is neglected and the concrete compression strength is properly modified through a suitable effectiveness factor (Nielsen 1999).

However, for concrete structures the structural performance must be considered as time-dependent, mainly because of the progressive deterioration of the mechanical properties of materials that makes the structural system less able to withstand the applied actions. Therefore, in order to ensure an adequate level of structural performance during the whole service life of the structure, the structural model must be also able to account for structural deterioration.

Starting from the previous considerations, a systematic approach to the limit analysis of plane framed structures, which considers axial force and bending moment as active and interacting generalized plastic stresses is considered (Biondini 2000). The structural system is considered to be exposed to an aggressive environment and the effects of the damaging process are described by the corresponding evolution of the axial force – bending moment resistance domains. In particular, such evolution is obtained by means of a proper methodology recently proposed for the durability analysis and lifetime assessment of concrete structures subjected to a diffusive attack from external aggressive agents (Biondini et al. 2004). The complete solution of the problem (i.e. the collapse loads, a stress distribution at the incipient collapse and a collapse mechanism) is then obtained at each time instant by linear programming.

The solution provided by the proposed time-variant limit analysis procedure makes the importance of environmental damage very clear from a qualitative point of view. However, due to the uncertainties in material and geometrical properties, in the magnitude and distribution of the loads, in the physical parameters that define the deterioration process, among others, the previous deterministic results cannot be used for reliable quantitative predictions. For this reason, the collapse loads must be considered as random variables or processes and the time-variant structural safety can be realistically assured only in probabilistic terms. This problem can be solved by using Monte Carlo simulations (Biondini 2000, Biondini *et al.* 2006). In this way, the probability density distribution of the collapse load multiplier is derived at each time instant and the corresponding time-variant probability of failure, as well as the expected structural lifetime associated with a prescribed reliability level, are evaluated. A final application to the probabilistic time-variant limit analysis and lifetime prediction of a reinforced concrete arch bridge is presented.

2. Limit analysis of framed structures

A systematic approach to the limit analysis of plane framed structures is presented here. The proposed approach neglects shear failures, which can be avoided by a proper capacity design, and considers the cross-sectional axial force n and bending moment m as active and interacting generalized plastic stresses. Following the general theory of limit analysis, a *rigid perfectly-plastic constitutive law* is adopted to relate these stresses to the correlative generalized plastic strains, represented by the cross-sectional axial elongation Δl and bending rotation θ, respectively. In this way, the behaviour of the discrete cross-sections where the plastic strains tend to develop can be represented by a *generalized plastic hinge* that allows a free axial-bending kinematic behaviour and, contemporaneously, fully transfers the corresponding plastic values of the axial force and bending moment. The plastic collapse under proportionally increasing loads is reached when the set of generalized plastic hinges is able to activate a kinematic mechanism for which the equilibrium can no longer be satisfied.

2.1 *Equilibrium and compatibility conditions*

Forces and generalized stresses are assumed in accordance with the conventions and with the reference systems shown in figure 1 (Biondini 2000). In the following, *equilibrium* and *compatibility* conditions are derived on the basis of the classical *small displacements hypothesis*.

The end forces and the internal generalized stress of a beam element can be posed as a function of the applied loads and of three independent statical quantities (Livesley

1975). Since it is reasonable to replace a distributed load with statically equivalent concentrated loads in an appropriate number of cross-sections, the element is considered subjected only to concentrated forces normal to the beam axis. When axial loads or applied moments are present, the beam element can be further subdivided. In this way, it is convenient to select as reference quantities the axial force n and the end moments $\bar{m}_1 = -m_{z'1}$ and $\bar{m}_2 = +m_{z'2}$ (figure 2). Thus, in the local coordinate system (x', y') the following relationships hold:

$$\mathbf{f}'_i = \mathbf{H}'_i \mathbf{r} - \mathbf{f}_i^{e'}, \quad i = 1, 2, \tag{1}$$

with $\mathbf{f}'_i = [n_{x'i} \ t_{y'i} \ m_{z'i}]^T$, $\mathbf{r} = [n \ \bar{m}_1 \ \bar{m}_2]^T$, and where the equivalent nodal force vectors $\mathbf{f}_i^{e'}$ and the equilibrium matrices $\mathbf{H}'_i$ are obtained from the equilibrium conditions (figure 2):

$$\mathbf{H}'_1 = \begin{bmatrix} -1 & 0 & 0 \\ 0 & -1/l & 1/l \\ 0 & -1 & 0 \end{bmatrix}, \quad \mathbf{H}'_2 = \begin{bmatrix} 1 & 0 & 0 \\ 0 & 1/l & -1/l \\ 0 & 0 & 1 \end{bmatrix}, \tag{2}$$

$$\mathbf{f}_1^{e'} = [0 \ 1 \ 0]^T \sum_j f_{y'j}(1 - l_j/l),$$

$$\mathbf{f}_2^{e'} = [0 \ 1 \ 0]^T \sum_j f_{y'j}(l_j/l). \tag{3}$$

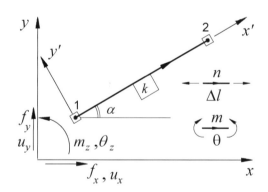

Figure 1. Reference systems.

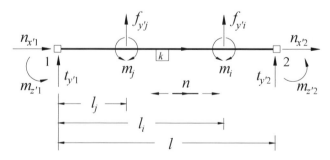

Figure 2. Reference statical quantities.

The previous equations can be written in the global reference system (x, y) by means of a coordinate transformation matrix $\mathbf{T} = \mathbf{T}(\alpha)$:

$$\mathbf{f}_i = \mathbf{T}\mathbf{f}'_i = \mathbf{T}(\mathbf{H}'_i\mathbf{r} - \mathbf{f}'^e_i) = \mathbf{H}_i\mathbf{r} - \mathbf{f}^e_i, \quad i = 1, 2, \quad (4)$$

$$\mathbf{T} = \begin{bmatrix} \cos\alpha & -\sin\alpha & 0 \\ \sin\alpha & \cos\alpha & 0 \\ 0 & 0 & 1 \end{bmatrix}. \quad (5)$$

In this way, by denoting with $\mathbf{f}^0_h$ the load vector applied at the node h of the structure in the global reference system, the nodal equilibrium conditions can be written as:

$$\mathbf{f}^0_h = \sum_{k \to h} \mathbf{f}_{ik}, \quad (6)$$

or,

$$\mathbf{f}^0_h + \sum_{k \to h} \mathbf{f}^e_{ik} = \sum_{k \to h} \mathbf{H}_{ik}\mathbf{r}_k, \quad (7)$$

where the sums are over all the elements k whose end i converges at the node h. Finally, by assembling over all the nodes h, the nodal equilibrium equations for the whole structure can be synthesized as:

$$\mathbf{f}_A = \mathbf{H}_A\mathbf{r}_A. \quad (8)$$

Apart from the ends, the more critical cross-sections of the element are those directly loaded. The bending moments in each cross-section i can also be expressed as a function of the applied loads:

$$m^0_i = \mathbf{h}^T_i\mathbf{r} + m_i, \quad (9)$$

where $\mathbf{h}_i$ and m^0_i are obtained again by simple equilibrium (figure 2):

$$\mathbf{h}^T_i = -[0 \quad 1 - l_i/l \quad l_i/l] \quad (10)$$

$$m^0_i = \sum_{j<i} f_{y'j}(l_i - l_j) - (l_i/l)\sum_j f_{y'j}(l - l_j). \quad (11)$$

The previous equilibrium equation does not depend on the reference system. After assembling over all the loaded cross-sections i, the internal equilibrium equations for the whole structure can be synthesized as:

$$\mathbf{f}_B = \mathbf{H}_B\mathbf{r}_A + \mathbf{r}_B. \quad (12)$$

In conclusion, the generalized stress vector $\mathbf{r} = [\mathbf{r}^T_A \quad \mathbf{r}^T_B]^T$ can be directly related to the load vector $\mathbf{f} = [\mathbf{f}^T_A \quad \mathbf{f}^T_B]^T$,

through the following equilibrium matrix $\mathbf{H}$:

$$\mathbf{H} = \begin{bmatrix} \mathbf{H}_A & \mathbf{0} \\ \mathbf{H}_B & \mathbf{I} \end{bmatrix}, \quad (13)$$

or,

$$\mathbf{f} = \mathbf{H}\mathbf{r}. \quad (14)$$

The generalized strains corresponding to the stresses n, $\bar{m}_1$, $\bar{m}_2$, m_i, are the elongation Δl, and the rotations $\bar{\theta}_1$, $\bar{\theta}_2$, θ_i, respectively. Therefore, the stress vector $\mathbf{r}$ can be associated with the strain vector $\mathbf{e} = [\mathbf{e}^T_A \quad \mathbf{e}^T_B]^T$, with $\mathbf{e}_A = [\mathbf{e}^T_1 \quad \mathbf{e}^T_2 \quad \ldots]^T$, $\mathbf{e}_k = [\Delta l \quad \bar{\theta}_1 \quad \bar{\theta}_2]^T$, $\mathbf{e}_B = [\theta_1 \theta_2 \ldots]^T$. In an analogous way, the load vector $\mathbf{f}$ is related to a vector of displacements $\mathbf{s} = [\mathbf{s}^T_A \quad \mathbf{s}^T_B]^T$, with $\mathbf{s}_A = [\mathbf{s}^T_1 \quad \mathbf{s}^T_2 \quad \ldots]^T$, $\mathbf{s}_k = [u_x \ u_y \ \theta_z]^T$, $\mathbf{s}_B = \mathbf{e}_B$. It can be verified that, between displacements $\mathbf{s}$ and strains $\mathbf{e}$, the following relationship holds (Livesley 1975):

$$\mathbf{e} = \mathbf{H}^T\mathbf{s}. \quad (15)$$

2.2 Yield conditions and flow rule

According to the hypothesis of the *rigid perfectly-plastic constitutive law*, (a) the *yielding criterion*, which defines the stress state corresponding to the start of the plastic flow, is convex, and (b) the *flow rule*, through which the increments of the plastic strains are correlated to the stress state, is associated with the yielding surface (normality rule).

By assuming the normal force n and the bending moment m as the only active generalized plastic stresses (figure 3(a)), the yielding criterion for the generic critical cross-section i can be written as $f_i(n_i, m_i) = 0$. Such a criterion defines, in the $n - m$ plane, a curve that can be reasonably idealized by a stepwise approximation, which is, for the sake of safety, inscribed within the convex domain $f_i(n_i, m_i) \leq 0$ (figure 3(b)). Therefore, by assuming a stepwise linearization with q_i sides, the yielding criterion for each critical cross-section i is rewritten as:

$$\boldsymbol{\phi}_i = \mathbf{N}_i\mathbf{r}_i - \mathbf{k}_i \leq \mathbf{0}, \quad (16)$$

where $\boldsymbol{\phi}_i = [\phi^i_1 \quad \phi^i_2 \quad \ldots \quad \phi^i_{q_i}]^T$, $\mathbf{N}_i = [\mathbf{n}^i_1 \quad \mathbf{n}^i_2 \quad \ldots \quad \mathbf{n}^i_{q_i}]^T$, $\mathbf{n}^i_j = [n^i_{j1} \quad n^i_{j2}]^T$, $\mathbf{k}_i = [k^i_1 \quad k^i_2 \quad \ldots \quad k^i_{qi}]^T \geq \mathbf{0}$, and $\mathbf{r}_i = [n_i \ m_i]^T$ (figure 3(c)). Finally, by assembling these conditions for the whole structure:

$$\boldsymbol{\phi} = \mathbf{N}\mathbf{r} - \mathbf{k} \leq \mathbf{0}. \quad (17)$$

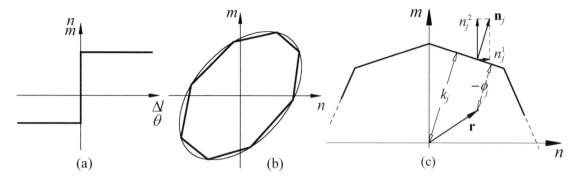

Figure 3. (a) Rigid plastic constitutive laws, (b) yielding curve with its stepwise linearization, and (c) flow rule.

The associated flow rule for each critical cross-section i is given by:

$$\Delta l_i = \mu_i \frac{\partial f_i}{\partial n_i}, \quad \theta_i = \mu_i \frac{\partial f_i}{\partial m_i}, \tag{18}$$

with the multiplier $\mu_i \geq 0$ that allows plastic flows only for the points lying on the yielding curve along the outside normal (figure 3(c)), that is $\mu_i f_i(n_i, m_i) = 0$. For the linearized case:

$$\mathbf{e}_i = \mathbf{N}_i^T \boldsymbol{\mu}_i, \tag{19}$$

with $\phi_j^i \mu_j^i = 0$ $(j = 1, \ldots, q_i)$ and where $\mathbf{e}_i = [\Delta l_i \ \theta_i]^T$, $\boldsymbol{\mu}_i = [\mu_1^i \ \mu_2^i \ \cdots \ \mu_{q_i}^i]^T \geq \mathbf{0}$. Summing up the axial strains Δl_i of all the critical cross-sections i of the element k:

$$\Delta l = \sum_{i \in k} \Delta l_i, \tag{20}$$

the plastic flow equations for the whole structure can be assembled as follows:

$$\mathbf{e} = \mathbf{N}^T \boldsymbol{\mu}. \tag{21}$$

2.3 Static and kinematic approach (duality)

Let $\mathbf{f}_0$ be a vector of constant loads and $\mathbf{f}$ a vector of loads whose intensity is proportional to a given scalar multiplier $\lambda \geq 0$. By assuming that the structure is safe for $\lambda = 0$, the *collapse multiplier* λ_c associated with its failure is derived from the two fundamental theorems of limit analysis. The *lower bound theorem* states that λ_c is the maximum of the multipliers associated with stress fields that satisfy both the equilibrium conditions and the yielding criterion. In a similar manner, the *upper bound theorem* states that λ_c is the minimum of the multipliers associated with plastic flows that satisfy both compatibility conditions and flow rule. Therefore, in mathematical terms, the fundamental theo-

rems of the limit analysis can be translated into the following dual linear programming problems:

$$\max \{\lambda | (\lambda \mathbf{f} - \mathbf{Hr}) = -\mathbf{f}_0, \mathbf{Nr} \leq \mathbf{k}, \lambda \geq 0\}, \tag{22}$$

$$\min \{\mathbf{k}^T \boldsymbol{\mu} - \mathbf{f}_0^T \mathbf{s} \,|\, (\mathbf{N}^T \boldsymbol{\mu} - \mathbf{H}^T \mathbf{s}) = \mathbf{0}, \mathbf{f}^T \mathbf{s} = 1, \boldsymbol{\mu} \geq \mathbf{0}\}. \tag{23}$$

It is worth noting that in the second case, the minimum condition is related to the work done by the proportional loads $\mathbf{f}$ for the displacements $\mathbf{s}$ associated with the collapse mechanism. Since this mechanism is associated with an arbitrary multiplier, it results in being univocally identified by the condition $\mathbf{f}^T \mathbf{s} = 1$.

The solution of the previous dual linear programs leads to the *complete solution* of the problem, i.e. the collapse multiplier, the stress distribution at the incipient collapse and the collapse mechanism. As it is the separator of two sets, λ_c is unique. However, the uniqueness of λ_c does not necessarily mean the uniqueness of the collapse mechanism, or that of the stress field at collapse.

2.4 Time-variant limit analysis

The previous limit analysis refers to prescribed states of the structure, where the cross-sectional performance is quantified by using fixed values of the quantities $\mathbf{N}$ and $\mathbf{k}$, which define the yielding criterion and the flow rule. However, due to the progressive deterioration of the mechanical properties of materials, such quantities, as well as the corresponding collapse multiplier λ_c, vary during time. In order to account for such variability, a structural analysis of the deteriorating cross-sections is firstly required to build the functions $\mathbf{N} = \mathbf{N}(t)$ and $\mathbf{k} = \mathbf{k}(t)$. In this way, a time-variant limit analysis leading to the time evolution of the collapse multiplier $\lambda_c = \lambda_c(t)$ can also be performed by solving the previous linear programs at several time instants.

3. Lifetime performance of deteriorating concrete cross-sections

Here, attention is focussed on the damaging process induced by the diffusive attack of environmental aggressive agents, like sulphate and chloride. In this context, the diffusion process may lead to the deterioration of the concrete and the corrosion of the reinforcement. In addition, damage induced by mechanical loading interacts with the environmental factors and accelerates the deterioration process (CEB 1992). For these reasons, a reliable tool for the assessment of the time-variant performance of concrete structures in aggressive environments should be capable of accounting for both the diffusion process and the corresponding mechanical damage, as well as for the coupling effects between diffusion, damage, and structural behaviour.

3.1 Simulation of the diffusion process

The simplest model to describe the kinetic process of the diffusion of chemical components in solids is represented by Fick's laws, which, in the case of a single component diffusion in isotropic, homogeneous and time-invariant media, can be reduced to the following second order partial differential linear equation:

$$D\nabla^2 C = \frac{\partial C}{\partial t}, \qquad (24)$$

where D is the diffusivity coefficient of the medium, $C = C(\mathbf{x}, t)$ is the concentration of the aggressive agent at point $\mathbf{x}$ and time t, $\nabla C = \text{grad } C(\mathbf{x}, t)$ and $\nabla^2 = \nabla \cdot \nabla$.

From the numerical point of view, the diffusion equation can be effectively simulated by using cellular automata that, in their basic form, consists of regular uniform grids of *sites* or *cells*, theoretically having an infinite extension, with a discrete variable in each cell that can take on a finite number of states (Wolfram 1994). In particular, Fick's laws in two dimensions can be accurately reproduced by adopting the following evolutionary rule (Biondini et al. 2004):

$$C_i^{k+1} = \phi_0 C_i^k + \frac{1 - \phi_0}{4} \sum_{j=1}^{2} (C_{i-1,j}^k + C_{i+1,j}^k), \qquad (25)$$

where the discrete variable $C_i^k = C(\mathbf{x}_i, t_k)$ represents the concentration of the aggressive agent at time t_k in the cell i of the automaton located at point $\mathbf{x}_i = (y_i', z_i')$ of the cross-section, $C_{i+1,i}^k$ is the concentration in the adjacent cells $i \pm 1$ in the direction $j = 1, 2$, and ϕ_0 is a suitable evolutionary

coefficient related to the rate of mass diffusion. Moreover, to regulate the process according to a given diffusivity D, a proper discretization in space and time should be chosen in such a way that the grid dimension Δx and the time step Δt satisfy the following relationship:

$$D = \frac{1 - \phi_0}{4} \frac{\Delta x^2}{\Delta t}. \qquad (26)$$

The deterministic value $\phi_0 = 1/2$ usually leads to good accuracy of the automaton. However, this evolutionary coefficient must be modelled as a random variable to take into account the stochastic effects in the diffusion process and the corresponding coupling effects between the diffusion process and the mechanical behaviour (Biondini et al. 2004).

3.2 Modelling of structural damage

Structural damage is modelled by introducing a degradation law of the effective resistant area for both the concrete matrix $A_c = A_c(t)$ and the steel bars $A_s = A_s(t)$:

$$dA_c(t) = [1 - \delta_c(t)]dA_{c0}, \quad dA_s(t) = [1 - \delta_s(t)]dA_{s0}, \qquad (27)$$

where the subscript '0' denotes the undamaged state at the initial time $t = t_0$, and the dimensionless functions $\delta_c = \delta_c(t)$ and $\delta_s = \delta_s(t)$ represent *damage indices* that provide a direct measure of the damage level within the range [0; 1]. The time evolution of these indices clearly depends on the corresponding evolution of the diffusion process.

The damaging processes in concrete structures undergoing diffusion are, in general, very complex. Moreover, the available information about environmental agents and material characteristics is usually not sufficient for detailed modelling. However, despite such complexities, very simple degradation models can often be successfully adopted. In the following, the damage indices $\delta_c = \delta_c(\mathbf{x}, t)$ and $\delta_s = \delta_s(\mathbf{x}, t)$ at a point $\mathbf{x} = (y', z')$ of the cross-section are correlated to the diffusion process by assuming, for both materials, a linear relationship between the rate of damage and the concentration $C = C(\mathbf{x}, t)$ of the aggressive agent (Biondini et al. 2004):

$$\frac{\partial \delta_c(\mathbf{x}, t)}{\partial t} = \frac{C(\mathbf{x}, t)}{C_c \Delta t_c}, \quad \frac{\partial \delta_s(\mathbf{x}, t)}{\partial t} = \frac{C(\mathbf{x}, t)}{C_s \Delta t_s}, \qquad (28)$$

where C_c and C_s represent the values of constant concentration $C(\mathbf{x}, t)$ that lead to complete damage of the concrete and steel after the time periods Δt_c and Δt_s respectively. In addition, the initial conditions $\delta_c(\mathbf{x}, t_{cr}) = \delta_s(\mathbf{x}, t_{cr}) = 0$ with $t_{cr} = \max\{t \mid C(\mathbf{x}, t) \le C_{cr}\}$ are assumed, where C_{cr} is a critical threshold of concentration.

3.3 Nonlinear structural analysis

The previous general criteria are now applied to the time-variant nonlinear analysis of deteriorating reinforced concrete cross-sections. By assuming the linearity of the concrete strain field and neglecting the bond-slip of reinforcement, the vectors of the stress resultants $\mathbf{r} = \mathbf{r}(t) = [n\ m_z\ m_y]^T$ and of the global strains $\mathbf{e} = \mathbf{e}(t) = [\varepsilon_0\ \chi_z\ \chi_y]^T$ are then related, at each time instant t, as follows:

$$\mathbf{r}(t) = \mathbf{S}(t)\mathbf{e}(t). \qquad (29)$$

The stiffness matrix $\mathbf{S}(t) = \mathbf{S}_c(t) + \mathbf{S}_s(t)$ is derived by integration over the area of the composite cross-section, or by assembling the following contributions of concrete and steel:

$$\mathbf{S}_c(t) = \int_{A_c} E_c(\mathbf{x}, t)\ \mathbf{b}(\mathbf{x})^T\ \mathbf{b}(\mathbf{x})[1 - \delta_c(\mathbf{x}, t)]\mathrm{d}A, \qquad (30)$$

$$\mathbf{S}_s(t) = \sum_m E_{sm}(t)\ \mathbf{b}_m^T \mathbf{b}_m[1 - \delta_{sm}(t)]A_{sm}, \qquad (31)$$

where the symbol 'm' refers to the mth steel bar located at $\mathbf{x}_m = (y'_m, z'_m)$, $E_c = E_c(\mathbf{x}, t)$ and $E_{sm} = E_{sm}(t)$ are the secant moduli of the materials, and $\mathbf{b}(\mathbf{x}) = [1\ -y'\ z']^T$ (Biondini *et al.* 2004). It is worth noting that the vectors $\mathbf{r}$ and $\mathbf{e}$ have to be considered as total or incremental quantities depending on the nature of the stiffness matrix $\mathbf{S}$, which depends on the type of formulation adopted (i.e. secant or tangent) for the generalized moduli of the materials.

Based on this model, the time-variant resistance domain $f(n, m_z, m_y) \leq 0$ of the cross-section can be evaluated with reference to a concrete compression strength properly modified through a suitable effectiveness factor (Nielsen 1999). In addition, the limited ductility of the materials can also be taken into account. For these reasons, these domains can be used as reliable yielding criterion in the limit analysis problem. In particular, the functions $\mathbf{N} = \mathbf{N}(t)$ and $\mathbf{k} = \mathbf{k}(t)$ can be defined by a stepwise approximation of the resistance curves $f(n, m_z, m_y) = 0$, with $m_z = m$ and $m_y = 0$ for the planar case.

4. Probabilistic prediction of structural lifetime

From a deterministic point of view, a structure is denoted as safe if the load multiplier λ is no larger than its collapse value $\lambda_c = \lambda_c(t)$. Because of the uncertainties involved in the problem, the quantity λ_c has to be considered as a random variable or process and a measure of structural safety is realistically possible only in probabilistic terms. In particular, by denoting $\hat{\lambda}_k$ an outcome of the random variable $\lambda_{ck} = \lambda_c(t_k)$, the probability of failure at prescribed time instants $t = t_k$ can be evaluated by the integration of

the density function $f_{\lambda_c}(\hat{\lambda}_k)$ within the failure domain $D_k = D(t_k) = \{\hat{\lambda}_k\ |\ \hat{\lambda}_k \leq \lambda\}$:

$$P_F(t_k) = P[\lambda(t) \geq \lambda_{ck}(t)] = \int_D f_{\lambda_c}(\hat{\lambda}_k)\mathrm{d}\hat{\lambda}. \qquad (32)$$

This formulation leads to a stochastic programming problem (Gavarini 1969), which is usually very expensive to solve. Moreover, in practice, the density distribution of λ_c is not known, and at most some information is available only about a set of N basic random variables $\mathbf{X} = [X_1\ X_2, \ldots, X_N]^T$ that defines the structural problem at the initial time $t = t_0$ (e.g. structural geometry $\mathbf{H}$, mechanical and geometrical properties of the cross-sections $\mathbf{N}$ and $\mathbf{k}$, dead $\mathbf{f}_0$ and live $\mathbf{f}$ loads). In addition, in concrete design the levels of verification are usually formulated in terms of functions of random variables $\mathbf{Y} = \mathbf{Y}(\mathbf{X})$ that describe the structural response at each time instant $t = t_k$ (e.g. stress resultants $\mathbf{r}$, global strains $\mathbf{e}$, etc.), and such derivation is generally only available in an implicit form. A numerical approach is then required and the reliability analysis can be performed by Monte Carlo simulations (Biondini 2000, Biondini *et al.* 2006).

Based on the probabilistic formulation of the time-variant limit analysis problem, the actual lifetime T of the structure associated with a prescribed target reliability level, for example expressed in terms of acceptable values of the probability of failure P_F^*, can finally be evaluated as:

$$T = \min\left\{(t - t_0)|P_F \leq P_F^*, \forall t \geq t_0\right\}, \qquad (33)$$

where t_0 is the time associated with the end of the construction phase.

5. Application to a concrete arch bridge

5.1 The arch bridge

The reinforced concrete arch bridge over the Corace river in Italy is now considered. The structural model, shown in figure 4, refers to the data presented in Galli and Franciosi (1955) and Ronca and Cohn (1979). The arch has a rectangular cross-section with nominal dimensions $d_y = 0.57$ m and $d_z = 6.00$ m, and it is reinforced with $45 + 45 = 90$ steel bars, each having a nominal diameter $\varnothing = 28$ mm (figure 5(b)). The beam has a two-cellular cross-section with main nominal dimensions $d_y = 2.00$ m and $d_z = 6.00$ m (figure 5(a)). The distribution of the reinforcement along the beam refers to the subdivision shown in figure 6, and is given in table 1. The structure is subjected to a set of dead loads g and to a live load p (see figure 4). These distributed loads are replaced by statically equivalent concentrated loads, 12 for each span of the girder and 6 for each span of the arch.

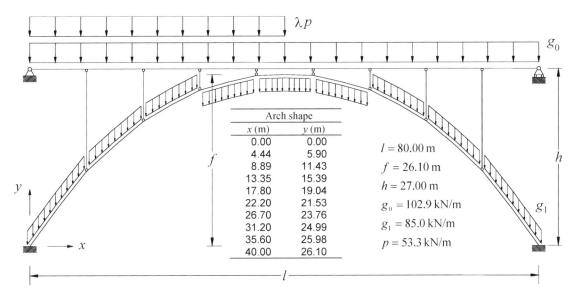

Arch shape	
x (m)	y (m)
0.00	0.00
4.44	5.90
8.89	11.43
13.35	15.39
17.80	19.04
22.20	21.53
26.70	23.76
31.20	24.99
35.60	25.98
40.00	26.10

$l = 80.00$ m
$f = 26.10$ m
$h = 27.00$ m
$g_0 = 102.9$ kN/m
$g_1 = 85.0$ kN/m
$p = 53.3$ kN/m

Figure 4. Arch bridge (nominal structure): overall dimensions and loading condition.

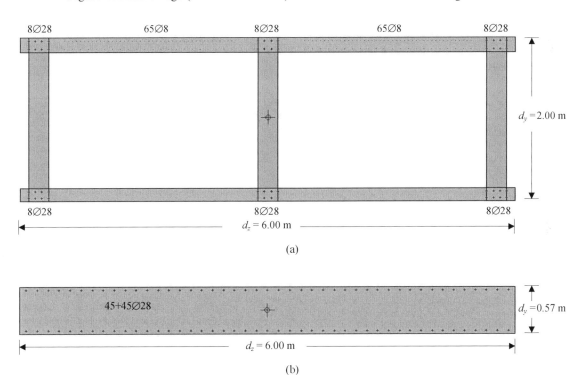

(a)

(b)

Figure 5. Details of the cross-sections: (a) beam (middle span), and (b) arch.

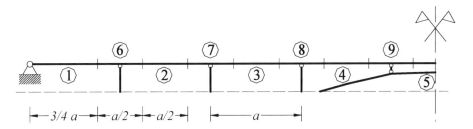

Figure 6. Beam segments having different reinforcement (see table 1).

149

Table 1. Distribution of the top A'_s and bottom A_s reinforcement along the beam (see figure 6). $n\varnothing d = n$ steel bars with diameter d [mm].

Span	1	2	3	4	5	6	7	8	9
A'_s	21⌀28	48⌀28	42⌀28	30⌀28	24⌀28	48⌀28	48⌀28	45⌀28	33⌀28
	130⌀8	130⌀8	130⌀8	130⌀8	130⌀8	130⌀8	130⌀8	130⌀8	130⌀8
A_s	21⌀28	30⌀28	42⌀28	24⌀28	24⌀28	21⌀28	36⌀28	27⌀28	24⌀28

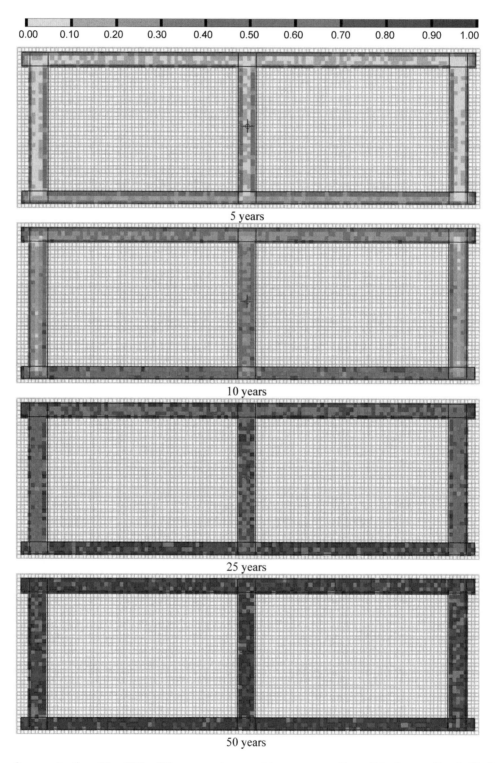

Figure 7. Maps of concentration $C(\mathbf{x}, t)/C_0$ of the aggressive agent in a cross-section of the beam after 5, 10, 25, and 50 years from the initial time of diffusion penetration.

For concrete, the stress–strain diagram is described by the Saenz law in compression and by an elastic perfectly plastic model in tension, with the following nominal parameters: effective compression strength $f_c = -30$ MPa; tension strength $f_{ct} = 0.25|f_c|^{2/3}$; initial modulus $E_{c0} = 9500|f_c|^{1/3}$; peak strain in compression $\varepsilon_{c0} = -0.20\%$; strain limit in compression $\varepsilon_{cu} = -0.35\%$; and strain limit in tension $\varepsilon_{ctu} = 2f_{ct}/E_{c0}$. For steel, the stress–strain diagram is described by an elastic perfectly plastic model in both tension and compression, with the following nominal parameters: yielding strength $f_{sy} = 300$ MPa; elastic modulus $E_s = 206$ GPa; and strain limit $\varepsilon_{su} = 1.00\%$. With reference to a nominal diffusivity coefficient $D = 10^{-11}$ m^2 s^{-1}, the cellular automaton is defined by a grid dimension $\Delta x = 50.2$ mm and a time step $\Delta t = 1$ year. The aggressive agent is assumed to be located with concentration $C(t) = C_0$ along the free edges of the cross-sections. Damage rates are assumed to be defined by the nominal values $C_{cr} = 0$, $C_c = C_s = C_0$, $\Delta t_c = 25$ years and $\Delta t_s = 50$ years.

5.2 Deterministic time-variant limit analysis

The diffusion process for the *nominal scenario* is highlighted in figures 7 and 8, which show the maps of concentration $C(\mathbf{x}, t)/C_0$ of the aggressive agent at different time instants in the beam and the arch respectively. The mechanical damage induced by diffusion can be evaluated from the diagrams in figures 9 and 10, which show the time evolution of the resistance bending moments of the axially unloaded beam (figure 9) and of the resistance curves $f(n, m) = 0$ of the arch (figure 10(a)), idealized at each time instant by a four-sides stepwise linearization (figure 10(b)). The deterioration of the five supporting walls, simply compressed, is not investigated since they are assumed as not critical with respect to collapse.

Figure 11 shows the results of the limit analysis carried out for the nominal scenario at the initial time and after 50 years of lifetime. These results highlight that the time evolution of damage leads to a significant variation of the collapse multiplier, which decreases from $\lambda_c = 4.28$ to $\lambda_c = 1.42$, as well as to a noteworthy redistribution of the internal stress resultants and a consequent modification of the collapse mechanism. Figure 12 shows the time evolution of the critical cross-sections where the generalized plastic hinges develop at the collapse. Despite the fact that the location of the plastic hinges tend to be nearly constant over time, local modifications of the plastic strain distribution arise after about 10 years in the arch, and after about 45 years in both the beam and the arch.

5.3 Probabilistic analysis and lifetime prediction

The probabilistic model assumes as random variables the location (x_h, y_h) of the nodal connections between the structural elements, the material strengths f_c and f_{sy}, the

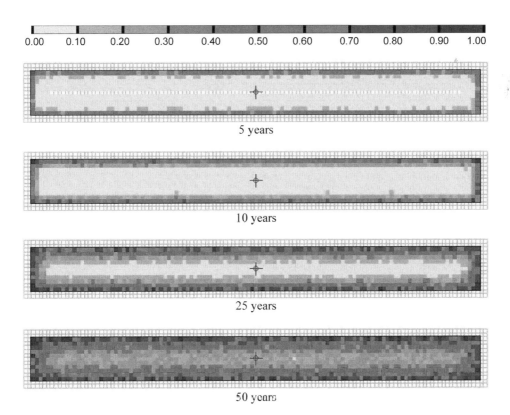

5 years

10 years

25 years

50 years

Figure 8. Maps of concentration $C(\mathbf{x}, t)/C_0$ of the aggressive agent in a cross-section of the arch after 5, 10, 25, and 50 years from the initial time of diffusion penetration.

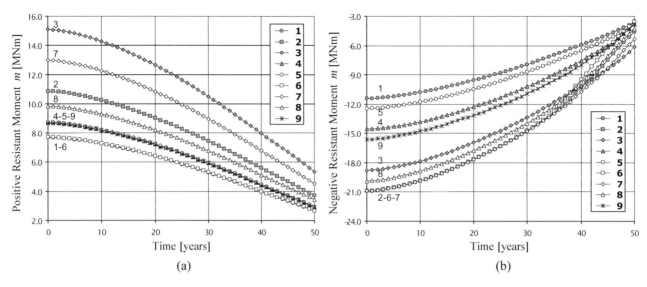

Figure 9. Time evolution of the (a) positive, and (b) negative resistance bending moments of the beam (see table 1).

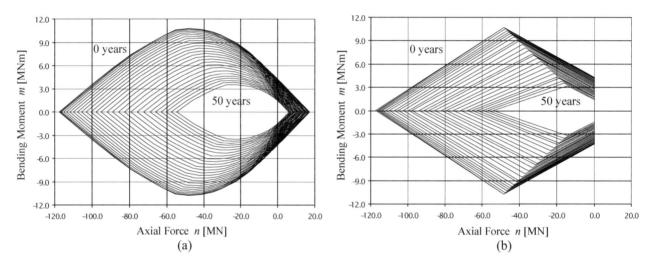

Figure 10. Time evolution of (a) the axial force – bending moment interaction diagram for the arch, and (b) its linearization.

coordinates (y'_p, z'_p) of the nodal points $p = 1, 2, \ldots$ that define the two-dimensional model of the concrete cross-sections, the coordinates (y'_m, z'_m) and the diameter $\varnothing_m$ of the steel bars $m = 1, 2, \ldots$, the diffusivity coefficient D, the damage rates $q_c = (C_c \Delta t_c)^{-1}$ and $q_s = (C_s \Delta t_s)^{-1}$, and the loads g and p in each beam element. A set of random variables is associated with each element of the structural model. These variables are considered to be statistically independent and are assumed to have the probabilistic distribution with the mean μ and standard deviation σ values listed in table 2.

Based on this model, a probabilistic time-variant limit analysis is carried out by using a set of about 28000 Monte Carlo simulations (1000 for each cross-section of

the beam, 1000 for the cross-section of each span of the arch, and 1000 for the global analysis of the structure). The time evolution of the collapse multiplier during the first 50 years of lifetime is shown in figure 13(a) for a selected sample of these simulations. In particular, figure 13(a) compares the nominal scenario with the scenarios associated with the minimum and maximum values of the collapse multiplier obtained at the initial time (min(0) and max(0)) and at the end of the structural lifetime (min(50) and max(50)), as well as with the scenarios characterized by the maximum and minimum variation of the collapse multiplier over the investigated lifetime period (max(0 – 50) and min(0 – 50)). With reference to the whole sample, figure 13(b) shows the time

0 years   50 years

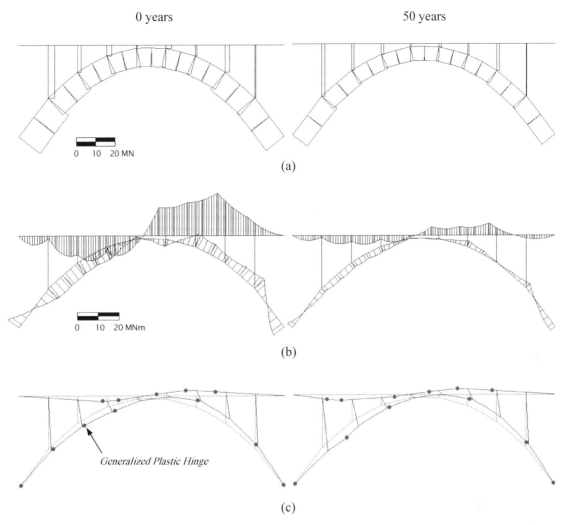

Figure 11. Limit analysis for the nominal scenario at the initial time of construction ($\lambda_c = 4.28$) and after 50 years of lifetime ($\lambda_c = 1.42$): (a) axial force, and (b) bending moment diagrams at the collapse; (c) collapse mechanism.

evolution of the statistical parameters (mean value μ, standard deviation σ, minimum and maximum values) of the collapse multiplier. The validity of the simulation results with respect to the sample size N is highlighted by figure 14, which allows the expected convergence towards stable values of both the mean μ and deviation σ/μ to be verified. A sensitivity analysis of the relative importance of each random variable in the probabilistic model has been presented in Biondini and Frangopol (2006).

The results of this simulation can be used to compute, at each point in time, the probability of failure for given deterministic target levels of the structural performance indicators, as shown by the probability curves in figure 15(a). These curves allow the time-variant reliability of the cross-section with respect to the required performance to be assessed. Moreover, based on these probability curves, the lifetime T associated with given

target reliability levels P^* can be computed as a function of the expected values of the live load multiplier, as shown in figure 15(b). These curves allow the remaining lifetime, which can be assured under prescribed reliability levels without maintenances, to be assessed.

6. Conclusions

The structural lifetime of deteriorating concrete frames with respect to the ultimate limit state of structural collapse has been investigated. The structural system is considered to be exposed to an aggressive environment and the effects of the damaging process induced by the diffusive attack of external agents are described by the corresponding time evolution of the axial force – bending moment resistance domains. Based on such domains, the time-variant collapse load is computed by means of limit analysis through a

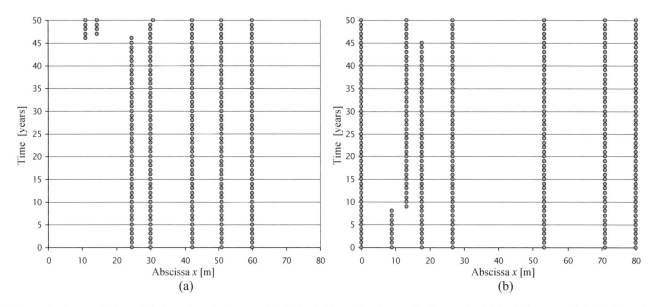

Figure 12. Time evolution of the location of the generalized plastic hinges for the nominal scenario: (a) in the beam, and (b) in the arch.

Table 2. Probability distributions and their parameters (truncated distributions provide non-negative outcomes).

Random Variable ($t = t_0$)	Distribution Type	μ	σ
Coordinates of the nodal points, (x_h, y_h)	Normal	$(x_h, y_h)_{nom}$	50 mm
Concrete strength, f_c	Lognormal	$f_{c,nom}$	5 MPa
Steel strength, f_{sy}	Lognormal	$f_{sy,nom}$	30 MPa
Coordinates of the nodal points, (y'_p, z'_p)	Normal	$(y'_p, z'_p)_{nom}$	5 mm
Coordinates of the steel bars, (y'_m, z'_m)	Normal	$(y'_m, z'_m)_{nom}$	5 mm
Diameter of the steel bars, $\varnothing_m$	Normal-truncated	$\varnothing_{m,nom}$	$0.10\varnothing_{m,nom}$
Diffusivity coefficient, D	Normal-truncated	D_{nom}	$0.10\,D_{nom}$
Concrete damage rate, $q_c = (C_c \Delta t_c)^{-1}$	Normal-truncated	$q_{c,nom}$	$0.30\,q_{c,nom}$
Steel damage rate, $q_s = (C_s \Delta t_s)^{-1}$	Normal-truncated	$q_{s,nom}$	$0.30\,q_{s,nom}$
Dead loads, g	Normal-truncated	g_{nom}	$0.10\,g_{nom}$
Live load, p	Normal-truncated	p_{nom}	$0.40\,p_{nom}$

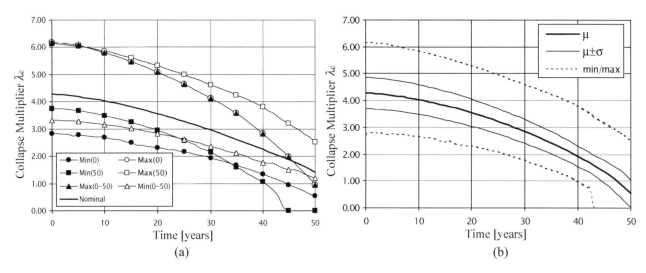

Figure 13. Time evolution of the collapse multiplier λ_c: (a) comparison among selected simulations, and (b) mean μ (thick line), standard deviation σ from the mean μ (thin lines), minimum and maximum values (dotted lines).

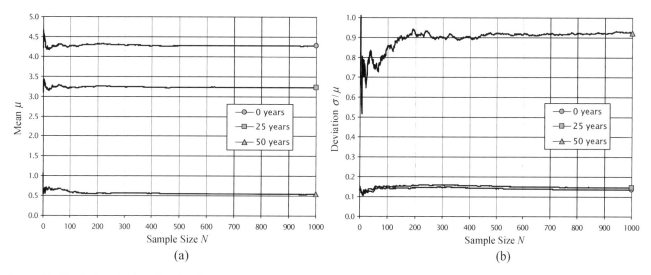

Figure 14. Evolution during the simulation of (a) mean value μ, and (b) deviation σ/μ of the collapse multiplier after 0, 25, and 50 years of lifetime.

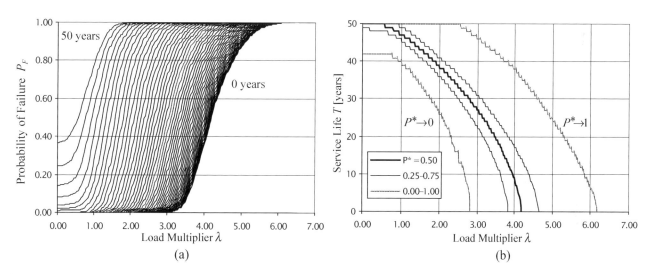

Figure 15. Time evolution of the collapse multiplier λ_c: (a) probability curves during the first 50 years of service life ($\Delta t = 2$ years), and (b) service life T associated with given values of the probability of failure P^* versus given target levels of the load multiplier λ.

systematic approach that considers axial force and bending moments as active and interacting generalized plastic stresses. The randomness of the main structural parameters is taken into account by Monte Carlo simulations, and both the time-variant probability of failure, as well as the expected structural lifetime associated to a prescribed reliability level are evaluated.

The application to the collapse reliability analysis and lifetime prediction of a reinforced concrete arch bridge shows that the proposed approach predicts the time-variant structural resistance with respect to a given demand or, conversely, the corresponding remaining lifetime that can

be assured under prescribed reliability levels without maintenance.

The accuracy of the results mainly depends on the values of the material parameters that define both the diffusive and damage processes. Further developments aimed to achieve a proper calibration of the material parameters and of their probabilistic distributions are required. However, despite the necessity of such developments, the proposed approach is proven to represent a powerful engineering tool for the reliability assessment and lifetime prediction of deteriorating reinforced concrete framed structures with respect to the collapse.

References

Biondini, F., Probabilistic limit analysis of framed structures. In *Proceedings of 8th ASCE Conference on Probabilistic Mechanics and Structural Reliability*, Paper 273, 2000.

Biondini, F. and Frangopol, D.M., Role of uncertainties in the lifetime performance of concrete structures. In *Proceedings of IFIP WG7.5 Working Conference on Reliability and Optimization of Structural Systems*, 2006.

Biondini, F., Bontempi, F., Frangopol, D.M. and Malerba, P.G., Cellular automata approach to durability analysis of concrete structures in aggressive environments. *J. Struct. Eng. ASCE*, 2004, **130**(11), 1724–1737.

Biondini, F., Bontempi, F., Frangopol, D.M. and Malerba, P.G., Probabilistic service life assessment and maintenance planning of deteriorating concrete structures. *J. Struct. Eng. ASCE*, 2006, **132**(5), 810–825.

CEB, *Durable Concrete Structures – Design Guide*, 1992 (Thomas Telford).

Galli, A. and Franciosi, V., Il calcolo a rottura dei ponti a volta sottile ed impalcato irrigidente. *Giorn. Genio Civile*, 1955, **11**, 686–700 (in Italian).

Gavarini, C., Concezione probabilistica del calcolo a rottura. *Giorn. Genio Civile*, 1969, **9**, 477–502 (in Italian).

Livesley, R.K., *Matrix Methods of Structural Analysis*, 1975 (Pergamon Press: New York).

Nielsen, M.P., *Limit Analysis and Concrete Plasticity*, 1999 (CRC Press).

Ronca, P. and Cohn, M.Z., Matrix-MP method for the analysis of inelastic arch structures. *Int. J. Numer. Methods Eng.*, 1979, **14**, 703–725.

Wolfram, S., *Cellular Automata and Complexity – Collected Papers*, 1994 (Addison-Wesley).

Integration of the effects of airborne chlorides into reliability-based durability design of reinforced concrete structures in a marine environment

Mitsuyoshi Akiyama, Dan M. Frangopol and Motoyuki Suzuki

In this paper, the hazard curve associated with airborne chlorides in a marine environment and the computational procedure to obtain the probability of occurrence of corrosion cracking in reinforced concrete (RC) structures are presented. A method for integration of the effects of airborne chloride into reliability-based durability design of RC structures in a marine environment is proposed. By using this method, it is possible to determine the probability of corrosion cracking due to airborne chlorides, regardless of region, distance from coastline, and the properties of the concrete controlled by the water to cement ratio.

1. Introduction

Reinforced concrete (RC) structures in a marine environment deteriorate with time due to chloride-induced corrosion of reinforcing bars. Corrosion is initiated by chloride contamination if the structures have poor quality concrete and/or inadequate concrete cover. Corrosion initiation could lead to cracking due to corrosion products and concrete cover spalling. Cracking and/or spalling accelerate the corrosion rate and finally lead to serviceability failure and a deterioration of structural performance. Therefore, it is necessary to determine adequate concrete cover and concrete quality for the design of new RC structures that will require less maintenance and repair over their lifetime. Design requirements for RC structures to improve the structural long-term performance (i.e. durability design) must include control of steel corrosion and concrete cracking (Ellingwood 2005).

Several marine environmental factors affect the degradation mechanisms of concrete structures. However, the exact influence of these factors is difficult to predict as they vary in time and space. In addition, the properties of concrete and steel bars are random variables. Because of the presence of uncertainties, it is necessary to formulate durability performance using probabilistic concepts and methods. Stochastic treatment of design problems takes into account the real nature of structural performance, making a reliable

design of RC structures possible (RILEM 1998). However, unlike the case associated with actions that are usually considered in structural design (e.g. probabilistic seismic hazard analysis), there is a lack of research on marine environmental hazard assessment. This is because the data on coastal atmospheric exposure is very limited.

Probabilistic methods for the evaluation of deterioration of RC bridges subjected to chloride attacks have been reported and a probabilistic framework for life-cycle analysis of RC structures has been established (Prezzi et al. 1996, Frangopol et al. 1997a, b, Val and Stewart 2003). Frangopol et al. (1997a) proposed a reliability-based design approach involving the optimisation of expected life-cycle cost. These reliability analyses have included examination of chloride threshold concentration, cracking and spalling of concrete cover, and/or deterioration of flexural and shear strength of concrete members. However, these prior studies have not taken into consideration the probabilistic assessment of the effects of de-icing salt and/or airborne chlorides.

The relationship between seismic intensity and the distance of the structure from the seismic source has been used in seismic probabilistic hazard assessment (SPHA) (Kameda and Nojima 1988). By applying the concept of SPHA to the assessment of the effect of airborne chlorides on durability design of RC structures, which is reduced with increasing the distance from the coastline, probabilistic hazard analysis can be performed and the effect of a marine environment can be quantified.

In this paper, an approach is proposed to establish the probabilistic hazard curve of airborne chlorides and provide a computational procedure to obtain the probability of occurrence of corrosion cracking. A performance function to avoid the occurrence of corrosion cracking within the lifetime of RC structures is also proposed. The flowchart describing the framework for computing the probabilities of occurrence of steel corrosion and corrosion cracking in RC structures in a marine environment is shown in Figure 1. This flowchart consists of four main parts: (a) defining the process of hazard assessment of airborne chlorides in a marine environment; (b) comparing the ratio of experimental to predicted values such as diffusion of chlorides; (c) computing the time-dependent probability of exceeding the limit states of steel corrosion and corrosion cracking; and (d) proposing the reliability-based durability design method to calculate the concrete cover and determine the concrete quality (water-cement ratio) such that the probability of

exceeding the limit state is at most equal to the target value.

2. Hazard assessment due to airborne chlorides

2.1. Attenuation of airborne chlorides

Environmental conditions should be quantitatively assessed and the evaluation results should be reflected in the rational durability design of RC structures in a marine environment. The effect of airborne chlorides should be modelled such that spatial-temporal variation is taken into account in the design of these structures.

Observed values in Japan (PWRI 1988) are used to obtain the attenuation relationship between the amount of airborne chlorides, C_{air}, and the distance from the coastline. There are 34 stations for collecting data on C_{air} located all over Japan, as shown in Figure 2. Since in Japan the wind usually blows from west to east, the structures near the Sea of Japan (East Sea) have suffered most from severe damage due to airborne chlorides. The time period for collecting

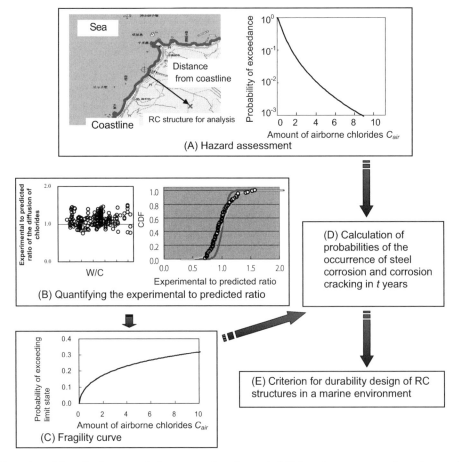

Figure 1. Flowchart describing the framework for computing the probabilities of occurrence of steel corrosion and corrosion cracking of RC structures in a marine environment.

airborne chloride samples used in the present study was three years. The speed of wind, the ratio of sea wind (defined as the percentage of time during one day when the wind is blowing from sea toward land) and the distance from the coastline, affect the amount of airborne chlorides, among other factors. The attenuation of C_{air} (mdd, i.e. 100mg/m^2/day) in the lateral direction can be expressed as:

$$C_{air} = 1.29 \cdot r \cdot u^{0.386} \cdot d^{-0.952} \qquad (1)$$

where r is the ratio of sea wind, u (m/sec) is the average wind speed during the observation period, and d (m) is the distance from the coastline. The comparison of observed values with predicted values by Equation (1) is shown in Figure 3. The ratio of sea wind r and the speed of wind u used in Equation (1) were obtained from the meteorological data collected.

Since the data on airborne chlorides are very limited, it is difficult to consider the effect of geological formation around structures, precipitation, and the differences in coastal topography (e.g. sand beach and reef) on the amount of airborne chlorides collected at each location. Also, Equation (1) cannot be applied for a splash zone. In order to reduce the uncertainty involved in the prediction of airborne chlorides, it is necessary to include additional parameters in the attenuation equation.

2.2. Hazard associated with airborne chlorides

The probability that C_{air} at a specific site will exceed an assigned value c_{air} is

$$q_s(c_{air}) = \int_0^\infty \int_0^\infty P(C_{air} > c_{air}|u,r) \cdot$$
$$f_u(u) \cdot f_r(r) \, du dr \qquad (2)$$

where $f_u(u)$ and $f_r(r)$ are the probability density functions (PDFs) of u and r, respectively, and $P(C_{air} > c_{air}|u, r)$ is the probability of $C_{air} > c_{air}$ given u and r.

The statistics of wind speed and the ratio of sea wind for the three years of C_{air} collection were obtained from meteorological data. Figure 4 shows the variation of average wind speed associated with two locations (Niigata and Uwajima) indicated in Figure 2. The uncertainty in r is small compared to that in u and, for this reason, it can be neglected.

Considering model uncertainty, the attenuation is

$$C_{air} = U_R \cdot \left(1.29 \cdot r_0 \cdot u^{0.386} \cdot d_0^{-0.952}\right) \qquad (3)$$

where U_R is a lognormal random variable representing model uncertainty shown in Figure 3.

Then, Equation (2) can be expressed as:

$$q_s(c_{air}) = \int_0^\infty P\left(U_R > \frac{C_{air}}{1.29 \cdot r_0 \cdot u^{0.386} \cdot d_0^{-0.952}}\right) \cdot$$
$$f_u(u) \, du \qquad (4)$$

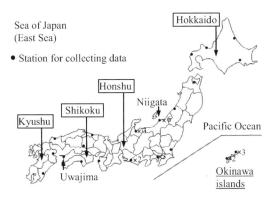

Figure 2. Data collection locations for airborne chlorides.

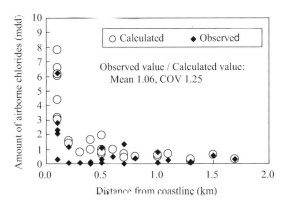

Figure 3. Amount of airborne chlorides versus distance from coastline.

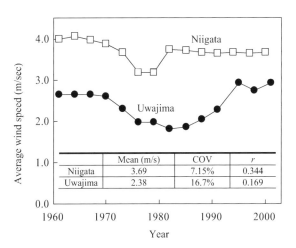

Figure 4. Variation of average wind speed in Niigata City and Uwajima City.

Hazard curves $q_s(c_{air})$ for Niigata City and Uwajima City are shown in Figure 5. The wind speeds and the ratios of sea wind are different in these two locations, resulting in different hazard levels. As shown, the probability of exceeding a prescribed amount of airborne chloride at Niigata City is much higher than that at Uwajima City. Based on the probabilities of exceeding a prescribed amount shown in Figure 5, the effect of marine environment can be quantified. In Niigata City many bridges are reported damaged by chloride attack, while in Uwajima City the sea wind did not blow often and there have been very few reports of damage by chloride attack. Figure 5 confirms these reports.

3. Evaluation of the probability of crack occurrence due to steel corrosion

3.1. *Performance function for steel corrosion*

When airborne chlorides were measured in each location shown in Figure 2, exposed concrete specimens were also placed at the same location to measure the surface chloride content, C_0. The exposed time period of concrete specimens was the same as that of airborne chlorides. The relationship between C_{air} (mdd) and C_0 (kg/m^3), shown in Figure 6, is

$$C_0 = 0.988 C_{air}^{0.379} \qquad (5)$$

The difference $C_{air} - C_0$ has a large dispersion. This dispersion is reflected by the fragility curve. The fragility curve shows the occurrence probability of the limit state under the condition that a specific value of airborne chloride, c_{air}, is given. The probability of occurrence of steel corrosion under the condition that c_{air} is given can be obtained by using the performance function:

$$g_{d,1} = C_T - \chi_1 \cdot C(c, \chi_2 D_c, \chi_3 C_0, t) \qquad (6)$$

where

$$\log D_c = -6.77(\text{W/C})^2 + 10.10(\text{W/C}) - 3.14 \qquad (7)$$

$$C(c, \chi_2 D_c, \chi_3 C_0, t) = \chi_3 C_0 \left\{ 1 - \text{erf}\left(\frac{0.1 \cdot c}{2\sqrt{\chi_2 D_c t}} \right) \right\} \qquad (8)$$

C_T is the critical threshold of chloride concentration (kg/m^3), c is the concrete cover (mm), t is the time after construction (years), W/C is the ratio of water to cement, erf is the error function, χ_1 is the model uncertainty associated with the estimation of C, D_c is the coefficient of diffusion of chloride, χ_2 is the model uncertainty associated with the estimation of D_c, and χ_3 is the model uncertainty related to the ratio of observed to predicted values (see Figure 6).

Based on the experimental results and survey of existing concrete structures in a marine environment (Val and Stewart 2003, Sasatani et al. 1997, Tsutsumi et al. 1996, Tanaka et al. 2001), it is assumed that the amount of surface chloride C_0 is determined by exposure for three years after the construction and the total amount of surface chloride is constant within a structure's lifetime. This assumption is consistent with the time period of collecting airborne chlorides in the present study. However, although many researchers have used Fick's law to predict the service life of a structure, the movement of chlorides in concrete is not a pure diffusion process from a physical point of view

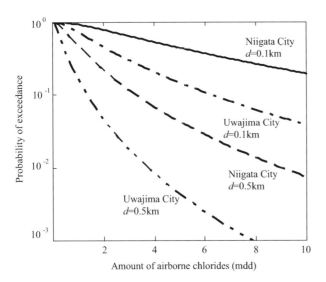

Figure 5. Hazard curves for amounts of airborne chlorides at two different locations at distance of 0.1 km and 0.5 km from coastline.

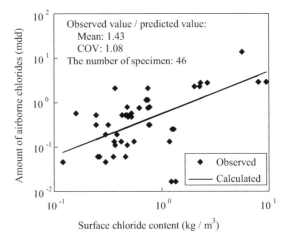

Figure 6. Amount of airborne chlorides versus the surface chloride content.

(Prezzi *et al.* 1996). There is need for additional investigation based on longer time periods of observation of the airborne chlorides and the amount of surface chlorides.

If ordinary Portland cement is used for concrete, the chloride coefficient of diffusion D_c is indicated in Equation (7). This equation was obtained using the data reported by Maeda *et al.* (2004). In Equations (6) to (8), the critical threshold chloride concentration, C_T, concrete cover, c, and χ_1, χ_2, and χ_3 are treated as random variables. It is well known that the values of the chloride threshold concentration vary widely. Based on Val and Stewart (2003), and Matsushima *et al.* (1998), C_T is treated as a normal random variable with a mean equal to 2.05 kg/m^3 and coefficient of variation (COV) of 37.5%. Kawamura *et al.* (2004) investigated the differences between the concrete cover specified in designs and the concrete cover measured in real structures. They reported that this difference can be modelled as a normal random variable with a mean of +8.5 mm and a standard deviation of 16.6 mm. The uncertainty in χ_1 is obtained by comparing observed values of the exposed specimens (PWRI 1988) with computed values by Fick's law. The uncertainty in χ_2 is obtained by comparing experimental results (Maeda *et al.* 2004) and predicted values by Equation (7). The uncertainty in χ_3 is obtained based on Figure 6. The PDFs of random variables χ_1, χ_2, and χ_3 and their parameters are shown in Table 1.

3.2. Performance function for crack occurrence due to steel corrosion

The probability of crack occurrence due to corrosion product on steel bars can be obtained based on the performance function:

$$g_{d,2} = \chi_4 Q_{cr}(c) - Q_b(V, T_{co}, t) \qquad (9)$$

where

$$Q_{cr}(c) = \eta(W_{c1} + W_{c2}) \qquad (10)$$

$$W_{c1} = \frac{\rho_s}{\pi(\gamma - 1)} \left[\alpha_0 \beta_0 \frac{0.22\left\{(2c + d)^2 + d^2\right\}}{E_c(c + d)} f_c'^{2/3} \right] \qquad (11)$$

$$W_{c2} = \alpha_1 \beta_1 \frac{\rho_s}{\pi(\gamma - 1)} \frac{c + d}{5c + 3d} w_c \qquad (12)$$

$$Q_b(V, T_{co}, t) = V(t - T_{co}) \qquad (13)$$

$$f_c' = -20.5 + 21.0/(\text{W/C}) \qquad (14)$$

$$\alpha_0 = (-0.0005d + 0.028)c + (-0.0292d + 1.27) \qquad (15)$$

$$\beta_0 = -0.0055f_c' + 1.07 \qquad (16)$$

$$\alpha_1 = (0.0007d - 0.04)c + 0.0663d + 5.92 \qquad (17)$$

$$\beta_1 = -0.0016f_c' + 1.04. \qquad (18)$$

ρ_s is the steel density (7.85 (mg/mm^3)), $\gamma = 3.0$ is the expansion rate of volume of corrosion product, T_{c0} is the time in years until steel corrosion occurrence obtained by using $g_{d,1} = 0$ for Equation (6), f_c' is the concrete strength (MPa), $w_c = 0.1$ mm is the crack width due to corrosion of the steel bar, E_c is the modulus of elasticity of concrete (MPa), d is the diameter of the steel bar (mm), V is the corrosion rate of the steel bar (mg/mm^2/year), α_0, β_0, α_1, and β_1 are coefficients taking into account the effects of concrete cover, steel bar diameter, and concrete strength (Qi and Seki 2001), η is the correction factor defined subsequently in this paper, and χ_4 is the model uncertainty associated with estimating Q_{cr}.

Table 1. Parameters of random variables.

Parameter		Distribution	Mean	COV	Reference
u	Niigata	Normal	3.69 m/s	7.15%	Figure 4
	Uwajima		2.38 m/s	16.7%	
U_R		Lognormal	1.06	125%	PWRI (1988)
χ_1		Lognormal	1.24	90.6%	PWRI (1988)
χ_2		Lognormal	1.89	184%	Maeda *et al.* (2004)
χ_3		Lognormal	1.43	108%	PWRI 1988
χ_4		Lognormal	1.00	33.0%	Matsushima *et al.* (2004), Nakagawa *et al.* (2004a)
Critical threshold chloride concentration at occurrence of steel corrosion		Normal	2.03 kg/m^3	37.5%	Val and Stewart (2003), Matsushima *et al.* (1998)
Construction errors of the concrete cover		Normal	Specified + 8.5 mm	16.6 mm/ (specified + 8.5 mm)	Kawamura *et al.* (2004)
Steel corrosion rate		Lognormal	0.061 mg/mm^2/year	58.0%	Nakagawa *et al.* (2004b)

Concrete structures crack due to corrosion if the amount of steel corrosion product is larger than the critical threshold of corrosion associated with crack initiation $\chi_4 Q_{cr}(c)$. Although there have been many reports on the critical threshold of corrosion amount at crack initiation, specific values have not been determined. In this study, the equation of Qi and Seki (2001) is used. This equation was obtained using a mechanical cylinder model and the finite element method.

In the original equation of Qi and Seki (2001), $\eta = 1.0$ and the coefficients α_0, β_0, α_1, and β_1 are based on comparing the computed results with the experimental results of specimens without stirrups. These experimental specimens were exposed to electric corrosion. However, stirrups can reduce the corrosion crack width, and the use of electric corrosion in the experiment caused the underestimation of corrosion amount at the crack initiation (Matsushima *et al.* 2004, Nakagawa *et al.* 2004a). In the present study, the correction factor η is introduced in the equation of Q_{cr} (Qi and Seki 2001) to consider the effects of stirrups and corrosion under drying/wetting conditions. Based on the experimental results of 25 specimens (Matsushima *et al.* 2004, Nakagawa *et al.* 2004a) with stirrups corroded under drying/wetting conditions, η was set to 3.78 and the model error was considered by using χ_4 (see Figure 7). Although Matsushima *et al.* (2004) reported that crack patterns due to steel corrosion are related to $(2c + d)/d$ (see Figure 7), χ_4 seems to be independent of this parameter.

The corrosion rate is determined based on Nakagawa *et al.* (2004b) who investigated many concrete structures in marine environments (in various regions and at various distances from the coastline) and reported the corrosion rate of steel bars before the occurrence of corrosion cracking. The corrosion rate is assumed as a lognormal random variable with a mean equal to 0.061 mg/mm^2/year and COV = 0.58. The use of these parameters is supported by data presented by Matsushima *et al.* (2001). All parameters of random variables used to calculate the probability of occurrence of corrosion cracking are summarised in Table 1.

3.3. Reliability assessment of RC structures in a marine environment

Monte Carlo simulation is used to evaluate conditional failure probability. Fragility curves, $F_r(c_{air})$, are shown in Figure 8. Such a curve is defined as the probability of occurrence of the limit state (see Equations (6) or (9)) under the condition that a specific value of airborne chlorides, c_{air}, is given. Figure 8 shows $F_r(c_{air})$ of concrete structures with W/C = 60% and concrete cover = 50 mm, and W/C = 45% and concrete cover = 100 mm, and the effect of concrete quality and concrete cover on the conditional limit state probability.

Using the fragility curve, $F_r(c_{air})$, probability of occurrence of corrosion cracking

$$p_f = \int_0^\infty \left(-\frac{dq_s(c_{air})}{dc_{air}} \right) \cdot F_r(c_{air})\, dc_{air} \qquad (19)$$

has to be evaluated. This probability is transformed into a reliability index β, as follows:

$$\beta = -\Phi^{-1}(p_f) \qquad (20)$$

where Φ is the cumulative distribution function of the standard normal variable.

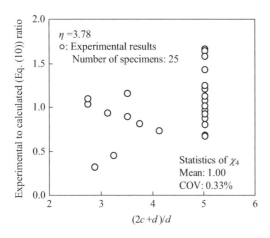

Figure 7. Comparison of calculated value with test results regarding the corrosion amount at the occurrence of cracking due to corrosion.

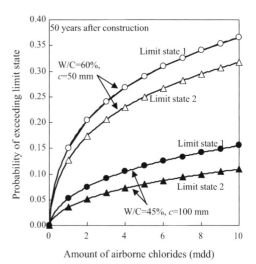

Figure 8. Fragility curve for limit states obtained by Equations (6) and (9) at 50 years after construction.

Using hazard curves of airborne chlorides in Niigata City (0.1 km from the coastline, see Figure 5) and fragility curves (see Figure 8), the relationship between the reliability index and time can be obtained by using Equations (19) and (20). As shown in Figure 9, the effect of the marine environment, concrete quality, and concrete cover on the durability performance of RC structures can be quantified by the proposed method.

4. Reliability-based design criterion

4.1. Proposed design criterion

Based on the proposed method, it is possible to ensure that the design criterion is satisfied ($\beta \approx \beta_{\text{target}}$, where β_{target} is target reliability index); however, practical design would require complex reliability computations. A design criterion is proposed herein, so that the reliability index for the occurrence of corrosion cracking will be very close to the target value without performing a complex reliability analysis by the designers. The designers determine the concrete cover and concrete quality (W/C) by only confirming that the time to the occurrence of crack, $T_{crack,d}$, is larger than the lifetime of the structure, T_d, in Equation (21). $T_{crack,d}$ is divided into the time to the occurrence of steel corrosion T_1 and the time from the occurrence of steel corrosion to the occurrence of crack T_2. The equation of $T_{crack,d}$ has the durability design factor φ taking into account the uncertainties in the computation of $T_{crack,d}$. Since this design criterion prevents the structure from cracking within its lifetime, the resulting structures will not

need any future maintenance. The proposed formulations are follows:

$$T_d \leq T_{crack,d} = \varphi\,(T_1 + T_2) \tag{21}$$

$$C_{0,d}\left\{1 - \text{erf}\left(\frac{0.1 \cdot c_d}{2\sqrt{D_c T_1}}\right)\right\} = C_{\text{lim},d} \tag{22}$$

$$T_2 = \frac{Q_{cr,d}}{V_d} \tag{23}$$

$$C_{0,d} = 4.2r^{0.25}u_a^{0.1}d^{-0.25} \tag{24}$$

where φ is durability design factor, c_d is the design concrete cover, $C_{\text{lim},d}$ is equal to C_T as the mean value, V_d is equal to V as the median value, $Q_{cr,d}$ is equal to Q_{cr} multiplied by χ_4 as the median value, u_a is the mean of wind speed and $C_{0,d}$ is the design amount of airborne chlorides.

The procedure to determine the durability design factor is based on code calibration. The steps are as follows:

(a) Set the target reliability index β_{target} and the lifetime of the structure, T_d.
(b) Calculate the design value of the surface chloride content using Equation (24).
(c) Assume the initial durability design factor φ.
(d) Determine the design concrete cover using Equation (21).
(e) Calculate β_i of structures that have the design concrete cover determined in step (d). In this study, β_i values for structures in 38 locations in Japan are calculated ($i = 1, 2, \ldots, 38$). Seven locations in Hokkaido, 16 locations in Honshu,

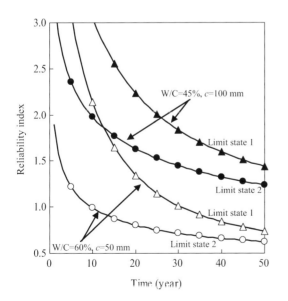

Figure 9. Reliability index for limit states obtained by Equations (6) and (9).

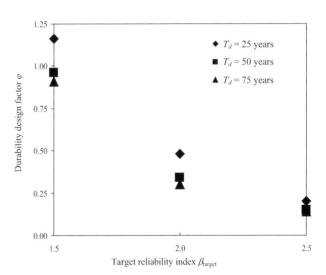

Figure 10. Relationship between lifetime of structure T_d, target reliability index β_{target}, and durability design factor φ.

Table 2. Reliability indices of reinforced concrete structures for different locations, distance from coastline, β_{target}, T_d, based on proposed design criterion and durability design factors.

Name of town or city	Island	$\beta_{\text{target}} = 2.0$, $T_d = 25$ years ($\varphi = 0.48$) Distance from coastline = 0.1 km			$\beta_{\text{target}} = 2.0$, $T_d = 50$ years ($\varphi = 0.34$) Distance from coastline = 0.1 km			$\beta_{\text{target}} = 1.5$, $T_d = 75$ years ($\varphi = 0.91$) Distance from coastline = 0.3 km			$\beta_{\text{target}} = 2.5$, $T_d = 25$ years ($\varphi = 0.20$) Distance from coastline = 0.3 km		
		W/C = 30%	W/C = 45%	W/C = 60%	W/C = 30%	W/C = 45%	W/C = 60%	W/C = 30%	W/C = 45%	W/C = 60%	W/C = 30%	W/C = 45%	W/C = 60%
Wakkanai	Hokkaido	1.91	1.97	2.00	2.00	2.01	2.13	1.51	1.44	1.44	2.44	2.51	2.58
Rumoi		1.91	1.97	1.99	2.00	2.00	2.13	1.51	1.43	1.43	2.44	2.50	2.58
Otaru		1.93	2.00	1.95	2.03	1.97	2.10	1.56	1.42	1.44	2.48	2.46	2.56
Esasi		2.08	1.95	1.98	1.98	2.06	2.17	1.46	1.39	1.39	2.40	2.46	2.62
Tomakomai		1.94	2.01	1.96	2.03	1.97	2.11	1.57	1.43	1.46	2.50	2.47	2.57
Kushiro		1.92	1.99	2.01	2.01	2.02	2.09	1.55	1.47	1.43	2.47	2.45	2.61
Nemuro		1.91	1.98	2.00	2.00	2.01	2.14	1.52	1.45	1.40	2.45	2.52	2.60
Hukaura	Honshu	2.09	1.97	1.97	2.06	2.00	2.05	1.64	1.52	1.54	2.49	2.38	2.47
Sakata		1.90	1.97	2.00	1.99	2.00	2.13	1.50	1.43	1.43	2.43	2.50	2.58
Miyako		2.01	1.99	1.96	1.98	1.99	2.08	1.54	1.48	1.48	2.41	2.47	2.52
Ishinomaki		1.98	1.96	2.01	1.94	1.95	2.05	1.50	1.51	1.48	2.38	2.45	2.55
Onahama		2.00	1.97	1.94	1.96	1.97	2.06	1.52	1.53	1.46	2.40	2.45	2.51
Niigata		1.93	2.00	1.95	2.02	1.97	2.10	1.56	1.42	1.45	2.49	2.46	2.56
Fushiki		1.98	1.95	2.00	1.94	2.02	2.10	1.49	1.50	1.48	2.37	2.44	2.55
Wajima		1.97	1.94	1.99	1.93	2.01	2.08	1.47	1.48	1.46	2.53	2.42	2.54
Shiga		2.11	1.99	1.99	1.94	2.02	2.02	1.66	1.55	1.52	2.51	2.41	2.42
Maizuru		1.95	2.03	1.95	1.98	1.99	2.06	1.57	1.60	1.52	2.36	2.44	2.45
Sakai		2.00	1.98	1.95	1.97	1.97	2.07	1.52	1.47	1.46	2.40	2.46	2.51
Hamada		2.04	2.02	1.99	2.01	2.02	2.06	1.58	1.53	1.48	2.44	2.42	2.49
Hagi		2.05	1.93	2.00	2.02	1.96	2.06	1.59	1.54	1.49	2.45	2.43	2.49
Shionomisaki		1.97	1.94	1.99	1.93	2.01	2.09	1.47	1.49	1.46	2.54	2.43	2.54
Owase		1.98	1.95	2.00	1.94	1.95	2.05	1.50	1.51	1.48	2.37	2.44	2.55
Katsuura		1.96	1.93	1.98	1.92	2.00	2.08	1.46	1.48	1.44	2.53	2.50	2.52
Uwajima	Shikoku	1.94	1.91	1.96	2.03	1.98	2.11	1.58	1.44	1.46	2.50	2.47	2.57
Muroto		2.08	1.95	1.95	2.05	1.98	2.03	1.62	1.57	1.52	2.48	2.46	2.45
Fukuoka	Kushu	2.08	1.96	1.96	2.05	1.99	2.04	1.63	1.50	1.53	2.47	2.46	2.46
Nagasaki		2.05	1.92	2.00	2.02	1.96	2.07	1.59	1.54	1.49	2.46	2.43	2.49
Ushibuka		2.10	1.97	1.97	2.07	2.01	2.06	1.65	1.53	1.51	2.50	2.38	2.47
Makurazaki		2.09	1.96	1.97	2.06	1.99	2.05	1.64	1.52	1.49	2.49	2.38	2.46
Aburatsu		1.93	1.99	1.94	2.02	2.03	2.09	1.55	1.48	1.43	2.47	2.45	2.61
Miyako	Okinawa Islands	1.97	1.94	1.99	1.93	2.01	2.09	1.47	1.48	1.45	2.53	2.51	2.53
Ishigaki		1.98	1.95	2.00	1.94	2.02	2.10	1.49	1.51	1.48	2.36	2.44	2.55
Iriomote		1.99	1.96	2.01	1.95	1.96	2.05	1.51	1.51	1.49	2.39	2.45	2.55
Minamidaito		1.99	1.97	1.94	1.95	1.96	2.06	1.51	1.52	1.50	2.39	2.45	2.56
Kume		2.00	1.96	1.94	1.96	1.96	2.06	1.51	1.53	1.45	2.39	2.46	2.51
Nago		2.00	1.97	1.94	1.96	1.97	2.06	1.52	1.53	1.46	2.39	2.46	2.50
Naha		2.01	1.99	1.96	1.98	1.98	2.08	1.54	1.48	1.48	2.41	2.47	2.52
Mean		1.99	1.97	1.98	1.99	1.99	2.08	1.54	1.49	1.47	2.45	2.45	2.53
Standard Deviation		0.06	0.03	0.02	0.04	0.03	0.03	0.06	0.05	0.03	0.05	0.03	0.05

two locations in Shikoku, five locations in Kyushu, and eight locations in Okinawa islands. Hokkaido, Honshu, Shikoku, Kyushu, and Okinawa islands are shown in Figure 2.

(f) Repeat steps (c) to (e) until

$$U = \sum_{i=1}^{38} \left(\beta_{\text{target}} - \beta_i(\varphi) \right)^2 \quad (25)$$

is minimised, and the durability design factor is found.

4.2. Durability design factors

RC structures with concrete cover specified by the JSCE Standard Specifications (2002) and Japan Road Specifications for Highway Bridges (2002) have reliability indices ranging from 1.5 to 2.5 for the occurrence of corrosion cracking. Since these specifications ignore the difference in the amount of airborne chlorides to determine the concrete cover, reliability index depends on the location. RILEM (1998) proposed that durability requirements for serviceability limit state have to use the target reliability indices $\beta_{\text{target}} = 2.5$, when the consequences of a durability failure are noticeable and the repair costs are high, and $\beta_{\text{target}} = 1.5$, when there are no noticeable consequences of a durability failure. In this study, β_{target} is set to be 1.5, 2.0 and 2.5, and the lifetime of the structure is set to be 25 years, 50 years and 75 years. Minimum concrete cover is assumed 10 mm.

Durability design factors φ minimising U (see Equation (25)) are calculated for each target reliability index and lifetime considered. Figure 10 shows the relationship between lifetime, T_d, durability design factor, φ, and target reliability index, β_{target}. It is confirmed in Figure 10 that φ is more sensitive to β_{target} than to T_d. The reliability indices for different locations, distances from coastline, target reliability indices and prescribed lifetime are indicated in Table 2. The results clearly indicate that the reliability indices are very close to the target values. Therefore, by using the design criterion and durability design factor proposed, structures having target durability reliability indices for prescribed lifetimes can be designed.

5. Conclusions

In this paper, integration of the effects of airborne chlorides into reliability-based durability design of RC structures in a marine environment was proposed.

(1) A computational procedure to establish the probabilistic hazard curve of airborne chlorides by applying the concept of seismic probabilistic

hazard assessment was proposed. This hazard curve can quantify the effect of marine environment.

(2) A method for calculating the occurrence probability of steel corrosion and corrosion cracking due to corrosion products in RC structures was proposed based on hazard curves of airborne chlorides and fragility curve.

(3) A discussion of durability design factors and criterion for designing RC structures that satisfy the target durability reliability level was presented.

(4) The proposed approach assumes that hazard of airborne chlorides depends only on the ratio of sea wind, average wind speed, and the distance from the coastline. The effects of other factors, such as the amount of rain and level of humidity on the hazard of airborne chlorides, need to be investigated. Also, temporal and spatial variations of corrosion rate (Marsh and Frangopol 2008) have to be incorporated.

(5) Further research is needed on the reliability of RC structures in a marine environment in connection with life-cycle maintenance and optimisation (Frangopol and Liu 2007).

Acknowledgements

The first author acknowledges support for this research by the Kajima Foundation. This study was carried out when the first author worked as a Visiting Research Associate at Lehigh University (October 2008–September 2009) in the research group of the second author.

References

Ellingwood, B.R., 2005. Risk-informed condition assessment of civil infrastructure: State of practice and research issues. *Structure and Infrastructure Engineering*, 1 (1), 7–18.

Frangopol, D.M., Lin, K.Y., and Estes, A.C., 1997a. Reliability of reinforced concrete girders under corrosion attack. *Journal of Structural Engineering, ASCE*, 123 (3), 286–297.

Frangopol, D.M., Lin, K.-Y., and Estes, A.C., 1997b. Life-cycle cost design of deteriorating structures. *Journal of Structural Engineering, ASCE*, 123 (10), 1390–1401.

Frangopol, D.M. and Liu, M., 2007. Maintenance and management of civil infrastructure based on condition, safety, optimisation, and life-cycle cost. *Structure and Infrastructure Engineering*, 3 (1), 29–41.

Japan Society of Civil Engineers (JSCE), 2002. *Standard specifications for concrete structures construction*. Tokyo, Japan: Maruzen.

Japan Road Association, 2002. *Specifications for highway bridges. III: Concrete bridges*. Tokyo, Japan: Maruzen.

Kameda, H. and Nojima, H., 1988. Simulation of risk-consistent earthquake motion. *Earthquake Engineering and Structural Dynamics*, 16, 1007–1019.

Kawamura, C., *et al.*, 2004. Investigation of construction errors of the cover for railway RC rigid frame viaducts. *Journal of Materials, Concrete Structures and Pavements, JSCE*, 64, 254–266 (in Japanese).

Maeda, S., Takewaka, K., and Yamaguchi, T., 2004. Quantification of chloride diffusion process into concrete under marine environment by analysis of salt damage data base. *Journal of Materials, Concrete Structures and Pavements, JSCE*, 63, 109–120 (in Japanese).

Marsh, P.S. and Frangopol, D.M., 2008. Reinforced concrete bridge deck reliability model incorporating temporal and spatial variations of probabilistic corrosion rate sensor data. *Reliability Engineering and System Safety*, 93, 394–409.

Matsushima, M., Nakagawa, T., and Tsutsumi, T., 2001. Study on estimation of deterioration of existing RC structures received chloride induced damage. *Journal of Construction Management and Engineering, JSCE*, 51, 93–100 (in Japanese).

Matsushima, M., *et al.*, 1998. A study of the application of reliability theory to the design of concrete cover. *Magazine of Concrete Research*, 50 (1), 5–16.

Matsushima, M., Yokota, M., and Seki, H., 2004. Corrosion products for cracking of surface concrete due to corrosion. *Proceedings of the Japan Concrete Institute*, 26 (2), 1669–1674 (in Japanese).

Nakagawa, T., *et al.*, 2004a. Assessment of corrosion speed of RC structure under the chloride deterioration environment. *In: Proceedings of JCI symposium on the analysis model supporting the verification of long-term performance of concrete structure in design*. October, 2004. Tokyo, Japan: JCI, 325–330 (in Japanese).

Nakagawa, H., *et al.*, 2004b. Experimental study on modes of concrete cracking and threshold amounts of corrosion for crack onset. *Doboku Gakkai Ronbunshuu E, JSCE*, 64 (1), 110–121 (in Japanese).

Prezzi, M., Geyskens, P., and Monterio, P.J.J., 1996. Reliability approach to service life prediction of concrete exposed to marine environments. *ACI Materials Journal*, 93 (6), 544–552.

Public Works Research Institute (PWRI), 1988. Survey on the geographical distribution of chloride particles carried by the sea wind. Technical note of Public Works Research Institute 3. Tsukuba, Japan (in Japanese).

Qi, L. and Seki, H., 2001. Analytical study on crack generation situation and crack width due to reinforcing steel corrosion. *Journal of Materials, Concrete Structures and Pavements, JSCE*, 50, 161–171 (in Japanese).

RILEM Technical Committee 130-CSL, 1998. *Durability design of concrete structures*. RILEM, Technical Research Center of Finland: E & FN SPON.

Sasatani, T., *et al.*, 1997. A study on the evaluation of chloride-ion penetration into the concretes under a marine environment. *Journal of Materials, Concrete Structures and Pavements, JSCE*, 36, 91–104 (in Japanese).

Tanaka, Y., *et al.*, 2001. Chloride penetration of high strength concrete. *Proceedings of the Japan Concrete Institute*, 23 (2), 517–522 (in Japanese).

Tsutsumi, T., *et al.*, 1996. Evaluation on parameters of chloride induced damage based on actual data. *Journal of Materials, Concrete Structures and Pavements, JSCE*, 32, 33–41 (in Japanese).

Val, D.V. and Stewart, M.G., 2003. Life-cycle cost analysis of reinforced concrete structures in marine environments. *Structural Safety*, 25, 343–362.

Fatigue system reliability analysis of riveted railway bridge connections

Boulent M. Imam, Marios K. Chryssanthopoulos and Dan M. Frangopol

A system-based model for fatigue assessment of riveted railway bridge connections, comprising a number of basic components, is presented in this article. Probabilistic fatigue load spectra are developed through Monte Carlo simulation of train passages over a finite element model of a typical, short-span bridge. Uncertainties arising from loading, resistance and modelling sources are taken into account. The riveted connection is treated through a set of generic sub-systems that capture potential damage in identifiable hot-spots, such as rivets, holes and angle fillets. The fatigue reliability over time is evaluated through system reliability methods by treating these hot-spots as the elements of a structural system. The results show that the probability of failure of the connection depends significantly on the form of the system adopted for the analysis and the rivet clamping force. Damage scenarios accounting for the potential loss of rivet clamping force are investigated, and it is shown that, in some cases, they can affect connection reliability considerably.

1. Introduction

Fatigue assessment of riveted bridges has received considerable attention during the past few decades. This is attributed to the fact that such bridges, which were not designed explicitly for fatigue, are believed to be approaching the end of their service life. Considering their large number, both in Europe and North America, replacement of all such bridges is extremely challenging from an economic point of view and would create severe network problems. In order to maintain these old bridges and plan repairs effectively, assessment of their remaining fatigue life is a vital requirement.

Riveted bridges were constructed using built-up sections, which comprise several typically flat-plated elements riveted together. The primary bridge members such as girders, stringers and floor-beams are joined through riveted connections consisting of a number of rivets and angle clips. The redundancy of these structures is generally high, since failure of an individual component does not usually lead to progressive collapse. This is because in a typical riveted member, a fatigue crack will propagate only within the component that has cracked and will not immediately transfer to other adjoining components since the interfaces between components act as crack stoppers (Sweeney 1979). However, following failure of the first

component and redistribution of the loads to the remaining components, a progressive type of failure may be initiated, depending on geometric configuration, acting load levels and component utilisation. It is, hence, both rational and desirable to extend methods to estimate the reliability of riveted bridges and connections beyond the level of individual component failure towards a systems reliability framework, which captures the relationships between individual component and overall system performance.

One of the appealing features of using a system approach to reliability is that it considers both the reliability of components and their relationship and importance to the entire system, thus allowing quantification of redundancy. In some cases, when a system is highly redundant, the reliability of individual components may be quite low while the system reliability remains high. In other cases, for weakest-link systems, although the reliability of every component may be satisfactory, the reliability of the system might be considered low. A system analysis is capable of capturing and quantifying such effects. On the other hand, considering individual components alone, which is still a common practice, may underestimate or overestimate the overall reliability of a structure. A number of recent studies have investigated the redundancy of structural systems from a system reliability point of view (Ghosn *et al.* 2010, Okasha and Frangopol 2010).

System reliability approaches have been commonly used in the past to estimate the service life of bridge structures (Hendawi and Frangopol 1994, Micic *et al.* 1995, Estes and Frangopol 1999, Nowak and Cho 2007). The deck, super-structure and sub-structure can be broken down to a varying degree and modelled as components of a higher level system representing the bridge. The reliability of different component failure modes can be estimated using methods such as First Order Reliability Method (FORM), Second Order Reliability Method (SORM) or through Monte Carlo simulations. A system representing the bridge is then assembled as a series-parallel combination of its components, and a system approach is employed to estimate the overall system reliability (Melchers 1999).

System reliability analysis of riveted bridges and connections has received limited attention, with only a few recent studies in that area. Wang *et al.* (2006, 2007) developed a system fatigue damage reliability model based on fracture mechanics principles investigating the behaviour of a riveted built-up girder. Probabilistic crack growth analysis of cracks initiating from the holes of angle clips was carried out and combined through systems analysis to estimate the reliability of the riveted girder. As a case study, their model was used to determine safe inspection intervals and maintenance strategies for an existing riveted bridge in China.

The objective of this article is to present a system-based model for fatigue assessment of riveted bridge connections building on previous work, which has quantified a number of important features exhibited by riveted grillage-type structures (Imam *et al.* 2006, 2008, Righiniotis *et al.* 2008). A finite element (FE) model of a short-span, wrought iron bridge, is used in order to convert train loading into fatigue load effects. Fatigue assessment is carried out using the recently developed theory of critical distances (TCD), which considers the entire distribution ahead of any given stress concentration. This theory (Taylor 2007) allows the criticality of different parts of a fatigue-sensitive detail to be independently assessed and lends itself to a system-based treatment of fatigue failure. A system approach is presented by treating each riveted connection through generic sub-systems that capture potential damage in identifiable hot-spots, such as rivets, holes and angle fillets. By treating these hot-spots as the elements of a structural system susceptible to fatigue failure, its reliability over time is evaluated using system reliability methods. The relative importance of various connection components, given their individual failure probability and their contribution to the system reliability, is discussed. The effect of different damage scenarios, in the form of loss of rivet clamping force, on the overall system reliability of the connection is also investigated through parametric studies. The results are compared with their S–N counterparts, which were obtained by the authors in earlier studies (Imam *et al.* 2008) by employing the traditional nominal stress approach, commonly adopted in bridge codes (BS5400 1980, AREA 1996), and modelling the connection as a single component.

2. Fatigue assessment using the TCD

The most popular method of fatigue analysis of structural components and connections is the well-known S–N method, which is employed by the majority of bridge codes (BS5400 1980, AREA 1996, AASHTO 1998, EC3 2005). The S–N method requires the classification of a detail into a fatigue class among typical standardised 'basic details', and its fatigue damage is estimated using the relevant S–N curve combined with remotely applied nominal stresses. Despite the fact that existing standards and recommendations cover a wide range of structural details and loading situations, there are a number of complex details that cannot be clearly classified, or even if they can, the definition of nominal stresses for these is far from straightforward. One such example is the deformation-induced fatigue that is caused by secondary stresses. In riveted bridges, such type of fatigue damage may take place at riveted connections such as, for example, stringer-to-cross-girder connections (Al-Emrani 2002). Conventional methods for fatigue analysis, such as the S–N nominal stress method, have in many cases proven themselves to be inappropriate for deformation-induced fatigue (Imam 2006). Assessment of existing bridges with such details using the conventional methods may render unrealistic results that can make a correct condition assessment and planning of maintenance strategies more difficult.

During the last decade or so, alternative fatigue design and assessment methods have been developed, such as the geometric (or hot-spot) stress approach (DNV 2005), the structural stress approach (Dong 2001) and the TCD (Taylor 2007). The hot-spot stress approach has been widely used in the offshore industry, and the TCD has been commonly employed in fatigue assessment of mechanical and automotive components. All of these methods are based on linear elastic FE analysis and can be used for fatigue assessment of complex details for which the S–N method is difficult to utilise. However, these approaches have so far rarely been used by bridge engineers and need to be developed and adapted for bridge assessment applications.

The hot-spot (or geometric) stress approach is based on extrapolation of stresses obtained at a distance from the stress concentration (notch) of the fatigue detail being investigated. On the other hand, the structural stress method is based on defining an equivalent stress state at some distance from the stress concentration and considered for fatigue calculations. Both of these methods have been found to give reasonable results for two-dimensional (2D) welded details but have been shown to have certain limitations for three-dimensional (3D) details and are difficult to apply in cases of complex 3D loadings (Doerk *et al.* 2003). In addition, their suitability for the fatigue assessment of non-welded details, such as riveted connections, has not been established so far.

The TCD is a recently developed methodology for fatigue assessment of notched components (Taylor 2007), in which the entire stress distribution ahead of the notch is considered for fatigue calculations. In other words, TCD does not rely on a single, remotely applied stress, as postulated by the S–N approach, which may be difficult to define, or on a stress at a single stress concentration point. This theory, which can be applied to any type of stress concentration, was developed so as to be used in conjunction with linear elastic FE analyses. These have advanced extensively during recent years and allow analyses of complex 3D geometries and reasonably accurate predictions of stress fields in the vicinity of stress concentrations.

One of the main advantages of the TCD is that it requires physical characterisation of the plain material only, rather than the extensive, but often lacking, physical testing on specific detail geometries. A fundamental aspect of the TCD is to determine a location or region within which all physical processes leading to fatigue crack initiation are assumed to take place. This size/location (or region) depends on fatigue material properties and more specifically on the fatigue limit of the plain material and its crack propagation threshold (threshold stress intensity factor). All stress calculations required for fatigue assessment are carried out within this region, the size/location of which can be characterised through a 'critical distance', L, defined as (Taylor 1999)

$$L = \frac{1}{\pi}\left(\frac{\Delta K_{\text{th}}}{\Delta \sigma_0}\right)^2 \qquad (1)$$

where ΔK_{th} is the threshold stress intensity (crack propagation threshold) and $\Delta \sigma_0$ is the fatigue limit of the plain (unnotched) material.

Following the definition of the critical distance L, there are different methods that may be employed for fatigue assessment based on this parameter. The point method suggests considering the maximum principal stress $\sigma_{\text{p(max)}}$ at a distance of $L/2$ from the tip of the notch (Taylor 1999). The line method suggests averaging the maximum principal stress along a line of length $2L$ starting from the notch tip (Taylor 1999). In the area method, the maximum principal stress is averaged over a semi-circular area, which is centred at the notch tip, having a radius of $1.32L$ (Bellett *et al.* 2005). For 3D applications, the volume method, which relies on averaging principal stresses over a hemisphere centred at the tip of the notch with a radius of $1.54L$, has been proposed (Bellett *et al.* 2005). After determining the 'characteristic stress' by one of these methods, this is then used for fatigue damage calculations.

Although the point and line methods may appear less complicated, analysis using the area and volume methods is not difficult to implement as part of post-processing after an FE analysis. The volume method lends itself to the analysis of 3D geometries where there are 3D stress fields around the investigated stress concentration, since the method can capture stress gradients in all directions.

The reliability of the TCD for fatigue assessment has been verified through comparisons of analytical and experimental results on a wide range of stress concentrations and materials, where it was shown that it can predict accurately both fatigue strength and fatigue life (Taylor 1999, 2005, Taylor and Wang 2000, Susmel and Taylor 2003).

A further appealing feature of the TCD is that it overcomes the mesh sensitivity problem that exists in FE analysis of stress concentrations where, in general, the maximum stress at notches does not converge with increasing mesh density since these points are usually associated with numerical singularities. As a result, the fatigue damage becomes unbounded. Since the TCD relies on the entire stress distribution ahead of the stress concentration and not on a single maximum stress value, it can overcome this problem. Previous studies carried out on a typical riveted bridge connection model have demonstrated the convergence characteristics of the TCD volume method (Righiniotis *et al.* 2008).

3. Reliability analysis framework

3.1. FE analysis

A probabilistic methodology has been employed in earlier studies by the authors for the fatigue assessment of a typical wrought iron riveted railway bridge (Imam *et al.* 2008). It is based on the S–N nominal stress method and accounts for uncertainties in loading, resistance and modelling. A global FE model of the bridge (shown in Figure 1) was used to convert present day and historical rail traffic into fatigue load effects.

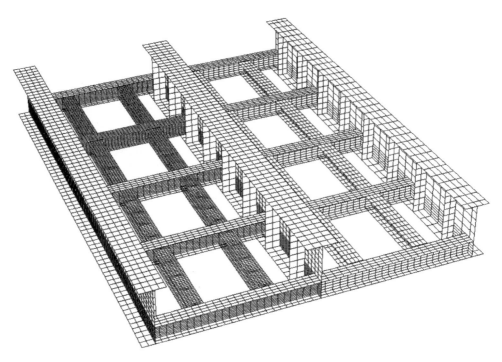

Figure 1. Global FE model of the investigated riveted bridge (Imam *et al.* 2007).

Annual fatigue response spectra were developed for different periods of the traffic models, spanning over 100 years, and were used for the estimation of the fatigue reliability of a stringer-to-cross-girder connection over time.

As the global model of the bridge treated the connection as a single entity and was unable to capture the fatigue behaviour of its individual elements, a detailed sub-model of the connection was later developed within the global model. This model, which has been fully described and validated in previous studies (Imam *et al.* 2007), is shown in Figures 2 and 3. As can be seen, all the individual elements of the connection, such as angle clips and rivets, and parts of the connected members in the vicinity of the connection were modelled explicitly to investigate their fatigue behaviour and assess their fatigue criticality. Fatigue damage calculations, which were carried out deterministically, were based on the TCD volume method, and the fatigue critical hot-spots on the connection were found to be located around the perimeter of the holes, around the circumference of the rivet head-to-shank intersection and along the depth of the angle fillet (Righiniotis *et al.* 2008).

In this article, the previous deterministic fatigue analyses using the TCD method (Righiniotis *et al.* 2008) are extended within a probabilistic framework. To this end, principal stress histories are obtained in the vicinity of the critical hot-spots of the connection (i.e. holes, rivets and angle clip fillets) by traversing

trains over the refined FE model of the bridge. The histories are obtained at the elements located within the critical volumes, which are in the form of a hemisphere centred at the notch and having a radius of 1.54 L. These stress histories are then combined in the context of the TCD volume method to obtain the characteristic (averaged within the volume) stress histories at each hot-spot (stress concentration), which are used with the rainflow counting method and Miner's rule (Miner 1945) to calculate fatigue damage. Uncertainties in loading, resistance and modelling are introduced at this stage.

3.2. Random variables

3.2.1. Resistance random variables

On the resistance side, uncertainties in the S–N curves and the cumulative damage model (Miner's rule) are incorporated within the methodology. For fatigue damage calculations, the S–N curve is treated probabilistically by assuming a fixed slope and a random fatigue limit. For wrought iron, a lognormal distribution with a mean value of 183 MPa defined at $N_{FL} = 2 \times 10^6$ cycles and a coefficient of variation (CoV) of 0.1 is assumed for the fatigue limit $\Delta\sigma_0$ (Righiniotis *et al.* 2008) together with a slope of 1/5 (Cullimore 1967). In graphical terms, the effect of random variability of the fatigue limit is to produce a parallel shifting of the S–N curve with the fatigue limit always defined at 2×10^6 cycles.

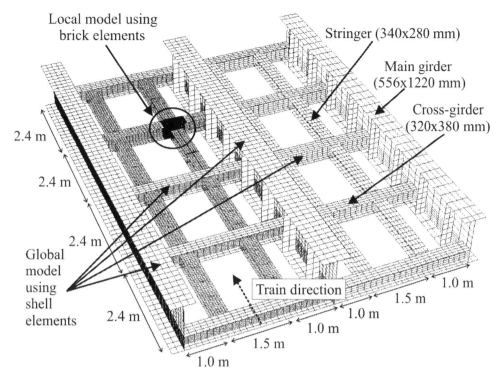

Figure 2.   Refined FE model of the investigated riveted bridge (Imam *et al.* 2007).

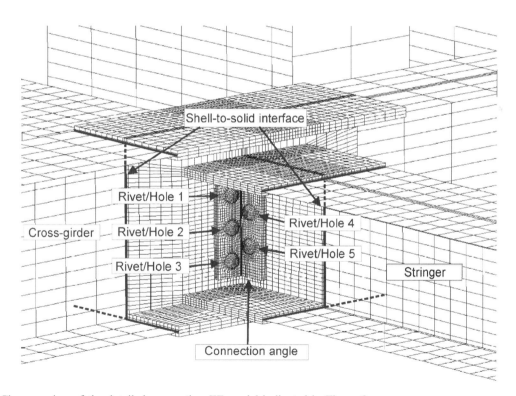

Figure 3.   Close-up view of the detailed connection FE model indicated in Figure 2.

The damage limit Δ in Miner's sum is also assumed to be a random variable. In deterministic analysis, this damage limit, which indicates fatigue failure, is assumed equal to one. Herein, a lognormal distribution with a mean value of 0.90 and a CoV of 0.30 is considered in accordance with earlier studies (Wirsching 1995).

Table 1. Characteristics of random variables for reliability analysis.

Random variable	Distribution	Mean	CoV	Description
$\Delta\sigma_0$	Lognormal	183 MPa	0.10	Fatigue limit
Δ	Lognormal	0.90	0.30	Damage limit in Miner's sum
DAF	Normal	1.10	0.14	Dynamic amplification factor
α	Normal	0.80	0.14	Ratio of actual-to-calculated stresses (modelling uncertainty)
f_{tj}	Lognormal	From BS5950 (1980)	0.14	Annual frequency for train j

The critical distance L, which is used to define the critical volume for carrying out the stress calculations for fatigue assessment, has been taken as deterministic due to the large amount of FE post-processing required by the TCD method. By using the mean value for the fatigue limit of wrought iron ($\Delta\sigma_0 = 183$ MPa) and a threshold stress intensity value of $\Delta K_{th} = 13.5$ MPa·m$^{1/2}$, the critical distance value is obtained as $L = 1.73$ mm. The value of the threshold stress intensity (crack propagation threshold) has been determined through crack growth experiments on wrought iron specimens reported by Helmerich et al. (2007).

3.2.2. Loading random variables

On the loading side, the problem is randomised by introducing uncertainties in train traffic volume and dynamic amplification. Live loading on the bridge is represented by the BS5400 (1980) medium traffic model, which is also adopted in the UK for present rail traffic. The annual frequency f_{tj} for each train type j crossing the bridge is obtained from BS5400 and is assumed to be lognormally distributed with a CoV of 0.14 (Imam et al. 2008). Randomness in train frequencies is, of course, directly linked to the annual number of applied stress cycles. For the dynamic amplification factor (DAF), a normal distribution with a mean value of 1.10 and a standard deviation of 0.15 is assumed. The latter is based on field measurements (Byers 1970, Tobias and Foutch 1997) on short-span steel railway bridges.

Further uncertainty in loading is incorporated by considering the epistemic or modelling uncertainty, represented by the ratio of actual-to-calculated stresses. Modelling uncertainty is here captured by the ratio of actual-to-calculated stresses, α, which is assumed to follow a normal distribution with a mean value of 0.80 and CoV of 0.14. These values are based on comparison of stresses obtained analytically or numerically with field measurements carried out in steel railway bridges (Byers 1976, Adamson and Kulak 1995, Sweeney et al. 1997).

A summary of all the random variables and their probabilistic modelling characteristics is shown in Table 1, whereas Table 2 presents the deterministic variables used in the analyses.

Table 2. Characteristics of deterministic variables for reliability analysis.

Deterministic variable	Mean	Description
ΔK_{th}	13.5 MPa·m$^{1/2}$	Threshold stress intensity
L	1.73 mm	Critical distance
m	3	Slope of S–N curve

3.3. Fatigue reliability calculations

The first step in the estimation of fatigue reliability is the development of annual fatigue load spectra (stress range histograms) for each critical hot-spot on the connection, which will subsequently be used for fatigue damage calculations. These spectra are developed by multiplying the static stress ranges (determined from the averaged, within the volume, maximum principal stress histories) obtained from previous deterministic analyses of the bridge (Righiniotis et al. 2008) with the DAF and the α factors. For each individual train crossing, a different value of DAF and α is sampled from their probability distributions in order to account for randomness. This process of calculating the stress ranges is carried out f_{tj} times, where f_{tj} is the annual frequency of train type j. In effect, a Monte Carlo simulation is carried out, with random variables DAF and α sampled from assumed distributions. Since the train frequencies f_{tj} are also taken to be random, the above process is repeated 10^3 times, in order to capture the uncertainty in the train frequencies.

The fatigue damage in each hot-spot is then calculated from its annual fatigue load spectrum (stress range histogram). The annual damage is given as

$$D_a = \frac{1}{N_{FL}(\Delta\sigma_0)^5} \sum_{i}^{k} n_i \left[(\Delta\sigma_{ave})_i\right]^5 \quad (2)$$

where k is the number of stress range blocks in the fatigue load spectrum, $(\Delta\sigma_{ave})_i$ is the ith stress range and n_i is the corresponding number of applied cycles. It should be recalled that $(\Delta\sigma_{ave})_i$ are the stress ranges obtained from the characteristic stress history, which is calculated by averaging the maximum principal stress histories within the critical volume. Since train loading is of variable amplitude, in fatigue damage calculations through Equation (2), the S–N curve is extended below $\Delta\sigma_0$ with the same slope (1/5).

The probability of failure is calculated by summing up the annual damages ($D_{a,l}$) over the years, where l represents any 1 year, and considering a limit state function given as

$$g = \Delta - \sum_l^T D_{a,l} \qquad (3)$$

where T is the number of years after 1970. The reason for performing the fatigue damage calculation from 1970 onwards is that up to that period fatigue damage was quite small, as demonstrated in earlier studies (Imam *et al.* 2006). For remaining fatigue life calculations, future load evolution is ignored, and the BS5400 medium traffic is extrapolated into the future up to the point of fatigue failure ($g \leq 0$). Thus, the remaining fatigue life T_r of any hot-spot is given as

$$T_r = \frac{\Delta - \sum_l^T D_{a,l}}{D_a} \qquad (4)$$

In Equation (4), variable D_a appearing in the denominator is the mean value of the annual damage, which is calculated from the annual fatigue load spectrum. For each Monte Carlo cycle, a single D_a value is generated and used.

Based on the limit state function of Equation (3), the probability of fatigue failure in any 1 year can be defined as

$$P_f = P[g(X) \leq 0] = \int_{g(X) \leq 0} f_X(x)dx$$
$$= P[T_r < t] \cong \frac{\Pi[g(X) \leq 0]}{\Pi_{\text{total}}} \qquad (5)$$

where $g \leq 0$ represents the 'failure' domain and $f_X(x)$ is the (unknown) joint distribution of the random variables entering the problem, and the integral is evaluated over the failure domain. In Monte Carlo simulation, the failure probability can be approximately evaluated as the ratio of $\Pi[g(X) \leq 0]$, which is the number of samples satisfying the event in brackets, over Π_{total}, which is the total number of samples generated in the simulations. In this article, the time-varying fatigue failure probability is estimated using Monte Carlo simulation with 10^6 samples, in accordance with a convergence study (Imam 2006).

3.4. System modelling

The connections in riveted bridges consist of a number of components; failure of an individual component may not lead to total connection failure. For example, a typical stringer-to-cross-girder connection, which is the focus of this investigation, consists of four angle clips and 10 rivets as shown in Figure 4. The approach described in Section 3.3 can be used to estimate the probability of fatigue failure of individual hot-spots on various connection components, e.g. of a single rivet, of a single hole, etc. These individual probabilities of failure can then be combined using a system reliability approach to estimate the overall reliability of the connection.

The connection can be considered as a system consisting of four sub-systems, each representing a single angle clip. Each angle clip, in turn, consists of various hot-spots as its individual elements (e.g. holes, rivets and fillets). For example, it can be seen in Figure 4a that there are three holes and three rivets on the leg of each angle clip that is connected to the cross-girder web (rivets/holes 1–3), whereas there are two holes and two rivets on its leg connected to the stringer web (rivets/holes 4–5). Furthermore, two hot-spots are considered along the fillet depth (fillets 1–2). Combined failure in hot-spots will result in angle clip failures, whereas combined angle clip failures will in turn imply total connection failure.

It is assumed that an individual angle clip can fail by one of the following five potential failure modes:

- Mode 1: failure of all three holes (holes 1–3) on the cross-girder leg of the angle clip
- Mode 2: failure of both holes (holes 4–5) on the stringer leg of the angle clip
- Mode 3: failure of all three rivets (rivets 1–3) on the cross-girder leg of the angle clip
- Mode 4: failure of both rivets (rivets 4–5) on the stringer leg of the angle clip
- Mode 5: failure of both hot-spots along the depth of the angle fillet (fillets 1–2).

The probability of failure of each of these five modes can be estimated from the probabilities of failure of the individual hot-spots that contribute towards its failure (i.e. holes, rivets and fillets). For instance, since failure in mode 1 is assumed to take place when all three holes 1–3 have failed, this can be considered as a parallel sub-system with the holes 1–3 being its individual elements. Thus,

$$P[\text{Mode 1}] = P[\text{Hole 1} \cap \text{Hole 2} \cap \text{Hole 3}] \qquad (6)$$

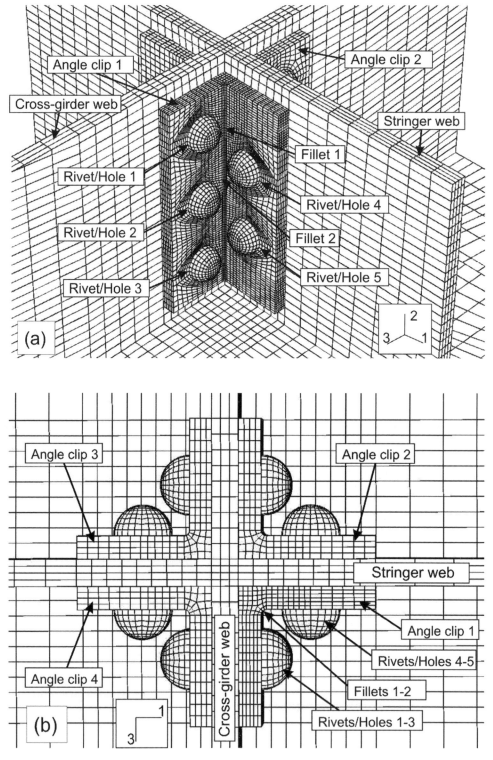

Figure 4. (a) Close-up view of the connection showing individual components (hot-spots). (b) Cross-sectional view of connection showing individual angle clips.

where the events in brackets imply failure at the named hot-spots. Assuming that hot-spot failures are statistically independent, the probability of mode 1 failure can be estimated as

$$P_{f(\text{Mode 1})} = P_{f(\text{Hole 1})} P_{f(\text{Hole 2})} P_{f(\text{Hole 3})} \qquad (7)$$

The failure probability in any one of the remaining modes can be estimated in a similar fashion. It is important to note that correlation between failure modes should be expected. Assuming statistical independence herein is a conservative assumption, i.e. it increases the probability associated with each mode of failure.

Since failure in any of the five modes given above postulates failure of an angle, an angle clip can be considered as a series system consisting of these individual elements contributing to each failure mode. A system representation of an angle is shown in Figure 5. Thus,

$$P[\text{Angle}] = P[\text{Mode 1} \cup \text{Mode 2} \cup \ldots \cup \text{Mode 5}] \quad (8)$$

and the probability of failure of an angle clip will be given as

$$P_{f(\text{Angle})} = 1 - \left[\left(1 - P_{f(\text{Mode 1})} \right) \cdots \left(1 - P_{f(\text{Mode 5})} \right) \right] \quad (9)$$

where each of the probabilities on the right-hand side can be calculated in a similar manner to Equation (7).

Furthermore, the entire stringer-to-cross-girder connection can be considered as a parallel system comprised of angles as individual elements (Figure 6). A number of assumptions can be made regarding the number of angle failures that would lead to total connection failure. The probability of failure of the connection assuming failure of one angle only is given by Equation (9). Assuming that any two angle failures would lead to total connection failure, the system probability of failure can be calculated as

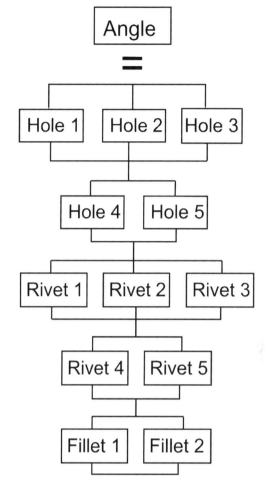

Figure 5. System representation of an angle clip subsystem.

is, therefore, not conservative. On the other hand, a conservative bound, which corresponds to the case of

$$P[\text{Connection}] = P\left[\begin{array}{l} (\text{Angle 1 and Angle 2}) \text{ or } (\text{Angle 1 and Angle 3}) \text{ or } (\text{Angle 1 and Angle 4}) \\ \text{or } (\text{Angle 2 and Angle 3}) \text{ or } (\text{Angle 2 and Angle 4}) \text{ or } (\text{Angle 3 and Angle 4}) \end{array} \right] \quad (10)$$

The bounds for the parallel system representation of the connection shown in Figure 6 can be expressed as

$$P_{f(\text{Angle 1})} P_{f(\text{Angle 2})} P_{f(\text{Angle 3})} P_{f(\text{Angle 4})} \leq P_{f(\text{Connection})}$$
$$\leq \min \left[P_{f(\text{Angle } i)} \right] \quad (11)$$

The lower bound, i.e. $P_{f(\text{Angle 1})} \ldots P_{f(\text{Angle 4})}$, corresponds to statistically independent failure modes and

perfectly correlated modes, is, in this case, given by the upper bound, i.e. $\min[P_{f(\text{Angle } i)}]$.

For the purposes of this investigation, only the stresses at different hot-spots on angle clip 1 were obtained and post-processed using the TCD. The stresses on the remaining angle clips (2–4) are assumed to be equal to those in angle clip 1, which in reality is not necessarily true. Consequently, in this article, the probability of failure of any angle is taken to be the same ($P_{f(\text{Angle 1})} = P_{f(\text{Angle 2})} = P_{f(\text{Angle 3})} = P_{f(\text{Angle 4})}$).

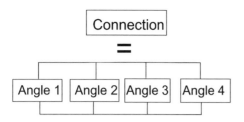

Figure 6. System representation of stringer-to-cross-girder connection.

3.5. *Damage scenarios*

The clamping force developed in a rivet is difficult to control, as it is affected by a number of factors such as the stiffness of the rivet and the connected plates, the driving and finishing temperatures, the method employed for riveting (Al-Emrani 2002) as well as the skill of individual workers. Field measurements have shown that there is a wide variation in the magnitude of the clamping stress developed with values ranging between 30 and 280 MPa, lower values being observed in rivets driven in the field. This variation is observed not only for connections with different geometric properties and driving conditions but also for rivets taken from the same riveted connections. In order to quantify the effect of rivet clamping force on the fatigue reliability of the investigated stringer-to-cross-girder connection, the analyses are carried out by assuming two levels of clamping for all rivets, i.e. a low-to-moderate clamping stress of 100 MPa and a high-clamping stress of 200 MPa.

Rivet defects may also affect the fatigue performance of a riveted connection, as shown in an earlier deterministic study (Imam *et al.* 2007). In this article, in addition to different rivet clamping stresses, the effect of a complete loss of rivet clamping force in each rivet is also investigated. Loss of clamping stress in a rivet may be gradual over a number of years through relaxation of the rivet due to vibrations and fretting between the connection components (Al-Emrani 1999). Moreover, tensile overloads in the rivets may also cause local yielding in the rivet head-to-shank junction leading to a partial or even complete loss of clamping force.

4. Results and discussion

4.1. *Critical volumes for TCD*

As presented earlier, the TCD method is based on determining a critical distance that defines the region within which stress averaging is to be performed for fatigue damage calculations. The critical volumes at the different hot-spots of the connection under investigation have been determined in earlier studies (Righiniotis *et al.* 2008). For presentation purposes,

some examples of these critical volumes, which are in the form of a hemisphere centred at the notch (stress concentration) and having a radius of $1.54\,L$, are also shown here in Figures 7 and 8. In this article, the stresses are estimated using a fixed $L = 1.73$ mm, as determined in Section 3.2.1. Within the TCD context, maximum principal stresses are averaged within these critical volumes, and the average maximum principal stress history is determined for each train that passes over the bridge. The rainflow counting method is used to convert these stress histories into stress range blocks, which are then updated, within a probabilistic framework (as described in Section 3) through Monte Carlo simulations, to generate annual response spectra for each hot-spot within the connection.

4.2. *Fatigue load spectra*

Figures 9 and 10 show typical probabilistic fatigue load spectra generated for hot-spots on the angle fillet 1 and hole 3 of the connection, respectively (see Figure 4a). The corresponding means and standard deviations of the stress range are also depicted in the figures. As mentioned previously, the spectra include the random effect of DAF, α and f_{tj}. The results pertain to the case of a 200 MPa clamping stress in all the rivets of the connection.

It is evident that most stress cycles experienced by these particular hot-spots are below the mean fatigue limit of the material (183 MPa). This was also observed in the fatigue load spectra developed in earlier studies using nominal, instead of local, stresses (Imam *et al.* 2008). Probability distributions may also be fitted to the fatigue load spectra as was done in the past by the authors in the case of the S–N nominal stress analysis of the connection (Imam *et al.* 2008). The challenge in using analytical distributions is to capture accurately the stress ranges that are close to the fatigue limit (right tail of the histograms in Figures 9 and 10) since these give the highest contribution to the overall fatigue damage.

Table 3 summarises the first and second order moment statistics for the annual fatigue load spectra of each connection hot-spot and for the two values of rivet clamping stress (100 and 200 MPa). The mean stress range at a detail can be considered as an important parameter in terms of fatigue reliability estimation and fatigue assessment. It can be seen that a higher clamping force in the rivets is beneficial in terms of reducing the mean stress range experienced by the majority of the hot-spots in the connection. The exception is at the angle fillets, where it can be seen that an increase in rivet clamping force results in increase in their mean stress range, which can be attributed to the higher restraint experienced by the angle clip.

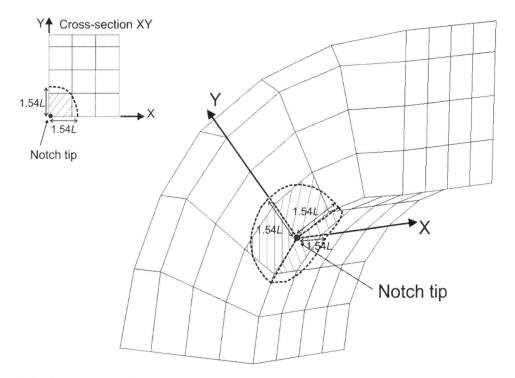

Figure 7.   Critical volume around a hole perimeter.

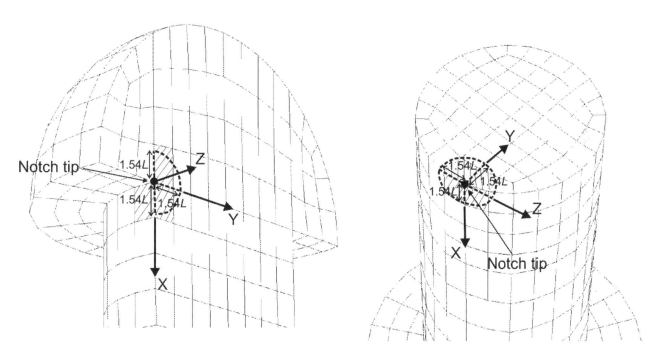

Figure 8.   Critical volume around the rivet perimeter at its head-to-shank intersection.

4.3.   *Fatigue reliability*

The fatigue damage in each hot-spot is calculated from its annual fatigue load spectrum, similar to Figures 9 and 10. The probability of failure and the remaining

fatigue life are, then, estimated through Monte Carlo simulations using Equations (4) and (5), respectively.

Figures 11 and 12 show the probability of fatigue failure from year 2010 onwards for the different connection hot-spots shown in Figure 4a. The results

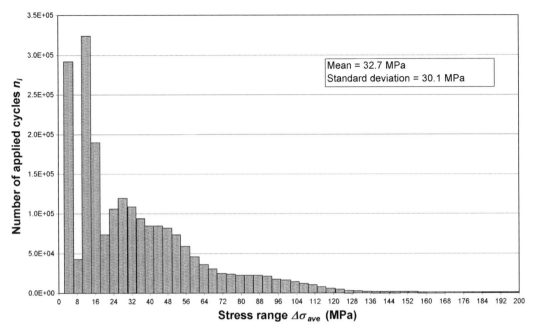

Figure 9. Annual fatigue load spectrum for angle fillet 1 (rivet clamping stress = 200 MPa).

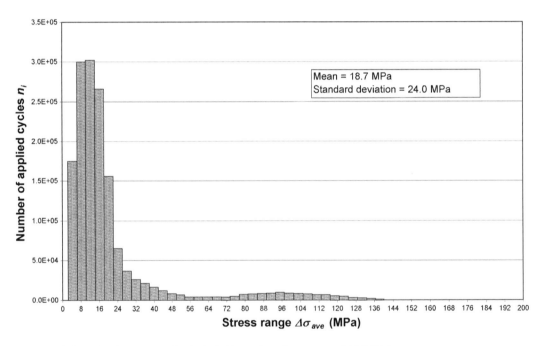

Figure 10. Annual fatigue load spectrum for hole 3 (rivet clamping stress = 200 MPa).

in Figure 11 are presented for a 100 MPa clamping stress in all rivets, whereas Figure 12 is for the case of 200 MPa clamping stress. As mentioned earlier, these hot-spots are then treated as the individual elements of a structural system representing an angle clip. The probability of failure of the system, comprising of one angle only and calculated using Equation (9), is also shown in these figures for comparison purposes.

It is evident from Figures 11 and 12 that there is a very wide range in the probabilities of failure of the different hot-spots. This overall range appears to be narrower in the case of the lower rivet clamping stress of 100 MPa, where the failure probabilities are generally higher as compared to the case of 200 MPa clamping stress. The remaining fatigue lives of the hot-spots range, for a 2.3% (design) probability of failure, from a couple of years to > 100 years. This illustrates the localised nature of the fatigue phenomenon with the damage being initiated from specific stress concentrations within a connection.

An increase in rivet clamping force can be seen to decrease the probability of failure for all holes and all rivets of the connection. Although higher clamping results in higher initial stresses being present in the rivets and, therefore, higher stress ratio ($\sigma_{min}/\sigma_{max}$) values, it reduces considerably the stress ranges experienced as a result of train loading, thus resulting in an overall reduction in fatigue damage

Table 3. First and second order statistics for stress range spectra at hot-spots.

Connection hot-spot	Clamping stress (100 MPa)		Clamping stress (200 MPa)	
	Mean (MPa)	Standard deviation (MPa)	Mean (MPa)	Standard deviation (MPa)
Hole 1	30.0	24.3	17.2	18.3
Hole 2	32.4	21.9	25.2	18.1
Hole 3	22.0	25.8	18.7	24.0
Hole 4	38.4	27.4	27.9	24.0
Hole 5	43.9	34.6	35.1	31.8
Rivet 1	36.8	28.2	32.1	20.2
Rivet 2	35.2	28.1	28.6	22.4
Rivet 3	40.8	32.6	31.8	27.2
Rivet 4	31.6	24.3	26.4	21.5
Rivet 5	28.7	23.3	27.8	19.5
Fillet 1	30.2	28.3	32.7	30.1
Fillet 2	28.3	26.5	29.8	26.9

accumulation. Different stress ratios would result in an upward or downward shift of the S–N curve of the material; however, in this article, the effect of the stress ratio is neglected. Higher clamping in the rivet also affects the stresses experienced around the holes by introducing compression, which has a beneficial effect on reducing the overall stress ranges at the hot-spots.

It can also be seen from the Figures 11 and 12 that the fatigue criticality ranking of the hot-spots is affected by the magnitude of the rivet clamping force. However, the highest failure probability (lowest remaining life) for both values of clamping force is obtained for hole 5 (see Figure 4a), which is found to be the most critical location in the connection. The figures also show that a higher rivet clamping stress has an adverse effect on the fatigue reliability of the angle fillets 1 and 2 due to the higher restraint provided to the connection and accompanying larger stresses and associated stress ranges (see also Table 3). Therefore, in high clamping connections, fatigue cracking may initiate either from hole 5 or from the fillet area near the top side of the angle clip. Such type of cracking has been observed in the field in damaged stringer-to-cross-girder connections (Al-Emrani 2002).

The system failure probabilities, assuming one-angle, two-angle and four-angle systems, calculated through Equation (11) are presented in Figure 13 for the two assumed clamping stress values. As mentioned earlier, the probability of failure for a one-angle system model represents the upper-bound solution (right-hand side of Equation (11)), whereas the two- and four-angle models represent the lower-bound solution (left-hand

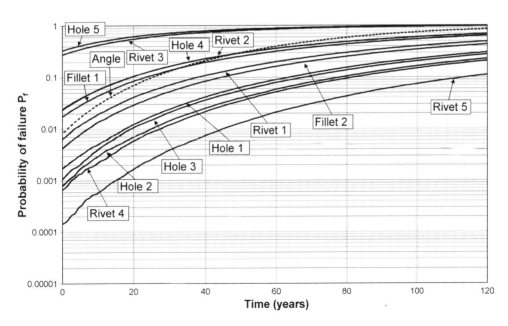

Figure 11. Probability of fatigue failure vs. time for different connection hot-spots and angle system (rivet clamping stress = 100 MPa).

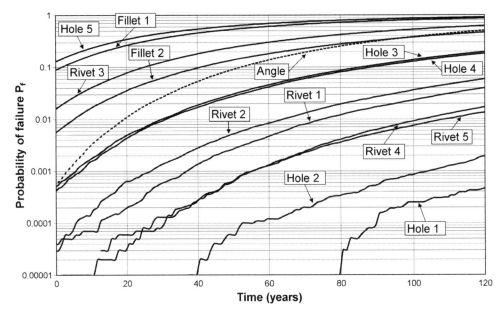

Figure 12. Probability of fatigue failure vs. time for different connection hot-spots and angle system (rivet clamping stress = 200 MPa).

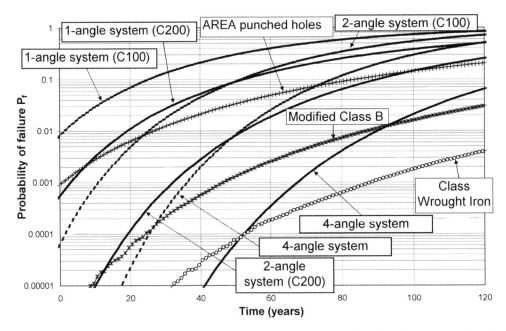

Figure 13. System probability of failure vs. time obtained using the TCD for two rivet clamping stress values (100 MPa (C100) and 200 MPa (C200)) compared to the probability of failure obtained using the S–N method (nominal stresses).

side of Equation (11)). In the same figure, the reliability profiles obtained in earlier studies by the authors using the nominal (S–N) stress method (Imam et al. 2008) and modelling the connection as a single component are also shown. The latter had been obtained for different fatigue detail S–N classifications according to British [Class Wrought Iron (Railtrack 2001), Modified Class B (BS5400 1980)] and American (AREA (1996) punched holes) bridge design/assessment codes. Class Wrought Iron refers to riveted wrought-iron connections at locations of rivets, whereas the S–N curve of Modified Class B has been shown to provide a good fit to available experimental fatigue results on riveted members (Imam 2006, Imam *et al.* 2008). The AREA punched holes classification refers to riveted details having normal clamping force and assembled by punching holes.

It can be seen in Figure 13 that, irrespective of the rivet clamping stress, the failure probability of the connection (system) consisting of one angle only is considerably higher than the case of assuming a two- or a four-angle system. In terms of remaining fatigue life, for a 2.3% probability of failure and 200 MPa clamping stress, a one-angle, two-angle and four-angle systems results in 28, 63 and > 100 years, whereas in the case of 100 MPa clamping stress in the rivets, the remaining life obtained is 8, 33 and 60 years, respectively. This demonstrates the significant effect of rivet clamping force on the fatigue reliability of the connection system, with higher clamping stresses reducing the probability of system failure of the connection for all system modelling assumptions (one, two and four angles).

By comparing the system reliability profiles with those obtained in previous studies through the traditional nominal stress (S–N) approach, it can be seen that a one-angle system assumption gives more conservative remaining life estimates than its nominal stress counterparts for both rivet clamping stress values. On the other hand, the remaining life estimates obtained from the four-angle system (for both clamping stress values) and the two-angle system (200 MPa clamping stress) can be seen to lie between its AREA (1996) punched holes and Class Wrought Iron (Railtrack 2001) counterparts. The TCD method employed in this article appears to result in a more rapid increase in the probability of failure of the connection with time as compared to the nominal stress method.

4.4. *Damage scenarios*

The effect of the different damage scenarios, in terms of complete loss of each rivet clamping force, on the fatigue reliability of the connection is shown in Figures 14, 15 and 16 for one-angle, two-angle and four-angle system representations of the riveted connection, respectively. The probability of failure of the undamaged connection over time is also shown in the figures for comparison purposes. The results are presented for the case of having a 200 MPa clamping stress in all the connection rivets.

It can be seen in Figures 14–16 that complete loss of the clamping force in rivet 3 has the most detrimental effect on the fatigue reliability of the connection. For a 2.3% probability of failure, loss of clamping in this rivet results in a 40–75% reduction in the remaining fatigue life, the highest reduction being observed in the case of a one-angle system representation of the connection. On the other hand, loss of clamping in rivet 1 results in a 10–30% reduction in the remaining fatigue life of the connection, which is notably lower than the first damage scenario. The damage scenarios associated with the other rivets (rivets 2, 4 and 5) were not found to be as highly critical as the previous cases, resulting in only a small reduction in the fatigue reliability of the connection in the case of rivet 4; for rivets 2 and 5, an increase in the overall connection reliability can be observed.

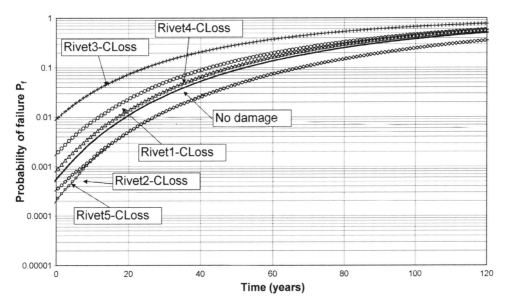

Figure 14.   Probability of fatigue failure vs. time for different rivet clamping force damage scenarios assuming a one-angle system (rivet clamping stress = 200 MPa).

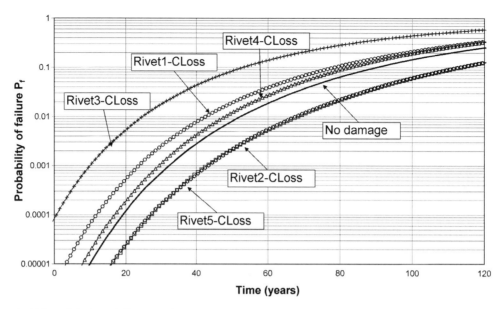

Figure 15. Probability of fatigue failure vs. time for different rivet clamping force damage scenarios assuming a two-angle system (rivet clamping stress = 200 MPa).

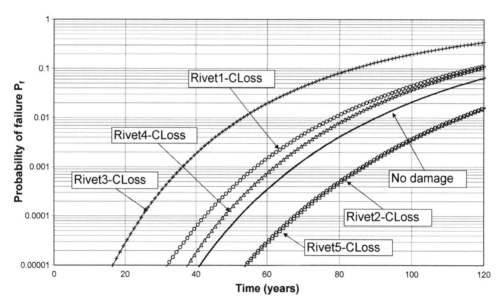

Figure 16. Probability of fatigue failure vs. time for different rivet clamping force damage scenarios assuming a four-angle system (rivet clamping stress = 200 MPa).

The results obtained in Figures 14–16 reveal that loss of clamping force in a connection rivet does not necessarily result in a detrimental effect on the fatigue reliability of the connection. For the particular stringer-to-cross-girder connection considered in this article, loss of clamping force in the top and bottom rivets connecting the leg of the angle to the cross-girder web was found to affect the fatigue performance of the connection adversely. It is not, therefore, surprising that the majority of the fatigue damage cases observed in the field in similar riveted connections were concentrated in either the top or bottom rivets (Al-Emrani 1999, 2002).

It should be noted that, for the purposes of the investigation, redistribution of stresses following a hot-spot (component) failure is not taken into account. The stresses on the remaining hot-spots are assumed to remain the same as in the undamaged state of the connection. In reality, the stresses will certainly change when a component has failed (e.g. a hole or a rivet), and the remaining hot-spots will carry the loads, which may lead to progressive failure of the connection. The current study has also adopted non-evolving loads from present day onwards, which is likely to be a non-conservative assumption given the current outlook on the development of rail freight in the UK and other countries. On the other hand, the assumption of independent failure modes in a weakest-link system, similar to the one investigated in this article, leads to conservative results for the failure probability.

5. Conclusions

This article presented a system-based model for fatigue assessment of riveted railway bridge connections building on previous work carried out by the authors, which considered the connection as a single component. An FE model of a typical, short-span, riveted railway bridge encompassing a detailed submodel of a stringer-to-cross-girder connection was used to convert the train loads to fatigue load effects. The criticality of the fatigue-sensitive details (hot-spots) was assessed through the TCD, a recently developed theory which considers the entire distribution ahead of any given stress concentration. Through Monte Carlo simulations, probabilistic fatigue load spectra for critical hot-spots on different components of the connection, such as holes, rivets and angle fillets, were developed. These spectra showed that the majority of the stress cycles experienced by the hot-spots are below the fatigue limit of the material.

By using the fatigue load spectra developed, reliability profiles and probabilistic remaining fatigue life estimates for the connection were determined. Loading, resistance and modelling uncertainties were taken into account. On the loading side, the problem was randomised through dynamic amplification and the annual train frequencies. On the response side, the S–N curve and the cumulative damage model were treated as random. Finally, on the modelling side, the ratio between actual and calculated stresses was represented as a random variable. The riveted connection was treated through generic sub-systems capturing potential damage in identifiable hot-spots, such as rivets, holes and angle fillets. These hot-spots were treated as the elements of a structural system susceptible to fatigue failure. The reliability profiles showed that the remaining life of the connection is sensitive to the assumptions made regarding the form of the system considered for the analysis. Low remaining lives were obtained for a number of connection's hot-spots indicating that fatigue cracking may be imminent or may have already initiated in a number of similar existing bridge connections. Reliability profiles obtained from the TCD method and the system representation have shown a more rapid increase in failure probability of the connection compared to similar results obtained using the traditional S–N approach. The rivet clamping force was found to have a considerable effect on the reliability of the connection for all system representations. Loss of clamping force in the top or bottom rivets connecting the angle leg to the cross-girder web was found to result in a considerable reduction in the remaining fatigue life of the connection. Loss of clamping in the remaining rivets was found not to have a significant adverse effect on the connection's reliability.

It is unwise to generalise the results and conclusions presented in this article to all types of riveted bridge connections with different configurations and geometries. For example, field observations from fatigue damage cases have shown that fatigue cracking occurs more commonly at the top and bottom parts of connections. This article is a first attempt at presenting a general system-based representation appropriate for fatigue assessment of complex connections made up of a number of components. As discussed, further work is required so that it can serve as a useful tool in planning inspections and/or repair actions in similar riveted bridge connections.

References

Adamson, D.E. and Kulak, G.L., 1995. *Fatigue tests of riveted bridge girders*. Edmonton, CA: Department of Civil Engineering, University of Alberta, Structural Engineering Report No 210.

Al-Emrani, M., 1999. *Stringer-to-floor-beam connections in riveted railway bridges – an introductory study of fatigue performance*. Publ. S99:4. Sweden: Chalmers University of Technology.

Al-Emrani, M., 2002. *Fatigue in riveted railway bridges – a study of the fatigue performance of riveted stringers and stringer-to-floor-beam connections*. Thesis (PhD). Chalmers University of Technology, Sweden.

American Association of State Highway Transportation Officials (AASHTO), 1998. *LRFD bridge design specifications*. 2nd ed. Washington, DC: American Association of State Highway Transportation Officials (AASHTO).

American Railway Engineering Association (AREA), 1996. *Manual for railway engineering: steel structures* [Chapter 15]. Washington, DC: American Railway Engineering Association (AREA).

Bellett, D., et al., 2005. The fatigue behaviour of three-dimensional stress concentrations. *International Journal of Fatigue*, 27 (3), 207–221.

BS5400, 1980. *Steel, concrete and composite bridges: Part 10: code of practice for fatigue*. London: British Standards Institute.

Byers, W.G., 1970. Impact from railway loading on steel girder spans. *Journal of Structural Division (ASCE)*, 96 (6), 1093–1103.

Byers, W.G., 1976. Rating and reliability of railway bridges. *In: Proceedings of the national structural engineering conference (ASCE)*, Madison, WI. Vol. 1. USA: ASCE, 153–170.

Cullimore, M.S.G., 1967. The fatigue strength of wrought iron after weathering in service. *The Structural Engineer*, 45 (5), 193–199.

DNV, 2005. *Fatigue design of offshore steel structures*. Norway: Det Norske Veritas. Recommended Practice, DNV-RP-C203.

Doerk, O., Fricke, W., and Weissenborn, C., 2003. Comparison of different calculation methods for structural stresses at welded joints. *International Journal of Fatigue*, 25 (5), 359–369.

Dong, P., 2001. A structural stress definition and numerical implementation for fatigue analysis of welded joints. *International Journal of Fatigue*, 23 (10), 865–876.

EC3, 2005. *EN 1993-1-9: Eurocode 3: design of steel structures – Part 1.9: Fatigue*. Brussels: CEN.

Estes, A.C. and Frangopol, D.M., 1999. Repair optimization of highway bridges using system reliability approach. *Journal of Structural Engineering (ASCE)*, 125 (7), 766–775.

Ghosn, M., Moses, F., and Frangopol, D.M., 2010. Redundancy and robustness of highway bridge superstructures and substructures. *Structure and Infrastructure Engineering*, 6 (1–2), 257–278.

Helmerich, R., Kühn, B., and Nussbaumer, A., 2007. Assessment of existing steel structures. A guideline for estimation of the remaining fatigue life. *Structure and Infrastructure Engineering*, 3 (3), 245–255.

Hendawi, S. and Frangopol, D.M., 1994. System reliability and redundancy in structural design and evaluation. *Structural Safety*, 16 (1–2), 47–71.

Imam, B., 2006. *Fatigue analysis of riveted railway bridges*. Thesis (PhD). University of Surrey, UK.

Imam, B., *et al.*, 2006. Analytical fatigue assessment of a typical riveted UK rail bridge. *Proceedings of the Institution of Civil Engineers (ICE) – Bridge Engineering*, 159 (3), 105–116.

Imam, B.M., Righiniotis, T.D., and Chryssanthopoulos, M.K., 2007. Numerical modelling of riveted railway bridge connections for fatigue evaluation. *Engineering Structures*, 29 (11), 3071–3081.

Imam, B.M., Righiniotis, T.D., and Chryssanthopoulos, M.K., 2008. Probabilistic fatigue evaluation of riveted railway bridges. *Journal of Bridge Engineering (ASCE)*, 13 (3), 237–244.

Melchers, R.E., 1999. *Structural reliability analysis and prediction*. Chichester: John Wiley & Sons.

Micic, T., Chryssanthopoulos, M.K., and Baker, M.J., 1995. Reliability analysis for highway bridge deck assessment. *Structural Safety*, 17, 135–150.

Miner, M.A., 1945. Cumulative damage in fatigue. *Journal of Applied Mechanics*, 12 (3), 159–164.

Nowak, A.S. and Cho, T., 2007. Prediction of the combination of failure modes for an arch bridge system. *Journal of Constructional Steel Research*, 63 (12), 1561–1569.

Okasha, N.M. and Frangopol, D.M., 2010. Time-variant redundancy of structural systems. *Structural and Infrastructure Engineering*, 6 (1–2), 279–301.

Railtrack, 2001. *Railtrack line code of practice: the structural assessment of underbridges*. Railtrack, UKRT/CE/C/025.

Righiniotis, T.D., Imam, B.M., and Chryssanthopoulos, M.K., 2008. Fatigue analysis of riveted railway bridge connections using the theory of critical distances. *Engineering Structures*, 30 (10), 2707–2715.

Susmel, L. and Taylor, D., 2003. Fatigue design in the presence of stress concentrations. *Journal of Strain Analysis for Engineering Design*, 38 (5), 443–452.

Sweeney, R.A.P., 1979. *Importance of redundancy in bridge-fracture control*. Washington, DC: Transportation Research Board, National Research Council, Transportation Research Record 711, 23–30.

Sweeney, R.A.P., Oommen, G., and Le, H., 1997. Impact of site measurements on the evaluation of steel railway bridges. *In: IABSE Workshop, Lausanne: evaluation of existing steel and composite bridges* [report], Lausanne, Zurich, 139–147.

Taylor, D., 1999. Geometrical effects in fatigue: a unifying theoretical model. *International Journal of Fatigue*, 21 (5), 413–420.

Taylor, D., 2005. Analysis of fatigue failures in components using the theory of critical distances. *Engineering Failure Analysis*, 12 (6), 906–914.

Taylor, D., 2007. *The theory of critical distances: a new perspective in fracture mechanics*. London: Elsevier.

Taylor, D. and Wang, G., 2000. The validation of some methods of notch fatigue analysis. *Fatigue and Fracture of Engineering Materials and Structures*, 23 (5), 387–394.

Tobias, D.H. and Foutch, D.A., 1997. Reliability-based method for fatigue evaluation of railway bridges. *Journal of Bridge Engineering (ASCE)*, 2 (2), 53–60.

Wang, C.S., *et al.*, 2006. Application of probabilistic fracture mechanics in evaluation of existing riveted bridges. *Bridge Structures: Assessment, Design and Construction*, 2 (4), 223–232.

Wang, C.S., *et al.*, 2007. System fatigue damage reliability assessment of railway riveted bridges. *Key Engineering Materials*, 347, 173–178.

Wirsching, P.H., 1995. Probabilistic fatigue analysis. *In*: C. Sundararajan, ed. *Probabilistic structural mechanics handbook*. New York: Chapman & Hall, 146–165.

Fatigue performance assessment and service life prediction of high-speed ship structures based on probabilistic lifetime sea loads

Kihyon Kwon, Dan M. Frangopol and Sunyong Kim

This article focuses on estimating probabilistic lifetime sea loads for high-speed ship structures with the aim of assessing fatigue performance and predicting service life from available data. Performance assessment and service life prediction for naval ship structures are extremely important issues. In particular, understanding the effect of sea loading on naval high-speed vessels is still a challenge. Potential lifetime load effects including low frequency wave-induced and high frequency slam-induced whipping loadings are investigated in this article by using a probabilistic approach. Clearly, integration of probabilistic sea loads into structural reliability assessment and service life prediction will provide a more reliable estimation of the long-term structural performance. Accordingly, this article presents an approach for fatigue reliability evaluation of ship structures based on the estimated lifetime sea loads. Loading information associated with sea states, ship speeds and relative wave headings is obtained from a joint high-speed sealift ship monohull structural seakeeping trials, while the S–N curves are established based on the British Standards.

1. Introduction

Ship structures subjected to various sea loads during operations experience strength degradation due to fatigue over their service life. For this reason, service life prediction for fatigue has to be carried out in design and assessment phases. In general, fatigue life can be assessed based on the stress–life (S–N) relationship (as a model of fatigue resistance) and the action of sea waves and the sea environment (as a model of fatigue loading), as suggested by Ayyub et al. (2002a). If the S–N category of the structural detail is correctly classified, the necessary information regarding fatigue resistance can be easily obtained. However, the accurate estimation of fatigue lifetime sea loads may be more challenging in time-dependent fatigue deterioration processes due to various uncertainties. These uncertainties include still water loading, wave-induced loading and transient impact-slamming, among others. Clearly, in fatigue design, experiments or simulations for predicting the potential lifetime sea loads are useful. Similarly, in fatigue assessment, structural health monitoring (SHM) during voyages provides real-time fatigue loadings that can be integrated into a time-dependent structural performance assessment. However, continuous monitoring up to the anticipated service life may not be feasible. This is because there

can be many restrictions due to budgetary, environmental and operational constraints. Alternatively, a probabilistic approach for fatigue life evaluation can be used to effectively estimate lifetime sea loads based on given information obtained from model tests, simulations or monitoring.

To date, the use of simulations, model tests and monitoring programs has been widely accepted for the estimation of lifetime sea loads. Kaplan et al. (1974) conducted a study with the computer program SCORES in order to estimate wave loads on the SL-7 container ship. The key factors of their study were ship speeds, wave lengths, headings and sea states. Similarly, Sikora et al. (1983) used the computer program SPECTRA for predicting primary load fatigue spectra for small waterplane area twin hull (SWATH) ships. Response amplitude operators for desired operating speeds and headings were used as input parameters as well as occurrence probabilities of sea state, heading and speed. As a result of these computer simulations, it was concluded that ship operational and wave conditions are important factors for the estimation of lifetime wave loads.

Ship model tests can be performed to provide various ship structural responses considering wave conditions, ship speeds and relative wave headings.

In general, performance measures obtained from model tests as well as monitoring can be used to provide more reliable structural responses, and to improve the decision-making process for ship maintenance management. The measured data from monitoring or model tests have been successfully used for structural performance assessment (Chiou and Chen 1990, Frangopol *et al.* 2008, Okasha *et al.* 2010a,b). Available sea loading information from model tests may allow not only the assessment of current ship structural performance but also the development of lifetime sea load prediction models using probabilistic methods.

This article focuses on estimating probabilistic lifetime sea loads based on model test and on integrating them into fatigue performance assessment and service life prediction. As illustrations, potential lifetime sea loads including low frequency wave-induced loading and high frequency slam-induced whipping loading are investigated, and a probabilistic approach for fatigue life evaluation is conducted. Occurrence probability associated with potential sea states is used to estimate probabilistic lifetime sea loads. Loading information is provided from the scaled test measurements of joint high-speed sealift ship (JHSS) monohull structural seaways loads test (Devine 2009). Based on all necessary information from the *S–N* approach for resistance and model test data for load effect, a fatigue reliability analysis is conducted by using the reliability software RELSYS (Estes and Frangopol 1998).

2. Fatigue resistance and loads

In many ship structures, the structural deterioration process due to fatigue significantly diminishes their service life. Typically, time-dependent fatigue strength can be assessed based on the *S–N* approach. Simultaneously, information on sea loadings, which is primarily associated with the action of sea waves and the sea environment, can be obtained from simulation programs, sea trial tests, segmented structural seakeeping model tests, and/or real-time SHM.

2.1. The S–N approach and Miner's rule

For fatigue life evaluation of steel structures, the *S–N* curve approach has been widely used and adopted by all standards and specifications. Fatigue strength of a structural detail is characterised in the relationship between stress range (nominal applied stresses) and cycles to failure for classified detail categories. The characteristic *S–N* curves are based on fatigue test data and correspond to the mean life of a detail which is shifted horizontally to the left by two standard

deviations (Fisher *et al.* 1998). The *S–N* curves are represented as sloping straight lines in logarithmic scale. The basic equation of fatigue strength is

$$S_r = \left(\frac{A}{N}\right)^{1/m} \qquad (1)$$

where S_r = nominal fatigue resistance (stress range), A = fatigue detail coefficient which can be treated as a random variable if uncertainty in fatigue strength is considered, N = number of cycles, and m = material constant. A typical set of *S–N* curves, as that shown in Figure 1, can be established based on the BS 5400 (1980).

Typically, fatigue damage is defined to be cumulative and the Palmgren–Miner rule is used to account for this damage accumulation. The linear damage rule proposed by Palmgren in 1924 was further investigated by Miner in 1945 (Fisher *et al.* 1998). It simply assumes that damage fraction at any particular stress range level is a linear function of the corresponding number of cycles. For a structural detail, the total damage can be expressed as the sum of damage occurrences that have taken place at individual stress range levels (i.e. Miner's rule)

$$D = \sum \frac{n_i}{N_i} \qquad (2)$$

where n_i = number of cycles at stress range level i and N_i = number of cycles to failure at stress range level i. Theoretically, the fatigue damage ratio, D, is equal to 1.0 at failure, while practically it may be less than 1.0 due to various uncertainties (Fisher *et al.* 1998, Ayyub *et al.* 2002a).

2.2. Estimation of sea loads based on simulation and monitoring

In the design phase, accurate estimates of potential sea loadings are important to ensure the desired structural

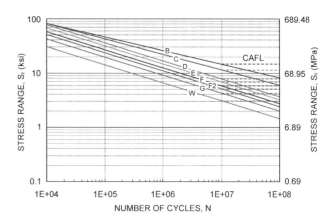

Figure 1. The *S–N* curves based on the BS 5400 (1980).

performance during the entire service life of ship structures, especially for high speed vessels. Primary structural loads on a ship result from its own weight, cargo, buoyancy and operation (Ayyub *et al.* 2002b). In assessing the reliability of ship structures, load effects may be estimated by finite element analysis, simulation and/or monitoring.

According to Paik and Frieze (2001), ship hull girder loads can be classified into three types: still water loads, low and high frequency wave-induced loads, and thermal loads. Still water loads are due to the difference between the weight and buoyancy distributions along the length of the ship. The low frequency wave-induced loads consist of vertical, horizontal and torsional wave loads, whereas the high frequency dynamic loads are due to slamming or whipping and springing (Devine 2009). Wave and dynamic loads are affected by many factors such as ship characteristics, ship speed, relative wave heading and sea states associated with significant wave heights (Ayyub *et al.* 2002b). Significant wave height is usually treated as a random variable that requires statistical analyses of ship response data collected from simulation, experiment or monitoring. For various sea states, efforts to estimate wave-induced load effects more accurately have been made (Glen *et al.* 1999, Wu and Moan 2006, Pedersen and Jensen 2009). For various ship speeds, Aalberts and Nieuwenhuijs (2006) analysed 1-year full scale measurements from a general cargo/container vessel in order to determine the effect of whipping (high frequency) and wave-induced (low frequency) loads on fatigue. Maximum wave-induced and dynamic bending moments that the ship may encounter during its service life should be taken into account in performance assessment and life prediction.

In recent years, the development of effective SHM systems for naval ships, especially for lightweight high speed ships, has been an important issue (Hess 2007, Salvino and Brady 2008). The SHM systems can be used to obtain prompt responses in terms of structural diagnosis and prognosis, and to offer possibilities for supporting operational and maintenance decisions. The use of available information from SHM is the most effective tool for the decision-making process. However, there are many restrictions to the adoption of this kind of SHM systems to high speed and high performance ships. In fact, these systems are still in an early stage of their development (Salvino and Brady 2008). Alternatively, ship model tests (e.g. segmented scaled model) or simulation analyses by using SPEC-TRA (Sikora *et al.* 1983) or LAMP (Lin and Yue 1990) can be employed to estimate lifetime sea loads considering various wave conditions. The simulation program SPECTRA developed by Sikora *et al.* (1983), is useful for computing vertical, lateral and torsional

moments applied to the hull girder of a monohull ship, and for creating a stress range bin histogram to evaluate fatigue life considering ship characteristics and wave conditions associated with specific sea routes (Michaelson 2000). In addition, ship model tests are useful for estimating various ship responses (e.g. stress, strain) in given sea states (e.g. moderate, high, hurricane), ship speeds and relative wave headings. Sea loads obtained from these model tests may be integrated into probabilistic lifetime sea loads prediction models. Consequently, probabilistic lifetime sea loads estimated from model tests can be used effectively for fatigue reliability evaluation.

2.3. Stress range bin histogram and probability density functions (PDFs)

As described previously, in terms of fatigue resistance, the S–N approach may be useful for estimating the total fatigue life including both crack initiation and crack propagation. On the other hand, in terms of fatigue load effects, variable amplitude loadings (i.e. stress range) must be appropriately taken into account for fatigue life evaluation. Cycle counting methods can be used to establish a stress range bin histogram (i.e. stress range vs. number of cycles). The ASTM Standard E 1049-85 (1997) addresses the following cycle counting techniques: level-crossing counting, peak counting, rain-flow counting, among others. In this article, the bending stress range bin histogram of a typical ship structure is computed by means of the peak counting technique. To consider the whole stress cycle (positive and negative), the values of the absolute peak stresses are doubled for the purpose of the histogram computation. This results in a conservative overestimation of the loads.

The procedure for creating a stress range bin histogram using peak counting is summarised as follows:

(1) determine the mean value of all time records
(2) filter all peak values (i.e. stresses) above the determined mean value
(3) set the stress range at two times the peak stress
(4) set the bin size (e.g. 0.5 ksi, 1.0 ksi) and count the assigned stress ranges
(5) establish a histogram of stress range occurrences.

Based on the established stress range bin histogram, effective stress range and number of cycles can be computed. Most importantly, an appropriate PDF for the prediction of sea loads should be determined. The probabilistic approach can be used to predict both resistance, R, and stress range, S, during fatigue life and eventually to perform fatigue reliability evaluation. The applicable PDFs associated with R and S are usually assumed to be lognormal and Weibull, respectively, for evaluating ship fatigue life. The PDFs of these distributions are .

(1) Lognormal distribution

$$f_R(r) = \frac{1}{r \cdot \zeta \cdot \sqrt{2 \cdot \pi}} \cdot \exp\left[-\frac{1}{2} \cdot \left(\frac{\ln(r) - \lambda}{\zeta}\right)^2\right]$$
$$\text{for } r > 0 \qquad (3)$$

where λ = mean of $\ln r$ (location parameter), and ζ = standard deviation of $\ln r$ (scale parameter).

(2) Weibull distribution

$$f_S(s) = \frac{\beta}{\alpha} \cdot \left(\frac{s}{\alpha}\right)^{\beta-1} \cdot \exp\left[-\left(\frac{s}{\alpha}\right)^\beta\right] \text{ for } s > 0 \quad (4)$$

where α = scale parameter, β = shape parameter and $\alpha > 0$, $\beta > 0$.

The parameters of the lognormal distribution can be easily obtained from fatigue resistance data, while those of the Weibull distribution are derived from the stress range bin histogram data. The effective stress range, S_{re}, could be derived as the qth moment of the Weibull PDF as follows:

$$S_{re} = \left[\int_0^\infty s^q \cdot f_S(s) \cdot ds\right]^{\frac{1}{q}} = [E(S^q)]^{\frac{1}{q}}. \qquad (5)$$

This can be also computed directly from the stress range bin histogram and Miner's rule (Miner 1945, Fisher et al. 1998):

$$S_{re} = \left[\sum \frac{n_i}{n_{total}} \cdot S_{ri}^m\right]^{\frac{1}{m}} \qquad (6)$$

where n_i = number of observations in the predefined stress range bin, S_{ri}, n_{total} = total number of observations during the monitoring period and m = slope of the S–N curve (material constant).

2.4. Probabilistic lifetime loads prediction for fatigue

A probabilistic approach to potential sea loads prediction for fatigue is herein addressed. This approach considers both effective stress range at a specified sea wave condition (e.g. sea state 7, ship speed of 35 knots, and heading of 0° for following seas) and number of cycles in its observed time period. As described previously, sea loads are function of ship characteristics, ship speed, relative wave heading and sea states associated with significant wave heights (wave condition). If ship model test data for certain wave conditions are provided, probabilistic lifetime sea loads can be estimated by considering both effective stress range and average daily number of cycles.

Based on given information (e.g. stress vs. time), wave-induced and whipping responses can be separately obtained by filtering. Wave-induced loadings are produced by the low-pass filtering, whereas wave impacts causing global hull girder whipping are collected using high-pass filtering (Brady 2004, Hildstrom 2007). Based on the filtering processes of raw data, individual stress range bin histograms for the given wave conditions are established using the peak counting method. Then, the effective stress range, S_{re}, and average daily number of cycles, N_{avg}, for an observed time period are calculated from the stress range histogram data. To estimate fatigue lifetime sea loads considering all possible wave conditions, the predicted effective stress range, S_{re}^*, can be derived under consideration of probabilistic ship operational profiles at a specific seaway. As an approximation, in this study it will be assumed that sea state, ship speed and relative wave heading are independent variables. The various probabilities of occurrence are considered to be the continuous representations of the relative frequencies n_i/n_{total} in Equation (6). Therefore, the resulting equation is:

$$S_{re}^* = \left[\sum_{i=1}^{ss} \sum_{j=1}^{sp} \sum_{k=1}^{wh} P_{SS,i} \cdot P_{SP,j} \cdot P_{WH,k} \cdot S_{re,ijk}^m\right]^{\frac{1}{m}} \qquad (7)$$

where S_{re} = effective stress range; m = material constant; $P_{SS,i}$ = probability of occurrence of the ith sea state ($i = 1, 2, \ldots, ss$), $P_{SP,j}$ = probability of occurrence of the jth ship speed ($j = 1, 2, \ldots, sp$) and $P_{WH,k}$ = probability of occurrence of the kth relative wave heading ($k = 1, 2, \ldots, wh$) for the applicable sea events. The corresponding schematic for estimating S_{re}^* is shown in Figure 2. As indicated, a new effective stress range bin histogram can be established by the individual effective stress ranges from each histogram and the occurrence probability associated with wave conditions.

Similarly, the predicted average daily number of cycles, N_{avg}^*, may be derived using the three occurrence probabilities which are associated with all potential sea wave conditions

$$N_{avg}^* = \sum_{i=1}^{ss} \sum_{j=1}^{sp} \sum_{k=1}^{wh} P_{SS,i} \cdot P_{SP,j} \cdot P_{WH,k} \cdot N_{avg,ijk}. \qquad (8)$$

The computed N_{avg}^* is used to estimate the accumulated number of stress cycles for future years, $N(y)$,

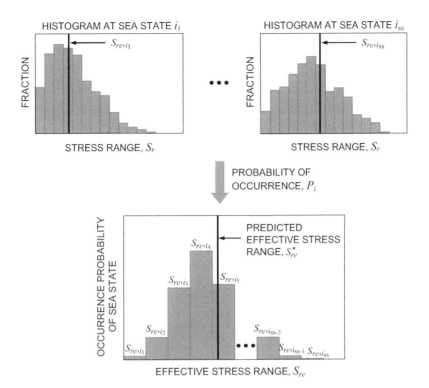

Figure 2. Schematic for estimation of the predicted effective stress range, S_{re}^*.

considering annual ship operation rate, α, in anticipated seaways. Therefore, $N(y)$ is estimated from the linear relationship to ship service life as

$$N(y) = 365 \cdot \alpha \cdot N_{avg}^* \cdot y. \qquad (9)$$

where y = number of years, and α = ship operation rate per year (e.g, $\alpha = 50\%$ for six months of operation, 75% or 90%).

3. Fatigue reliability analysis

Performance assessment and service life prediction for fatigue are herein addressed. As mentioned previously, ship fatigue life can be assessed more reliably based on both the S–N curve for ship capacity and the test data for load effects under uncertainties. It is noted that the predicted effective stress range, S_{re}^*, derived from Equation (7) is used for the prediction of lifetime load effect for fatigue.

3.1. Fatigue limit state

Under the repeated or fluctuating application of stresses, ship performance assessment and service life prediction for fatigue can be performed by fatigue reliability analysis with a well-defined fatigue limit-state function consisting of fatigue resistance, R, and

load effect, S. This is important because maintenance-management actions including inspection, monitoring and repair can be better planned if based on the well-quantified ship reliability. For the fatigue reliability evaluation, the limit-state functions of structural details are established, and PDFs for resistance and stress range are assumed. Typically, the safety of any structure would be preserved when its resistance, R, is larger than the predicted effective stress range, S_{re}^*.

The limit-state function used in fatigue reliability analysis is defined based on the S–N approach and Miner's rule (1945) as follows:

$$g(\mathbf{X}) = \Delta - D = 0 \text{ for } D = \sum n_i/N_i = (N/A) \cdot (eS_{re}^*)^m \qquad (10)$$

where Δ is Miner's critical damage accumulation index in terms of resistance and is assumed as lognormal with mean value of 1.0 and coefficient of variation (COV) of 0.3 for metallic materials (Wirsching 1984); D is Miner's damage accumulation index, e is a typical measurement error factor and m is a constant defined in the BS 5400 (1980). The number of cycles, N, which is obtained from Equation (9), is treated as random with COV of 0.2 and A is also considered random.

Complete details for all random variables are presented in Table 1.

3.2. Fatigue reliability evaluation

Based on the function $g(X)$, the fatigue reliability analysis is performed by using the reliability software RELSYS (Estes and Frangopol 1998). S_{re}^* is treated as Weibull PDF with COV of 0.2, while other random variables (i.e. Δ, A, N and e) are Lognormal (see Table 1).

The flowchart for the fatigue reliability evaluation is shown in Figure 3, and the corresponding steps are summarised as follows:

3.2.1. Step 1. Details of structural members based on the S–N approach

The S–N approach in terms of fatigue resistance, R, provides relevant information including the S–N category, material constant, m, constant amplitude fatigue limit (CAFL), and fatigue detail coefficient, A.

Table 1. Random variables for fatigue reliability evaluation.

Random variables	Notation	Distribution	Source
Critical damage accumulation index	Δ	Lognormal, $E(\Delta) = 1.0$, $COV(\Delta) = 0.3$	Wirsching (1984)
Fatigue detail coefficient	A	Lognormal, $E(A)^a = 6.29E + 11$ MPa^3, $(1.92E + 09$ ksi$^3)$, $COV(A)^a = 0.54$	BS 5400 (1980)
Measurement error factor	e	Lognormal, $E(e) = 1.0$, $COV(e) = 0.1$	Ayyub et al. (2002) and Frangopol et al. (2008)
Predicted effective stress range	S_{re}^*	Weibull (see Table 2), $COV(S_{re}^*) = 0.2$	Based on model test data
Predicted average daily number of cycles	N_{avg}^*	Lognormal (see Table 2), $COV(N_{avg}^*) = 0.2$	Based on model test data

Note: ^{a}The values $E(A)$ and $COV(A)$ assigned by the S–N category F.

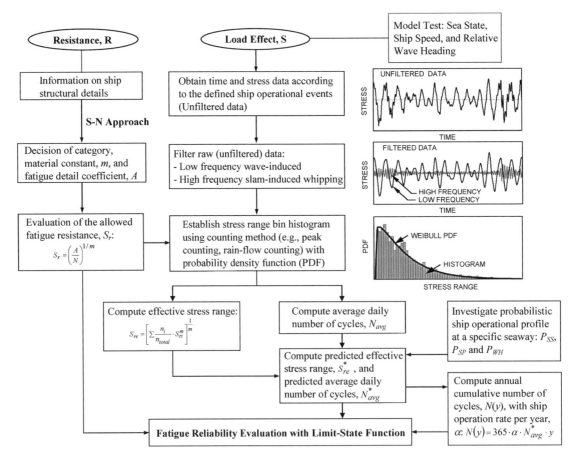

Figure 3. Flowchart for the fatigue reliability evaluation.

3.2.2. Step 2. Low-pass and high-pass filtering based on the collected unfiltered data

From the unfiltered (raw) data, wave-induced and slamming-induced whipping responses are obtained by filtering at low and high frequency levels, respectively, in order to provide separately useful responses for ship fatigue life evaluation.

3.2.3. Step 3. Stress range bin histogram and PDFs

The stress range bin histograms are established by using peak counting method from the unfiltered or filtered data at the selected locations (stations) of structural members. Based on the stress range bin histogram, effective stress range, S_{re}, (see Equations (5) and (6)) and the average daily number of cycles, N_{avg}, from a monitoring time period, T_{mon}, can be computed. Mean modal wave period, T_w, which is different at each sea state, is used to estimate N_{avg} by

multiplying the ratio (i.e. T_{mon}/T_w) by the counted number of occurrences during T_{mon}. An appropriate PDF for predicting sea loads is used considering uncertainty during fatigue lifetime. In ship fatigue reliability evaluation, lognormal and Weibull PDFs can be used for resistance and load effects, respectively.

3.2.4. Step 4. Probabilistic lifetime sea loads prediction

The probabilistic approach to potential sea loads prediction for fatigue evaluation is developed considering ship speeds, relative wave headings, and sea states associated with wave heights. The calculated S_{re} and N_{avg} according to the sea states (e.g. $0 \sim 9$) or applicable sea events are used to estimate both the predicted effective stress range, S_{re}^*, and the predicted average daily number of cycles, N_{avg}^*. All possible ship operational conditions through anticipated seaways are taken into account.

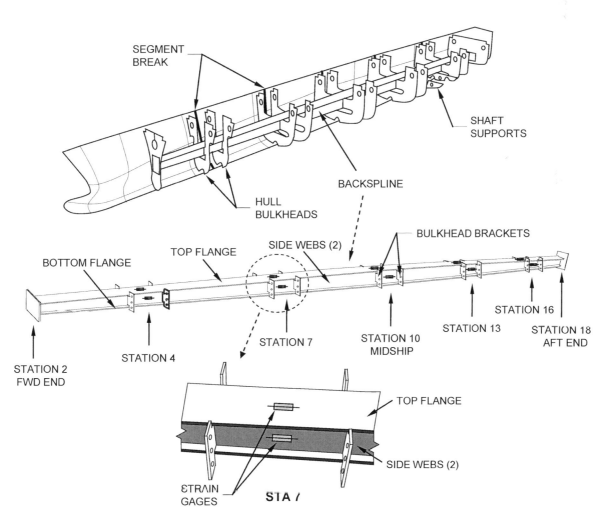

Figure 4. JHSS model (adapted from Devine 2009).

3.2.5. Step 5. Cumulative number of cycles, $N(y)$

By using Equation (9), $N(y)$ is estimated for time-dependent fatigue reliability evaluation. In this study, $N(y)$ does not reflect instantaneous but progressive time effect for fatigue life of ship, meaning that the number of cycles up to a specific year has been accumulated since the first ship operation year.

3.2.6. Step 6. Fatigue reliability analysis

For a given service year, the fatigue reliability analysis is performed with all necessary information from steps 1–5. For the assumed PDFs (lognormal and Weibull), the reliability software RELSYS (Estes and Frangopol 1998) is used to compute the fatigue reliability index. This program uses the first-order reliability method (FORM) to compute the reliability index.

4. Application

As an illustration, probabilistic lifetime sea loads of the JHSS for fatigue are estimated based on model test data and integrated into the fatigue performance assessment and service life prediction. Potential lifetime load effects, which are associated with low frequency wave-induced and high frequency slam-induced whipping loadings due to vertical bending moment, are investigated. For fatigue reliability analysis, the collected sea loadings from the scaled test measurements of a JHSS monohull structural seaways loads test (Devine 2009) are used together with the S–N curve provided by the BS 5400 (1980).

4.1. Segmented model

A full-scaled JHSS monohull length was scaled down to reach the value of 6.1 m (20 ft) in the segmented model (Devine 2009). It is noted that appropriate scale factors for the involved quantities (e.g. length, time, moment of inertia, bending moment) were obtained based on Froude scaling laws.

The segmented model approach was used to measure detailed hull response using a simple internal backspline (see Figure 4). The vertical, lateral and torsional stiffness and vibrational characteristics of the hull were modelled by using the internal backspline (Devine 2009). During each test run, realistic vibrational response, including hull primary and secondary

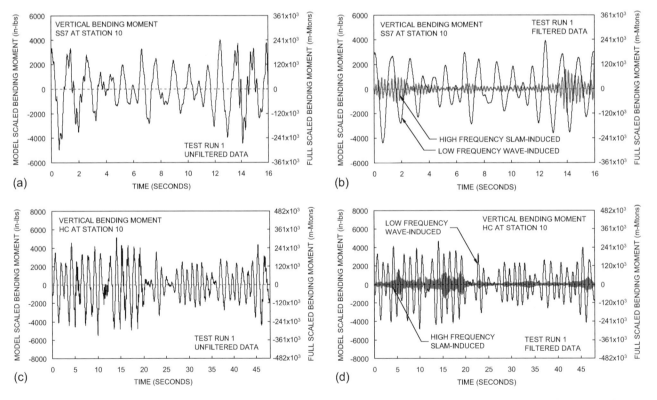

Figure 5. JHSS primary vertical bending moment. (a) unfiltered data at Station 10, 35 knots, SS 7 and heading of 0°. (b) low and high frequency filtered data at Station 10, 35 knots, SS 7 and heading of 0°. (c) unfiltered data at Station 10, 15 knots, HC and heading of 0°. (d) low and high frequency filtered data at Station 10, 15 knots, HC and heading of 0°.

loads, was collected from the installed strain gages on the Froude-scaled structural component at Stations 4, 7, 10, 13 and 16 (see Figure 4). As shown in Figure 4, the shell sections were connected with a continuous backspline beam and strain gages were installed at each segment cut to measure the vertical, lateral and torsional bending moments and vertical/lateral shear forces. It is noted that section modulus at the identified stations on the backspline varies along the beam length. Description of the JHSS segmented model tests and further details can be found in Devine (2009).

4.2. Fatigue resistance and load effects

Details of fatigue resistance and the scaled test data, which are associated with the strain gages installed on the top flanges of the backspline at five stations (i.e. Stations 4, 7, 10, 13 and 16 in Figure 4), are used to illustrate the fatigue reliability assessment and service life prediction based on the estimated probabilistic lifetime sea loads. For fatigue resistance, the S–N curves based on the BS 5400 (1980) are used and the

corresponding S–N parameters (i.e. category, CAFL and fatigue detail coefficient, A) are investigated at the respective structural details. Typically, the rational procedure to find the S–N parameters is to identify the worst weld detail in the design and assessment phases. In this study, for illustrative purposes, the S–N category F, which may be the worst case, is assumed for all the details, for illustrative purposes. The material constant, m, is 3.0, while the mean value of A is 6.29E+11 MPa^3 (1.92E+09 ksi^3) with coefficient of variation COV$(A) = 0.54$. The corresponding constant amplitude fatigue limit is CAFL = 39.78 MPa (5.77 ksi).

In this study, two sets of test data provided by Devine (2009) are used: (i) sea state 7 (SS7); 35 knots and heading of 0°; (ii) Hurricane Camille (HC), 15 knots and heading of 0°. Based on the given model test data, primary vertical hull-girder bending moments are investigated at the gage stations. At midship (i.e. Station 10), vertical bending moments due to SS7 and HC are presented in Figure 5. Hogging moment is positive and sagging is negative. Ship speeds in SS7

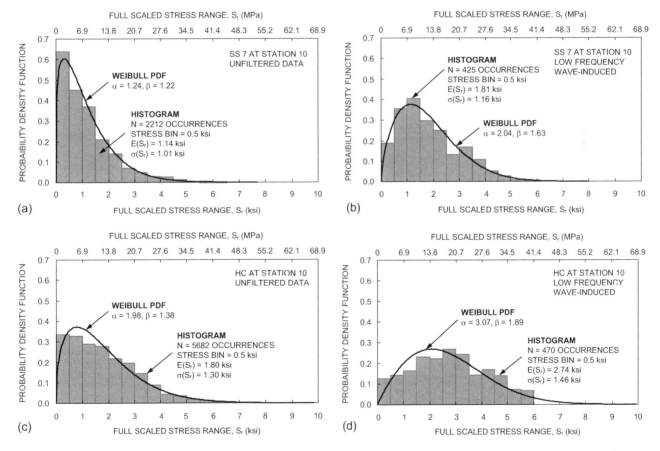

Figure 6. Stress range bin histogram and Weibull PDF. (a) unfiltered data at Station 10, 35 knots, SS 7 and heading of 0°. (b) low frequency wave-induced data at Station 10, 35 knots, SS 7 and heading of 0°. (c) unfiltered data at Station 10, 15 knots, HC and heading of 0°. (d) low frequency wave-induced data at Station 10, 15 knots, HC and heading of 0°.

and HC were 35 knots and 15 knots, respectively, in the same heading of 0° (i.e. following seas). It is noted that the Froude scale factor with respect to the bending moment is $1.025\lambda_F^4$ where $\lambda_F = 47.5255$ (Devine 2009). In both wave conditions, the filtering procedure has been applied to data, using low-pass and high-pass filtering to extract separately wave-induced moment and slamming-induced whipping moment (see Figure 5b and d).

For the wave conditions SS7 and HC, stress range bin histograms using peak counting are established based on unfiltered (wave-induced and slam-induced) and filtered (wave-induced) data. To convert bending moment, M, to stress, σ (i.e. $\sigma = M/S_m$), the Froude scale factor $0.346\lambda_F^4$ for section modulus, S_m, was used (Devine 2009). Weibull PDF, which is widely accepted for lifetime sea loads prediction, is used for the probabilistic approach. As shown in Figure 6a–d, Weibull PDFs of full scaled stress range are fitted on the established stress range bin histograms, for illustrative purposes. The parameters α and β indicate scale and shape of the Weibull PDF, respectively, while $E(S_r)$ and $\sigma(S_r)$ denote the mean value and standard deviation of the stress range, respectively. It is found that the $E(S_r)$ from the filtered data (i.e. neglecting high frequency load effect) is larger than that from the unfiltered data (i.e. including high frequency) at both loading conditions (see Figure 6). This is because the contribution of lower stress ranges to fatigue damage is diminished in the filtered data, as shown in Figure 5b and d. However, since the number of cycles for high frequency can be large, the cumulative effect of these numbers can be important.

For each test run of SS7 and HC at Stations 10 and 13, effective stress range, S_{re}, and average daily number of cycles, N_{avg}, in the observed time period are computed and presented in Figure 7a–d. With the sampling rate for this primary hull response data of 200 Hz, full scaled observed time periods for the total concatenated runs of SS7 and HC are about 42.4 min. and 66.6 min., respectively. As shown in Figure 7, S_{re} and N_{avg} are fluctuating through each test run. For lifetime fatigue assessment and prediction, these two parameters are herein treated as random variables considering loading uncertainty associated with the limited test runs.

4.3. Fatigue reliability analysis by using probabilistic lifetime sea loads

As described previously, under uncertainty associated with wave loading, a probabilistic approach for

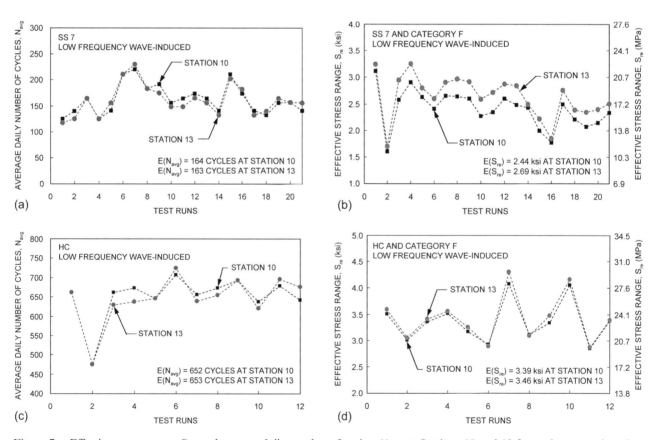

Figure 7. Effective stress range, S_{re}, and average daily number of cycles, N_{avg}, at Stations 10 and 13 for each test run based on low frequency wave-induced moment: (a) N_{avg} of SS 7, (b) S_{re} of SS7, (c) N_{avg} of HC, and (d) S_{re} of HC.

Table 2. Lifetime prediction of sea loads for fatigue at each station.

| Station | Sea state 7 | | | Hurricane Camille | | | Predicted value | | |
| | S_{re}, MPa (ksi) | | N_{avg} (cycles) | S_{re}, MPa (ksi) | | N_{avg} (cycles) | S_{re}^*, MPa (ksi) | | N_{avg}^* (cycles) |
	Weibull	Miner		Weibull	Miner		Weibull	Miner	
Station 4	5.83 (0.85)	5.93 (0.86)	187	10.99 (1.59)	10.70 (1.55)	703	5.43 (0.79)	5.48 (0.79)	143
Station 7	12.43 (1.80)	12.34 (1.79)	175	19.88 (2.88)	19.41 (2.81)	656	11.40 (1.65)	11.29 (1.64)	134
Station 10	16.93 (2.45)	16.82 (2.44)	164	23.81 (3.45)	23.39 (3.39)	652	15.40 (2.23)	15.27 (2.21)	126
Station 13	18.77 (2.72)	18.55 (2.69)	163	24.17 (3.50)	23.85 (3.46)	653	17.01 (2.47)	16.83 (2.44)	125
Station 16	13.19 (1.91)	13.10 (1.90)	162	15.43 (2.23)	15.29 (2.22)	668	11.91 (1.73)	11.86 (1.72)	125

Note: Equations (5) and (6) are used in the calculation of S_{re} by Weibull PDF and Miner's rule, respectively.

potential sea loads prediction is necessary to be developed based on given information (e.g. model tests, simulations, monitoring). In particular, if model test data for each sea state is available, lifetime sea loads for fatigue life evaluation can be reliably estimated, using occurrence probability of sea states in a seaway, and the computed S_{re} and N_{avg} from applicable operational conditions. As a result, probabilistic lifetime sea loads of JHSS monohull from model test data can be computed by using the proposed approach.

The established stress range bin histograms from low frequency wave-induced data of SS7 and HC, which are filtered from total concatenated runs, are used to estimate S_{re} and N_{avg} at the five stations. In the calculation of S_{re}, Equations (5) and (6) are employed considering Weibull PDF and Miner's rule, respectively. The calculated S_{re} and N_{avg} at the five stations are presented in Table 2. The maximum value of effective stress range was observed at Station 13, not at midship (i.e. Station 10) for both SS7 and HC, whereas the maximum bending moment was recorded at Station 10 (see Figure 8). This is because the section modulus on the backspline varies along the length of JHSS monohull. By using Equations (7) and (8), the predicted effective stress range, S_{re}^*, and predicted average daily number of cycles, N_{avg}^*, considering potential sea states at the worst area (i.e. North Atlantic Ocean) as presented in Table 3 (Brady et al. 2004), are estimated to perform the fatigue reliability assessment. Due to the lack of information, occurrence probability of sea state is only considered in order to estimate probabilistic lifetime sea loads. Occurrence probabilities of ship speed and relative wave heading are ignored in this application.

All necessary information for the probabilistic fatigue reliability analysis is obtained from steps 1 to 5 (see also Figure 3), and fatigue reliability analyses are conducted using reliability software KELSYS (Estes and Frangopol 1998). The established S–N curve based on the BS 5400 (1980) is herein used. Predicted lifetime

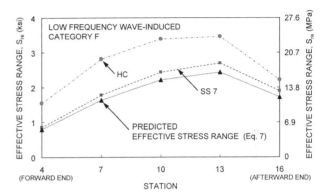

Figure 8. Predicted effective stress range, S_{re}^*, at the five stations of the JHSS.

Table 3. Modal wave period and probability of occurrence according to sea states of North Atlantic Ocean (Brady et al. 2004).

| Sea state | Mean value of significant wave height | | Mean modal wave period (s) | Probability of occurrence[a] (%) |
	(ft)	(m)		
0 – 1	0.16	0.05	–	1.0
2	0.98	0.30	6.9	6.6
3	2.87	0.87	7.5	19.6
4	6.15	1.87	8.8	29.7
5	10.66	3.25	9.7	20.8
6	16.40	5.00	12.4	14.1
7	24.61	7.50	15.0	6.8
8	37.73	11.50	16.4	1.3
> 8	> 45.90	> 13.99	20.0	0.1

Note: [a]Probabilities reported for the North Atlantic annual.

loads are estimated based on the low frequency wave-induced data filtered. Fatigue reliability evaluation at the identified critical location is performed considering (i) annual ship operation rate, α, of 50, 75 and 90% and (ii) low frequency wave-induced moment and complete history including high frequency slam-induced

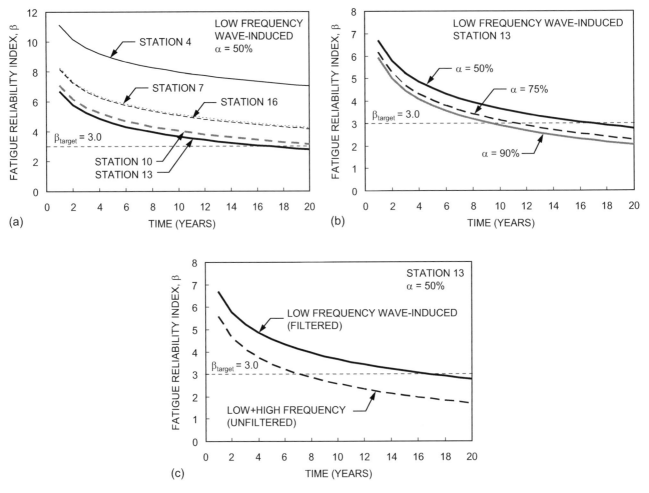

Figure 9. Fatigue performance assessment and service life prediction of the JHSS based on the predicted lifetime sea loads. (a) at the five stations with ship annual operation rate, α = 50%. (b) at Station 13 according to α = 50%, 75% and 90%. (c) using low frequency wave-induced filtered data and unfiltered data at Station 13.

whipping moment. Target reliability, β_{target}, is assumed to be 3.0. This target is in the range of target reliability indices for fatigue (i.e. $2.0 \le \beta_{target} \le 4.0$) recommended in Mansour *et al.* (1996).

The critical location of JHSS monohull is first identified. As shown in Figure 9a, at Station 13 fatigue reliability attains its lower bound, whereas the upper bound is at Station 4. Fatigue reliability analyses at the critical Station 13 are performed for both cases (i) and (ii). The result for case (i) is shown in Figure 9b. As expected, fatigue life of JHSS decreases significantly when the ship operation rate increases. For the predefined β_{target} of 3.0, the predicted fatigue life was only about 9 years in the case of α of 90%, whereas it was 16 years in the case of α = 50%. The result of the fatigue reliability analysis for case (ii) is presented in Figure 9c. It is found that the effect of high frequency slam-induced whipping moment on fatigue life could not be neglected when considering operations in the worst areas.

5. Conclusions

This article presents an approach for fatigue reliability assessment and service life prediction of high-speed ship structures based on the probabilistic lifetime sea loads estimated from model test data. The *S–N* approach applied to the identified structural details was used to estimate structural capacity in the fatigue reliability evaluation, whereas model test data were used to estimate probabilistic lifetime sea loads in terms of load effects. Under uncertainties associated with fatigue resistance and loading history, two PDFs (i.e. Lognormal, Weibull) were used. The unfiltered (raw) data collected on a scaled JHSS monohull was used to establish stress range bin histogram using peak counting method and to illustrate the proposed approach.

The following conclusions are drawn:

(1) The model test data can be used for estimating probabilistic lifetime sea loads associated with effective stress range and number of cycles.

(2) Using filtering process, low frequency wave-induced and high frequency slam-induced whipping moments can be extracted from unfiltered test data in order to identify structural responses separately.

(3) Based on the established stress range bin histogram, individual effective stress ranges for given wave conditions (which are related to ship characteristics, ship speeds, relative wave headings and sea states) can be computed and used to estimate the predicted effective stress range, S_{re}^*, considering all possible ship operational conditions.

(4) Based on the estimated probabilistic lifetime sea loads and the S–N approach, fatigue performance assessment and service life prediction of ship structures can be performed. Therefore, the remaining fatigue life can be rationally estimated by using the proposed probabilistic approach.

Acknowledgements

The support from the Office of Naval Research to Lehigh University under award N-00014-08-0188 is gratefully acknowledged. The authors greatly appreciate the technical discussions with Dr. Edward Devine and Dr. Liming Salvino, Naval Surface Warfare Center, Carderock Division (NSWCCD) and thank them for providing the data used in this article. The opinions and conclusions presented in this article are those of the authors and do not necessarily reflect the views of the sponsoring organisation.

References

Aalberts, P.J. and Nieuwenhuijs, M., 2006. Full scale wave and whipping induced hull girder loads. *Proceedings of the Fourth International Conference on Hydroelasticity in Marine Technology*, 10–14 September Wuxi, China. Beijing, China: National Defence Industry Press, 65–78.

ASTM Standard E 1049-85, 1997 (reapproved). Standard practices for cycle counting in fatigue analysis. *In: Annual Book of ASTM Standards,* Vol. 03.01, 710-718, Philadelphia.

Ayyub, B.M., *et al.*, 2002a. Reliability-based design guidelines for fatigue of ship structures. *Naval Engineers Journal (ASNE)*, 114 (2), 113–138.

Ayyub, B.M., *et al.*, 2002b. Reliability-based load and resistance factor design (LRFD) guidelines for hull girder bending. *Naval Engineers Journal (ASNE)*, 114 (2), 43–68.

Brady, T.F., 2004. *Global structural response measurement of Swift (HSV-2) from JLOTS and Blue Game rough water trials*. West Bethesda, MD: Naval Surface Warfare Center, Carderock Division, NSWCCD-65-TR-2004/33.

Brady, T.F., *et al.*, 2004 *HSV-2 swift instrumentation and technical trials plan*. West Bethesda, MD: Naval Surface Warfare Center, Carderock Division, NSWCCD-65-TR-2004/18.

BS 5400, Part 10, 1980. *Steel, concrete, and composite bridges: code of practice for fatigue*. London, England: British Standards Institute.

Chiou, J.W. and Chen, Y.K., 1990. *Fatigue prediction analysis validation from Sl-7 hatch corner strain data*. Washington, DC, Ship Structure Committee, Report No. SSC-338.

Devine, E.A., 2009. *An overview of the recently-completed JHSS Monohull and Trimaran structural seaways loads test program*. West Bethesda, MD: Naval Surface Warfare Center, Carderock Division (NSWCCD) Power-Point Briefing.

Estes, A.C. and Frangopol, D.M., 1998. RELSYS: a computer program for structural system reliability analysis. *Structural Engineering and Mechanics*, 6 (8), 901–919.

Fisher, J.W., Kulak, G.L., and Smith, I.F., 1998. *A fatigue primer for structural engineers*. Chicago, IL: National Steel Bridge Alliance.

Frangopol, D.M., Strauss, A., and Kim, S., 2008. Bridge reliability assessment based on monitoring. *Journal of Bridge Engineering (ASCE)*, 13 (3), 258–270.

Glen, I.F., Paterson, R.B., and Luznik, L., 1999. *Sea operational profiles for structural reliability assessment*. Washington, DC, Ship Structure Committee, Report No. SSC-406.

Hess, P.E. III, 2007. Structural health monitoring for high-speed naval ships. *Proceedings of the 6th International Workshop on Structural Health Monitoring, Inc.*, Lancaster, PA: DEStech Publications (keynote paper).

Hildstrom, G.A., 2007. *JHSV analysis engine*. West Bethesda, MD: NSWCCD-65-TR-2006/15, Naval Surface Warfare Center, Carderock Division.

Kaplan, P., Sargent, T.P., and Cilmi, J., 1974. *Theoretical estimates of wave loads on the SL-7 container ship in regular and irregular seas*. Washington, DC, Ship Structure Committee, Report No. SSC-246.

Lin, W.M. and Yue, D.K.P., 1990. Numerical solutions for large-amplitude ship motions in the time-domain. *Proceedings of the 18th Symposium Naval Hydrodynamics*, 20–22 August, University of Michigan, Ann Arbor, MI. Washington, DC: National Academy Press, 41–66.

Mansour, A.E., *et al.*, 1996. *Probability-based ship design: implementation of design guidelines*. Washington, DC, Ship Structure Committee, Report No. SSC-392.

Michaelson, R.W., 2000. *User's guide for SPECTRA: Version 8.3*. West Bethesda, MD: Naval Surface Warfare Center, Carderock Division, NSWCCD-65-TR-2000/07.

Miner, M.A., 1945. Cumulative damage in fatigue. *Journal of Applied Mechanics*, 12 (3), 159–164.

Okasha, N.M., Frangopol, D.M., and Decò, A., 2010a. Integration of structural health monitoring in life-cycle performance assessment of ship structures under uncertainty. *Marine Structures*, 23 (3), 303–321.

Okasha, N.M., *et al.*, 2010b. Reliability analysis and damage detection in high-speed naval craft based on structural health monitoring data. *Structural Health Monitoring*. doi: 10.1177/1475921710379516.

Paik, J.K. and Frieze, P.A., 2001. Ship structural safety and reliability. *Progress in Structural Engineering and Materials*, 3 (2), 198–210.

Palmgren, A., 1924. The service life of ball bearings. *Zeitschrift des Vereines Deutscher Ingenieure*, 68 (14), 339–341.

Pedersen, P.T. and Jensen, J.J., 2009. Estimation of hull girder vertical bending moments including non-linear and flexibility effects using closed form expressions. *Proccedings of IMechE*, 223 (3), 377–390.

Salvino, L.W. and Brady, T.F., 2008. Hull monitoring system development using a hierarchical framework for data and information management. *In*: *Proceedings of the 7th International Conference on Computer and IT Applications in the Maritime Industries (COMPIT'08)*, 21–23 April, Liège, Belgium, 589–602.

Sikora, J.P., Dinsenbacher, A., and Beach, J.E., 1983. A method for estimating lifetime loads and fatigue lives for swath and conventional monohull ships. *Naval Engineers Journal*, 95 (3), 63–85.

Wirsching, P.H., 1984. Fatigue reliability for offshore structures. *Journal of Structural Engineering (ASCE)*, 110 (10), 2340–2356.

Wu, M.K. and Moan, T., 2006. Numerical prediction of wave-induced long-term extreme load effects in a flexible high-speed pentamaran. *Journal of Marine Science and Technology*, 11 (1), 39–51.

Experimental investigation of the spatial variability of the steel weight loss and corrosion cracking of reinforced concrete members: novel X-ray and digital image processing techniques

Sopokhem Lim, Mitsuyoshi Akiyama, Dan M. Frangopol and Haitao Jiang

ABSTRACT

The material properties of concrete structures and their structural dimensions are known to be random due to the spatial variability associated with workmanship and various other factors. This randomness produces spatially variable corrosion damages, such as steel weight loss and corrosion cracks. The structural capacity of reinforced concrete (RC) members strongly depends on the local conditions of their reinforcements. Modelling the spatial variability of steel corrosion is important, but steel corrosion in RC members can only be observed after severely damaging the concrete members. To understand the steel corrosion growth process and the change in the spatial variability of steel corrosion with time, continuous monitoring is necessary. In this study, X-ray photography is applied to observe steel corrosion in RC beams. The steel weight loss is estimated by the digital image processing of the X-ray photograms. The non-uniform distribution of steel weight loss along rebars inside RC beams determined using X-ray radiography and its correlation with longitudinal crack widths are experimentally investigated.

1. Introduction

The corrosion of embedded rebars due to chloride attack is a major cause of reductions in the service life of reinforced concrete (RC) structures in marine environments. The rebar inside the RC element is protected from corrosion by the passive alkaline nature of the surrounding concrete cover. However, this protected barrier can be damaged or destroyed as a result of gradual chloride attack. The active corrosion of reinforcement is initiated when the chloride reaches the reinforcement surface (Ann & Song, 2007; Trejo & Pillai, 2004). The built-up corrosion product occupies a greater volume than the original rebar, leading to cracking in the concrete cover, followed by surface cracking.

Previous studies (Andrade, Alonso, & Molina, 1993; Otsuki, Miyazato, Diola, & Suzuki, 2000; Vu, Stewart, & Mullard, 2005) have reported that the corrosion-induced cracking time is far shorter than the propagation and widening time. In addition, the residual rebar cross section loss, which has been associated with the crack width during propagation, does not significantly affect the ultimate limit state; however, it does affect the serviceability state and the long-term performance of the structure. According to Palsson and Mirza (2002), at the time the Dickson Bridge was decommissioned in 1994, the bridge was structurally deficient but did not collapse even its deck was severely delaminated and damaged by chloride contamination, showing exposed corroded rebars, concrete cracking and spalling, leeching of the cement paste from the exposed piers, etc. Consequently, deteriorated RC structures are associated with a considerable economic loss in terms of maintenance and repair costs. Therefore, the development of a reliability model for predicting the long-term performance of deteriorated RC structures is important for minimising repair and maintenance during their life-cycle.

Recently, prediction models based on probabilistic concepts have been widely used to estimate the long-term structural performance of corrosion-affected RC structures (Marsh & Frangopol, 2008; Mori & Ellingwood, 1993; Stewart & Mullard, 2007). In assessing the probabilistic structural performance, Stewart (2004) emphasises the need to consider the spatial variability of steel corrosion. Ignoring the distribution of local steel corrosion can lead to the underestimation of the failure probability. Hence, the localised corrosion damage of rebar is an important input parameter for estimating the remaining service life of corroding RC structures.

However, the limited experimental data on the relationship between the spatial variations in steel weight loss and the width of the surface cracks have been reported to hinder the improvement of the accuracy of the prediction models (Akiyama & Frangopol, 2014; Akiyama, Frangopol, & Yoshida, 2010). The scarcity of experimental data is due to the difficulty in continuously observing the non-uniform spatial corrosion of steel weight loss during various stages of corrosion. Although it is possible to use

Table 1. Details of test specimens.

Notation	Cross section (mm)	Span length (mm)	Bar diameter (mm)	Cover (mm)	W/C (%)	Stirrup
I-1	140 × 80	1460	13	20	50	DB6@100[a]
II-1	140 × 80	1460	13	20	40	DB6@100
II-2	140 × 80	1460	13	20	65	DB6@100
III-1	140 × 80	1460	13	20	50	–
III-2	140 × 80	1460	13	20	50	DB6@165

[a]Deformed bars with a diameter of 6 mm arranged in intervals of 100 mm.

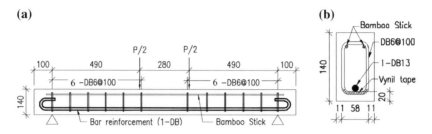

Figure 1. (a) Front view of the specimen for a 100-mm stirrup spacing and (b) typical cross section for all the specimens (all dimensions are in mm).

destructive methods (e.g. Alonso, Andrade, Rodriguez, & Diez, 1998; Vidal, Castel, & François, 2004; Zhang, Castel, & François, 2010) to study the relationship between the two parameters by repeatedly breaking specimens to weigh the rebar at various stages of corrosion, this method can suffer from uncertainties and errors associated with the different conditions encountered when remaking the specimens and the different growth patterns and measurements of the corrosion crack widths.

The use of X-ray technology as a non-destructive method is a promising means of avoiding these problems. This technique enables a continuous investigation of steel corrosion throughout the corrosion process. In concrete engineering, X-ray technology has been used as a non-destructive method to visualise and investigate steel corrosion in concrete structures. Beck, Goebbels, Burkert, Isecke, and Babler (2010) used X-ray computer tomography to examine the surface of a very small steel cylinder (9 mm in diameter and 10 mm long) at various stages of corrosion inside a mortar specimen. The reported difference between the mass loss, which was determined using the constructed 3D X-ray tomography image, and the actual mass loss after breaking the specimen was 40–60%. Akiyama and Frangopol (2013) demonstrated that the X-ray apparatus was a suitable tool for continuously investigating the weight loss of a corroded rebar that was embedded in a cylinder (100 mm × 200 mm) and a prism (100 mm × 100 mm × 400 mm). A digital image analysis based on the X-ray radiography of the shape of a corroded rebar from different viewing angles was used to determine the steel weight loss. The difference between the calculated steel weight loss based on the digital image analysis of the X-ray photogram and the measured value after damaging the specimen is only about 10%.

This paper presents an experimental study aiming to comprehensively investigate the continuous increase in the spatial variability of the steel weight loss along corroded rebars and the associated longitudinal crack widths of RC beams at various stages of corrosion, using novel X-ray and digital image processing techniques. The effects of various water-to-cement (W/C) ratios and stirrups on these two main parameters are discussed. Although the method of investigation is similar to that of Akiyama and Frangopol (2013), larger specimens with longer

rebars are studied. Moreover, using a new, upgraded X-ray apparatus and image intensifier, higher resolution X-ray photograms can be obtained, allowing the detection of the corrosion product. The more advanced digital image analyser used also provides more accurate estimates of the steel weight loss. The experimental outcomes are expected to provide a fundamental understanding of the non-uniform spatial growth of the steel weight loss and the corresponding corrosion cracking. These experimental results can help provide the stochastic fields associated with steel weight loss in RC components and estimate the steel weight loss based on the measured corrosion cracking widths.

2. Experimental programme

2.1. Overview of experimental plan

An experimental plan is established to study the effects of W/C ratio and stirrups on the spatial steel weight loss and corrosion cracking. To achieve these goals, five RC beams were fabricated and divided into two groups. The details of the specimens are shown in Table 1 and Figure 1. For the first group, specimens I-1, II-1 and II-2 were produced with W/C ratios of 50, 40 and 65%, respectively. The second group consists of specimens I-1, III-1 and III-2. Specimens I-1 and III-2 have stirrup spacings of 100 and 165 mm, respectively, whereas specimen III-1 has no stirrups. For the compressive strength test, six cylinders (100 mm × 200 mm) were also produced for each concrete mixture.

The experimental procedure is as follows. The corrosion of embedded longitudinal rebar was accelerated via the electrochemical technique. At specific time intervals prior to performing the X-ray radiography, the crack widths were recorded by obtaining images of surface cracking on the bottom of the beams. X-ray radiography was performed once before the steel corrosion initiated and several times during the corrosion process to capture photograms of the developing morphology of the non-corroded and corroded rebars from different viewing angles. These photograms were used in the digital image processing to estimate the steel weight loss.

Table 2. Mixing proportions.

Notation	G_{max} (mm)	W/C (%)	s/a[a] (%)	Water (kg/m³)	Cement (kg/m³)	FA[c] (kg/m³)	CA[d] (kg/m³)	AE[b] (ml/m³)
I-1	20	50	44.3	181	362	754	961	2715
II-1	20	40	42.3	178	445	694	961	3338
II-2	20	65	47.3	185	285	829	940	2138
III-1 & III-2	20	50	44.3	181	362	754	961	2715

[a]Fine aggregate ratio.
[b]Air entranced agent.
[c]Fine aggregate.
[d]Coarse aggregate.

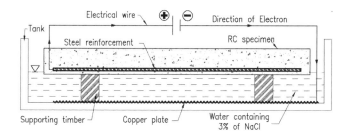

Figure 2. Electrolytic experimental test set-up.

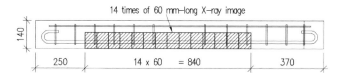

Figure 3. Total steel bar length captured by the X-ray apparatus (all dimensions are in mm).

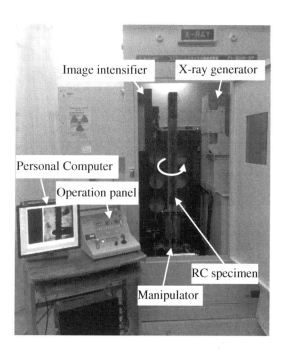

Figure 4. X-ray imaging set-up at Material Engineering Laboratory, Waseda University.

2.2. Materials and concrete mix proportion

All specimens were fabricated using identical material constituents. Ordinary Portland cement with a specific density of 3.16 g/cm³ was used. The fine aggregate has a fineness modulus of 2.64 and a specific density of 2.60 g/cm³. The coarse aggregate has a maximum size of 20 mm ($G_{max} \leq 20$ mm) and a specific density of 2.64 g/cm³. A deformed rebar with a diameter of 13 mm (DB13) was used as the longitudinal rebar, and deformed rebars with a diameter of 6 mm (DB6) were used as stirrups. All rebars were of the same steel quality grade, SD345. The details of the concrete mixing proportions are shown in Table 2.

2.3. Specimen fabrication procedure

The same fabrication procedure was performed for all the specimens. When used, the stirrups were wrapped with vinyl tape to prevent direct contact with the longitudinal rebar. The stirrups were arranged in specified intervals at the shear span to prevent abrupt shear failure of the concrete during the corrosion process. Before pouring the concrete, electrical wire was tied to one end of the steel reinforcement. Two days after fabrication, the mould was stripped off from the specimens, and the specimens were cured in water in a 23–25 °C room for 28 days.

2.4. Electrolytic experiment

After the specimens were cured, the steel corrosion process was initiated using the electrolytic technique. The detailed assembly of the electrolytic experimental test is shown in Figure 2. The RC specimen was placed on two pieces of supporting timber and partially immersed in a 3% sodium chloride (NaCl) solution in a tank in a controlled environment at 23–25 °C. The external copper plate, which is placed under the supporting timbers below the specimen, served as the cathode; the embedded rebar inside the specimen served as the anode. To ensure that the tests could be completed within a reasonable timeframe, the total impressed current was adjusted for each specimen to maintain the same current density (i.e. 1000 µA/cm²) to pass over the surface of the rebar. The accelerated corrosion process proceeded until the accumulated current time reached approximately 620 h (i.e. about 25 days).

2.5. Surface crack width measurement

The external surface cracks that occurred along the bottom of the specimens at various steel weight losses were imaged

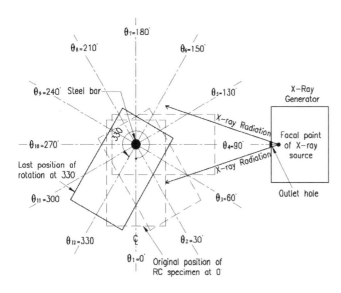

Figure 5. Top view of the specimen set-up and views of the steel bar at different angles associated with the rotation of the RC specimen.

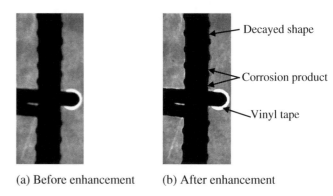

(a) Before enhancement (b) After enhancement

Figure 7. X-ray images of a corroded steel bar at 0° with a mean steel weight loss of 8.79% (a) before and (b) after enhancement.

by a digital camera before X-ray radiography was performed. The location of the captured images corresponds to that of the captured X-ray image, i.e. 250–1090 mm from the left side of the specimens, as shown in Figure 3. This required seventeen 50-mm-long images to be continuously obtained along the bottom of the specimens. Note that in this experimental study, the visual longitudinal corrosion cracking occurred only at the bottom of the corroded beams. Crack width measurements on the photographs were continuously performed every 5 mm until a distance of 1090 mm was reached using an advanced image analysis programme.

2.6. X-ray photogram acquisition procedure

Using the X-ray configuration in Figure 4, images of the non-corroded and corroded areas of the rebar inside the specimen were captured from different viewing angles once before the initiation of corrosion and several times during the corrosion process at various steel weight losses. Images of the rebar from 12 viewing angles (i.e. 0°, 30°, 60°, 90°,120°, 150°, 180°, 210°, 240°, 270°, 300° and 330°) were recorded. At each angle, the total length of the rebar imaged by the X-ray apparatus was 840 mm. Fourteen 60-mm-long images were consecutively captured (see Figure 3). The X-ray radiography procedure used to acquire the photograms consists of two main steps, as described below.

The first step is setting the specimen in an appropriate position. The specimen was placed on a manipulator located between the X-ray intensifier and generator in the X-ray chamber. This

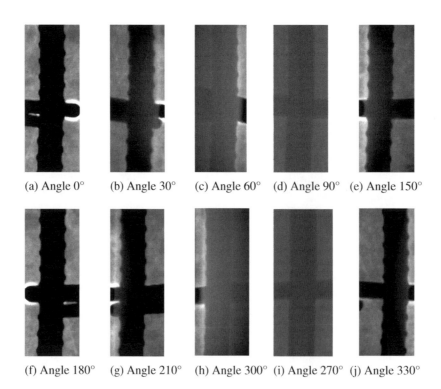

(a) Angle 0° (b) Angle 30° (c) Angle 60° (d) Angle 90° (e) Angle 150°

(f) Angle 180° (g) Angle 210° (h) Angle 300° (i) Angle 270° (j) Angle 330°

Figure 6. Ten X-ray images of the original steel bar obtained at different viewing angles.

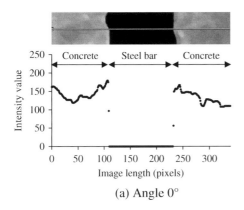

(a) Angle 0°

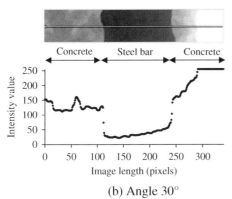

(b) Angle 30°

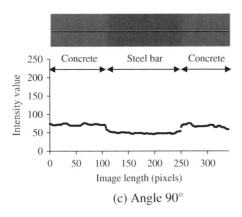

(c) Angle 90°

Figure 8. Profiles of the intensity values for a row of pixels in the X-ray photograms at three different viewing angles.

(a) Non-corroded steel bar at $MRw = 0.00\%$

(b) Corroded steel bar at $MRw = 6.05\%$

Figure 9. Sliced 5-mm-high X-ray photograms in (a) non-corroded steel bar at $MRw = .00\%$ and (b) corroded steel bar at $MRw = 6.05\%$.

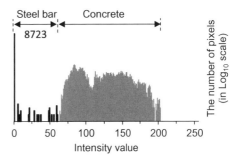

(a) Non-corroded steel bar at $MRw = 0.0\%$

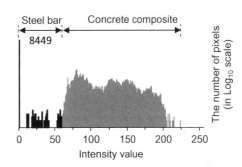

(b) Corroded steel bar at $MRw = 6.05\%$

Figure 10. Histograms of the accumulated number of pixels classified by intensity values for the steel bars in Figures 9(a) and (b).

Table 3. Estimated weight loss vs. actual measured weight loss.

Specimens	Estimated weight loss (%)	Actual weight loss (%)	Absolute difference (%)
I-1	19.65	16.63	3.02
II-1	16.97	13.71	3.26
II-2	23.18	20.62	2.56
III-1	27.23	24.48	2.75
III-2	25.54	22.51	3.03

manipulator was used to translate and rotate the specimen into the desired positions via the operation panel. The specimen was first rotated to $\theta_1 = 0°$ and horizontally adjusted into the position where the centre of the embedded rebar was aligned with the middle point of the X-ray source outlet. This setting fixes the centre of the rebar as the centre point of rotation of the specimen, as indicated in Figure 5. In the vertical direction, the specimen was then shifted up or down relative to the X-ray radiation source to obtain the starting position of 370 mm from the base of the specimen on the computer screen.

After setting up the specimen, radiography images were acquired by attenuating the primary X-ray beam with materials of different densities and thicknesses. The chosen power and current settings for the X-ray radiography are functions of the source-to-specimen distances and concrete thicknesses when the specimen is rotated to different viewing angles: 120 kV, 1.2 mA for viewing angles of 0°, 30°, 60°, 120°, 150°, 180°, 210°, 240°, 300°, and 330° and 145 kV, 1.2 mA for viewing angles of 90° and 270°. Starting from a known position at 370 mm from the base of the specimen and a viewing angle of $\theta_1 = 0°$, the RC specimen is vertically translated 14 times in increments of 60 mm relative to the radiation cone beam supplied by the X-ray generator. The radiation penetrates the RC specimen for visualisation, and

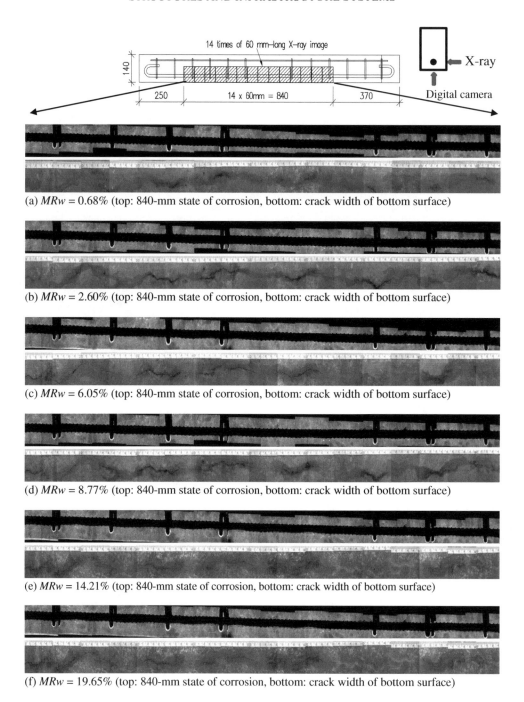

(a) *MRw* = 0.68% (top: 840-mm state of corrosion, bottom: crack width of bottom surface)

(b) *MRw* = 2.60% (top: 840-mm state of corrosion, bottom: crack width of bottom surface)

(c) *MRw* = 6.05% (top: 840-mm state of corrosion, bottom: crack width of bottom surface)

(d) *MRw* = 8.77% (top: 840-mm state of corrosion, bottom: crack width of bottom surface)

(e) *MRw* = 14.21% (top: 840-mm state of corrosion, bottom: crack width of bottom surface)

(f) *MRw* = 19.65% (top: 840-mm state of corrosion, bottom: crack width of bottom surface)

Figure 11. Spatial growth of the steel weight loss and corrosion cracking of specimen I-1 for six different values of *MRw*.

the attenuated X-ray radiation detected by the image intensifier reveals the composition details of various densities. After the attenuated X-rays are converted into visible light on a fluorescent screen, an equipped charge-coupled device camera unit, whose capture command is linked to and controlled by a software program, is used to capture and store the light intensity as digital values. These digital values comprise a 1024×768-pixel greyscale image.

The same process was performed repeatedly to capture images from the remaining viewing angles. The viewing angles of 0°, 30°, 60°, 90°,120°, 150°, 180°, 210°, 240°, 270°, 300° and 330° refer to those rebar views at which the specimen was rotated to angles θ_1, θ_2, θ_3, θ_4, θ_5, θ_6, θ_7, θ_8, θ_9, θ_{10}, θ_{11} and θ_{12}, respectively (see Figure 5).

3. Procedure for estimating the steel weight loss

3.1. Image enhancement before analysis

Various 60-mm-long images of the original rebar that were captured using the X-ray apparatus before corrosion are shown in Figure 6. In general, the original images at 0° and 180° provide the clearest views, followed by images at 30°, 150°, 210° and 330° and then by images at 90° and 270°, respectively. The worst images are those at 60° and 330°. This ordering is a result of the differences in the thickness of the concrete composite penetrated by the X-ray radiation during the image capturing for a particular specimen rotation angle. For example, at 0° and 180°, the specimen is in a favourable position, as the concrete thickness

(a)

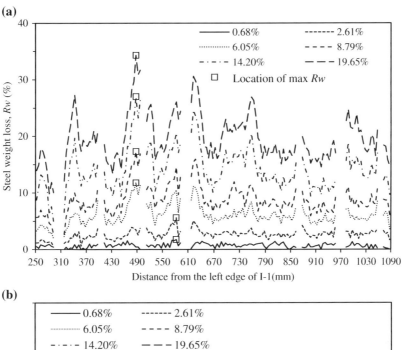

(b)

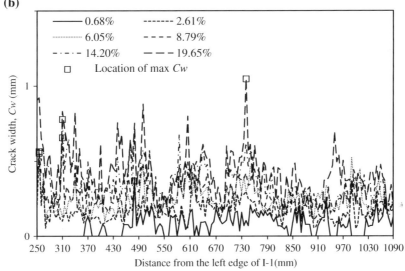

Figure 12. Spatial distribution of the (a) steel weight loss and (b) surface crack width of beam I-1.

encountered by the X-ray radiation is only 80 mm, providing a notably clear image. In contrast, at 90° and 270°, the angle is unfavourable, as 140 mm of concrete is penetrated by the X-ray radiation.

Furthermore, the post-corrosion image at the viewing angle of 0° in Figure 7(a) illustrates the shape of the corroded rebar at a mean steel weight loss of 8.79%, which corresponds to that of non-corroded rebar in Figure 6(a). However, although the image at this angle provides the clearest view, it remains difficult to carefully examine the corrosion products or decayed shape of the rebar. In Figure 7(b), for the image after enhancement, the corrosion products and decayed shape of the corroded rebar can be more easily identified.

Therefore, it is necessary to enhance the image before the analysis to readily obtain detailed information from the image. In the enhancement process, the fine details of the image were revealed or the blurred regions were reduced using Image-Pro Plus software version 7.0 of Media Cybernetics, Inc. (2012) to accentuate the intensity changes and make the high-contrast edges visible. Visualising the high-contrast edges between the rebar and concrete composite allows the area shapes of the

concrete composite and rebar to be easily distinguished, which is important for determining the area of the corroded rebar to estimate the steel weight loss.

3.2. Steel weight loss estimated by digital image analysis

To estimate the steel weight loss using the X-ray photograms, the area of the original rebar before corrosion and that of the corroded rebar at a given time during the corrosion process need to be determined. At each of the eight viewing angles, the two types of rebar areas are determined through the digital image analysis of X-ray photograms (i.e. a manipulation of the stored digital data of the image in terms of numerical representations of pixels). Note that only X-ray images from 8 viewing angles (i.e. $\theta_1 = 0°$, $\theta_2 = 30°$, $\theta_4 = 90°$, $\theta_6 = 150°$, $\theta_7 = 180°$, $\theta_8 = 210°$, $\theta_{10} = 270°$ and $\theta_{12} = 330°$) were used for estimating the steel weight loss along the corroded bars because: (1) the time-consuming and laborious works involved in digital image analysis of considerably large amount of data and (2) a good accuracy of estimated steel weight loss could be obtained which will be presented in the next section.

(a)

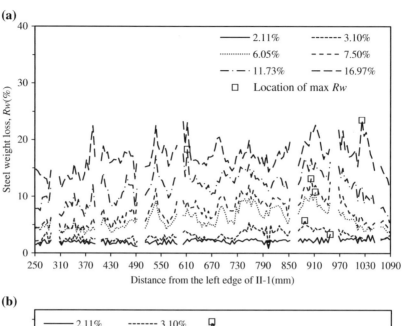

(b)

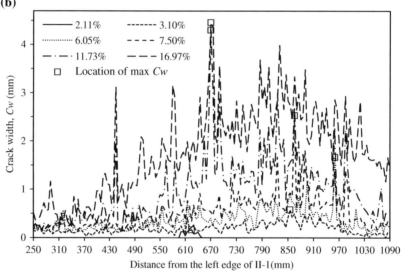

Figure 13. Spatial distribution of the (a) steel weight loss and (b) surface crack width of beam II-1.

In this paper, the acquired X-ray photograms are 8-bit grey-scale images with 1024×768 pixels. The greyness levels of the pixels are numerically represented by 256 intensity values ranging from 0 for completely black to 255 for completely white. This numeric representation enables the image software to distinguish pixels of different colours. Because the reinforcing rebar is denser than the concrete composite or corrosion product, it absorbs the X-ray radiation most efficiently. Thus, the rebar always produces the darkest pixels with the lowest intensity range compared to the concrete composite and corrosion product in the images. Figure 8 illustrates the profiles of the intensity values for a row of pixels that correspond to a line drawn on each of the above X-ray photograms. All the profile lines show that the intensity values of the rebar at 0°, 30° and 90° in the middle part of each graph are always lower than those of the corrosion products located close to the corroded rebar and the concrete in the left- and right-hand sides of the graphs. This finding also holds for the images of other viewing angles, as the X-ray images captured at 0° and 180°; 30°, 150°, 210° and 330°; and 90° and 270° are very similar (see Figure 6).

Therefore, by manipulating the intensity values of the pixels of the rebar, the total number of pixels below a minimum threshold of intensity values that represent the area of rebar alone can be selected, counted and classified using the image processing software. To facilitate the analysis of the digital data of the image, the 60-mm-high image of all viewing angles is sliced into twelve 5-mm-high images (see Figures 9(a) and (b)). Figures 10(a) and (b) present histograms of the accumulated numbers of pixels classified by intensity values for the 5-mm images of the non-corroded and corroded rebars in Figures 9(a) and (b), respectively. The histograms explicitly show that the total number of pixels of the selected intensity threshold of 0–62 is 8723 for the non-corroded rebar area, greater than the corresponding value of 8449 pixels of the selected intensity threshold of 0–59 for the corroded rebar area.

After the number of pixels of non-corroded and corroded rebar was determined, the total area was obtained by multiplying the number of pixels by the unit area per pixel as follows:

$$A_{\theta_n} = P_{\theta_n} \times A_p \tag{1}$$

$$A'_{\theta_n} = P'_{\theta_n} \times A_p \tag{2}$$

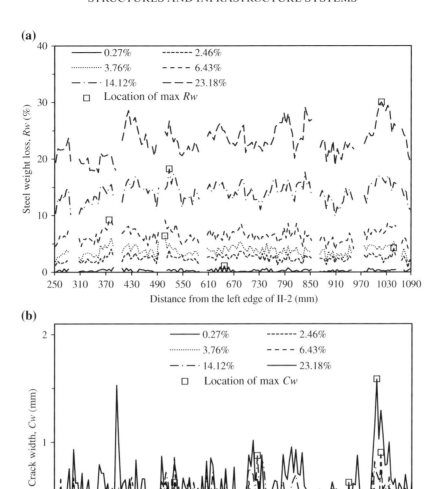

Figure 14. Spatial distribution of the (a) steel weight loss and (b) surface crack width of beam II-2.

where A_{θ_n} and A'_{θ_n} are the areas of the original rebar before corrosion and the corroded rebar at each viewing angles, respectively; θ_n denotes the viewing angles in which the subscript $n = \{1, 2, 4, 6, 7, 8, 10, 12\}$; P_{θ_n} and P'_{θ_n} are the number of pixels of the area of the original rebar and corroded rebar, respectively, at any of n viewing angles; and A_p is the unit area per pixel in the image.

The volume of the rebar before and after corrosion can then be calculated as follows:

$$V_{\theta_n} = \frac{\pi \left(A_{\theta_n} \right)^2}{4L} \tag{3}$$

$$V'_{\theta_n} = \frac{\pi \left(A'_{\theta_n} \right)^2}{4L} \tag{4}$$

where V_{θ_n} and V'_{θ_n} are the volumes of the original and corroded rebars, respectively, at a viewing angle θ_n; and L is the length of the rebar, which is 5 mm herein.

The steel weight loss per length L (mm) of the rebar is determined by taking the average of each value of the steel weight loss for each viewing angle as follows:

$$Rw = \frac{1}{k} \sum \frac{\left(V_{\theta_n} - V'_{\theta_n} \right)}{V_{\theta_n}} \times 100 \tag{5}$$

where Rw is the steel weight loss in percentage (%) per length L (mm) of rebar, θ_n denotes the viewing angles in which $n = \{1, 2, 4, 6, 7, 8, 10, 12\}$, and k is the number of viewing angles ($k = 8$ for the eight different viewing angles considered).

4. Results and discussion

4.1. Accuracy of the estimation method

After the completion of all the tests, the concrete specimens were demolished to remove the embedded rebars from them. All the rebars were immersed in a water tank containing 10% diammonium hydrogen citrate solution for 24 h to remove the corrosion products. Next, the weights of the corroded rebars

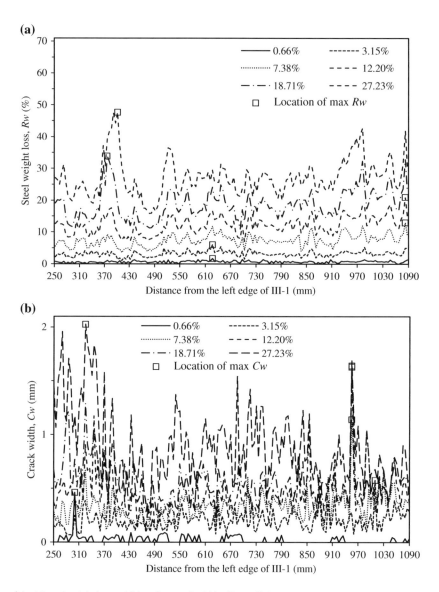

Figure 15. Spatial distribution of the (a) steel weight loss and (b) surface crack width of beam III-1.

were measured using a laboratory digital scale. The measured steel weights and the estimated weight loss of rebars calculated using digital image processing are compared in Table 3.

Note that the differences between the steel weight losses estimated using the digital image analysis and the actual measured steel weight losses are approximately 3%. This result demonstrates the good accuracy of the estimation method used for steel weight loss. The weight losses quantified via the X-ray images appear to be higher than the actual measured amounts, indicating that the employed X-ray method marginally overestimates the actual measured weight loss of embedded rebars in RC members. One possible cause of the overestimation is the inability of the projected 2D X-ray images to provide information about pit corrosion on the rebar surface. Consequently, the estimated cross section areas based on the projected areas of the rebar using the X-ray image are slightly smaller than those based on the actual areas of rebars embedded in the RC beams.

4.2. Spatial variability of the steel weight loss and crack widths

Figure 11 illustrates the continuous spatial growth of the steel corrosion visualised using the X-ray technique and the propagation of longitudinal corrosion cracking at the bottom of the beams at different stages of corrosion. Comparing the X-ray images and photographs, it is found that the steel weight loss Rw occurs gradually and at locations near the areas exacerbated by the longitudinal surface crack widths Cw. These weight loss locations are noticeable at some of the locations of the arranged stirrups and in the centre of the span.

Figures 12–16 show the spatial distribution of the steel weight loss Rw and surface crack width Cw at their corresponding locations along beams I-1, II-1, II-2, III-1, and III-2, at various mean steel weight losses MRw. Note that because Rw could not be obtained at the locations of the arranged stirrups, there are regular gaps in the graphs at both of the shear spans, from 250

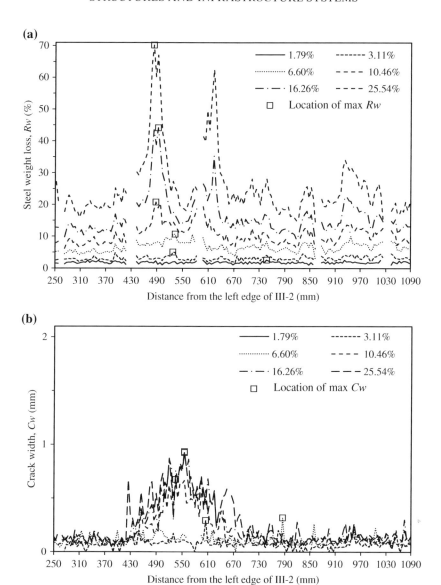

Figure 16. Spatial distribution of the (a) steel weight loss and (b) surface crack width of beam III-2.

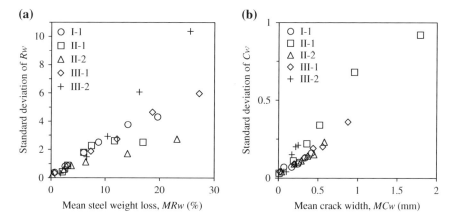

Figure 17. Relationship between the standard deviations and mean values of Rw and Cw.

to 610 mm and from 850 to 1090 mm, except for specimen III-1, which had no stirrups. In general, it can be observed that Cw increases with Rw. The distribution of Rw and Cw at various stages of steel corrosion is spatially non-uniform because Rw and Cw fluctuate erratically along the specimens. The non-uniformity of Rw and Cw becomes increasingly prominent with increasing MRw and MCw. In Figure 17, the standard deviations of Rw and Cw indicate that the distributions of Rw and Cw increasingly

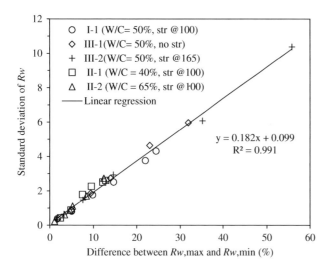

Figure 18. Relationship between the standard deviation and the difference between Rw, max and Rw, min.

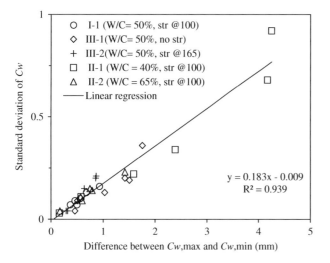

Figure 19. Relationship between the standard deviation and the difference between Cw, max and Cw, min.

diverge from their means as MRw and MCw increase. Therefore, the spatial variability in Rw and Cw is becoming larger as the steel weight loss and crack width increase. Yamamoto, Oyado, Mikata,

Kobayashi, and Shimomura (2011) studied the distribution of steel weight loss along the corroded bar using a great number of corroded RC beams. They also reported that the non-uniform degree of steel cross section losses of the corroded reinforcement increases with the corrosion loss. The cause of this larger spatial variability might result from the greater contribution of the larger surface cracks to the increase of Rw. When chloride penetrated from outside, corrosion cracking allowed much higher concentration of chloride to reach the rebar surface more easily, according to Andrade et al. (1993).

On the other hand, the maximum steel weight loss Rw, max and crack width Cw, max, which are denoted by square symbols in the graphs, do not usually occur at the corresponding locations. Several maxima of Rw, denoted by square symbols in the graphs, and other peaks (not maximum points) of Rw occur close to the locations of the stirrups (i.e. approximately at 300, 400, 500, 600, 860, 960 and 1060 mm) for specimens I-1, II-1 and II-2 in Figures 12(a), 13(a) and 14(a). This is probably due to the fact that the thinner concrete covers at the locations of the installed stirrups expose the longitudinal rebar to chloride and thus cause it to corrode more quickly than the rebar at other locations. The locations of the peaks often shift depending on the mean values. For example, the Rw, max of specimen I-1 is located at approximately 580 mm at MRw of .68% and 2.60% but shifts to approximately 490 mm at larger values of MRw. However, Figure 12(a) and (b) shows that some peaks of Rw and Cw also occur at approximately the same locations, namely 310, 490, 610 and 730 mm. This finding emphasises the influence of steel weight loss on the increase in crack widths at corresponding locations along the specimen.

4.3. Trend of steel weight loss and crack widths

In addition to the previously mentioned non-uniform spatial distributions of Rw and Cw, a trend is consistently observed among the graphs in Figures 12–16. Although the spatial variabilities of Rw and Cw increase as their mean values increase, it is worth noting that their erratic shapes tend to have a clear trend as MRw exceeds approximately 5%. For example, in Figure 12(a) for specimen I-1, the spatial distributions of Rw at MRw of 8.79, 14.20 and 19.65% seem to increase, following a very similar fluctuating pattern to the previous distributions of Rw at

(a)

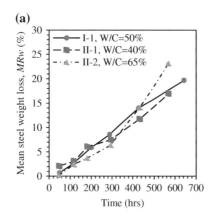

(b)

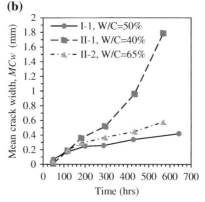

Figure 20. Effects of W/C ratios on the development of steel corrosion and crack width.

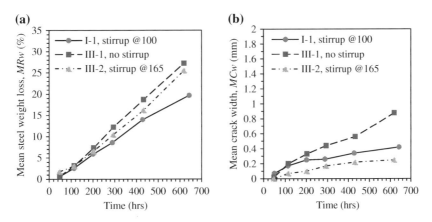

Figure 21. Effects of stirrups on the development of steel corrosion and crack width.

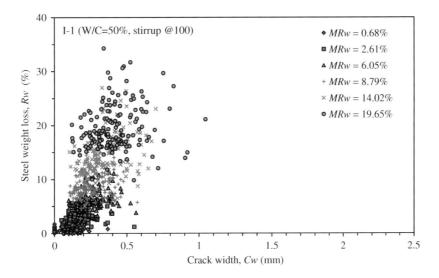

Figure 22. Relationship between the steel weight loss and crack width of specimen I-1.

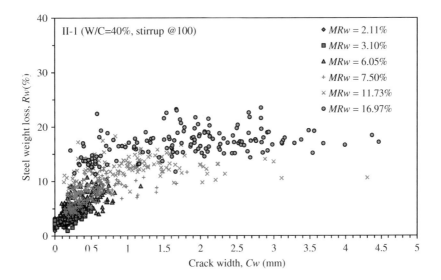

Figure 23. Relationship between the steel weight loss and crack width of specimen II-1.

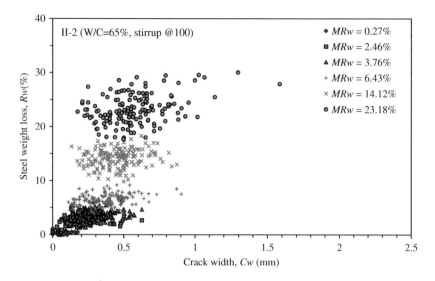

Figure 24. Relationship between the steel weight loss and crack width of specimen II-2.

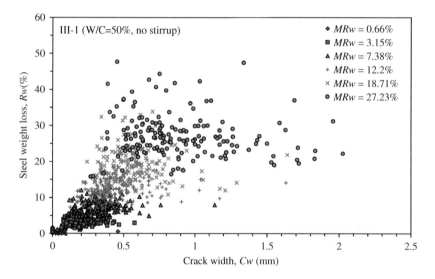

Figure 25. Relationship between the steel weight loss and crack width of specimen III-1.

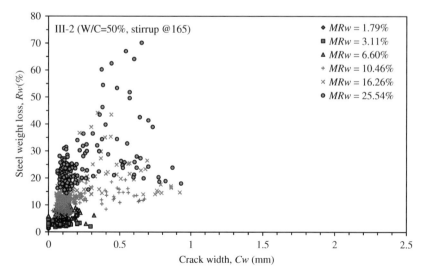

Figure 26. Relationship between the steel weight loss and crack width of specimen III-2.

MRw = 6.05%. The spatial distributions of Rw in Figures 13(a), 14(a), 15(a) and 16(a) also exhibit a similar spatial growth behaviour as MRw increases beyond 5%. Similarly, the fluctuating pattern of the distribution of crack widths in Figures 12(b), 13(b), 14(b), 15(b) and 16(b) seems to follow the same trend as MRw exceeds 5%; however, the trend of the spatial distribution of crack widths appears to be weaker than that of Rw.

With respect to this behaviour, a strong relationship can be found between the standard deviation and differences between the maximum and minimum values of steel corrosion (Rw, max − Rw, min) and crack widths (Cw, max − Cw, min), as shown in Figures 18 and 19. The significant merit of this relationship is that it might be possible to estimate the spatial variability of Rw and Cw between two different points if the maximum and minimum values of Rw or Cw between these points can be determined via *in situ* inspection. Due to this consistent spatial growth trend, it might be preferable to estimate the distribution of steel corrosion of RC members using the inspection results of steel corrosion in the *in situ* structure when the mean steel weight loss exceeds 5%.

4.4. Effect of water-to-cement ratio

Figure 20 shows the increase in the means of Rw and Cw as a function of corrosion time for specimens with different W/C ratios. From Figure 20(a), because the trend lines of the three specimens exhibit similar behaviour, there appears to be no clear effect of W/C ratios on the growth of steel corrosion. Figure 20(b) shows the effects of different W/C ratios on the surface crack widths. It can be seen that the crack widths of the specimens with W/C ratios of 40% grew more quickly than those with W/C ratios of 50% and 65%. Therefore, the crack widths of specimens with lower W/C ratios tend to grow more quickly than those with higher W/C ratios. These results are in agreement with the findings of Alonso et al. (1998). There is a delay in the increase of the crack widths for specimens with higher W/C ratios, which have greater porosity to accommodate the corrosion product and reduce the internal pressure.

4.5. Effect of stirrups

Figure 21 shows the effects of stirrups on the development of steel corrosion and crack widths for the beams exposed to approximately the same amount of accumulated current. Figures 21(a) and (b) indicates that the steel corrosion and surface crack width of the specimen having no stirrups (i.e. III-1) increase more quickly than those of other specimens having stirrups (i.e. I-1 and III-2). For the steel corrosion of the specimens having stirrups, Figure 21(a) shows that the longitudinal rebar of specimen III-2, with a stirrup spacing of 165 mm, is corroded more quickly than that of specimen I-1, with a stirrup spacing of 100 mm. Thus, decreasing the stirrup spacing or increasing the number of stirrups slows the steel corrosion. In Figure 21(b), because the mean crack width of specimen I-1 increases more quickly than that of specimen III-2, one can conclude that increasing the number of stirrups might accelerate the growth of crack widths.

4.6. Relationship between steel weight loss and crack widths

The graphs in Figures 22–26 show the relationships between the steel weight loss and the surface crack widths at their corresponding locations for all the specimens. Generally, the scattered points are increasingly highly dispersed with increasing MRw. Therefore, the relationship between Rw and Cw weakens significantly at higher values MRw.

However, it is also found that the steel weight loss increases with increasing crack width. In particular, in the initial part of the corrosion process, up to a crack width of about .3 mm, both the steel weight loss and crack width appear to be linearly related for all the specimens. This result is inconsistent with previous findings reported in the literature. For example, Vidal et al. (2004) found a linear approximation between the two parameters for the RC specimens with corrosion crack widths of about 1 mm, which is similar to the finding by Alonso et al. (1998) although these authors reported that the scatter of their relationships increased with the corrosion crack width and became significant as the crack width was over 1 mm. This inconsistency might be caused by the differences in the experimental procedure such as the magnitude of current density to corrode the specimens. Further research is needed to investigate the effects of the current densities, structural details and concrete qualities on the relationship between the steel weight loss and corrosion crack width.

5. Conclusions

In this paper, a procedure using an X-ray technique to continuously investigate the spatial growth of corroded rebars inside RC beams has been illustrated, and a method for estimating steel weight loss via the digital image analysis of X-ray photograms has been presented. The accuracy of the estimation method was also discussed. The following conclusions can be drawn:

(1) The estimated rebar weight loss calculated using the digital image analysis of X-ray photograms was found to be only 3% higher than that of the actual measured steel weight loss. This demonstrated the good accuracy of the application of the X-ray technique for investigating the spatial growth of corroded rebars in RC members.

(2) The distributions of the steel weight loss and crack width are spatially non-uniform, and their degree of non-uniformity significantly increases as the steel weight loss and corrosion cracking increase. The cause of this larger spatial variability might result from the greater contribution of the larger surface cracks to the increase in steel weight loss (i.e. the larger corrosion cracking allows the much higher chloride concentration to penetrate through and reach the reinforcement more easily).

(3) The maximum steel weight loss usually does not occur at the locations corresponding to the maximum crack width and often occurs close to the stirrups. Additionally, these locations vary depending on the mean steel weight loss.

(4) A strong relationship was found between the standard deviations of the steel weight loss or crack width and the differences between their maximum and minimum values when the mean steel weight loss exceeded 5%. This relationship may allow the estimation of the variability between two points after data from the *in situ* inspection of steel corrosion are obtained.

(5) The effect of W/C ratio on the increased steel weight loss is not obvious. Meanwhile, the crack width of the specimens with low W/C ratios increased faster than those with high W/C ratios.

(6) The crack width and steel corrosion of the specimen having no stirrups increased more quickly than those of specimens having stirrups. Increasing the number of stirrups slows the steel corrosion but accelerates the growth of the crack widths.

The parameters used to produce the stochastic field associated with the steel weight loss in RC structures could be determined using the experimental results presented in this paper. This in turn can facilitate the life-cycle reliability assessment of ageing RC structures by incorporating the spatial variability of steel weight loss (as presented in Akiyama & Frangopol, 2014). In addition, a relationship between the steel weight loss and width of the longitudinal surface cracks might be established with corrosion cracks smaller than .3 mm. This relationship might be incorporated into an estimation model of the long-term structural performance of RC elements in a chloride-contaminated environment. At larger crack widths, a weaker relationship was found. This suggests that it is desirable to estimate the steel corrosion based on the inspected data of the steel corrosion levels of the *in situ* structures when the crack width is above .3 mm.

On the other hand, the difficulty in quantifying the non-uniform distribution of corrosion pits along the corroded rebars has been reported as the key issue in assessing the performance and structural reliability of corrosion-affected RC members (e.g. Coronelli & Gambarova, 2004; Shimomura, Saito, Takahashi, & Shiba, 2011; Stewart, 2009; Stewart & Al-Harthy, 2008; Val, 2007) and also in assessing their residual life (e.g. Torres-Acosta, Navarro-Gutierrez, & Terán-Guillén, 2007). Although the derivation of residual cross-sectional loss according to time is not presented in this paper, one can imagine that, by employing a similar estimating method to that reported in this paper, the X-ray images of non-corroded and corroded bars from the 12 different viewing angles can be used to derive the pitting corrosion and residual cross-sectional loss at any specific location along the corroded bars. The research related to the derivation of corrosion pits using X-ray images and assessment of corrosion-affected RC members using a finite element model with the consideration of the spatial variability of steel weight loss is under development and will be presented in the near future.

Disclosure statement

No potential conflict of interest was reported by the authors.

References

Akiyama, M., & Frangopol, D. M. (2013). Estimation of steel weight loss due to corrosion in RC members used on digital image processing of X-ray photogram. In A. Strauss, D. M. Frangopol, & K. Bergmeister (Eds.), *Life-cycle and sustainability of civil infrastructure systems* (pp. 1885–1891). London: CRC Press/Balkema, Taylor & Francis Group plc.

Akiyama, M., & Frangopol, D. M. (2014). Long-term seismic performance of RC structures in an aggressive environment: Emphasis on bridge piers. *Structure and Infrastructure Engineering, 10,* 865–879.

Akiyama, M., Frangopol, D. M., & Yoshida, I. (2010). Time-dependent reliability analysis of existing RC structures in a marine environment using hazard associated with airborne chlorides. *Engineering Structures, 32,* 3768–3779.

Alonso, C., Andrade, C., Rodriguez, J., & Diez, J. M. (1998). Factors controlling cracking of concrete affected by reinforcement corrosion. *Materials and Structures, 31,* 435–441.

Andrade, C., Alonso, C., & Molina, F. J. (1993). Cover cracking as a function of bar corrosion: Part I – Experimental test. *Materials and Structures, 26,* 453–464.

Ann, K. Y., & Song, H.-W. (2007). Chloride threshold level for corrosion of steel in concrete. *Corrosion Science., 49,* 4113–4133.

Beck, M., Goebbels, J., Burkert, A., Isecke, B., & Babler, R. (2010). Monitoring of corrosion processes in chloride contaminated mortar by electrochemical measurement and X-ray tomography. *Materials and Corrosion, 61,* 475–479.

Coronelli, D., & Gambarova, P. (2004). Structural Assessment of corroded reinforced concrete beams: Modeling guidelines. *Journal of Structural Engineering, 130,* 1214–1224.

Marsh, P. S., & Frangopol, D. M. (2008). Reinforced concrete bridge deck reliability model incorporating temporal and spatial variations of probabilistic corrosion rate sensor data. *Reliability Engineering & System Safety, 93,* 394–409.

Media Cybernetics, Inc. (2012). Image-Pro Plus version 7.0 for Windows, start-up guide. Rockville, MD: Media Cybernetics.

Mori, Y., & Ellingwood, B. R. (1993). Reliability-based service-life assessment of aging concrete structures. *Journal of Structural Engineering, 119,* 1600–1621.

Otsuki, N., Miyazato, S.-I., Diola, N., & Suzuki, H. (2000). Influences of bending crack and water–cement ratio on chloride-induced corrosion of main reinforcing bars and stirrups. *ACI Material Journal, 97,* 454–464.

Palsson, R., & Mirza, M. S. (2002). Mechanical response of corroded steel reinforcement of abandoned concrete bridge. *ACI Structural Journal, 99,* 157–162.

Shimomura, T., Saito, S., Takahashi, R., & Shiba, A. (2011). Modelling and nonlinear FE analysis of deteriorated existing concrete structures based on inspection. In C. Andrade & G. Mancini (Eds.), *Modelling of corroding concrete structures*, RILEM Bookseries 5 (pp. 259–272). Dordrecht, Heidelberg, London, New York: Springer.

Stewart, M. G. (2004). Spatial variability of pitting corrosion and its influence on structural fragility and reliability of RC beams in flexure. *Structural Safety, 26,* 453–470.

Stewart, M. G. (2009). Mechanical behaviour of pitting corrosion of flexural and shear reinforcement and its effect on structural reliability of corroding RC beams. *Structural Safety, 31,* 19–30.

Stewart, M. G., & Al-Harthy, A. (2008). Pitting corrosion and structural reliability of corroding RC structures, experimental data and probabilistic analysis. *Reliability Engineering System and Safety, 93,* 273–382.

Stewart, M. G., & Mullard, J. A. (2007). Spatial time-dependent reliability analysis of corrosion damage and the timing of first repair for RC structures. *Engineering Structures, 29,* 1457–1464.

Torres-Acosta, A. A., Navarro-Gutierrez, S., & Terán-Guillén, J. (2007). Residual flexure capacity of corroded reinforced concrete beams. *Engineering Structures, 29,* 1145–1152.

Trejo, D. & Pillai, R. G. (2004). Accelerated chloride threshold testing – Part II: Corrosion resistant reinforcement. *ACI Materials Journal, 101,* 57–64.

Val, D. V. (2007). Deterioration of strength of RC beams due to corrosion and its influence on beam reliability. *Journal of Structural Engineering, 133*, 1297–1306.

Vidal, T., Castel, A., & François, R. (2004). Analyzing crack width to predict corrosion in reinforced concrete. *Cement and Concrete Research, 34*, 165–174.

Vu, K., Stewart, M. G., & Mullard, J. (2005). Corrosion-induced cracking: Experimental data and predictive models. *ACI Structural Journal, 102*, 719–726.

Yamamoto, T., Oyado, M., Mikata, Y., Kobayashi, K., and Shimomura, T. (2011). Systematic laboratory test on structural performance of corroded reinforced concrete and its utilization in practice. In C. Andrade & G. Mancini (Eds.), *Modelling of corroding concrete structures*, RILEM Bookseries 5 (pp. 113–124). Dordrecht, Heidelberg, London, New York: Springer.

Zhang, R., Castel, A., & François, R. (2010). Concrete cover cracking with reinforcement corrosion of RC beam during chloride-induced corrosion process. *Cement and Concrete Research, 40*, 415–425.

Reliability-based durability design and service life assessment of reinforced concrete deck slab of jetty structures

Mitsuyoshi Akiyama, Dan M. Frangopol and Koshin Takenaka

ABSTRACT

Reinforced concrete (RC) structures in a marine environment deteriorate with time due to chloride-induced corrosion of reinforcing bars. Since the RC deck slabs of jetty structures are exposed to a very aggressive environment, higher deterioration rates can develop. In this paper, a reliability-based durability design and service life assessment of RC jetty structures are presented. For new RC jetty structures, the concrete quality and concrete cover necessary to prevent the chloride-induced reinforcement corrosion causing the deterioration of structural performance during the whole lifetime could be determined. Based on the airborne chloride hazard depending on the vertical distance from the sea level surface to the RC deck slab, the probability associated with the steel corrosion initiation is estimated. The water to cement ratio and concrete cover to satisfy the target reliability level are provided. For evaluating the service life of existing structures, the condition state based on the visual inspection of RC structure can be provided. The deterioration process of the RC jetty structure can be modelled as a Markov process. Therefore, the transition probability matrix at time t after construction can be updated by visual inspection results. A procedure to update the transition probability matrix by the Sequential Monte Carlo Simulation method is indicated. In an illustrative example, the effect of the updating on the life-cycle reliability estimate of existing RC deck slab in a jetty structure subjected to the chloride attack is presented.

1. Introduction

For structures located in a moderately or highly aggressive environment, multiple environmental and mechanical stressors lead to deterioration of structural performance. Despite the fact that concrete is a reliable structural material with good durability performance, exposure to severe environments makes it vulnerable (Moradi-Marani, Shekarchi, Dousti, & Mobasher, 2010). Reinforcement corrosion in concrete is the dominant cause of premature deterioration of reinforced concrete (RC) structures. The corrosion-induced structural deterioration is a gradual process consisting of a few phases during the service life of RC structures, including corrosion initiation, cover concrete cracking, serviceability loss, and collapse due to loss of structural strength. Such deterioration will reduce the service life of structures and increase their life-cycle cost due to maintenance actions. Whole-life performance prediction of RC structures is gradually becoming a requirement for the design of these structures and a necessity for optimal decision-making with respect to their inspection, repair, strengthening, replacement and demolition.

Despite extensive research in this field, a number of issues still remain unclear. One of the main intricacies is the uncertainty associated with the physical parameters involved in the problem.

Because of the presence of many kinds of uncertainties associated with the prediction of deterioration process of RC structures, various theoretical frameworks have been developed to assess the service life performance of structures, including advanced reliability theories (Ellingwood, 2005; Frangopol, 2011; Li, 2003, 2004; Mori & Ellingwood, 1993).

In this paper, the reliability-based durability design and service life assessment of RC deck slab of jetty structures are presented. Figure 1 shows a typical RC jetty in Tokyo port of Japan. Dousti, Moradian, Taheri, Rashetnia and Shekarchi (2013) reported, based on their field investigation, that the assessment of RC jetty structures showed that, when the structures are exposed to aggressive environments, very high deterioration rates can develop, leading to serious damage conditions in very short time periods. The highest deterioration rates were observed in zones where the concrete surface is subjected to perpetual wetting and drying cycles of salt water.

For new RC deck slab in a jetty structure, it is possible to determine the water to cement ratio and concrete cover to prevent the steel corrosion initiation during the whole lifetime of the structure under various uncertainties by conducting an adequate time-dependent reliability analysis. Several marine environmental factors affect the degradation mechanisms of concrete structures. The exact influence of these factors is difficult to predict

RC jetty structure

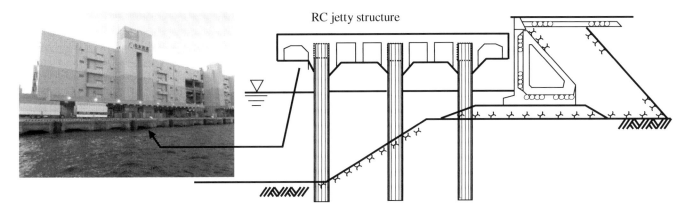

Figure 1. RC jetty structure in Tokyo port of Japan (Source: Author).

as they vary in time and space. Unlike the case associated with actions that are usually considered in structural design (e.g. probabilistic seismic hazard analysis and traffic load simulation), there is a lack of research on marine environmental hazard assessment. This is because the data on coastal atmospheric exposure is very limited compared with the array measurements of seismic ground motion. In the seismic probabilistic hazard analysis (SPHA), which determine the relation between an earthquake motion intensity (e.g. PGA, PGV and response acceleration) and the corresponding annual probability of exceedance, it is necessary to evaluate the attenuation of seismic intensity with the distance from the seismic source (Cornell, 1968). By applying the concept of SPHA to the assessment of the effect of airborne chlorides on durability design of RC structures, which is reduced with increasing the distance from the coastline, probabilistic hazard analysis can be performed and the effect of a marine environment can be quantified. Akiyama et al. (Akiyama, Frangopol, & Suzuki, 2012; Akiyama, Frangopol, & Matsuzaki, 2014) proposed an approach to establish the probabilistic hazard curve associated with airborne chlorides depending on the horizontal distance from the coastline and provided a computational procedure to obtain the probability of occurrence of steel corrosion and corrosion cracking. Since RC jetty structures are located above the sea, it is necessary to evaluate the hazard associated with airborne chlorides depending on the vertical distance from the sea-level surface. The procedure to estimate the failure probability associated with the steel corrosion initiation for RC deck slab in the jetty structure is presented in this paper.

Because of the lack of adequate knowledge to ensure the durability of concrete structures, existing RC jetty structures have significantly deteriorated. Investigation of corrosion damage in RC jetty structures is mainly focused on the RC deck slab. A systematic visual inspection is conducted to get a realistic data about the deterioration and distress of the concrete deck slabs and to take the extensive photographs of the deck slabs for determining the extent of investigation (Dousti et al. 2013). Based on these visual inspections, the probabilities of transition between condition states of RC deck slab are provided. These probabilities could be incorporated into the Markov chain.

Markov chains are commonly used for performance assessment of deteriorating components and systems. The rate of transition from one state to another is constant over time for homogeneous Markov chains. Madanat (1993) presented a methodology for planning the maintenance and rehabilitation activities of transportation facilities based on the latent Markov decision process. Kato, Iwanami, Yokota and Yamaji (2002) applied the Markov model to the life-cycle reliability estimate of the RC jetty structures. Recently, Saydam, Frangopol and Dong (2013) presented a methodology for quantifying lifetime risk of bridge superstructures. In their model, a scenario-based approach integrating the Pontis element condition rating system into risk assessment procedure was used to identify expected losses, and a Markov process was applied to estimate the deterioration level of bridge components regarding the transition between the condition states. Reliability estimation of RC structures under hazard associated with environmental stressors and material deterioration models requires complex computations including hazard and fragility analyses (e.g. Akiyama, Frangopol, & Yoshida, 2010). Since in Markov model, deterioration process can be expressed as the transition probability, it can provide significant time efficiency in life-cycle reliability analysis.

The transition probability matrix used in Markov model depends on the evaluation of environment, corrosion process of steel bars and deterioration of structural performance. The model uncertainties associated with the estimation of the transition probability matrix could be very large if there is only information on design condition of existing RC structures analysed. Meanwhile, for existing structures it is possible to reduce epistemic uncertainties using inspection results (Akiyama et al. 2010; Yoshida, 2009). In this study, a computational procedure for estimating the life-cycle reliability of existing RC jetty structures subjected to the chloride attack is presented. The Markov model and Sequential Monte Carlo Simulation (SMCS) are both used in conjunction with time-dependent condition state probability for existing RC jetty structures. An illustrative example is presented.

2. Reliability-based durability design of RC deck slab of a jetty structure

2.1. Hazard assessment associated with airborne chloride

Environmental conditions should be quantitatively assessed and the evaluation results should be reflected in the rational durability design of RC structures in a marine environment. The difference in the amount of airborne chlorides among structure locations

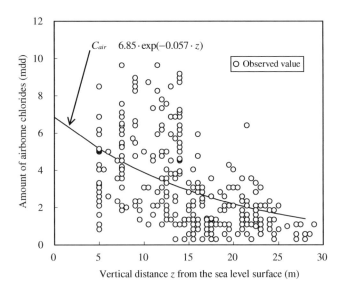

$$C_{air} = 6.85 \cdot \exp(-0.057 \cdot z)$$

Figure 2. Amount of airborne chlorides vs. vertical distance from the sea level surface (based on the observed values reported by Aoyama, Torii and Matsuda, 2003).

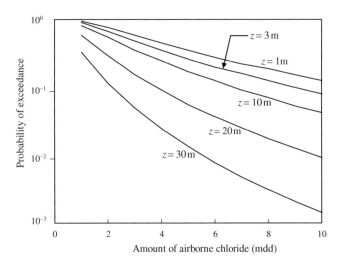

Figure 3. Probability profiles for amounts of airborne chlorides depending on the vertical distance z from the sea level surface.

should be considered taking into account the spatial-temporal variation.

Akiyama et al. (2012, 2014) proposed the attenuation relation between the amount of observed airborne chlorides, C_{air}, and the horizontal distance from the coastline. The speed of wind, the ratio of sea wind (defined as the percentage of time during one day when the wind is blowing from sea toward land) and the distance from the coastline affect the amount of airborne chlorides. C_{air} (mdd = 100 mg/m²/day) in the lateral direction can be expressed as:

$$C_{air} = 1.29 \cdot r \cdot u^{0.386} \cdot d^{-0.952} \qquad (1)$$

where r is the ratio of sea wind, u (m/s) is the average wind speed during the observation period and d (m) is the distance from the coastline. The statistics of wind speed and the ratio of sea wind of C_{air} collection could be obtained from meteorological data.

Since RC jetty structures are located above the sea, Equation (1) cannot be applied to estimate the probability associated with

the occurrence of steel corrosion. Aoyama, Torii and Matsuda (2003) reported that the amount of airborne chloride attenuates with the increase in the vertical distance from the sea level surface based on the measurements of chloride concentration as shown in Figure 2. The attenuation of C_{air} (mdd) in the vertical direction can be expressed as:

$$C_{air} = 6.85 \cdot \exp(-0.057 \cdot z) \qquad (2)$$

where z (m) is the vertical distance of the bottom of RC deck slab from the sea level surface.

The mean and coefficient of variation (COV) of the ratio of observed to calculated airborne chloride is 0.96 and 0.646, respectively. Since the effect of the speed of wind and the ratio of sea wind on the amount of airborne chloride is negligible, the attenuation of C_{air} in Equation (2) depends only on the distance from the sea level surface. Because of large uncertainty involved in the prediction of airborne chloride spread in the vertical direction, larger concrete cover and smaller water to cement ratio are needed to ensure the prescribed target reliability level. In order to reduce this uncertainty, further research is needed to include additional parameters in the attenuation equation.

The probability that C_{air} at a specific site will exceed a prescribed value c_{air} is:

$$\begin{aligned} q_s(c_{air}) &= P(C_{air} > c_{air}) \\ &= P\left(U_R > \frac{c_{air}}{6.85 \cdot \exp(-0.057z)} \right) \end{aligned} \qquad (3)$$

where U_R is the lognormal random variable representing attenuation uncertainty associated with Equation (2).

Figure 3 shows the probability profiles (i.e. hazard curve) associated with the airborne chloride. Unlike the airborne chloride probability profiles depending on the horizontal distance from the coastline (Akiyama et al. 2012, 2014), those shown in Figure 3 are independent of the location of the structure.

2.2. Reliability assessment of RC deck slab in a jetty structure

When airborne chlorides were measured in each location, exposed concrete specimens were also placed at the same location to measure the surface chloride content, C_0. Based on the measurements of chloride concentration (Aoyama 2003), the relationship between C_{air} (mdd) and C_0 (kg/m³) is:

$$C_{air} = 0.563 \cdot C_0^{0.948} \qquad (4)$$

The fragility curve shows the occurrence probability of the limit state under the condition that a specific value of airborne chloride, c_{air} is given. The probability of occurrence of steel corrosion given c_{air} can be obtained using the performance function (Akiyama et al. 2012):

$$g_{d,1} = C_T - \chi_1 \cdot C(c, \chi_2 D_c, \chi_3 C_0, t) \qquad (5)$$

where $\quad \log D_c = -6.77(W/C)^2 + 10.10(W/C) - 3.14 \quad (6)$

$$C(c, \chi_2 D_c, \chi_3 C_0, t) = \chi_3 C_0 \left\{ 1 - \mathrm{erf}\left(\frac{0.1 \cdot c}{2\sqrt{\chi_2 D_c t}} \right) \right\} \qquad (7)$$

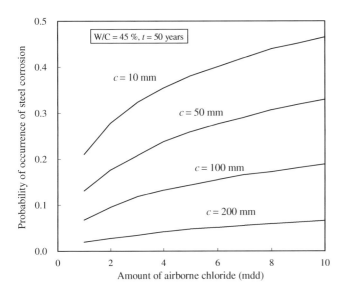

Figure 4. Effect of concrete cover on the probability profiles associated with the steel corrosion assuming W/C = 50% and t = 50 years after construction.

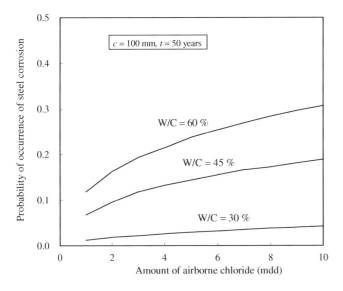

Figure 5. Effect of water to cement ratio on the probability profiles associated with the steel corrosion assuming c = 100 mm and t = 50 years after construction.

where C_T is the critical threshold of chloride concentration (kg/m^3), c is the concrete cover (mm), t is the time after construction (years), W/C is the ratio of water to cement, erf is the error function, χ_1 is the model uncertainty associated with the estimation of C, D_c is the coefficient of diffusion of chloride, χ_2 is the model uncertainty associated with the estimation of D_c and χ_3 is the model uncertainty associated with Equation (6).

The parameters of random variables were provided in Akiyama et al. (2012). Figures 4 and 5 show the fragility curves $F_r(c_{air})$. The effects of concrete cover and water to cement ratio on the probabilities of occurrence of steel corrosion given c_{air} are presented in Figures 4 and 5, respectively. Using the fragility curve, $F_r(c_{air})$, the probability of steel corrosion initiation is estimated by:

$$p_f = \int_0^\infty \left(-\frac{dq_s(c_{air})}{dc_{air}} \right) \cdot F_r(c_{air}) \, dc_{air} \tag{8}$$

Figures 6 and 7 illustrate the relationship between the probability of steel corrosion initiation and distance z from the sea level surface assuming RC deck slab with W/C = 45% and t = 10 years, and W/C = 45% and t = 50 years, respectively. Using Equation (8), the concrete cover and water to cement ratio to ensure the target reliability could be determined; however, this would require the structural designer to conduct the reliability computations.

2.3. Reliability-based durability design

A design criterion to ensure that the reliability index for the occurrence of steel corrosion will be close to the target value, without performing a complex reliability analysis by the designers, is proposed herein. The proposed formulation is:

$$\gamma \frac{C_d}{C_{T,d}} \leq 1.0 \tag{9}$$

$$C_d = C_{0,d} \left(1 - \text{erf} \frac{0.1 \cdot c}{2 \sqrt{D_{c,d} \cdot T}} \right) \tag{10}$$

$$C_{0,d} = 4.5 \cdot \exp(-0.05 \cdot z) \tag{11}$$

where γ is the durability design factor, $C_{T,d}$ is the design critical threshold of chloride concentration (C_T in Equation (5) as the mean value), and T is the lifetime of RC jetty structure.

The procedure to determine the durability design factor is based on code calibration. The steps are as follows (Akiyama et al. 2012, 2014):

(a) Set the target reliability index β_{target} and the lifetime of the structure T.
(b) Calculate the design value of the surface chloride content using Equation (11).
(c) Assume the initial durability design factor γ and the distance z from the sea level surface.
(d) Select the design concrete cover and water to cement ratio to satisfy the design criterion (i.e. Equation (9)).
(e) Calculate the probability using Equation (8) and transform it into reliability index.
(f) Repeat steps (c) to (e) until:

$$U = \sum_{i=1} \left(\beta_{target} - \beta_i(\gamma) \right)^2 \tag{12}$$

is minimised, and the durability design factor is found.

When T = 50 years, the durability design factors associated with the target reliability indices 1.0, 1.5 and 2.0 are 1.07, 3.45 and 12.9, respectively. Figure 8 indicates the reliability index associated with the steel corrosion initiation assuming that the concrete with W/C = 30% is used for the RC jetty structure. The results indicate that the reliability indices are close to the target values. Using the design criterion and durability design factor proposed, RC jetty structures having target durability reliability indices for prescribed lifetime T can be designed.

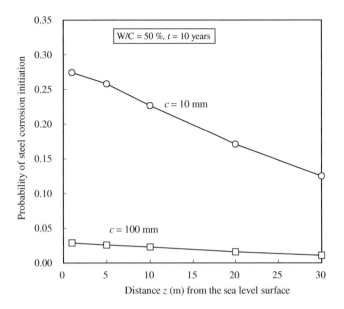

Figure 6. Relationship between probability of steel corrosion initiation and distance z from the sea level surface assuming W/C = 50% and t = 10 years after construction for two concrete covers, c = 10 mm and c = 100 mm.

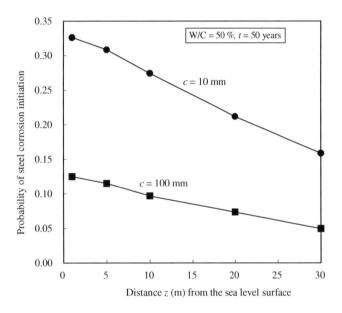

Figure 7. Relationship between probability of steel corrosion initiation and distance z from the sea level surface assuming W/C = 50% and t = 50 years after construction for two concrete covers, c = 10 mm and c = 100 mm.

3. Life-cycle reliability estimation of existing RC deck slab of a jetty structure

RC deck slab of a jetty structure with high quality concrete and adequate concrete cover designed by Equation (9) prevents the chloride-induced reinforcement corrosion causing the deterioration of structural performance during its whole lifetime. Based on the life-cycle cost analysis, corrosion-resistant stainless steel reinforcing bars would be used to protect RC jetty structure from the chloride attack, even if it is much more expensive than conventional carbon steel (Val & Stewart, 2003). Although cost estimation is outside of the scope of this paper, life-cycle reliability analysis can be applied to select optimal strategies for improving durability of new RC jetty structures.

Meanwhile, some existing structures designed without adequate durability detailing deteriorate severely. To confirm whether these deteriorated structures still conform to the safety and/or serviceability requirements, it is necessary to investigate the effect of the chloride-induced reinforcement corrosion on the structural capacity and stiffness. Although the life-cycle reliability assessment of existing corroded RC structures has been developed, it requires complex reliability computations. For existing RC jetty structures, it is important to identify when the maintenance actions including the repair and/or strengthening are needed. Markov model is an efficient tool to estimate the time-variant performance of structures. The accuracy of Markov model in the life-cycle reliability analysis depends strongly on the transition probability which is the probability of moving from any given state to the subsequent state on the next time interval. The uncertainty associated with the estimation of the transition probability becomes large if the transition probability is determined based on the information on the design condition of RC jetty structures. Using inspection results, the epistemic uncertainties associated with the service life reliability prediction of existing RC structures can be reduced compared with new RC structures.

In this section, the deterioration process of a RC jetty structure transitioning between condition states is modelled as a Markov process. A procedure to obtain the transition probability matrix at time t after construction updated by SMCS is indicated. Transition probability matrix could be updated in order to be consistent with the observational information.

3.1. Markov model

The rate of transition from one state to another is assumed to be constant over time for homogeneous Markov chains. The probability of moving from any given state $(k-1)$ to state k on the next time interval is called the transition probability p_k. This probability is the core of a Markov chain. In this paper, a RC jetty structure in a marine environment with six condition states (i.e. k = 0, 1, 2, 3, 4 and 5) is considered. The deterioration states of RC jetty structures due to chloride-induced corrosion of reinforcing bars are classified according to the criteria defined by the Japan Society of Civil Engineers (JSCE, 2001) and Komure, Hamada, Yokota, and Yamaji (2002) as listed in Table 1. The relationship between the state probability vector $X(t)$ and the transition probability matrix that can be used in the prediction of the structural performance is:

$$
\begin{Bmatrix} X_0 \\ X_1 \\ X_2 \\ X_3 \\ X_4 \\ X_5 \end{Bmatrix} = \begin{bmatrix} 1-p_1 & 0 & 0 & 0 & 0 & 0 \\ p_1 & 1-p_2 & 0 & 0 & 0 & 0 \\ 0 & p_2 & 1-p_3 & 0 & 0 & 0 \\ 0 & 0 & p_3 & 1-p_4 & 0 & 0 \\ 0 & 0 & 0 & p_4 & 1-p_5 & 0 \\ 0 & 0 & 0 & 0 & p_5 & 1 \end{bmatrix}^t \begin{Bmatrix} 1 \\ 0 \\ 0 \\ 0 \\ 0 \\ 0 \end{Bmatrix}
$$

$$
= \begin{Bmatrix} f_0(p_1,t) \\ f_1(p_1,p_2,t) \\ f_2(p_1,p_2,p_3,t) \\ f_3(p_1,p_2,p_3,p_4,t) \\ f_4(p_1,p_2,p_3,p_4,p_5,t) \\ f_5(p_1,p_2,p_3,p_4,p_5,t) \end{Bmatrix}
\qquad (13)
$$

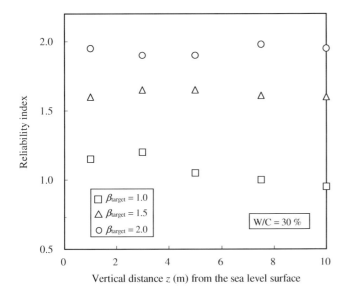

Figure 8. Reliability index of RC deck slab in a jetty structure designed by Equation (10) assuming W/C = 30%.

Table 1. Deterioration states of RC jetty defined by Japan Society of Civil Engineers (JSCE, 2001) and Komure, Hamada, Yokota and Yamaji (2002).

Condition state	Deterioration status
0	Sound
1	Some rusts are observed on the concrete surface. Crack width is less than 0.3 mm
2	Minor corrosion cracks are shown. Crack width is larger than 0.3 mm
3	Major corrosion cracks are shown. Crack width is larger than 1.0 mm
4	Deteriorated seriously. More than 10% of cover concrete of RC deck slab are spalled. Structural performance declined
5	Deteriorated totally. More than 40% of cover concrete of RC deck slab are spalled. Obvious decline of structural performance

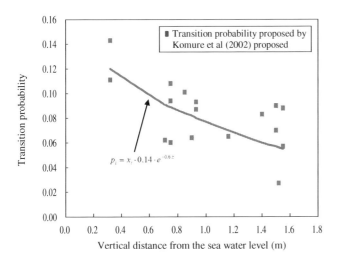

Figure 9. Relationship between transition probability and distance from the sea water level to the RC deck slab in the jetty structure.

where $X(t) = [X_0, X_1, X_2, X_3, X_4, X_5]$.

Komure et al. (2002) proposed the annual transition probability to quantify the actual chloride-induced damage state based on the survey results of many existing RC jetty structures around

Japan. To prevent moving from any given state $(k - 1)$ to state $k + 1$, the time interval in Equation (13) was assumed to be one year. Figure 9 shows the relationship between transition probability and the vertical distance from the seawater level to the RC deck slab. Assuming that transition probability is independent of conditional states (i.e. $p_1 = p_2 = \ldots = p_5$), p_i is provided by:

$$p_1 = p_2 = p_3 = p_4 = p_5 = 0.14 \cdot e^{-0.6 \cdot z} \qquad (14)$$

When model uncertainty associated with the estimation of transition probability p_i is taken into consideration, p_i is provided by

$$p_i = x_i \cdot 0.14 \cdot e^{-0.6 \cdot z} \qquad (15)$$

where x_i is the model uncertainty, assumed to follow a lognormal distribution, associated with the estimation of p_i.

3.2. Sequential Monte Carlo Simulation

For existing structures, the uncertainties associated with predictions can be reduced by the effective use of information obtained from visual inspections, field test data regarding structural performance, and/or monitoring (Enright & Frangopol, 1999; Estes & Frangopol, 2003; Estes, Frangopol, & Foltz, 2004; Frangopol, 2011; Frangopol & Liu, 2007; Frangopol, Saydam, & Kim, 2012; Frangopol, Strauss, & Kim, 2008a, 2008b, Okasha & Frangopol, 2009; Okasha, Frangopol, & Orcesi, 2012). This information helps engineers to improve accuracy of long-term structural performance estimation. In this paper it is assumed that the state probability vector at t years is given based on the survey of existing RC jetty structures. Since the relationship between the state probability vector and related random variables is nonlinear as described in Equations (13–15), it is impossible to perform the updating of these random variables by a theoretical closed-form solution such as a Kalman Filter algorithm.

In this paper, SMCS is applied to update the random variables x_i. The state space model consists of two processes, the time updating process and the observation updating process. The time

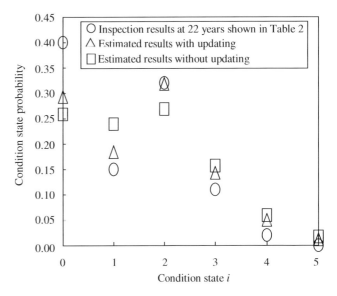

Figure 10. Comparison of inspection results at 22 years, with estimated results with updating at 16 years and estimated results without updating at 16 years.

Table 2. Probabilities of condition state based on inspection of an existing RC jetty structure (adapted from Taniguchi, Tamura, Sano, and Hamada 2004).

Condition state	Probability of condition state					
	0	1	2	3	4	5
16 years after construction	0.42	0.18	0.30	0.08	0.02	0.00
22 years after construction	0.40	0.15	0.32	0.11	0.02	0.00

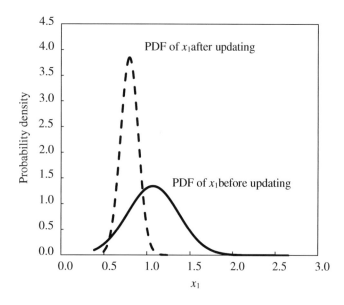

Figure 11. Comparison of the PDFs of x_1 with updating at 16 years and without updating.

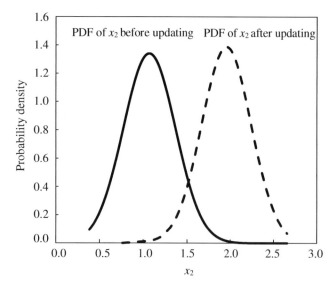

Figure 12. Comparison of the PDFs of x_2 with updating at 16 years and without updating.

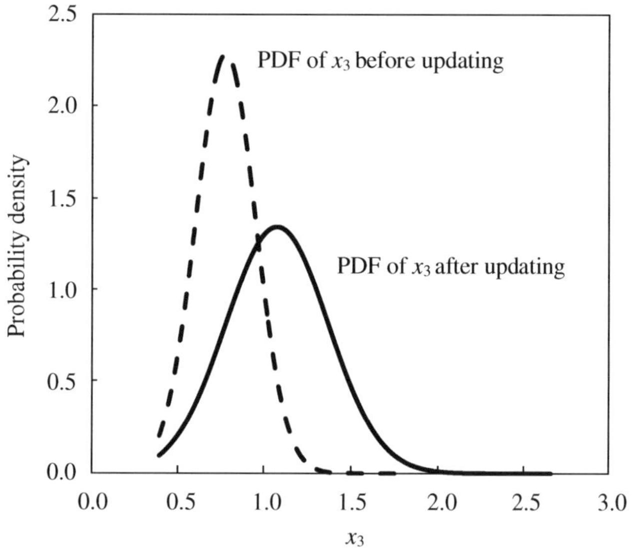

Figure 13. Comparison of the PDFs of x_3 with updating at 16 years and without updating.

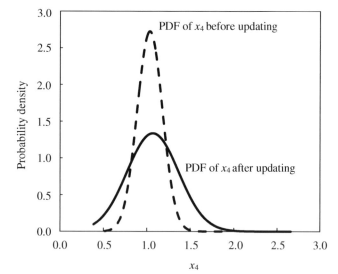

Figure 14. Comparison of the PDFs of x_4 with updating at 16 years and without updating.

updating process is the one step ahead prediction based on the information at the $(k-1)$-th step. The predicted vector is:

$$x_{k/k-1} = F(x_{k-1/k-1}, w_k) \qquad (16)$$

where w_k is the system noise represented by the noise involved in the prediction process. It is assumed that observation information z_k is a function H of state vector $x_{k/k}$ and observation noise v_k as:

$$z_k = H(x_{k/k}, v_k) \qquad (17)$$

The probability density functions (PDFs) of these noises are assumed known and independent. The algorithm based on MCS starts by assuming samples drawn from the distribution at $(k-1)$-th step:

$$x_{k-1/k-1}^{(j)} \sim p(x_{k-1}|Z_{k-1}), \quad j = 1, \dots, n \qquad (18)$$

$$Z_{k-1} = (z_1, z_2, \dots, z_{k-1}) \qquad (19)$$

The superscript (j) denotes the generated j-th sample realisation. The PDF is approximately expressed by the samples with Dirac delta function δ as:

$$p(x_{k-1}|Z_{k-1}) \cong \frac{1}{n} \sum_{j=1}^{n} \delta\left(x_{k-1} - x_{k-1/k-1}^{(j)}\right) \quad (20)$$

The above approximation form of PDF is called empirical PDF. The samples of the k-th step before observation updating are obtained by simply substituting them into Equation (16) to become:

$$x_{k/k-1}^{(j)} = F\left(x_{k-1/k-1}^{(j)}, w_k^{(j)}\right) \quad (21)$$

The empirical PDF of the k-th step before updating is similarly estimated by the sample realisation:

$$p(x_k|Z_{k-1}) \cong \frac{1}{n} \sum_{j=1}^{n} \delta\left(x_k - x_{k/k-1}^{(j)}\right) \quad (22)$$

The PDF after updating is:

$$\begin{aligned} p(x_k|Z_k) &= p(x_k|z_k, Z_{k-1}) \\ &= \frac{p(x_k, z_k|Z_{k-1})}{p(z_k|Z_{k-1})} \\ &= \frac{p(z_k|x_k, Z_{k-1}) \cdot p(x_k|Z_{k-1})}{\int p(z_k|x_k, Z_{k-1}) \cdot p(x_k|Z_{k-1}) \cdot dx_k} \end{aligned} \quad (23)$$

Substituting Equations (22) into (23) and using the property of a delta function results in:

$$\begin{aligned} p(x_k|Z_k) &= \sum_{j=1}^{n} \left[\frac{q_k^{(j)}}{\sum_{i=1}^{n} q_k^{(j)}}\right] \cdot \delta(x_k - x_{k/k-1}^{(j)}) \\ &= \sum_{j=1}^{n} a_k^{(j)} \cdot \delta(x_k - x_{k/k-1}^{(j)}) \end{aligned} \quad (24)$$

where:

$$q_k^{(j)} = p(z_k|x_{k/k-1}^{(j)}) \quad (25)$$

$$a_k^{(j)} = \frac{q_k^{(j)}}{\sum_{i=1}^{n} q_k^{(i)}} \quad (26)$$

The term $a_k^{(j)}$ is the weight (likelihood ratio) of sample j. When a new observation is available, the weights are recalculated and the approximate posterior PDF is sequentially updated. The detailed procedure of SMCS applied to reliability analysis of concrete structure in a marine environment was given by Yoshida (2009).

3.3. Illustrative example

Based on the visual inspections, Taniguchi, Tamura, Sano and Hamada (2004) reported the difference of condition state probabilities of an existing RC deck slab in a jetty structure subjected to the chloride attack between 16 and 22 years after construction as listed in Table 2. RC jetty structures deteriorate due to the chloride attack, and, therefore, the probability of severe condition

states increases with time. Based on the inspection result at 16 years, the condition state at 22 years is predicted using the Markov model and SMCS.

Since computational results are almost the same if the number of samples N is more than 10,000, N in SMCS is set to 10,000. Based on the observation data, Equation (17) becomes:

$$\begin{aligned} \begin{pmatrix} Z_0 \\ Z_1 \\ Z_2 \\ Z_3 \\ Z_4 \\ Z_5 \end{pmatrix} &= \begin{pmatrix} X_0 \\ X_1 \\ X_2 \\ X_3 \\ X_4 \\ X_5 \end{pmatrix} + \begin{pmatrix} v_0 \\ v_1 \\ v_2 \\ v_3 \\ v_4 \\ v_5 \end{pmatrix} \\[2mm] &= \begin{pmatrix} f_0(p_1, t) \\ f_1(p_1, p_2, t) \\ f_2(p_1, p_2, p_3, t) \\ f_3(p_1, p_2, p_3, p_4, t) \\ f_4(p_1, p_2, p_3, p_4, p_5, t) \\ f_5(p_1, p_2, p_3, p_4, p_5, t) \end{pmatrix} + \begin{pmatrix} v_0 \\ v_1 \\ v_2 \\ v_3 \\ v_4 \\ v_5 \end{pmatrix} \end{aligned} \quad (27)$$

where Z_i represents the the inspection results of condition state i, X_i is the prediction of each condition state i and v_i is the observation noises at condition state i.

Figure 10 compares the probabilities of the six condition states i ($i = 0, 1, 2, 3, 4, 5$) reported in Table 2, with estimated results without and with updating at 16 years. The difference between inspection results and predictions from the proposed model

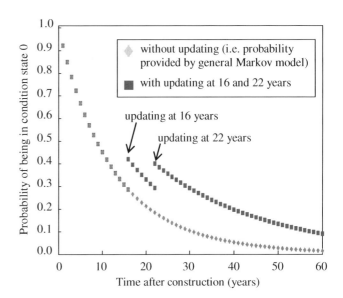

Figure 15. Time-dependent probability of being in condition state 0 with and without updating.

Table 3. List of assumed observation data.

X ($t - 20$ year)	Probability	
	Case 1	Case 2
X_0	0.44	0.19
X_1	0.37	0.33
X_2	0.14	0.27
X_3	0.04	0.14
X_4	0.01	0.05
X_5	0.00	0.02

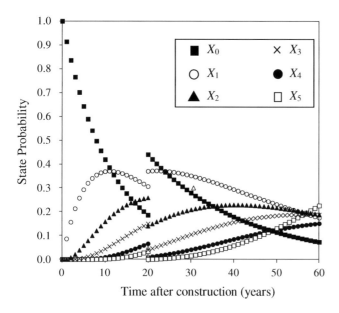

Figure 16. Time-dependent state probability (Case 1).

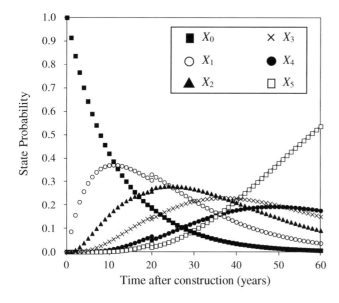

Figure 17. Time-dependent state probability (Case 2).

using SMCS is smaller than that between inspection results and Equation (13) at $t = 22$ years. The proposed model using SMCS could perform a more accurate life-cycle reliability estimate of the RC jetty structure. Figures 11–14 show the effect of updating on the PDF of model uncertainty associated with the estimation of transition probability. The COV after updating is smaller than that without updating.

Figure 15 shows the time-dependent probability of condition state 0 with updating at 16 and 22 years. To examine the effect of updating on the probability of condition state 0, the time-dependent probability estimated by the general Markov model is also shown in Figure 15. At 60 years, the difference between the probabilities of condition state 0 with and without updating is not negligible. It is important to update the conditional probabilities using inspection results for establishing the rational maintenance strategy.

To examine the effect of observation information on the time-dependent condition state probability, two cases of state probability vector $X(t)$ determined by the visual inspection as listed in Table 3 are considered. It is assumed that the observation information is provided at 20 years after construction. Figures 16 and 17 show time-dependent condition state probabilities using Markov model and SMCS. As shown in Table 3, since the assumed probability of condition state 0 of RC jetty structure in Case 1 is larger than that in Case 2, X_5 at 60 years in Case 1 is smaller than that in Case 2. The condition state probability after updating depends on the additional information. It is updated in order to be consistent with inspection results.

At the design stage, life-cycle reliability assessment of new RC jetty structure is conducted using the transition probability p_k. This probability could be determined based on the survey results acquired from many structures over time, for example, Equation (14). The general the trend in the time-dependent reliability of RC jetty structures could be estimated. However, a different value of p_k needs to be used to represent the deterioration of different classes of structures under various environmental conditions. In the case of available adequate inspection data from the specific structure, p_k has to be updated for consistency with the inspection results.

4. Conclusions

The findings of the present study can be summarised as follows:

(1) Based on the attenuation of the amount of observed airborne chloride in the vertical direction, probability profiles associated with airborne chloride for the life-cycle reliability assessment of RC deck slab in the jetty structure were proposed. These profiles can quantify the effect of marine environment.

(2) A reliability-based method for durability design of RC jetty structures using a durability design factor was introduced. For new RC jetty structures, concrete quality and concrete cover to prevent the chloride-induced reinforcement corrosion during the whole lifetime of these structures could be determined using the proposed durability design method. RC structures could be designed so that the time taken for the occurrence of the steel corrosion provided by the partial factor and design criterion is longer than the design lifetime. Using this design method, RC jetty structures can satisfy the target reliability level without any reliability computations performed by structural designers.

(3) A life-cycle reliability estimation method using Markov chain and SMCS was presented for existing RC deck slab in a jetty structure. The structural deterioration is modelled using condition state probabilities (i.e. Markov chain state probabilities) and the transition probability matrix is updated by SMCS. It is important to have inspection results in order to reduce the epistemic uncertainties. This can help in performing a more accurate life-cycle reliability estimate of existing RC jetty structures.

(4) Further research is needed on the reliability of RC jetty structures in a marine environment in connection

with probabilistic nonlinear FE modeling (Teigen et al. 1991a, 1991b), reliability-based inspection optimization (Onoufriou & Frangopol 2002), and life-cycle maintenance and optimisation (Frangopol & Kong 2001, Frangopol & Tsompankis 2014). Further research is necessary on life-cycle performance of RC jetty structures by improving durability, reducing maintenance costs and extending service life (e.g. use of high-performance concrete, admixtures and/or corrosion-resistant stainless steel rebars).

Acknowledgements

The opinions and conclusions presented in this paper are those of the authors and do not reflect the views of the sponsoring organisations.

Disclosure statement

No potential conflict of interest was reported by the authors.

Funding

This work was supported by JSPS Grant-in-Aid for Scientific Research (B) [grant number 24360185], and the Exploratory Research [grant number 25630198].

References

Akiyama, M., Frangopol, D. M., & Matsuzaki, H. (2014). Reliability-based durability design and service life assessment of concrete structures in an aggressive environment. Chapter 1. In D.M. Frangopol & Y. Tsompanakis (Eds.), *Maintenance and safety of aging infrastructure* (pp. 1–26). London: CRC Press/Balkema, Taylor & Francis Group.

Akiyama, M., Frangopol, D. M., & Suzuki, M. (2012). Integration of the effects of airborne chlorides into reliability-based durability design of reinforced concrete structures in a marine environment. *Structure and Infrastructure Engineering, 8*, 125–134.

Akiyama, M., Frangopol, D. M., & Yoshida, I. (2010). Time-dependent reliability analysis of existing RC structures in a marine environment using hazard associated with airborne chlorides. *Engineering Structures, 32*, 3768–3779.

Aoyama, M., Torii, K., & Matsuda, T. (2003). A study on the chloride ion penetrability into concrete in concrete structures in a severe saline environment. *JSCE Journal of Materials, Concrete Structures and Pavements, 61*, 251–264 (in Japanese).

Cornell, C. A. (1968). Engineering seismic risk analysis. *Bulletin of the Seismological Society of America, 58*, 1583–1606.

Dousti, A., Moradian, M., Taheri, S. R., Rashetnia, R., & Shekarchi, M. (2013). Corrosion assessment of RC deck in a jetty structure damaged by chloride attack. *Journal of Performance of Constructed Facilities, 27*, 519–528.

Ellingwood, B. R. (2005). Risk-informed condition assessment of civil infrastructure: State of practice and research issues. *Structure and Infrastructure Engineering, 1*, 7–18.

Enright, M. P., & Frangopol, D. M. (1999). Condition prediction of deteriorating concrete bridges using Bayesian updating. *Journal of Structural Engineering, 125*, 1118–1125.

Estes, A. C., & Frangopol, D. M. (2003). updating bridge reliability based on bridge management systems visual inspection results. *Journal of Bridge Engineering, 8*, 374–382.

Estes, A. C., Frangopol, D. M., & Foltz, S. D. (2004). Updating reliability of steel miter gates on locks and dams using visual inspection results. *Engineering Structures, 26*, 319–333.

Frangopol, D. M. (2011). Life-cycle performance, management, and optimization of structural safety under uncertainty: Accomplishments and challenges. *Structure and Infrastructure Engineering, 7*, 389–413.

Frangopol, D. M., & Liu, M. (2007). Maintenance and management of civil infrastructure based on condition, safety, optimisation, and life-cycle cost. *Structure and Infrastructure Engineering, 3*, 29–41.

Frangopol, D. M., & Kong, J. S. (2001). Expected maintenance cost of deteriorating civil infrastructures. In D. M. Frangopol & H. Furuta (Eds.), *Life-Cycle Cost Analysis and Design of Civil Infrastructure Systems* (pp. 22–47). Reston, VA: ASCE.

Frangopol, D. M., Saydam, D., & Kim, S. (2012). Maintenance, management, life-cycle design and performance of structures and infrastructures: A brief review. *Structure and Infrastructure Engineering, 8*(1), 1–25.

Frangopol, D. M., Strauss, A., & Kim, S. (2008a). Use of monitoring extreme data for the performance prediction of structures: General approach. *Engineering Structures, 30*, 3644–3653.

Frangopol, D. M., Strauss, A., & Kim, S. (2008b). Bridge reliability assessment based on monitoring. *Journal of Bridge Engineering, 13*, 258–270.

Frangopol, D. M., & Tsompanakis, Y. (Eds.). (2014). *Maintenance and Safety of Aging Infrastructure*, Structures & Infrastructures Book Series, Vol. 10. London: CRC Press / Balkema - Taylor & Francis Group, 746 p.

Japan Society of Civil Engineers (JSCE). (2001). *Standard specifications for concrete structures*. Tokyo: Maruzen.

Kato, E., Iwanami, M., Yokota, H., & Yamaji, T. (2002). Development of life-cycle management system for open type wharf. *Annual Report by Port and Airport Research Institute, 48*, 3–36 (in Japanese).

Komure, K., Hamada, H., Yokota, H., & Yamaji, T. (2002). Development of a model on deterioration progress for RC deck of open type wharf. *Annual Report by Port and Airport Research Institute, 41*, 3–38 (in Japanese).

Li, C. Q. (2003). Life-cycle modeling of corrosion-affected concrete structures: Propagation. *Journal of Structural Engineering, 129*, 753–761.

Li, C. Q. (2004). Reliability based service life prediction of corrosion affected concrete structures. *ASCE Journal of Structural Engineering, 130*, 1570–1577.

Madanat, S. (1993). Optimal infrastructure management decisions under uncertainty. *Transportation Research Part C: Emerging Technologies, 1*, 77–88.

Moradi-Marani, F., Shekarchi, M., Dousti, A., & Mobasher, B. (2010). Investigation of corrosion damage and repair system in a concrete jetty structure. *Journal of Performance of Constructed Facilities, 24*, 294–301.

Mori, Y., & Ellingwood, B. R. (1993). Reliability-based service-life assessment of aging concrete structures. *Journal of Structural Engineering, 119*, 1600–1621.

Okasha, N. M., & Frangopol, D. M. (2009). Lifetime-oriented multi-objective optimization of structural maintenance, considering system reliability, redundancy, and life-cycle cost using GA. *Structural Safety, 31*, 460–474.

Okasha, N. M., Frangopol, D. M., & Orcesi, A. D. (2012). Automated finite element updating using strain data for the life-time reliability assessment of bridges. *Reliability Engineering & System Safety, 99*, 139–150.

Onoufriou, T., & Frangopol, D. M. (2002). Reliability-based inspection optimization of complex structures: A brief retrospective. *Computers & Structures, 80*, 1133–1144.

Saydam, D., Frangopol, D. M., & Dong, Y. (2013). Assessment of risk using bridge element condition ratings. *Journal of Infrastructure Systems, 19*, 252–265.

Taniguchi, O., Tamura, T., Sano, K., & Hamada, H. (2004). Condition state estimate of deteriorated RC superstructure of wharves using inspection results. *Proceedings of Japan Concrete Institute, 26*, 2049–2054 (in Japanese).

Teigen, J. G., Frangopol, D. M., Sture, S., & Felippa, C. A. (1991a). Probabilistic FEM for nonlinear concrete structures. I: Theory. *Journal of Structural Engineering, 117*, 2674–2689.

Teigen, J. G., Frangopol, D. M., Sture, S., & Felippa, C. A. (1991b). Probabilistic FEM for nonlinear concrete structures. II: Applications. *Journal of Structural Engineering, 117*, 2690–2707.

Val, D. V., & Stewart, M. G. (2003). Life-cycle cost analysis of reinforced concrete structures in marine environments. *Structural Safety, 25*, 343–362.

Yoshida, I. (2009). Data assimilation and reliability estimation of existing structure. *COMPDYN 2009 international conference*. Greece: Rhodes.

Part IV

Life-cycle performance under earthquakes

Life-cycle cost of civil infrastructure with emphasis on balancing structural performance and seismic risk of road network

Hitoshi Furuta, Dan M. Frangopol and Koichiro Nakatsu

The life-cycle cost (LCC) concepts and methods have impacted remarkably on the field of civil infrastructure. In this paper, LCC design concepts and methods are discussed with emphasis on bridges and road networks. The relationships among several performance measures are discussed. An attempt is made to provide rational balances of these measures by using a multi-objective genetic algorithm. Furthermore, LCC is evaluated focusing on the effects of earthquakes. At first, LCC analysis is formulated to consider the socio-economic effects due to the collapse of structures as well as the minimisation of lifetime maintenance cost. A stochastic model of structural response under earthquake effects is proposed. Then, the probability of failure of an individual bridge due to the earthquake excitation is calculated based on the reliability theory. In addition, LCC evaluations are performed not only for a single bridge but also for groups of bridges forming road networks.

1. Introduction

The life-cycle cost (LCC) concepts and methods have impacted remarkably on the field of civil infrastructure (Frangopol and Furuta 2001). Many international symposia and workshops have been held on this topic all over the world. The basic concept and methodology of LCC itself are not new. They were adopted several decades ago in the fields of electrical and mechanical engineering. In the field of civil engineering, seismic risk analysis was established on the basis of LCC. However, it is noted that the seismic risk analysis has not devoted enough attention to the structural maintenance activities.

LCC design is formulated as an optimisation problem, which aims to implement an optimal inspection/repair strategy for the minimum expected total life-cycle cost including initial, preventive maintenance, inspection, repair and failure costs, while the structure maintains the target safety. As a representative civil infrastructure, highway bridges have their life-cycle consisting of design, construction, inspection, repair and replacement. At present, it is necessary to develop an optimal strategy for the bridge management through the lifetime in order to reduce the overall LCC. An efficient method that provides adequate inspection/repair strategies was proposed by Frangopol *et al.*

(1997). This method can determine how many inspections are appropriate for lifetime, and at what time inspections and repairs should be done, while taking into account all bridge repair possibilities based on an event tree.

However, if the number of design variables increases, it is difficult to solve the problem. Therefore, an attempt was made to extend and improve the work by Frangopol *et al.* (1997) using genetic algorithm (GA) (Furuta *et al.* 1998). Using GA, it is possible to easily decide the number of lifetime inspections, the time of each inspection, and which inspection has to be used. LCC is a useful indicator for the lifetime structural performance of deteriorating structures. The strategy obtained by LCC optimisation can be different according to the prescribed level of structural performance and required service life. The relationships among several performance measures are discussed and rational balances of these measures are provided by using the multi-objective genetic algorithm (MOGA). So far the authors have discussed the relationships among the minimisation of LCC, the optimal extension of structural service life, and the target durability level by using MOGA (Furuta *et al.* 2004). By introducing MOGA, it is possible to obtain several available solutions that have different structural life spans, durability levels, and LCC values.

Furthermore, LCC is evaluated focusing on the effects of earthquakes that are major natural disasters in Japan (Furuta and Koyama 2003). At first, LCC analysis is formulated to consider the socio-economic effects due to the collapse of structures as well as the minimisation of maintenance cost. The loss by the collapse of structures due to earthquake can be defined in terms of an expected cost and introduced into the calculation of LCC. A stochastic model of structural response under earthquake effects is proposed. Then, the probability of failure due to the earthquake excitation is calculated based on the reliability theory. In addition, LCC evaluations are performed not only for a single bridge but also for groups of bridges forming road networks (Furuta *et al.* 2002, Liu and Frangopol 2004a, b).

2. Application of genetic algorithm to life-cycle cost optimisation

In LCC optimisation, the cost of repair is calculated by taking into account the effects of aging and corrosion, and all possibilities of repair based on an event tree. Failure cost is calculated based on the consequences of failure and the probability of failure occurrence. Moreover, expected total cost is calculated by using the net discount rate of money. The details of costs computations are provided in Frangopol *et al.* (1997).

LCC optimisation is a non-linear problem that includes integer and discrete variables. Therefore, it is necessary to apply a combinatorial optimisation method to solve it. The purpose of LCC optimisation is to find the most economical plan for inspection/repair. It was demonstrated that a non-uniform interval of inspection/repair strategy is more economical than a uniform one (Frangopol *et al.* 1997). It is easily understood that the combination of inspection techniques with different detection capabilities in a strategy is desirable in order to reduce the cost. Discrete variables are useful in determining when and how inspections and repairs have to be performed and what methods have to be used. GA is a representative algorithm of combinatorial optimisation methods (Goldberg 1989). Using GA, it is possible to decide the number of lifetime inspections, the time of each inspection, and which inspection has to be selected. Then, the time of repair is decided based on an event tree analysis.

The LCC optimisation is reduced to the following mathematical programming problem:

Minimise

$$C_{ET} \tag{1a}$$

subject to

$$P_{f,life} \leq P^*_{f,life} \tag{1b}$$

where C_{ET} is the expected total cost, $P_{f,life}$ is the lifetime (i.e. cumulative time) probability of failure, and $P^*_{f,life}$ is the maximum acceptable lifetime probability of failure.

3. Multi-objective genetic algorithm and structural performance

3.1. Structural performance

In order to establish a rational maintenance programme for bridge structures, it is necessary to evaluate the structural performance of existing bridges in a quantitative manner. So far, several structural performance indicators have been developed, some of which are reliability, durability, and damage indices (Furuta *et al.* 2003, Neves *et al.* 2003). However, it is often necessary to discuss the structural performance from the socio-economic point of view.

LCC is one of the useful measures for evaluating the structural performance from another standpoint by reducing the overall cost and achieving an appropriate allocation of resources. In general, LCC optimisation consists in minimising the expected total cost which includes the initial cost involving design and construction, routine or preventive maintenance cost, inspection, repair and failure cost. The optimal strategy obtained by LCC optimisation can be different according to the prescribed level of structural performance and required service life. In this paper, an attempt is made to discuss the relationships among several performance measures and provide appropriate balances among them by using MOGA.

3.2. Formulation using multi-objective genetic algorithm

A multi-objective optimisation problem has several objectives, and Pareto solutions are obtained as a set of solutions. GA evaluates the optimal solution by random and multiple-point searches. The MOGA is performed according to the following five steps (Goldberg 1989):

Step 1: Generation of initial populations.
Step 2: Crossover. Two-point crossover is used herein.
Step 3: Mutation.
Step 4: Evaluation of fitness function.
Step 5: Selection.

Although several performance measures have been so far developed, the following five measures are considered here (Furuta *et al.* 2003, Neves *et al.* 2003, 2006a, b):

(1) Durability level, defined in terms of lifetime safety.

(2) Service life, defined as the prescribed expected period for a structure to be used safely.

(3) Life-cycle cost, defined as the total cost including initial construction cost, maintenance cost, repairing cost, and replacement cost.

(4) Condition rate, defined as the damage state found as a result of an inspection and assigned as 0, 1, 2, 3, 4, or 5 representing excellent, no damaged, slightly damaged, moderately damaged, severely damaged, and collapse, respectively.

(5) Reliability index, obtained by reliability analysis.

The relationships among these performance measures are discussed by using MOGA. The following two problems with three conflicting objectives can be formulated as follows:

Problem 1
Objective function 1: Durability level should be maximised.
Objective function 2: Service life should be maximised.
Objective function 3: Life-cycle cost should be minimised.

Problem 2
Objective function 1: Condition rate should be minimised.
Objective function 2: Durability level should be maximised.
Objective function 3: Life-cycle cost should be minimised.

By introducing MOGA, it is possible to obtain several available solutions that have different durability levels, service lives and LCC values, or condition rates, reliability indices and LCC values. This was illustrated in Furuta *et al.* (2004), where 244 Pareto solutions were obtained by using MOGA. As an example, Figure 1 shows Pareto solutions associated with LCC and maximum service life when the allowable durability level is 0.6.

4. New multi-objective genetic algorithm for life-cycle cost optimisation

Bridge maintenance planning has several constraints. In general, it is not easy to solve multi objective optimisation problems with constraints by applying the traditional MOGA.

4.1. New MOGA

A new MOGA was developed by introducing the sorting technique into the selection process (Furuta *et al.* 2006). The selection is performed using the so-called sorting rules which arrange the list of individuals in a prescribed order of evaluation values. Then, the fitness values are assigned to them by using the linear normalisation technique. In general, if the fitness values are calculated directly according to the evaluation values, the differences among individuals decrease so that the effective selection cannot be done. In this study, the selection procedure is constructed by coupling the linear normalisation technique and the sorting technique. Using the evaluation values, the individuals are re-ordered and given the new fitness values. Figure 2 presents the process of the selection proposed herein. The individuals satisfying the constraints are arranged first according to the evaluation

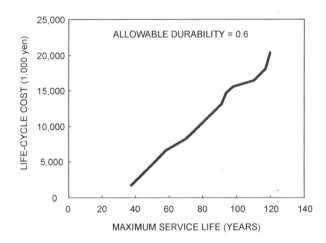

Figure 1. Pareto solutions.

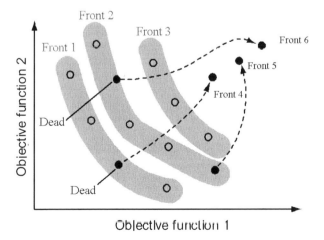

Figure 2. New sorting rules.

values and further the individuals non-satisfying the constraints are arranged according to the degree of violating the constraints. Accordingly, all the individuals are given the fitness values using the linear normalisation technique.

In order to apply the sorting rules to the multi-objective optimisation problems, the non-dominated sorting method is used (Kitano 1995). In the non-dominated sorting method, the Pareto solutions are defined as *Front1*. Then, *Front2* is determined by eliminating the *Front1* from the set of solution candidates. Repeating the process, the new *Front* is pursued until the solution candidates diminish. Further, the *Fronts* are stored in the pool of the next generation. If the pool is full, the individuals in the *Front* are divided into the solutions to survive or die based on the degree of congestion. Then, the individuals are divided into the group satisfying the constraints and the group without satisfying the constraints. The former is called an 'alive individual', and the latter a 'dead individual'. While alive individuals are given the fitness values according to the evaluation values after the non-dominated sorting, the dead individuals are given the same fitness value. When implementing the non-dominated sorting, the Pareto *Front* may not exist at the initial generation, because a lot of dead individuals remain after the non-dominated sorting. Then, the dead individuals are arranged in the order of degree of violating the constraints and some of them are selected for the next generation. Thus, the multi-objective optimisation problems with constraints are transformed into the minimisation problem of violation of constraints. The elite preserve strategy is employed for the selection of survival individuals (Kitano 1995).

When the generation progresses, alive individuals appear and then both the alive individuals forming the Pareto front and the dead individuals arranged in the order of violation degree exist together. In this case, appropriate numbers of alive and dead individuals are selected for the next generation.

It is desirable to determine an appropriate life-cycle maintenance plan by comparing several solutions for various conditions. A new decision support system is developed herein from the viewpoint of multi-objective optimisation, in order to provide various solutions needed for decision-making.

4.2. Formulation using new MOGA

In this study, LCC, durability level and service life are used as objective functions. LCC is minimised, durability level is maximised, and service life is maximised (Furuta *et al*. 2004). There are trade-off relations among the three objective functions. For example, LCC increases when service life is extended, and durability level and service life decrease due to the reduction of LCC. Then, multi-objective optimisation can provide a set of Pareto solutions that cannot improve an objective function without making other objective functions worse.

A group of 10 concrete highway bridges are considered in this study. Maintenance management planning for 10 consecutive piers and floor slabs (composite structure of steel girders and reinforced concrete (RC) slabs) is considered (Furuta *et al*. 2006).

In this study, environmental corrosion due to carbonation of concrete, chloride attack, frost damage, chemical corrosion, or alkali-aggregate reaction are considered as major aggressive agents. Deterioration of a bridge due to corrosion depends on the concrete cover of its components and environmental conditions, among other factors. For each component, the major degradation mechanism is assumed corresponding to associated environmental conditions. In this study, four environmental conditions are considered: one typical carbonation (corrosion induced by carbonation) condition and three chloride attack conditions (mild, medium and severe).

4.3. Numerical example

In the implementation of the proposed new MOGA, the GA parameters considered are as follows: number of individuals = 1000, crossover rate = 0.60, mutation rate = 0.05 and number of generations = 3000.

Figure 3 presents the results obtained by the new MOGA. This figure shows the evolution of the results from the 1st generation (iteration number) to the 100th generation, indicating that the proposed new MOGA could obtain various optimal solutions with different LCC values and durability levels. From Figure 4, it is observed that the new MOGA can find good solutions and that the final solutions are sparse and have discontinuity. In other words, the surfaces associated with the trade-off relations are not smooth. This implies that an appropriate long term maintenance plan cannot be created by the repetition of the short term plans.

The new MOGA can provide many useful solutions with different characteristics for determining appropriate maintenance plans available for practical use. It is clear that LCC can be reduced by adopting simultaneous repair works. Finally, it is confirmed that the proposed method using linear normalisation technique and sorting technique can provide many near-optimal maintenance plans with various reasonable LCC values, durability levels, and service lives. Note that it is quite difficult to obtain such near-optimal solutions by the traditional MOGA.

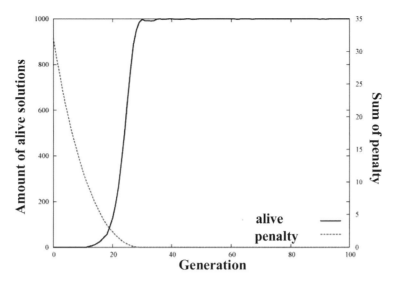

Figure 3.   Evolution of new MOGA under carbonation environment (alive solutions and penalty).

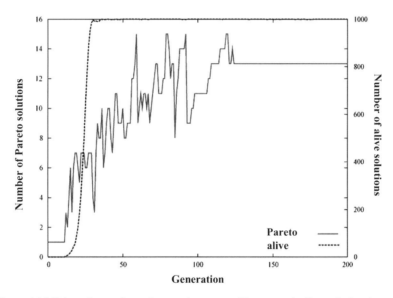

Figure 4.   Evolution of new MOGA under carbonation environment (Pareto and alive solutions).

5.   Life-cycle cost optimisation considering seismic risk

In general, LCC analysis considers the damage and deterioration of materials and structures under aggressive agents (Ellingwood 2005). However, in regions exposed to frequent natural hazards such as typhoons and earthquakes, it is necessary to account for the effects of these hazards (Chang and Shinozuka 1996).

In this paper, based on the seismic risk analysis, LCC is evaluated focusing on the effects of earthquakes that are major natural disasters in Japan. At first, LCC analysis is formulated to consider the socio-economic effects due to the collapse of structures as well as the minimisation of maintenance cost. The loss by the collapse of structures due to the earthquake can

be defined in terms of an expected cost and accounted for in LCC. A stochastic model of structural response under earthquake effects is proposed. Then, the probability of failure due to the earthquake excitation is calculated based on the reliability theory.

LCC is defined herein as the sum of initial construction cost and seismic risk. Seismic risk includes both loss due to earthquake and user cost. LCC considering seismic risk is calculated as

$$LCC = C_i + \sum P_d(a) \cdot C_d(a) \qquad (2)$$

$$P_d(a) = P_h(a) \cdot P(DI|a) \qquad (3)$$

where C_i = initial construction cost, $P_d(a)$ = probability of seismic damage occurrence, $C_d(a)$ = seismic loss, $P_h(a)$ = earthquake occurrence probability, $P(DI|a)$ = seismic damage probability given maximum acceleration a, DI = damage index.

Equation (3) provides the probability of damage occurrence due to the earthquake, which is the multiplication of earthquake occurrence probability with seismic damage probability. In this study, the earthquake occurrence probability is calculated by using the earthquake hazard curve shown in Figure 5 and the damage probability is calculated by using the damage curve.

5.1. Calculation of damage curve and definition of damage degree

Seismic damage probability is defined in terms of the probability that a bridge pier shows each damage degree among the prescribed damage ranges. The damage curve is calculated here by using the dynamic analysis of the reinforced concrete (RC) bridge pier. The RC pier is designed according to *Design Specifications of Highway Bridges*, Earthquake Version (MLIT 2002). Then, it is assumed that the ground condition is Type II and the importance of the bridge is B (MLIT 2002). Dynamic analysis is performed for bridges, in which a single mass and single degree of freedom model is used and the Newmark method is used. It is assumed that the compressive strength of concrete is $f_c = 21$ N/mm^2 and the reinforcing bars are SD295. As input earthquake wave, the ground condition of Type II is used, and Type I and II earthquakes are used. Table 1 presents six input earthquakes. Using these conditions, the dynamic analysis is performed 600 times for a RC

pier. Several damage indices have been proposed so far (Park and Ang 1985, Shoji *et al.* 1997). However, in this research, damage degree is defined in terms of the maximum response displacement, horizontal force, and horizontal displacement of pier. The damage is categorised into five degrees As, A, B, C, and D representing collapse, severe damage, partial buckling and deformation, deformation, and no damage or minor damage, respectively. In order to calculate the damage probability, it is necessary to determine the distribution function of damage degree (damage index) corresponding to the maximum earthquake acceleration. In this study, the log-normal distribution is assumed. When the distribution of damage degree is determined, the damage probability can be calculated as

$$P(DI|a) = \int_a^b f_{DI}(x, a)dx \qquad (4)$$

where $[a, b]$ is the interval of each damage degree.

The damage probability is calculated for each damage degree and the results are plotted on a graph with the exceedance probability as the vertical axis and the maximum acceleration as the horizontal axis (JSCE 1996). Then, the damage probability curve can be obtained by combining them. Figure 6 shows the computed damage probability for several damage degrees.

Table 1. Input data of six earthquake waves.

Earthquake type	Name	M	Direction
Type I	1968, Hyuga-Nada earthquake	7.5	Longitudinal Transversal
	1994, Hokkaido-Toho earthquake	8.1	Transversal
Type II	1995, Hyogo-Nanbu earthquake	7.2	N-S E-W N27W

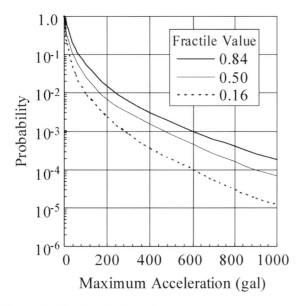

Figure 5. Earthquake hazard curve.

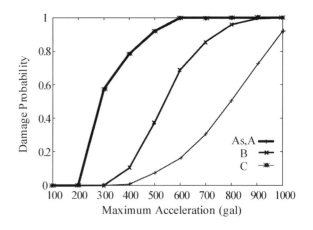

Figure 6. Damage probability versus maximum acceleration for damage degrees As, A, B, and C.

6. LCC optimisation for road network

Practically, it is necessary to consider the effects of network to calculate LCC of bridge systems. It can be expected that although the effects of seismic risk are not large in the case of a single bridge, they are large and important in the case of multiple bridges in a road network, because the user cost becomes quite large for the entire road network.

6.1. *Road network model*

For road networks, three network models (Model 1, Model 2, and Model 3) are employed, which are presented in Figure 7. In these models, it is assumed that each network model includes a road passing over a river, and that traffics reach the destination through detours when some bridges cannot be used. Moreover, it is assumed that the traffic volume and the speed have the relation shown in Figure 8 (Kinki Branch, MLIT1999).

6.2. *User cost*

The user cost (UC) is defined as the sum of the time cost (UC_T) and energy consumption cost (UC_C) due to the detour or closure of road. The cost associated with increasing driving time is calculated as the difference between (a) the cost associated with detour and road closure and (b) the usual cost (without detour and road closure).

$$UC_T = \alpha \cdot \{(Q \cdot T) - (Q_0 \cdot T_0)\} \qquad (5)$$

$$UC_C = \beta \cdot \{(Q \cdot L) - (Q_0 \cdot L_0)\} \qquad (6)$$

where α = unit time cost, β = unit driving cost, Q, T, L represent detour traffic volume, detour driving time, and detour link length at the time of road closure, respectively, and Q_0, T_0, L_0 are initial traffic volume, initial driving time, and initial link length, respectively.

Using the data given by Nihon Sogo Research Institute (1998) and assuming the ratio of small and medium trucks to be 10%, the unit time cost α is estimated as 82 Yen/car/min., and β is assumed in the interval 18 to 35 Yen/car/km. The restoring periods are assumed to be two months and two weeks for the damage (As, A) and B, respectively.

6.3. *Calculation of LCC*

6.3.1. *Calculation of LCC and repair cost*

Taking into account the discount rate of money, LCC is calculated as

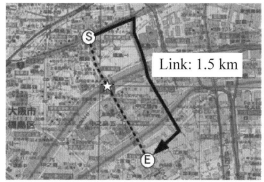

Model 1. Tamae-bashi
Fukushima-ku, Osaka city

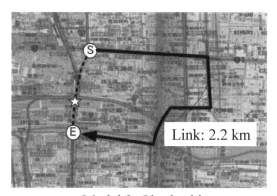

Model 2. Ohe-bashi
Kita-ku, Osaka city

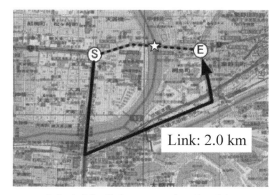

Model 3. Sakuranomiya-bashi
Tyuo-ku, Osaka city

Figure 7. Three road network models.

$$LCC = C_i + \sum P_f(a) \cdot P(DI|a)$$
$$\times \left\{ \frac{C_m(DI|a) + UC(DI|a)}{(1+i)^T} \right\} \qquad (7)$$

where $C_m(DI|a)$ = repair cost for each damage degree, $UC(DI|a)$ = UC for each damage degree, i = discount rate, T = service life.

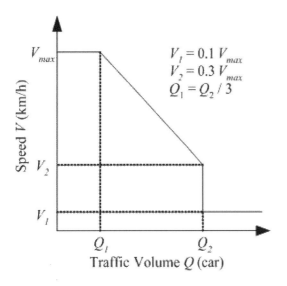

Figure 8. Relation between traffic volume and speed.

For each damage degree, restoring method and cost are presented in Table 2.

6.3.2. LCC results for three network models

For the three road networks in Figure 7, LCC is calculated by assuming that the fractile value in the hazard curve is 0.5, discount rate is 0, and service life is 100 years. Figure 9 shows the calculated results, which indicate that there are important differences among the three networks in Figure 7, because of the differences in distances of detour and the initial traffic volumes. In the network with high traffics, the seismic risk is 104,559,000 Yen representing about 10 times the initial construction cost and 26 times the maintenance/repair cost.

Comparing the case involving the user cost in the seismic risk with that without involving it, the seismic risk is only 40% of the initial cost when the user cost is not considered.

6.3.3. Relation between LCC and maximum acceleration

Paying attention to the damage probability curve in Figure 6, it is evident that there is some difference in the damage probabilities according to the earthquake intensity. Figure 10 shows the relation between the seismic risk and the maximum acceleration indicating that the seismic risk decreases as the maximum acceleration increases. This is due to the fact that the bridge piers were designed to satisfy the requirement that the damage should be minor and the bridge function can be recovered in a short period.

Finally, the effect of damage degree on the seismic risk is also examined. Figure 11 presents the percentage

Table 2. Restoring method and cost.

Damage index	Restoring method	Repair cost	Repair time
As, A	Rebuild	120% of initial construction cost	2 months
B	Repair	73,000 yen/m^2	1 month
C	Repair	35,000 yen/m^2	2 weeks
D	No repair	–	–

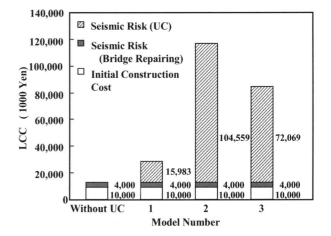

Figure 9. LCC of each model.

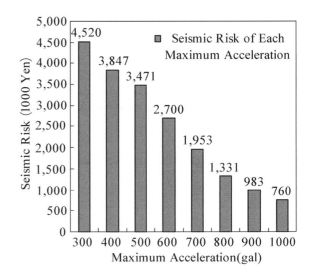

Figure 10. Seismic risk associated with each maximum acceleration.

of seismic risk corresponding to each damage degree. As indicated, the damage degree C is associated with the largest percentage of seismic risk, and the severe damage degrees. As and A are associated with a relatively small percentage of seismic risk. This is due to the fact that while the occurrence probabilities of As and A increase with the maximum acceleration, the earthquake occurrence probability decreases.

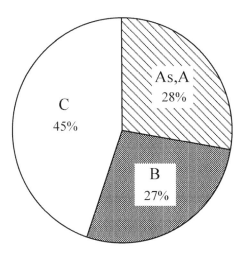

Figure 11. Relation between seismic risk and damage index.

7. Conclusions

In this paper, an attempt was made to formulate LCC optimisation based on GA and to discuss the relationships among several structural performance measures. In addition, seismic risk was introduced into the LCC optimisation. The MOGA was adopted to successfully solve the large and complex combinatorial scheduling problems for the maintenance of groups of damaged RC bridges in road networks.

By considering LCC, durability level and service life as objective functions, it is possible to obtain the relationships among these three performance indicators and provide bridge maintenance management engineers with various cost-effective maintenance plans balancing several conflicting objectives.

Based on the results presented in this paper, the following conclusions are drawn:

(1) Since the optimal maintenance problem is a very complex combinatorial problem, it is difficult to obtain multi-objective solutions by the current optimisation techniques.

(2) Although GA is applicable to solve multi-objective problems, it is difficult to apply it to large and very complex bridge network maintenance problems. By introducing the technique of Non-Dominated Sorting GA-2 (NSGA2), it is possible to obtain efficient near-optimal solutions for the maintenance planning of a group of bridge structures in a road network.

(3) The Pareto solutions obtained by the proposed method show discontinuity. This means that the surfaces constructed by the trade-off relationships are not smooth. Therefore, an appropriate long term maintenance plan

cannot be created by the simple repetition of short term plans.

(4) In the examples presented, the relation between durability level and LCC is non-linear. The increase of LCC does not affect significantly the improvement of durability level.

(5) LCC can be reduced by adopting simultaneous repair works. The proposed method using linear normalisation technique and sorting technique can provide many near-optimal maintenance plans with various reasonable LCC values and durability levels.

(6) The seismic damage probability is calculated for each damage degree of RC bridge piers.

(7) Through the LCC calculation of several representative road networks, it is concluded that the topology of road network greatly affects the seismic risk.

(8) The effect of seismic risk is increasing with user cost.

References

Chang, S.E. and Shinozuka, M., 1996. Life-cycle cost analysis with natural hazard risk. *Journal of Infrastructure Systems*, 2 (3), 118–126.

Ellingwood, B.R., 2005. Risk-informed condition assessment of civil infrastructure: state of practice and research issues. *Structure and Infrastructure Engineering*, 1 (1), 7–18.

Frangopol, D.M. and Furuta, H., eds. 2001. *Life-cycle cost analysis and design of civil infrastructure systems*. Reston, Virginia: ASCE.

Frangopol, D.M. and Liu, M., 2007. Maintenance and management of civil infrastructure based on condition, safety, optimisation, and life-cycle cost. *Structure and Infrastructure Engineering*, 3 (1), 29–41.

Frangopol, D.M., Lin, K.-Y., and Estes, A.C., 1997. Life-cycle cost design of deteriorating structures. *Journal of Structural Engineering*, 123 (10), 1390–1401.

Furuta, H., Frangopol, D.M., and Saito, M., 1999. Application of genetic algorithm to life-cycle cost maintenance of bridges. *In: Proceedings of 10th KKNN symposium*, 16–20 August, Taejon, Korea, Taejon: KAIST.

Furuta, H., Nose, Y., Dogaki, M., and Frangopol, D.M., 2002. Bridge maintenance system of road network using life-cycle cost and benefit. *In: Proceedings of IABMAS*, July, Barcelona, Spain, Barcelona: CIMNE Publications. CD-ROM.

Furuta, H., Kameda, T., Fukuda, Y., and Frangopol, D.M., 2004. Life-cycle cost analysis for infrastructure systems: Life cycle cost vs. safety level vs. service life. Keynote Paper. *In: D.M. Frangopol, E. Brühwiler, M.H. Faber, and B. Adey, eds. Life-cycle performance of deteriorating structures: Assessment, design and management*. Reston, Virginia: ASCE, 19–25.

Furuta, H. and Koyama, K., 2003. Optimal maintenance planning of bridge structures considering earthquake effects. *In: Proceedings of IFIP TC7 conference*, Sophia Antipolis, France. Boston: Springer.

Furuta, H., Kameda, T., and Frangopol, D.M., 2004. Balance of structural performance measures. *In: Proceedings of structures congress*, May, Nashville, Tennessee. Reston, Virginia: ASCE, CD-ROM.

Furuta, H., Kameda, T., Nakahara, K., Takahashi, Y., and Frangopol, D.M., 2006. Optimal bridge maintenance planning using multi-objective genetic algorithm. *Structure and Infrastructure Engineering*, 2 (1), 33–41.

Goldberg, D.E., 1989. *Genetic algorithms in search, optimisation and machine learning*. Toronto: Addison-Wesley Publishing Company, Inc.

JSCE (Japan Society of Civil Engineers), 1996. *Report on damage by Hanshin Awaji earthquake* (in Japanese).

MLIT (Ministry of Land, Infrastructure and Transportation), Kinki Branch, 1999. *Road traffic census* (in Japanese).

Kitano, H. ed. 1995. *Genetic algorithm 3*. Tokyo: Sangyotosho (in Japanese).

Liu, M. and Frangopol, D.M., 2004a. Probabilistic maintenance prioritisation for deteriorating bridges using a multi-objective genetic algorithm. *In: Proceedings of the 9th ASCE joint specialty conference on probabilistic mechanics and structural reliability*, 26–28 July, Sandia National Laboratories, Omnipress, Albuquerque, New Mexico. Six pages on CD-ROM.

Liu, M. and Frangopol, D.M., 2004b. Optimal bridge maintenance planning based on probabilistic performance prediction. *Engineering Structures*, 26 (7), 991–1002.

MLIT (Ministry of Land, Infrastructure and Transportation), 2002. *Design Specifications for Highway Bridges*. Tokyo: Maruzen (in Japanese).

Neves, L.C., Frangopol, D.M., and Hogg, V., 2003. Condition–reliability–cost interaction in bridge maintenance. *In: Proceedings of the 9th international conference on applications of statistics and probability in civil engineering (ICASP9)*, 6–9 July, San Francisco, California. Rotterdam: Millpress, 2, 1117–1122.

Neves, L.A.C., Frangopol, D.M., and Cruz, P.J.S., 2006a. Probabilistic lifetime-oriented multi-objective optimisation of bridge maintenance: single maintenance type. *Journal of Structural Engineering*, 132 (6), 991–1005.

Neves, L.A.C., Frangopol, D.M., and Petcherdchoo, A., 2006b. Probabilistic lifetime-oriented multi-objective optimisation of bridge maintenance: combination of maintenance types. *Journal of Structural Engineering*, 132 (11), 1821–1834.

Nihon Sogo Research Institute, 1998. *Draft of guideline for evaluation of investment on road*. Tokyo: NSRI (in Japanese).

Park, Y.J. and Ang, A.H.S., 1985. Mechanistic seismic damage model for reinforced concrete. *Journal of Structural Engineering*, 111 (4), 722–739.

Shoji, M., Fujino, Y., and Abe, M., 1997. Optimisation of seismic damage allocation of viaduct systems. *Journal of Japan Society of Civil Engineers*, No. 563, I-39, 79–94 (in Japanese).

Long-term seismic performance of RC structures in an aggressive environment: emphasis on bridge piers

Mitsuyoshi Akiyama and Dan M. Frangopol

Even though accurate structural models have been developed for the performance of corroded structures subjected to monotonic flexure and/or shear, studies on seismic performance that include corrosion damage are scarce. For the lifetime assessment of structures in aggressive environments and earthquake-prone regions, the effects of corrosion on seismic performance need to be taken into consideration. Whereas the seismic demand depends on the results of seismic hazard assessment, the deterioration of seismic capacity depends on the environmental hazard assessment. The analysis of the life-cycle reliability of corroded reinforced concrete (RC) structures under earthquake excitations is the topic of this paper. It includes (a) estimation of the seismic capacity of corroded RC components; (b) seismic and airborne chloride hazard assessment and (c) life-cycle seismic reliability of bridges with corrosion damage. In particular, this paper introduces the visualisation of corrosion process in RC members using X-ray for modelling the spatial variability of rebar corrosion. A novel computational procedure to integrate the probabilistic hazard associated with airborne chlorides into life-cycle seismic reliability of bridge piers is presented.

1. Introduction

For structures located in a moderately or highly aggressive environment, multiple environmental and mechanical stressors lead to deterioration of structural performance. Such deterioration will reduce the service life of structures and increase the life-cycle cost of maintenance actions. Various environmental stressors affect the degradation mechanisms of structures. However, the effects of these stressors are difficult to predict, as they vary in time and space. Because of the presence of such uncertainties, long-term structural performance must be predicted based on probabilistic concepts and methods, and life-cycle reliability assessment methodologies must be established (Ellingwood, 2005; Frangopol, 2011; Mori, 1992; Mori & Ellingwood, 1993).

Many researchers have reported theoretical predictions of structural performance as a function of time. Even though some accurate structural models have been developed for corroded reinforced concrete (RC) structures subjected to monotonic flexure and/or shear (Coronelli & Gambarova, 2004), studies on seismic performance including corrosion damage are scarce. Earthquakes are still a dominant hazard to structures in many parts of the world. For the lifetime assessment of structures in aggressive environments and earthquake-prone regions, the effects of corrosion on seismic performance need to be taken into consideration. Within this context, the life-cycle seismic reliability of RC structures in a marine environment is discussed.

Figure 1 shows the flowchart describing the framework for computing the life-cycle seismic reliability of RC bridge structures in a marine environment. When estimating the life-cycle seismic reliability based on the flowchart shown in Figure 1, a given performance limit state related to predefined level of damage of RC bridge structures should be considered. This flowchart consists of three main parts: (a) determining the seismic capacity of corroded RC structures based on inspection results and monitoring, and predicting the seismic fragility curve associated with a given performance limit state (part (a)); (b) considering simultaneously two types of hazards including seismic hazard and the hazard associated with airborne chlorides (part (b)) and (c) estimating the time-dependent structural reliability including updating (part (c)).

In part (a), experimental data on the seismic performance of corroded RC structures are collected from the existing literature and the relationship between material deterioration and structural capacity is

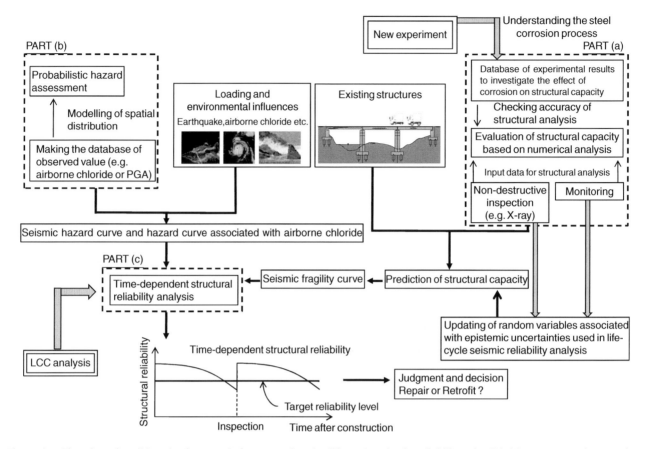

Figure 1. Flowchart describing the framework for computing the life-cycle seismic reliability of RC bridge structures in a marine environment.

experimentally investigated. Then, based on the experimental results, numerical models for evaluating the performance of corroded RC structures are presented (Shimomura et al., 2006). These models must predict the stiffness, strength, ductility and cyclic behaviour of RC members. In a capacity design procedure, these models could be used to obtain the hierarchy of resistance of the various structural components and failure modes necessary to ensure a suitable plastic mechanism and avoid the brittle modes (Priestley, Seible, & Calvi, 1996). For existing structures, inspection results obtained from non-destructive testing methods can become input data for numerical analysis to evaluate their seismic capacity. For example, X-ray photography is a suitable tool to visualise the spatial distribution of steel corrosion. The spatial distribution of cross-sectional area of corroded rebars provided by X-ray photography could be used as input data. This makes the improvement of the accuracy of predicting the seismic capacity of corroded structural components possible. This novel technique is introduced in this paper.

Regarding part (b), in seismic reliability assessment and prediction of RC structures located in an aggressive environment, two kinds of hazards must be accounted for: seismic hazard and hazards associated with other environmental stressors. Unlike conventional probabilistic seismic hazard analysis (PSHA; Kameda & Nojima, 1988), environmental hazard assessment lacks significant research. Therefore, the appropriate methodology for quantifying environmental effects must be established. Whereas the seismic demand depends on the results of seismic hazard assessment, the deterioration of seismic capacity depends on the environmental hazard assessment (Akiyama, Frangopol, & Matsuzaki, 2011a). Based on the seismic capacity and hazard assessment obtained from parts (a) and (b), an analysis of the life-cycle reliability of structures under earthquake excitations, including corrosion damage, is presented in part (c). Results provided by visual inspection, non-destructive inspection and/or monitoring can update the random variables used in time-dependent reliability analysis and improve the accuracy of the present and future failure probability.

This paper aims to highlight recent accomplishments associated with the life-cycle reliability of corroded RC structures under earthquake excitations in a marine environment.

2. Seismic capacity of RC columns in a marine environment

The probabilistic assessment of structures under seismic hazard has developed rapidly over the last two decades. A probabilistic methodology developed by Cornell and his co-workers (Cornell, Jalayer, Hamburger, & Foutch, 2002; Yun, Hamburger, Cornell, & Foutch, 2002) for seismic risk assessment of moment-resisting steel frames has been widely used. The potential of this methodology to provide seismic risk assessments of RC bridge structures has been investigated by Lupoi, Franchin, and Schotaunus (2003). In addition, several researchers have discussed the fragility curve, which quantifies the probability of a structure reaching a certain damage state under a given ground motion intensity. This fragility curve plays an important role in the overall seismic risk assessment of structures (Nielson & DesRoches, 2007; Padgett & DesRoches, 2007). The purpose of these past studies was to present efficient computational methods for estimating the seismic reliability of bridges.

However, little attention has been devoted to the assessment of the seismic reliability of corroded structures. Since corrosion accelerates the vulnerability of bridges subjected to seismic hazard, it is important to consider the effect of corrosion on the lifetime seismic reliability of bridges in earthquake-prone regions and aggressive environments. Kumar, Gardoni, and Sanchez-Silva (2009) presented a probabilistic approach to compute the life-cycle cost of corroding RC bridges in earthquake-prone regions. However, they used a very simple numerical model to consider the effect of steel corrosion on seismic capacity and demand. Choe, Gardoni, Rosowsky, and Haukaas (2008) also developed probabilistic capacity models for corroding RC columns by using a reduced diameter of reinforcement steel. Simon, Bracci,

and Gardoni (2010) presented the effects of corrosion on the seismic response of a typical RC bridge based on realistic lifetime deterioration in strength due to the reduction in cross-sectional area of the reinforcement. Performing the analysis at the section level using the reduced steel rebar cross section may be an over-simplification to evaluate the seismic capacity such as ductility capacity. Since ductility loss frequently occurs when considering the response of corroding structures even under monotonic loads, the intensity of corrosion process could have a considerable influence on the behaviour of RC columns under cyclic loads (Li, Gong, & Wang, 2009). Biondini, Palermo, and Toniolo (2011) investigated the lifetime seismic performance of concrete structures under diffusive attack of aggressive agents, assuming that structural damage induced by diffusion can be modelled by introducing a degradation law of the effective resistant areas for concrete matrix and steel bars. Berto, Vitaliani, Saetta, and Simioni (2009) assessed the seismic performance of existing RC structures affected by degradation phenomena, taking into consideration the reduction of rebar sections and rebar ultimate deformation, and degradation of concrete cover. Although Berto et al. (2009) introduced the effect of corrosion on the mechanical properties of concrete and rebar, the interaction between concrete and rebar degradations was not considered in the evaluation of seismic capacity of RC corroded structures. In addition, their models need to be validated in quantitative terms by means of a proper calibration of the deterioration model with experimental data.

As a result, it is difficult to determine whether the seismic safety of deteriorated bridge pier with corrosion cracks in a marine environment as shown in Figure 2 is compromised, or to explain how long this bridge pier will have a safety level above a prescribed threshold.

Figure 2.   Corroded RC bridge pier in a marine environment (Photos taken by the first author).

Improving the understanding of the influence that corrosion has on the seismic performance of structures is needed.

Recently, there have been experimental studies on corroded columns subjected to cyclic loading. Yamamoto and Kobayashi (2006) investigated the structural performance of deteriorated concrete members. Saito, Oyado, Kanakubo, and Yamamoto (2007) presented the fundamental properties of strength and deformation capacity of corroded RC columns subjected to cyclic load. They observed that the corrosion of longitudinal bars caused reduction in the column deformation capacity. Fang, Gylltoft, Lundgren, and Plos (2006) examined the relationship between bond stress and slip response of corroded reinforcement with concrete under cyclic loading. Their results revealed that bond behaviour was significantly reduced under cyclic loading. Kobayashi (2006) investigated the effect of steel corrosion on the seismic capacity and hysteresis loop of RC beams as shown in Figures 3 and 4. Figure 3 shows the cracks of RC beams with and without corrosion after monotonic and cyclic loading. Kobayashi (2006) showed that after the reversed cyclic loading, RC beams suffered from severe spalling of concrete cover within constant-moment region. This spalling was caused by the steel corrosion and corrosion cracks that led to debonding between the concrete cover and rebar. Figure 4 shows the relationship between applied force and displacement of RC beams. As the amount of corrosion of the longitudinal rebar increases, the displacement ductility capacities and flexural capacities of beams decrease. All specimens tested by Kobayashi (2006) exhibited flexural failure even under cyclic loading, since sufficient shear reinforcement was present to prevent brittle failure. However, it should be noted that the failure mode might change from flexure failure, largely owing to buckling of the longitudinal reinforcement, to flexural-shear failure, which is mainly caused by fracturing of the shear reinforcement (Ou, Tsai, & Chen, 2011).

It is necessary in seismic reliability analysis to relate deformation demands placed on structural components with the probability of reaching specific levels of damage.

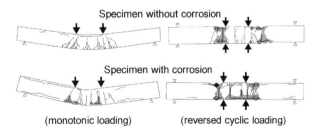

Figure 3. Appearance of specimens with and without corrosion after loading (based on information provided by Prof. Kobayashi (2006) to the first author).

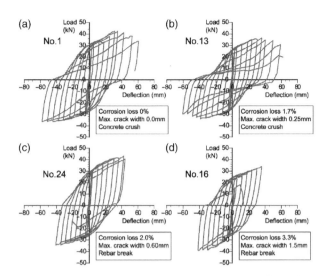

Figure 4. Hysteresis curves of RC specimens subjected to cyclic loading (based on information provided by Prof. Kobayashi (2006) to the first author).

The onset of buckling of longitudinal rebars in RC component is a key damage state (Berry & Eberhard, 2005; Dhakal & Maekawa, 2002; Naito, Akiyama, & Suzuki, 2011). This is because rebar buckling requires extensive repairs. Figure 5 shows an example of plastic hinge analysis to evaluate the plastic rotation and displacement ductility at the onset of rebar buckling of corroded RC bridge piers. Based on the comparison of experimental results of corroded RC columns with small axial load subjected to cyclic loading, the reduction of confining pressure applied to the concrete core could be ignored when evaluating the ductility capacity of RC columns. To establish a numerical method for analysing corroded RC members, it is necessary in the plastic hinge analysis to consider the effects of the cross-section reduction of steel bar, cracking of concrete cover due to expansion of corrosion products and bond degradation between concrete and steel. These lead to the reduction of the prevention force by ties and concrete cover for lateral deformation of longitudinal rebars and cause the decline of ductility capacity. Based on the buckling analysis of longitudinal rebars, for a lumped plasticity model, Akiyama et al. (2011a) presented an analytical method to evaluate the relationship between the moment and curvature and to predict the curvature at the onset of buckling of the longitudinal rebars of corroded RC columns taking into consideration the reduction of the prevention force by ties and concrete cover due to corrosion. Details of structural analysis shown in Figure 5 are given in Akiyama et al. (2011a). Using the steel corrosion amount and concrete deterioration derived from measurements of corroded specimens, computational results are in good agreement with the experimental results as shown in Figure 6. A combined experiment and

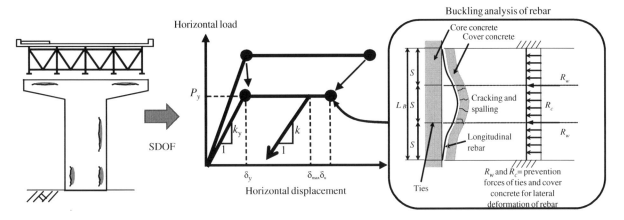

Figure 5. An example of seismic analysis model for corroded RC bridge piers to evaluate the structural capacity and demand (adapted from Frangopol and Akiyama (2009)).

analysis assessment procedure can provide reliable techniques to assess the seismic capacity of corroding RC structures.

To reach satisfactory computational results on the capacity of an existing RC structure, the key problem is the collection of reliable data on the corrosion level and concrete deterioration in the field. What is actually important and difficult in numerical simulation is how to accurately predict the degree and location of material deterioration in a real structure and how to adequately represent them in terms of input data in the structural analysis (Shimomura et al., 2010). In particular, since ductility capacity and energy dissipation depend strongly on the localised condition of reinforcements in the plastic hinge, further research is needed on the integration of modelling of spatial variability of steel corrosion into the estimation of seismic capacity such as the buckling analysis of longitudinal rebars (Akiyama et al.) even though some researchers reported spatial time-dependent reliability analysis (Marsh & Frangopol, 2007, 2008; Stewart, 2004; Stewart & Suo, 2009). Recently, X-ray technology has been applied to the visualisation of concrete cracking to investigate the behaviour of fracture

process zone in concrete. Otsuka and Date (2000) reported that microcracks in a complex configuration near the tip of the notch of concrete specimens could be observed by X-ray technology. Akiyama, Nakajima, and Komoriya (2011b) established the digital picture processing method for X-ray photography to estimate the amount of corrosion products in RC components. Figure 7 shows pictures of corroded tensile rebar in a single RC beam using X-ray. In addition, corrosion cracks on the bottom surface using digital camera and the geometry of RC beam are shown in Figure 7. This beam was corroded by electrical corrosion. The pictures using X-ray were taken at $W_{xray,g} = 0.0\%$, 2.9%, 9.7%, 19.7%, 25.6% and 27.3%, where $W_{xray,g}$ is the averaged steel weight loss of tensile rebar within the constant-moment region. After taking the pictures associated with $W_{xray,g} = 27.3\%$, the beam was tested under a monotonically increasing load until failure by using a four-point bending set-up. The steel weight loss was estimated based on pictures using X-ray taken from various angles. As shown in Figure 7, as the steel weight loss increases, the diameter of tensile rebar decreases and the crack width on the bottom surface increases. Figure 8 shows the relationship between distance from left loading

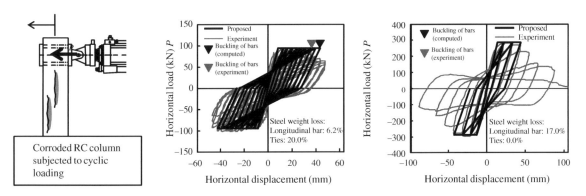

Figure 6. Comparison of the relationship between horizontal load and displacement of corroded columns (adapted from Frangopol and Akiyama (2009)).

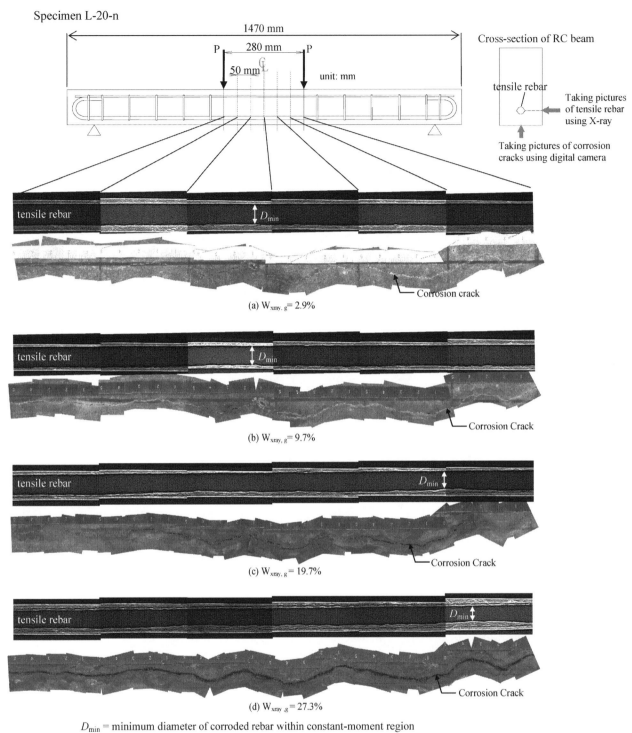

D_{min} = minimum diameter of corroded rebar within constant-moment region
$W_{xray, g}$ = averaged steel weight loss of tensile rebar within constant-moment region

Figure 7.   Visualisation of steel corrosion in RC member by X-ray.

point in Figure 7 and local steel weight loss of tensile rebar along the rebar, $W_{xray,l}$. Figures 7 and 8 indicate that the location with the maximum $W_{xray,l}$ depends on $W_{xray,g}$. Also, the difference between the maximum and minimum $W_{xray,l}$ ranges from 10% to 15%, independent of $W_{xray,g}$, when $W_{xray,g}$ is larger than 9.7%. Figure 9 shows the relationship between corrosion crack width and steel weight loss of tensile rebar. Corrosion crack width and steel weight loss of tensile rebar were measured at the same location. Even though previous researches show that there is large variability in the steel weight loss evaluated based on the visual inspection of corrosion

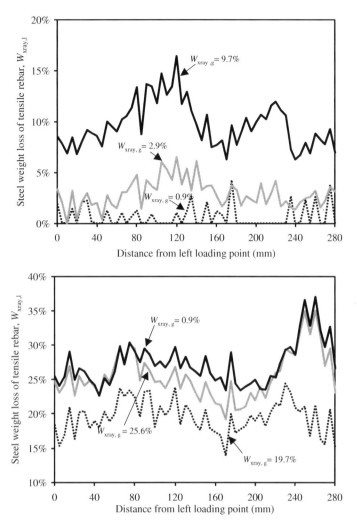

Figure 8. Distribution of steel weight loss of tensile rebar within constant-moment region ($W_{xray,g}$ is the averaged steel weight loss of tensile rebar within the constant-moment region and $W_{xray,l}$ is the local steel weight loss of tensile rebar along the rebar).

crack width (Vidal, Castel, & François, 2004, 2007), a linear relationship between corrosion crack width $C_{w,l}$ and steel weight loss $W_{xray,l}$ can be found as shown in Figure 9 if $W_{xray,g}$ is less than 9.7%. It should be noted that this relationship needs to be validated using additional experimental results. Using X-ray technology, corrosion process in RC components can be observed continuously. These experiments will help improve the accuracy of estimation of the corrosion distribution and corrosion amount of rebar. In addition, if the steel weight loss $W_{xray,l}$ in the RC member could be estimated accurately by the corrosion crack width $C_{w,l}$ provided from visual inspection, it is not necessary to adopt X-ray photography on site. Further research is needed in the laboratory test of X-ray photography to improve the accuracy of the relationship between $W_{xray,l}$ and $C_{w,l}$. Once $W_{xray,l}$ is provided by X-ray photography on site or $C_{w,l}$ based on visual inspection, random variables associated with the estimation of steel weight loss can

be updated as described later. Having the inspection results helps in performing an accurate life-cycle reliability assessment.

When a computational model to evaluate the seismic capacity of corroded RC structures is established, the effect of corrosion amount of rebars on the fragility curve can be examined. Figure 10 depicts an example of fragility curve of RC bridge pier shown in Figure 11 with steel corrosion amount in plastic hinge $C_w = 0\%$, 5%, 20% and 40%. The bridge pier was originally designed according to the Design Specifications for Highway Bridges by the Japan Road Association (2002). In Figure 10, an RC bridge pier was modelled as a single degree of freedom and the longitudinal rebar buckling of this pier was considered as the sole limit state when estimating the fragility. The vertical axis in Figure 10 represents the conditional failure probability of seismic demand D_e exceeding the seismic capacity C_a given seismic intensity (i.e. maximum velocity at the bedrock). As described later,

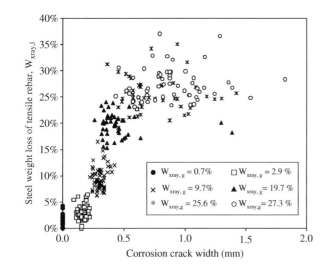

Figure 9. Relationship between steel weight loss and corrosion crack width.

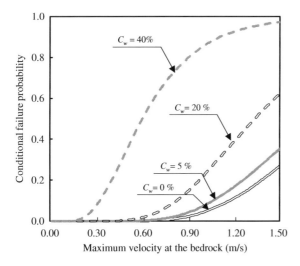

Figure 10. Comparison of fragility estimates with different amounts of steel weight loss of the RC bridge pier shown in Figure 11.

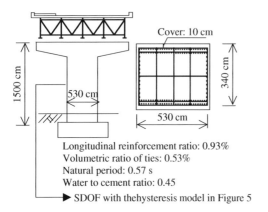

Figure 11. RC bridge pier analysed.

the horizontal axis of seismic hazard curves used in this study is the maximum velocity of the ground motion. This is used as seismic intensity in the fragility analysis. The fragility is estimated from the ratio of the number of times D_e exceeds C_a to the total number of Monte Carlo simulations (MCS). D_e is calculated by nonlinear dynamic analysis using a number of ground motions. In Figure 10, artificial ground motions are used. Since the fragility curves are almost the same if the number of samples in MCS is larger than 2000, the number of samples and artificial ground motions is set to 2000. The details associated with the generation of the artificial ground motions are provided in Akiyama, Matsuzaki, Dang, and Suzuki (2012c). To capture the true demands on RC bridges, the nonlinear dynamic analytical model of RC bridge used in seismic fragility analysis must be enhanced [e.g. three-dimensional finite element model (Ghosh & Padgett, 2010)]. Also, to evaluate the failure probability of the RC bridge, additional limit states need to be considered (e.g. limit state for corroded steel bearing). Fragility curves as shown in Figure 10 are provided as the conditional probabilities given levels of maximum velocity of the ground motion and airborne chloride.

The relationship between time, amount of corrosion and seismic performance could be established as shown in Figure 12. Using the relationship between time and amount of steel corrosion provided by the analysis of material transport and material deterioration, the prediction of seismic fragility at any time after construction could be provided. For future prediction of seismic capacity, some parameters used in the analysis of material transport and material deterioration have to be updated based on information provided by inspection and monitoring on site (see Figure 1), and the effect of the location of structures in a marine environment on their seismic capacity must be taken into consideration. In addition, to predict the seismic performance accurately, further studies are needed on the interaction between microscopic chloride and water diffusion and internal damage of concrete component due to seismic loading (Melchers, Li, & Lawanwisut, 2008).

3. Hazard assessment

In conventional seismic hazard analysis, it is assumed that earthquakes occur randomly and independently. The annual probability that the random intensity Γ at a specific site will exceed a value γ is (Kameda & Nojima, 1988)

$$p_0(\gamma) = 1 - \exp\left\{-\sum_{k=1}^{n} \nu_k q_k(\gamma)\right\} \cong \sum_{k=1}^{n} \nu_k q_k(\gamma) \quad (1)$$

where n is the number of potential earthquake sources

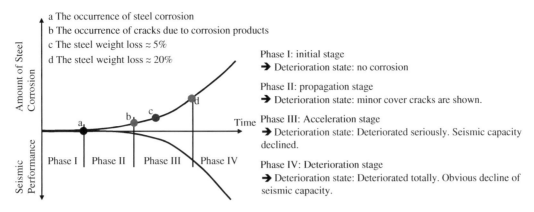

a The occurrence of steel corrosion
b The occurrence of cracks due to corrosion products
c The steel weight loss ≈ 5%
d The steel weight loss ≈ 20%

Phase I: initial stage
➔ Deterioration state: no corrosion

Phase II: propagation stage
➔ Deterioration state: minor cover cracks are shown.

Phase III: Acceleration stage
➔ Deterioration state: Deteriorated seriously. Seismic capacity declined.

Phase IV: Deterioration stage
➔ Deterioration state: Deteriorated totally. Obvious decline of seismic capacity.

Figure 12.   Effect of time on the double amount of steel corrosion and associated seismic performance.

around the site, ν_k is the earthquake occurrence rate at source k with upper and lower bound magnitudes m_{uk} and m_{lk}, respectively, and $q_k(\gamma)$ is the probability of $\Gamma > \gamma$ given that an earthquake occurs at source k. This probability is

$$q_k(\gamma) = \int_{m_{lk}}^{m_{uk}} \int_{r_{lk}}^{r_{uk}} P(\Gamma > \gamma|m,r) f_{Mk}(m) f_{Rk}(r) \, dm \, dr \quad (2)$$

where $f_{Mk}(m)$ is the probability density function (PDF) of magnitude M of an earthquake which could occur at source k, $f_{Rk}(r)$ is the PDF of distance R (upper and lower values are r_{uk} and r_{lk}) from the site to the rupturing fault at source k and $P(\Gamma > \gamma|m,r)$ is the probability of $\Gamma > \gamma$ given $M = m$ and $R = r$.

Attenuation relationships, also called ground motion prediction equations, play a key role in probabilistic seismic hazard assessment (PSHA). When the attenuation

uncertainty is involved, the random intensity is

$$\Gamma = U \hat{\gamma}(m,r) \quad (3)$$

where $\hat{\gamma}(m,r)$ is the attenuation relation represented as a function of m and r, and U is the lognormal random variate representing attenuation uncertainty with median of 1.0 and coefficient of variation δ_γ. Then $P(\Gamma > \gamma|m,r)$ in Equation (2) can be expressed as

$$P(\Gamma > \gamma|m,r) = P(U \hat{\gamma}(m,r) > \gamma)$$
$$= P\left(U > \frac{\gamma}{\hat{\gamma}(m,r)}\right) \quad (4)$$

The seismic hazard curve is obtained from Equation (1) for various values of γ. Figure 13 shows three seismic hazard curves for Fukuoka City, Maizuru City and Sakata City in Japan.

Earthquake environment is quantified by seismic hazard curve as indicated in Figure 13. A marine environment should be quantitatively assessed and the evaluation results should be reflected in the estimation of life-cycle performance of RC structures. Even though many probabilistic methods for the evaluation of deterioration of RC bridges subjected to chloride attacks have been reported and a probabilistic framework for life-cycle analysis of RC structures has been established, prior studies have not taken into consideration the hazard associated with marine environment.

The probabilistic model of hazard associated with airborne chlorides taking into account the spatiotemporal variation was proposed by Akiyama, Frangopol, and Suzuki (2012b). As shown in Figure 14, the amount of airborne chlorides decreases with increasing the distance from the coastline, as does the seismic intensity (e.g. the peak ground acceleration attenuates with distance from the seismic source). The observed airborne chlorides in Figure 14 were collected in the stations near the coastline

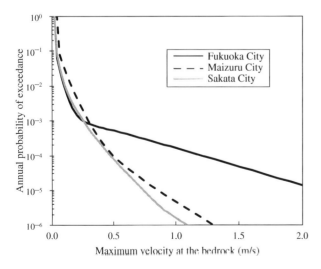

Figure 13.   Seismic hazard curves.

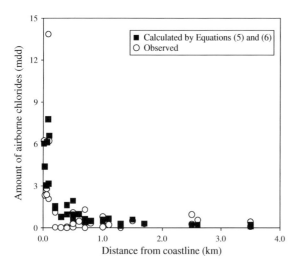

Figure 14. Attenuation of airborne chloride with distance from the coastline.

all over Japan. Since in Japan the wind usually blows from west to east, the structures near the Sea of Japan (East Sea) have suffered most from severe damages due to airborne chlorides. The speed of wind, the ratio of sea wind (defined as the percentage of time during 1 day when the wind is blowing from sea towards land), significant wave height and the distance from the coastline affect the amount of airborne chlorides. The attenuation relation associated with airborne chloride can be expressed as

$$C_{air} = 0.003 \cdot r \cdot u^{2.52} \cdot h^{11.5} \cdot d^{-2.34} \quad \text{if } d \geq 0.1 \text{ km} \quad (5)$$

$$C_{air} = 0.656 \cdot r \cdot u^{2.52} \cdot h^{11.5} \quad \text{if } d < 0.1 \text{ km} \quad (6)$$

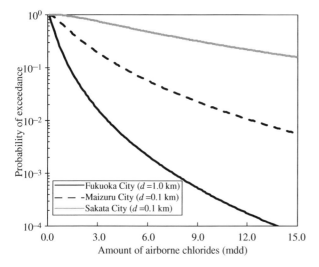

Figure 15. Hazard curve associated with airborne chlorides.

where u is the average wind speed (m/s) during the observation period, r is the ratio of sea wind, h is the significant wave height (m) and d is the distance from the coastline (km). Since the measured amounts of airborne chloride are mostly independent of distance from the coastline if d is less than 0.1 km, the attenuation relation provided by Equation (6) does not include d. Equation (6) can be applied for a splash zone.

The comparison of observed values with predicted values by Equations (5) and (6) is shown in Figure 14. The ratio of sea wind r, average wind speed u and significant wave height h are obtained from the meteorological data collected. Then, to obtain the hazard curve associated with airborne chlorides, equations similar to those used to obtain seismic hazard curves can be used. The probability that C_{air} at a specific site will exceed a prescribed value c_{air} is

$$q_s(c_{air}) = \int_0^\infty \int_0^\infty \int_0^\infty P(C_{air} > c_{air}|u,r,h)$$
$$\times f_u(u) \cdot f_r(r) \cdot f_h(h) \, du \, dr \, dh \quad (7)$$

where $f_u(u)$, $f_r(r)$ and $f_h(h)$ are the PDFs of u, r and h, respectively, and $P(C_{air} > c_{air}|u,r,h)$ is the probability of $C_{air} > c_{air}$ given u, r and h.

Figure 15 shows the probability of exceedance of various amounts of airborne chlorides in Fukuoka City at $d = 1.0$ km, Maizuru City at $d = 0.1$ km and Sakata City at $d = 0.1$ km. Figure 16 shows the relationships between the mean of steel weight loss and time for the RC bridge pier shown in Figure 10, assuming that the RC bridge pier is located in Fukuoka City, Maizuru City and Sakata City. The mean of steel weight loss in Figure 16 depends on the hazard curve associated with airborne chloride in Figure 15. Based on the hazard assessment associated with airborne chlorides, the effect of different marine environments and distance from the coastline on the bridge performance can be quantified.

The hazard curves associated with airborne chlorides indicate the large uncertainties involved in the evaluation of marine environment. These uncertainties affect the seismic reliability of RC structures. In order to develop a rational maintenance strategy, it is necessary to try to reduce the epistemic uncertainties associated with hazard assessment, in particular the scatter on attenuation relations. In PSHA, the issue of avoiding physically unrealisable ground motion amplitudes becomes important (Bommer, 2002). From a physical point of view, for hazard assessment associated with airborne chlorides, spatial spread has to be modelled instead of simply attenuation relation. Also, the effects of precipitation and the differences in coastal topography (e.g. sand beach and reef) on the amount of airborne chlorides need to be considered.

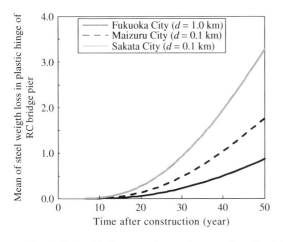

Figure 16. Relationship between time and mean of steel weight loss.

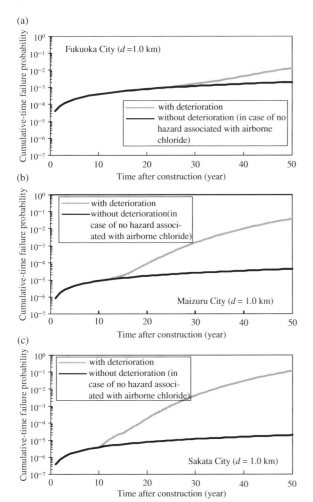

Figure 17. Relationship between time and cumulative-time failure probability.

4. Life-cycle seismic reliability including corrosion damage

When only seismic hazard and one limit state are considered, the expected risk is (Frangopol & Akiyama, 2009)

$$R = C(S) \int_{\alpha} \left[-\frac{dH(\alpha)}{d\alpha} \right] P[S|\Gamma = \alpha] \, d\alpha \qquad (8)$$

where $C(S)$ is the consequence associated with limit state S, $P[S|\Gamma = \alpha]$ is the conditional probability that the limit state S is achieved given that ground motion intensity Γ (such as peak ground acceleration or velocity) is equal to α (fragility curve), and $H(\alpha)$ is the probability that the ground motion intensity α is exceeded at least once during a time interval T (seismic hazard curve).

In the calculation of the conditional probability $P[S|\Gamma = \alpha]$ of structures subjected to aggressive environment (e.g. marine environment), the effect of corrosion on this probability has to be taken into consideration. In seismic reliability analysis of structures in marine environment, the structural capacity such as the displacement at the occurrence of buckling of longitudinal rebars of RC bridge pier (e.g. Figure 5) depends on the results of hazard associated with airborne chlorides; however, the demand depends on the results of seismic hazard assessment. By assuming that seismic capacity and demand are statistically independent and that the occurrence of earthquake is modelled as a Poisson process, the annual probability of exceedance of seismic capacity C_a under earthquake excitation at t years after construction can be expressed as

$$p_{fa}(t) = \int_0^{100} \int_0^{\infty} -\frac{dp_0(\gamma)}{d\gamma} \cdot P[D_e \geq C_a|\Gamma = \gamma,$$
$$C_w = c_w(t)] \cdot f(c_w(t)) \, d\gamma dc_w \qquad (9)$$

where $P[D_e \geq C_a|\Gamma = \gamma, C_w = c_w(t)]$ is the conditional probability of the seismic demand D_e exceeding the seismic capacity C_a conditioned upon the seismic intensity γ and steel weight loss $c_w(t)$; and $f(c_w(t))$ is the PDF of $c_w(t)$.

During a given time interval T, the cumulative-time failure probability p_f of bridge pier subjected to both seismic hazard and hazard associated with airborne chlorides is

$$p_f = 1 - (1 - p_{fa}(1)) \cdot (1 - p_{fa}(2)) \cdots (1 - p_{fa}(t = T)) \qquad (10)$$

It is assumed in Equation (10) that the events associated with the failure probabilities at different times t are statistically independent. However, in reality, they are dependent through the strength. There have been studies providing seismic hazard estimates based on non-Poissonian seismicity (Bommer, 2002). Since the lifetime of the structure is much shorter than the return period of the earthquake, the assumption of using the same seismic hazard curve $p_0(\gamma)$ for the time interval T could be acceptable. However, further research on the effect of the degree of correlation among the random variables at different times, and time-dependent seismic hazard used in Equation (9) on life-cycle reliability of structures is needed.

The relationships between cumulative-time failure probability and time after construction of RC bridge pier are shown in Figure 17, using the seismic hazard curves in Figure 13 and the airborne chloride hazard curves in Figure 15. The elevation and cross section of the bridge pier are shown in Figure 11. Only the limit state associated with the longitudinal rebar buckling was considered, assuming that the RC bridge pier has sufficient ties to prevent shear failure. To examine the effect of corrosion on the safety of the bridge pier, cumulative-time failure probabilities without deterioration (i.e. annual failure probability $pf_a(t)$ in Equation (9) does not depend on time after construction) are also shown in Figure 17. In the beginning, the failure probability of the bridge piers only depends on the seismic hazard, and the RC bridge pier located in Fukuoka City has the highest failure probability. However, as shown in Figure 15, the probability of exceedance of a prescribed amount of airborne chloride in Sakata City is much higher than that in Fukuoka City. The cumulative-time failure probability of RC bridge pier in Sakata City increases with time due to chloride attack. The difference between the cumulative-time failure probability associated with both seismic and airborne chloride hazards and that associated with seismic hazard alone is the largest for Sakata City. Finally, 50 years after construction, the cumulative-time failure probability of RC bridge pier in Sakata City is the highest, although the seismicity in Sakata City is the lowest among the three cities. Regarding RC structures in an earthquake region and marine environment, the decision on seismic retrofit and/or repair due to material deterioration should be supported by the

multiple hazard assessments of environments surrounding the structure analysed.

As previously indicated, when assessing the seismic reliability of existing and corroding RC components with light transverse reinforcement, the change of failure mode from ductile to brittle has to be taken into consideration. Priestley, Verma, and Xiao (1994) pointed out that shear strength is a function of the flexural ductility. As plastic hinge rotations increase, the widening of the flexure-shear cracks reduces the capacity of shear transfer by aggregate interlock. Further research is needed on the modelling of shear strength of corroded RC columns subjected to cyclic lateral shear force.

Even though aleatory uncertainty cannot be reduced, improvement in our knowledge or in the accuracy of predictive models will reduce the epistemic uncertainty (Ang & De Leon, 2005). This means that for existing structures, the uncertainties associated with predictions can be reduced by the effective use of information obtained from visual inspections, field test data regarding structural performance and/or monitoring (Frangopol, 2011; Frangopol & Liu, 2007, Frangopol, Strauss, & Kim, 2008a, 2008b; Frangopol, Saydam, & Kim, 2012; Okasha, Frangopol, & Orcersi, 2012). This information helps engineers to improve the accuracy of structural condition prediction. However, the updated random variables do not follow, in general, widely used PDFs (such as normal and lognormal). The difficulties of the solution in Bayesian updating depend on the relationships between observed physical quantities, such as inspection results, and the PDFs of associated random variables. For problems in which all these relationships are linear and all random variables are modelled as Gaussian, a closed form solution exists. When nonlinear relations or non-Gaussian variables are involved, a rigorous theoretical approach is generally impossible to implement in realistic cases. An approximate solution can, however, be found by using several approaches. Monte Carlo (MC) approach is, in general, used because of its versatility. MC-based methods for nonlinear filtering technique have been developed since the 1990s. These methods include MC filter, Bootstrap filter, recursive MCS, sequential MCS, the sampling importance resampling method and the sequential importance sampling (Gordon, Salmond, & Smith, 1993; Kitagawa, 1996; Ristic et al., 2004). Figure 18 shows the general flowchart of MC-based reliability estimation (Yoshida, 2009). Akiyama, Frangopol, and Yoshida (2010a) reported the time-dependent reliability analysis of existing RC structures using MC-based reliability analysis. They show how multiple random variables related to steel weight loss estimated by the corrosion crack width could be updated simultaneously based on inspection results. The information obtained from inspection and/or monitoring can be used to reduce epistemic uncertainties and, consequently, to better estimate the

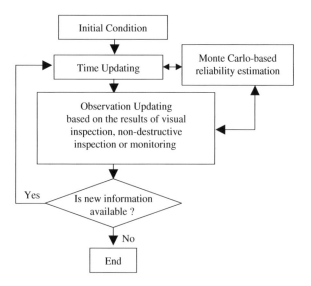

Figure 18.   Flowchart of MC-based reliability estimation.

seismic reliability of RC structures in an aggressive environment.

5. Conclusions

(1) This paper presents the procedure for estimating the life-cycle seismic reliability of RC structures in earthquake-prone region and an aggressive environment. This consists of three main components: (a) corrosion, namely determining the seismic capacity of structures and evaluating the seismic fragility curve; (b) multiple hazard assessments of environments surrounding the structure and (c) lifetime seismic reliability analysis with updating.

(2) The corrosion process of rebar has not been investigated thoroughly due to the difficulties with observing it. X-ray digital picture processing method for corroded RC members is introduced in this study. This can visualise the corrosion process in the RC member and can contribute to understanding of how spatial distribution of rebar corrosion is established. From the experiments using X-ray, a model of spatial variability of rebar corrosion was presented for estimating the reliability of RC structures.

(3) A novel computational procedure to integrate the probabilistic hazard associated with airborne chlorides into life-cycle seismic reliability assessment of RC bridge piers is presented. The seismic demand depends on the results of seismic hazard assessment, whereas the deterioration of seismic capacity depends on the airborne chloride hazard. In seismic reliability estimation of RC bridge piers in a marine environment, it should be noted that even if the failure probability is relatively low at the beginning of time after construction, it will increase with time due to the marine environment.

(4) Further research is needed to establish the methodology for life-cycle management of RC bridges. Especially, cost analysis in a life-cycle perspective and the effects of insufficient data and knowledge associated with long-term structural performance of existing bridges under multi-hazards have to be examined. To estimate the seismic reliability of bridges accurately, the effects of corroded bearing, reduction of shear strength due to corrosion and soil–structure interaction on the bridge safety need to be investigated. In addition, life-cycle bridge maintenance optimisation and resilience of bridges under multiple hazards including earthquake, tsunami, corrosion and traffic are topics for further investigation. Efforts in this direction are already underway (Akiyama, Frangopol, Arai, & Koshimura, 2012a; Bocchini & Frangopol, 2011, 2012; Decò & Frangopol, 2011, Frangopol & Bocchini, 2012; Okasha & Frangopol, 2010).

Acknowledgements

The authors would like to acknowledge the work by Prof. Takumi Shimomura (Nagaoka University of Technology), Prof. Koichi Kobayashi (Gifu University) and members contributing to research activities in the Japan Society of Civil Engineers Committee 331 on 'Structural performance of deteriorated concrete structures'. The authors also thank Prof. Ikumasa Yoshida (Tokyo City University) for many suggestions on the topic of MC-based reliability analysis.

References

Akiyama, M., Frangopol, D.M., Arai, M., & Koshimura, S. (2012a). Probabilistic assessment of structural performance of bridges under tsunami hazards. ASCE Structures Congress 2012, Chicago, USA, pp. 1919–1928.

Akiyama, M., Frangopol, D.M., & Matsuzaki, H. (2011a). Life-cycle reliability of RC bridge piers under seismic and airborne chloride hazards. *Earthquake Engineering and Structural Dynamics*, 40(15), 1671–1687.

Akiyama, M., Frangopol, D.M., & Suzuki, M. (2012b). Integration of the effects of airborne chlorides into reliability-based durability design of R/C structures in a marine environment. *Structure and Infrastructure Engineering*, 8(2), 125–134.

Akiyama, M., Frangopol, D.M., & Yoshida, I. (2010a). Time-dependent reliability analysis of existing RC structures in a marine environment using hazard associated with airborne chlorides. *Engineering Structures*, 32(11), 3768–3779.

Akiyama, M., Matsuzaki, H., Dang, T.H., & Suzuki, M. (2012c). Reliability-based capacity design for reinforced concrete bridge structures. *Structure and Infrastructure Engineering*, 8(12), 1096–1107.

Akiyama, M., Nakajima, K., & Komoriya, T. (2011b). Visualization of corrosion process in RC component using X-Ray. *Concrete Research and Technology*, 22(3), 35–45, (in Japanese).

Ang, A.H-S., & De Leon, D. (2005). Modeling and analysis of uncertainties for risk-informed decisions in infrastructures engineering. *Structure and Infrastructure Engineering*, 1(1), 19–21.

Berry, M.P., & Eberhard, M.O. (2005). Practical performance model for bar buckling. *Journal of Structural Engineering*, 131(7), 1060–1070.

Berto, L., Vitaliani, R., Saetta, A., & Simioni, P. (2009). Seismic assessment of existing RC structures affected by degradation phenomena. *Structural Safety*, 31, 284–297.

Biondini, F., Palermo, A., & Toniolo, G. (2011). Seismic performance of concrete structures exposed to corrosion: Case studies of low-rise precast buildings. *Structure and Infrastructure Engineering*, 7(1-2), 109–119.

Bocchini, P., & Frangopol, D.M. (2011). Generalized bridge network performance analysis with correlation and time-variant reliability. *Structural Safety*, 33(2), 155–164.

Bocchini, P., & Frangopol, D.M. (2012). Restoration of bridge networks after an earthquake: Multi-criteria intervention optimization. *Earthquake Spectra*, 28(2), 1–25.

Bommer, J.J. (2002). Deterministic vs. probabilistic seismic hazard assessment: An exaggerated and obstructive dichotomy. *Journal of Earthquake Engineering*, 6(1), 43–73.

Choe, D-E., Gardoni, P., Rosowsky, D., & Haukaas, T. (2008). Probabilistic capacity models and seismic fragility estimates for RC columns subject to corrosion. *Reliability Engineering and System Safety*, 93, 383–393.

Cornell, C.A., Jalayer, F., Hamburger, R.O., & Foutch, D.A. (2000). Probabilistic basis for 2000 SAC Federal Emergency Management Agency steel moment frame guidelines. *Journal of Structural Engineering*, 128(4), 526–533.

Coronelli, D., & Gambarova, P. (2004). Structural assessment of corroded reinforced concrete beam: Modeling guidelines. *Journal of Structural Engineering*, 130(8), 1214–1224.

Decò, A., & Frangopol, D.M. (2011). Risk assessment of highway bridges under multiple hazards. *Journal of Risk Research*, 14(9), 1057–1089.

Dhakal, R.P., & Maekawa, K. (2002). Reinforcement stability and fracture of cover concrete in reinforced concrete members. *Journal of Structural Engineering*, 128(10), 1253–1262.

Ellingwood, B.R. (2005). Risk-informed condition assessment of civil infrastructure: State of practice and research issues. *Structure and Infrastructure Engineering*, 1(1), 7–18.

Fang, C., Gylltoft, K., Lundgren, K., & Plos, M. (2006). Effect of corrosion on bond in reinforced concrete under cyclic loading. *Cement and Concrete Research*, 36, 548–555.

Frangopol, D.M. (2011). Life-cycle performance, management, and optimization of structural systems under uncertainty: Accomplishments and challenges. *Structure and Infrastructure Engineering*, 7(6), 389–413.

Frangopol, D.M., & Akiyama, M. (2009). Lifetime seismic reliability analysis of corroded reinforced concrete bridge piers. In M. Papadrakakis, M. Fragiadakis, & N.D. Lagaros (Eds.), *Computational methods in earthquake engineering*. Dordrecht-Heidelberg-London-New York: Springer (Chapter 23).

Frangopol, D.M., & Bocchini, P. (2012). Bridge network performance, maintenance, and optimization under uncertainty: Accomplishments and challenges. *Structure and Infrastructure Engineering*, 8(4), 341–356.

Frangopol, D.M., & Liu, M. (2007). Maintenance and management of civil infrastructure based on condition, safety, optimization, and life-cycle cost. *Structure and Infrastructure Engineering*, 3(1), 29–41.

Frangopol, D.M., Saydam, D., & Kim, S. (2012). Maintenance, management, life-cycle design and performance of structures and infrastructures: A brief review. *Structure and Infrastructure Engineering*, 8(1), 1–25.

Frangopol, D.M., Strauss, A., & Kim, S. (2008a). Use of monitoring extreme data for the performance prediction of structures: General approach. *Engineering Structures*, 30(12), 3644–3653.

Frangopol, D.M., Strauss, A., & Kim, S. (2008b). Bridge reliability assessment based on monitoring. *Journal of Bridge Engineering*, 13(3), 258–270.

Ghosh, J., & Padgett, J.E. (2010). Aging consideration in the development of time-dependent seismic fragility curves. *Journal of Structural Engineering*, 136(12), 1497–1511.

Gordon, N., Salmond, D., & Smith, A. (1993). A novel approach to nonlinear/non-Gaussian Bayesian state estimation. *Proceedings of IEEE on Radar and Signal Processing*, 140, 107–113.

Japan Road Association (2002). Design specification for highway bridges; Part V; seismic design, Maruzen, Japan.

Kameda, H., & Nojima, H. (1988). Simulation of risk-consistent earthquake motion. *Earthquake Engineering and Structural Dynamics*, 16, 1007–1019.

Kitagawa, G. (1996). Monte Carlo filter and smoother for non-Gaussian state space models. *Journal of Computational and Graphical Statistics*, 5(1), 1–25.

Kobayashi, K. (2006). The seismic behavior of RC member suffering from chloride-induced corrosion. *Proceedings of the second fib Congress*, Paper ID. 19-3.

Kumar, R., Gardoni, P., & Sanchez-Silva, M. (2009). Effect of cumulative seismic damage and corrosion on the life-cycle cost of reinforced concrete bridges. *Earthquake Engineering and Structural Dynamics*, 38, 887–905.

Li, J., Gong, J., & Wang, L. (2009). Seismic behavior of corrosion-damaged reinforced concrete columns strengthened using combined carbon fiber-reinforced polymer and steel jacket. *Construction and Building Material*, 23(7), 2653–2663.

Lupoi, A., Franchin, P., & Schotaunus, M. (2003). Seismic risk evaluation of RC bridge structures. *Earthquake Engineering and Structural Dynamics*, 32, 1275–1290.

Marsh, P.S., & Frangopol, D.M. (2007). Lifetime multi-objective optimization of cost and spacing of corrosion rate sensors embedded in a deteriorating reinforced concrete bridge deck. *Journal of Structural Engineering, ASCE*, 133(6), 777–787.

Marsh, P.S., & Frangopol, D.M. (2008). Reinforced concrete bridge deck reliability model incorporating temporal and spatial variations of probabilistic corrosion rate sensor data. *Reliability Engineering & System Safety*, 93(3), 394–409.

Melchers, R.E., Li, C.Q., & Lawanwisut, W. (2008). Probabilistic modeling of structural deterioration of reinforced concrete beams under saline environment corrosion. *Structural Safety*, 30, 447–460.

Mori, Y. (1992). Reliability-based condition assessment and life prediction of concrete structures. Ph.D. thesis, Johns and Hopkins University, Baltimore, MD.

Mori, Y., & Ellingwood, B.R. (1993). Reliability-based service life assessment of aging concrete structures. *Journal of Structural Engineering*, 119(5), 1600–1621.

Naito, H., Akiyama, M., & Suzuki, M. (2011). Ductility evaluation of concrete-encased steel bridge piers subjected to lateral cyclic loading. *Journal of Bridge Engineering*, 16(1), 72–81.

Nielson, B.G., & DesRoches, R. (2007). Seismic fragility methodology for highway bridges using a component level approach. *Earthquake Engineering and Structural Dynamics*, 36, 823–839.

Okasha, N.M., & Frangopol, D.M. (2010). Novel approach for multi-criteria optimization of life-cycle preventive and essential maintenance of deteriorating structures. *Journal of Structural Engineering*, 136(8), 1009–1022.

Okasha, N.M., Frangopol, D.M., & Orcesi, A.D. (2012). Automated finite element updating using strain data for the lifetime reliability assessment of bridges. *Reliability Engineering & System Safety*, 99, 139–150.

Otsuka, K., & Date, H. (2000). Fracture process zone in concrete tension specimen. *Engineering Fracture Mechanics*, 65, 111–131.

Ou, Y.-C., Tsai, L.-L., & Chen, H.-H. (2011). Cyclic performance of large-scale corroded reinforced concrete beams. *Earthquake Engineering and Structural Dynamics*, DOI: 10.1002/eqe.1145

Padgett, J.E., & DesRoches, R. (2007). Sensitivity of seismic response and fragility to parameter uncertainty. *Journal of Structural Engineering*, 133(12), 1710–1718.

Priestley, M.J.N., Seible, F., & Calvi, G.M. (1996). *Seismic design and retrofit of bridges*. New York: Wiley.

Priestley, M.J.N., Verma, R., & Xiao, Y. (1994). Seismic shear strength of reinforced concrete columns. *Journal of Structural Engineering, 120*(8), 2310–2329.

Ristic, B., Arulampalam, S., & Gordon, N. (2004). Beyond the Kalman filter. *Particle filters for tracking applications*. Norwood, Massachusetts, USA: Artech House.

Yamamoto, T., & Kobayashi, K. (2006). Report of research project on structural performance of deteriorated concrete structures by JSCE 331-review of experimental study. *Proceedings of the International Workshop on Life Cycle Management of Coastal Concrete Structures, Nagaoka, Japan* (pp. 171–180).

Yoshida, I. (2009). Data assimilation and reliability estimation of existing RC structure. *COMPDYM 2009,* Rhodes, Greece, CD281.

Yun, S.-Y., Hamburger, R.O., Cornell, C.A., & Foutch, D.A. (2002). Seismic performance evaluation for steel moment frames. *Journal of Structural Engineering, 128*(4), 534–545.

Saito, Y., Oyado, M., Kanakubo, T., & Yamamoto, Y. (2007). Structural performance of corroded RC column under uniaxial compression load. *First International Workshop on Performance, Protection & Strengthening of Structures under Extreme Loading*. Canada: Whistler.

Shimomura, T., Miyazato, S., Yamamoto, T., Sato, S., Kato, Y., & Tsuruta, H. (2010). Systematic research on structural performance of deteriorated concrete structures in Japan. 2nd International Symposium on Service Life Design for Infrastructures, Delft, The Netherlands.

Shimomura, T., Sato, T., Miyazato, S., Saito, S., Yamamoto, T., Kamiharako, A., Kobayashi, K., & Akiyama, M. (2006). Report of research project on structural performance of deteriorated concrete structures by JSCE-331., Life Cycle Management of Coastal Concrete Structures, Nagaoka, Japan.

Simon, J., Bracci, J.M., & Gardoni, P. (2010). Seismic response and fragility of deteriorated reinforced concrete bridges. *Journal of Structural Engineering, 136*(10), 1273–1281.

Stewart, M.G. (2004). Spatial variability of pitting corrosion and its influence on structural fragility and reliability of RC beams in flexure. *Structural Safety, 26*, 453–470.

Stewart, M.G., & Suo, Q. (2009). Extent of spatially variable corrosion damage as an indicator of strength and time-dependent reliability of RC beams. *Engineering Structures, 31*(1), 198–207.

Vidal, T., Castel, A., & François, R. (2004). Analyzing crack width to predict corrosion in reinforced concrete. *Cement and Concrete Research, 34*, 165–174.

Vidal, T., Castel, A., & François, R. (2007). Corrosion process and structural performance of a 17 year old reinforced concrete beam stored in chloride environment. *Cement and Concrete Research, 37*, 1551–1561.

Performance analysis of Tohoku-Shinkansen viaducts affected by the 2011 Great East Japan earthquake

Mitsuyoshi Akiyama, Dan M. Frangopol and Keita Mizuno

Tohoku-Shinkansen viaducts in eastern Japan were designed in accordance with specifications published in the 1970s. They have less shear reinforcement than required by current design specifications. Consequently, the No. 5 Inohana viaducts of Tohoku-Shinkansen failed in shear during the 2003 Sanriku-Minami earthquake. The viaducts were retrofitted by means of steel jacketing so that they had sufficient shear capacity. The retrofitted columns of the viaducts performed well during the 2011 Great East Japan earthquake. However, there is a lack of understanding of the impact of these retrofitting interventions on the vulnerability of the viaduct. It is important to recognise the relationship between the damage to the Shinkansen viaduct with retrofitted reinforced concrete columns and the ground motion intensity. After a brief review of the history of the performance of Shinkansen viaducts to several earthquakes prior to the 2011 Great East Japan earthquake, this paper presents fragility curves of the retrofitted and non-retrofitted Tohoku-Shinkansen viaducts based on nonlinear dynamic analyses and Monte Carlo simulation. It was found that the median of the fragility curve associated with the ultimate limit state for retrofitted viaduct is at least five times larger than that for as-built viaduct.

1. Introduction

Bridges are susceptible to damage during an earthquake event, particularly if they are designed without adequate seismic detailing. Reinforced concrete (RC) columns built using earlier seismic design specifications often lack enough flexural strength, flexural ductility capacity and/or shear strength. When these columns are subjected to strong ground motions, they have the potential to exhibit brittle failure. Several recent destructive earthquakes in Japan (e.g. 1995 Hyogoken-Nanbu earthquake, 2003 Sanriku-Minami earthquake and 2004 Niigata-ken-Chuetsu earthquake) inflicted various levels of damage on the Shinkansen viaducts. The investigation of these negative consequences gave rise to serious discussions about seismic design philosophy and to extensive research activity on the retrofit of existing bridges. The seismic design methodology for new bridges was also improved. General column retrofits often include some type of encasement to improve the shear or flexural strength, flexural confinement and ductility capacity. Steel jackets are common for this purpose.

The Tohoku-Shinkansen viaducts in eastern Japan, which were designed using specifications published in the 1970s, were damaged during the 2003 Sanriku-Minami earthquake because of the inadequacies of the shear design of RC components. After that earthquake, the RC bridge piers of the Tohoku-Shinkansen viaducts were retrofitted with steel jacketing to prevent brittle failure from a severe earthquake and to enhance the ductility capacity. This work was completed before the 2011 Great East Japan earthquake. As reported in a damage investigation (Akiyama & Frangopol, 2012; Kawashima et al., 2011), the retrofitted Shinkansen viaducts experienced no damage, and the effectiveness of the seismic retrofit involving the application of steel jacketing to RC columns was demonstrated. However, there is a lack of understanding of the impact of these retrofits on the vulnerability of the viaduct. It is important to recognise the relationship between the damage to the Shinkansen viaduct with retrofitted RC columns and the ground motion intensity.

Seismic fragility curves are essential tools for assessing the vulnerability of viaducts. These curves describe the probability that the actual damage to a viaduct exceeds the damage thresholds when the structure is subjected to a specific ground motion intensity. Fragility curves can offer a means of communicating the risk of damage over a range of potential earthquake ground motion intensities. This information is essential for seismic risk management and decision-making on retrofit and mitigation strategies. Empirical fragility curves based on bridge damage data from past earthquakes have been developed for as-built bridges (e.g. Shinozuka, Feng, Kim, & Kim, 2000; Shinozuka, Feng, Lee, & Naganuma, 2000). However, as a result of the limited empirical data available, developing the fragility curves for retrofitted bridges based on damage investigation is impossible.

In the absence of adequate empirical data, analytical methods have been used to develop the fragility curves (e.g. Karim & Yamazaki, 2003). Analytically derived fragility curves for retrofitted road bridges were developed by Kim and Shinozuka (2004) and Padgett and DesRoches (2007, 2008). Demand and capacity are compared in a fragility analysis. In previous studies on analytically computed fragility curves, the demands were evaluated by a nonlinear dynamic analysis of the retrofitted bridge, and the capacities of each component were defined based on the results from past experiments and the results of an expert opinion survey (Padgett & DesRoches, 2007). In their fragility analysis, Padgett and DesRoches (2008) used the curvature ductility capacities of steel-jacketed columns with median values of 9.35 for slight damage, 17.7 for moderate damage, 26.1 for extensive damage and 30.2 for complete damage. However, as the ductility capacities of retrofitted columns depend on (a) the structural details, (b) failure modes of as-built column and (c) seismic specifications used in the retrofit design, these values could not be applied to develop the fragility curves of the Shinkansen viaducts with steel-jacketed RC columns.

In this paper, first, damage investigations of Shinkansen viaducts before and after retrofit are briefly presented. By comparing the damage state before and after the seismic retrofits during the 2011 Great East Japan earthquake, the effectiveness of the seismic retrofit against the strong ground motions are investigated. Then, enhancement of seismic capacity of Shinkansen viaducts is analytically examined. The moment–curvature relationships of RC columns with steel jacketing based on experimental results are established. The limit states for the fragility analysis of the as-built and retrofitted viaducts are defined. Finally, the improvement due to steel jacketing is quantified by comparing the fragility curves of the viaduct before and after retrofit.

2. Seismic design and retrofit in Japan

In Japan, seismic design specifications for structures and infrastructures have been significantly revised. The first seismic design code for road bridges, which included the seismic analysis considering the inelastic bridge behaviour and the level 2 ground motions (i.e. seismic design actions for the verification of the no-collapse requirement), was

issued in Japan in 1990 (Japan Road Association, 1990). For railway bridges, limit-state design methodology was introduced in 1992 (Railway Technical Research Institute, 1992). Railway bridges were designed to ensure the ductile behaviour under the ground motions with a return period of approximately 100 years. Hyogo-Ken Nanbu earthquake of 17 January 1995 caused destructive damage to many highway and railway bridges, buildings and harbour facilities. For this reason, many seismic design codes in Japan were revised after Hyogo-Ken Nanbu earthquake. The evolution of seismic design code including the seismic analysis method, seismic design actions and structural details is reported in Kawashima (2012).

Since the 1995 Hyogoken-Nanbu earthquake, seismic retrofits have been conducted for RC bridge piers that have insufficient shear reinforcement and/or have cut-offs of longitudinal rebars without adequate anchorage length at the midpoint of bridge pier. However, there were still a number of bridge piers that required retrofitting before the 2011 Great East Japan earthquake.

2.1 Damage to bridges due to the recent earthquakes in Japan

2.1.1 2003 Sanriku-Minami earthquake

The Sanriku-Minami earthquake with the magnitude of 7.0 occurred on 26 May 2003. The epicentre of this earthquake was located near that of the 2011 Great East Japan earthquake. Concrete structures in Miyagi-ken (i.e. Miyagi Prefecture) and Iwate-ken (i.e. Iwate Prefecture) were severely damaged due to the strong ground motion as shown in Figure 1. Especially, Tohoku-Shinkansen viaducts were damaged during the 2003 Sanriku-Minami earthquake. There were no reports on damage due to the tsunami.

Tohoku-Shinkansen entered service in 1982 between Omiya near Tokyo Metro and Morioka Stations. As the Shinkansen viaducts were designed prior to the occurrence of the 1978 Miyagiken-Oki earthquake, they have less shear reinforcement than required by current seismic specifications. An allowable stress design approach (i.e. seismic coefficient method) based on static and elastic analyses was used to design the Tohoku-Shinkansen viaducts. Design response acceleration was approximately 20% of gravity, independent of natural period of viaducts. Some viaducts of the Tohoku-Shinkansen between the Morioka and Mizusawa-Esashi stations in Iwate-ken were extensively damaged by shear during the 2003 Sanriku-Minami earthquake (Japan Society of Civil Engineers, 2004). Figure 1 shows that RC columns of Shinkansen viaducts failed in shear. The Shinkansen viaducts in only Tokyo and Sendai areas had been retrofitted since the 1995 Hyogoken-Nanbu earthquake. There were no retrofitted Shinkansen viaducts in the area with high seismic intensity during the 2003 Sanriku-Minami earthquake. Most

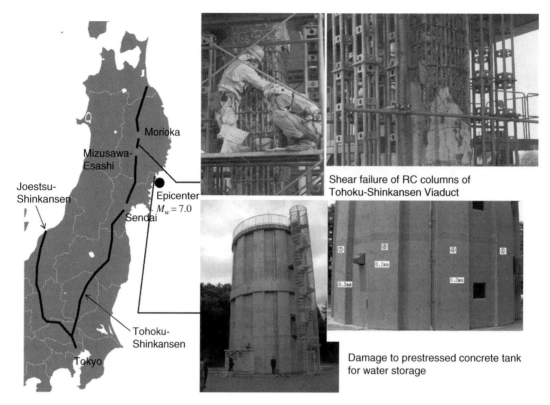

Figure 1.   Damage to infrastructures during the 2003 Sanriku-Minami earthquake.

viaducts in Iwate-ken were single-story RC moment-resisting frames with a Gerber girder on both sides, as shown in Figure 2. Damage was concentrated at the side columns of the viaduct during the 2003 Sanriku-Minami earthquake. Because the side columns are shorter than the centre columns to fit under the Gerber girder, the ratio of the shear strength to the shear force corresponding to maximum flexural strength was smaller in the side columns than in the centre columns. This led to shear failure in the side columns.

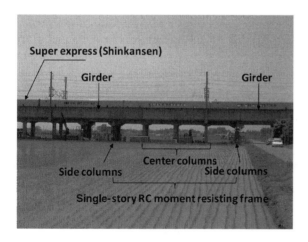

Figure 2.   Single-story RC moment-resisting frame pier with a Gerber girder on both sides.

2.1.2   2004 Niigataken-Chuetsu earthquake

The Niigataken-Chuetsu earthquake with the magnitude of 6.6 occurred on 23 October 2004. As shown in Figure 3, this earthquake caused damage to Joetsu-Shinkansen viaducts that began the service between Omiya and Niigata stations in 1982. Joetsu and Tohoku-Shinkansen viaducts were designed according to the same seismic specification. There were a few retrofitted RC columns of viaducts using steel jacketing before the 2004 Niigataken-Chuetsu earthquake. Retrofitted columns had no damage during the 2004 Niigataken-Chuetsu earthquake.

A bullet train of Joetsu-Shinkansen running near the epicentre derailed due to the 2004 Niigataken-Chuetsu earthquake. This was the first derailment in the 40-year history of the Shinkansen. The columns of a single-story RC rigid frame without retrofitting were seriously damaged by shear as shown in Figure 3. In addition, RC bridge piers had damages due to the insufficient anchorage length of longitudinal rebars as shown in Figure 3.

After the 2003 Sanriku-Minami earthquake, the first seismic retrofit programme was initiated for the Tohoku and Joetsu-Shinkansen viaducts. The objectives of the programme were to prevent shear failure by ensuring that the shear strength of steel-jacketed RC columns exceeds the shear corresponding to maximum flexural strength and/or to prevent the damage of bridge piers that have the cut-off of the longitudinal rebars by improving the flexure and shear resistance using steel jacketing, RC jacketing or

Derailment of Joetsu-Shinkansen   Shear failure of RC columns

Damage to Joetsu-Shinkansen Viaduct which has the cut-off of the longitudinal rebars

Figure 3.   Damage to Joetsu-Shinkansen viaducts during the 2004 Niigataken-Chuetsu earthquake.

carbon fibre-reinforced polymer sheet. After retrofitting the as-built RC columns, these columns can be the prime source of energy dissipation responding to strong ground motion. The programme was completed in May 2008 after retrofitting 15,400 columns of single-story RC moment-resisting frames and 2340 bridge piers. In 2009, the second retrofit programme for 6700 columns of single-story RC moment-resisting frames to enhance the shear and flexural strengths and ductility capacity of the RC columns was initiated and is still in progress.

2.2   Seismic design and retrofit of bridges

Figure 4 presents a brief overview of seismic specifications for new bridges and seismic retrofit for existing bridges in Japan. Based on the experience of seismic damage to bridges and research progress on earthquake resistance, shear and ductility designs of RC components were modified and rubber bearings were introduced. The bridges designed according to the current seismic specification can exhibit ductile behaviour to secure the hierarchy of strengths of the various structural components necessary for leading to the intended configuration of plastic hinges and for avoiding brittle failure modes (i.e. capacity design).

As evidenced by the past earthquakes, such as the 2003 Sanriku-Minami earthquake and 2004 Niigataken-Chuetsu earthquake, existing RC structures that may exhibit brittle failure need retrofit as shown in Figure 4. However, there are still a very large number of bridges without adequate seismic design detailing. Even though it is very important that rapid progress is made in the seismic retrofit programme, it will take more time to eliminate the number of existing bridges without adequate seismic design detailing. It is necessary to establish a seismic retrofit strategy. To determine the priorities for seismic retrofit, the seismic risk of individual structures needs to be compared with the structures, and the effect of the seismic retrofit on the risk reduction has to be quantified. To do this, it is necessary to obtain the fragility curves of individual structures with and without seismic retrofit as will be described later.

3.   Damage investigation of the Shinkansen viaducts subjected to the 2011 Great East Japan earthquake

3.1   Comparison of damage states of the Shinkansen viaducts before and after seismic retrofit

Figures 5 and 6 show a comparison of the acceleration time history and 5% damping response accelerations of the 2003

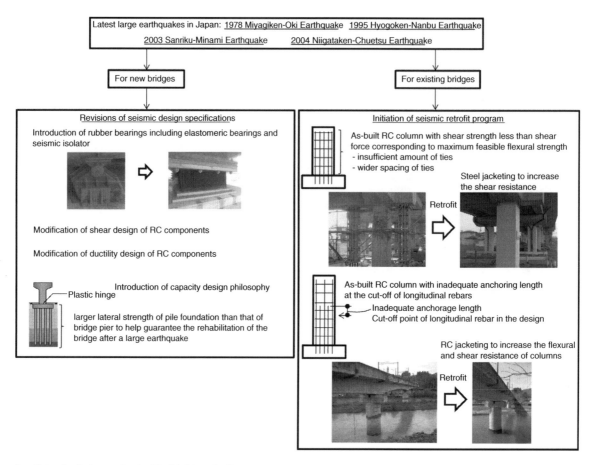

Figure 4. Seismic design and retrofit of bridges in Japan.

Sanriku-Minami earthquake with those of the 2011 Great East Japan earthquake. The ground motions were measured at the seismic stations IWT014 and IWT011 near the No. 5 Inohana viaducts and the No. 3 Odaki viaducts, respectively, as shown in Figure 7. As shown in Figure 5, the duration of the ground motion was very different in the two earthquakes. However, the peak ground accelerations (PGAs) were almost the same. In addition, as presented in Figure 6, because the fundamental natural period of a single-story RC rigid frame ranges from 0.4 to 0.6 s depending on the soil condition and column height, it is reasonable to assume that the response accelerations of the No. 5 Inohana viaducts and the No. 3 Odaki viaducts were nearly the same in the 2003 Sanriku-Minami earthquake and the 2011 Great East Japan earthquake.

Figures 8 and 9 show the damaged states of the No. 5 Inohana viaducts and the No. 3 Odaki viaducts taken after the 2003 Sanriku-Minami earthquake and the 2011 Great East Japan earthquake, respectively. Because of deficiencies in the number of ties used to prevent brittle failure, the RC columns of the No. 5 Inohana viaducts and the No. 3 Odaki viaducts failed in shear during the 2003 Sanriku-Minami earthquake. Under the first seismic retrofit programme described earlier, the viaducts were retrofitted by means of steel jackets for the RC columns as

shown in Figure 10(a), so that they had sufficient shear capacity. Steel jackets with 6-mm thickness were used, and the gap between the as-built RC column and the steel jacket was approximately 30 mm. The gap was filled with a low-strength and self-compacting material. For road bridges in Japan, a steel jacket is used to increase the shear and flexural strengths and the ductility capacity (Kawashima, 2000). To increase the flexural strength of road bridge, anchor bolts were provided to connect the bottom of the steel jacket to the footing as shown in Figure 10(b). However, under the first seismic retrofit programme for the Shinkansen viaducts, the steel jacket was only used to increase the shear strength and ductility capacity, and for this reason, it was not anchored to other components.

The retrofitted columns of the viaducts performed well, with almost no damage during the 2011 Great East Japan earthquake (Akiyama & Frangopol, 2012). There have been no reports on the damage of steel-jacketed RC columns or on the undesirable consequences elsewhere in the bridge produced by the steel jacketing. The effectiveness of seismic retrofitting using steel jacketing to prevent significant damage to the Shinkansen viaducts was demonstrated.

Under the first retrofit programme for Shinkansen viaducts, RC columns were retrofitted if the ratio α of

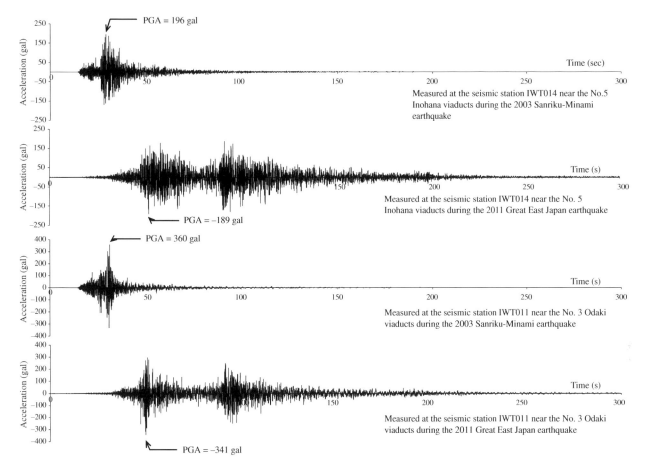

Figure 5. Comparison of the acceleration time history measured during the 2003 Sanriku-Minami earthquake and the 2011 Great East Japan earthquake based on the data from the National Research Institute for Earth Science and Disaster Prevention (2012).

shear strength to shear force corresponding to maximum flexural strength is less than threshold C_t. The inequality $\alpha < C_t$ represents that RC column has a high probability of exhibiting the brittle failure. Even if $\alpha \geq C_t$, RC columns may exhibit brittle behaviour because of the

variability of concrete and reinforcement rebar strength, and uncertainties associated with the estimation of shear and flexure strength. Figure 11 shows the damage to RC columns of Tohoku-Shinkansen viaducts in Miyagi-ken and Iwate-ken during the 2011 Great East Japan

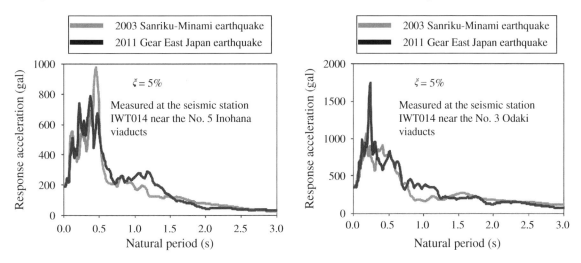

Figure 6. Response accelerations during the 2003 Sanriku Minami earthquake and the 2011 Great East Japan earthquake based on the data from the National Research Institute for Earth Science and Disaster Prevention (2012).

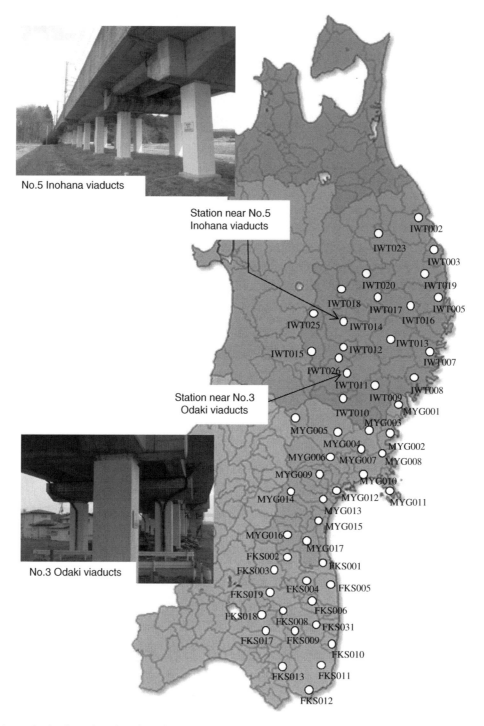

Figure 7. Locations of seismic stations based on the National Research Institute for Earth Science and Disaster Prevention (2012).

earthquake (Akiyama & Frangopol, 2012). It should be noted that all the viaducts that suffered damage during the 2011 Great East Japan earthquake had not yet been retrofitted. The investigation after the 2011 Great East Japan earthquake confirmed that RC bridge piers were damaged due to the insufficient anchorage length at the cut-off point of longitudinal rebars, or the RC columns of the single-story RC moment-resisting frame failed in shear (Akiyama & Frangopol, 2012). These failure modes of RC members observed in the 2011 Great East Japan earthquake are the same as those observed in the past earthquakes as shown in Figures 1 and 3. As seismic activity in Tohoku region is very high, it is necessary to rapidly enhance the seismic performance of RC columns of Shinkansen viaducts with a low value of α in the second seismic retrofit programme.

Figure 8.   R13–R15 of the No. 5 Inohana viaduct taken after the 2003 Sanriku-Minami earthquake and the 2011 Great East Japan earthquake (Note: the viaducts were retrofitted before the 2011 Great East Japan earthquake) (photos taken by the first author).

3.2   Damage investigations of the railway viaduct designed according to latest seismic specification and the Shinkansen viaduct subjected to strong aftershock

Although this study places emphasis on the damage investigation and vulnerability evaluation of Shinkansen viaducts before and after seismic retrofit, it is important to present the effectiveness of upgrade of seismic specifications against the strong ground motions comparing the performance of railway viaducts designed according to old and latest seismic specifications during the 2011 Great

Figure 9.   R2 of the No. 3 Odaki viaducts taken after the 2003 Sanriku-Minami earthquake and the 2011 Great East Japan earthquake (Note: the viaducts were retrofitted before the 2011 Great East Japan earthquake) (photos taken by the first author).

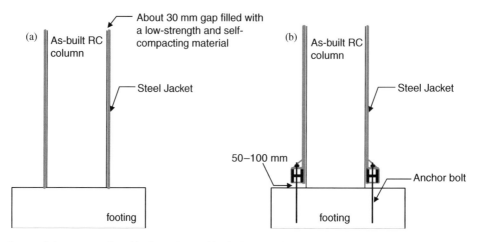

Figure 10. Seismic retrofit by means of steel jackets: (a) steel jacketing to enhance the shear strength and ductility capacity and (b) steel jacketing to enhance the flexural strength as well as shear strength and ductility capacity.

East Japan earthquake. In addition, the impact of aftershock on the damage to Shinkansen viaducts is described briefly in this section.

In the Nagamachi Area, 5 km away from the Sendai station, local trains run parallel to bullet trains of Tohoku-Shinkansen as shown in Figure 12. The viaducts of both local line and Tohoku-Shinkansen are single-story RC moment-resisting frames. However, the local line viaducts were designed according to the seismic specification revised after the 1995 Hyogoken-Nanbu earthquake. Although Shinkansen viaducts without seismic retrofit were severely damaged and had large diagonal cracks as shown in Figure 12, local line viaducts had minor flexural cracks. This proved that the revisions of seismic specifications contributed to improve the seismic performance of bridges.

Since 11 March 2011, there have been many aftershocks in the east Japan. Especially, on 7 April 2011, an aftershock with the magnitude of 7.2 caused damages to many structures and infrastructures in the

Tohoku Region. Although Shinkansen viaducts had been repaired since 11 March 2011, some of repaired RC columns were damaged again due to this aftershock. Figure 13 shows the RC bridge pier damaged due to the aftershock on 7 April under re-repair work, as the RC bridge pier repaired after the 2011 Great East Japan earthquake had large diagonal cracks. According to the railway company, the number of Shinkansen viaducts requiring repair due to the aftershock on 7 April is larger than that due to the mainshock on 11 March. Further research is needed to investigate the effect of aftershock on the seismic fragility (Li & Ellingwood, 2007).

4. Fragility analysis of the Shinkansen viaducts

4.1 Basic equations for developing fragility curves for as-built and retrofitted viaducts

Fragility is defined as the conditional probability of occurrence of the event $g_i < 0$, representing that the

Figure 11. Example of damages to Tohoku Shinkansen viaducts before retrofit during the 2011 Great East Japan earthquake (left and right photos were provided by East Japan Railway Company and Dr Takahashi, Kyoto University, respectively).

Tohoku Shinkansen viaduct constructed in the 1970s.

Viaduct for local line constructed in the 2000s.

RC columns of viaducts for local line

RC columns of Shinkansen viaducts

Figure 12. Comparison of damage states of RC columns constructed in the 1970s and the 2000s (photos taken by the first author).

seismic demand placed on the structure exceeds its capacity, for a given level of seismic intensity Γ, such as PGA and response acceleration. The limiting performance requirement is defined as $g_i = 0$, which is the limit sate and g_i is the performance function. The fragility can be expressed as follows:

$$\text{Fragility} = P[g_i < 0|\Gamma = \gamma] \qquad (1)$$

In this study, the performance functions g_1, g_2 and g_3 used in Equation (1) are determined for the as-built viaduct, as indicated in the following equations:

$$g_1 = \chi_1(V_c + V_s) - V_{D,b} \qquad (2)$$

$$g_2 = \chi_2 \theta'_{y,b} - \theta_{D,b} \qquad (3)$$

$$g_3 = \chi_3 \theta_{u,b} - \theta_{D,b} \qquad (4)$$

where V_c and V_s are the shear strengths contributed by the concrete and the shear reinforcement, respectively; $V_{D,b}$ and $\theta_{D,b}$ are the peak shear forces of the RC column of the as-built viaduct and the rotation demand of the plastic hinge of the RC column of the as-built viaduct obtained by nonlinear dynamic analysis using a number of measured ground motions, respectively; $\theta'_{y,b}$ and $\theta_{u,b}$ are the yielding and ultimate rotations of the plastic hinge of the RC column of the as-built viaduct, respectively and χ_1, χ_2 and χ_3 are the lognormal random variables representing the model uncertainties associated with the estimation of $V_c + V_s$, $\theta_{y,b}$ and $\theta_{u,b}$, respectively. The shear strength $V_c + V_s$ in Equation (2) was proposed by Akiyama, Matsuzaki, Dang, and Suzuki (2012), and the statistical descriptors of χ_1 are shown in Table 1.

Figure 13. Damage to RC columns repaired after the 2011 Great East Japan earthquake on 11 March due to the large aftershock on 7 April (photos taken by the first author).

Table 1. Parameters of random variables.

	Mean	Covariance	Distribution	Reference
χ_1	1.38	0.185	Lognormal	Akiyama et al. (2012)
χ_2	0.936	0.298	Lognormal	Figure 17, Tamai and Sato (1998)
χ_3	1.08	0.167	Lognormal	Figure 17, Tamai and Sato (1998)
χ_4	1.27	0.161	Lognormal	Figure 17, Tamai and Sato (1998)
Concrete compressive strength	28.2 MPa	0.100	Normal	Akiyama et al. (2012)
Yield strength of rebar	354 MPa	0.070	Normal	Akiyama et al. (2012)
Young's modulus of rebar	2.05×10^5 MPa	0.010	Normal	Akiyama et al. (2012)

The Shinkansen viaducts were retrofitted by means of steel jacketing of the as-built RC columns. Because the viaduct has sufficient shear resistance after retrofitting, the performance function associated with shear failure was not considered for the retrofitted viaduct. The performance functions g_4 and g_5 for the retrofitted viaduct are

$$g_4 = \chi_2 \theta'_{y,a} - \theta_{D,a} \qquad (5)$$

$$g_5 = \chi_3 \theta_{u,a} - \theta_{D,a} \qquad (6)$$

where $\theta'_{y,a}$ and $\theta_{u,a}$ are the yielding and ultimate rotation capacity of the plastic hinge of the steel-jacketed RC column, respectively, and $\theta_{D,a}$ is the rotation demand on the plastic hinge of the steel-jacketed RC column obtained by nonlinear dynamic analysis using a number of measured ground motions.

The fragility defined in Equation (1) is estimated using the Monte Carlo simulation (MCS) and determined from the ratio of the number of times $g_i < 0$ (i.e. demand exceeds capacity), where $i = 1, 2, 3, 4, 5$, to the total number of MCSs. If the relationship between the fragility $P[g_i < 0 | \Gamma = \gamma]$ and seismic intensity γ is modelled as a lognormal distribution based on the fragility analysis using MCSs, then the fragility curve is represented by

$$F(\gamma) = \Phi \left[\frac{\ln(\gamma/\lambda)}{\zeta} \right] \qquad (7)$$

where λ and ζ are the median and log-standard deviation of the fragility curve and $\Phi[\cdot]$ is the standard-normal distribution function. The parameters used to determine the lognormal distribution (median λ and log-standard deviation ζ) are estimated by the method of maximum likelihood (Shinozuka, Feng, Lee, & Naganuma, 2000).

The difference between $P[g_i < 0 | \Gamma = \gamma]$ provided by MSC and $F(\gamma)$ in Equation (7) is negligible at each seismic intensity γ. The parameters of lognormal distribution for the seismic fragility are almost the same if the number of samples in MCS is > 1000. Therefore, the number of samples in MCS is set to 1000.

4.2 Dynamic response analysis of the Shinkansen viaducts

The side columns of the single-story RC moment-resisting frame R15 of the No. 5 Inohana viaducts, as shown in Figure 14, are analysed in this study (hereafter referred to as the 'frame pier'). The seismic behaviour of the frame pier is captured in the transverse direction. Figure 14 presents a two-dimensional (2D) response model to develop the fragility curves before and after the column retrofitting with a steel jacket. Within the assumption of a stiff soil, the bases of the frame piers are assumed to be fully fixed. If the frame pier is constructed on the soft soil, the effect of the dynamic interaction between the foundation and the soil on the dynamic behaviour and fragility estimation needs to be investigated. The column is modelled as an elastic zone with a pair of plastic hinges at each end of the column. Each plastic hinge is modelled as a nonlinear rotational spring, as shown in Figure 14. In Figure 15, the relationship between the moment and the rotation of the plastic hinge is represented by the two lines, and the stiffness beyond the yielding point is assumed to be zero. The rotation at the yielding point is

$$\theta_y' = \left(\frac{\chi_4 M_u}{M_y} \right) \theta_y \qquad (8)$$

where M_u is the maximum moment of the plastic hinge of the as-built or retrofitted column, M_y and θ_y are the moment and rotation of the as-built or retrofitted column when the strain in the extreme tensile rebar reaches its yield point, respectively, and χ_4 are the lognormal random variables representing the model uncertainties associated with the estimation of M_u.

As described previously, steel jacketing has been used as a retrofit measure to enhance the flexural ductility and shear strength in RC columns. Experiments on steel jacketed bridge columns were conducted in several studies (e.g. Chai, Priestley, & Seible, 1991; Priestley, Seible, Xiao, & Verma, 1994). In an earthquake prone region, steel jacketing is the most common column retrofit. To estimate the improvement of seismic capacity of circular columns of a common bridge type in the USA due to steel jacketing, material models for the concrete fibres were modified (Kim & Shinozuka, 2004; Padgett & DesRoches,

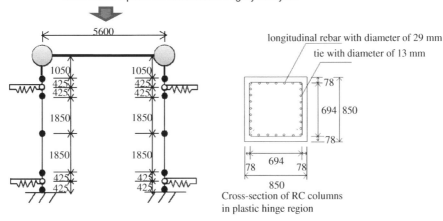

Figure 14.   Two-dimensional modelling for a single-story RC moment-resisting frame pier (photo taken by the first author) (unit: mm).

2008). In this study, the dynamic behaviour of steel-jacketed columns with square cross section of Shinkansen viaduct is evaluated. In addition, details of steel jacketing for Shinkansen viaduct are different from those for bridges in the USA. Especially, the typical jacket thickness for Shinkansen viaduct retrofit is 6 mm, whereas that for bridge retrofit in the USA is 16 mm (Padgett, 2007). In this study, the seismic capacity estimation method for steel-jacketed RC column of Shinkansen viaduct is established using the experimental results of square columns with a thinner jacket thickness.

The rotation capacity $\theta_{u,a}$ proposed by Tamai, Takiguchi, and Hattori (1996) is as follows:

$$\theta_{u,a} = \frac{4.66t}{b} + 0.024 \qquad (9)$$

where t is the thickness of the steel jacket and b the cross-sectional width of the as-built column. Tamai, Takiguchi, and Hattori (1996) reported that the lateral force acting on a steel-jacketed RC column decreases sharply beyond the rotation capacity $\theta_{u,a}$.

Based on the experimental results of steel-jacketed columns subjected to cyclic loading, Tamai and Sato (1998) reported that the hysteretic model of these columns is based on that proposed by Takeda, Sozen, and Nielse (1970). In this paper, the unloading stiffness k_r is assumed (Tamai & Sato, 1998)

$$k_r = k_y \left(\frac{\theta_{max}}{\chi_2 \theta'_y} \right)^{-0.5} \qquad (10)$$

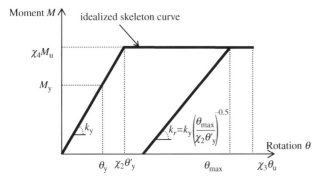

Figure 15.   Moment and rotation relationship of a plastic hinge for an RC column with a steel jacket.

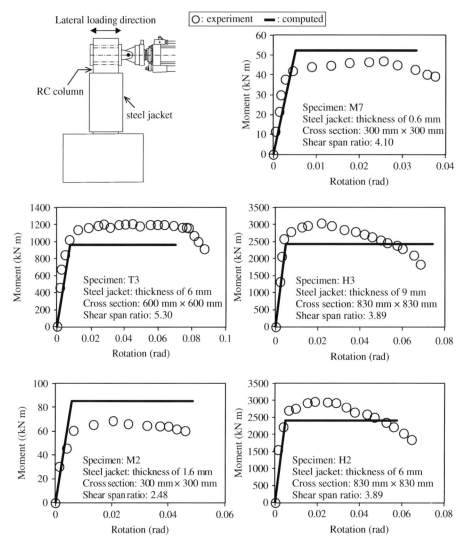

Figure 16. Comparisons of skeleton curves between tested and computed results of steel-jacketed RC column subjected to cyclic loading.

where k_y is the yielding stiffness and θ_{max} is the maximum curvature attained in the direction of loading.

The statistics of χ_2, χ_3 and χ_4 can be derived by the ratio of the experimental results of steel-jacketed RC columns subjected to cyclic loading to computed values. Experimental results reported by Tamai and Sato (1998) are used in this paper. The number of specimens used was 12, the thickness of the steel jacket ranged from 1.6 to 9.0 mm, the cross section of the specimens ranged from 300 mm × 300 mm to 860 mm × 860 mm and the shear span ratio defined as the ratio of shear span to the effective depth of cross section ranges from 2.48 to 5.30. Figure 16 illustrates an example of envelope curves between experiment and computed results of steel-jacketed RC columns. As experimental results reported by Tamai and Sato (1998) did not include the hysteresis loops of steel-jacketed RC columns subjected to cyclic loading, only envelope curves of experiment results are

compared with those of computed results as shown in Figure 16.

Figure 17 depicts the comparison between the computed and experimental results of $\theta'_{y,a}$, M_u and $\theta_{u,a}$. The accuracy of estimation of $\theta_{u,a}$ for steel-jacketed RC columns is almost the same as that for typical RC columns which exhibit flexural failure (Akiyama, Matsuzaki, Dang, & Suzuki, 2012). Although the estimation method of $\theta'_{y,a}$ needs to be improved to reduce the coefficient of variation, dispersion in the yielding point on the moment and rotation relationship do not affect the nonlinear response of frame pier. It could be expected that the effect of $\theta'_{y,a}$ on $\theta_{D,a}$ is not significant in the fragility estimation. As the steel jacket was not anchored to the other components, the maximum moment M_u of the retrofitted RC columns was calculated without considering the existence of the steel jacket. The steel-jacketing increases not only the ductility and shear capacity, but also the flexural strength

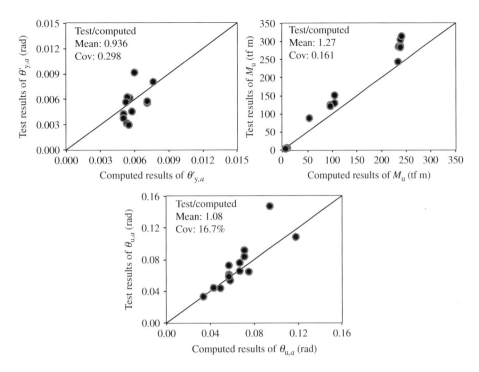

Figure 17. Comparison of computed and experiment results.

(Yashinsky, 1998). The experimental maximum moments of 10 steel-jacketed RC columns are underestimated except two RC columns with smaller cross section (i.e. specimens M7 and M2). In the fragility analysis, χ_2, χ_3 and χ_4 are assumed to be lognormal variables. The statistics listed in Table 1 are used as the model uncertainties.

Although χ_2 and χ_3 are evaluated based on the comparison of experimental and computed results of steel-jacketed RC columns subjected to cyclic loading, the statistics of model uncertainties associated with the evaluation of $\theta'_{y,b}$ and $\theta_{u,b}$ are assumed to be the same as those associated with the evaluation of $\theta'_{y,a}$ and $\theta_{u,a}$. The model uncertainties associated with the capacity evaluation of as-built RC column can be expected to be smaller. However, as the variability in demand is very large in comparison with the variability in structural capacity (Li & Ellingwood, 2009), this assumption may not affect the fragility estimation. The uncertainties associated with material strength are considered when M_y, M_u, θ_y and θ_u are calculated. The parameters associated with material strength are also indicated in Table 1.

When estimating the fragility of as-built viaducts, 1000 samples for each of the random variables are created in MCS from their probabilistic characteristics shown in Table 1. Failure modes of RC columns of as-built viaducts are determined as

$$\eta = \frac{\chi_1(V_c + V_s)a}{\chi_4 M_u}, \tag{11}$$

where a is the shear span of the RC column (i.e. half the column height). Because as-built RC columns with $\eta < 1.0$ exhibit shear failure, fragility can be estimated using the performance function g_1. If η is > 1.0, as-built RC columns exhibit flexural failure or shear failure after flexural yielding. Fragility can be estimated using the performance functions g_2 and g_3. Figure 18 shows the flowchart for estimating the fragility of the as-built viaduct. As this paper aims to investigate the effect of seismic retrofit on the improvement in fragility, shear failure after flexural yielding is not considered when estimating $\theta_{u,b}$ in Equation (4) for as-built RC column. To estimate the fragility of as-built viaducts designed according to an old specification, shear–flexure inter-action model is needed in the seismic response assessment of viaducts (e.g. Xu & Zhang, 2011).

Fifty-four seismic stations in Tohoku region of Japan including IWT011 and IWT014, as shown in Figure 7, were selected for the nonlinear dynamic analysis performed on the frame pier. At each seismic station, three ground motions (i.e. north–south, east–west and up–down orientations) are measured. As the effect of the ground motions of vertical component on the variation in axial force on RC columns could be ignored, only the horizontal component of ground motions measured during the 2003 Sanriku-Minami earthquake, 2008 Iwate-Miyagi Nairiku earthquake and 2011 Great East Japan earthquake was used to develop the fragility curves. Distance R to the causative fault is different among seismic stations shown in Figure 7. However, Bazzurro and Cornell (1994a,

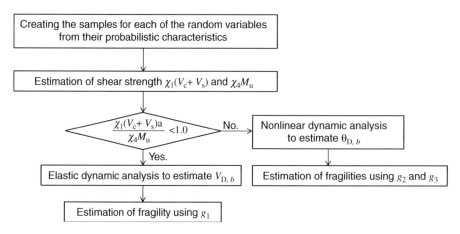

Figure 18.   Flowchart for estimating the fragility of as-built viaduct for a given level of seismic intensity Γ.

1994b) reported that nonlinear response of structure shows no systematic dependence on R.

In the MCS, one horizontal component from 108 ground motion records during the 2003 Sanriku-Minami earthquake, 2008 Iwate-Miyagi Nairiku earthquake or 2011 Great East Japan earthquake is randomly selected. The seismic intensity Γ in the fragility analysis is the PGA or $S_a(T_1, \xi)$, where $S_a(T_1, \xi)$ is the ξ-damped spectra acceleration at the fundamental frequency T_1 (≈ 0.43 s) of the analysed frame pier for the considered earthquake record. Each ground motion is amplified such that the PGA or $S_a(T_1, \xi)$ is equal to a specified seismic intensity. Figure 19 presents the response acceleration of 108 ground motions that are amplified such that PGA $= 1000$ gal or $S_a(T_1, \xi) = 1000$ gal. There is large variability in response accelerations of ground motions measured during three earthquakes. Structures with shorter natural period tend to have larger lateral forces due to these ground motions.

4.3   Fragility curves of the as-built and retrofitted Shinkansen viaducts

In the MCS analyses, g_i from Equations (2) to (6) is calculated using the maximum value of the peak shear forces of the RC columns and the ductility demands sustained by the plastic hinges. Figure 20 shows the fragility curves illustrating the vulnerability of the as-built and retrofitted frame pier over a range of the seismic intensities PGA or $S_a(T_1, \xi)$ using 108 ground motions measured during the 2011 Great East Japan earthquake. It shows the median λ in Equation (7) for each performance function.

Because RC columns of the as-built frame pier do not have sufficient shear reinforcement and the shear strength is less than the flexural strength, the fragility curve associated with g_1 is located to the left of that associated with g_2. From Figure 5, it might be roughly estimated that

the as-built frame pier of No. 5 Inohana viaducts was subjected to ground motion with PGA ≈ 200 gal during the 2011 Great East Japan earthquake. If the frame pier had not been retrofitted, there is a high probability that it would have suffered severe damage due to shear.

Comparing the fragility curve associated with the performance function g_1 for the as-built frame pier to that associated with g_5 for the retrofitted frame pier, the improvement to seismic vulnerability from steel jacketing can be quantified when enhanced fragility curves are plotted as a function of PGA or $S_a(T_1, \xi)$. For example, the median λ associated with g_5 is about 10 times larger than that associated with g_1 as shown in Figure 20 when response acceleration is selected as seismic intensity. Seismic retrofit by means of steel jacketing is very significant to satisfy the non-collapse requirement for the viaducts.

The enhanced curves shift to the right relative to those associated with the frame piers before retrofit. Kim and Shinozuka (2004) presented the fragility curve of a typical California-type multiframe concrete bridge based on a 2D response analysis and an MCS. They developed the fragility curves of a bridge with steel jacketing, indicating the states of damage as none, slight, moderate, extensive and complete collapse. It is interesting to note that the fragility curve associated with g_5 in Figure 20 is similar to that associated with the complete damage state presented by Kim and Shinozuka (2004) as shown in Figure 20, even though the details of steel jacketing, bridge types and ground motions used were different.

Figures 21 and 22 depict the fragility curves of as-built and retrofitted frame pier using 108 ground motions measured during the 2003 Sanriku-Minami earthquake and 2008 Iwate-Miyagi Nairiku earthquake, respectively. Despite the fact that ground motions are amplified such that Γ(PGA) $= \gamma$, response accelerations at the fundamental natural period of frame pier are very different as shown in Figure 19. Fragility

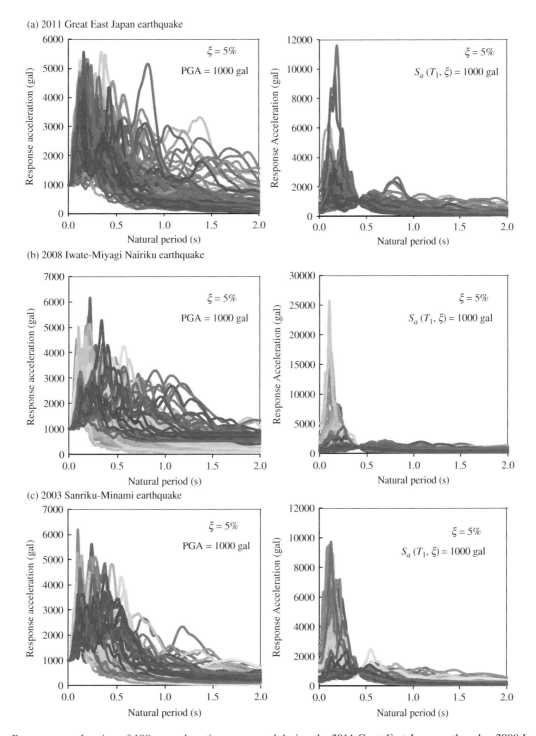

(a) 2011 Great East Japan earthquake

(b) 2008 Iwate-Miyagi Nairiku earthquake

(c) 2003 Sanriku-Minami earthquake

Figure 19. Response acceleration of 108 ground motions measured during the 2011 Great East Japan earthquake, 2008 Iwate-Miyagi Nairiku earthquake and 2003 Sanriku Minami earthquake based on the data from the National Research Institute for Earth Science and Disaster Prevention (2012).

curves using PGA as seismic intensity depend strongly on earthquake. When $S_a(T_1, \xi)$ is used as seismic intensity, differences in medians of fragility curves associated with the performance functions g_1, g_2 and g_4 as shown in Figures 20–22 are not so large, compared with using PGA as seismic intensity. Although

nonlinear response depends on phase characteristics of ground motion and fragility curves associated with the ultimate rotation (i.e. g_3 and g_5) are different among earthquakes as shown in Figures 20–22, $S_a(T_1, \xi)$ is a more appropriate indicator for fragility curves of viaducts than the PGA.

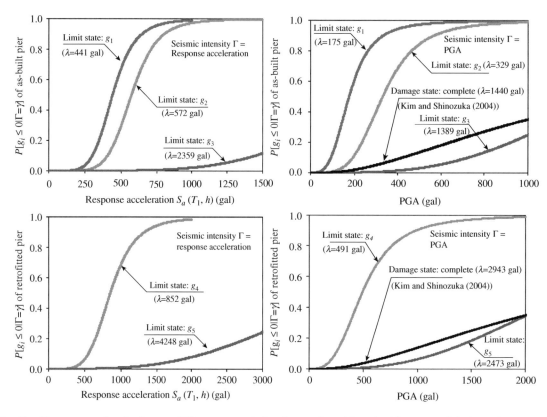

Figure 20. Fragility curves of a single-story RC moment-resisting frame pier with and without a steel jacket using ground motions measured during the 2011 Great East Japan earthquake.

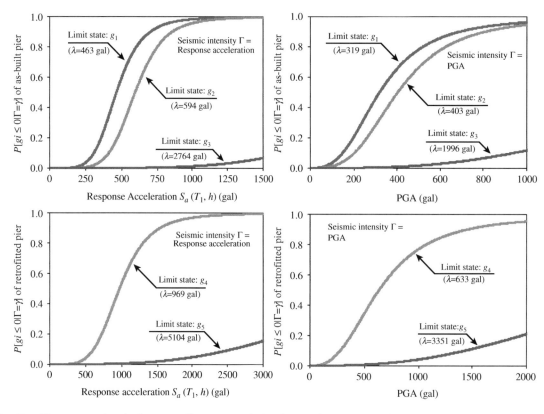

Figure 21. Fragility curves of a single-story RC moment-resisting frame pier with and without a steel jacket using ground motions measured during the 2003 Sanriku-Minami earthquake.

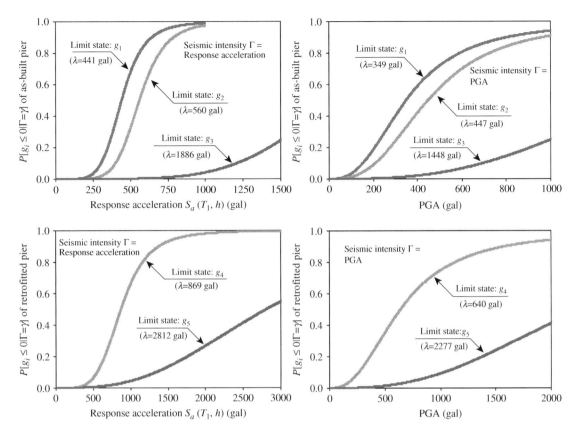

Figure 22. Fragility curves of a single-story RC moment-resisting frame pier with and without a steel jacket using ground motions measured during 2008 Iwate-Miyagi Nairiku earthquake.

5. Conclusions

The effectiveness of the seismic retrofit to the Shinkansen viaduct was demonstrated during the 2011 Great East Japan earthquake. The RC columns retrofitted by means of steel jacketing experienced no damage during this earthquake. Some of the Shinkansen viaducts were not retrofitted in the first seismic retrofit programme because they were identified as having sufficient shear reinforcement to prevent brittle failure. They were supposed to be retrofitted in the second seismic retrofit programme. Some of the non-retrofitted viaducts were severely damaged during the 2011 Great East Japan earthquake. Satisfactory seismic response of RC structures requires non-occurrence of brittle failure modes. It is important that rapid progress is made in the seismic retrofit programme for the Shinkansen viaducts to assure their long-term safe performance.

The viaduct with retrofitted RC columns sustained no damage during the 2011 Great East Japan earthquake. However, it should be noted that the ground motions at the site of the Shinkansen viaducts due to the 2011 Great East Japan earthquake were not strong compared with those prescribed by the current seismic design code. To evaluate the likelihood of damage of the as-built and retrofitted viaduct over a range of potential earthquake ground motion intensities, seismic fragility curves were developed. Lognormal distribution functions were derived by a nonlinear dynamic analysis and using MCS. The simulated fragility curves of the retrofitted viaduct show a great improvement in seismic performance compared with those of the as-built viaduct. The median of the fragility curve associated with the ultimate limit state for retrofitted viaduct is at least five times larger than that for as-built viaduct depending on the characteristics of ground motion.

It is necessary to establish a seismic retrofit strategy to improve the seismic performance of existing viaducts in order to minimise the difference between their seismic performance and the performance required in the latest seismic specifications. Based on the comparison of risk among existing viaducts, the high-priority viaducts should be identified. In addition, when determining priorities for seismic retrofit, other performance indicators should be investigated such as resilience (Bocchini & Frangopol, 2012a, 2012b; Decò, Bocchini, & Frangopol, 2013; Frangopol & Bocchini, 2011, 2012) and sustainability (Dong, Frangopol, & Saydam, 2013), and the effect of the aggressive environment (Akiyama & Frangopol, 2013; Akiyama, Frangopol, & Matsuzaki, 2011; Akiyama, Frangopol, & Suzuki, 2012; Decò & Frangopol, 2013; Frangopol & Akiyama, 2011) using a life-cycle performance optimisation context (Frangopol, 2011) under tsunami (Akiyama, Frangopol, Arai, & Koshimura, 2013) and multi-hazards (Decò & Frangopol, 2011; Decò & Frangopol, 2013; Zhu & Frangopol, 2013).

Acknowledgements

The first author acknowledges the support provided by the J-RAPID programme of the Japan Science and Technology Agency (project leader: Prof. Kazuhiko Kawashima, Tokyo Institute of Technology). The author thank the National Research Institute for Earth Science and Disaster Prevention, Japan, for providing the strong-motion data recorded by the K-NET. The opinions and conclusions presented in this paper are those of the authors and do not necessarily reflect the views of the sponsoring organisation.

References

Akiyama, M., & Frangopol, D.M. (2012). Lessons from the 2011 Great East Japan Earthquake: Emphasis on life-cycle structural performance. *Proceedings of the third international symposium on life-cycle civil engineering*, Vienna, Austria.

Akiyama, M., & Frangopol, D.M. (in press). Long-term seismic performance of RC structures in an aggressive environment: Emphasis on bridge piers. *Structure and Infrastructure Engineering*. doi: 10.1080/15732479.2012.761246

Akiyama, M., Frangopol, D.M., Arai, M., & Koshimura, S. (2013). Reliability of bridges under tsunami hazards: Emphasis on the 2011 Tohoku-Oki Earthquake. *Earthquake Spectra, 29*(S1), S295–S314.

Akiyama, M., Frangopol, D.M., & Matsuzaki, H. (2011). Life-cycle reliability of RC bridge piers under seismic and airborne chloride hazards. *Earthquake Engineering and Structural Dynamics, 40*(15), 1671–1687.

Akiyama, M., Frangopol, D.M., & Suzuki, M. (2012). Integration of the effects of airborne chlorides into reliability-based durability design of reinforced concrete structures in a marine environment. *Structure and Infrastructure Engineering, 8*(2), 125–134.

Akiyama, M., Matsuzaki, M., Dang, D.H., & Suzuki, M. (2012). Reliability-based capacity design for reinforced concrete bridge structures. *Structure and Infrastructure Engineering, 8*(12), 1096–1107.

Bazzurro, P., & Cornell, C.A. (1994a). Seismic hazard analysis of nonlinear structures. I: Methodology. *ASCE Journal of Structural Engineering, 120*, 3320–3344.

Bazzurro, P., & Cornell, C.A. (1994b). Seismic hazard analysis of nonlinear structures. II: Applications. *ASCE Journal of Structural Engineering, 120*, 3345–3365.

Bocchini, P., & Frangopol, D.M. (2012a). Optimal resilience- and cost-based post-disaster intervention prioritization for bridges along a highway segment. *ASCE Journal of Bridge Engineering, 17*(1), 117–129.

Bocchini, P., & Frangopol, D.M. (2012b). Restoration of bridge networks after an earthquake: Multi-criteria intervention optimisation. *Earthquake Spectra, 28*(2), 426–455.

Chai, Y.H., Priestley, M.J.N., & Seible, F. (1991). Seismic retrofit of circular bridge columns for enhanced flexural performance. *ACI Structural Journal, 88*, 572–584.

Decò, A., Bocchini, P., & Frangopol, D.M. (in press). A probabilistic approach for the prediction of seismic resilience of bridges. *Earthquake Engineering and Structural Dynamics*. doi: 10.1002/eqe.2282

Decò, A., & Frangopol, D.M. (2011). Risk assessment of highway bridges under multiple hazards. *Journal of Risk Research, 14*(9), 1057–1089.

Decò, A., & Frangopol, D.M. (2013). Life-cycle risk assessment of spatially distributed aging bridges under seismic and traffic hazards. *Earthquake Spectra, 29*(1), 1–27.

Dong, Y., Frangopol, D.M., & Saydam, D. (in press). Time-variant sustainability assessment of seismically vulnerable bridges subjected to multiple hazards. *Earthquake Engineering and Structural Dynamics*. doi: 10.1002/eqe.2281

Frangopol, D.M. (2011). Life-cycle performance, management, and optimization of structural systems under uncertainty: Accomplishments and challenges. *Structure and Infrastructure Engineering, 7*(6), 389–413.

Frangopol, D.M., & Akiyama, M. (2011). Lifetime seismic reliability analysis of corroded reinforced concrete bridge piers. In M. Papadrakakis, M. Fragiadakis, & N.D. Lagaros, (Eds.), *Computational Methods in Earthquake Engineering, Vol. 21 in Computational Methods in Applied Sciences* (pp. 527–537). Springer.

Frangopol, D.M., & Bocchini, P. (2011). Resilience as optimization criterion for the bridge retrofit of a transportation network subject to earthquake. *Proceedings of the ASCE structures congress* (pp. 2044–2055). Las Vegas, NV.

Frangopol, D.M., & Bocchini, P. (2012). Bridge network performance, maintenance, and optimization under uncertainty: Accomplishments and challenges. *Structure and Infrastructure Engineering, 8*(4), 341–356.

Japan Road Association. (1990). *Design specification for highway bridges. Part V: Seismic design*. Tokyo, Japan: Maruzen.

Japan Society of Civil Engineers. (2004). Damage analysis of concrete structures due to the earthquakes in 2003. Concrete Library, 114, Tokyo (in Japanese).

Karim, K.R., & Yamazaki, F. (2003). A simplified method of constructing fragility curves for highway bridges. *Earthquake Engineering and Structural Dynamics, 32*, 1603–1626.

Kim, S.-H., & Shinozuka, M. (2004). Development of fragility curves of bridges retrofitted by column jacketing. *Probabilistic Engineering Mechanics, 19*, 105–112.

Kawashima, K. (2000). Seismic design and retrofit of bridges. *Proceedings of 12th World Conference on Earthquake Engineering (WCEE)*, Keynote paper. Auckland, New Zealand.

Kawashima, K. (2012). Damage of bridges due to the 2011 Great East Japan earthquake. *Proceedings of the international symposium on engineering lessons learned from the 2011 Great East Japan Earthquake* (pp. 82–101). Tokyo, Japan.

Kawashima, K., Kosa, K., Takahashi, Y., Akiyama, M., Nishioka, T., Watanabe, G., Koga, H., & Matsuzaki, H. (2011). Damages of Bridges during 2011 Great East Japan Earthquake. *Proceedings of 43rd joint meeting, US–Japan panel on wind and seismic effects*, UJNR, Tsukuba Science City, Japan.

Li, Q., & Ellingwood, B.R. (2007). Performance evaluation and damage assessment of steel frame buildings under main-shock-aftershock earthquake sequences. *Earthquake Engineering and Structural Dynamics, 36*, 405–427.

Li, Y., & Ellingwood, B.R. (2009). Framework for multihazard risk assessment and mitigation for wood-frame residential construction. *Journal of Structural Engineering, 135*, 159–168.

National Research Institute for Earth Science and Disaster Prevention. (2012). Strong-motion seismograph networks (K-net and Kik-net). Retrieved from www.kyoshin.bosai.go.jp.

Padgett, J.E. (2007). *Seismic vulnerability assessment of retrofitted bridges using probabilistic methods.* (Ph.D. thesis). Atlanta, GA: Georgia Institute of Technology.

Padgett, J.E., & DesRoches, R. (2007). Sensitivity of seismic response and fragility to parameter uncertainty. *ASCE Journal of Structural Engineering, 133*, 1710–1718.

Padgett, J.E., & DesRoches, R. (2008). Methodology for the development of analytical fragility curves for retrofitted bridges. *Earthquake Engineering and Structural Dynamics, 37*, 1157–1174.

Priestley, M.J.N., Seible, F., Xiao, Y., & Verma, R. (1994). Steel jacket retrofitting of reinforced concrete bridge columns for enhanced shear strength. Part 2: Test results and comparison with theory. *ACI Structural Journal, 91*, 537–551.

Railway Technical Research Institute (1992). *Seismic design standards for railway structures.* Tokyo, Japan: Maruzen.

Shinozuka, M., Feng, M.Q., Kim, H.-K., & Kim, S.-H. (2000). Nonlinear static procedure for fragility curve development. *ASCE Journal of Engineering Mechanics, 126*, 1287–1295.

Shinozuka, M., Feng, M.Q., Lee, J., & Naganuma, T. (2000). Statistical analysis of fragility curves. *ASCE Journal of Engineering Mechanics, 126*, 1224–1231.

Takeda, T., Sozen, M., & Nielse, N. (1970). Reinforced concrete response to simulated earthquake. *ASCE Journal of Structural Engineering, 96*, 2557–2573.

Tamai, S., Takiguchi, M., & Hattori, N. (1996). Ductility capacity of retrofitted RC columns. *Proceedings of the Japan Concrete Institute, 20*(3), 1111–1116 (in Japanese).

Tamai, S., & Sato, T. (1998). Ductility of steel jacketed RC columns. *RTRI Report, 12*(9), 39–44 (in Japanese).

Xu, S.-Y., & Zhang, J. (2011). Hysteretic shear–flexure interaction model of reinforced concrete columns for seismic response assessment of bridges. *Earthquake Engineering and Structural Dynamics, 40*, 315–337.

Yashinsky, M. (1998). Performance of bridge seismic retrofits during Northridge earthquake. *ASCE Journal of Bridge Engineering, 3*, 1–14.

Zhu, B., & Frangopol, D.M. (2013). Risk-based approach for optimum maintenance of bridges under traffic and earthquake loads. *ASCE Journal of Structural Engineering, 139*(3), 422–434.

Probabilistic assessment of an interdependent healthcare–bridge network system under seismic hazard

You Dong and Dan M. Frangopol

ABSTRACT

Strong earthquakes can destroy infrastructure systems and cause injuries and/or fatalities. Therefore, it is important to investigate seismic performance of interdependent infrastructure systems and guarantee their abilities to cope with earthquakes. This paper presents an integrated probabilistic framework for the healthcare–bridge network system performance analysis considering spatial seismic hazard, vulnerability of bridges and links in the network, and damage condition of a hospital at component and system levels. The system level performance is evaluated considering travel and waiting time based on the damage conditions of the components. The effects of correlation among the seismic intensities at different locations are investigated. Additionally, the correlations associated with damage of the investigated structures are also incorporated within the probabilistic assessment process. The conditional seismic performance of the hospital given the damage conditions of the bridge network and the effect of bridge retrofit actions are also investigated. The approach is illustrated on a healthcare system located near a bridge network in Alameda, California.

1. Introduction

Strong earthquakes can destroy infrastructure systems and cause injuries and/or fatalities. Therefore, it is important to investigate seismic performance of healthcare systems to guarantee immediate medical treatment after earthquakes. The 1971 San Fernando, California, earthquake caused a severe damage to hospitals associated with about 53 million US$ loss in building damage and more than 1 million US$ loss in equipment damage (Murphy, 1973). The World Health Organization (WHO, 2007) stated that healthcare systems "must be physically resilient and able to remain operational and continue providing vital health services" after disasters. Thus, healthcare systems need to be resilient enough to cope with earthquakes and to provide timely medical treatment. In this paper, the seismic performance assessment of a healthcare system located near a bridge network is investigated considering both component and system performance levels.

The assessment of healthcare–bridge network system performance depends on the seismic vulnerability of bridges located in a bridge network and hospital, as well as on the ground motion intensity. After a destructive earthquake, the functionality of a highway network can be affected significantly; this, in turn, may lead to hinder the emergency management. Additional travel time would result due to the damaged bridges and links; consequently, injured persons may not receive treatment in time. Thus, it is important to account for the effects of damage condition associated with highway bridge network on the healthcare system performance. In this paper, the extra travel time of injured persons through the damaged bridge network to a hospital under the seismic hazard is investigated.

Myrtle, Masri, Nigbor, and Caffrey (2005) carried out a series of surveys on performance of hospitals during several earthquakes to identify the important components; Yavari, Chang, and Elwood (2010) investigated performance levels for interacting components (i.e. structural, nonstructural, lifeline and personnel) using data from past earthquakes; Achour, Miyajima, Kitaura, and Price (2011) investigated the physical damage of structural and nonstructural components of an hospital under seismic hazard; and Cimellaro, Reinhorn, and Bruneau (2011) introduced a model to describe the hospital performance under earthquake considering waiting time. However, the damage conditions associated with bridge networks have not been incorporated within the healthcare system performance assessment process. Additionally, the correlation effects have also not been addressed in these studies.

After an earthquake, it is common to experience a sudden increase in the number of patients for a period of time, which in turn can bring delay in treating them. The estimation of hospital capacity after an extreme event is of vital importance to determine the waiting time of the injured persons. Hospital functionality may be disrupted by damage associated with structural and nonstructural components or medical equipment. A proposed approach considering both structural and nonstructural components (e.g. medical equipment) is presented in this paper to investigate the hospital performance under a given seismic scenario. The relationship between structural and nonstructural seismic demands (e.g. peak inter story drift ratio and peak floor acceleration) is considered. Additionally, the correlations among the damages of structural and nonstructural components are also considered in the hospital functionality assessment process. Finally, the effects of correlation on the healthcare–bridge network system performance at a system level are investigated.

The performance of interdependent infrastructure systems under seismic hazard could be incorporated within the framework of life-cycle analysis of structural systems under extreme events. By considering the probability of occurrence of seismic hazard and structural deteriorations, the seismic performance of interdependent systems in a life-cycle context could be investigated. This will help to (a) promote the concept of interdependency in life-cycle design and assessment of structural systems under extreme events and (b) improve the risk-informed decision-making process for maintenance actions on interdependent infrastructures. It is clear that a life-cycle framework considering interdependent systems under seismic hazard is more realistic than the approach based on independent systems, since it involves the consideration of complex interaction processes among deteriorating structures during their lifetime. Overall, this study provides a novel approach to compute the seismic performance of interdependent healthcare–bridge network systems.

This paper aims to assess probabilistically an interdependent healthcare–bridge network system under seismic hazard and to aid the emergency preparedness to cope with the sudden increase of patients. The damage conditions of the bridges, links and hospital are considered in the overall system performance assessment. Fragility curves are employed to identify the components vulnerability under seismic hazard. The effects of disruption associated with transportation system on the emergency management are investigated. Additionally, the correlations among structural damages and the effect of bridge retrofit actions are considered in the assessment process. The system level performance indicators are expressed in terms of the extra travel and waiting time of the injured persons from the damaged region to the hospital given the occurrence of the earthquake. The approach is illustrated on a healthcare system located near a bridge network in Alameda, California.

2. Earthquake scenarios

The first step in seismic performance assessment of a healthcare system located near a bridge network is to identify the seismic scenarios at the location of the system. The seismic scenarios associated with an earthquake fault are introduced herein. The earthquake rupture is given as a characteristic magnitude distribution, modelled as a Gaussian distribution using the mean, a coefficient of variation of .12, and a truncation at ±2 standard deviations of magnitude above and below the mean (USGS, 2003). Generally, the mean magnitude M associated with an earthquake rupture can be computed as (Hanks & Bakun, 2002):

$$M = \begin{cases} 3.98 + \log_{10}(A_F) & A_F \leq 468 \text{ km}^2 \\ 3.09 + \frac{4}{3}\log_{10}(A_F) & A_F > 468 \text{ km}^2 \end{cases} \quad (1)$$

where A_F is the total area of the fault segment (km^2). Given the seismic scenario, the following step is to predict the ground motion intensity at the location of the structure. The attenuation relation, expressed in a logarithmic form, is used to predict the ground motion intensity at a certain site (Campbell & Bozorgnia, 2008).

The seismic intensities at different sites caused by the same earthquake are correlated. It is necessary to consider the spatial correlation of ground motion intensities within the seismic performance assessment of interdependent hospital – bridge network system. Several studies (e.g. Jayaram & Baker, 2009; Wang & Takada, 2005) revealed that the peak ground acceleration (PGA) associated with a given seismic scenario at different sites is spatially correlated and this correlation is higher for closer sites. Accordingly, the correlation among the seismic intensities at different locations is modelled as an exponential decay function (Wang & Takada, 2005):

$$\rho\left(IM_i, IM_j\right) = \exp\left(-h/l_{co}\right) \quad (2)$$

where h is the distance between two sites i and j (km); IM is the ground motion intensity measure (e.g. PGA); and l_{co} is the correlation length (km). The value of l_{co} can be estimated based on the statistical analysis of the past earthquake data. Given the distribution types associated with the ground motion intensities and correlation coefficients in Equation (2), the correlated ground motion intensities could be generated using straightforward numerical procedures, such as Monte Carlo simulation. In this paper, the effects of ground motion correlation are accounted for within the seismic assessment of spatially distributed interdependent healthcare–bridge network system.

3. Bridge, link and hospital seismic damage assessment

3.1. Bridge seismic vulnerability

A transportation network is composed of nodes, links and bridges. Nodes describe the locations of highway intersections, while links represent the highway segments connecting two nodes. Generally, bridges are considered to be the most vulnerable components in a transportation network (Dong, Frangopol, & Saydam, 2014; Lee & Kiremidjian, 2007; Liu & Frangopol, 2006). Fragility curves are commonly used to quantify structural performance under seismic hazard and are defined by the exceedance probability of a damage state under a given ground motion intensity (Dong, Frangopol, & Saydam, 2013; Mander, 1999). The fragility curve of a structural system under a specific ground motion intensity can be expressed as (Mander, 1999):

$$P_{S \geq DS_i \mid IM} = \Phi\left(\frac{\ln(IM) - \ln(m_i)}{\beta_i}\right) \quad (3)$$

where $\Phi(.)$ is the standard normal cumulative distribution function; β_i is the standard deviation of the logarithm of ground motion intensity associated with damage state i; and m_i is the median value of the intensity measure of damage state i. For the seismic performance assessment of a transportation network, it is necessary to develop the fragilities of all bridges. The correlations

275

among the seismic performance of bridges in terms of fragility have also to be considered. Specifically, bridges in a transportation network can be classified into different subgroups to characterise their fragilities considering structural characteristics, such as number of spans and material types (e.g. steel, concrete). The median values of ground motion intensities with different damage states of highway bridges have been investigated by Mander (1999) and Shinozuka, Feng, Kim, Uzawa, and Ueda (2001). For a given ground motion intensity, the probability $P_{DSi|IM}$ of a bridge being in a damage state i is given by the difference between the probabilities of exceedance of damage states i and $i+1$.

The expected bridge damage index BDI is obtained by multiplying the probability of a bridge being in each damage state with the corresponding bridge damage state index $BDDI$. Accordingly, the BDI of bridge k under a certain ground motion intensity is (Shiraki et al., 2007):

$$BDI_k = \sum_{i=1}^{n_{BDS}} BDDI_i \cdot P_{DSki|IM} \qquad (4)$$

where $BDDI_i$ is the damage index associated with the damage state i; n_{BDS} is the number of damage states; and $P_{DSki|IM}$ is the probability of a bridge k being in a damage state i. In this paper, the considered damage states for a highway bridge are none, slight, moderate, major and complete.

3.2. Link seismic damage assessment

A link is considered as an element connecting the nodes of a network. The performance of network links is related to individual bridge located on the link. The performance of the link after an earthquake can be expressed in terms of link damage index LDI, which depends on the $BDIs$ of the bridges on the link, as (Chang, Shinozuka, & Moore, 2000):

$$LDI = \sqrt{\sum_{j=1}^{n_b} \left(BDI_j \right)^2} \qquad (5)$$

where n_b is the number of the bridges located on the link. Traffic flow in the damaged link can be different and speed limits might be reduced. The level of link traffic flow capacity and speed for a damaged link depends on damage states of the link. The intact state LDS_1, slight LDS_2, moderate LDS_3 and major damage LDS_4 states represent $LDI \leq .5$, $.5 < LDI \leq 1.0$, $1.0 < LDI \leq 1.5$ and $LDI > 1.5$, respectively (Chang et al., 2000). The correlation coefficient $\rho(BDDI_{Bk}, BDDI_{Bj})$ between the damage indices of bridge k and j is considered in the assessment process.

3.3. Hospital functionality assessment

The effects of damage states associated with structural and nonstructural components on the damage performance of a hospital are introduced in this section. The damage assessment of the hospital should determine the capacity of how many patients it can handle to provide timely treatment to the injured people (Cimellaro et al., 2011). The functionality of a hospital could be assessed based on its components (e.g. structural, nonstructural) (Yavari et al., 2010). Building components can be categorised into structural, nonstructural and content (Mitrani-Reiser, 2007). The

performance associated with different components of a hospital system should be investigated using different seismic demand indicators (e.g. story drift, floor acceleration). As beams and columns can provide support for the structure's weight and the strength needed to resist lateral forces, the beams and columns are usually classified as structural components. Nonstructural components usually include architectural, mechanical and electrical components within a building (FEMA, 2009; HAZUS, 2003). The damage to structural components is usually related to structural drift and ground motion acceleration, while the damage of nonstructural components is generally associated with the floor acceleration and structural drift (Dong & Frangopol, 2016; Mitrani-Reiser, 2007).

Given the ground motion intensity, the seismic performance of a hospital is investigated using fragility curves. Based on HAZUS (2003), the PGA is adopted to predict the performance of structural components under earthquakes. The probability of the structural components being in different damage states could be computed accordingly. The peak floor acceleration (PFA) acts as seismic demand for the damage assessment of nonstructural components. The peak floor acceleration amplification Ω (i.e. PFA/PGA) is adopted herein to compute the PFA, which in turn could be utilised for the vulnerability analysis of nonstructural components. The peak floor acceleration amplification factor Ω is (Chaudhuri & Hutchinson, 2004):

$$\Omega = \left(1.0 + \alpha_1 \sqrt{h_{nor}} \right) \left(1.0 - h_{nor} \right) + \left(\alpha_2 h_{nor}^2 \right) h_{nor} \qquad (6)$$

where α_1 and α_2 are empirical constants and h_{nor} is normalised height computed as the floor height divided by the total building height. Given the detailed information of the investigated hospital and seismic inputs, the PFA could also be obtained using nonlinear time history or incremental dynamic analysis (Dong & Frangopol, 2016).

The expected damage indices associated with structural and nonstructural components can be expressed, respectively, as follows:

$$D_{SC} = \sum_{i=1}^{n_{SD}} HCDI_{SC,i} \cdot P_{SCi|IM} \qquad (7)$$

$$D_{NSC} = \sum_{i=1}^{n_{NSD}} HCDI_{NSC,i} \cdot P_{NSCi|IM} \qquad (8)$$

where $HCDI_{SC,i}$ and $HCDI_{NSC,i}$ are the damage indices associated with state i of structural and nonstructural components of the hospital, respectively; n_{SD} and n_{NSD} are the numbers of damage states associated with structural and nonstructural components, respectively; and $P_{SCi|IM}$ and $P_{NSCi|IM}$ are the probabilities of the structural and nonstructural components being in damage state i, respectively. These probabilities are obtained based on the fragility curves considering different seismic demands. The correlation coefficient between the damage indices of structural and nonstructural components $\rho(HCDI_{SC}, HCDI_{NSC})$ is considered in the assessment process.

A comprehensive performance assessment of a hospital should be conducted based on a system level approach. Herein, a composite damage index of a hospital based on the structural and nonstructural damages is developed. Given the weighting factors associated with structural and nonstructural components, the composite expected hospital damage index HDI is

$$HDI = r_{SC} \cdot D_{SC} + r_{NSC} \cdot D_{NSC} \qquad (9)$$

where r_{SC} and r_{NSC} are the weighting factors associated with structural D_{SC} and nonstructural D_{NSC} damage indices, respectively. Given the probability density function (PDF) associated with hospital damage index HDI and threshold values (i.e. lower and upper bounds), the probability of a hospital being in different functionality levels $HFLs$ can be identified. Holmes and Burkett (2006) suggested classifying structural and nonstructural damages into four levels: none, minor, affecting hospital operations and temporary closure. Yavari et al. (2010) presented an approach considering the overall facility as fully functional, functional, affected functionality and not functional. In this paper, the four functionality levels for a hospital are none, slight, moderate and major (Yavari et al., 2010).

The capacity of a hospital depends on the classified hospital functionality levels $HFLs$. The waiting time is an important parameter to evaluate the capacity of a hospital during normal and extreme event conditions (Yi, 2005). When the number of injured persons is larger than the number of patients treated, additional waiting time is necessary. The waiting time associated with hospital functional level HFL_i under daily patient rate λ is (Paul, George, Yi, & Lin, 2006):

$$WT_i(\lambda) = \exp\left(A_i + B_i\lambda\right) \qquad (10)$$

$$A_i = \frac{\lambda_U \ln\left(WT_L\right) - \lambda_L \ln\left(WT_{Ui}\right)}{\lambda_U - \lambda_L} \qquad (11)$$

$$B_i = \frac{\ln\left(WT_{Ui}\right) - \ln\left(WT_L\right)}{\lambda_U - \lambda_L} \qquad (12)$$

where A_i and B_i are constants associated with the hospital functionality level i; λ_L is the pre-disaster average daily patient arrival number; WT_L is the waiting time associated with the normal hospital operation; λ_U is the maximum daily arrival number; and WT_{Ui} is the waiting time associated with functionality level i under the maximum arrival rate λ_U. Equation (10) is used herein to compute the waiting time; given additional data, other functions could also be adopted.

4. System level performance assessment

In this section, the effects of the damage states associated with a bridge network and a hospital on system level performance of an interdependent healthcare–bridge network are investigated. The extra travel and waiting time at the system level are computed. Based on the configuration of the investigated bridge network, travel time is defined as the time it takes to transfer the injured people to the hospital immediately after the earthquake. With respect to waiting time analysis, the healthcare system performance is measured by the waiting time needed to get the injured persons treated.

The extra travel time experienced by an injured person is due to the damages of bridges and links in a bridge network. The travel time is representative of the functionality of a bridge network; large travel time reveals a high reduction of functionality associated with a bridge network (Bocchini & Frangopol, 2011; Dong, Frangopol, & Sabatino, 2015; Frangopol & Bocchini, 2011). The extra daily travel time $EDTT$ for the injured persons in a bridge network can be expressed as (Dong et al., 2015):

$$EDTT = \sum_{j=1}^{n_l} \sum_{i=1}^{n_{LD}} P_{LDSj,i|IM} \left[ADR_{ij} \cdot \left(\frac{l_j}{S_{Dj,i}} - \frac{l_j}{S_{0j}} \right) + ADT_{ij} \cdot \frac{D_j}{S} \right] \qquad (13)$$

where n_l is the number of links in a bridge network; n_{LD} is the number of damage states associated with link damage; $P_{LDSj,i|IM}$ is the conditional probability of the jth link being in damage state i; ADT_{ij} is average daily number of injured persons that follow detour due to damage state i of the jth link; D_j is the length of the extra detour of jth link (km); S is the detour speed (km/h); ADR_{ij} is the average daily number of injured persons that remain on the jth link under damage state i; l_j is the length of link j (km); S_{0j} is the traffic speed on intact link j (km/h); and $S_{Dj,i}$ is the traffic speed on link j associated with damage state i (km/h).

The waiting time is related to the hospital functionality levels under a given seismic scenario. Given the limited functionality associated with the hospital, the extra waiting time of the injured people could be computed. Based on the theorem of total probability, the extra daily waiting time $EDWT$ can be computed as:

$$EDWT = \sum_{i=1}^{n_H} \left[P_{HFi|IM} \cdot WT_i(ATV) \right] \cdot ATV - ATV \cdot WT_0(ATV) \qquad (14)$$

where ATV is the total number of injured persons transferred though a bridge network to a hospital; WT_i is the waiting time associated with functionality level i given ATV; n_H is the number of functionality levels of a hospital under investigation; $WT_0(ATV)$ is the waiting time associated with the intact hospital under ATV; and $P_{HFi|IM}$ is the conditional probability of hospital being in functionality level i under IM. The flowchart of the computation associated with the extra travel and waiting time of a healthcare system under seismic hazard is shown in Figure 1. As indicated, the seismic scenarios should be identified first. Then, the vulnerability analyses of a bridge network and a hospital are conducted. Finally, based on the damage conditions and functionality levels, the extra travel and waiting time are computed as system level performance indicators.

During the system level performance assessment process, the correlations associated with the ground motion intensities and seismic damage indices are considered. For example, the correlations among the IMs at different locations are computed using Equation (2). Then, using Monte Carlo simulation, these correlated IMs could be generated. Overall, given the correlation coefficients, the correlated random variables used in the functionality assessment procedure could be generated. The flowchart of generating these random variables using Monte Carlo simulation is shown in Figure 2. Finally, the system level performance

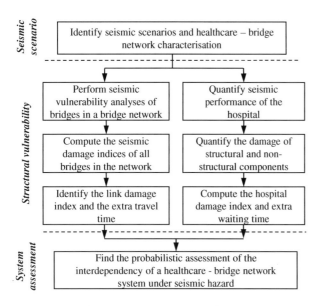

Figure 1. Flowchart of component and system levels functionality assessment of an interdependent healthcare–bridge network system under seismic hazard.

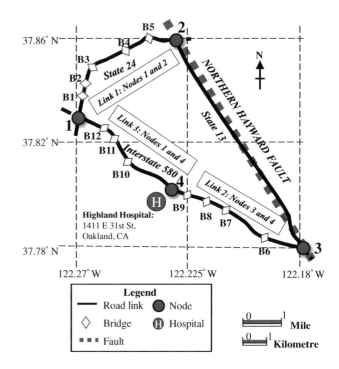

Figure 3. Layout of the healthcare–bridge network system under investigation.

indicators (e.g. extra travel and waiting time) could be computed using Equations (13) and (14).

Due to the correlations among the random variables of ground motion intensities and seismic damage indices associated with bridges and hospital, the seismic performance of a bridge network and a hospital is correlated. The correlation effects on the conditional seismic performance of a hospital given the damage state of a link are investigated herein. The conditional probability of a hospital being in functionality level j given the link in damage state i is:

$$P\left(HFL_{j|IM}\middle|LDS_{i|IM}\right) = \frac{P\left(HFL_{j|IM}\cap LDS_{i|IM}\right)}{P\left(LDS_{i|IM}\right)} \quad (15)$$

where $HFL_{j|IM}$ is the event that a hospital is in functionality level j given IM and $LDS_{i|IM}$ is the event that a link is in damage state i given IM. The probability $P(HFL_{j|IM}\cap LDS_{i|IM})$ could be computed by considering the events $HFL_{j|IM}$ and $LDS_{i|IM}$ as a parallel system. Then, given the correlation coefficients among the random variables (e.g. ground motion intensities, seismic damage indices), the probability $P(HFL_{j|IM}\cap LDS_{i|IM})$ is computed. When the events $HFL_{j|IM}$ and $LDS_{i|IM}$ are

independent, $P(HFL_{j|IM}\cap LDS_{i|IM}) = P(HFL_{j|IM})\times P(LDS_{i|IM})$. Finally, given $P(HFL_{j|IM}\cap LDS_{i|IM})$ and $P(LDS_{i|IM})$, the conditional probability of the hospital being in functionality level j is computed according to Equation (15).

5. Illustrative example

The proposed approach is illustrated on an interdependent healthcare–bridge network system located in Alameda, California. The schematic layout of the hospital and bridge network is shown in Figure 3. The Highland Hospital, constructed in 1954, is investigated herein. This is a public hospital and is operated by the Alameda Health system. A set of locations is connected to the hospital via a bridge network composed of 12 bridges. Link 1 connects the nodes 1 and 2, link 2 connects nodes 3 and 4 and link 3 connects nodes 1 and 4. The hospital investigated herein is treated as a mid-rise steel moment frame building and designed based on moderate-code seismic provision.

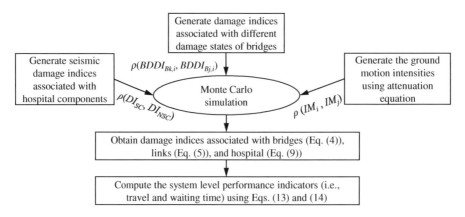

Figure 2. Flowchart of generating correlated random variables using Monte Carlo simulation to compute the system level performance indicators.

5.1. Seismic performance of bridges and links

In order to quantify the seismic performance of the bridges and links, the probabilistic earthquake scenarios should be identified. The seismic scenarios are selected based on the seismic rupture sources in the San Francisco bay area (USGS, 2003). The investigated earthquake magnitudes are associated with the Northern Hayward Fault as the healthcare–bridge network system is located in this area. The segment length and width of the fault are 50 and 14 km, respectively. Using Equation (1), the expected magnitude associated with the investigated rupture is 6.88. Using Monte Carlo Simulation, 100,000 samples of seismic scenarios are generated. Then, based on the attenuation equation (Campbell & Bozorgnia, 2008), the ground motion intensity at the location of the healthcare–bridge network system is predicted.

The PGA is utilised as ground motion intensity measure and other ground motion intensities (e.g. spectral acceleration) could also be used (Campbell & Bozorgnia, 2008). The PGA is assumed lognormal. Its expected value at the location of the hospital is .865 g, and the standard deviation is .51 g using the attenuation equation (Campbell & Bozorgnia, 2008). The probabilistic PGAs at the locations of the bridges and hospital are generated using Monte Carlo simulation. The correlation among the PGAs at the locations of the hospital and bridges is computed based on Equation (2). The exponential decay function has been widely adopted in the assessment of the spatial correlation of PGA (Esposito & Iervolino, 2011; Goda & Hong, 2008; Jayaram & Baker, 2009). Then, the correlated random variables associated with ground motion intensities are generated using Monte Carlo simulation as indicated in Figure 2.

The parameters of the fragility curves associated with the bridges located on the network are based on Shinozuka et al. (2001). Bridges 2, 7 and 9 are single-span concrete bridges; the remaining bridges are multiple-span continuous concrete bridges. The fragility curves of the basic single and multiple-span

continuous concrete bridges are shown in Figures 4(a) and (b). Given the skew angle and soil condition of the specific bridges, the fragility curves could be updated accordingly (Shinozuka et al., 2001).

Given the ground motion intensity and fragility curves, the probabilities of the bridges being in different damage states are computed. Then, using Equation (4), the bridge damage index BDI under the investigated ground motion intensity is obtained. The damage state index BDDI is considered lognormal with a coefficient of variation of .5 (HAZUS, 2003; Shinozuka et al., 2008). The expected values of the damage index associated with slight, moderate, major and complete damage states are .1, .3, .75 and 1, respectively (Shiraki et al., 2007). Monte Carlo simulation is adopted to generate these random variables. The bridge damage states associated with bridges 1 (multiple-span continuous concrete) and 2 (single-span concrete bridge) are shown in Figure 4(c) and (d), respectively. The bridge damage indices associated with bridges 1 and 2 are best fitted by a gamma distribution with mean values .338 and .196, and standard deviations .272 and .187, respectively. Subsequently, the damage indices of links are computed using Equation (5).

The correlation among the random damage indices BDDI of different seismic damage states is now considered. Herein, the correlation coefficients are assumed to be 0, .5 and 1, representing uncorrelated, partially and fully correlated random variables, respectively. These values are adopted to investigate the correlation effects on the network performance under seismic hazard. Additionally, the damage indices of different bridges are also correlated. The probabilistic damage index associated with link 1 under different correlation coefficients among the damage indices BDDI is shown in Figure 5(a). As indicated, the standard deviation of the link damage index increases as the correlation coefficient among the random variables increases. Given the threshold associated with link performance, the probabilities of the link 1 being in different performance levels (i.e. from no

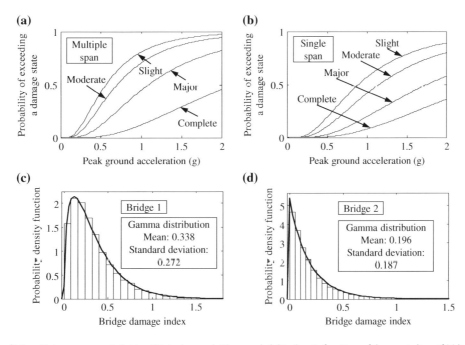

Figure 4. Fragility curves of (a) multiple-span concrete bridge, (b) single-span bridges; probability density functions of damage indices of (c) bridge 1 (B1) and (d) bridge 2 (B2).

(a)

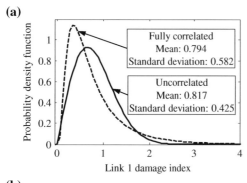

(b)

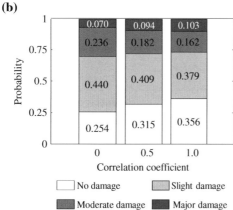

Figure 5. (a) Effects of correlation among the damage indices *BDDI* on the link 1 damage index and (b) probabilities of link 1 being in different damage states using 0, .5 and 1.0 correlation coefficients.

damage to major damage) are shown in Figure 5(b). As indicated, without considering the correlation effect, the probabilities of link being in none and major damage states would be underestimated. Consequently, the correlation coefficients among the damage indices *BDDI* have a significant effect and should be carefully evaluated. The probabilistic damage indices associated with links 2 and 3 could also be obtained. The probabilities of

Table 1. Probabilities of links 2 and 3 in Figure 3 being in different damage states considering correlation among the random damage indices *BDDI*.

Link	Correlation coefficient	No damage	Slight damage	Moderate damage	Major damage
Link 2	.0	.3514	.4509	.1643	.0334
	.5	.4136	.4001	.1356	.0507
	1.0	.4498	.3573	.1277	.0652
Link 3	.0	.4472	.4223	.1111	.0194
	.5	.4963	.3666	.1051	.0320
	1.0	.5327	.3239	.0977	.0457

the links 2 and 3 being in different damage states are shown in Table 1.

5.2. Hospital damage assessment

In order to quantify the hospital damage index *HDI*, the seismic intensity and vulnerability of structural and nonstructural components should be identified. The PGA is adopted as seismic demand indicator for the structural components, while the PFA is used to investigate the seismic performance of nonstructural components (HAZUS, 2003). The relationship between PFA and PGA is indicated in Equation (6). For the mid-rise hospital building, α_1 and α_2 are assumed 1.63 and 1.53, respectively (Chaudhuri & Hutchinson, 2004). The maximum PFA occurs at $h_{nor} = 1$ for the investigated hospital. The parameters of the fragility curves associated with structural and nonstructural components are based on HAZUS (2003) and shown in Table 2. Given moderate-code seismic provision, the median values and standard deviations of fragility curves associated with structural and nonstructural components are obtained. Subsequently, given the PGA and PFA, the seismic vulnerabilities of the structural and nonstructural components are computed.

Herein, the damage indices of damage states associated with structural and nonstructural components are based on Aslani and Miranda (2005). The mean values of slight, moderate, major and complete damage states of structural components are .025, .12, .6 and 1.2, respectively. The mean values of slight, moderate, major and complete damage states of nonstructural components are .02, .12, .36 and 1.2, respectively. The coefficients of variation of these random variables are .7. Then, based on Equations (7) and (8), the damage indices of structural and nonstructural components are computed. The PDFs of damage indices of structural and nonstructural components are indicated in Figure 6. As indicated, the expected damage index of the structural component is almost 2.05 times of that associated with the nonstructural component.

The correlation coefficient among the damage indices of structural and nonstructural components $\rho(HCDI_{SC}, HCDI_{NSC})$ is considered in the assessment process. The three correlation coefficients 0, .5 and 1 are considered. Monte Caro simulation is used to generate the random variables considering correlation as indicated in Figure 2. Given the damage index of structural and nonstructural components, the composite building damage index is computed using Equation (9). This equation is used to compute the hospital damage index. Herein, r_{SC} and r_{NSC} are assumed .5. The expected value and standard deviation of *HDI*

Table 2. Median and standard deviation associated with fragility curves of structural and nonstructural components (adapted from HAZUS, 2003).

Design level	Slight damage		Moderate damage		Major damage		Complete damage	
	Median	Standard deviation	Median	Standard deviation	Median	Standard deviation	Median	Standard deviation
Structural component								
Moderate code	.16	.64	.28	.64	.6	.64	1.27	.64
Low code	.15	.64	.23	.64	.42	.64	.73	.64
High code	.17	.64	.34	.64	.85	.64	2.1	.64
Nonstructural component								
Moderate code	.38	.67	.75	.67	1.5	.67	3	.67
Low code	.3	.65	.6	.67	1.2	.67	2.4	.67
High code	.45	.66	.9	.67	1.8	.68	3.6	.66

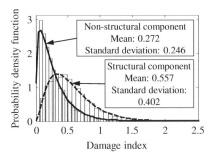

Figure 6. Probability density functions of structural and nonstructural components damage indices.

Table 3. Probabilities of the hospital having different functionality levels under different correlation coefficients among the damage indices of structural and nonstructural components.

Correlation coefficient	No damage	Slight damage	Moderate damage	Major damage
.0	.3323	.4946	.1526	.0205
.5	.4116	.3903	.1569	.0412
1.0	.4625	.3234	.1527	.0614

associated with $\rho(HCDI_{SC}, HCDI_{NSC}) = 0$ are .416 and .221, respectively. Given $\rho(HCDI_{SC}, HCDI_{NSC}) = 1$, the expected value and standard deviation of HDI are .416 and .319, respectively. As indicated, the correlation has a significant effect on the standard deviation of the hospital damage index.

For the hospital functionality level analysis, none, slight, moderate and major damage are represented by the values of the hospital damage index $HDI \leq .3$, $.3 < HDI \leq .6$, $.6 < HDI \leq 1$ and $HDI > 1$, respectively. Given the hospital functionality criterion, the probabilities of the hospital being in different functionality levels are identified and shown in Table 3. As revealed, the probabilities of being in none and major damaged functionality levels increase when considering correlation effects. Given the threshold values associated with hospital damage indices determined by a decision-maker, the probabilities of the hospital being in different functionality levels could be updated.

5.3. System level performance

The seismic performance of a healthcare–bridge network system is investigated considering two indicators: (a) travel time and (b) waiting time. After the earthquake, the injured persons from nodes 1, 2 and 3 in Figure 3 are transferred to node 4, as the hospital is near this node. The extra travel and waiting time associated with the daily patient volume are investigated. If the damage state of the link is slight, the remaining patient volume and the flow speed are 100% and 75% of those for the intact link, respectively. In moderate damage state, the remaining volume and the flow speed are 75% and 50% of those for the intact link. In major damage state, the remaining volume and the flow speed are 50% and 50% of those for the intact link (Chang et al., 2000; Dong et al., 2014). The extra detour length of links is 3.5 km. The daily patient volumes after the earthquake from node 1 to 4, 2 to 4 and 3 to 4 are 60, 120 and 60, respectively. The total number of daily injured persons transferred to the hospital is 240 (i.e. 60 + 120 + 60).

Then, given the probabilities of links being in different damage states as computed previously, the extra daily travel time

considering the number of injured persons through the damaged transportation network is computed using Equation (13). The correlation coefficients $\rho(BDDI_{Bk,i}, BDDI_{Bj,i})$ and $\rho(PGA_i, PGA_j)$ are denoted as ρ_1 and ρ_2, respectively. Given $\rho_1 = \rho_2 = 0$, the extra daily travel time is 15.13 h; this value reduces to 12.02 h given $\rho_1 = \rho_2 = 1$. As indicated, the correlation among the random variables could affect the extra travel time significantly. Compared with the uncorrelated case (i.e. $\rho_1 = \rho_2 = 0$), the extra travel time associated with fully correlated random variables ($\rho_1 = \rho_2 = 1$) decreases by almost 26%. Furthermore, the correlation among the ground motion intensities has a larger effect on the extra travel time. Given $\rho_1 = \rho_2 = 0$, the extra daily travel time is 15.13 h; this value reduces to 13.22 h when $\rho_1 = 0$ and $\rho_2 = 1$ and to 14.09 h when $\rho_1 = 1$ and $\rho_2 = 0$.

Given the hospital being in the performance levels as shown in Table 3, the extra daily waiting time is evaluated using Equation (14). The pre-earthquake average patient arrival per day λ_L is 80. The waiting time WT_L associated with the normal operation condition is 20 min. Given more information (e.g. number of beds, number of operating rooms) of the investigated hospital, the arrival rate and waiting time could be updated. The maximum arrival per day λ_U is assumed 450. The waiting times associated with none, slight, moderate and major damage levels under maximum arrival rate are 40, 60, 80 and 120 min, respectively. Then, using Equations (10) – (12), the waiting times associated with different hospital functional levels are shown in Figure 7(a). The waiting time is expressed using the exponential function (Paul et al., 2006). With additional data, other models could also be incorporated within the assessment process.

The correlation effects are considered in the extra waiting time assessment process. Herein, $\rho(HCDI_{SC}, HCDI_{NSC})$ is denoted as ρ_3. Given $\rho_3 = 0$, the extra daily waiting time of the injured person

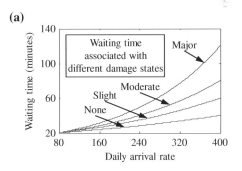

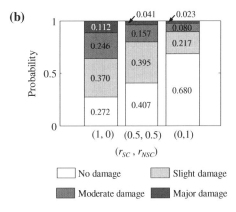

Figure 7. (a) Waiting time associated with hospital having different functionality levels (b) effects of r_{SC} and r_{NSC} on the probabilities of hospital being in different functionality levels.

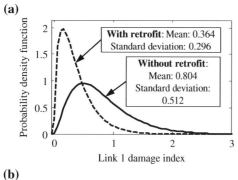

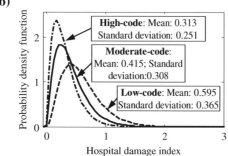

Figure 8. Effects of (a) retrofit actions on link 1 damage index and (b) high- and low-design codes on the hospital–bridge network system under seismic hazard.

is 16.84 h; this value reduces to 15.46 h when $\rho_3 = 1$. The extra daily waiting time of the injured people is slightly larger than the extra travel time. The effects of r_{SC} and r_{NSC} on the hospital functionality levels are indicated in Figure 7(b). The two parameters (i.e. r_{SC} and r_{NSC}) have a significant effect on the hospital functionality assessment and should be carefully assessed.

The effects of retrofit actions associated with the bridges on extra travel time are also studied. The fragility enhancement of bridges retrofitted by steel jacketing is investigated based on the approach presented in Shinozuka et al. (2008). The enhancement ratios associated with fragility median values are 55%, 75%, 104% and 145% considering slight, moderate, major and complete damage states, respectively (Shinozuka et al., 2008). Given $\rho(BDDI_{Bk,i}, BDDI_{Bj,i}) = .5$ and Equation (2), the extra daily travel time with and without retrofit actions are 3.12 and 13.01 h, respectively. As revealed, the seismic retrofit actions have a profound effect on the extra travel time and can improve the healthcare–bridge network performance significantly. The probabilistic damage index associated with link 1 under these two scenarios (i.e. with and without retrofit actions) is shown in Figure 8(a).

Additionally, the effects of seismic vulnerability of hospital on the system level performance are investigated. Medians and standard deviations associated with fragility curves of structural

and nonstructural components under low- and high-code design provisions are shown in Table 2. Given the fragility curves associated with these design provisions, the extra daily waiting time with low-, moderate- and high-design codes are 25.41, 16.06 and 10.59 h, respectively. The PDF of the hospital damage index is shown in Figure 8(b). As revealed, the seismic design code has a significant effect on the hospital functionality levels.

Moreover, the effects of correlation on the conditional probabilities of hospital being in different functionality levels given the seismic damage of the link are investigated using Equation (15). Given $\rho(BDDI, HCDI) = \rho(PGA_i, PGA_j) = .5$, the conditional probabilities of the hospital being in different functionality levels are shown in Table 4. The case without considering the correlation effects is also shown in this table. As indicated, the correlation effects could affect the conditional performance of the hospital significantly. For example, given the link 1 being in damage state 3 (i.e. moderate damage state), the conditional probability of the hospital being in moderate damaged functionality level is .2555, while this value is reduced to .1526 without considering the correlation effects.

6. Conclusions

This paper presents an approach for assessment of an interdependent healthcare–bridge network system under seismic hazard considering uncertainties and correlation effects. The functionalities associated with the individual bridges, bridge networks and hospital are investigated and combined for the system performance assessment. Rather than focusing only on structural damage, the extra travel and waiting time are investigated. The correlation among the damage indices is considered in the performance assessment process in addition to the correlation among the spatial seismic intensities. The approach is illustrated on a healthcare system near a bridge network in Alameda, California. The approach presented can be extended to systems with multiple hospitals, in which the interdependencies among hospitals and the interdependencies healthcare facilities–bridge network systems could be considered.

The following conclusions are drawn:

(1) The proposed performance assessment of an interdependent healthcare–bridge network system under seismic hazard provides system level probabilistic measures that can aid the emergency management process. In addition to the consideration of bridge damage assessment, the damages associated with structural and nonstructural components of a hospital are considered in the functionality assessment.

Table 4. Conditional probabilities of the hospital being in different functionality levels given the seismic damage of the link 1 considering the correlations among the ground motion intensities and damage indices of bridges and hospital.

Correlation coefficient	Hospital functional level	$P(HFL_{i|iM}|LDS_{1|iM})$	$P(HFL_{i|iM}|LDS_{2|iM})$	$P(HFL_{i|iM}|LDS_{3|iM})$	$P(HFL_{i|iM}|LDS_{4|iM})$
$\rho = .5$	No	.6460	.4032	.2115	.0823
	Slight	.2978	.4397	.4681	.3273
	Moderate	.0520	.1367	.2555	.3906
	Major	.0042	.0204	.0649	.1998
$\rho = 0$	No	.3323	.3323	.3323	.3323
	Slight	.4946	.4946	.4946	.4946
	Moderate	.1526	.1526	.1526	.1526
	Major	.0205	.0205	.0205	.0205

(2) The correlation among the random damage indices has a significant effect on the probabilities of links being in different damage states and should be carefully evaluated. Without considering the correlation effect, the probabilities of link being in none and major damage states would be underestimated.

(3) It is necessary to consider the correlation coefficients among the spatial ground motion intensities and component-to-component damage indices for the healthcare–bridge network system performance assessment. The correlation coefficients have a significant effect on the standard deviations of the damage indices of bridges, links and hospital.

(4) Regarding the system level performance assessment, the extra travel and waiting time decrease when the correlation coefficients (e.g. correlations among the ground motion intensities and seismic damage indices) are accounted for. Additionally, the correlation among the ground motion intensities has a larger effect on the extra travel time than that among the damage indices.

(5) The seismic performance of interdependent healthcare–bridge network system depends highly on the seismic performance of both bridge network and hospital. The effects of retrofit and seismic strengthening associated with bridges and hospital are significant. Bridge retrofit actions could result in a profound improvement of the performance of healthcare–bridge network system. Given the specific bridge retrofit actions and correlation coefficients considered in this paper, the extra travel time of the injured persons associated with the case when bridges are retrofitted is about 24% (i.e. 3.12/13.01) of the case when bridges are not retrofitted.

(6) Based on the damage conditions of the link, the functionality of the hospital under the investigated seismic scenarios could be predicted considering correlation effects. Overall, it is of vital importance to incorporate correlation effects within the seismic performance assessment of interdependent infrastructure systems.

Acknowledgements

The support from the National Science Foundation, the Commonwealth of Pennsylvania, Department of Community and Economic Development, through the Pennsylvania Infrastructure Technology Alliance (PITA) and the U.S. Federal Highway Administration Cooperative Agreement is gratefully acknowledged. The opinions and conclusions presented in this paper are those of the authors and do not necessarily reflect the views of the sponsoring organisations.

Disclosure statement

No potential conflict of interest was reported by the authors.

Funding

This work was supported by the National Science Foundation [grant number CMS-0639428], [grant number CMMI-1537926]; the Commonwealth of Pennsylvania, Department of Community and Economic Development, through the Pennsylvania Infrastructure Technology Alliance (PITA); the U.S. Federal Highway Administration Cooperative Agreement [award number DTFH61-07-H-00040].

References

Achour, N., Miyajima, M., Kitaura, M., & Price, A. (2011). Earthquake-induced structural and nonstructural damage in hospital. *Earthquake Spectra, 27*, 617–634.

Aslani, H., & Miranda, E. (2005). *Probabilistic earthquake loss estimation and loss disaggregation in building* (Report No. 157). Stanford, CA: Department of Civil and Environmental Engineering, Stanford University.

Bocchini, P., & Frangopol, D. M. (2011). A stochastic computational framework for the joint transportation network fragility analysis and traffic flow distribution under extreme events. *Probabilistic Engineering Mechanics, 26*, 182–193.

Campbell, K. W., & Bozorgnia, Y. (2008). NGA ground motion model for the geometric mean horizontal component of PGA, PGV, PGD and 5% damped linear elastic response spectra for periods ranging from 0.01 to 10 s. *Earthquake Spectra, 24*, 139–171.

Chang, S. E., Shinozuka, M., & Moore, J. E. (2000). Probabilistic earthquake scenarios: Extending risk analysis methodologies to spatially distributed systems. *Earthquake Spectra, 16*, 557–572.

Chaudhuri, S., & Hutchinson, T. (2004). *Distribution of peak horizontal floor acceleration for estimating nonstructural element vulnerability.* Proceedings of the 13th World Conference on Earthquake Engineering, Vancouver, BC, Canada, paper No. 1721.

Cimellaro, G. P., Reinhorn, A. M., & Bruneau, M. (2011). Performance-based metamodel for healthcare facilities. *Earthquake Engineering and Structural Dynamics, 40*, 1197–1217.

Dong, Y., Frangopol, D. M., & Saydam, D. (2013). Time-variant sustainability assessment of seismically vulnerable bridges subjected to multiple hazards. *Earthquake Engineering and Structural Dynamics, 42*, 1451–1467.

Dong, Y., Frangopol, D. M., & Saydam, D. (2014). Sustainability of highway bridge networks under seismic hazard. *Journal of Earthquake Engineering, 18*, 41–66.

Dong, Y., Frangopol, D. M., & Sabatino, S. (2015). Optimizing bridge network retrofit planning based on cost-benefit evaluation and multi-attribute utility associated with sustainability. *Earthquake Spectra, 31*, 2255–2280.

Dong, Y., & Frangopol, D. M. (2016). Performance-based seismic assessment of conventional and base-isolated steel buildings including environmental impact and resilience. *Earthquake Engineering and Structural Dynamics, 45*, 739–756.

Esposito, S., & Iervolino, I. (2011). PGA and PGV spatial correlation models based on European multievent datasets. *Bulletin of the Seismological Society of America, 101*, 2532–2541.

FEMA. (2009). *NEHRP recommended provisions for seismic regulations for new buildings and other structures, FEMA P-750*. Washington, DC: Federal Emergency Management Agency.

Frangopol, D. M., & Bocchini, P. (2011, April, 14–16). *Resilience as optimization criterion for the rehabilitation of bridges belonging to a transportation network subjected to earthquake.* Proceedings of the SEI-ASCE 2011 Structures Congress, Las Vegas, NV, USA.

Goda, K., & Hong, H. P. (2008). Spatial correlation of peak ground motions and response spectra. *Bulletin of the Seismological Society of America, 98*, 354–365.

Hanks, T. C., & Bakun, W. H. (2002). A bilinear source-scaling model for M-log observations of continental earthquakes. *Bulletin of the Seismological Society of America, 92*, 1841–1846.

HAZUS. (2003). *Multi-hazard loss estimation methodology, earthquake model*. Technical Manual. Washington, DC: Department of Homeland Security Emergency Preparedness and Response Directorate, FEMA, Mitigation Division.

Holmes, W. T., & Burkett, L. (2006). *Seismic vulnerability of hospitals based on historical performance in California.* Proceedings of the 8th U.S. National Conference on Earthquake Engineering, San Francisco, CA, USA.

Jayaram, N., & Baker, J. W. (2009). Correlation model for spatially distributed ground-motion intensities. *Earthquake Engineering and Structural Dynamics, 38*, 1687–1708.

Lee, R. G., & Kiremidjian, A. S. (2007). *Uncertainty and correlation in seismic risk assessment of transportation systems* (PEER Report 2007/05). Berkeley, CA: Pacific Earthquake Engineering Research Center College of Engineering, University of California.

Liu, M., & Frangopol, D. M. (2006). Probability-based bridge network performance evaluation. *Journal of Bridge Engineering, 11*, 633–641.

Mander, J. B. (1999). *Fragility curve development for assessing the seismic vulnerability of highway bridges* (Technical Report). Buffalo, NY: University at Buffalo, State University of New York.

Mitrani-Reiser, J. (2007). *Probabilistic loss estimation for performance-based earthquake engineering* (PhD Dissertation) (pp. 1–155). Pasadena, CA: California Institute Technology.

Murphy, L. (1973). *San Fernando, California, Earthquake of February 9 1971*. Washington, DC: U.S Department of Commerce.

Myrtle, R. C., Masri, S. E., Nigbor, R. L., & Caffrey, J. P. (2005). Classification and prioritization of essential systems in hospitals under extreme events. *Earthquake Spectra, 21*, 779–802.

Paul, J. A., George, S. K., Yi, P., & Lin, L. (2006). Transient modelling in simulation of hospital operations for emergency response. *Prehospital and Disaster Medicine, 21*, 223–236.

Shinozuka, M., Feng, M. Q., Kim, H., Uzawa, T., & Ueda, T. (2001). *Statistical analysis of fragility curves* (Technical Report MCEER). Los Angeles, CA: Department of Civil and Environmental Engineering, University of South California.

Shinozuka, M., Zhou, Y., Kim, S. H., Murachi, Y., Banerjee, S., Cho, S., & Chung, H. (2008). *Social-economic effect of seismic retrofit implemented on bridges in Los Angeles highway network* (Report CA06-0145). Irvine, CA: Department of Civil and Environmental Engineering, University of California.

Shiraki, N., Shinozuka, M., Moore, J. E., Chang, S. E., Kameda, H., & Tanaka, S. (2007). System risk curves: probabilistic performance scenarios for highway networks subject to earthquake damage. *Journal of Infrastructure Systems, 13*, 43–54.

USGS. (2003). *Earthquake probabilities in the San Francisco Bay Region: 2002–2031* (Open File Report 03-214). Menlo Park, CA: United States Geological Survey.

Wang, M., & Takada, T. (2005). Macrospatial correlation model of seismic ground motions. *Earthquake Spectra, 21*, 1137–1156.

World Health Organization (WHO). (2007). *Risk Reduction in the Health Sector and Status of Progress*. Proceedings of Disaster Risk Reduction in the Healthcare Sector- Thematic Workshop, World Health Organization (WHO), Geneva, Switzerland.

Yavari, S., Chang, S., & Elwood, K. J. (2010). Modeling post-earthquake functionality of regional health care facilities. *Earthquake Spectra, 26*, 869–892.

Yi, P. (2005). *Real-time generic hospital capacity estimation under emergency situations*. Buffalo, NY: State University of New York at Buffalo.

Part V

Inspection and monitoring

Application of the statistics of extremes to the reliability assessment and performance prediction of monitored highway bridges

Thomas B. Messervey, Dan M. Frangopol and Sara Casciati

The present paper examines how statistics of extremes can be used to enhance the assessment and performance prediction of monitored highway bridges. This is achieved by proposing an approach to obtain a monitoring-based live-load for use in a reliability analysis. Time effects that correlate observed data to code required return periods are considered. Additionally, a method to identify, minimise and properly account for the epistemic uncertainty inherent to any monitoring record within a reliability analysis is presented. The approach can be utilised to plan data collection efforts or to maximise the utility of a limited amount of data. The extension of extreme value statistics to the monitoring of highway bridges is developed first using simulations and is then demonstrated on a case study using 90 days of in-service data collected from a bridge located in Pennsylvania, USA.

Introduction

The deteriorated condition of many existing highway bridges is well documented and research towards the development of structural health monitoring (SHM) enabling bridge management programs to optimally maintain and repair these structures in a life-cycle context is ongoing (FHWA 2007, Frangopol and Liu 2007, Frangopol and Messervey 2009a). In short, many bridges built in the early to mid part of the twentieth century are approaching or have passed their original intended service lives. Rehabilitation and maintenance needs are greater than funds available and as such bridge service lives must be extended. In urban areas, disrupting normal traffic flow is also a major concern. Further aggravating this problem, research has shown that traffic volumes and vehicle weights are increasing. In Europe, vehicular kilometres driven have generally doubled since 1990 and are projected to double again by the year 2030 putting additional strain on already congested networks (Enevoldsen 2008). Such a scenario of aged and deteriorating bridges subjected to progressively in-creasing traffic demands has created the need for methods to assess and manage these structures optimally with respect to cost, condition, and safety.

Recent advances in SHM and sensor technologies offer a potential solution to this problem. Through the collection of in-service data and data processing

algorithms, when changes in structural performance violate safety thresholds, alerts could be sent to a manager for decision. With a sufficient amount of data, repair and maintenance strategies could be planned and optimised for the service life of deteriorating structures. These are in fact the goals of many research efforts worldwide (Frangopol and Messervey 2007, 2009b). Although the goals are clear, how to achieve them is not as easily determined. What to monitor, across how many members, with what sensors, and for how long are all challenging questions. A large bridge could have hundreds of sensors making the acts of separating noise, minimising power consumption, data reduction, data processing, and data storage non-trivial issues. Furthermore, it should be expected that a bridge manager would not elect to employ monitoring solutions unless they are shown to be simple, practical, and cost effective.

The collection of in-service monitoring data and the characterisation of a monitoring based distribution poses several unique challenges. Unlike recording a singular structural response to a specific load condition (e.g. a park test) or the detection of the presence or absence of a particular indicator (e.g. corrosive agents), defining a probability distribution is sensitive to the amount of data collected raising two important questions. First, how does the observed information relate to safety or code requirements? To answer this question, another dimension, time, must be addressed. For example, if a structure is monitored for a short period and only light load demands are recorded, one cannot conclude that the structure has enough safety. One could conclude that the structure *was safe* for the loads encountered during the period monitored, but does not adequately convey the safety of the structure over its intended service life. The second question that arises using in-service data involves determining the appropriate balance between the amounts of data collected, the accuracy of the characterised distribution, and the uncertainty associated with the parameter point estimates.

In this paper, statistics of extremes are utilised to address several of these challenges for the application of bridge monitoring. By describing the distribution of the largest observed value in a specified observation timeframe, the statistics of extremes are well suited for structural safety assessment. Because the approach utilises only maximum values, it is efficient in terms of data reduction. An important characteristic of extreme value distributions (EVDs) is their asymptotic behaviour. Assuming no change in the underlying behaviour, once defined for a specific sample size or for a given timeframe, a simple transformation defines the distribution for a larger sample size or a longer desired timeframe. Leveraging this property, information observed in reasonably collectable period of time (days, weeks, or months) can be related to the much longer code specified return periods for use in a reliability analysis. Successive changes in the characterisation of this data with respect to a specific timeframe could indicate a change in the underlying behaviour or damage. After presenting a brief theoretical overview, this paper proposes an approach to characterise monitoring based live loads for use in a reliability analysis. Time effects, the reduction of uncertainty, maximising the utility of a limited amount of data, and the incorporation of parameter uncertainty into a reliability analysis are discussed. Simulations are used to illustrate method development which is then demonstrated using data from an in-service monitoring program.

Theoretical background and its application to highway bridges

The design and assessment of civil infrastructure is often concerned with the largest or smallest (extreme values) of a number of random variables. For example, buildings must withstand maximum wind loads, dams maximum flood levels, and bridges maximum traffic loads for a given time period. With respect to monitoring, this concept can be used as the selection criteria for what data to keep. In permanent monitoring approaches, the magnitude of a continuous data stream across hundreds of sensors becomes a management problem in itself. Selecting, logging, and maintaining peak values is one way to efficiently manage data (Frangopol and Messervey 2009a).

As explained in Ang and Tang (1984), the largest value from samples of size n taken from a population of an initial random variable X with known initial probability density function (PDF), $f_X(x)$, is also a random variable

$$Y_n = \max (X_1, X_2, X_3, \ldots, X_n) \qquad (1)$$

whose PDF may be derived from that of X. Assuming in agreement with random sample theory that the sample random variables $(X_1, X_2, \ldots, X_n)$ are statistically independent and identically distributed as the initial variable X, the cumulative distribution function (CDF) and the PDF of Y_n take, respectively, the forms (Ang and Tang 1984)

$$F_{Y_n}(y) = [F_X(y)]^n \qquad (2)$$

$$f_{Y_n}(y) = n[F_X(y)]^{n-1} f_X(y) \qquad (3)$$

Therefore, the final distribution of the extreme values is a function only of the initial distribution of X and the sample of size n. Depending on how the tail of the underlying distribution decays in the direction of the extreme, Equations (2) and (3) can lead to one of three well known asymptotic forms: the Type I double exponential (Gumbel), the Type II single exponential (Fisher-Tippett), or the Type III bounded exponential (Weibull). Figure 1 shows the the distributions of the largest value from a Gaussian distribution (Figure 1(a)) and a Gumbel distribution (Figure 1(b)), where n is the sample size (i.e. number of occurrences or time). It is noted that for the Gaussian distribution, the PDF of the largest value is shifted to the right and the shape of the distribution changes. For the Gumbel distribution, the PDF of the largest value is of the same shape as that of the initial value, except that the distribution is shifted as indicated in Figure 1(b). Researchers conducting a reliability analysis can leverage this asymptotic behaviour to consider the time effect of any recurring live load with the transformation defining the distribution of the most likely maximum value. The equations that govern the transformation of several common distributions into these asymptotic forms can be found in Ang and Tang (1984).

The use of extreme value statistics is already well established in the design and assessment of highway bridges for the calibration of load factors, resistance factors, load combination factors and the treatment of design trucks (Ghosn *et al.* 2003, Ghosn and Moses

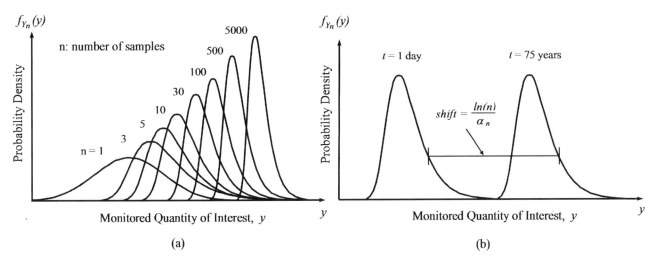

Figure 1. Distributions of the largest value from (a) a Gaussian initial distribution, and (b) a Gumbel initial distribution.

1986). The application of extreme value statistics utilising structure specific data in a probabilistic analysis can consistently account for uncertainties at the local level and develop essentially what could be termed a 'bridge-specific' code (Enevoldsen 2008). It should be noted that the extension of coupling extreme value statistics with bridge monitoring data is slightly different than the transformations depicted in Figure 1. By peak picking maximum values from a monitoring record, the extreme value distribution is observed and characterised directly instead of beginning with the underlying (non-transformed) distribution. Goodness of fit can be investigated to determine which EVD model is most appropriate for any given monitoring record (Liu and Frangopol 2009). However, the Type I Gumbel distribution offers two significant advantages. First, the EVD parameters can be estimated directly from recorded maxima without requiring numerical methods or tables. Second, the transformation to the PDF of the maximum live load during 75 years results in a simple shift of the distribution with no change in shape (Figure 1(b)). For these reasons, this paper develops a method specific to the Type I EVD. Advantages of characterising a monitoring based live load include ensuring that live load parameters are reflective of modern truck weights, passing from a general to a structure specific analysis, and replacing the difficult modelling of a traffic simulation that incorporates vehicle speeds, vehicle spacing, and multiple lane side-by-side truck occurrences.

Characterising monitoring-based distributions and their uncertainty

The characterisation of a monitoring-based distribution is sensitive to the amount of data available. In some cases, the researcher or manager will have the option to collect additional data and in other cases only a limited dataset may be available. Differences in the statistical characterisation of distribution parameters can be addressed in terms of error, confidence, or uncertainty where more information decreases error, increases confidence, or reduces uncertainty. In this study, an approach using uncertainty is developed adopting the Ang and De Leon (2005) classification of uncertainty as either aleatory (data-based uncertainty associated with natural randomness) or epistemic (knowledge-based uncertainty associated with lack of information and imperfect models).

Observation of the transformation of a single known distribution

Method development begins with the investigation of an idealised simulation to quantify how quickly and how accurately the transformation of a Gumbel distribution can be identified via peak picking. In Messervey and Frangopol (2007), a Gumbel distribution was added to the results of the Gindy and Nassif (2006) WIM study which included millions of recent highway truck records in the state of New Jersey. The resulting parameters for the best fit Gumbel were a characteristic value of $u_n = 156$ kN and a shape factor of $\alpha = 0.015$ corresponding to an expected value of $\mu_X = 194.4$ kN and standard deviation of $\sigma_X = 85.5$ kN. The shape factor α is the inverse of the commonly utilised scale factor. For this first simulation, this distribution is randomly sampled 300 times corresponding to a hypothetical average daily truck volume of 300 and the daily maximum value Y_n (Y_{300} if related to the sample size and Y_1 if related to time) is selected. The process is repeated to form a vector of maximum values Y_n which defines the observed extreme value distribution. The mean μ_{Y_n} and standard deviation σ_{Y_n} of this vector of maximum values are related to the Type I Gumbel EVD parameters as (Ang and Tang 1984)

$$\alpha_n = \frac{\pi}{\sqrt{6}\sigma_{Yn}} \text{ and } u_n = \mu_{Y_n} - \frac{\gamma}{\alpha_n} \quad (4)$$

where $\gamma = 0.5772$ is Euler's number. If the considered number of samples is infinitely large, the distribution obtained via 'peak picking' will match the asymptotic transformation of the initial distribution calculated by (Ang and Tang 1984)

$$\mu_{Yn} = \mu_X + \frac{\ln(n)}{\alpha} = 194.4 + \frac{\ln(300)}{0.015} = 574.65 \, \text{kN} \quad (5)$$

Specific to the Gumbel distribution, the standard deviation and shape factor are invariant and no equations are needed to transform these quantities.

The concept of observing a transformed distribution from a known underlying distribution can be visualised in Figure 1(b) where the distribution on left side of this figure is being sampled, the distribution on the right is being observed via the 'peak picking' of maximum values, and Equation (5) provides the theoretical result which enables an evaluation of the peak picking process. The simulation is repeated 1500 times. Each simulation provides a point estimate for μ_{Y_n} and σ_{Y_n}. Repeating the simulation allows the determination of the variability of these point estimates with respect to the number of observed maximum values. 1500 iterations of the simulation for each number of investigated maximum values were enough to ensure that the average of the point estimates converged closely to the value predicted by Equation (5) (e.g. $\mu_{Y_n} = 574.65$ kN) and that the standard deviation also converged closely to its invariant value (e.g. $\sigma_{Y_n} = 85.5$ kN). As such, the standard deviation of the mean and the standard deviation of the standard deviation of the maximum values are determined for increasing sets of maximum

values observed. Table 1 reports the results of this experiment. As expected, the consideration of more maximum values provides less uncertain parameter estimates. Minimum and maximum values of the parameters show the range of possible values and in particular to highlight the sensitivity of a point estimate of σ_{Y_n} to a small number of observations. This variability is noteworthy because short monitoring records of one or two weeks are much more common than long-term monitoring records.

Expanding the results of this particular simulation to a more generalised approach produces some interesting findings. First, comparing the results in the first and third rows of Table 1 reveals that, at least for considered case of a Gumbel distribution, the standard deviation of the mean and the standard deviation of the standard deviation are approximately equal, that is

$$\sigma_{\mu_{Yn}} \approx \sigma_{\sigma_{Yn}} \quad (6)$$

Second, it is noted that the standard deviation of the mean decreases with the number of observations of the maximum value as follows:

$$\sigma_{\mu_{Yn}} = \frac{\sigma_{Yn}}{\sqrt{m}} \quad (7)$$

where m is the number of observations of the maximum value. For $m = 10$ and $\sigma_{Y_n} = 85.5$ kN

$$\sigma_{\mu_{Yn}} = \frac{85.5 \, \text{kN}}{\sqrt{10}} = 27.0 \, \text{kN} \approx \sigma_{\sigma_{Yn}} \quad (8)$$

which is approximately equal to the simulated results for $m = 10$ in Table 1. Figure 2 shows the results of similar calculations from $m = 5$ to $m = 100$ and plots the coefficient of variation vs. the number of observed maximum values of both μ_{Y_n} and σ_{Y_n}. Each plot consists of two curves, one simulated and the other predicted using Equation (7). The curves overlap nearly perfectly except for small values of m where small deviations are present. The last step in generalising the approach is to address that the standard

Table 1. Simulation results describing the variability of monitoring-based parameters for a Type I Gumbel Extreme Value Distribution based on the number of maximum values observed.

Number of maximum values observed	5	10	20	50	100
Standard deviation of μ_{Y_n} (kN)	38.1	26.8	19.1	12.0	8.5
Minimum/maximum observed μ_{Y_n} (kN)	474.7/735.0	500.7/661.2	523.7/651.2	537.8/621.3	550.3/604.6
Standard deviation of σ_{Y_n} (kN)	35.2	26.2	18.8	12.4	8.7
Minimum/maximum observed σ_{Y_n} (kN)	4.8/291.2	29.3/212.7	30.9/179.9	50.8/128.1	57.0/113.6

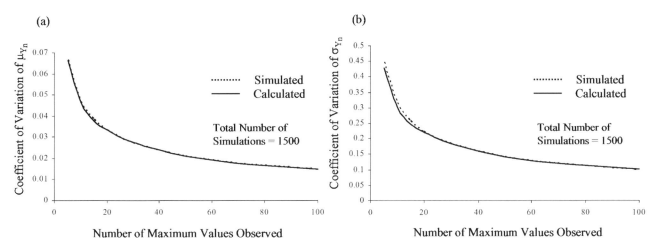

Figure 2. Coefficient of variation vs. the number of maximum values observed for (a) the mean μ_{Y_n} and (b) the standard deviation σ_{Y_n} of a Type I EVD.

deviation of the underlying distribution will be unknown for monitoring applications.

Observation on the transformation of multiple distributions into a single EVD

Characterising the underlying distribution of the live loads upon a bridge structure is more complicated than the simulation of a single known distribution. Changes in daily traffic, side-by-side truck occurrences, the effects of vehicle speeds, wind, and temperature results in a much more uncertain phenomenon. The application of extreme statistics to this data must consider that the process is a convolution of unknown distributions, each characterised by different and potentially varying sampling frequencies. For example, daily truck traffic may be 300 trucks today and 500 trucks tomorrow. Wind events may be fairly infrequent whereas temperature fluctuates daily and so on. For such a process, the transformation into an extreme value distribution for maximum values may be impossible to predict theoretically and may only be observable experimentally.

Based on these observations, the possibility of measuring instead of modelling the live loads is highly desirable. However, such a scenario raises several important questions. Do the maximum value descriptors converge to stable values in a reasonable amount of observations? Can the process still be modelled by an extreme value distribution? And lastly, what is the appropriate timeframe in which to observe and select the maximum values? Although a longer observation timeframe is generally more desirable, it must be balanced against the fact that it reduces the number of maximum values available to define the EVD using the same amount of data.

To begin answering these questions, three additive and independent processes are simulated as detailed in Table 2. Each process has a different frequency of sampling within each unit time of the simulation. Process 1 follows a Gamma distribution. Its frequency of sampling each unit time is determined by a uniform random variable ranging between 400 and 1600. Process 2 follows a Gaussian distribution. Its frequency of sampling is controlled by a random variable with a probability of sampling 5% during unit time. Process 3 also follows a Gaussian distribution. This distribution is sampled exactly once per unit time. The convolution of these three processes is conducted according to the following rules. Each unit time the maximum value from Process 1 is selected. If Process 2 occurs, this value is added directly to Process 1. Process 3 does occur each unit time and its value is added directly to the summation of Processes 1 and 2.

This additive process is motivated by, but does not accurately model, the live-load demand on a generic member of a highway bridge where Process 1 can represent the weight of truck crossings, Process 2 models an occasional wind effect that results in increasing the load effect associated with Process 1, and Process 3 models thermal effects during unit time that result in increasing or decreasing the load effect associated with Processes 1 and 2. Simulation unit times notionally correspond to daily monitoring records where data processing results in one recorded maximum value per day. The Gamma distribution for Process 1 is again fit to the Gindy and Nassif (2006) WIM study. The magnitude of the wind effect (Process 2) and temperature effect (Process 3) are both small compared to the truck weight effect. Of course, this scenario could be improved for any specific structure of interest. However, assuming a linear elastic

Table 2. Simulation distribution types, parameters and number of samples per unit time.

Process 1		Process 2		Process 3	
Distribution	Number of samples per unit time	Distribution	Number of samples per unit time	Distribution	Number of samples per unit time
Gamma	Uniform distribution	Normal		Normal	
$\alpha = 5.69, \beta = 35.83$	$a = 400, b = 1600$	$\mu = 50, \sigma = 10$	0.05 (unit time)	$\mu = 0, \sigma = 20$	One

behaviour, it is reasonable to assume that the load effects of trucks, wind, and temperature on structural members can be superposed. As such, the objective of the simulation is not to accurately model the live load demand upon any particular structure or member, but instead to model and capture the mathematical aspects of a random process in order to determine how long this process must be observed to define its characteristics.

The simulation is run for 45,000 unit times enabling an equal comparison of 1500 observations of the daily, weekly, and monthly timeframes. Therefore, the first 1500 unit times of the simulation characterise the daily observation timeframe, the first 10,500 unit times the weekly observation timeframe, and the full 45,000 unit times the monthly observation timeframe. Starting on the second unit time (or day) of the simulation, the average, μ_{Y_n}, and the standard deviation, σ_{Y_n}, of the maximum are calculated to investigate how many observations are required before these parameters converge on a particular value. The same process is repeated using the same data, but weekly and monthly maxima are selected. The parameters μ_{Y1}, μ_{Y7}, μ_{Y30} and σ_{Y1}, σ_{Y7}, σ_{Y30} (where 1, 7, and 30 denote one day, one week, and one month, respectively) are calculated. These parameters are representative of the same underlying distribution characterised in different observation timeframes using extreme value statistics. Figure 3 depicts the results of the simulation. The observations are recorded on the abscissa and magnitude of the mean load demand or standard deviation is recorded on the ordinate. In each case, excellent convergence is generally achieved after 500 observations, but a sufficiently close convergence is observed to occur much sooner with a small margin of error. As expected with extreme values, longer timeframes (more samples) result in larger mean values. The average daily maximum converges to a value of 598 for the daily maximums, 699 for the weekly maximums, and 769 for the monthly maximums. With respect to the standard deviations, it is noted that longer timeframes decrease variability. The standard deviation of the daily maxima converges to a value of 70, the weekly maxima to 62, the monthly maxima to 59.

Having demonstrated that the simulation results converge to stable values, it is desirable to investigate if an EVD is an appropriate model for this convolution of independent random processes. Using the daily observation timeframe and all data generated by the simulation, Figure 4(a) illustrates the histogram of the 45,000 maxima which has the characteristic shape of an EVD. An empirical cumulative distribution function (ECDF) is created for this data and is plotted in Figure 4(b) with the best fit Type I, Type II, and Type III extreme value distribution CDFs which are found by minimising the sum squared error (SSE). Each EVD model provides a reasonable fit to the ECDF. The Type III (Weibull) fits the data nearly exactly with a slight deviation at the upper tail, the Type I (Gumbel) fits the data very well with a slight deviation at the lower tail, and the Type II (Fisher-Tippett) has slightly larger deviations at both tails.

The Gumbel distribution is selected as the preferred model and it is assumed that this distribution characterises the underlying behaviour and that no further changes in shape occur if the distribution is observed within or transformed to a longer reference timeframe. This simplifying assumption allows for the direct characterisation of the EVD from the monitoring data and allows its simple transformation to longer reference timeframes. The consideration of different types of EVDs and the sensitivity of parameter estimates using monitoring records can be found in Liu et al. (2009). The error introduced by assuming the Gumbel model characterises the underlying behaviour is left for further investigation across the convolution of different types of processes, sampling frequencies, and distributions.

Next, the selection of the best possible observation timeframe is investigated and discussed. To compare alternatives, the distributions must be transformed to a common reference timeframe. Because truck loads dominate the response of a highway bridge it is most appropriate to map the distribution defined by each observation timeframe to the 75 year design truck live load return period consistent with existing codes (Ghosn et al. 2003, Ghosn and Moses 1986). Using the mean and standard deviation obtained in Figure 3 for 1500 observations in a daily timeframe ($\mu_{Y_n} = 598$

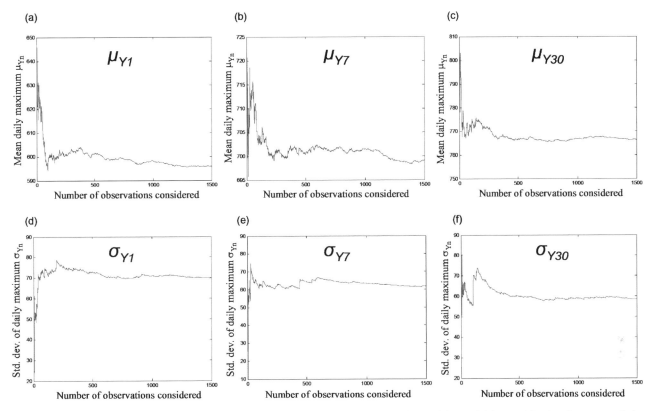

Figure 3.   1500 observations of the mean of daily maxima, μ_{Y_n}, in (a) daily, (b) weekly, and (c) monthly timeframes, and the standard deviation of daily maximums in (d) daily, (e) weekly, and (f) monthly timeframes.

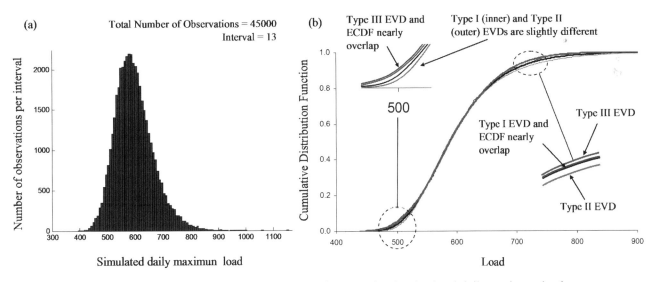

Figure 4.   (a) Histogram and (b) simulation ECDF and best fit EVDs for the simulated daily maximum loads.

and $\sigma_{Y_n} = 70$) the shape parameter of the daily extreme value distribution is calculated using Equation (4) as

$$\alpha_1 = \frac{\pi}{\sqrt{6}(70)} = 0.01833 \qquad (9)$$

Equation (5) can then be utilised to transform the distribution of the daily maximums to the 75-year return period as 1155.45. The standard deviation and the shape parameter remain invariant.

Table 3 reports these values together with the ones obtained when considering the weekly and monthly

Table 3. Extreme value distribution (EVD) parameters characterized in different observations timeframes and transformed to 75-year EVDs.

EVD parameters	Daily, Y_1 $n = 1$	75-year $n = 27375$	Weekly, Y_7 $n = 1$	75-year $n = 3900$	Monthly, Y_{30} $n = 1$	75-year $n = 900$
α	0.0183	0.0183	0.0207	0.0207	0.0217	0.0217
μ (kN)	598	1155	699	1099	769	1082
σ (kN)	70	70	62	62	59	59

observation timeframes. The results show that each observed timeframe (daily, weekly, and monthly) maps to similar, but different 75-year distributions as depicted in Figure 5. Ideally, the observed distributions should be identical in shape (e.g. have the same standard deviation) and should successively shift to the right (larger mean values) on the abscissa. Instead, the transformed distributions should overlap (e.g. be the same distribution). For this simulation, there are some differences in shape among the observed distributions corresponding to the different observed standard deviations. For the transformed distributions, the weekly and monthly 75-year transformed distributions nearly overlap whereas the daily 75-year transformed distribution is shifted further to the right. From Equation (5), it is noted that the shape factor directly correlates to the magnitude of the transformation with higher standard deviations corresponding to larger shifts. For the application of monitoring highway bridges, higher standard deviations in part correspond to less monitoring information.

In themselves, the results from this simulation or from Figure 5 cannot be utilised to state which observation timeframe is best suited for the reliability analysis of any particular structure.

Case study: Lehigh River Bridge

The Lehigh River Bridge SR-33 (Figure 6) was constructed in 2001 and is situated in Bethlehem, PA, USA. The bridge is a four-span continuous weathering steel deck truss with a main span of 181.05 metres. The depth of the truss varies from 10.97 metres (at midspans) to 21.95 metres (over the supports). The structure is subjected to light to medium truck traffic. The Lehigh River Bridge is unique because the reinforced concrete deck is not only composite with the longitudinal steel stringers and transverse floor beams, but also with the upper chord members of the truss through the use of shear studs connecting the upper chords directly to the bridge deck. This structure is the only composite truss in the State of Pennsylvania and possibly the United States (Connor and McCarthy 2006). The main truss members (i.e. upper chords, lower chords and diagonals) are fabricated from structural steel plates into box or 'H' shapes. The steel

stringers, sway bracing, and cross bracing members are all rolled 'W' shapes. The bridge was monitored during construction, for controlled load tests using test trucks, over time for temperature measurements, and for several short periods of in-service usage (Connor and McCarthy 2006, Connor and Santosuosso 2002). The objective of the monitoring program was to measure mechanical and thermal strains during construction and while in-service in order to better understand the performance of the structure. Data is available from representative periods of time that include all seasons. Instrumentation and testing were conducted by personnel from Lehigh University's Center for Advanced Technology for Large Structural Systems (ATLSS). Complete descriptions of the bridge layout as well as the field and instrumentation programs are available in ATLSS reports prepared by Connor and Santosuosso (2002) and by Connor and McCarthy (2006).

Frangopol et al. (2008) used data collected from this monitoring program to assess the reliability of the truss upper chord, truss lower chord, and deck stringers at each construction or testing milestone. The end result of this work with respect to the reliability assessment of these members is reported in Frangopol et al. (2008) and determined that the structure had a high reliability during construction, for the controlled crawl load test, and for the one day of in-service live load monitoring. The next logical step for the assessment of this structure is an extrapolation of the future reliability of the bridge based upon live-loads monitored over time. For this purpose, periodic in-service data is available from 24 sensors located on various members during dates that range from June 2004 to February 2005 (Connor and McCarthy 2006). For each sensor, approximately 90 days of data are available from two control units (one for each direction of traffic) that collected strain measurements when predetermined strain thresholds were exceeded. The selection of the critical members for this study was aided significantly by dynamic tests conducted and reported by Connor and McCarthy (2006). The responses of a lower chord, upper chord, and two stringers to a series of vehicles are all reported in Connor and McCarthy (2006). These tests consistently identify the upper truss chords and four (of nine) stringers as those having the most significant responses

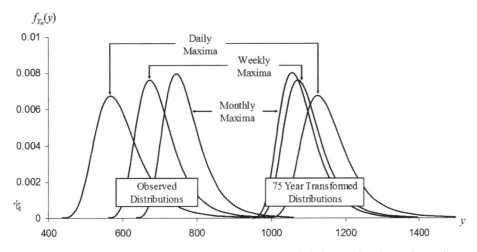

Figure 5. Simulation based observed daily, weekly, and monthly EVDs and their associated transformations to a 75-year EVD.

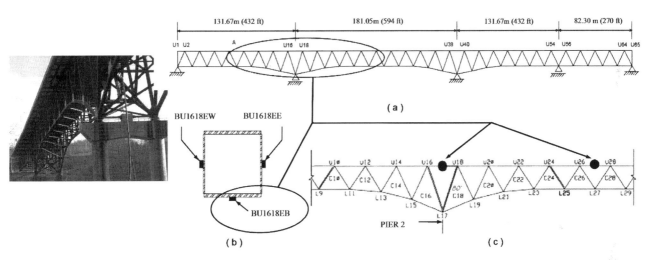

Figure 6. Lehigh River Bridge (a) structural schematic (b) sensor placement on members of interest, and (c) location of the members of interest selected for the reliability analysis (photo taken by Sunyong Kim on 8 January 2008).

during traffic loading. Using the results reported in Connor and McCarthy (2006), and considering member importance, the two upper truss chords, located above a support and at bridge midspan respectively, were selected as the critical members for this structure. Although the stringers have slightly lower reliability indexes as reported in Frangopol *et al.* (2008), the failure of one stringer does not typically lead to global structural failure. In contrast, the failure of a truss component may result in a global collapse and as such these members are deemed more important. The structural scheme and the critical locations selected for analysis are shown in Figure 6.

EVD parameter estimation and selection of the observation timeframe

Daily maximum stresses were computed from the measurements of four sensors centrally placed

on the bottom flange of their respective upper chord truss members (Figure 6(c)). Two sensors were located above the main support (one in each direction of traffic) and two sensors were located at the centre of the main span of the bridge (again one in each direction of traffic). Both locations were of interest due to the differences in member sizes and the deck cross section affecting both the resistance capacity and the dead loads. The data showed that the two eastbound sensors consistently had greater stress demands and as such these were selected for the reliability analysis. Statistical descriptors of the daily maximum stresses for the eastbound sensors were calculated and are reported in Table 4. It is important to note that the bridge experiences very small live-load stresses compared to the average yield strength of 380 MPa for the M270 Grade 50W steel (Frangopol *et al.* 2008).

Although the data in Table 4 could be utilised directly to characterise the daily EVDs for both sensor locations, it is first desirable to investigate if longer observation timeframes provide a more accurate (less uncertain) characterisation of the EVD. For this purpose, maxima are selected from the same data sets by picking the maximum value from timeframes of increasing length (e.g. every two days, every three days, and so on) creating a new vector of maximum values for each timeframe. Once each vector is established, the mean and standard deviation are computed and Equation (4) is utilised to define the respective EVDs. The results from these calculations are shown in Figure 7 for the eastbound sensor at bridge midspan.

Ideally, each successive distribution should shift slightly to the right on the abscissa as maxima are

selected from longer observation timeframes and no change in shape should be present. While the first aspect is observed as expected, a large change in shape between the 1 day, 2 day and 3 day EVDs is present implying that the 1 and 2 day EVDs are likely affected by short term recurring fluctuations in the data. For the same data, the 3, 4, and 5 day distributions are nearly identical in shape indicating that the EVD is well defined by these timeframes. Because this analysis was for a fixed and limited amount of data, the standard deviations begin to increase for the 6 through 12 day observation timeframes as there are fewer and fewer maximums available to define the EVD parameters. From these results, the 3 day distribution is selected as the optimal timeframe from which to select the maximum values because it minimises the standard deviation of the EVD and maximises the number of available maxima once a stable shape is obtained. The impact of this choice is illustrated in Figure 7, which shows the transformation of both the 1 day, 3 day, and 5 day EVDs to a 75-year EVD. It is seen that a selection of the daily observation timeframe for this data would result in a significantly greater mean value and standard deviation for the 75-year EVD and that the 5-day observation timeframe is also not optimal. Further investigating the 3-day observation timeframe, Figure 8 shows its ECDF overlayed with the best fit Type I EVD. Although a limited amount of data is available (30 observations), the model provides a reasonable approximation.

Selecting the 3-day observation timeframe and its corresponding 3-day EVD, the variability of the 3-day mean and standard deviation parameters is now

Table 4. Daily maximum live load stress descriptors for the two selected sensors of interest.

Descriptor	Span 16–18 over support (eastbound)	Span 26–28 midspan (eastbound
Number of days monitored	92	93
Average daily maximum stress, μ_Y (MPa)	8.33	12.75
Standard deviation of the daily maximum stress, σ_Y (MPa)	1.13	2.16
Maximum recorded daily maximum stress (MPa)	10.75	16.98
Minimum recorded daily maximum stress (MPa)	3.63	5.61

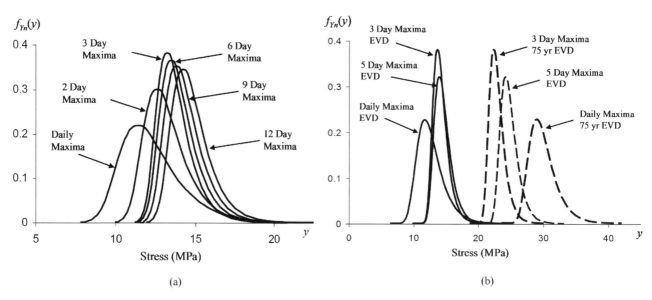

(a) (b)

Figure 7. (a) Selection of the optimal observation timeframe and (b) a comparison of the transformation of the same data to a 75-year EVD observed in three separate timeframes.

investigated. Although this particular 90-day dataset is best observed using 3-day maxima, a rolling average of both parameters reveals that neither converges to a stable value within the 30 available observations as shown in Figure 9.

The uncertainty of each parameter is estimated by applying Equation (7) to the mean and standard deviation of the maximum values using $m = 31$ for the sensor at midspan and $m = 30$ for the sensor above the support. Additionally, the 3-day EVD is transformed to the 75-year EVD using Equation (5) where $n = 9131$ reflects the number of 3-day periods in 75 years. The parameter variability calculated by Equation (7) is identical for both the 3-day and 75-year EVDs due to the invariant shape of the Gumbel distribution when transformed. The results of these calculations are reported in Table 5.

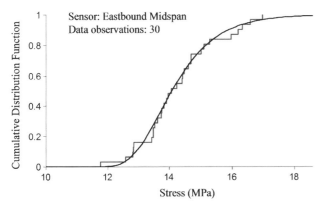

Figure 8. Empirical cumulative distribution function and the best fit Type I EVD for the eastbound sensor at midspan.

The 75-year EVDs are now used for the reliability analysis of these members. Since the calculation of the reliability index requires the mean and standard deviation of the Type I EVD, incorporating the results of Table 5 into a reliability analysis requires multiple iterations of the analysis using each time different realisations of the EVD parameters. To perform the analysis, elastic behaviour is assumed and a simple component analysis is conducted utilising the performance function

$$g = R - L_D - L_L \qquad (10)$$

where R is the member resistance, L_D the dead load, and L_L the 75 year monitoring-based live load reported in Table 5. Table 6 summarises the random variable descriptors for this analysis. The resistance and dead load parameters are adopted from Frangopol et al. (2008). All parameters are reported as stresses and each distribution is characterised by its mean and standard deviation.

Conducting the analysis, the mean reliability index for the upper truss chord at bridge midspan is $\beta = 9.297$ and for the upper chord above the support is $\beta = 8.994$. These results are consistent with those of the park and crawl tests conducted after construction (Frangopol et al. 2008) and indicate that the bridge maintains a high level of in-service safety. In addition, the distribution of the reliability index was also investigated and is reported in Messervey (2009). Because this particular structure experiences such small live load stresses relative to both its capacity and dead load, the reliability index is large in magnitude and the dispersion of the EVD parameters

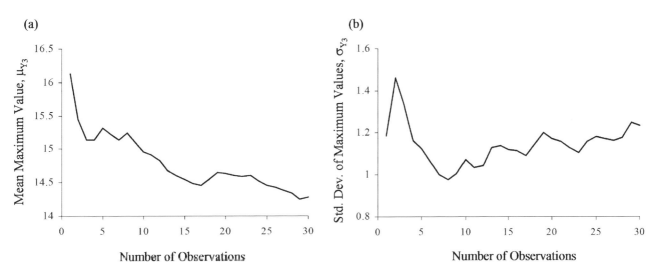

Figure 9. Beginning with the second observation, (a) the mean and (b) the standard deviation of the maximum values as each addition 3-day maximum is recorded.

Table 5. Based on approximately 90 days of monitoring data, the 3-day EVD, the transformed 75-year EVD and the parameters of the EVD mean and standard deviation parameters.

Sensor	Descriptor	3-day EVD	75-year EVD	75-year EVD parameters	
	Number of observations	$m = 31$	$n = 9131$		
Span 26–28 Eastbound sensor at bridge midspan	μ_Y (MPa)	14.26	23.03	Mean of mean	23.03
				Standard deviation of mean	0.22
	σ_Y (MPa)	1.23	1.23	Mean of standard deviation	1.23
				Standard deviation of standard deviation	0.22
	Number of observations	$m = 30$	$n = 9131$		
Span 16–18 Eastbound sensor above support	μ_Y (MPa)	9.13	14.00	Mean of mean	14.00
				Standard deviation of mean	0.12
	σ_Y (MPa)	0.684	0.684	Mean of standard deviation	0.68
				Standard deviation of standard deviation	0.12

Table 6. Reliability analysis distributions and parameters.

Member	Random variable	Distribution and parameters (MPa)
Span 26–28	R	Normal (380, 28)
Eastbound upper	L_D	Normal (92.7, 4.9)
Chord at midspan	L_L	Gumbel (23.03, 1.23)*
Span 16–18	R	Normal (380, 28)
Eastbound upper	L_D	Normal (109, 5.7)
Chord at midspan	L_L	Gumbel (14.0, 0.68)*

*These parameters are random variables themselves as indicated in Table 5.

has little impact on the reliability index although a clear distribution is present. This result is not to be expected for all structures, and especially not for short span bridges typically dominated by the live load response. Regardless, the methodology provides a rational approach for the inclusion of monitoring data into a reliability analysis that considers both time effects and the treatment of the uncertainties inherent to any monitoring data set.

Conclusions

The major findings are summarised as follows:

(1) Statistics of extremes, provided that a number of assumptions are verified, can be applied to bridge monitoring data to characterise monitoring-based live load distributions for use in a reliability analysis. The application of extreme value statistics in this manner is advantageous because it provides a simple and efficient data processing approach and prevents the difficult modelling of vehicle weights, speeds, configurations, side-by-side truck crossings, wind

effects, temperature effects, etc. Furthermore, the asymptotic behaviour of extreme value distributions provides a mechanism to incorporate time effects into the analysis.

(2) This paper developed an approach specific to the Type I Gumbel EVD and furthermore assumed this distribution was characteristic of the underlying phenomenon. Using this assumption, it was possible to characterise EVD parameters directly from the monitoring data. Extending the ideas developed in this paper to other types of EVDs and a more detailed study of the error introduced by the simplifying assumption utilised in this paper are left for further research.

(3) The characterisation of EVD parameters will always occur with less than perfect information. Because the live load on highway bridges is highly uncertain and because monitoring based assessments will occur in short intervals with respect to the lifetime of the structure, it is necessary to optimise the timeframe from which maximum values are selected. The optimal timeframe can be obtained by creating the vectors of maximum values for timeframes of increasing length and by selecting the timeframe with the lowest standard deviation.

(4) Any monitoring-based distribution will be sensitive to the amount of data utilised for its characterisation. A rational method to account for parameter uncertainty based on the amount of available data is presented. The method can serve as a design aid for monitoring program development or can assist the analysis of, in particular, short monitoring records.

(5) The parameter variability of a monitoring-based distribution can be addressed within a

reliability analysis by running multiple iterations of the analysis where the distribution parameters are randomly selected. The end result is a distribution of the reliability index which can be utilised for maintenance and management decisions.

(6) While the proposed approach has a number of strengths, it also has some limitations. Because only maximum values are considered, a significant amount of information is discarded which likely eliminates a more detailed study of the underlying phenomenon and contributing sources of uncertainty. The method is best suited for a top-down analysis where a simplified model of resistance and demand are employed to calculate structural reliability or as a tool to specifically determine the live load distribution within a more detailed bottom-up analysis. Future work on this method has to investigate the development of metrics that indicate changes in structural performance over time based on changes in the EVD parameters.

Acknowledgements

The support from (a) the National Science Foundation through grants CMS-0638728 and CMS-0639428, (b) the Commonwealth of Pennsylvania, Department of Community and Economic Development, through the Pennsylvania Infrastructure Technology Alliance (PITA), (c) the US Federal Highway Administration Cooperative Agreement Award DTFH61-07-H-00040 and (d) the US Office of Naval Research Contract Number N00014-08-1-0188 is gratefully acknowledged. The opinions and conclusions presented in this paper are those of the authors and do not necessarily reflect the views of the sponsoring organisations.

References

Ang, A.H.-S. and De Leon, D., 2005. Modeling and analysis of uncertainties for risk-informed decision in infrastructures engineering. *Structure and Infrastructure Engineering*, 1 (1), 19–31.

Ang, A.H.-S. and Tang, W.H., 1984. *Probability concepts in engineering planning and design*. Volume II. New York: John Wiley & Sons.

Connor, R.J. and McCarthy, J.R., 2006. Report on field measurements and uncontrolled load testing of the Lehigh River Bridge (SR-33). Lehigh University's Center for Advanced Technology for Large Structural Systems (ATLSS), ATLSS Phase II Final Report No. 06–12.

Connor, R.J. and Santosuosso, B.J., 2002. Field measurements and controlled load testing on the Lehigh River Bridge (SR-33). Lehigh University's Center for Advanced Technology for Large Structural Systems (ATLSS), ATLSS Report 02–07.

Enevoldsen, I., 2008. Practical implementation of probability based assessment methods for bridges. *In*: H.-M. Koh and D.M. Frangopol, eds. *Bridge maintenance, safety, management, health monitoring and informatics*. London: Taylor & Francis Group/CRC Press.

FHWA, Federal Highway Administration, 2007. *National bridge inventory* [online]. Available from: http://www.fhwa.dot.gov/bridge/nbi.htm [Accessed 22 May 2007].

Frangopol, D.M. and Liu, M., 2007. Maintenance and management of civil infrastructure based on condition, safety, optimisation, and life-cycle cost. *Structure and Infrastructure Engineering*, 3 (1), 29–41.

Frangopol, D.M. and Messervey, T.B., 2007. Integrated life-cycle health monitoring, maintenance, management and cost of civil infrastructure. *In*: F. Lichu, S. Limin, and S. Zhi, eds. *Proceedings of the international symposium on integrated life-cycle design and management of infrastructures*, 16–18 May 2007, Tongji University, Shanghai, China (keynote paper), Shanghai: Tongji University Press, 216–218; and full 12 page paper on CD-ROM.

Frangopol, D.M. and Messervey, T.B., 2009a. Maintenance principles for civil structures. Chapter 89. *In*: C. Boller, F.K. Chang, and Y. Fujino, eds. *Encyclopedia of structural health monitoring*, Volume 4. UK: John Wiley & Sons, 1533–1562.

Frangopol, D.M. and Messervey, T.B., 2009b. Life-cycle cost and performance prediction: Role of structural health monitoring. Chapter 16. *In*: S.-S. Chen and A.H.-S. Ang, eds. *Frontier technologies for infrastructure engineering*. Boca Raton, London, New York, Leiden: CRC Press/Balkema, 361–381.

Frangopol, D.M., Strauss, A., and Kim, S., 2008. Bridge reliability assessment based on monitoring. *Journal of Bridge Engineering, ASCE*, 13 (3), 258–270.

Ghosn, M., Moses, F., and Wang, J., 2003. Design of highway bridges for extreme events. NCHRP TRB Report 489, Washington, DC.

Ghosn, M. and Moses, F., 1986. Reliability calibration of a bridge design code. *Journal of Structural Engineering*, 112 (4), 745–763.

Gindy, M. and Nassif, H., 2006. Effect of bridge live load based on 10 years of WIM data. *In*: *Proceedings of the 3rd international conference on bridge maintenance and safety*, IABMAS'08, Porto, Portugal. Taylor and Francis. 9 pages on CD-ROM.

Gumbel, E.J., 1958. *Statistics of extremes*. New York: Columbia University Press.

Liu, M., Frangopol, D.M., and Kim, S., 2009. Bridge system performance assessment from structural health monitoring: a case study. *Journal of Structural Engineering, ASCE*, 135 (6), 733–742.

Messervey, T.B., 2009. Integration of structural health monitoring into the design, assessment, and management of civil infrastructure. Thesis (PhD). University of Pavia, Pavia, Italy.

Messervey, T.B. and Frangopol, D.M., 2007. Bridge live load effects based on statistics of extremes using on–site load monitoring. *In*: D.M. Frangopol, M. Kawatani, and C.-W. Kim, eds. *Reliability and optimisation of structural systems: Assessment, design, and life-cycle performance*. London: Taylor & Francis Group, 173–180.

Probabilistic bicriterion optimum inspection/monitoring planning: applications to naval ships and bridges under fatigue

Sunyong Kim and Dan M. Frangopol

Initiation and propagation of fatigue cracks in steel structures induced by repetitive actions are highly random due to both aleatory and epistemic uncertainties related to material properties, loads, damage, modelling and other factors. For this reason, a probabilistic approach is necessary to predict the fatigue crack growth damage. This study presents a probabilistic approach for combined inspection/monitoring planning for fatigue-sensitive structures considering uncertainties associated with fatigue crack initiation, propagation and damage detection. This combined inspection/monitoring planning is the solution of an optimisation formulation, where the objective is minimising the expected damage detection delay. Furthermore, this formulation is extended to a bicriterion optimisation considering the conflicting relation between expected damage detection delay and cost. A set of Pareto solutions is obtained by solving this bicriterion optimisation problem. From this set, a solution can be selected balancing in an optimum manner inspection and monitoring times, quality of inspections, monitoring duration, and number of inspections and monitorings. The proposed approach is applied to a naval ship and a bridge subjected to fatigue.

Introduction

One of main deterioration processes of steel structures is fatigue, defined as the process of initiation and growth of cracks under repetitive loads. In general, the fatigue evolution is affected by uncertainties associated with the location and size of initial crack, stress range near the initial crack, number of cycles, and material and geometric properties (Fisher *et al.* 1998). For this reason, a probabilistic approach is necessary to predict the fatigue crack growth damage for inspection and monitoring planning. During the last decades, several probabilistic approaches have been developed and applied to steel structures including ships and bridges subjected to fatigue (Madsen and Sørensen 1990, Madsen *et al.* 1991, Ayyub *et al.* 2002, Moan 2005, Kwon and Frangopol 2010). These studies were extended into cost-effective inspection and maintenance planning considering probability of fatigue damage detection (Garbatov and Soares 2001, Chung *et al.* 2006, Moan 2011).

The probability of fatigue damage detection has been generally formulated by including the uncertainties in crack size and inspection quality. In order to increase the probability of fatigue damage detection,

advanced damage detection techniques, including structural health monitoring (SHM), have been developed and applied. The objectives of these developments include effective and timely repair actions. The probability of crack detection was defined as the conditional probability that the crack is detected when it has a specific size (Chung *et al.* 2006). Ideally, the probability of damage detection has to be 1.0. However, even in this case, there will be still time lapse from the damage occurrence to the time when the damage is detected (Kim and Frangopol 2011a). Therefore, in order to reduce this time lapse and repair delay, inspection planning should consider, in a rational way, the uncertainties associated with both inspection quality and prediction of damage occurrence. Kim and Frangopol (2011b) proposed a probabilistic approach to establish the optimum inspection planning of ship structures based on minimisation of the expected damage detection delay. However, the effect of SHM on the expected damage detection delay was not investigated.

In this study, a probabilistic approach to establish optimum combined inspection/monitoring planning for fatigue-sensitive structures is presented. In order to compute the expected damage detection delay, the probabilistic approach considers uncertainties associated with fatigue crack initiation, propagation and damage detection. The combined inspection/monitoring planning is the solution of an optimisation formulation, where the objective is minimising the expected damage detection delay. Furthermore, this formulation is extended to a bicriterion optimisation consisting of minimisation of both (a) expected damage detection delay and (b) expected total inspection and monitoring costs. The cost estimation includes costs associated with the type of inspection (i.e. inspection quality), monitoring duration, number of inspections and monitorings, and discount rate of money. For a given number of inspections and monitorings, all the possible combinations of inspection and monitoring are considered, and the associated bicriterion optimisation problems are formulated and solved. Each bicriterion problem has its own Pareto solution set. Based on these Pareto sets, the final Pareto set is obtained. This procedure is extended to determine the optimum-balanced number of inspections and monitorings. The proposed approach is applied to a naval ship and a bridge subjected to fatigue.

Prediction of crack growth

Various types of steel structures including bridges, offshore structures and naval structures are sensitive to fatigue cracking induced by repetitive loads (Fisher *et al.* 1998). It may not be possible to avoid initial fatigue cracks because the cracks may be pre-existing from fabrication. These initial cracks may be propagated into macro-cracks resulting in structural failure. The rate of crack growth depends on the size of initial crack, stress range near the initial crack, number of cycles associated with the stress range, and material and geometry properties of the steel detail (Fisher 1984). Paris' equation (Paris and Erdogan 1963), among other empirical- and phenomenological-based crack propagation models, has been generally used (Fatemi and Yang 1998, Schijve 2003, Mohanty *et al.* 2009). The ratio of the crack size increment to cycle increment is (Paris and Erdogan 1963)

$$\frac{\mathrm{d}a}{\mathrm{d}N} = C(\Delta K)^m \qquad (1)$$

where a = crack size, N = number of cycles and ΔK = stress intensity factor. C and m are material crack growth parameters. The stress intensity factor ΔK is expressed in terms of crack size a and stress range S_{re} as (Irwin 1958)

$$\Delta K = S_{\mathrm{re}} \cdot G(a) \cdot \sqrt{\pi a} \qquad (2)$$

where $G(a)$ = geometry function. Based on Equations (1) and (2), the cumulative number of cycles N associated with crack size a_N can be predicted as (Fisher 1984)

$$N = \frac{1}{C \cdot S_{\mathrm{re}}^m} \cdot \int_{a_o}^{a_N} \frac{1}{(G(a)\sqrt{\pi \cdot a})^m} \mathrm{d}a \qquad (3)$$

where a_o = initial crack size. Furthermore, the time t (years) associated with the occurrence of the crack size a_N is predicted by considering the annual number of cycles N_{an} and the annual increase rate of number of cycles r_c as (Madsen *et al.* 1987)

$$t = \frac{\ln\left[1 + \frac{1}{N_{\mathrm{an}} \cdot C \cdot S_{re}^m} \cdot \ln(1 + r_c) \cdot \int_{a_o}^{a_N} \frac{1}{(G(a)\sqrt{\pi \cdot a})^m} \mathrm{d}a\right]}{\ln(1 + r_c)}$$
$$\text{for } r_c > 0 \qquad (4a)$$

$$t = \frac{1}{N_{\mathrm{an}} \cdot C \cdot S_{re}^m} \cdot \int_{a_o}^{a_N} \frac{1}{(G(a)\sqrt{\pi \cdot a})^m} \mathrm{d}a \quad \text{for } r_c = 0$$
$$(4b)$$

Probability of fatigue damage detection

Probability of fatigue damage detection is defined as the conditional probability that the crack is detected by an inspection method, when the crack exists with a specific size (Chung *et al.* 2006). Probability of damage detection associated with an inspection method has been quantified in terms of crack size (or defect size) and inspection quality (Packman *et al.* 1969, Berens and Hovey 1981, Madsen *et al.* 1991, Mori and Ellingwood 1994a, Frangopol *et al.* 1997, Chung *et al.* 2006). The representative relations between probability of detection P_{ins} and crack size a (or defect size) are:

(a) Shifted exponential form (Packman *et al.* 1969):

$$P_{\mathrm{ins}} = 1 - \exp\left(-\frac{a - a_{\min}}{\lambda}\right) \quad \text{for } a > a_{\min} \qquad (5)$$

where $a_{\min}$ = smallest detectable crack size and λ = characteristic parameter for inspection quality. The value of this parameter ranges from 0 to ∞, and λ decreases with increasing the quality of inspection.

(b) Log-logistic form (Berens and Hovey 1981):

$$P_{\mathrm{ins}} = \frac{\exp[\chi + \kappa \ln(a)]}{1 + \exp[\chi + \kappa \ln(a)]} \qquad (6)$$

where χ and κ are statistical parameters. These parameters can be estimated using the maximum likelihood method for a specific inspection method (Chung *et al.* 2006).

(c) Normal cumulative distribution function (CDF) form (Frangopol *et al.* 1997):

$$P_{\text{ins}} = \Phi\left(\frac{\delta - \delta_{0.5}}{\sigma_\delta}\right) \quad (7)$$

where $\Phi(\cdot)$ = standard normal CDF; δ = damage intensity; $\delta_{0.5}$ = damage intensity at which the inspection method has a probability of detection of 0.5 and σ_δ = standard deviation of $\delta_{0.5}$. The value of $\delta_{0.5}$ represents the quality of inspection. A higher quality of inspection is associated with a smaller value of $\delta_{0.5}$. In this article, the normal CDF form in Equation (7) is used, and the coefficient of variation of $\delta_{0.5}$ is assumed as 0.1 (i.e. $\sigma_\delta = 0.1 \cdot \delta_{0.5}$). The damage intensity δ is defined as (Kim and Frangopol 2011b)

$$\delta = 0 \quad \text{for } a < a_{\min} \quad (8a)$$

$$\delta = \frac{a - a_{\min}}{a_{\max} - a_{\min}} \quad \text{for } a_{\min} \leq a < a_{\max} \quad (8b)$$

$$\delta = 1 \quad \text{for } a \geq a_{\max} \quad (8c)$$

where $a_{\min}$ and $a_{\max}$ are the minimum and maximum detectable crack sizes when the result of the detection is

uncertain (i.e. if $a < a_{\min}$ and $a \geq a_{\max}$ the probability of detection is 0 and 1, respectively).

Expected damage detection delay

Expected damage detection delay when inspection is used

The time lapse from the damage occurrence to the time for the damage to be detected by an inspection method is referred as damage detection delay (Huang and Chiu 1995). When the damage occurs at time t and is detected at time t_{ins} by an inspection after time t, the damage detection delay t_{del} is $t_{\text{ins}} - t$. The formulation of t_{del} considering probability of detection and number of inspections is based on an event tree model (Kim and Frangopol 2011a, 2011b). For example, if inspections are used at time $t_{\text{ins},1}$ and $t_{\text{ins},2}$, and the damage occurs in the time interval t_s to t_e, there will be three possible cases according to damage occurrence time as follows: (a) case 1: $t_s \leq t < t_{\text{ins},1}$; (b) case 2: $t_{\text{ins},1} \leq t < t_{\text{ins},2}$; and (c) case 3: $t_{\text{ins},2} \leq t < t_e$. For case 1 (i.e. $t_s \leq t < t_{\text{ins},1}$), there are three branches as shown in Figure 1. The gray circle node in Figure 1 is a chance node at every inspection where there are two mutually exclusive events (i.e. detection and no detection). The probabilities associated with these two events are P_{ins} (i.e. detection) and $1 - P_{\text{ins}}$ (i.e. no detection), respectively. Branch 1 in Figure 1 represents the event of damage detection at the first inspection. The

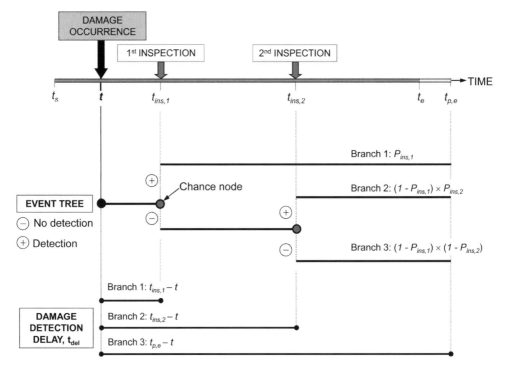

Figure 1. Damage detection delay when damage occurs in the time interval between t_s and $t_{\text{ins},1}$ (i.e. $t_s \leq t < t_{\text{ins},1}$).

corresponding expected damage detection delay is $P_{\text{ins},1} \times (t_{\text{ins},1} - t)$. The expected t_{del} for branch 2 is $(1 - P_{\text{ins},1}) \times P_{\text{ins},2} \times (t_{\text{ins},2} - t)$. In this manner, the expected damage detection delay for each case can be formulated. Furthermore, considering the time t for damage to occur as a continuous random variable, the expected damage detection delay $E(t_{\text{del}})$ for N_i inspections is formulated as (Kim and Frangopol 2011b)

$$E(t_{\text{del}}) = \sum_{j=1}^{N_i+1} \left\{ \int_{t_{\text{ins},j-1}}^{t_{\text{ins},j}} \left[E(t_{\text{del}})_{\text{case},j} \cdot f_T(t) \right] dt \right\} \quad (9)$$

where $f_T(t) = $ probability density function (PDF) of damage occurrence time and $E(t_{\text{del}})_{\text{case},j} = $ expected damage detection delay for case j (i.e. $t_{\text{ins},j-1} \leq t < t_{\text{ins},j}$).

$t_{\text{ins},0}$ for $j = 1$ and t_{ins,N_i+1} for $j = N_i + 1$ are t_s and t_e, respectively. $t_{\text{p,e}}$ in Figure 1 is associated with the time when the damage can be detected with perfect detectability (i.e. probability of damage detection is 1.0). In this study, $t_{\text{p,e}}$ is defined as $t_{\text{p,e}} = t_e + t_p$, where $t_e = $ upper bound of the damage occurrence time, and $t_p = $ time interval during which the expected probability of damage detection is at least 99.9%. The lower and upper bounds of damage occurrence time (i.e. t_s and t_e) are defined as (Kim and Frangopol 2011b)

$$t_s = F_T^{-1}[\Phi(-q)] \quad (10a)$$

$$t_e = F_T^{-1}[\Phi(q)] \quad (10b)$$

where $F_T^{-1}(\cdot) = $ the inverse CDF of the damage occurrence time t. The value of q is assumed 3.0 herein.

Expected damage detection delay when monitoring is used

SHM data allow updating the information on the structural performance. The quality of the information is related to the monitoring duration, the location of sensors and the number of sensors installed. If there is no damage detection delay during monitoring duration t_{md}, and monitoring is applied N_m times with the same duration t_{md}, the expected damage detection delay $E(t_{\text{del}})$ based on Equation (9) becomes

$$E(t_{\text{del}}) = \sum_{j=1}^{N_m+1} \left[\int_{t_{\text{mon},j-1}+t_{\text{md}}}^{t_{\text{mon},j}} (t_{\text{mon},j} - t) \cdot f_T(t) dt \right] \quad (11)$$

where $t_{\text{mon},j} = j$th monitoring starting time. $t_{\text{mon},0} + t_{\text{md}}$ for $j = 1$ and t_{mon,N_m+1} for $j = N_m + 1$ are t_s and t_e, respectively.

Expected damage detection delay when combined inspection/monitoring is used

When combined inspection/monitoring is used to detect damage, the expected damage detection delay $E(t_{\text{del}})$ can be formulated using Equations (9) and (11).

For instance, if one inspection and one monitoring are used, and the inspection is applied *before* monitoring (i.e. $t_{\text{mon},1} > t_{\text{ins},1}$) as shown in Figure 2, there will be four possible cases according to damage occurrence time: (a) case 1: $t_s \leq t < t_{\text{ins},1}$; (b) case 2: $t_{\text{ins},1} \leq t < t_{\text{mon},1}$; (c) case 3: $t_{\text{mon},1} \leq t < t_{\text{mon},1} + t_{\text{md}}$; (d) case 4: $t_{\text{mon},1} + t_{\text{md}} \leq t \leq t_e$. The associated expected damage detection delay is formulated as

$$E(t_{\text{del}}) = \int_{t_s}^{t_{\text{ins},1}} \left[P_{\text{ins},1} \cdot (t_{\text{ins},1} - t) + (1 - P_{\text{ins},1}) \right.$$
$$\left. \cdot (t_{\text{mon},1} - t) \right] \cdot f_T(t) dt + \int_{t_{\text{ins},1}}^{t_{\text{mon},1}} (t_{\text{mon},1} - t)$$
$$\cdot f_T(t) dt + \int_{t_{\text{mon},1}+t_{\text{md}}}^{t_e} (t_{\text{p,e}} - t) \cdot f_T(t) dt \quad (12)$$

It should be noted that case 3 is not considered in Equation (12) because it is assumed that there is no detection delay during monitoring duration t_{md} (i.e. probability of damage detection is 1.0).

On the contrary, when the inspection is used to detect damage *after* monitoring (i.e. $t_{\text{mon},1} + t_{\text{md}} < t_{\text{ins},1}$), the expected damage detection delay is

$$E(t_{\text{del}}) = \int_{t_s}^{t_{\text{mon},1}} (t_{\text{mon},1} - t) \cdot f_T(t) dt$$
$$+ \int_{t_{\text{mon},1}+t_{\text{md}}}^{t_{\text{ins},1}} \left[P_{\text{ins},1} \cdot (t_{\text{ins},1} - t) + (1 - P_{\text{ins},1}) \right.$$
$$\left. \cdot (t_{\text{p,e}} - t) \right] \cdot f_T(t) dt + \int_{t_{\text{ins},1}}^{t_e} (t_{\text{p,e}} - t) \cdot f_T(t) dt$$
$$(13)$$

Inspection and monitoring cost

The inspection cost is related to the quality of an inspection method. In general, inspection methods associated with a higher quality (i.e. higher probability of damage detection) are more expensive (Frangopol et al. 1997). In this study, the cost C_{ins} associated with an inspection method is expressed using $\delta_{0.5}$ in Equation (7) (i.e. damage intensity at which the inspection method has a probability of detection of 0.5) as (Mori and Ellingwood 1994b)

$$C_{\text{ins}} = \alpha_{\text{ins}}(1 - 0.7\delta_{0.5})^{20} \quad (14)$$

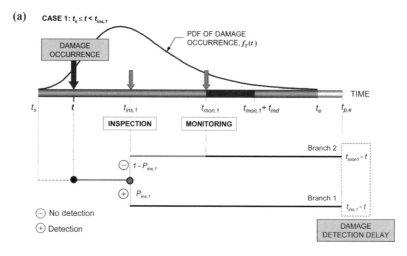

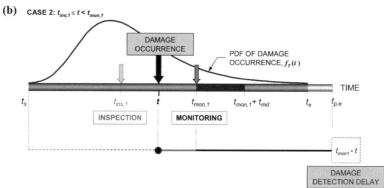

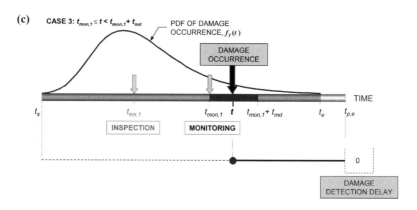

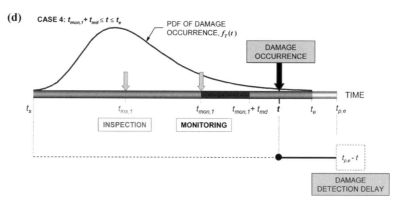

Figure 2. Damage detection delay when inspection and monitoring are used. (a) Case 1: $t_s \leq t < t_{\text{ins},1}$; (b) Case 2: $t_{\text{ins},1} \leq t < t_{\text{mon},1}$; (c) Case 3: $t_{\text{mon},1} \leq t < t_{\text{mon},1} + t_{\text{md}}$ and (d) Case 4: $t_{\text{mon},1} + t_{\text{md}} \leq t \leq t_{\text{e}}$.

where α_{ins} is a constant (i.e. $\alpha_{ins} = 5$). The total inspection cost C_{Tins} for N_i inspections is computed as

$$C_{Tins} = \sum_{j=1}^{Ni} \frac{C_{ins}}{(1+r)^{t_{ins,j}}} \qquad (15)$$

where r = discount rate of money, $t_{ins,j}$ = jth inspection time.

The monitoring cost includes initial design, installation, operation and repair cost of the monitoring system (Frangopol and Messervey 2009). The monitoring cost C_{mon} can be estimated as

$$C_{mon} = C_{mon,ini} + t_{md} \times C_{mon,ann} \qquad (16)$$

where t_{md} = monitoring duration (years), $C_{mon,ini}$ = initial cost of monitoring system consisting of design and installation cost of the monitoring system and $C_{mon,ann}$ = annual cost related to operation and repair of the monitoring system. In this article, $C_{mon,ini}$ and $C_{mon,ann}$ are assumed 10 and 20, respectively.

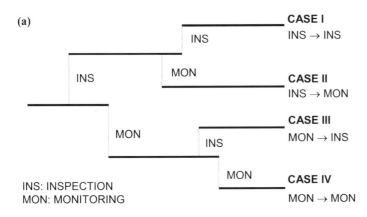

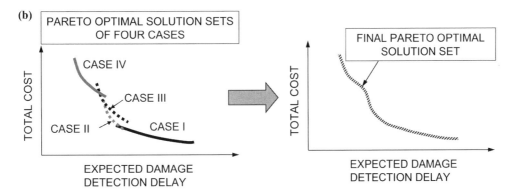

Figure 3. (a) Four possible cases for number of inspections and/or monitorings $N = 2$; (b) Pareto optimal solution sets associated with four possible cases, and final Pareto solution set for $N = 2$.

Table 1. Design variables associated with each case in Figure 3(a); $N_i + N_m = 2$.

Case		Number of inspections N_i	Number of monitorings N_m	Objective functions	Design variables			
I	INS → INS	2	0	$E(t_{del})$ (see Equation (9)); C_{Tins}	$t_{ins,1}$	$t_{ins,2}$	$\delta_{0.5}$	–
II	INS → MON	1	1	$E(t_{del})$ (see Equation (12)); $C_{Tins} + C_{Tmon}$	$t_{ins,1}$	$t_{mon,1}$	$\delta_{0.5}$	t_{md}
III	MON → INS	1	1	$E(t_{del})$ (see Equation (13)); $C_{Tmon} + C_{Tins}$	$t_{ins,1}$	$t_{mon,1}$	$\delta_{0.5}$	t_{md}
IV	MON → MON	0	2	$E(t_{del})$ (see Equation (11)); C_{Tmon}	$t_{mon,1}$	$t_{mon,2}$	–	t_{md}

Note: INS, inspection; MON, monitoring.

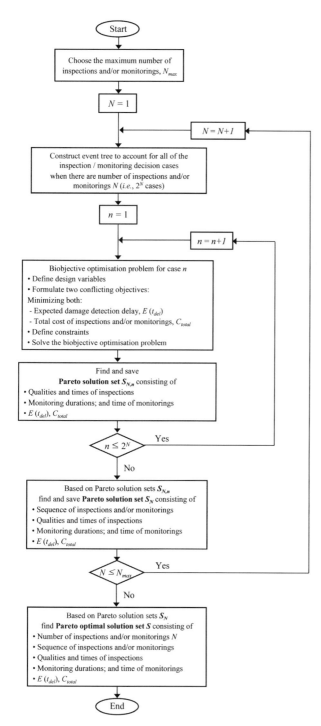

Figure 4. Flow chart to find the final Pareto optimal solution set S.

Furthermore, when a structure is monitored N_m times with the same monitoring duration t_{md}, the total monitoring cost is

$$C_{Tmon} = C_{mon,ini} + \sum_{j=1}^{N_m} \frac{t_{md} \cdot C_{mon,ann}}{(1+r)^{t_{mon,j}}} \qquad (17)$$

where $t_{mon,j} = j$th monitoring starting time.

Bicriterion optimisation

Damage detection with less delay can lead to timely repair actions. In order to reduce damage detection delay, the quality and/or the number of inspections should increase. Furthermore, when monitoring is used to detect damage, damage detection delay depends on monitoring duration and/or number of monitorings. Increasing the monitoring duration and/or number of monitorings can lead to reduction of the damage detection delay (Kim and Frangopol 2011a). However, in general, the limited financial resources constrain the selection of the quality and/or number of inspections and monitorings. Therefore, well-balanced inspection and monitoring planning should be formulated as a bicriterion optimisation with two conflicting objectives by minimising both (a) the expected damage detection delay and (b) the inspection and/or monitoring cost.

If both inspection and monitoring are used to detect damage, and the available number N of inspection N_i and/or monitorings N_m is equal to 2 (i.e. $N = N_i + N_m = 2$), then there will be four possible cases (inspection followed by inspection (case I), inspection followed by monitoring (case II), monitoring followed by inspection (case III) and monitoring followed by monitoring (case IV)) as shown in Figure 3a. The event tree in Figure 3a is used to consider all possible cases (I, II, III and IV). Every case is associated with its own bicriterion optimisation. Each bicriterion optimisation has its own design variables (see Table 1) and produces a Pareto solution set. For example, the design variables of case I in Figure 3a are inspection times (i.e. $t_{ins,1}$ and $t_{ins,2}$) and inspection quality represented by $\delta_{0.5}$ as indicated in Table 1. The objective functions associated with this

Table 2. Random variables used for crack growth model of a joint between bottom plate and longitudinal plate.

	Notation (units)	Mean	COV	Type of distribution
Initial crack size	a_o (mm)	0.5	0.2	Lognormal
Annual number of cycles	N_{an} (cycles/year)	1.0×10^6	0.2	Lognormal
Stress range	S_{re} (MPa)	40	0.1	Weibull
Material crack growth parameter	C	3.54×10^{-11}	0.3	Lognormal

Note: COV, coefficient of variation.

case are the expected damage detection delay $E(t_{del})$ of Equation (9) and the total inspection cost C_{Tins} of Equation (15), when the number of inspections $N_i = 2$. For this case, the total cost C_{total} (i.e. $C_{Tins} + C_{Tmon}$) is equal to C_{Tins}, since there is no monitoring (i.e. $C_{Tmon} = 0$). For case IV in Figure 3a, the bicriterion optimisation problem is formulated by selecting the design variables as monitoring times (i.e. $t_{mon,1}$ and $t_{mon,2}$) and monitoring duration t_{md} (see Table 1). The associated objective functions are indicated in Equations (11) and (17) for $N_m = 2$. Pareto fronts corresponding to the four cases can be obtained after solving bicriterion optimisation problems as shown in Figure 3b. Based on these four Pareto solution sets, the final Pareto solution set can be determined. This Pareto solution set S_N for $N = 2$ will provide the sequence of inspections and monitorings (i.e. inspection followed by inspection, inspection followed by monitoring, monitoring followed by inspection or monitoring followed by monitoring) as well as inspection and/or monitoring times, inspection quality and monitoring durations. This procedure to determine the Pareto solution set S_N for given number of inspections and/or monitorings N can be extended to find the final Pareto solution set S when the available number of inspections and/or monitorings N ranges from 1 to N_{max}. Figure 4 provides a flow chart to find the final Pareto solution set S. The final Pareto solution set S will provide the number of inspections and/or monitorings, the sequence of inspections and monitorings, the inspection and/or monitoring times, inspection quality and monitoring duration.

Application to a naval ship

Description of ship hull structure subjected to fatigue

The proposed approach is applied to a naval ship hull structure. A critical location subjected to fatigue is assumed to be the joint between longitudinal plate and bottom plate. The fatigue crack in the bottom plate can initiate on the edge connected to the stiffener in the transverse direction under repeated loading due to the action of sea water waves. In order to predict the crack length size, Equation (4) is used assuming $m = 2.54$ and using the random variables defined in Table 2. The stress range S_{re} is assumed to be a random variable with a Weibull PDF (Madsen et al. 1991). Initial crack length a_o, annual number of cycles N_{an} and material crack growth parameter C are treated as log-normally distributed random variables. The mean value of material crack growth parameter C is assumed as 3.54×10^{-11} for high-yield steel (HY80) (Dobson et al. 1983). It should be noted that the geometry function $G(a)$ is assumed to be 1.0 (Madsen et al. 1991, Akpan et al. 2002), and there

is no annual increase rate of the number of cycles r_c (i.e. $r_c = 0$).

In this study, the crack size of 1.0 mm is referred to the fatigue crack damage criterion. In other words, the minimum crack size a_{min} for damage intensity δ in Equation (8) is 1.0 mm. The maximum crack size a_{max} in Equation (8) is assumed to be 50 mm. Figure 5 shows the PDF of fatigue damage occurrence (i.e. $a_{min} = 1.0$ mm) time obtained from Monte Carlo simulation with 100,000 samples and its best fitted PDF (i.e. generalised extreme value (GEV) PDF). The GEV PDF is defined as indicated in a study by Kim and Frangopol (2011b). The lower and upper bounds of damage occurrence time (i.e. t_s and t_e in Equation (10)) are 0.41 and 17.56 years, respectively.

Optimum balance of cost and expected damage detection delay

When the available number of inspections and/or monitorings is $N = 2$, there will be four cases. Each case will have its own bicriterion optimisation formulation as mentioned previously (see Figure 3 and Table 1). The bicriterion optimisation formulations of these four cases are formulated as

$$\text{Find} \quad t_{ins,1}, t_{ins,2} \text{ and } \delta_{0.5} \quad \text{for case I} \qquad (18a)$$

$$t_{ins,1}, t_{mon,1}, \delta_{0.5} \text{ and } t_{md} \quad \text{for cases II and III} \qquad (18b)$$

$$t_{mon,1}, t_{mon,2} \text{ and } t_{md} \quad \text{for case IV} \qquad (18c)$$

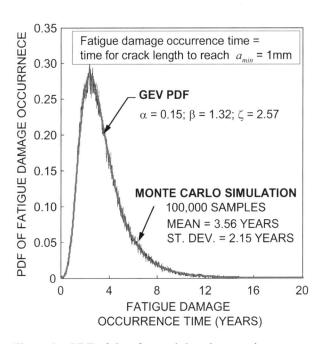

Figure 5.   PDF of time for crack length to reach a_{min}.

to minimise both $E(t_{\text{del}})$ and C_{total} (19)

such that $t_{\text{ins,2}} - t_{\text{ins,1}} \geq 1.0$ year, and

$\quad 0.01 \leq \delta_{0.5} \leq 0.1$ for case I (20a)

$t_{\text{mon,1}} - t_{\text{ins,1}} \geq 1.0$ year, $0.01 \leq \delta_{0.5} \leq 0.1$ and

$\quad 0.3$ year $\leq t_{\text{md}} \leq 1.0$ year for case II (20b)

$t_{\text{ins,1}} - t_{\text{mon,1}} \geq 1.0$ year, $0.01 \leq \delta_{0.5} \leq 0.1$ and

$\quad 0.3$ year $\leq t_{\text{md}} \leq 1.0$ year for case III (20c)

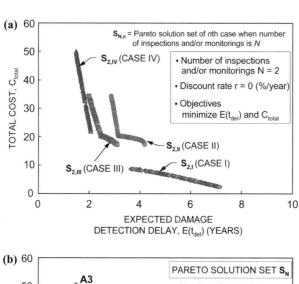

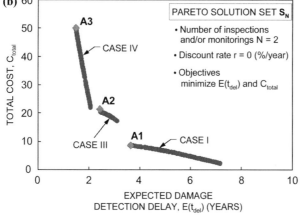

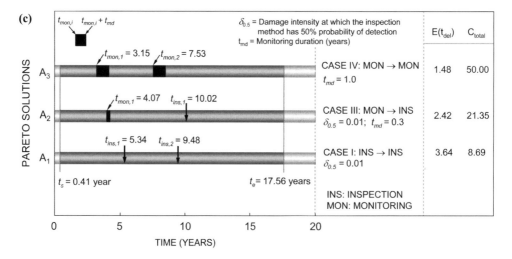

Figure 6. Number of inspections and/or monitorings $N = 2$. (a) Pareto solution sets $S_{N,n}$ for cases I, II, III and IV; (b) Pareto solution set S_N; (c) combined inspection/monitoring plans for solutions A_1, A_2 and A_3 in (b).

$t_{mon,2} - t_{mon,1} \geq 1.0$ year and 0.3 year $\leq t_{md} \leq 1.0$ year

for case IV   (20d)

given $N = N_i + N_m = 2$, and $f_T(t)$ (21)

The design variables and constraints of the bicriterion optimisation formulations for cases I, II,

III and IV are indicated in Equations (18) and (20), respectively. The objectives are to minimise both expected damage detection delay $E(t_{del})$ and total cost C_{total}. The time interval between inspections and/or monitorings has to be at least 1 year, and $\delta_{0.5}$ representing the quality of inspection should be in the interval 0.01–0.1. Monitoring duration t_{md} has to be in the time interval 0.3–1.0 year. The GEV PDF $f_T(t)$

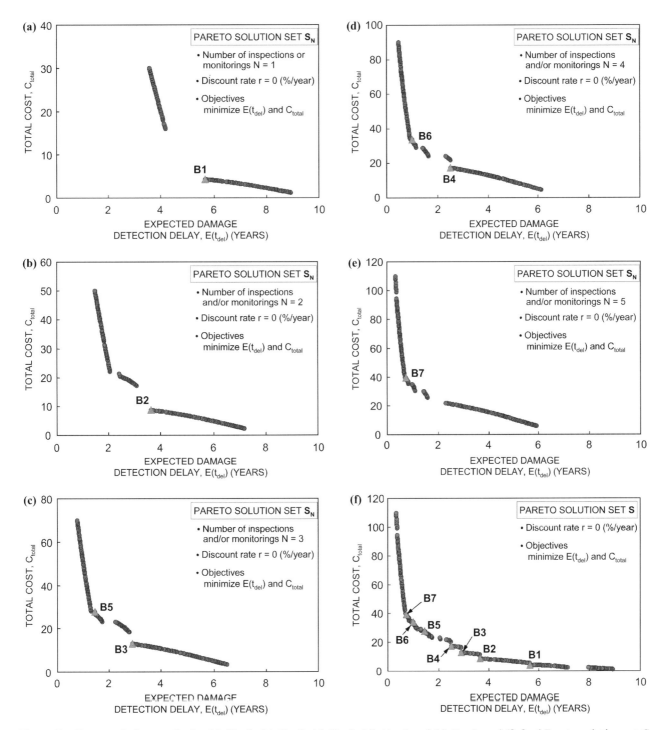

Figure 7.   Pareto solution set S_N for (a) $N=1$; (b) $N=2$; (c) $N=3$; (d) $N=4$ and (e) $N=5$, and (f) final Pareto solution set S.

in Figure 5 indicated in Equation (21) is used to formulate $E(t_{del})$. Non-dominated sorting in genetic algorithms (NSGA-II) programme developed by Deb *et al.* (2002) is used to find the Pareto solution set of the bicriterion optimisation formulations in Equations (19)–(21).

The genetic algorithm process with 500 generations provides the Pareto sets for cases I, II, III and IV shown in Figure 6a. $S_{N,n}$ denotes a Pareto set of nth case when available number of inspections and/or monitorings is N. For example, $S_{2,I}$ in Figure 6a is the Pareto solution set of case I (INS → INS case in

Table 3. Pareto optimum solutions in Figure 7: values of objectives and design variables.

| Pareto optimum solution | Objectives | | Design variables | | | | | | | |
	$E(t_{del})$ (years)	C_{total}	N	Optimum inspection or monitoring times (years)					$\delta_{0.5}$	Monitoring duration t_{md} (years)
B_1	5.67	4.35	1	$t_{ins,1}$ 7.47					0.01	–
B_2	3.64	8.69	2	$t_{ins,1}$ 5.34	$t_{ins,2}$ 9.48				0.01	–
B_3	2.90	13.03	3	$t_{ins,1}$ 4.51	$t_{ins,2}$ 6.82	$t_{ins,3}$ 10.82			0.01	–
B_4	2.49	17.38	4	$t_{ins,1}$ 4.11	$t_{ins,2}$ 5.61	$t_{ins,3}$ 7.73	$t_{ins,4}$ 11.81		0.01	–
B_5	1.45	27.69	3	$t_{mon,1}$ 3.03	$t_{mon,2}$ 5.68	$t_{ins,1}$ 11.49			0.01	0.33
B_6	0.99	33.90	4	$t_{mon,1}$ 2.58	$t_{mon,2}$ 4.21	$t_{mon,3}$ 6.68	$t_{ins,1}$ 12.59		0.01	0.33
B_7	0.74	39.44	5	$t_{mon,1}$ 2.19	$t_{mon,2}$ 3.50	$t_{mon,3}$ 5.10	$t_{mon,4}$ 7.79	$t_{ins,1}$ 13.46	0.01	0.31

Note: N, total number of inspections and/or monitorings.

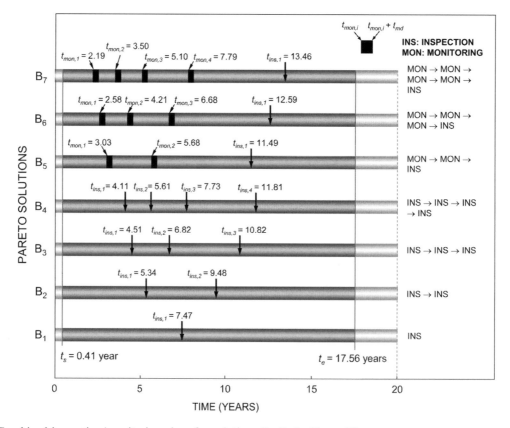

Figure 8. Combined inspection/monitoring plans for solutions B_1–B_7 in Figure 7f.

Table 4. Random variables used for crack growth model of a cover plate.

Random variables	Notation (units)	Mean	COV	Type of distribution
Initial crack size	a_o (mm)	0.5	0.2	Lognormal
Annual number of cycles	N_{an} (cycles/year)	1.62×10^6	0.2	Lognormal
Annual increase rate of number of cycles	r_c (%)	2	0.1	Lognormal
Stress range	S_{re} (Mpa)	13.78	0.1	Weibull
Material crack growth parameter	C	2.024×10^{-13}	0.25	Lognormal
Weld size	Z (mm)	16	0.1	Lognormal

Note: COV, coefficient of variation.

Figure 3 and Table 1). A Pareto set $S_{N,n}$ consists of 100 populations. The final Pareto solution set S_2, based on the Pareto solution sets for $n = $ I–IV in Figure 6a, is obtained using the ε-constraint approach by minimising the selected single objective function, while other objective functions are treated as constraints (Haimes et al. 1971) as follows

$$\text{Minimise} \quad f_i \quad (22)$$

$$\text{subject to} \quad f_j \leq \varepsilon_j \text{ for all } j = 1, 2, \ldots, q; j \neq i \quad (23)$$

where $i \in \{1, 2, \ldots, q\}$ and $q = $ number of objective functions. The final Pareto solution set S_2 is shown in Figure 6b. Combined inspection/monitoring plans of the three representative solutions A_1, A_2 and A_3 in Figure 6b are illustrated in Figure 6c. The inspection and monitoring plan for solution A_1 requires two inspections (case I) applied at time $t_{ins,1} = 5.34$ years and $t_{ins,2} = 9.48$ years with $\delta_{0.5} = 0.01$ (see Figure 6c), and the associated $E(t_{del})$ and C_{total} are 3.64 years and 8.69, respectively (see Figure 6b). If Pareto solution A_2 is selected instead of A_1, the expected damage detection delay $E(t_{del})$ will be reduced from 3.64 years to 2.42 years, but an additional cost of 12.66 (i.e. 21.35–8.69) is needed as shown in Figure 6b. The inspection and monitoring plan associated with A_2 (case III) consists of the monitoring starting time $t_{mon,1} = 4.07$ years with monitoring duration $t_{md} = 0.3$ year and the inspection at time $t_{ins,1} = 10.02$ years with $\delta_{0.5} = 0.01$ (see Figure 6c). It should be noted that the discount rate of money was not considered (i.e. $r = 0$); the value of $\delta_{0.5}$ is assumed to be the same for the first and second inspections associated with case I, and the same monitoring duration t_{md} is used for the first and second monitorings associated with case IV.

In a similar way, the Pareto sets S_N for $N = 1$–5 are obtained as shown in Figure 7a–e. The final Pareto set S considering N as a design variable is also found by using the ε-constraint approach based on the Pareto solution sets S_N. The detailed procedure to find the final Pareto set S is provided in Figure 4. Figure 7f shows the Pareto set S. The optimum values of design

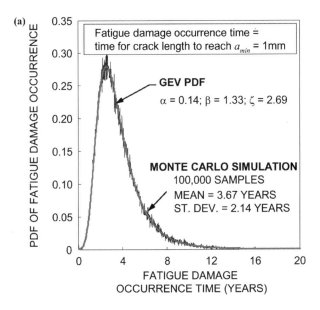

(a)

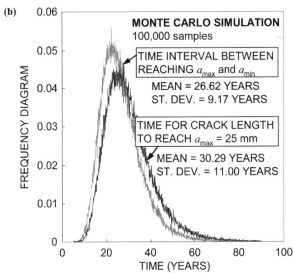

(b)

Figure 9. PDF: (a) time for crack length to reach a_{min}; (b) time for crack length to reach a_{max} and time interval between reaching a_{max} and a_{min}.

variables and objective functions of the seven representative solutions B_1–B_7 in Figure 7 are provided in Table 3. Combined inspection/monitoring plans for

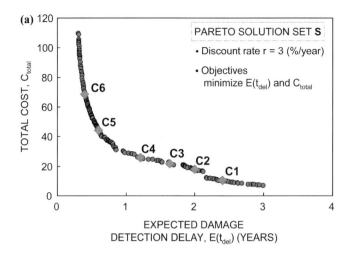

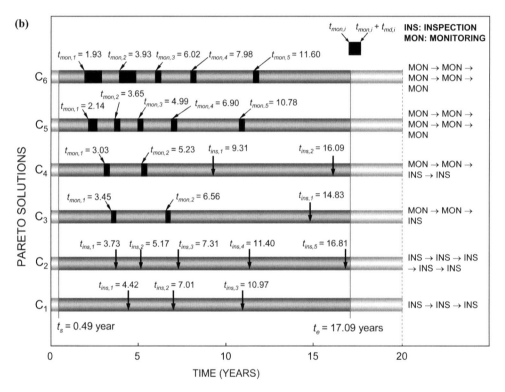

Figure 10. (a) Final Pareto solution set S and (b) combined inspection/monitoring plans for solutions C_1–C_6 in (a).

solutions B_1–B_7 are illustrated in Figure 8. Solutions B_3 and B_5 in Figure 7f are found in the Pareto solution set S_3 in Figure 7c. Solutions B_4 and B_6 in Figure 7f are associated with the Pareto set S_4 in Figure 7d. Solution B_6 requires three monitorings with the same duration $t_{md} = 0.33$ year and one inspection with $\delta_{0.5} = 0.01$, and the corresponding $E(t_{del})$ and C_{total} are 0.99 year and 33.90, respectively (see Table 3). Monitoring times $t_{mon,1}$, $t_{mon,2}$, $t_{mon,3}$ are 2.58, 4.21, 6.68 years, and inspection time $t_{insp,1}$ is 12.59 years as shown in Figure 8. In order to reduce the total cost C_{total}, solution B_4 consisting of four inspections with

$\delta_{0.5} = 0.01$ can be selected. As a result, C_{total} can be reduced from 33.90 to 17.38, but $E(t_{del})$ will increase from 0.99 to 2.49 years.

Application to an existing bridge

Description of an existing highway bridge subjected to fatigue

The proposed approach is applied to an existing highway bridge, the Yellow Mill Pond Bridge located in Bridgeport, Connecticut, USA. In this application, critical location is assumed to be the end of the cover plate weld. Detailed information is available in a study by Fisher (1984). In order to predict the time for the occurrence of a given crack size at this critical location, Equation (4) is used. The geometry function $G(a)$ in Equation (4) is defined as (Fisher 1984)

$$G(a) = G_e(a) \cdot G_s(a) \cdot G_w(a) \cdot G_g(a) \quad (24)$$

where $G_e(a)$ = crack shape factor = 0.952; $G_s(a)$ = front face factor = $1.211 - 0.186 \cdot \sqrt{a/c}$; $G_w(a)$ = finite width factor = 1.0 and $G_g(a)$ = stress gradient factor = $K_{tm} \cdot [1 + 6.79 \cdot (a/t_f)^{0.435}]^{-1}$, where a = depth crack size; c = width crack size; t_f = flange thickness; K_{tm} = stress concentration factor = $-3.54 \cdot \ln(Z/t_f) + 1.98 \cdot \ln(t_{cp}/t_f) + 5.80$; Z = weld size; t_{cp} = cover plate thickness. The relation between depth crack size a and width crack size c is assumed as $c = 5.462 \times a^{1.133}$. The deterministic parameters used are the flange thickness $t_f = 32.0$ mm, the cover plate thickness $t_{cp} = 31.8$ mm and the material parameter $m = 3.0$ (Shetty and Baker 1990). All random variables necessary to predict crack growth of this critical location are provided in Table 4.

Figure 9a shows the PDF of time for the crack size to reach $a_{min} = 1.0$ mm assumed as the fatigue crack damage criterion. Through comparison with Monte Carlo simulation with 100,000 samples, best fitting PDF (i.e. GEV PDF) with $\alpha = 0.14$, $\beta = 1.33$ and $\zeta = 2.69$ is obtained as shown in Figure 9a. If the maximum crack size a_{max} for damage intensity defined in Equation (8) is assumed to be 25 mm, the time for damage intensity to reach 1.0 will have the mean value of 30.29 years and the standard deviation of 11.00 years as shown in Figure 9b. Furthermore, the PDF associated with the time interval between damage occurrence (i.e. crack size $a = a_{min}$) and full damage (i.e. crack size $a = a_{max}$) is shown in Figure 9b. In general, damage should be detected and repaired before the time when the crack size reaches a_{max}. Since crack size will increase from a_{min} to a (see Equation (8b)) during the damage detection delay, the damage detection delay has to be less than the time associated with $a_{max} - a_{min}$. Therefore, the time interval between damage occurrence and full damage in Figure 9b can provide an upper bound of the damage detection delay.

Optimum balance of cost and expected damage detection delay

The general formulation of the bicriterion optimisation problem for a given number of inspections and/or monitorings N is

$$\text{Find} \quad \boldsymbol{t_{ins}} = \{t_{ins,1}, t_{ins,2}, \ldots, t_{ins,N_i}\};$$
$$\boldsymbol{t_{mon}} = \{t_{mon,1}, t_{mon,2}, \ldots, t_{mon,N_m}\};$$
$$\boldsymbol{t_{md}} = \{t_{md,1}, t_{md,2}, \ldots, t_{md,N_m}\} \text{ and}$$
$$\boldsymbol{\delta_{0.5}} = \{\delta_{0.5,1}, \delta_{0.5,2}, \ldots, \delta_{0.5,N_i}\} \quad (25)$$

$$\text{to minimise both} \quad E(t_{del}) \text{ and } C_{total} \quad (26)$$

such that $t_{ins,j} - t_{ins,j-1} \geq 1.0$ year;
$0.01 \leq \delta_{0.5,j} \leq 0.1$; $t_{mon,j} - t_{mon,j-1}$
≥ 1.0 year; 0.3 year $\leq t_{md,j} \leq 1.0$ year and
$|\boldsymbol{t_{mon}} - \boldsymbol{t_{ins}}| \geq 1.0$ year $\quad (27)$

$$\text{given} \quad N = N_i + N_m; \text{ and } f_T(t) \quad (28)$$

The design variables are the vectors of inspection times $\boldsymbol{t_{ins}}$, monitoring times $\boldsymbol{t_{mon}}$, monitoring durations $\boldsymbol{t_{md}}$ and quality of inspections $\boldsymbol{\delta_{0.5}}$. The time intervals between inspections and/or monitorings have to be at

Table 5. Pareto optimum solutions in Figure 10(a): values of objectives and design variables.

Pareto optimum solution	$E(t_{del})$ (years)	C_{total}	N	Optimum inspection and/or monitoring times (years)					$\delta_{0.5}$ and/or monitoring duration t_{md} (years)				
C_1	2.40	10.41	3	$t_{ins,1}$ 4.42	$t_{ins,2}$ 7.01	$t_{ins,3}$ 10.97			$\delta_{0.5,1}$ 0.01	$\delta_{0.5,2}$ 0.01	$\delta_{0.5,3}$ 0.01		
C_2	2.00	17.95	5	$t_{ins,1}$ 3.73	$t_{ins,2}$ 5.17	$t_{ins,3}$ 7.31	$t_{ins,4}$ 11.40	$t_{ins,5}$ 16.81	$\delta_{0.5,1}$ 0.01	$\delta_{0.5,2}$ 0.01	$\delta_{0.5,3}$ 0.01	$\delta_{0.5,4}$ 0.02	$\delta_{0.5,5}$ 0.10
C_3	1.64	21.61	3	$t_{mon,1}$ 3.45	$t_{mon,2}$ 6.56	$t_{ins,1}$ 14.83			$t_{md,1}$ 0.32	$t_{md,2}$ 0.30	$\delta_{0.5,1}$ 0.08		
C_4	1.22	26.02	4	$t_{mon,1}$ 3.03	$t_{mon,2}$ 5.23	$t_{ins,1}$ 9.31	$t_{ins,2}$ 16.09		$t_{md,1}$ 0.36	$t_{md,2}$ 0.30	$\delta_{0.5,1}$ 0.01	$\delta_{0.5,2}$ 0.07	
C_5	0.60	17.95	5	$t_{mon,1}$ 2.14	$t_{mon,2}$ 3.65	$t_{mon,3}$ 4.99	$t_{mon,4}$ 6.90	$t_{mon,5}$ 10.78	$t_{md,1}$ 0.51	$t_{md,2}$ 0.30	$t_{md,3}$ 0.30	$t_{md,4}$ 0.30	$t_{md,5}$ 0.30
C_6	0.40	68.70	5	$t_{mon,1}$ 1.93	$t_{mon,2}$ 3.93	$t_{mon,3}$ 6.02	$t_{mon,4}$ 7.98	$t_{mon,5}$ 11.60	$t_{md,1}$ 1.00	$t_{md,2}$ 0.98	$t_{md,3}$ 0.34	$t_{md,4}$ 0.30	$t_{md,5}$ 0.31

Note: N, total number of inspections and/or monitorings.

least 1 year. Constraints for t_{md} and $\delta_{0.5}$ are indicated in Equation (27). The GEV PDF $f_T(t)$ in Figure 9a is used to formulate $E(t_{del})$. For given N, the total number 2^N of Pareto sets $S_{N,n}$ can be obtained by solving the bicriterion optimisation problems in Equations (25)–(28). Finally, the Pareto solution set S can be obtained through the procedure given in Figure 4. Figure 10a shows this final Pareto set S and six representative solutions C_1–C_6. Values of design variables (i.e. N, t_{ins}, t_{mon}, t_{md} and $\delta_{0.5}$) and objective functions (i.e. $E(t_{del})$ and C_{total}) are given in Table 5. It should be noted that the annual discount rate r is assumed 3%. The combined inspection/monitoring plans corresponding to solutions C_1–C_6 are illustrated in Figure 10b.

Conclusions

This article presented a probabilistic approach to establish an optimum combined inspection/monitoring planning for ship and bridge structures subjected to fatigue based on bicriterion approach. For given number of inspections and monitorings, all possible combinations of inspection and monitoring were considered. Each combination was associated with a bicriterion optimisation formulation consisting of two conflicting objectives by simultaneously minimising both the expected damage detection delay and the total inspection and monitoring cost. Based on the Pareto solution sets of all combinations of inspection and monitoring for the given number of inspections and monitorings, the final Pareto set was obtained. Finally, this procedure was extended to determine the optimum number of inspections and monitorings. The following conclusions can be drawn from this study:

(1) Due to the scarcity of financial resources allocated for maintaining and/or improving the reliability of the deteriorating civil and marine infrastructure systems, an optimisation considering damage detection delay and inspection/monitoring is crucial. Along these lines, the proposed bicriterion approach provides powerful means to optimise inspection/monitoring planning for fatigue-sensitive structures under uncertainty.

(2) In addition, performance measures such as system reliability, robustness and redundancy may be integrated in the proposed approach, by extending it to a multicriterion optimisation formulation under uncertainty (Frangopol and Liu 2007, Frangopol 2011). Moreover, the optimum combined inspection/monitoring planning approach proposed in this article can be extended to life-cycle cost design of fatigue-sensitive structures including bridges and naval ships by considering initial, inspection, monitoring, maintenance, repair and failure costs.

(3) In general, fatigue damage can be detected with less delay by using monitoring than inspection. However, monitoring is usually more expensive than inspection. Therefore, combined inspection/monitoring planning provides an optimal-balanced solution. Damage detection delay leads to repair delay. This delay increases the probability of failure.

(4) The fatigue damage occurrence and propagation are random processes involving intermittent growths and dormant periods. In order to consider these evolutionary features, Markov chains, jump process models and stochastic differential equations have been developed (Sobczyk 1987). The scheduling of inspection and monitoring can be affected by the time evolution model of fatigue cracks. Therefore, further studies are needed to incorporate such advanced stochastic modellings into the approach proposed in this article.

Acknowledgements

The support from grants by (a) the National Science Foundation through CMS-0639428, (b) the Commonwealth of Pennsylvania, Department of Community and Economic Development, through the Pennsylvania Infrastructure Technology Alliance (PITA), (c) the US Federal Highway Administration Cooperative Agreement Award DTFH61-07-H-00040 and (d) the US Office of Naval Research Contract Number N00014-08-1-0188 is gratefully acknowledged. The opinions and conclusions presented in this article are those of the authors, and do not necessarily reflect the views of the sponsoring organisations.

References

Akpan, U.O., et al., 2002. Risk assessment of aging ship hull structures in the presence of corrosion and fatigue. *Marine Structures*, 15 (3), 211–231.

Ayyub, B.M., et al., 2002. Reliability-based design guidelines for fatigue of ship structures. *Naval Engineers Journal*, 114 (2), 113–138.

Berens, A.P. and Hovey, P.W., 1981. *Evaluation of NDE reliability characterization*. Dayton: Air Force Wright-Aeronautical Laboratory, Wright-Patterson Air Force Base.

Chung, H.-Y., Manuel, L., and Frank, K.H., 2006. Optimal inspection scheduling of steel bridges using nondestructive testing techniques. *Journal of Bridge Engineering*, 11 (3), 305–319.

Deb, K., et al., 2002. A fast and elitist multiobjective genetic algorithm: NSGA-II. *IEEE Transactions on Evolutionary Computation*, 6 (2), 182–197.

Dobson, W.G., et al., 1983. *Fatigue considerations in view of measured load spectra*. Washington, DC: SSC-315, Ship Structure Committee.

Fatemi, A. and Yang, L., 1998. Cumulative fatigue damage and life prediction theories: a survey of the state of the art for homogeneous materials. *International Journal of Fatigue*, 20 (1), 9–34.

Fisher, J.W., 1984. *Fatigue and fracture in steel bridges*. New York, NY: John Willey & Sons.

Fisher, J.W., Kulak, G.L., and Smith, I.F., 1998. *A fatigue primer for structural engineers*. Chicago, IL: National Steel Bridge Alliance.

Frangopol, D.M., 2011. Life-cycle performance, management, and optimization of structural systems under uncertainty: accomplishments and challenges. *Structure and Infrastructure Engineering*, 7 (6), 389–413.

Frangopol, D.M., Lin, K.Y., and Estes, A.C., 1997. Life-cycle cost design of deteriorating structures. *Journal of Structural Engineering*, 123 (10), 1390–1401.

Frangopol, D.M. and Liu, M., 2007. Maintenance and management of civil infrastructure based on condition, safety, optimization, and life-cycle cost. *Structure & Infrastructure Engineering: Maintenance, Management, Life-Cycle*, 3 (1), 29–41.

Frangopol, D.M. and Messervey, T.B., 2009. Life-cycle cost and performance prediction: role of structural health monitoring [Chapter 16]. *In*: S.-S. Chen and A.H.-S. Ang, eds. *Frontier technologies for infrastructures engineering*. Leiden, The Netherlands: CRC Press-Balkema-Taylor & Francis Group, 361–381.

Garbatov, Y. and Soares, C.G., 2001. Cost and reliability based strategies for fatigue maintenance planning of floating structures. *Reliability Engineering and System Safety*, 73 (3), 293–301.

Haimes, Y.Y., Lasdon, L.S., and Wismer, D.A., 1971. On a bicriterion formulation of the problems of integrated system identification and system optimization. *IEEE Transactions on Systems, Man, and Cybernetics*, 1 (3), 296–297.

Huang, B.-S. and Chiu, H.-N., 1995. The quality management of the imperfect production process under two monitoring policies. *International Journal of Quality & Reliability Management*, 12 (3), 19–31.

Irwin, G.R., 1958. *The crack-extension-force for a crack at a free surface boundary*. NRL report 5120. Washington, DC: Naval Research Laboratory.

Kim, S. and Frangopol, D.M., 2011a. Inspection and monitoring planning for RC structures based on minimization of expected damage detection delay. *Probabilistic Engineering Mechanics*, 26 (2), 308–320.

Kim, S. and Frangopol, D.M., 2011b. Optimum inspection planning for minimizing fatigue damage detection delay of ship hull structures. *International Journal of Fatigue*, 33 (3), 448–459.

Kwon, K. and Frangopol, D.M., 2010. Bridge fatigue reliability assessment using probability density functions of equivalent stress range based on field monitoring data. *International Journal of Fatigue*, 32 (8), 1221–1232.

Madsen, H.O., *et al.*, 1987. Probabilistic fatigue crack growth analysis of offshore structures, with reliability updating through inspection. *In*: *Proceedings of the marine structural reliability symposium*, 5–8 October 1987. Arlington, VA: SSC/SNAME, 45–55.

Madsen, H.O. and Sørensen, J.D., 1990. Probability-based optimization of fatigue design, inspection and maintenance. *In*: *Proceedings of the 4th international symposium on integrity of offshore structures*, 2–3 July 1990. London: Elsevier, 421–432.

Madsen, H.O., Torhaug, R., and Cramer, E.H., 1991. Probability-based cost benefit analysis of fatigue design, inspection and maintenance. *In*: *Proceedings of the marine structural inspection, maintenance and monitoring symposium*, 18–19 March 1991. Arlington, VA: SSC/SNAME, II.E.1–II.E.12.

Moan, T., 2005. Reliability-based management of inspection, maintenance and repair of offshore structures. *Structure and Infrastructure Engineering*, 1 (1), 33–62.

Moan, T., 2011. Life-cycle assessment of marine civil engineering structures. *Structure and Infrastructure Engineering*, 7 (1–2), 11–32.

Mohanty, J.R., Verma, B.B., and Ray, P.K., 2009. Prediction of fatigue crack growth and residual life using an exponential model: part I (constant amplitude loading). *International Journal of Fatigue*, 31 (3), 418–424.

Mori, Y. and Ellingwood, B.R., 1994a. Maintaining reliability of concrete structures. I: role of inspection/repair. *Journal of Structural Engineering*, 120 (3), 824–845.

Mori, Y. and Ellingwood, B.R., 1994b. Maintaining reliability of concrete structures. II: optimum inspection/repair. *Journal of Structural Engineering*, 120 (3), 846–862.

Packman, P.F., *et al.*, 1969. Definition of fatigue cracks through nondestructive testing. *Journal of Materials*, 4 (3), 666–700.

Paris, P.C. and Erdogan, F.A., 1963. Critical analysis of crack propagation laws. *Journal of Basic Engineering*, 85 (4), 528–534.

Schijve, J., 2003. Fatigue of structures and materials in the 20th century and the state of the art. *International Journal of Fatigue*, 25 (8), 679–702.

Shetty, N.K. and Baker, M.J., 1990. Fatigue reliability of tubular joints in offshore structures: reliability analysis. *In*: *Proceedings of the 9th international conference on offshore mechanics and arctic engineering*, 18–23 February 1990. Vol. 2. Houston, USA: ASME, 231–239.

Sobczyk, K., 1987. Stochastic models for fatigue damage of materials. *Advances in Applied Probability*, 19 (3), 652–673.

Integration of structural health monitoring in a system performance based life-cycle bridge management framework

Nader M. Okasha and Dan M. Frangopol

In this paper, an approach for integrating the information obtained from structural health monitoring in a life-cycle bridge management framework is proposed. The framework is developed on the basis of life-cycle system performance concepts that are also presented in this paper. The performance of the bridge is quantified by incorporating prior knowledge and information obtained from structural health monitoring using Bayesian updating concepts. This performance is predicted in the future using extreme value statistics. Advanced modelling tools and techniques are used for the lifetime reliability computations, including incremental nonlinear finite element analyses, quadratic response surface modelling using design of experiments concepts, and Latin hypercube sampling, among other techniques. The methodology is illustrated on an existing bridge in the state of Wisconsin.

Introduction

At Lehigh University, a research programme under the leadership of the second author has led to the development of a framework for the life-cycle management of bridges under uncertainty. This framework is presented herein. In essence, the objective of this framework is to provide a practical computational methodology for the assessment and maintenance of bridge structures. Inherent uncertainties in load and resistance parameters impose a necessity for using probabilistic tools to quantify the structural performance. Lifetime reliability has been a pillar in many of the proposed approaches for bridge management. The use of advanced tools can greatly improve the quality and increase the accuracy of these frameworks.

Reduction of epistemic uncertainty and validation of present information is a key step in a good life-cycle management framework. In the past, researchers have used visual inspection results in reducing uncertainties in the performance quantification process. However, visual inspection may not distinguish structural from cosmetic deterioration, nor identify damage states that threaten public safety or significant economic loss (Ellingwood 2005). This issue may be resolved by using destructive and/or non-destructive testing results. However, these testing techniques are often time consuming, expensive, and access to all critical locations is not always possible (Chang et al. 2003). Also, destructive testing may introduce local damage to the structure. Proof load testing has been used for updating the lifetime reliability (Faber et al. 2000). However, proof load testing carries with it a number of research issues, most of which centre around determining the appropriate load intensity for the test to be informative (Ellingwood 2005).

Advances in structural health monitoring (SHM) have paved the road for SHM's practical utility. SHM can provide accurate information about both the resistance parameters of the structure and load effects acting on its components. Further, SHM can be performed over discrete time intervals or throughout the service-life of the structure. The former approach provides information about the changes in structural performance between SHM assessments and the latter provides knowledge of these changes as they happen.

One challenge in dealing with SHM has been the processing of the large amount of data extracted and interpreting it into meaningful information that can be used for decision making. Accordingly, SHM has enjoyed significant research efforts aimed at the development of algorithms and methodologies for system identification, damage detection, and updating of finite element models. Only recently, research that treats SHM under uncertainty has emerged (Frangopol et al. 2008a, b, Strauss et al. 2008a, b, Catbas et al. 2008, Liu et al. 2009a, b). However, these studies have mainly focused on information related to the load effects that SHM provides. Furthermore, these studies have primarily used classical inference concepts in which prior information cannot be easily incorporated and, in fact, is neglected. Okasha *et al.* (2010) have proposed an approach in which the SHM information can be, in fact, used to update the structural parameters of the structure that are in turn used in updating the lifetime reliability of the structure.

This paper uses Bayesian inference concepts which enable the inclusion of necessary prior information and it focuses on updating the load effects. The procedures in this paper and in Okasha *et al.* (2010) together are intended to provide the means to bridge the gap between SHM and the life-cycle management of bridge structures under uncertainty. These procedures are a crucial part of the framework presented herein.

In short, an approach is proposed in this paper in which information obtained from SHM is integrated in a life-cycle bridge management framework. The framework is developed on the basis of life-cycle system performance concepts that are also presented. Probabilistic concepts are used for treating uncertainties, such as Bayesian updating, extreme value statistics, and lifetime reliability evaluation. Advanced modelling tools and techniques are used for the lifetime reliability computations, including incremental nonlinear finite element analyses, quadratic response surface modelling using design of experiments concepts, and Latin hypercube sampling, among other techniques. The methodology is illustrated on an existing bridge in the state of Wisconsin.

Bridge management framework

Ultimately, optimal decisions are needed for ensuring the continuous safety of structural systems under multiple objectives and constraints and only a proper integrated framework would yield such decisions (Frangopol 2010). Figure 1 shows a flowchart of the integrated life-cycle management framework proposed in this paper. This figure shows the main features of the framework. Figure 2 shows the framework with extended details. Uncertainty is an integral component in all aspects of this (or any) life-cycle management framework (Frangopol and Okasha 2008). Therefore, each of the steps specified in this approach is to be conducted with careful treatment of the associated uncertainties. The following is a description of the framework.

Build FE models

For a given bridge structure, and following this framework, the first step is to build finite element (FE) models of its system and components. These models can be used for at least two purposes:

(a) To perform the life-cycle performance (reliability) analysis. Even though, performance

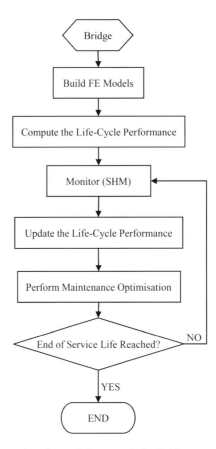

Figure 1. Flowchart of framework for bridge management.

indicators can be developed without using FE analysis (FEA), this framework takes advantage of the advances in such analysis in obtaining performance assessment with higher accuracy. More details on this point are given in the next subsection.

(b) To be used with the SHM to update the resistance parameters. Traditionally, SHM data has been used to update FE models of structures for the purpose of obtaining a FE model that captures the performance of the structure more accurately. Okasha *et al.* (2010) have used the finite element updating to update the parameters of the structure that are in turn used in updating the lifetime reliability of the structure.

Compute the life-cycle performance

Indicators that can be used in representing the behaviour of the structure under uncertainty are numerous (Frangopol and Okasha 2008). A major factor that is present when deciding the choice of a performance indicator is its compatibility with the framework used. For instance, it is important that the

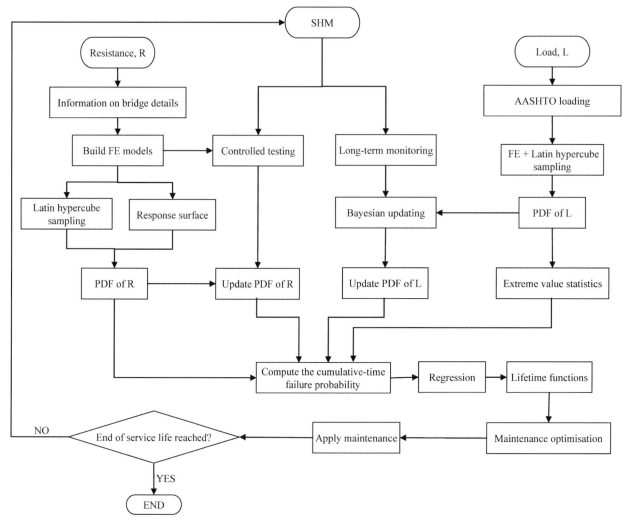

Figure 2. Flowchart of framework for bridge management with extended details.

performance indicator provides the ability to be accurately updated by SHM data. Also, depending on the maintenance optimisation method, the performance indicator should be computationally affordable. Frangopol and Okasha (2009a) compared the computational requirements of various performance indicators in a genetic algorithm based maintenance optimisation. It was found that the lifetime functions are the most computationally affordable, but least accurate. In order to improve their accuracy Okasha and Frangopol (2010a) developed lifetime functions of the system performance of a bridge using FEA and response surface methods. Also, Okasha et al. (2010) developed lifetime functions for the components of the same bridge using FEA and Latin hypercube sampling. In addition, Okasha and Frangopol (2010c, d) have discussed the importance of including the system redundancy as a performance indicator.

Performance indicators for both the components of the structure and its overall system are necessary for the optimisation of maintenance. On one hand, it is necessary to know the effect of the maintenance actions to a component on its performance. On the other hand, the advantages of using the system performance as criterion are well established.

System performance has traditionally been established as a function of the component performance. This function, however, has mainly been established based solely on engineering judgment. For example, component-system relationships have been established based on assumptions of series-parallel models (Estes and Frangopol 1999). The results have been found to be quite sensitive to the system model assumed (Liu et al. 2009b). Alternatively, the system performance can be found directly without the need to quantify component performance (Okasha and Frangopol 2010a). However, in this case, the component performance indicators are still needed for performing a practical maintenance optimisation.

In this framework, both component and system performance indicators are obtained using advanced computational tools. The component-system performance relationship is established by correlating the component performance to the true system performance by means of multiple nonlinear regression. This generates an equation that accurately relates the changes in the system performance over time as a function of the changes in the component performance over time. This avoids the need for assumptions in obtaining this important relationship.

Perform SHM

Monitoring of structural health can be beneficial for bridge management in various aspects, and many response quantities can be measured by a large number of devices. Cost and accuracy is a trade-off in the choice of the SHM system and monitored quantities. Details of the various SHM systems, measured quantities and techniques can be found elsewhere (Chang *et al.* 2003, Glaser *et al.* 2007, Aktan *et al.* 2002, Liu *et al.* 2009b).

Update the life-cycle performance

To date, the primary monitored response quantity measured by SHM that has been used in reliability analysis is strain (Frangopol *et al.* 2008a, b, Strauss *et al.* 2008a, b, Liu *et al.* 2009a, b). Temperature and temperature-induced strains have also been used (Catbas *et al.* 2008). In each of these investigations, the structural response due to the applied loads (load effect) has been the target for integration of SHM in reliability analysis. However, prior information has not been considered. Furthermore, only the instantaneous reliability index has been adopted.

In this framework, both the structural resistance and the structural response are updated with SHM. Lifetime functions are used and prior information is incorporated. Details of the approach for updating the lifetime functions and instantaneous reliability index by updating the structural resistance parameters can be found in Okasha *et al.* (2010). In this approach, static strain measurements are used to update the finite element model of the bridge and, in turn, update its lifetime reliability. The finite element updating is conducted using an optimisation technique known as Particle Swarm (Okasha *et al.* 2010). A methodology for updating the structural response to traffic load is presented in the next section of this paper.

Perform maintenance optimisation

In a maintenance optimisation problem, one must carefully formulate the problem and select the solution tool. Various techniques can be found in the literature. If regular time interval preventive maintenance with one essential maintenance is employed, the approach developed by Neves *et al.* (2006a,b) can be used. When different parts of the structure can be maintained at different times, the approach by Okasha and Frangopol (2009) can be used, as well as the approach developed by Okasha and Frangopol (2010b). In the latter approach, regular or irregular-time interval preventive maintenance combined with any number of appropriate essential maintenance types applied at the optimum times are found by multi-objective optimisation and solved by genetic algorithms. Lifetime functions, namely the unavailability and redundancy, are used as performance indicators. As previously mentioned, advanced techniques for developing these lifetime functions are described in Okasha and Frangopol (2010a) and Okasha *et al.* (2010). The maintenance optimisation is formulated and solved as a nested problem in which an algorithm for finding the best essential maintenance plan is imbedded into a genetic algorithm for finding the best preventive maintenance schedule. The essential maintenance algorithm uses an intelligent event tree search through all potentially optimum plans. For the genetic algorithm, each timing of preventive maintenance application is represented by a continuous design variable. The number of times of application of each preventive maintenance type is represented with an integer design variable and its value determines the number of continuous design variables to consider for each preventive maintenance type (Okasha and Frangopol 2010b).

End of service life

A maintenance plan is selected from the results of the optimisation and is put in action. It is possible that SHM is applied in a continuous fashion throughout the structure and the framework is updated in real-time. However, this may not be economically feasible. In such a case, and after a certain duration of time, the application of SHM may be repeated and the predicted results of the framework can be validated and updated (i.e. the process is repeated from the 'Monitor (SHM)' step in Figure 1). This is done until the specified service life of the structure is consumed. At this point, the structure is deemed unsafe and not fit for service.

SHM and Bayesian updating

In essence, the main motivation for devising ways to incorporate SHM data into a life-cycle framework is to find an estimation of a certain quantity of interest, such as the load effects and structural resistance parameters. Therefore, updating is a mathematical estimation problem. The quantities of interest are inherently random. Accordingly, mathematical estimation consists of two steps: (a) mathematically model the data (i.e. describe it by its probability density function (PDF)), and (b) estimate its parameters (Kay 1993). The classical estimation approach treats the parameters of the PDF deterministically, and they can be found by methods such as the maximum likelihood estimate (Bucher 2009). To account for estimation errors, the classical estimation approach uses confidence intervals to express the degree of these errors (Ang and Tang 2007). However, the classical estimation approach provides no room for incorporating prior information. Instead, the Bayesian estimation approach treats the parameters of the PDF as random variables, and makes it possible to use prior knowledge (Kay 1993). Judgment based solely on SHM data obtained over a period of time may lack information on extreme events that are encountered outside this period. Therefore, it is crucial to combine this SHM data with established estimates of these quantities, for example the AASHTO specified live load (AASHTO 2002) to be combined with the load effects obtained from SHM. Therefore, the classical estimation is inappropriate and a systematic methodology for using the Bayesian estimation approach in integrating SHM data is needed. This is provided in this section.

Basic concepts

When dealing with the Bayesian approach for estimation of the properties of a random variable X (e.g. the maximum applied bending moment), it is of extreme importance to keep in mind that the Bayesian updating targets the random variables θ_i that represent the parameters of the PDF of X and not directly X itself. This is different from the classical approach where the parameters θ_i are treated as deterministic. In the Bayesian approach, and before updating, the θ_i parameters are described by a PDF called the prior PDF $f'(\theta)$, which constitutes the prior knowledge that is intended for combining with the SHM data. In addition, the prior PDF has parameters of its own. To distinguish them from the parameters of the PDF of the underlying random variable X, the parameters of the prior are called hyper-parameters (Fink 1995).

The new information obtained from SHM, is in a form of discrete sample values. For instance, the maximum daily bending moment encountered at a section over a period of N days may constitute the sample set to be used. This data is used to construct the likelihood function $L(\theta)$ as (Ang and Tang 2007)

$$L(\theta) = \prod_{i=1}^{N} f_X(x_i|\theta) \quad (1)$$

where $f_X(x_i|\theta)$ is the PDF of X evaluated at the SHM data value x_i, given that the parameter of the PDF is θ. The likelihood function $L(\theta)$ may be formally defined as the likelihood of observing the SHM data values x_i assuming a given parameter θ.

Combining the prior with the likelihood provides the new updated information about the parameter θ, which is in the form of the posterior PDF $f''(\theta)$ given as (Ang and Tang 2007)

$$f''(\theta) = kL(\theta)f'(\theta) \quad (2)$$

where k is a normalising constant required to make $f''(\theta)$ a proper PDF and is calculated independently of θ as

$$k = \left[\int_{-\infty}^{\infty} L(\theta)f'(\theta)d\theta \right]^{-1} \quad (3)$$

Updating X

The result of the Bayesian approach is evidently the updated PDF of the random parameter of the underlying random variable X. However, in life-cycle management, it is most likely that a reliability-based performance indicator is used. Methods and software that compute these performance indicators are built to treat random variables with deterministic parameters. It is therefore necessary to combine the inherent variability in X with the variability in the posterior PDF of its parameter in order to obtain the updated PDF of X whose parameters become determinestic. By virtue of the total probability theorem, the updated PDF of X is

$$f'_X(x) = \int_{-\infty}^{\infty} f_X(x|\theta)f''(\theta)d\theta \quad (4)$$

Closed-form solutions for Equation (4) are not always possible to obtain, as will be seen with SHM data treatment. Alternatively, an approximate updated distribution can be found by first performing numerical integration to obtain the cumulative distribution function (CDF)

$$F'_X(y_i) = \int_{-\infty}^{y_i} \int_{-\infty}^{\infty} f_X(u|\theta)f''(\theta)d\theta du \quad i = 1, 2, ..., k \quad (5)$$

where $Y = [y_1, y_2, ..., y_k]$ is an array of values large enough to cover the range of probable values of X and with small enough interval, and then, an appropriate CDF is fitted to the obtained values using the method

of least squares, where the parameters $\boldsymbol{p}$ of the distribution are determined by solving the following optimisation problem (Bucher 2009).
Find:

the parameters vector $\boldsymbol{p} = \{p_1, p_2, \ldots, p_n\}$ (6a)

To minimise:

$$s = \sum_{i=1}^{k} [F(y_i) - F^*(y_i, \boldsymbol{p})]^2 \quad (6b)$$

where $F^*(y_i, \boldsymbol{p})$ is the fitted CDF at y_i with parameters $\boldsymbol{p}$. The optimisation problem associated with Equation (6) is solved for several distributions. The goodness of fit is determined using the coefficient of determination R^2 statistic (Bucher 2009)

$$R^2 = 1 - \frac{s}{\sum_{i=1}^{k} [F(y_i)]^2} \quad (7)$$

The R^2 statistic takes values between 0 and 1, where the closer to 1 the better the fit. The distribution with the highest R^2 value is the one chosen for fitting the cumulative-time failure probability.

It is advised to avoid sampling-based numerical integration techniques, such as the adaptive Simpson quadrature, to compute Equation (5). These methods may find that the function is essentially zero at every point that it sampled, not knowing that the function is an essential delta function with a spike at some location it may not sample. Alternatively, the trapezoidal rule with a fine enough interval can be used.

Conjugate Priors

A prior that is conjugate of the underlying random variable is a prior distribution when combined with the likelihood function; a posterior function in the same mathematical form of the prior is obtained. For example, if the underlying random variable X is Gaussian (with unknown mean μ and known standard deviation σ), and the prior of its parameter μ (i.e. its mean) is also Gaussian (with hyper-parameters μ' and σ'), the posterior of μ is obtained as Gaussian with hyper parameters

$$\mu'' = \frac{\bar{x}(\sigma')^2 + \mu'(\sigma^2/n)}{(\sigma')^2 + (\sigma^2/n)} \quad (8a)$$

$$\sigma'' = \sqrt{\frac{(\sigma')^2(\sigma^2/n)}{(\sigma')^2 + (\sigma^2/n)}} \quad (8b)$$

where $\bar{x}$ is the sample mean.

Conjugate prior distributions can be constructed if the data set of an arbitrary size can be completely characterised by a fixed number of summaries with respect to the likelihood function, i.e. it depends on the existence of sufficient statistics of fixed dimension for the given likelihood function (Fink 1995). For example, the sufficient statistics in the case of Equation (8) are $\bar{x}$ and n. Hence, all the information that is needed from the data set in order to perform Bayesian updating is contained in the number of points in that data set and its mean. A comprehensive collection of conjugate priors can be found in Fink (1995).

Non-conjugate priors

The basic justification for the use of conjugate prior distributions is that it is easy to understand the results, which can often be put in analytic form, and they simplify computations (Gelman et al. 2003). Unfortunately, it is mostly the case currently that conjugate priors are not used and thus special techniques are used for Bayesian computations (Ghosh et al. 2006). Except in simple cases, explicit evaluation of the integrals in Bayesian analysis is rarely possible (Smith and Gelfand 1992). In fact, the recent popularity of the Bayesian approach to statistical applications is mainly due to advances in statistical computing (Ghosh et al. 2006).

In this paper, treatment of SHM data is associated with monitoring of extreme events (i.e. load effects of very heavy trucks). The literature of papers linking the themes of Bayesian updating and extreme value modelling is sparse, in part due to computational difficulties, some of which have recently been overcome by techniques such as Markov chain Monte Carlo (McMC) (Coles and Powell 1996). Extreme value distributions do not lend themselves easily to Bayesian updating; the main problem is that there is no conjugate distribution; the Weibull distribution does not belong to the exponential family (Bedford and Cooke 2001). Thus, Bayesian-updating of extreme value distributions cannot lead to explicit posterior distributions (Robert 1994). Hence, a simulation procedure is the best way to determine the posterior distribution (Thodi et al. 2009). The Metropolis-Hastings (MH) algorithm has been suggested for this purpose (Robert 1994, Robert and Cassella 1999). In fact, Thodi et al. (2009) have observed that the MH algorithm produced better results than those obtained with an alternative Laplace approximation method.

Metropolis-Hasting algorithm

The basic idea of the Metrapolis-Hasting (MH) algorithm, whose development led to very considerable progress in Bayesian analysis, is not to directly simulate from the target density (which may be computationally very difficult) at all, but to simulate an easy Markov chain that has this density as its stationary distribution (Ghosh *et al.* 2006). Instead of finding the closed-form solution of a posterior PDF, which may not be available, samples from this PDF are generated. However, this requires evaluating the normalising factor k in Equation (2) (see Equation (3)). In many cases, this is a formidable task. The power of McMC methods, particularly the MH algorithm is their ability to generate the samples without resorting to computing the integration required in the normalising factor k (see Equation (3)). Once the samples are generated, the PDF of the posterior can be approximated (Smith and Gelfand 1992).

According to Tierney and Mira (1999), the MH algorithm dates back to Metropolis *et al.* (1953), was extended by Hasting (1970) and came into widespread use as a tool for Bayesian analysis by the work of Tanner and Wong (1987) and Gelfand and Smith (1990). The following is a brief description of the algorithm based on MathWorks (2009):

(1) Assume an initial value $x(t)$.
(2) Draw a sample, $y(t)$, from a proposal distribution $q(y \mid x(t))$.
(3) Accept $y(t)$ as the next sample $x(t + 1)$ with probability $r(x(t), y(t))$, and keep $x(t)$ as the next sample $x(t + 1)$ with probability $1 - r(x(t), y(t))$, where:

$$r(x, y) = \min \left\{ \frac{f(y)}{f(x)} \frac{q(x|y)}{q(y|x)}, 1 \right\} \qquad (9)$$

(4) Increment $t \rightarrow t + 1$, and repeat steps 2 and 3 until the desired number of samples are obtained.

As in any McMC method, the draws are regarded as a sample from the target density only after the chain has passed the transient stage and the effect of the fixed starting value has become so small that it can be ignored (Chib and Greenberg 1995).

The slice sampling algorithm

To produce quality samples efficiently with MH algorithm, it is crucial to select a good proposal distribution. If it is difficult to find an efficient proposal distribution, the slice sampling algorithm without explicitly specifying a proposal distribution can be used (MathWorks 2009). Slice sampling originates with the observation that to sample from a univariate

distribution, points can be sampled uniformly from the region under the curve of its PDF and then only the horizontal coordinates of the sample points are looked at (Neal 2003). The following is a brief description of the algorithm based on MathWorks (2009):

(1) Assume an initial value $x(t)$ within the domain of $f(x)$.
(2) Draw a real value y uniformly from $(0, f(x(t)))$, thereby defining a horizontal 'slice' as $S = \{x: y < f(x)\}$.
(3) Find an interval $I = (L, R)$ around $x(t)$ that contains all, or much of the 'slice' S.
(4) Draw the new point $x(t + 1)$ within this interval.
(5) Increment $t \rightarrow t + 1$ and repeat steps 2 through 4 until the desired number of samples are obtained.

Test example

Example 9.8 of Ang and Tang (2007) is considered in this paper to illustrate some of the above concepts. In this example, both the underlying and prior densities are Gaussian and thus the problem is solved in Ang and Tang (2007) using Equation (8). Herein, the problem is solved by constructing the original likelihood function which is used with the MH and slice sampling algorithms in MATLAB (MathWorks 2009) to find the estimates of the posterior and its PDF. Briefly, the prior of the mean of the underlying distribution is N(20.420 m, 0.020 m) and the underlying random variable X is also Gaussian with a known standard deviation 0.08 and unknown mean (i.e. to be estimated). A sample of five points is used $x_i = \{20.45, 20.38, 20.51, 20.42, 20.46\}$ m, giving a sample mean $\bar{x} = 20.444$ m. Hence, using Equation (8), the posterior mean and standard deviations of the mean of the posterior are found in closed-form as 20.426 m and 0.017 m, respectively. Accordingly, the accuracy of the simulation methods can be tested.

The prior PDF of the mean of the underlying distribution and the likelihood function are constructed, respectively, as

$$f'(\mu) = \frac{1}{\sigma'\sqrt{2\pi}} e^{\left[-\frac{1}{2}\left(\frac{\mu - \mu'}{\sigma'}\right)^2\right]} \qquad (10)$$

$$L(\mu) = \left(\frac{1}{\sigma\sqrt{2\pi}}\right)^5 e^{\left[-\frac{1}{2}\left(\frac{20.45 - \mu}{\sigma}\right)^2\right]} e^{\left[-\frac{1}{2}\left(\frac{20.38 - \mu}{\sigma}\right)^2\right]}$$
$$\times e^{\left[-\frac{1}{2}\left(\frac{20.51 - \mu}{\sigma}\right)^2\right]} e^{\left[-\frac{1}{2}\left(\frac{20.42 - \mu}{\sigma}\right)^2\right]} e^{\left[-\frac{1}{2}\left(\frac{20.46 - \mu}{\sigma}\right)^2\right]} \qquad (11)$$

where $\sigma = 0.08$ m, $\mu' = 20.42$ m, and $\mu' = 0.02$ m. Recalling that the sampling techniques generate samples 'from' the posterior PDF without computing the normalising factor k (see Equation (3)), the target density is proportional to the function

$$\pi(\mu) = L(\mu)f'(\mu). \qquad (12)$$

In order to use the MH algorithm, the proposal density is used as the uniform PDF

$$q(y|x(t)) = \begin{cases} 1 & -0.5 \le x(t) - y \le 0.5 \\ 0 & \text{otherwise} \end{cases} \qquad (13)$$

For both the MH and slice sampling algorithms, the initial value $x(t) = 21$ m, and 10000 samples are used. The mean and standard deviation of the posterior according to both methods and the closed-form solution are given in Table 1. Also, the frequency histograms and PDFs generated from both algorithms are shown in Figure 3. Clearly, both methods provide good approximate solutions to the closed-form. In the remainder of this paper, the slice sampling algorithm is used for the main computations, while the MH algorithm (with a normal proposal distribution whose parameters are chosen based on the problem) is used for checking some of the slice sampler results.

Updating with large SHM samples

In long-term SHM, very large amounts of data are usually generated. The treatment of data sets of this size in Bayesian analysis requires special considerations. First, in the absence of conjugate priors, which is mostly the case in SHM, Equation (2) has to be computed directly. Construction of the likelihood function (see Equation (1)) requires multiplying all the $f_X(x_i|\theta)$ functions at all SHM data points. With only five observations, the above test example resulted in a long equation for the likelihood function (see Equation (11)). When hundreds of SHM data points are at hand, construction of such equation is impractical. Alternatively, the likelihood function at a given parameter value can be computed sequentially in algorithmic form as follows:

- Initiate $L(\theta) = 1$
- Repeat for $i = 1, \ldots, N$, where N = number of SHM points.
- $L(\theta) = L(\theta) f_X(x_i|\theta)$.

Second, $f_X(x_i|\theta)$ is a PDF whose value is most likely less than 1.0. In this case, as N increases, $L(\theta)$ is decreasing, creating numerical difficulties in the Bayesian computations. To study how $L(\theta)$ decreases with the sample size N (see Equation (1)), consider again the test example above. Assume that 200 samples are available. For this illustration, these samples are drawn from the PDF N(20.4, 1.0), and the underlying standard deviation is assumed as $\sigma = 0.3$. Figure 4 shows the likelihood function evaluated at $\mu = 20.4$ with increasing the sample size N. It is observed that with $N > 135$, the values are so small that MATLAB returned a value of zero (cannot be shown on logarithmic scale).

Clearly, a large SHM data set associated with extreme events cannot be accommodated at once in Bayesian analysis. Luckily, a key aspect of Bayesian analysis is the ease with which sequential analysis can be performed (Gelmen et al. 2003). In sequential Bayesian analysis, each posterior acts as the prior in the subsequent analysis (Ang and Tang 2007). Thus, the whole SHM set can be divided into sets of sizes

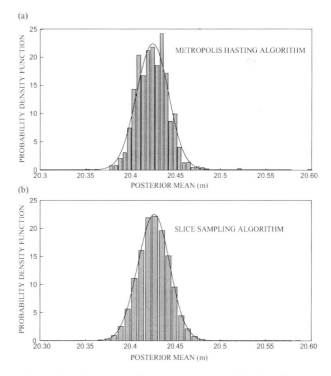

Figure 3. Frequency histograms and PDFs for the test example generated using: (a) Metropolis-Hasting algorithm, and (b) slice sampling algorithm.

Table 1. Results of the Bayesian updating example.

	Exact	MH algorithm	Slice sampling
Posterior mean (m)	20.426	20.4240	20.4260
Posterior standard deviation (m)	0.017	0.0178	0.0178

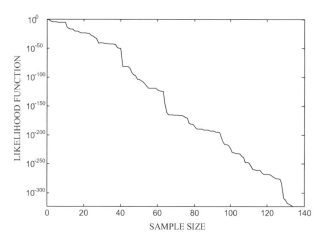

Figure 4. Change in likelihood function with increase in sample size.

determined on the basis of a curve similar to that in Figure 4 and the analysis is conducted sequentially. In fact, this sequential approach also allows the updating in real-time with the SHM data as they are obtained, where the analysis is conducted with one SHM data point at a time, or in near real-time, where a set of points obtained at regular/irregular-time intervals are used.

Modelling of live load

Whether it is the instantaneous or cumulative probability of failure that is being calculated, the two components to be known are the structural strength R and the load effect L. The load effect is the structural response to various types of loads, dead and live. SHM instrumented on a completely constructed bridge is only able to detect variations in live load. In this paper, the SHM data treated is that representing the structural response due to traffic. In particular, the strain produced by passing traffic and measured using strain gages is converted herein to bending moments at the critical section.

The strain gauges are able to record the strain produced by passing of all light and heavy vehicles. However, for reliability purposes, the safety of the structure is quantified with respect to extreme events produced by the load effect L_n. Therefore, the extreme values obtained from the sensors, and resulting from the passing of heavy trucks, are considered to constitute the load effect. Appropriate probabilistic tools to model this type of loading are the extreme value distributions.

Exact modelling of extreme value distributions is usually a formidable task. Alternatively, one of three asymptotic extreme value distributions are assigned based on the tail behaviour of the original distribution (Ang and Tang 1984). The distribution of loads produced by traffic, heavy and light (i.e. the original distribution L), is modelled as Gaussian (Nowak 1999). The Gaussian distribution is known to have an exponential tail. Therefore, an appropriate extreme value distribution to be used for modelling heavy vehicle loads is the type I extreme value distribution L_n and is given as (Ang and Tang 1984)

$$f_{L_n}(l) = \alpha_n e^{-\alpha_n(l-u_n)} \exp\left[-e^{-\alpha_n(l-u_n)}\right] \qquad (14)$$

where u_n is the characteristic largest value of the initial variate L, and α_n is an inverse measure of the dispersion of L_n. These values are related to the mean μ_{L_n} and standard deviation σ_{L_n} of L_n according to

$$\alpha_n = \frac{\pi}{\sqrt{6}\sigma_{L_n}} \qquad (15a)$$

$$u_n = \mu_{L_n} - \frac{\gamma}{\alpha_n} \qquad (15b)$$

As previously mentioned, the goal of the methodology presented in this paper is to incorporate prior information in addition to what is obtained from SHM in the life-cycle management framework. This prior information can be obtained from code values (AASHTO 2002) or statistical studies of traffic loads (Nowak 1999). In this paper, it is considered that the parameter u_n has a greater impact on $f_{L_n}(l)$ than the parameter α_n (see Equation (15)) and thus α_n is assumed known while u_n is obtained from SHM using Bayesian updating. Also, from the prior information an initial estimate of u_n is used, i.e. the prior PDF of u_n, where it is assumed that the u_n is lognormal and given as (Ang and Tang 2007)

$$f'(u_n) = \frac{1}{\zeta u_n \sqrt{2\pi}} e^{\left[-\frac{1}{2}\left(\frac{\ln u_n - \lambda}{\zeta}\right)^2\right]} \qquad (16)$$

where ζ and λ are the hyper-parameters of the variate u_n and are related to the mean and standard deviation of the variate u_n according to

$$\zeta = \sqrt{\ln(1 + \frac{\sigma^2}{\mu^2})} \qquad (17a)$$

$$\lambda = \ln \mu - \frac{1}{2}\zeta^2 \qquad (17b)$$

The mean μ is obtained from the prior information and σ is calculated based on an assumed coefficient of variation (σ / μ) of 10%. Hence, the posterior for this problem is proportional to the target density

$$\pi(u_n) = L(u_n)f'(u_n)$$

$$= \left[\prod_{i=1}^{N} \alpha_n e^{-\alpha_n(l_i - u_n)} \exp\left[-e^{-\alpha_n(l_i - u_n)}\right] \right]$$

$$\times \left[\frac{1}{\zeta u_n \sqrt{2\pi}} e^{\left[-\frac{1}{2}\left(\frac{\ln u_n - \lambda}{\zeta}\right)^2\right]} \right] \qquad (18)$$

where N = number of SHM data points considered.

Extreme value prediction

Ideally, a SHM system would be designed for fulfilling objectives of the life-cycle management framework. The type of SHM devices used and the locations of these devices would be established to obtain the optimum details needed for the enhancement of the life-cycle management framework. For instance, appropriate sensors are installed at locations that can be determined by means of the reliability importance factor. However, many SHM investigations have already been conducted on various bridges for purposes that are different from what a life-cycle management framework would intend. With proper treatment, the results of these investigations may also be integrated into the life-cycle management framework. One treatment is to spatially extrapolate the results of SHM from one location to another.

The primary structural response in this framework is the bending moment caused by the moving traffic. The SHM response quantity measured is the strain. The strain is converted to stress assuming Hooke's law. The stress is then converted to bending moment acting on the monitored section M_m. However, the monitored section may not be a critical one and thus the moments at the critical section M_c are not readily available. Using the bending moment envelope, the moment M_c can be found as

$$M_c = \frac{\alpha_c}{\alpha_m} M_m \qquad (19)$$

where α_c and α_m are the ordinates of the bending moment envelope at the critical and monitored sections, respectively.

Next, Bayesian updating is implemented to incorporate prior information, while the values of M_c are used to construct the likelihood function. As a result, the updated bending moment M_u, in a form of an extreme value distribution, is found.

In order to account for the changes inflicted upon the updated response due to increase in traffic or deterioration of the section over time, the obtained moment M_u is extrapolated using extreme value statistical concepts. For this purpose, an approach proposed by Liu et al. (2009a) is implemented herein.

In this approach, the prediction function $\zeta(t = T)$ is introduced as

$$\zeta(t = T) = \max\left[\frac{\varepsilon_{\max}(T)}{\max(\varepsilon_1, \varepsilon_2, \ldots, \varepsilon_N)}, 1 \right] \qquad (20)$$

where ε_i ($i = 1, 2, 3, \ldots, N$) are the strain measurements from SHM, and

$$\varepsilon_{\max}(T) = u_n - \frac{1}{\alpha_n} \cdot \ln\left[-\ln\left(1 - \frac{1}{N_T}\right) \right] \qquad (21)$$

where N_T is total number of the passages of the heavy vehicles in the next T years (see Liu et al. (2009a)).

Case study: the I-39 Bridge

Description of the bridge

The I-39 Bridge is located near Wausau, Wisconsin, USA, and carries US-51 and I-39 north bound over the Wisconsin River. It is a five span continuous steel girder bridge and was opened to traffic in 1961. It crosses the Wisconsin River from the village of Rothschild on the southeast side, to the town of Weston, Marathon County, on the Northwest side. The bridge is symmetrical about the mid point of the third span and has a total length of 196 m. The I-39 Bridge underwent a controlled testing and long-term monitoring programme in 2004 by the National Center for Advanced Technology for Large Structural Systems (ATLSS), at Lehigh University, Bethlehem, Pennsylvania, and is considered in this paper to illustrate the concepts introduced.

System performance analysis

The main performance indicator used in this study is the cumulative-time probability of failure. This probability is computed using an incremental non-linear finite element analysis in which the resistance is the load increment at which the structure collapses or a deflection limit is reached. For the components of this bridge, the performance indicators were computed in Okasha et al. (2010) using an incremental nonlinear finite element method in conjunction with a Latin hypercube sampling. For the system, this performance indicator was computed in Okasha and Frangopol (2010a) using also an incremental nonlinear finite element method but in conjunction with the response surface method and design of experiments. In both studies, the cumulative-time probabilities of failure were fitted to Weibull distributions that represent the lifetime functions of the system $F_s(t)$, interior girders $F_i(t)$

and exterior girders $F_e(t)$. The CDF of the Weibull distribution is given as

$$F(t) = \begin{cases} 1 - e^{-(\lambda t)^\kappa} & \text{for} \quad t > 0 \\ 0 & \text{otherwise} \end{cases} \qquad (22)$$

where λ is a scale parameter and κ is a shape parameter (i.e. $p_1 = \lambda$ and $p_2 = \kappa$). The scale and shape parameters of the girders and system Weibull functions are given in Table 2.

In this paper, a component-system performance relationship is established using multiple nonlinear regression. This relationship is assumed as

$$Y = A X_1^{P_1} X_2^{P_2} X_3^{P_3} X_4^{P_4} \qquad (23)$$

where $Y = F_s(t)$ for all values of t considered, and $X_k = F_k(t)$ for all values of t considered, and $k = 1, 2, 3,$ and 4 represents girders 1, 2, 3, and 4, respectively. However, owing to the symmetry of the bridge, the external girders 1 and 4 are assumed to have the same lifetime function $F_e(t)$, and the internal girders 2 and 3 are assumed to have the same lifetime function $F_i(t)$. Thus, Equation (23) is reduced to

$$Y = A X_i^{P_i} X_e^{P_e} \qquad (24)$$

It is worth noting that the assumption of symmetry giving the same lifetime functions to the internal girders, and also the exterior girders, can be easily removed by analysing all girders in the bridge without affecting the methodology. However, the bridge example considered herein only aims to illustrate the concepts proposed in this paper and thus two different girders are considered. By taking the natural logarithm of both sides of Equation (24), the nonlinear equation becomes linear

$$\ln(Y) = B + P_i \ln(X_i) + P_e \ln(X_e) \qquad (25)$$

where B, P_i and P_e are constants obtained by multiple linear regression and $A = \exp(B)$.

In order to compare with the traditional series-parallel method of establishing the component-system relationship, the four systems shown in Figure 5 are investigated. In these systems it is assumed that the

Table 2. Scale and shape parameters for the Weibull distribution of the girders and system of the I-39 Bridge.

	Scale parameter (1/year)	Shape parameter
Interior girder	1/171.095	23.066
Exterior girder	1/88.704	10.979
System	1/109.579	16.197

system fails if (a) any girder fails, (b) any two adjacent girders fail (c) any three adjacent girders fail and (d) all four girders fail. Accordingly, the component-system performance relationships for these four models, respectively, are found using the minimum-cut set method (Frangopol and Okasha 2009b) as

$$Y_1 = 1 - (1 - X_i)^2 (1 - X_e)^2 \qquad (26a)$$

$$Y_2 = 1 - (1 - X_i^2)(1 - X_i X_e)^2 \qquad (26b)$$

$$Y_3 = 1 - (1 - X_i^2 X_e)^2 \qquad (26c)$$

$$Y_4 = X_i^2 X_e^2 \qquad (26d)$$

The results from Equations (24) and (26) and the component and system cumulative-time failure probabilities are plotted in Figure 6. The constants of Equation (24) obtained from the multiple linear regression are $A = \exp(B) = \exp(6.62) = 748.759$, $P_i = 0.662$, and $P_e = 0.088$. Clearly, the predicted model in Equation (24) produced results that are in excellent agreement with the true system performance. It is

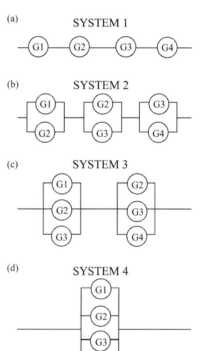

Figure 5. Series-parallel models considered, where the system fails if any: (a) any one girder fails, (b) any two adjacent girders fail, (c) any three adjacent girders fail, and (d) all four adjacent girders fail.

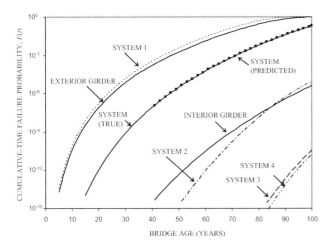

Figure 6.   System analysis for the lifetime reliability of the I-39 Bridge.

Description of the monitoring program and data extraction

The instrumentation of the bridge consisted of installation of weldable resistance strain gauges at key locations both to understand the response of the bridge to controlled load tests (crawl tests and dynamic tests) and quantify the stress-range histograms at critical details (long-term monitoring) (see Mahmoud *et al.* 2005 for further details).

Static results from crawl tests were used in Okasha *et al.* (2010) to update the finite element model and, in turn, update the lifetime reliability of the structure. In this paper, the illustrated methodology targets the long-term monitoring data obtained from Channels 15 and 16 (denoted as CH15 and CH16, respectively) of the I-39 Bridge (see Figure 7) which measured and recorded the structural responses of the east exterior girder (G4) and east interior girder (G3), respectively (Figure 7). The corresponding sensors were installed at the lower surface of the bottom flanges in the section S_1 (Figure 7), which is located about 41.8 m north of the central line of the south abutment of the bridge.

The monitored strain data from the uncontrolled load tests were extensively collected and investigated during the period of 95 days from 29 July to 3 November 2004. In order to minimise the volume of monitoring data and to consider only the heavy vehicles, recording of the data in Channels 15 and 16 was triggered when a vehicle induced a strain larger

noted that, in this particular example, the Weibull function of the interior girder returns zero for the years prior to year 40. Therefore, Equation (24) returns a value of zero at these years as well.

The first system (see Figure 5a and Equation (26a)) is the only model that provides conservative (although quite different) estimates of the true system performance. The true system performance lies between that captured by systems 1 and 2. However, the range is considerably too large to be of practical use.

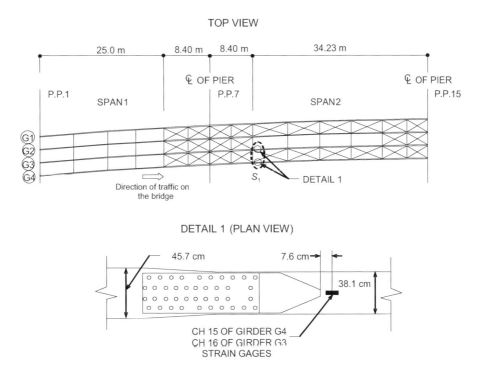

Figure 7.   Sensor locations on the I-39 Wisconsin Bridge (adapted from Mahmoud *et al.* 2005).

than the predefined strain (Mahmoud *et al.* 2005). There were a total of 630 events captured during the monitoring period of 95 days in which heavy vehicles crossed the bridge on the right (east) lane. The bending moment corresponding to these strains were calculated. However, section S_1 (see Figure 7) is not the critical location for the bending moment limit state, which is in the middle of the second span (12.92 metres away from section S_1). Using the bending moment envelope and Equation (19), the bending moments at the critical locations are obtained.

Bayesian updating

Prior information about the applied bending moment is obtained by applying an HS-20 AASHTO design truck to the finite element model of the girder and using the AASHTO distribution factors to compute the moment applied at both exterior and interior girders. Using these values as the mean of the applied bending moment, and with an assumed coefficient of variation (COV) of 10%, the hyper-parameters of the prior of the parameter u_n are calculated using Equation (17) for the exterior and interior girders. The parameter α_n is calculated based on a COV of 19% (Nowak 1999) and using Equation (15a). The SHM data are used to update the hyper-parameters of the parameter u_n.

Based on a similar investigation to that conducted using Figure 4 in the test example, it was decided that the 630 sample size of SHM is applied sequentially in increments of 100 points in the Bayesian updating. Figure 8 shows the results of the Bayesian analysis using the slice sampler. In this figure, the change in both the mean and standard deviation of the parameter u_n is shown for the interior and exterior girders. It is clear that the prior information is in better agreement with the Bayesian results in the interior girder case than in the exterior girder case. This is mainly contributed to the results of the AASTHO distribution factors used to establish the prior information. In fact, Orcesi and Frangopol (2010) illustrated how different these distribution factors are from those obtained from testing of the bridge. In addition, it is seen that the largest change in the prior mean and standard deviation occurs within the updating using the first 100 samples and then the change stabilises. This is also shown in Figure 9, where the change in COV of the parameter u_n is shown for the interior and exterior girders as more SHM data is used. In order to shed more light on this issue, a smaller sample size is used within this interval. This is shown in Figure 10. Indeed, the first sample SHM data point produced a large effect on the prior information. This was followed by a decreasing effect as the sample size grows.

The implications of the observations made from Figures 8, 9 and 10 are threefold. First, it is clear that

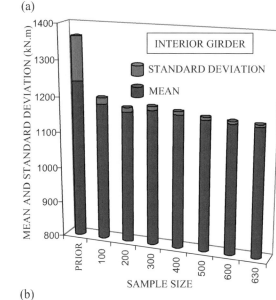

(a)

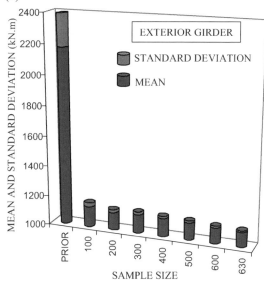

(b)

Figure 8. Changes in mean and standard deviation of the extreme value distribution parameter u_n with increase in SHM information for an: (a) interior girder, and (b) exterior girder.

the influence of the prior information tends to diminish with the growth of the SHM sample size. Thus, and in consistency with Ghosh *et al.* (2006), for a large sample size of SHM data, which is typically the case in long-term monitoring, a precise mathematical specification of the prior is not necessary. Accordingly, the above assumption of COV = 10% for the parameter u_n is washed away in this analysis. Second, the reduction of the uncertainty due to incorporing the SHM data is evident. Third, results such as those in Figures 8, 9 and 10 provide indicators for deciding the total sample size of the SHM data, and the appropriate timing for terminating the long-term monitoring. Clearly, the

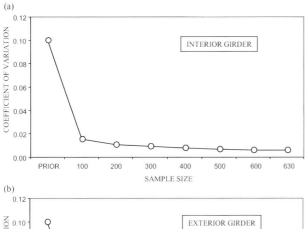

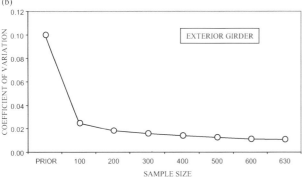

Figure 9. Changes in the coefficient of variation of the extreme value distribution parameter u_n with increase in SHM information for an: (a) interior girder, and (b) exterior girder.

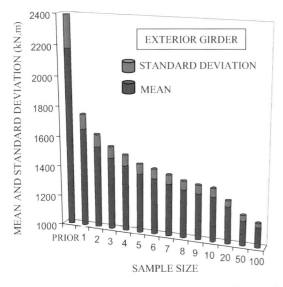

Figure 10. Changes in mean and standard deviation of the extreme value distribution parameter u_n with increase in SHM information for an exterior girder (smaller sample size).

period in which the ATLSS center conducted the long-term monitoring of this bridge was long enough such that the reduction in the uncertainty became insignificant towards the end.

The prior and posterior PDFs of the parameter u_n are shown in Figure 11. Also shown are the frequency

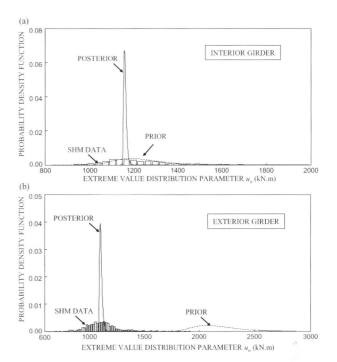

Figure 11. Bayesian updating results of the extreme value distribution parameter u_n for an: (a) interior girder, and (b) exterior girder.

histograms of SHM data obtained from CH15 (the exterior girder) and CH16 (the interior girder). It is emphasised that the prior and posterior PDFs are not intended to model the SHM data. They are intended to model the parameter u_n of the extreme value distribution that models the behaviour which the SHM data is considered to follow. Therefore, the variability of the prior and posterior only reflects the uncertainty in this parameter and not the uncertainty in the underlying distribution. Accordingly, the uncertainty in this parameter has evidently been reduced considerably using with the introduction of the SHM information and using Bayesian updating, as seen in Figure 11. Thus, the positive impact of introducing the SHM data is clear.

Using the obtained posterior, the CDF of the underlying distribution, i.e. the applied bending moment due to heavy trucks, (see Equation (5)) is obtained by numerical integration and the results are fitted to the extreme value distribution type I by optimisation (see Equation (6)). The results are shown in Figure 12 for the interior and exterior girders. Clearly, the extreme value distribution type I is an excellent candidate for fitting this random variable. The PDF of the obtained updated distribution is shown with the original distribution and the SHM data in Figure 13 for the interior and exterior girders. As shown in the figure, the updated distribution migrated to embrace the SHM data in both girders. Figure 14

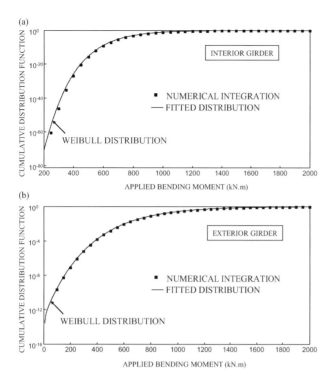

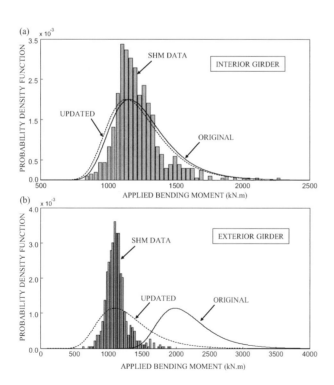

Figure 12. Computed and fitted extreme-value type I CDF of the applied bending moment for an: (a) interior girder, and (b) exterior girder.

Figure 13. Original and updated PDF of the applied bending moment for an: (a) interior girder, and (b) exterior girder.

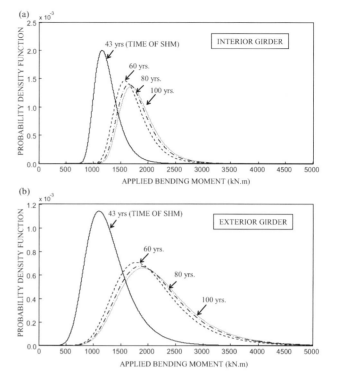

Figure 14. PDF of the extrapolated applied bending moments for an: (a) interior girder, and (b) exterior girder.

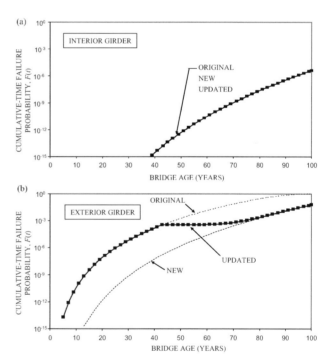

Figure 15. Lifetime reliability profile before and after updating at year 43 (2004) for an: (a) interior girder, and (b) exterior girder.

shows the extrapolated PDF from the updated distribution of the applied bending moment due to heavy trucks in the interior and exterior girders.

Updating the lifetime reliability

Using the approach proposed in Okasha *et al.* (2010), the cumulative-time failure probability is computed and shown in Figure 15 for the interior and exterior girders. In this figure, the curves labeled 'original' and 'new' represent the CDF of the distributions fitted to the cumulative-time failure probability computed without and with updating using the SHM data, respectively, and the curve labelled 'updated' represents the CDF of the final distribution obtained by combining these two (Okasha *et al.* 2010). The results in this figure reflect the updating of the live load effect performed above as well as the updating of the resistance parameters of the bridge conducted in Okasha *et al.* (2010). It is clear in the figure that the updated reliability of the interior girder aligns with the original reliability. This is because the reduction in the load effect for this girder is met with reduction in strength found in Okasha *et al.* (2010), such that the two effects almost cancel out. Even though a reduction in strength was also found for the exterior girder in Okasha *et al.* (2010), this reduction was overcome by the large reduction in the load effect (see Figure 13). Accordingly, the deterioration in the lifetime reliability is reduced significantly.

Conclusions

In this paper, an approach is proposed to integrate the information provided by SHM in a life-cycle management framework for bridge structures. The life-cycle management framework is presented. In addition, a new approach for establishing the relationship between the component and system performance is proposed. The concepts presented are illustrated on a bridge in Wisconsin, USA.

The life-cycle management framework presented in this paper is the culmination of a research programme conducted at Lehigh University. The details of several parts of the framework have been introduced in previous publications (Frangopol and Okasha 2008, 2009a, b, Okasha and Frangopol 2009, 2010a, b, c, d, Okasha *et al.* 2010).

This paper aims to provide details on the framework step associated with integrating the information obtained from SHM data. In particular, these details target the extraction of useful information from SHM data related to the load effect caused by the traffic, especially by heavy trucks. Then, this information is combined with prior knowledge, extrapolated, and used to compute a new profile of performance. The performance indicators used in this framework are based on lifetime functions that are derived using advanced computational tools. Prior knowledge and SHM data are combined using Bayesian analysis. This paper illustrates the systematic application of Bayesian analysis for integrating the SHM information. Since the underlying distribution of the load effect is an extreme value distribution, a closed-form solution for the Bayesian updating of the problem does not exist. Therefore, advanced simulation algorithms are used to provide an alternative. As the SHM data sample size is growing, the influence of the prior is found to wash away. Thus a precise mathematical specification of the prior is not necessary with large SHM data samples. Also, this observation provides indicators for deciding the total sample size of the SHM data, and the appropriate timing for terminating the long-term monitoring.

The traditional technique for determining the crucial relationship between the component and system performance is to establish a series-parallel model that supposedly captures the behaviour of the system and relates the system failure cause to the failures of the individual components. In this paper, this technique is reviewed. It is found that for the bridge investigated, the only conservative model among a group of series-parallel models considered is the series model in which the system failure is assumed to occur by the failure of any component. The true system performance is found to lie between two of the assumed system models. However, the range between these two system models is considerably large. Alternatively, it is proposed that presently available advanced computational tools are used to compute the true system performance as well as the component performance. These are used to establish an accurate nonlinear equation that represents the component-system relationship. Using logarithmic transformation, the equation becomes linear and the coefficients of which are found by multiple linear regression. The results of this approach were found to be in excellent agreement with the true system performance.

The framework presented in this paper is believed to be a practical tool for establishing rational, continuous and accurate decision support for decision makers in charge of managing bridge structures.

Acknowledgements

The support from (a) the National Science Foundation through grants CMS-0638728 and CMS-0639428, (b) the Commonwealth of Pennsylvania, Department of Community and Economic Development, through the Pennsylvania Infrastructure Technology Alliance (PITA), (c) the US Federal Highway Administration Cooperative Agreement Award DTFH61-07-H-00040, and (d) the US Office of Naval Research Contract Number N00014-08-1-0188 is gratefully acknowledged.

References

American Association of State Highway and Transportation Officials (AASHTO), 2002. *Standard specifications for highway bridges.* 16th edn. Washington, DC: AASHTO.

Aktan, A.E., Catbas, F.N., Grimmelsman, K.A. and Pervizpour, M., 2002. Development of a model health monitoring guide for major bridges. Report submitted to Federal Highway Administration Research and Development, No. DTFH61-01-P-00347, Drexel Intelligent Infrastructure and Transportation Safety Institute.

Ang, A.H.-S. and Tang, W.H., 1984. *Probability concepts in engineering planning and design. Vol. II. Decision, risk and reliability.* New York: Wiley & Sons.

Ang, A.H.-S. and Tang, W.H., 2007. *Probability concepts in engineering. Emphasis on applications to civil and environmental engineering.* 2nd edn. New York: Wiley & Sons.

Bedford, T. and Cooke, R., 2001. *Probabilistic risk analysis.* The Edinburgh Building, Cambridge: Cambridge University Press.

Bucher, C., 2009. . D.M. Frangopol, ed. *Computational analysis of randomness in structural mechanics: Vol. 3. Structures and infrastructures series.* Leiden, The Netherlands: CRC Press/Balkema/Taylor & Francis.

Catbas, F.N., Susoy, M. and Frangopol, D.M., 2008. Structural health monitoring and reliability estimation: Long span truss bridge application with environmental monitoring data. *Engineering Structures,* 30 (9), 2347–2359.

Chang, P.C., Flatau, A. and Liu, S.C., 2003. Review paper: health monitoring of civil infrastructure. *Structural Health Monitoring,* 2 (3), 257–267.

Chib, S. and Greenberg, E., 1995. Understanding the Metropolis-Hastings algorithm. *The American Statistician,* 49 (4), 327–335.

Coles, S.G. and Powell, E.A., 1996. Bayesian methods in extreme value modelling: a review and new developments. *International Statistical Review,* 64, 119–136.

Ellingwood, B.R., 2005. Risk-informed condition assessment of civil infrastructure: state of practice and research issues. *Structure and Infrastructure Engineering,* 1 (1), 7–18.

Estes, A.C. and Frangopol, D.M., 1999. Repair optimisation of highway bridges using system reliability approach. *Journal of Structural Engineering,* 125 (7), 766–775.

Faber, M.H., Val, D.V. and Stewart, M.G., 2000. Proof load testing for bridge assessment and upgrading. *Engineering Structures,* 22 (12), 1677–1689.

Fink, D., 1995. A compendium of conjugate priors. In progress report: Extension and enhancement of methods for setting data quality objectives. DOE contract 95–831.

Frangopol, D.M., 2010. Life-cycle performance, management, and optimisation of structural systems under uncertainty: Accomplishments and challenges. *Structure and Infrastructure Engineering,* DOI: 10.1080/15732471003594427.

Frangopol, D.M. and Okasha, N.M., 2008. Life-cycle performance and redundancy of structures. *In: Proceedings of the 6th international probabilistic workshop,* 26–27 November 2008, Darmstadt, Germany (keynote lecture). Darmstadt, Germany: Technische Universität Darmstadt, 1–14.

Frangopol, D.M. and Okasha, N.M., 2009a. Multi-criteria optimisation of life-cycle maintenance programs using advanced modelling and computational tools. Chapter 1 in. B.H.V. Topping, L.F. Costa Neves, and C. Barros, eds. *Trends in civil and structural computing.* Stirlingshire, Scotland: Saxe-Coburg Publications, 1–26.

Frangopol, D.M. and Okasha, N.M., 2009b. Redundancy of structural systems based on survivor functions. *In:* L. Griffiths, T. Helwig, M. Waggoner, and M. Hoit, eds. *Proceedings of the 2009 structures congress: Don't mess with structural engineers,* 30 April–2 May, SEI-ASCE, Austin, Texas. ASCE, CD-ROM, 1781–1790.

Frangopol, D.M., Strauss, A. and Kim, S., 2008a. Bridge reliability assessment based on monitoring. *Journal of Bridge Engineering,* 13 (3), 258–270.

Frangopol, D.M., Strauss, A. and Kim, S., 2008b. Use of monitoring extreme data for the performance prediction of structures: General approach. *Engineering Structures,* 30 (12), 3644–3653.

Gelfand, A.E. and Smith, A.F.M., 1990. Sampling-based approaches to calculating marginal densities. *Journal of the American Statistical Association,* 85, 398–409.

Gelman, A., Carlin, J., Stern, H. and Rubin, D.B., 2003. *Bayesian data analysis.* 2nd edn. Boca Raton, Florida: Chapman & Hall/CRC.

Ghosh, J.K., Delampady, M. and Samanta, T., 2006. *An introduction to Bayesian analysis, theory and methods.* New York, NY: Springer.

Glaser, S.D., Li, H., Wang, M.L., Ou, J. and Lynch, J., 2007. Sensor technology innovation for the advancement of structural health monitoring: a strategic program of US–China research for the next decade. *Smart Structures and Systems,* 3 (2), 221–244.

Hastings, W.K., 1970. Monte Carlo sampling methods using Markov chains and their applications. *Biometrika,* 57 (1), 97–109.

Kay, S.M. 1993. *Fundamentals of statistical signal processing.* Upper Saddle River, NJ: Prentice Hall.

Liu, M., Frangopol, D.M. and Kim, S., 2009a. Bridge safety evaluation based on monitored live load effects. *Journal of Bridge Engineering,* 14 (4), 257–269.

Liu, M., Frangopol, D.M. and Kim, S., 2009b. Bridge system performance assessment from structural health monitoring: A case study. *Journal of Structural Engineering,* 135 (6), 733–744.

Mahmoud, H.N., Connor, R.J. and Bowman, C.A., 2005. Results of the fatigue evaluation and field monitoring of the I-39 Northbound Bridge over the Wisconsin RiverLehigh University, Bethlehem, PA, USA: Lehigh University. ATLSS Report 05–04. .

MathWorks, 2009. *Statistics Toolbox^{TM} 7 user's guide.* Natick, MA: The MathWorks, Inc., 2112 pp.

Metropolis, N., Rosenbluth, A.W., Rosenbluth, M.N., Teller, A.H. and Teller, E., 1953. Equations of state calculations by fast computing machines. *Journal of Chemical Physics,* 21, 1087–1092.

Neal, R.M., 2003. Slice sampling. *The Annals of Statistics,* 31 (3), 705–741.

Neves, L.C., Frangopol, D.M. and Cruz, P.J., 2006a. Probabilistic lifetime-oriented multiobjective optimisation of bridge maintenance: Single maintenance type. *Journal of Structural Engineering,* 132 (6), 991–1005.

Neves, L.C., Frangopol, D.M. and Petcherdchoo, A., 2006b. Probabilistic lifetime-oriented multi-objective optimisation of bridge maintenance: Combination of maintenance types. *Journal of Structural Engineering,* 132 (11), 1821–1834.

Nowak, A.S., 1999. Calibration of LRFD Bridge Design Code. Transportation Research Council, Washington, DC. NCHRP Report 368.

Okasha, N.M. and Frangopol, D.M., 2009. Lifetime-oriented multi-objective optimisation of structural maintenance considering system reliability, redundancy and life-cycle cost using GA. *Structural Safety*, 31 (6), 460–474.

Okasha, N.M. and Frangopol, D.M., 2010a. Advanced modelling for the life-cycle performance prediction and service-life estimation of bridges. *Journal Computing in Civil Engineering*, in press.

Okasha, N.M. and Frangopol, D.M., 2010b. Novel approach for multi-criteria optimisation of life-cycle preventive and essential maintenance of deteriorating structures. *Journal of Structural Engineering*, 136 (8), in press.

Okasha, N.M. and Frangopol, D.M., 2010c. Redundancy of structural systems with and without maintenance: An approach based on lifetime functions. *Reliability Engineering & System Safety*, 95 (5), 520–533.

Okasha, N.M. and Frangopol, D.M., 2010d. Time-variant redundancy of structural systems. *Structure and Infrastructure Engineering*, 6 (1/2), 279–301.

Okasha, N.M., Frangopol, D.M. and Orcesi, A.D., 2010. Automated finite element updating using strain data for the lifetime reliability assessment of bridges. *Reliability Engineering & System Safety*, submitted.

Orcesi, A.D. and Frangopol, D.M., 2010. Inclusion of crawl tests and long-term health monitoring in bridge serviceability analysis. *Journal of Bridge Engineering*, 15 (3), 312–326.

Robert, C.P., 1994. *The Bayesian choice*. Springer Texts in Statistics. New York, NY: Springer-Verlag.

Robert, C.P. and Cassella, G., 1999. *Monte Carlo statistical methods*. Springer Texts in Statistics. New York, NY: Springer-Verlag.

Smith, A.F.M. and Gelfand, A.E., 1992. Bayesian statistics without tears: A sampling–resampling perspective. *The American Statistician*, 46 (2), 84–88.

Strauss, A., Frangopol, D.M. and Kim, S., 2008a. Statistical, probabilistic and decision analysis aspects related to the efficient use of structural monitoring systems. *Beton- und Stahlbetonbau*, 103, 23–28.

Strauss, A., Frangopol, D.M. and Kim, S., 2008b. Use of monitoring extreme data for the performance prediction of structures: Bayesian updating. *Engineering Structures*, 30 (12), 3654–3666.

Tanner, M.A. and Wong, W.H., 1987. The calculation of posterior distributions by data augmentation. *Journal of the American Statistical Association*, 82, 528–540.

Thodi, P., Khan, F. and Haddara, M., 2009. Risk based integrity modelling of gasp facilities using Bayesian analysis. *In: Proceedings of the 1st annual gas processing symposium*, 10–12 January, Qatar. Qatar: Elsevier, 297–306.

Tierney, L. and Mira, A., 1999. Some adaptive Monte Carlo methods for Bayesian inference. *Statistics in Medicine*, 18, 2507–2515.

Critical issues, condition assessment and monitoring of heavy movable structures: emphasis on movable bridges

F. Necati Catbas, Mustafa Gul, H. Burak Gokce, Ricardo Zaurin, Dan M. Frangopol and Kirk A. Grimmelsman

In this paper, a relatively less studied class of structures is presented based on the research conducted on Florida's movable bridges over the last several years. Movable bridges consist of complex structural, mechanical and electrical systems that provide versatility to these bridges, but at the same time, create intermittent operational and maintenance challenges. Movable bridges have been designed and constructed for some time; however, there are fewer studies in the literature on movable bridges as compared to other bridge types. In addition, none of these studies provide a comprehensive documentation of issues related to the condition of movable bridge populations in conjunction with possible monitoring applications specific to these bridges. This paper characterises and documents these issues related to movable bridges considering both the mechanical and structural components. Considerations for designing a monitoring system for movable bridges are also presented based on inspection reports and expert opinions. The design and implementation of a monitoring system for a representative bascule bridge are presented along with long-term monitoring data. Various movable bridge characteristics such as opening/closing torque, bridge balance and friction are shown since these are critical for maintenance applications on mechanical components. Finally, the impact of environmental effects (such as wind and temperature) on bridge mechanical characteristics is demonstrated by analysing monitoring data for more than 1000 opening/closing events.

1. Introduction

Heavy movable structures such as retractable stadium roofs, stands, pitches, navigation locks, dams, flap gates, flood control equipment, movable bridges and others go beyond conventional civil engineering structures with respect to the complexity of their designs and operation. Heavy movable structures have large machinery for which most operational speeds are low and critical forces are significant. Mechanical and electrical components are integrated with the structural elements creating a very unique type of structure often referred as 'kinetic architecture', which provides the flexibility to increase the usage of these structures with different configurations. The focus of this paper is on movable bridges. These are one of the most common types of movable transportation structures. The main advantage of this type of structure is that the bridge can be constructed with little vertical clearance, avoiding the expense of high piers and long approaches. Moving components of these bridges are operated by various types of machinery to open the passageway for the waterborne traffic.

While movable bridges are a viable alternative to fixed bridges, they also present significant problems associated with their long-term operation and performance. Many movable bridges are located in coastal areas or in areas where de-icing treatments are applied to roadways exposing them to conditions favourable to corrosion and rapid deterioration through cross-sectional loss of the structural members. Mechanical component failures due to friction and wear caused by the repeated movements during openings and closings are also deemed very critical by bridge owners. Even with regular maintenance, continuous downgrading of all parts of such complex bridges is inevitable, leading to breakdowns causing problems for both land and maritime traffic. These problems result in high maintenance costs for these bridges.

Timely and effective implementations of repair programmes for movable bridges are critical management issues since (a) traffic disruptions will occur when the bridge is not operable, and (b) the problems associated with movable bridges can be difficult and time consuming to fix. In Florida, which has the second highest movable bridge population in the USA, it is estimated that the unit maintenance cost per square foot of the deck of a movable bridge can be up to 100 times that of a fixed bridge, based on the private communication with state engineers. Consequently, a small movable bridge population owned by an agency may require a considerable maintenance budget.

Movable bridges have been constructed and their designs have been studied since as early as 1882 by Fränkel (1882). Additional documentation for some early movable bridges can be found in Waddell (1895), Greene and McKeen (1938) and Quade (1954). Furthermore, some examples of movable bridges in the USA can be found in Hardesty, Christie, and Fischer (1975a, 1975b), Wengenroth, Hardesty, and Mix (1975) and Ecale and Lu (1983). In 1998, the American Association of State Highway and Transportation Officials (AASHTO) published a manual that provides uniform guidelines and procedures for inspecting, evaluating and maintaining movable bridges (AASHTO, 1998). This manual also includes information related to the structural, mechanical, hydraulic and electrical components, and the operational characteristics that are unique to movable bridges. Although there are specific operational and maintenance issues associated with movable bridges, there are very few studies in the literature which focus on these bridges. Most of the bridge maintenance studies focused on fixed bridges (Okasha & Frangopol, 2010, 2012). In addition, there are few papers in the archival journals which focus on movable bridges, in general, or specifically on bascule-type movable bridges. Papers describing the replacement of movable spans (Fisher & Robitaille, 2011) and the kinematics of movable bridges (Byers, 2008; Wallner & Pircher, 2007) are examples of studies for movable bridges in general. For bascule bridges, a brief review of the Chicago-type bascule bridge design (Ecale & Lu 1983) and full-scale testing of procedures for assembling trunnion–hub–girder interactions by Besterfield, Nichani, Kaw, and Eason (2003) are important examples. Other than these examples, there are also some studies on lift bridges, mainly related to construction by Ramey (1983), Griggs (2006) and Zhao, Wang, and Pang (2011). The authors have investigated the load rating of movable bridges, modelling for various possible damage simulations, identification of the most common maintenance issues and implementation of novel sensing and image technologies on movable bridges (Catbas, Gokce, Gul, & Frangopol, 2011; Catbas, Gokce, & Gul, 2012; Catbas, Zaurın, Gul, & Gokce, 2012). However, there are no comprehensive studies in the literature which outline all the critical damage, deterioration considerations, con-

dition assessment methods for maintenance for movable bridge populations along with an extensive monitoring system for both structural and mechanical components.

The objective of this paper is to present and characterise the operation and maintenance issues specific to movable bridges by considering both the mechanical and the structural components. In addition, condition monitoring strategies appropriate for movable bridges based on an analysis of inspection reports and expert opinions are presented along with a structural health monitoring (SHM) system that has been implemented on an in-service, bascule-type movable bridge. Finally, some representative long-term measurements obtained from the monitoring system are presented and evaluated.

This paper first describes the most common types of movable bridges and provides a detailed analysis of the entire movable bridge inventory in Florida. Inspection reports for a sample of 51 movable bridges from Florida's inventory were evaluated to identify and characterise critical damage, deterioration, malfunction issues related to their operation and maintenance. Subsequently, two different case studies related to bascule bridges are presented. The first case study is an in-depth evaluation of the performance of a bascule bridge in Florida over a 35-year period. Condition state data were tracked for different components to illustrate the deterioration patterns and rehabilitation effects. The second case study describes the development of a SHM system for a representative, in-service bascule bridge. The pilot design of the SHM system is presented in the context of tracking and evaluating the previously identified critical operational and maintenance issues. Long-term monitoring data that describe the bridge balance and friction characteristics of the bridge are also presented and evaluated. Bridge balance and friction are important characteristics of bascule bridges since they can directly affect the performance of the mechanical and structural components. The impact of preventative maintenance and environmental effects, such as wind and temperature, on the long-term mechanical operation of the bridge is discussed by analysing bridge monitoring data from about 1000 opening/closing events. Although some basic data analyses are presented in the paper as case studies, a detailed analysis of the monitoring data from the specific mechanical and structural components of the bridge is not in the scope of this study. These studies have been and will be reported in other publications (see for example Catbas, Gokce, et al., 2012; Catbas, Zaurin, et al., 2012).

2. Movable bridges and condition assessment

Movable bridges can be found in many parts of the world, especially crossing navigable waterways in urban and coastal regions. In the USA, the state of Florida has a large

Figure 1.   Main types of movable bridges.

population of movable bridges, most of which are owned by Florida Department of Transportation (FDOT). The bridge population analysed and the case studies presented in this paper are specific to Florida's inventory of 146 movable bridges, which includes 3 lift-type, 133 bascule-type (the highest in the nation) and 10 swing-type bridges [National Bridge Inventory (NBI), 2009]. Examples of these different types of movable bridges were discussed in Catbas, Zaurin, Susoy, and Gul (2007) and are also shown in Figure 1. The majority of these bridges have main spans between 20 and 111 m, with a mean span length of about 40 m. Almost half of this bridge population is 40–50 years old, and the mean age of the movable bridge inventory is 43 years. Figure 2 summarises Florida's inventory of movable bridges according to the type of structure, span length and year built (Catbas et al., 2007).

The distributions of the condition ratings for Florida's movable bridges are shown in Figure 3, and the mean values of these condition ratings (computed from data given in Figure 3) are 6.54, 6.33 and 6.53 for the deck, superstructure and substructure, respectively. The NBI Condition Guide of Federal Highway Administration (FHWA, 1995) defines numerical condition rating values ranging from 0 to 9 which are used to describe the condition of a bridge element based on visual inspection. A condition rating value of 0 corresponds to a failed condition, a rating value of 6 corresponds to a satisfactory condition and a rating value of 9 corresponds to excellent condition. Based on the interviews with bridge engineers from the FDOT, it was determined that the interactions of the mechanical components with structural components were generally a more critical consideration for the day-to-day operation and maintenance of these bridges than the condition ratings established through biennial visual inspections.

3.   Analysis of inspection data for a movable bridge population

The reports generated from the biennial bridge inspections include information related to the general characteristics of the bridge such as location, age, dimensions, type and date of inspection in addition to element condition ratings. The inspection reports also include recommended maintenance or repairs for each element that has been identified as deficient. Although it is acknowledged that condition ratings given in bridge inspection reports are subjective and limited (Turner-Fairbank Highway

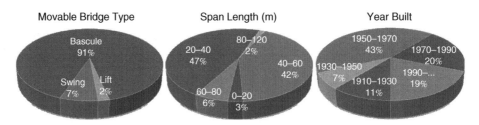

Figure 2.   Florida's inventory of movable bridges.

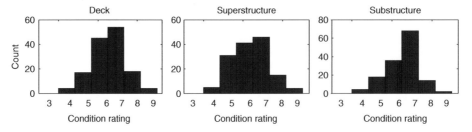

Figure 3.   Condition rating of deck, superstructure and substructure.

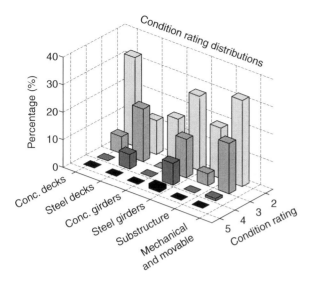

Figure 4.   Distribution of condition ratings for movable bridge elements.

Research Center, 2005), an analysis of current and past inspection reports can provide valuable representative information for evaluating the general condition of the movable bridges and for identifying the components most susceptible to deterioration and damage.

Consequently, a sample of 51 movable bridges in FDOT Districts 4 and 5 was analysed by examining the data from bridges' inspection reports to identify the most common problems, the components that experience these problems and their number of occurrences. Figure 4 illustrates the condition rating distribution of the critical components obtained from the analysis using the data made available to the authors. The bridge elements are shown with the percentage of bridges in the sample associated with each element category having a condition rating equal to or worse than 2. The condition rating system used by state engineers in Florida is different from the NBI condition rating scale, and condition rating values of 1 and 5 correspond to the best and worst conditions, respectively.

Although problems with steel girders are the most common source of low condition ratings for the sample of movable bridges evaluated, the highest maintenance costs are associated with the movable components. The mechanical and movable element category shown in Figure 4 includes all of the mechanical parts and machinery that are required for opening and closing the bridge. Some examples of the mechanical parts include shafts, trunnions, gearboxes and open gears. As noted previously, proper functioning of these components is very critical for efficient operation of a movable bridge. For the sample population of 51 Florida bridges that was analysed, 58% of the bridges had a condition rating of 2 or worse for their mechanical and movable elements, and 22% of

bridges had a condition rating of 3 or worse for these elements. A large portion of the movable bridge malfunctions is due to failures of these mechanical parts and systems.

Further analysis was conducted on the most commonly observed problems, which were then evaluated to identify the number of occurrences of improper functioning of components related to the mechanical and movable elements. Based on this analysis, the six most common types of problems associated with the mechanical and movable elements were identified as missing fasteners, cracking, leakage of lubricant from components such as the gearbox and trunnion, misalignments, section loss and inadequate lubrication. The total number of occurrences of each problem for the various mechanical components is summarised in Figure 5. Leakage of lubricant is observed to be the most frequent issue for these components.

A major source for most of the problems listed in Figure 5 is a change in the centre of gravity of the bridge, which affects the dead load distribution, balance and opening/closing torque of the bridge. Increased friction and changes in the balance state are critical conditions as they create mechanical problems for the drive motors, gearboxes and gears, hydraulic systems, shafts, open gear racks, bearings, pinions and the trunnions. The bridge balance and friction characteristics of a bascule bridge are investigated in conjunction with environmental effects, such as temperature and wind, through analysis of long-term monitoring data in a following section of this paper. Currently, movable bridges are not monitored continuously over the long term; however, inspection and maintenance are carried out according to regularly scheduled intervals. In Section 4, the inspection results for a single bascule bridge are evaluated for tracking the changes in the condition of the different bridge components and investigating the impacts of rehabilitation over a 35-year period.

4.   Case study I: long-term analysis of inspection results for Christa Mcauliffe Bridge

As shown in Section 3, bascule bridges are by far the most numerous type of movable bridge in Florida's inventory. A bascule bridge over Florida SR-3, known as the Christa McAuliffe Bridge (Figure 6), was selected as a representative bridge from the sample of 51 bridges described in the previous section to illustrate the deterioration patterns and effects of rehabilitation.

The structure studied is the southbound span of the two parallel spans on SR-3, crossing the Barge Canal in Merritt Island, FL. This span was constructed in 1961 and underwent extensive rehabilitation during the period 1999–2003. The bridge has double bascule leaves, each 21 m long and 12 m wide and carries two lanes of traffic.

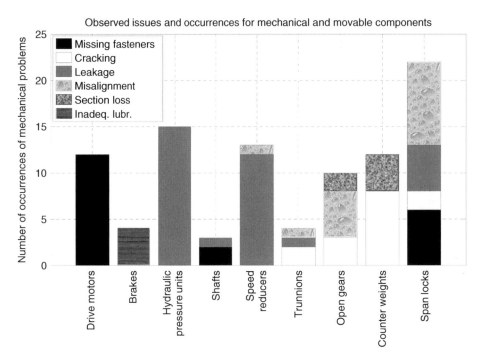

Figure 5.   Most commonly observed movable component issues and their number of occurrences identified from the inspection reports of the studied sample set.

The bridge spans are opened about six to seven times a day.

The Christa McAuliffe Bridge is a bascule-type bridge with a rack-and-pinion mechanism. The bascule leaves are rotated at the trunnions, which are the pivot points on the main girders. The weight of the span is balanced with a cast-in-place concrete counterweight that minimises the torque required to lift the leaf. The counterweight for

Figure 6.   Christa McAuliffe Bridge.

the main girders stays below the approach span deck in the closed position. When the bridge is opened, the leaves rotate upwards and the counterweights rotate downwards. An electrical motor generates the driving torque, which is then distributed to the drive shafts through a gearbox. The gearbox consists of an assembly of gears that operate similar to automobile differentials, providing an equal lifting torque to both sides of the leaf. The drive shafts transmit the torque to the final gear called the pinion that engages the rack assembly, which is directly attached to the main girder. In the closed position, the girders rest on bearing devices called live load shoes. These bearings limit the transfer of live load forces to the mechanical system by instead transmitting them to a pier located in front of the trunnion. In addition, double-leaf bascule bridges have a locking device at the tip of each leaf called a span lock and are arranged to act as cantilevers when closed. The span locks transfer live load shear forces between the ends of the two opposing leaves and also prevent the ends from bouncing as traffic passes over the bridge (Koglin, 2003). The bascule bridge also includes fixed components, such as reinforced concrete piers and approach spans.

Inspection reports for the Christa McAuliffe Bridge were analysed to evaluate the changes in condition ratings from 1969 to 2005. Examples of the analysis results are shown in Figure 7 for the condition ratings of the structural (Figure 7(a)) and the mechanical and movable elements (Figure 7(b)) over this time period. It should be noted that prior to the 1980s, the mechanical and movable elements were not assigned a condition rating, so these ratings were only analysed for the years 1982 through 2005.

The condition states indicate the trend of the bridge's condition over its lifetime. The deterioration patterns can be identified and modelled, making it possible to predict the future condition and the effects of rehabilitation. It is seen that the condition of steel girder elements starts deteriorating after 7 or 8 years due to the effects of surface corrosion, section loss and operating impacts causing dents and misalignment (Figure 7(a)). A rehabilitation period, shown in the figure, produces a positive effect on the condition ratings of the bridge components. The downward trend of the condition ratings is much more rapid for the case of movable elements since they are subjected to repeated opening/closing events and are very sensitive to alignment, lubrication and proper mainten- ance. Figure 7(b) shows that some movable components can change rapidly, changing the condition rating from 1 to 3 in < 1 year after rehabilitation. This rapid change in rating may also be due to poor maintenance. It is also observed that the rating of some of the components increased although a major rehabilitation is not applied. Since movable bridges undergo reactive maintenance based on visual inspection, these components may have been maintained/repaired individually during these times.

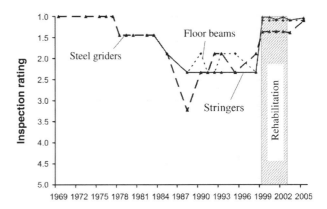

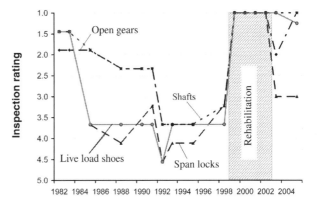

Figure 7. Christa McAuliffe Bridge: (a) structural and (b) mechanical/movable element condition ratings.

Moreover, the subjective nature of the visual inspections, which is one of the main motivations of this study, may have also caused these increases. The movable com- ponents must be continuously maintained in order to avoid unexpected failures. For effective condition-based main- tenance, continuous monitoring can be considered as an enabling technology.

5. Case study II: continuous monitoring of Sunrise Boulevard Bridge

A representative movable bridge was made available for designing and demonstrating a continuous SHM system. The representative bridge was identified based on the bridge population analysis described earlier, input from movable bridge engineers and expert practitioners, considering the characteristics that included structure type, span length, age, opening frequency, type of traffic and accessibility. It was determined that a movable bridge on Florida SR-838 (the Sunrise Boulevard Bridge shown in Figure 8) would be representative of Florida's bascule bridges. The particular bascule span selected for monitoring was constructed in 1989 and consisted of the westbound span of the two parallel bascule bridges on the Sunrise Boulevard (SR-838) in Ft. Lauderdale, Florida.

Figure 8. Sunrise Boulevard Bridge.

The bridge is a double-leaf bascule structure with a total span length of 35.7 m and a width of 16.3 m. The westbound span carries three traffic lanes over the Atlantic Intercoastal Waterway and can be opened for maritime traffic on the hour and half-hour when requested. Depending on the boat traffic, the bridge is usually opened about 10–15 times a day. Factors that were considered in designing the monitoring programme for the structural, mechanical and electrical components and the overall SHM system are discussed in the following sections.

5.1. Design of the monitoring system

The inspection information evaluated previously was combined with the expert opinions of bridge engineers, FDOT officials and consultants to identify specific issues and possible monitoring needs. The kinematic behaviour of a movable bridge is the main reason for most of the problems and high maintenance costs associated with its operation and performance. The mechanical and movable elements that form the complex operating system for the bridge also require special expertise for maintenance and repairs.

Consequently, monitoring of components that frequently require inspection and maintenance becomes very critical. For static structures, monitoring of structural components is usually the only concern for maintenance, safety and operations; however, for movable bridges, monitoring of the mechanical and electrical components is equally important. With these considerations, the components that need to be monitored are defined based on the

issues given in the inspection reports, preliminary finite element model analysis and also based on extensive discussions and feedback provided by movable bridge experts (such as experienced engineers who have been working with the Sunrise Boulevard Bridge and this type of bridges for a long time) and bridge owners (including the decision-makers in different districts and the main office of FDOT). Tables 1–3 provide an overview of some of the issues, components and corresponding instrumentation.

Data analysis methods and the expected outcomes for structural, mechanical, electrical components as well as operational and environmental effects were developed based on the issues and the related measurement needs. Such a monitoring system can provide information about the root causes of the structural and mechanical problems, as exemplified later in this paper, and future designs can be improved using this information.

A comprehensive instrumentation plan was developed to monitor the most critical structural, mechanical and electrical components for the bascule bridge. Since the two leaves of the bascule bridge are physically separated from each other, a wireless communication system was established to permit seamless data transmission between the opposing leaves of the bridge. Two GPS units were also employed to synchronise the timing of the measurements recorded from the opposing spans of the bridge. Data from the monitoring system can be collected during rush hours, during opening/closing events and at other times depending on demand. Details of the Sunrise

Table 1. SHM considerations for structural components.

Structural components	Issues	Installed sensors	Data analysis methods	Expected outcome
Live load shoes	• Loss of contact, pounding impact loading on the shoes • Corrective action in case of high-impact loads Fully/particularly seated girders	• Strain rosette • Accelerometer • Dynamic strain gage • Vibrating wire strain gage	• Dynamic analysis in time and frequency domain	• Calculate the shear at the live load shoe for load rating • Track condition of live load shoe and shims
Girders and floor beams and stringers	• Section loss due to corrosion • Bending, deformation and mis alignment • Missing fasteners	• Accelerometer • Dynamic strain gage • Vibrating wire strain gage • Srain rosette	• Dynamic analysis in time and frequency domain • Strain correlation • Time series analysis	• Monitor the stress levels • Determine the structural condition at critical and vulnerable locations • Track shear and moment for load rating

Table 2. SHM considerations for mechanical and electrical components.

Mechanical and electrical components	Issues	Installed sensors	Data analysis methods	Expected outcome
Electrical motors	• Unexpected interruption, schedule maintenance, replacement ahead of time • Predictive maintenance for replacement	• Accelerometer • Infrared temperature • Amp-meter	• Frequency domain analysis (FDA) • Root mean square (RMS) analysis	• Define baseline and track changes of acceleration, temperature and electric current • Correlate various electrical motor monitoring parameters for predictive maintenance
Gearboxes	• Lubrication, wearing of the gears, load transfer within the system • Maintenance of the gearbox	• Accelerometer • Microphone	• FDA • RMS analysis • Artificial neutral networks (ANNs)	• Track acoustic print and the vibration characteristics to determine if the gearbox lubrication is normal and the gear condition is satisfactory

Table 2 – *continued*

Mechanical and electrical components		Issues	Installed sensors	Data analysis methods	Expected outcome
Shafts		• Load transfer within the system, differential working of shafts and loading on the system • Bridge balance, changes in friction numbers	• Strain rosette	• Friction, torque, balance calculation • Correlation analysis	• Track torque and friction values for balance problems of the counterweight • Correlate wind and temperature with torque and friction characteristics
Open gears and backs		• Lubrication, corrosion, cracks and missing bolts • Determine if maintenance needed based on measurement history	• Accelerometer • Video camera	• Image processing • ANN	• Detect lubrication level and corrosion • Track bolt conditions of the racks
Trunnions		• Maintenance for lubrication measures from torque data • Stress monitoring around this critical area	• Tiltmeter • Microphone • Strain rosette	• Balance calculation • Shear rotation • Pressure calculation	• Track friction over time for lubrication in the trunnion • Track shear force on the trunnion area • Track rotation of two main girders for balance
Span locks		• Alignment problems, fatigue effects and dynamic loading leading to failure, excessive shear • Hydraulic pressure problems	• Pressure gage • Strain rosette • Tiltmeter	• Shear rotation • Pressure calculation • Strain correlation	• Check for excessive stress on the lock bar • Tiltmeters on the lock bars for horizontal and vertical alignment • Track condition of span lock shims and track hydraulic pressure

Table 3. Considerations for operational and environmental monitoring.

Monitoring of operational and environmental effects	Issues	Installed sensors	Data analysis methods	Expected outcome
	• Evaluation of environmental inputs on mechanical and structural responses for maintenance purposes • Evaluation of operational inputs on structural components	• Weather station • Video camera	• Time domain and FDA • Image processing and image sensor correlation	• Track the environmental effects on the components • Track critical traffic inputs to the structure

Boulevard Bridge monitoring system (Figure 9) can be found in Catbas et al. (2010).

5.2. *Bridge balance and friction for movable bridge operations*

A critical operational consideration for movable bridge owners is the mechanical component failures due to friction and wear caused by the movement of the span during openings and closings. The current practice to mitigate failures and assess maintenance effectiveness is to conduct bridge balance tests intermittently. Technicians travel to different bridges, conduct bridge balance tests and determine whether there is a considerable change in bridge friction from the previous measurement. If there are significant changes in the results, further inspections may be conducted to investigate whether an immediate maintenance is required to readjust the bridge balance or not. The open gears, bearings of the gearbox and trunnions are maintained according to a predetermined schedule to minimise this issue. The maintenance of these components also has an impact on the alignment of mechanical and structural elements. Even with regular maintenance, continuous downgrading of all parts of these bridges is inevitable, leading to breakdowns. The engineers mainly compare consecutive bridge friction numbers to make decisions.

Operating the bridge under balance conditions other than slight leaf heaviness may place excessive wear on the structural and mechanical components of the bridge leading to costly repairs or possible safety concerns. In the unbalanced condition, excessive stresses can be directly transferred to mechanical components of the bridge. The immediate maintenance measures performed for an unbalanced operating condition are typical lubrication of the mechanical system components or adjusting the balance of the bascule leaf with weights. The friction and balance problems during openings/closings are monitored with torque values calculated from strain rosettes on the shafts and the tiltmeters on the trunnions. After obtaining strain rosette-based torque values with fundamental mechanics of material formulations, the bridge balance and friction equations can be derived as follows (Malvern, Lu, & Jenkins, 1982). If the distance between centre of gravity and trunnion is L, then the horizontal moment arm X of the leaf weight W is

$$X = L\cos(\alpha + \theta), \qquad (1)$$

where α is the angle of centre of gravity elevation below the horizontal axis and θ is the opening angle. Consequently, the imbalanced moment (M_i) is

$$M_i = WL\cos(\alpha + \theta). \qquad (2)$$

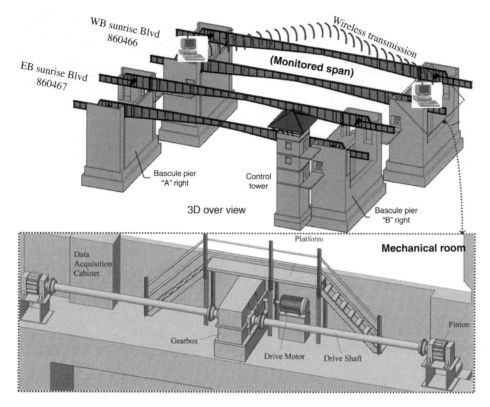

Figure 9.   Sunrise Boulevard Bridge wireless communication scheme and mechanical room 3D overview.

Subsequently, by taking the friction into account, the required opening (T_O) and closing (T_C) bridge torques are

$$T_O = WL\cos(\alpha + \theta) + T_F, \qquad (3)$$

$$T_C = WL\cos(\alpha + \theta) - T_F. \qquad (4)$$

By adding and subtracting Equations (3) and (4), friction torque (T_F) and average torque (AVT) are

$$T_F = 0.5(T_O - T_C), \qquad (5)$$

$$AVT = 0.5(T_O + T_C) = WL\cos(\alpha + \theta). \qquad (6)$$

In monitoring applications, friction torque (T_F) is calculated in segments and the final average of these values yields an average torque friction (AVTF):

$$AVTF = \frac{\sum T_F}{\text{Number of values}} \qquad (7)$$

Determination of the WL and α parameters, which characterise the imbalance state, is as follows (Figure 10 (a)). If the cosine factor in Equation (6) is expanded, then

the components of WL are

$$A = WL\cos(\alpha), \qquad (8)$$

$$B = WL\sin(\alpha). \qquad (9)$$

By inserting these values into Equation (6):

$$AVT = A\cos(\theta) - B\sin(\theta). \qquad (10)$$

By performing a least-squares regression analysis of AVT data changing for each opening degree, horizontal (A) and vertical (B) components of WL can be obtained:

$$WL = \sqrt{A^2 + B^2}. \qquad (11)$$

Typical plots of AVT values with respect to opening angle θ are given in Figure 11. An AVT plot created from actual measurements is presented in Figure 11. The AVT changes with the horizontal distance between the trunnion and the centre of gravity. The positive region of this plot is unbalanced towards the leaf side, and the negative region corresponds to unbalance towards the counterweight side, indicating that counterweight is too heavy. It is desirable to have the centre of gravity located on the leaf side in

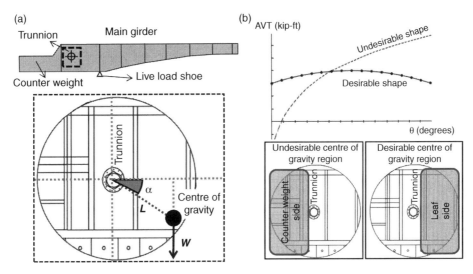

Figure 10. Trunnion (coordinate centre) and centre of gravity location of the leaf (a), desired and undesired AVT shapes and centre of gravity regions (b).

bascule bridges because the bridge should stay in the closed position when there is a malfunction.

Figure 11 illustrates typical bridge opening–closing data as recorded by the monitoring system. This figure summarises the analysis of tiltmeter and strain rosette measurements to determine the bridge balance/friction for a single opening and closing cycle. For this particular opening/closing cycle, adjusted opening torque is 122 kNm (90 kip-ft), and this torque reduces to around 68 kNm (50 kip-ft) when the leaf angle is 65° from horizontal. While closing the bridge, the measured torque is about 27 kNm (20 kip-ft) and as the span is about to be closed, the torque increases up to about 81 kNm (60 kip-ft). This indicates that the initial opening requires the highest torque, and closing from the vertical position requires the lowest torque. For this case, the average friction is found to be around 22.2 kNm (16.4 kip-ft). The relative increase in bridge friction is typically attributed to a lack of proper maintenance such as lubrication of the gearboxes and the open gears. While this may be true, it is also imperative to explore other possible factors that might also have an impact on the behaviour observed from the continuous monitoring system.

5.3. Wind and temperature effects on bridge balance

In this section, wind and temperature effects on the bridge balance are investigated. Wind speed and α (the angle of centre of gravity elevation below the horizontal axis as shown in Figure 10) computed from measurements recorded during 462 opening/closing events over a 1-month period are presented in Figure 12. This figure indicates that a clear relationship exists between these parameters. An increase in wind speed directly affects α, which is a characteristic of the centre of gravity. This observation also illustrates the effects of the wind on bridge balance parameters A and B. It shows that taking discrete bridge balance measurements as applied in current practice may reveal biased values depending on the wind speed. Tracking these values using a continuous monitoring system may mitigate such problems. Of course, it should be mentioned that wind is not the only parameter and a complex coupling of other parameters such as ambient temperature with wind speed and directions may also impact the bridge balance parameters.

While balance adjustment decisions are generally based on the history of friction numbers, there is no formula that defines the load amounts, which should be added to or removed from the counterweights. In addition, there is no consideration or formulation for the impact of ambient temperature on the bridge balance and friction characteristics. As a result, tracking the bridge balance and friction becomes important especially for understanding the possible correlation between friction and temperature. For this reason, temperature data from a weather station at the bridge and the friction numbers determined from the monitoring system measurements were analysed in detail for 1047 opening and closing events over a 3-month period (Figure 13). The October 2009–January 2010 period is especially selected due to changes in temperature.

When the ambient temperature and bridge friction numbers were compared, an inverse correlation with a correlation coefficient of 0.72 was computed. The high correlation between these parameters is also observable in Figure 13. This is an indication of the effect of ambient

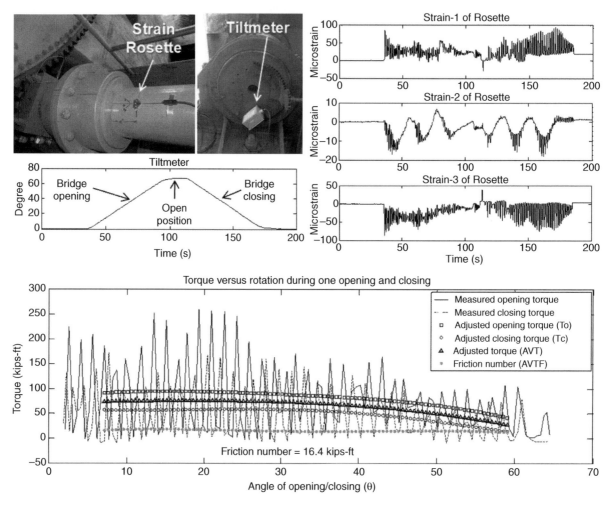

Figure 11. Sample tiltmeter data, sample strain rosette data and balance/friction analysis based on SHM during one opening and closing (1 kip-ft = 1.36 kNm).

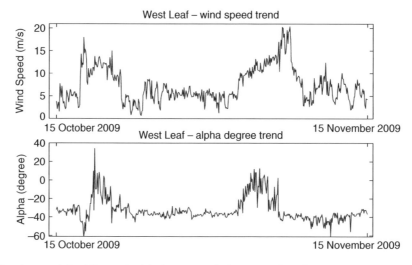

Figure 12. Wind speed and α trend for 462 opening/closing events during 1-month period.

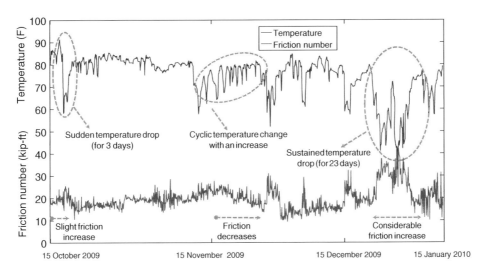

Figure 13. Monitoring friction and temperature over time from West leaf (1 kip-ft = 1.36 kNm and $C = (F - 32) \times 5/9$) for 1047 opening/closing events.

temperature on bridge friction. The effect is particularly clear when sudden and sustained temperature changes are observed, such as occurred in late December 2009 and early January 2010 from Figure 13. The temperature decrease (from $\sim 26.7°C$ to $\sim 4.4°C$) that lasted for 23 days resulted in a considerable increase in the friction number. Figure 13 also shows that the bridge had a friction number of 26.6 kNm (19.6 kip-ft) obtained at 26.8°C (80.3°F) on 15 October 2009. When monitored on 4 January 2010, the friction number is 54.8 kNm (40.3 kip-ft) at 4.9°C (40.8°F). This also suggests that the ambient temperature has a significant effect on the bridge friction.

Although temperature is a non-deterministic phenomenon, daily, monthly and seasonal temperature cycles can be observed with intermittent temperature shocks and changes (Catbas, Susoy, & Frangopol, 2008). Because of the cyclic nature of the ambient temperature and its observed correlation with friction, it would be appropriate to investigate the ambient temperature and friction responses by means of Fourier series with the expansion form given as

$$f = a_0 + \sum_{n=1}^{N} a_n \sin(n\omega t) + b_n \cos(n\omega t). \quad (12)$$

The temperature and friction data are characterised using an eighth-order Fourier series ($N = 8$) as shown in Figure 14. The seasonal characteristics of the signals along with the major and sustained temperature drop in late December are captured. It should be noted that the curve fit here might change slightly as more data are included in such an analysis. It is also observed that the main contributors to the Fourier series are the low-frequency ($n\omega$) terms (a_n and b_n), indicating the correlation of higher

period-sustained temperature changes. This is an indication that for the same lubrication conditions, the friction values can be expected to be different in hot summer months compared to cold winter months.

6. Summary and conclusions

6.1. Summary

Movable bridges have been used and their designs have been studied for some time; however, the condition, performance and continuous monitoring of movable bridges, and more specifically bascule bridges, are not extensively documented as compared to other bridge types. In this paper, the characteristics of the movable bridge inventory in Florida were presented along with the evaluation of their condition based on the inspection database and also based on detailed discussions with movable bridge experts. It is seen that mechanical and movable components are the main concern for the routine operation of these bridges. The bridge owners are mostly concerned with the mechanical failures, which require immediate action and maintenance since mechanical malfunctions lead to disruptions to land and/or marine traffic. As a result, mechanical components are maintained more frequently.

Issues such as misalignment, inadequate lubrication and leakage at mechanical components are analysed along with the number of occurrences. In addition, two different case studies are presented for an in-depth evaluation of the performance of a bridge over a 35-year period and for the development of a monitoring system on a representative movable bridge. For the first bridge, condition-rating data were tracked for different components to illustrate the

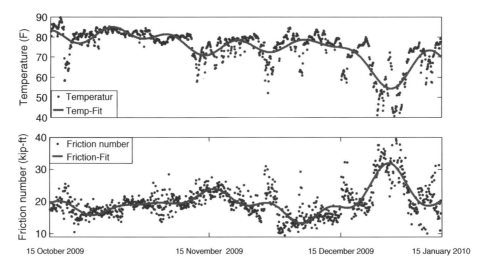

Figure 14. Fourier series fit to monitored friction and temperature over time from West leaf (1 kip-ft = 1.36 kNm and $C = (F - 32) \times 5/9$).

deterioration patterns and rehabilitation effects. The erratic and sudden changes in condition of mechanical components are observed over long term. Such sudden failures make monitoring systems very desirable for continuously tracking the condition and performance of the mechanical components. An SHM system design is presented for tracking previously identified common issues on a representative bascule-type bridge as a second case study. The design of the system and its justification, which is based on the evaluations of inspection reports, preliminary finite element models and expert opinions, was discussed.

6.2. Conclusions

The monitoring system is utilised to track a movable bridge-specific issue 'bridge balance and friction', which can directly affect all of the mechanical/movable components. It is shown that wind and temperature have impact on the bridge balance and friction. In addition to bridge maintenance, the bridge friction is also correlated to temperature (especially during sudden and sustained temperature changes) and wind speed. Based on the analysis of monitoring data, the correlation between ambient temperature and bridge friction was determined. The correlation of the ambient temperature and bridge friction is further emphasised through the characterisation of these signals using a low-order Fourier series analysis. This proved to be a useful approach due to the periodicities observed for the ambient temperature and the corresponding friction data.

Additional studies are recommended for developing models that characterise the bridge friction in terms of time-dependent environmental effects for estimating/scheduling optimum maintenance for movable bridges under uncertainty and in a life cycle context. At the

beginning, these studies can be based on results reported for fixed bridges by Frangopol and Liu (2007), Frangopol (2011) and Frangopol and Bocchini (2012) and on the performance indicators reported by Saydam and Frangopol (2011) and Zhu and Frangopol (2012).

Acknowledgements

The research project described in this paper is supported by the Florida Department of Transportation (FDOT) Contract No. BD548/RPWO 23 and FHWA Cooperative Agreement Award DTFH61-07-H-00040. The authors thank Mr Alberto Sardinas of FDOT for his support and feedback throughout the project. The authors express their profound gratitude to Dr Hamid Ghasemi of FHWA for his support of this research. The support of both agencies and their engineers is greatly recognised and appreciated. Several graduate, undergraduate students, post-doctoral research associates and research engineers also made significant contributions to the design, installation and implementation of the SHM system; we thank all these individuals. The opinions, findings and conclusions expressed in this publication are those of the authors and do not necessarily reflect the views of the sponsoring organisations.

References

American Association of State Highway and Transportation Officials (AASHTO). (1998). *Movable bridge inspection, evaluation and maintenance manual* (1st ed.). Washington, DC: Author.

Besterfield, G., Nichani, S., Kaw, A.K., & Eason, T. (2003). Full-scale testing of procedures for assembling trunnion-hub-girder in bascule bridges. *Journal of Bridge Engineering, 8*(4), 204–211.

Byers, W. (2008). Discussion of 'Kinematics of movable bridges' by Markus Wallner and Martin Pircher. *Journal of Bridge Engineering, 13*(2), 211.

Catbas, F.N., Gokce, H.B., & Gul, M. (2012). Non-parametric analysis of structural health monitoring data for identification and localization of changes: Concept, lab and real life studies. *Structural Health Monitoring, 11*(5), 613–626.

Catbas, F.N., Gokce, H.B., Gul, M., & Frangopol, D.M. (2011). Condition and performance considerations for movable bridges. *ICE: Bridge Engineering, 164*(3), 145–155.

Catbas, F.N., Gul, M., Zaurin, R., Gokce, H.B., Terrell, T., Dumlupinar, T., & Maier, D. (2010). *Long term bridge maintenance monitoring demonstration on a movable bridge. A framework for structural health monitoring of movable bridges (Report).* Tallahassee, FL: Florida Department of Transportation (FDOT).

Catbas, F.N., Susoy, M., & Frangopol, D.M. (2008). Structural health monitoring and reliability estimation: Long span truss bridge application with environmental monitoring data. *Engineering Structures, 30*(9), 2347–2359.

Catbas, F.N., Zaurin, R., Gul, M., & Gokce, H.B. (2012). Sensor networks, computer imaging and unit influence lines for structural health monitoring: A case study for bridge load rating. *Journal of Bridge Engineering, ASCE, 17*(4), 662–670.

Catbas, F.N., Zaurin, R., Susoy, M., & Gul, M. (2007). *Integrative information system design for Florida department of transportation – A framework for structural health monitoring of movable bridges (Report, Contract No. BD548-RPWO#11).* Tallahassee, FL: Florida Department of Transportation, 231 pages.

Ecale, H., & Lu, T.-H. (1983). New Chicago-type bascule bridge. *Journal of Structural Engineering, 109*(10), 2340–2354.

Federal Highway Administration (FHWA). (1995). *Recording and coding guide for the structural inventory and appraisal of the nation's bridges.* Washington, DC.

Fisher, A.D., & Robitaille, A.M. (2011). The replacement of the movable span of the Thames River Bridge: How 19th century technology impacted 21st century construction. *Journal of Construction Engineering and Management, ASCE, 137*(10), 895–900.

Frangopol, D.M. (2011). Life-cycle performance, management, and optimization of structural systems under uncertainty: Accomplishments and challenges. *Structure and Infrastructure Engineering, 7*(6), 389–413.

Frangopol, D.M., & Bocchini, P. (2012). Bridge network performance, maintenance, and optimization under uncertainty: Accomplishments and challenges. *Structure and Infrastructure Engineering, 8*(4), 341–356.

Frangopol, D.M., & Liu, M. (2007). Maintenance and management of civil infrastructure based on condition, safety, optimization, and life-cycle cost. *Structure and Infrastructure Engineering, 3*(1), 29–41.

Fränkel, W. (1882). Der brückenbau – Bewegliche brücken. in T. Schäffer & E. Sonne, eds. Leipzig: Wilhelm Engelmann.

Greene, W.K., & McKeen, E.E. (1938). Erecting the marine parkway bridge. *Engineering News-Record,* 371–374.

Griggs, J.F.E. (2006). Development of the vertical lift bridge: Squire whipple to J.A.L. Waddell, 1872–1917. *Journal of Bridge Engineering, 11*(5), 642–654.

Hardesty, E.R., Christie, R.W., & Fischer, H.W. (1975a). Fifty-year history of movable bridge construction – Part I. *Journal of the Construction Division, 101*(3), 511–527.

Hardesty, E.R., Christie, R.W., & Fischer, H.W. (1975b). Fifty-year history of movable bridge construction – Part II. *Journal of the Construction Division, 101*(3), 529–543.

Koglin, T.L. (2003). *Movable bridge engineering.* New Jersey, USA: Wiley.

Malvern, L.E., Lu, S.Y., & Jenkins, D.A. (1982). *Handbook of bascule bridge balance procedures (Report).* Tallahassee, FL, USA: Florida Department of Transportation.

National Bridge Inventory (NBI). (2009). *Public disclosure of national bridge inventory (NBI) data.* Washington, DC: Federal Highway Administration.

Okasha, N.M., & Frangopol, D.M. (2010). Novel approach for multi-criteria optimization of life-cycle preventive and essential maintenance of deteriorating structures. *Journal of Structural Engineering, ASCE, 136*(8), 1009–1022.

Okasha, N., & Frangopol, D.M. (2012). Integration of structural health monitoring in a system performance based life-cycle bridge management framework. *Structure and Infrastructure Engineering, 8*(11), 999–1016.

Quade, M.N. (1954). Special design features of the Yorktown bridge. *Transactions of the American Society of Civil Engineers, 119*(1), 109–123.

Ramey, G.E. (1983). Lift system for raising continuous concrete bridges. *Journal of Transportation Engineering, 109*(5), 733–746.

Saydam, D., & Frangopol, D.M. (2011). Time-dependent performance indicators of damaged bridge superstructures. *Engineering Structures, 33*(9), 2458–2471.

Turner-Fairbank Highway Research Center. (2005). *Bridge study analyzes accuracy of visual inspections.* Retrieved from http://www.tfhrc.gov/focus/jan01/bridge_study.htm

Waddell, J.A.L. (1895). The Halstead street lift bridge. *Transactions of the American Society of Civil Engineers, 33*(1), 1–60.

Wallner, M., & Pircher, M. (2007). Kinematics of movable bridges. *ASCE Journal of Bridge Engineering, 12*(2), 147–153.

Wengenroth, R.H., Hardesty, E.R., & Mix, H.A. (1975). Fifty-year history of movable bridge construction – Part III. *Journal of the Construction Division, 101*(3), 545–557.

Zhao, Y., Wang, J.F., & Pang, M. (2011). The integral lifting project of the qifeng bridge: Case study. *Journal of Performance of Constructed Facilities, ASCE, 25*(3), 353–361.

Zhu, B., & Frangopol, D.M. (2012). Reliability, redundancy and risk as performance indicators of structural systems during their life-cycle. *Engineering Structures, 41*, 34–49s.

Part VI

Redundancy as life-cycle performance indicator

Time-variant redundancy of structural systems

Nader M. Okasha and Dan M. Frangopol

Structural redundancy is expected to change over time due to time-variant loading and damage under uncertainties. The objective of this paper is to investigate the time-variant redundancy of structural systems. Analyses of structural reliability and redundancy affected by deterioration in structural resistance and increase in applied loads are conducted by using numerical examples. It is shown that the structural system redundancy is influenced by several factors, such as the material type, the resistance correlation structure and deterioration rate and the rate of increase in applied loads. The results show the importance of including the time factor in the quantification of redundancy. Such results are useful for identifying the best measures to take in order to maintain a satisfactory level of redundancy throughout the life of structural systems and to incorporate the time-variant redundancy in a lifetime-oriented multi-objective optimisation framework of risk-based management of structural systems.

1. Introduction

Uncertainties are inherent in all aspects of the assessment process of structural reliability and redundancy. Due to this inherent nature, probabilistic methods have to be used (Ang and De Leon 2005, Ellingwood 2005, Moan 2005, Sorensen and Frangopol 2008). During the past three decades, researchers have been investigating the application of system reliability concepts and techniques in structural design and evaluation (Frangopol 1989, 1992, De *et al.* 1990, Paliou *et al.* 1990, Frangopol and Nakib 1991, Frangopol and Iizuka 1992, Ghosn and Moses 1998, Bertero and Bertero 1999, Liu *et al.* 2000, Ghosn and Frangopol 2007a,b, Ghosn *et al.* 2010). Probability-based system redundancy concepts have been introduced and used to asses the condition of structural systems and seek warnings of partial or total collapse (Frangopol 1987, Frangopol and Curley 1987, Fu 1987, De *et al.* 1990, Paliou *et al.* 1990, Frangopol *et al.* 1992, 1998, Gharaibeh and Frangopol 2000, Gharaibeh *et al.* 2000a,b).

To the best of our knowledge, redundancy measures have been studied and sometimes implemented with no regard to the aspect of time. Structural redundancy, however, is expected to change over time due to various time-variant mechanical and environmental stressors. Studying the time-variant redundancy of a structure provides a better perspective of the structural performance over time and helps to improve the design and maintenance of structural systems. By incorporating the time-variant redundancy in a lifetime-oriented multi-objective optimisation framework of civil infrastructure management, optimised cost-effective structures can maintain a satisfactory level of lifetime reliability and redundancy.

The objective of this study is to investigate the effects of the resistance deterioration and load increase on the redundancy of structural systems. Analyses of structural reliability and redundancy affected by structural deterioration and load increase are conducted via numerical examples. The effects of various factors on the system redundancy, such as the material type, the rate of increase in loading, the member resistance correlation structure and the deterioration rate are also investigated. The results show the importance of including the time factor in the quantification of redundancy of structural systems.

2. Time-invariant system reliability and redundancy

2.1. System reliability

Civil engineering structures are to be designed for loads associated with environmental, mechanical and human actions that are uncertain in their manifestations. Several materials used in civil engineering also display a wide scatter in their properties. The topic of structural reliability offers a rational framework to quantify these uncertainties mathematically. This topic combines theories of probability, statistics and random processes with principles of structural mechanics, and forms the basis on which modern structural design and assessment codes are developed and calibrated.

Structural system reliability defines safety as the condition in which system failure will not occur. The reliability of a structural system $P_{s(sys)}$ is usually addressed by dealing with the probability of occurrence of the complement of the survival event, i.e. the probability of system failure $P_{f(sys)} = 1 - P_{s(sys)}$. The probability of failure of a system is defined as the probability of violating any of the limit state functions that define its failure modes. Limit states of structural systems, as an example, are expressed by equations relating the resistances of the structural components to the load effects acting on these components. These limit states are violated when the value of the respective performance function is less than zero. Once a limit state is violated, the structure fails in the mode defined by that limit state. Performance functions can be expressed as:

$$g_i = (R_s)_i - (Q)_i, \qquad (1)$$

where g_i is the system performance function with respect to failure mode i; $(R_s)_i$ is the system resistance associated with failure mode i; and $(Q)_i$ is the load effect associated with failure mode i ($i = 1, 2, \ldots, n$). The probability of occurrence of failure mode i is thus the probability of the load effect exceeding the resistance, which, accordingly, causes g_i to fall below zero. Due to uncertainties in the component resistance and in the applied loads, a structure may fail in any one of its possible failure modes. For this reason, all possible failure modes are taken into account when the probability of system failure is calculated. Therefore, the probability of system failure of a structure with n possible failure modes is:

$$P_{f(sys)} = P[\text{any } g_i < 0], \quad i = 1, 2, \ldots, n. \qquad (2)$$

An exact calculation of $P_{f(sys)}$ can only be carried out by performing an integration of the joint probability density function $f_{X_1, X_2, \ldots, X_k}(x_1, \ldots, x_k)$ of the random variables $\mathbf{X} = \{X_1, X_2, \ldots, X_k\}$ involved in the problem over the failure region defined by the aforementioned performance functions (Ang and Tang 1984). Let E_i be the event of occurrence of failure mode i, i.e. $E_i = [g_i < 0]$. The probability of system failure is:

$$P_{f(sys)} = \int_{(E_1 \cup \cdots \cup E_n)} \cdots \int f_{X_1, X_2, \cdots, X_k}(x_1, \ldots, x_k) dx_1 \ldots dx_k, \qquad (3)$$

where $\cup$ is the union. The exact calculation of $P_{f(sys)}$ however, is generally a formidable task. For practical purposes, approximate methods of finding solutions to Equation (3) are implemented. Probabilities of occurrence of the individual limit states are usually calculated by using first-order (FORM) or second-order (SORM) moment methods. Since the different failure modes of a system are generally correlated through the loading and resistances, first-order (Cornell 1967), or, more accurately, second-order bounds (Ditlevsen 1979) can be used to account for these correlations and estimate upper and lower bounds on $P_{f(sys)}$. Alternatively, Monte Carlo simulations can be performed to directly estimate the probability of system failure. The drawback of this method, however, is the high computational cost of the simulation process due to the usually very large number of simulations needed to accurately compute the required probability of failure of a realistic structural system.

The incremental method (Moses 1982, Rashedi and Moses 1983, 1988) is used to formulate the performance functions used in this paper. Mathematically, the goal of the incremental method is to find an expression for the system resistance in terms of the component resistances. In the case of a ductile system, for instance, the form of this expression is as follows (Moses 1982):

$$(R_s)_i = \sum_{j=1}^{m} (C_j)_i R_j, \qquad (4)$$

where $(R_s)_i$ is the system resistance associated with failure mode i; $(C_j)_i$ is the coefficient representing the participation of component j in resisting failure mode i; R_j is the resistance of component j; and m is the number of components that need to fail in order to cause the occurrence of failure mode i. It is assumed, throughout this paper, that the resistances are all lognormally distributed random variables. The lognormal distribution has been assumed for its exclusion of negative values, which is appropriate in most structural engineering applications.

Consider, for example, the two parallel bar system shown in Figure 1. It is assumed that (a) the horizontal member linking the two bars is perfectly rigid and constrained to remain horizontal and (b) the bars have the same cross-sectional area and modulus of elasticity. Consider first the case where the system is ductile. The system resistance assuming that bar 1 fails before bar 2 is found by the incremental method as:

$$R_s = R_1 + R_2 \qquad (5)$$

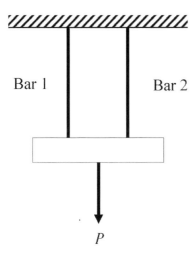

Figure 1. Structural system of two parallel bars.

and the performance function is

$$g = R_1 + R_2 - P. \tag{6}$$

If bar 2 was assumed to fail first, the same performance function is obtained. This is because failure modes of ductile systems are independent of the failure sequence. However, this is not the case in brittle systems. Different failure sequences lead to different load redistributions and thus different failure modes (Rashedi and Moses 1983). All these failure modes must be anticipated and accounted for in a correct reliability analysis. Two failure modes can be anticipated if the two parallel bar system is brittle, namely, mode 1 caused by failure of bar 1 followed by bar 2 and mode 2 caused by failure of bar 2 followed by bar 1. The system resistance associated with mode 1 is:

$$(R_s)_1 = \max\{2R_1, R_2\} \tag{7}$$

and the system resistance associated with mode 2 is:

$$(R_s)_2 = \max\{2R_2, R_1\}. \tag{8}$$

The load effect that causes the occurrence of both failure modes is the same, i.e. $(Q)_1 = (Q)_2 = P$. Hence, the performance functions associated with modes 1 and 2, respectively, are:

$$g_1 = \max\{2R_1, R_2\} - P \tag{9}$$

and

$$g_2 = \max\{2R_2, R_1\} - P. \tag{10}$$

The probability of damage occurrence to the system is also required for redundancy analyses. The occurrence of damage is usually described as being the result of the failure of the first component in the system. An axial-force component fails if its axial force reaches its resistance level. A ductile component continues to support a load equal to its resistance after its failure, while a brittle component redistributes the loads it supported prior to its failure to other components in the system (Hendawi and Frangopol 1994). The load that causes the first component failure, however, is the same, whether the system is ductile or brittle. Therefore, the probability of damage occurrence to a ductile system is equal to the probability of damage occurrence to its brittle counterpart. The probability of damage occurrence to the system can be defined as follows. Let h_j be a performance function with respect to failure of component j, and V_j be the event of failure of component j, i.e. $V_j = [h_j < 0]$. The probability of damage occurrence to the system $P_{f(dmg)}$ is:

$$P_{f(dmg)} = \int_{(V_1 \cup \cdots \cup V_k)} \cdots \int f_{X_1, X_2, \cdots, X_k}(x_1, \ldots, x_k) \mathrm{d}x_1 \ldots \mathrm{d}x_k. \tag{11}$$

The performance functions h_j for both brittle and ductile systems are merely the internal load carried by component j subtracted from the resistance of this component. It can be shown for the two parallel bar system in Figure 1 that:

$$F_1 = \frac{A_1}{A_1 + A_2} P = l_1 P = \frac{1}{2} P \tag{12}$$

and

$$F_2 = \frac{A_2}{A_1 + A_2} P = l_2 P = \frac{1}{2} P, \tag{13}$$

where F_i, A_i and l_i are the internal force, cross-sectional area and load sharing factor of bar i, respectively. The performance functions with respect to failure of bars 1 and 2, respectively, are thus:

$$h_1 = R_1 - l_1 P = R_1 - \frac{1}{2} P \tag{14}$$

and

$$h_2 = R_2 - l_2 P = R_2 - \frac{1}{2} P. \tag{15}$$

2.2. System redundancy

System redundancy has been defined as the availability of system warning before the occurrence of structural collapse. Structural redundancy depends on many factors, such as the configuration, number of components, component size, material properties, connection types, secondary systems location, intensity of applied loads, statistical parameters and correlation among basic random variables (Frangopol *et al.* 1992). At present, the literature is rich with studies presenting measures of quantifying redundancy in design and evaluation of structures. However, no universal measure has yet been agreed on. Most efforts in this field have been directed towards defining probabilistic system redundancy measures (Frangopol 1987, Frangopol and Curley 1987, Fu 1987, Frangopol *et al.* 1992). The redundancy index RI for the probabilistic representation of system redundancy, defined as (Fu 1987):

$$\text{RI} = \frac{P_{\text{f(dmg)}} - P_{\text{f(sys)}}}{P_{\text{f(sys)}}}, \tag{16}$$

is used in this study, where $P_{\text{f(sys)}}$ is the probability of system failure and $P_{\text{f(dmg)}}$ is the probability of damage occurrence (e.g. first component failure) to the system. The difference between the probability of damage occurrence to the system $P_{\text{f(dmg)}}$ and the probability of system failure $P_{\text{f(sys)}}$ can be interpreted as the availability of system warning. A structure is redundant if the loads that cause its failure (i.e. collapse) are higher than those that trigger a sign of warning, which is usually the failure of the first component. The larger the difference between $P_{\text{f(dmg)}}$ and $P_{\text{f(sys)}}$, the larger the redundancy index RI. An increase in the value of RI indicates a higher system redundancy and vice versa. On the other hand, a structural system is considered non-redundant if $P_{\text{f(sys)}} = P_{\text{f(dmg)}}$ or, equivalently, RI $= 0$.

3. Time-variant system reliability and redundancy

In most studies of reliability and redundancy of structural systems, it is implicitly assumed that both the loads and resistances are time-independent random variables. That is, the probability density functions of the random variables are kept unchanged during the lifetime of a structure. This assumption is usually resorted to because of the complexity involved in dealing with time-dependent systems. However, including the system time-dependent effects, as demonstrated in this study, may create significant impacts on the reliability and redundancy analyses and leads to better modelling of the true performance of structural systems.

As hinted by the preceding paragraph, including the time aspect into the reliability model is performed by establishing time-variant prediction models for both the loads and resistances. The resistances of structural components and systems experience degradation as time goes by. Degradation can be caused by various mechanisms, such as corrosion and fatigue. On the other hand, loading may increase over time, especially as in traffic loads over bridge systems. The decrease in resistance accompanied by an increase in load may lead to significant drops in the levels of reliability and redundancy expected and originally designed for.

3.1. Resistance degradation

In general, a structure will begin to deteriorate the day it is placed in service. A predictive model is needed to estimate how the resistance changes over time. The resistance degradation model is usually derived theoretically, obtained from laboratory data or extrapolated from the behaviour of similar structures under the same conditions (Estes and Frangopol 2005). For illustrative purposes, the resistance degradation model used in this study is deterministic (Biondini *et al.* 2008). It is assumed that the degradation in resistance is due primarily to a continuous section loss over time. The original resistance probability density function (PDF) is updated by modifying its mean and standard deviation to account for the loss in the cross-section. The deterministic remaining cross-section of component i at time t is computed as:

$$A_i(t) = [1 - \text{DR}_i] \times A_i(t-1) = [1 - \text{DR}_i]^t \times A_i(0), \tag{17}$$

where $A_i(t)$, $A_i(t-1)$ and $A_i(0)$ are the cross-sectional areas of component i at time t, $t-1$ and 0, respectively, and DR_i is the deterioration rate, or the fraction of section loss, of component i. The mean of the time-variant random resistance of component i at time t, $\mu_{\text{R}i}(t)$, is related to its cross-section as follows:

$$\mu_{\text{R}i}(t) = A_i(t) \times \left(\mu_{\text{Fy}}\right)_i, \tag{18}$$

where $\left(\mu_{\text{Fy}}\right)_i$ is the mean of the random yield stress Fy of a component i, where Fy is assumed to be lognormal and constant over time. Once the yield stress of a component is reached, it may yield but still support loads equal to its resistance if ductile, or fail completely and stop contributing to the system if brittle.

For the system in Figure 1, the initial cross-section of bar i, $A_i(0)$, is computed based on a chosen initial central safety factor $\text{SF}_i(0)$, a chosen initial load sharing factor $l_i(0)$ and a given initial mean applied load $\mu_{\text{p}}(0)$. The initial load sharing factor determines the portion of the applied load the bar initially supports, i.e. the ratio of the initial mean internal force in the bar $\mu_{\text{F}i}(0)$ to the mean initial applied load $\mu_{\text{p}}(0)$. The initial central safety factor of bar i, $\text{SF}_i(0)$, is the ratio of its initial mean resistance $\mu_{Ri}(0)$ to its initial mean internal force $\mu_{Fi}(0)$. Therefore:

$$A_i(0) = \frac{SF_i(0) \times \mu_{Fi}(0)}{(\mu_{Fy})_i} = \frac{SF_i(0) \times l_i(0) \times \mu_P(0)}{(\mu_{Fy})_i}.$$

(19)

The initial load sharing factors $l_1(0)$ and $l_2(0)$ are determined based on the desired cross-section distribution among the two bars while maintaining the sum of initial load sharing factors of both bars to 1.0. For instance, $l_i(0) = 0.5$ results in equal initial bar cross-sections.

The standard deviation of the resistance is assumed to increase over time because present uncertainties in properties of random variables tend to increase further in the future. The standard deviation of the resistance of component i at time t, $\sigma_{Ri}(t)$, is thus given by:

$$\sigma_{Ri}(t) = [1 + DR_i] \times \sigma_{Ri}(t-1) = [1 + DR_i]^t \times \sigma_{Ri}(0),$$

(20)

where $\sigma_i(t)$, $\sigma_i(t-1)$ and $\sigma_i(0)$ are the standard deviations of the resistance of component i at times t, $t-1$ and 0, respectively. The initial standard deviation of the resistance of component i, $\sigma_{Ri}(0)$, is given as:

$$\sigma_{Ri}(0) = A_i(0) \times (\sigma_{Fy})_i,$$

(21)

where $(\sigma_{Fy})_i$ is the standard deviation of the material yield stress of component i. The parameters of the lognormally distributed resistance of component i at time t are:

$$\zeta_{Ri}(t) = \sqrt{\ln\left(1 + \frac{\sigma_{Ri}^2(t)}{\mu_{Ri}^2(t)}\right)}$$

(22)

and

$$\lambda_{Ri}(t) = \ln(\mu_{Ri}(t)) - \frac{1}{2}\zeta_{Ri}^2(t).$$

(23)

3.2. Load amplification

Live loads on some types of structures are expected to gradually increase with time. In the case of highway bridges, for instance, the annual increase in traffic has a significant effect on the fatigue life of the bridge component and the reliability of the structure (Frangopol *et al.* 2008). A model that assumes a constant annual increase rate, LIR, in live load is considered. The mean of the time-variant load is given by:

$$\mu_P(t) = [1 + LIR] \times \mu_P(t-1) = [1 + LIR]^t \times \mu_P(0),$$

(24)

where $\mu_P(t)$, $\mu_P(t-1)$ and $\mu_P(0)$ are the means of the time-variant loads at times t, $t-1$ and 0, respectively.

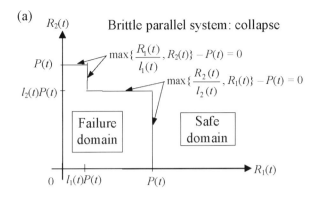

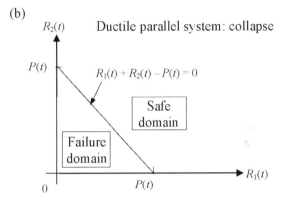

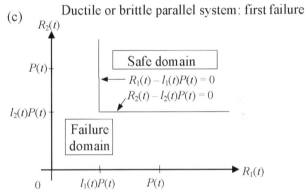

Figure 2. Time varying failure domain of the two parallel bar system for: (a) brittle system failure, (b) ductile system failure and (c) first yield (ductile or brittle).

3.3. Time-variant reliability

One approach of assessing the time-variant reliability is the point-in-time method. This approach is used in the present paper. The reliability in this method is computed at various points in time to establish a lifetime reliability profile. Alternatively, a different approach, yet increasingly complex, is to asses the reliability and redundancy during specified periods of time (Yang *et al.* 2004, 2005, Estes and Frangopol 2005). This method makes use of specified survivor functions. The survivor function defines the probability that an element is safe during a time interval.

The point-in-time method requires the updating of the performance functions at every point in time with the new information obtained by the resistance predictive models. It is worth noting that the changes in the load effects over time have no bearing on the performance functions. The performance function and probability of system failure equations at time t are obtained by updating Equations (1) and (2), respectively, as:

$$g_i(t) = (R_s(t))_i - (Q(t))_i \qquad (25)$$

and

$$P_{f(sys)}(t) = P[\text{any } g_i(t) < 0], \qquad (26)$$

where $g_i(t)$, $(R_s(t))_i$ and $(Q(t))_i$ are the performance function, system resistance and load effects associated

with the ith failure mode at time t, respectively, and $P_{f(sys)}(t)$ is the probability of system failure at time t. The incremental method is used herein to obtain general forms of the time-variant performance functions.

Consider again the case of the two parallel bar system in Figure 1. Recall that the performance functions developed previously were obtained under the assumption of equal areas and the same moduli of elasticity. Deterioration of a bar at a point in time might not be the same as the other, which leads to the violation of the above assumption of equal areas in the future. It is assumed that the modulus of elasticity is time-invariant. The performance functions are, therefore, reformulated next for a general case of two bars with two different time-variant areas and the same time-invariant modulus of elasticity.

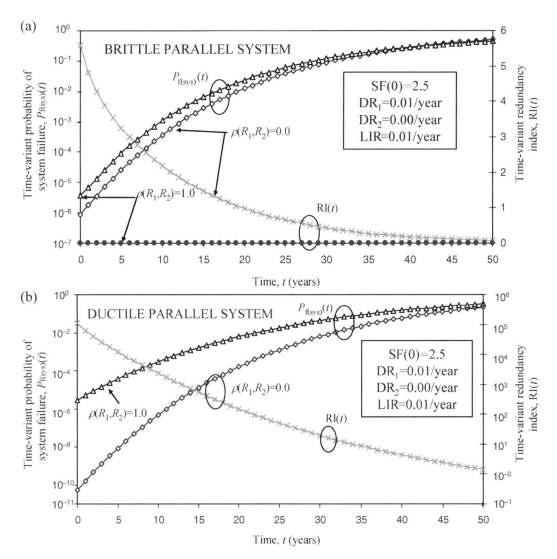

Figure 3. Effect of the coefficient of correlation, $\rho(R_1, R_2)$, on the time-variant probability of system failure, $P_{f(sys)}(t)$, and redundancy index, RI(t), of: (a) the brittle system and (b) the ductile system.

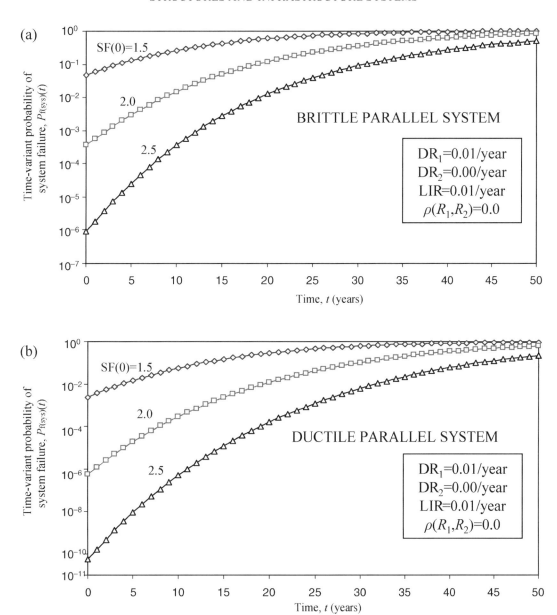

Figure 4. Effect of the initial central safety factor, SF(0), on the time-variant probability of system failure, $P_{f(sys)}(t)$, of: (a) the brittle system and (b) the ductile system.

The time-variant load sharing factors for both bars are found from Equations (12) and (13):

$$l_1(t) = \frac{A_1(t)}{A_1(t) + A_2(t)} \qquad (27)$$

and

$$l_2(t) = \frac{A_2(t)}{A_1(t) + A_2(t)} = 1 - l_1(t), \qquad (28)$$

where $l_1(t)$ and $l_2(t)$ are the load sharing factors for bars 1 and 2 at time t, respectively. Interestingly enough, the expression for the performance function

with respect to failure of the ductile system at time t is found to be:

$$g(t) = R_1(t) + R_2(t) - P(t), \qquad (29)$$

which is similar to Equation (6), at a given time t. This is consistent with the fact that the collapse modes of ductile systems are independent of the material distribution in the system, i.e. the cross-section of the components of the system (Fu and Frangopol 1990). For the case of the brittle system, the performance functions are formed as:

$$g_1(t) = \max\left\{ \frac{R_1(t)}{l_1(t)}, R_2(t) \right\} - P(t) \qquad (30)$$

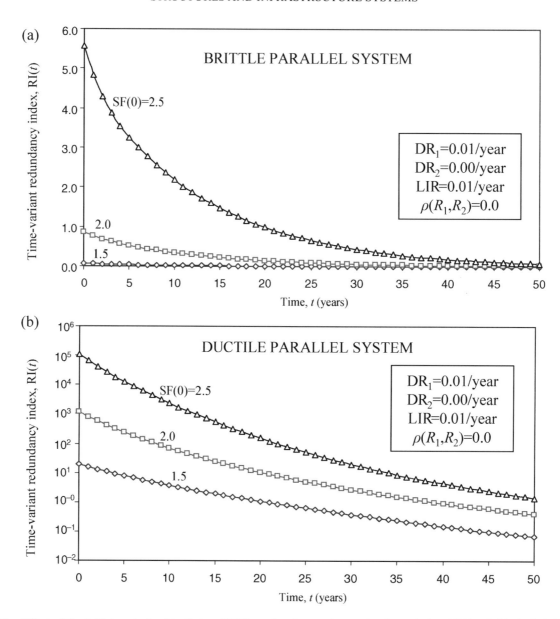

Figure 5. Effect of the initial central safety factor, SF(0), on the time-variant redundancy index, RI(t), of: (a) the brittle system and (b) the ductile system.

and

$$g_2(t) = \max\left\{ \frac{R_2(t)}{l_2(t)}, R_1(t) \right\} - P(t). \qquad (31)$$

Clearly, for the special case of equal load sharing, $l_1(t) = l_2(t) = 0.5$, Equation (30) becomes Equation (9) and Equation (31) becomes Equation (10) at a given time t. The performance functions with respect to first failure of the bars are:

$$h_1(t) = R_1(t) - l_1(t)P(t) \qquad (32)$$

and

$$h_2(t) = R_2(t) - l_2(t)P(t). \qquad (33)$$

Limit state functions are obtained by setting the value of the performance functions to zero. They set the boundaries between the failure and safe domains. The limit state functions associated with Equations (29–33) are shown in Figure 2. The corresponding probabilities of failure are found by integrating the bivariate lognormal (PDF) over areas enclosed by the respective failure domains shown in Figure 2.

3.4. Time-variant redundancy

Now that the probabilities of system failure $P_{f(sys)}(t)$ and damage occurrence to the system $P_{f(dmg)}(t)$ at time t can be obtained, the time-variant

(a)

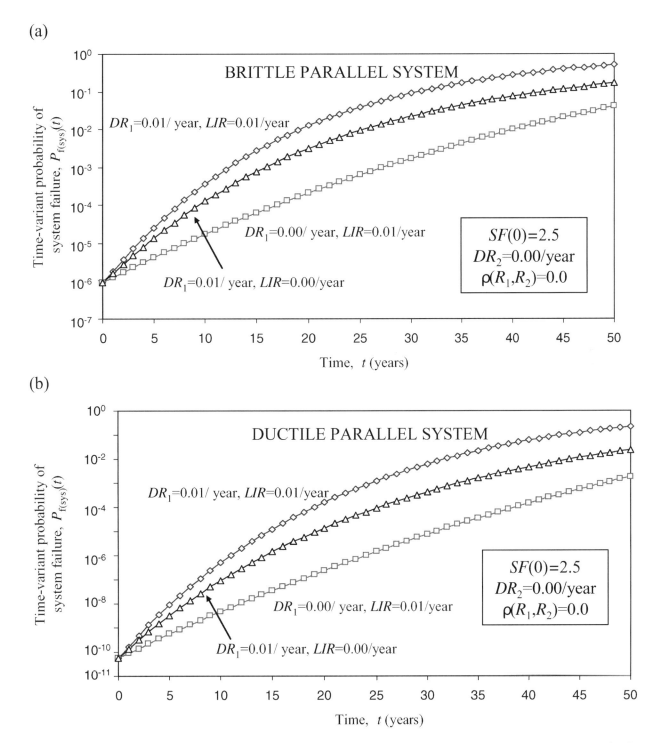

Figure 6. Effect of the deterioration rate of bar 1, DR_1, while bar 2 is kept intact, and load increase rate, LIR, on the time-variant probability of system failure, $P_{f(sys)}(t)$, of: (a) the brittle system and (b) the ductile system.

redundancy index given in Equation (16) can be rewritten as:

$$RI(t) = \frac{P_{f(dmg)}(t) - P_{f(sys)}(t)}{P_{f(sys)}(t)}, \qquad (34)$$

where $RI(t)$, $P_{f(dmg)}(t)$ and $P_{f(sys)}(t)$ are the redundancy index, probability of damage occurrence to the system, and probability of system failure at time t, respectively.

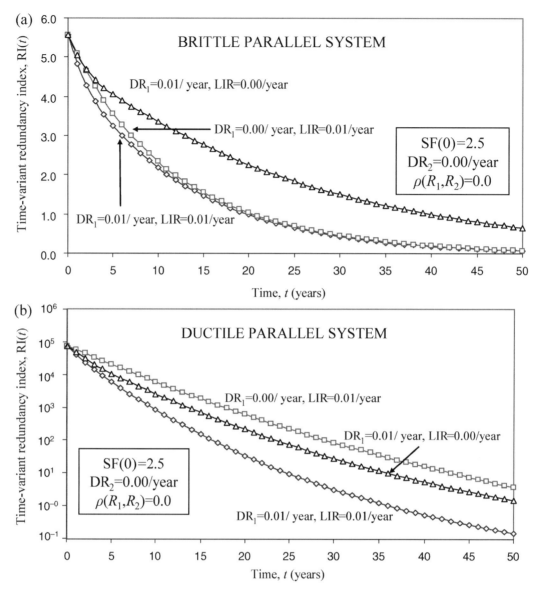

Figure 7. Effect of the deterioration rate of bar 1, DR_1, while bar 2 is kept intact, and load increase rate, LIR, on the time-variant redundancy index, $RI(t)$, of: (a) the brittle system and (b) the ductile system.

4. Numerical examples

4.1. Example 1: parallel system

The two parallel bar system shown in Figure 1 is further discussed here to illustrate the application of time-variant system reliability and redundancy and their computations. The yield stresses of the bars are assumed to be lognormal random variables with the same mean and standard deviation of 10 and 2 kN/cm^2, respectively. The applied load is assumed to be deterministic. An initial applied load of 10 kN is assumed. The two bars have equal initial load sharing factors, i.e. $l_1(0) = l_2(0) = 0.5$, and the same initial central safety factor $SF_1(0) = SF_2(0) = SF(0)$, and hence equal initial mean and equal initial standard

deviation of their resistances, i.e. $\mu_{R1}(0) = \mu_{R2}(0)$ and $\sigma_{R1}(0) = \sigma_{R2}(0)$. Except where specified otherwise, an initial safety factor $SF(0) = 2.5$ is used, which provides initial mean resistances $\mu_{R1}(0) = \mu_{R2}(0) = 12.5$ kN and initial standard deviations of resistances $\sigma_{R1}(0) = \sigma_{R2}(0) = 2.5$ kN.

The load at time t is computed using Equation (24). The mean and standard deviation of the resistances of the two bars at time t and their corresponding parameters are given by the degradation models in Equations (17–23). The time-variant load sharing factors of the bars are computed using Equations (27) and (28). Several deterioration rates are considered to study the sensitivity of the results to the deterioration rate. The probabilities of failure and damage occurrence are computed by integration of the bivariate lognormal PDF. The following results are obtained:

1. Figure 3 plots the time-variant probability of system failure and time-variant redundancy index for the brittle (Figure 3a) and ductile (Figure 3b) systems considering statistical independence, i.e. $\rho(R_1, R_2) = 0$, and perfect correlation, i.e. $\rho(R_1, R_2) = 1$, between the resistances of the bars, where ρ is the coefficient of correlation. The load is increased by 1% per year and the resistance of bar 1 is degrading by 1% per year, while bar 2 remains intact. An initial central safety factor of 2.5 is used.

Clearly, the time-variant reliability and redundancy of both systems degrade rapidly with time. It is obvious from the figures that the ductile system provides higher time-variant reliability and redundancy than its brittle counterpart. Furthermore, given the assumed initial central safety factor, the plots show a tremendous advantage in using components with statistically independent bar resistances in ensuring greater reliability and redundancy of the parallel system. In fact, both systems have no redundancy when the resistances of their components are perfectly correlated. Note that zero values cannot be shown on the logarithmic scale of the time-variant redundancy in Figure 3b. These observations may be justified as follows. In a deterministic sense, two parallel bars with equal resistances and load sharing factors definitely fail simultaneously, regardless of whether the system is ductile or not. Consider first the case where the time is $t = 0$, the two parallel bars considered have equal initial means and standard deviations of their resistances and equal initial load sharing factors. Due to the uncertainties in the material yield stress, however, the initial resistances of the bars may take values different from their initial means. Whether the two bars take the same resistance values or not is determined by the resistance correlation. The more they are correlated, the higher the chances that they are equal. Therefore, with low correlation, it is very likely that the two initial resistances are different, as they may probabilistically vary. Accordingly, one of the bars may fail before the other, providing redundancy to the system. On the other hand, perfect positive resistance correlation causes that, for any value, the initial resistance of one bar may take, the other will take the same initial resistance value. For the case where $t > 0$, the previous argument holds if both bars have the same deterioration characteristics. However, if the bars deteriorate differently, their means will differ, and according to the deterioration

model used, the coefficient of variation will differ too. This will give little rise to the redundancy. It was found, however, that the resulting increase in redundancy is insignificant. Ductile systems have higher redundancy than brittle ones, since the failed component continues to support loading.

It is evident that the time-variant reliability and/or redundancy can reach intolerable limits at a certain point in time (sometimes not too far from construction). Depending on these limits, the reliability may or may not reach its pre-described threshold before the redundancy of the structure does. This urges the need for incorporating both the time-variant reliability and redundancy in any structural time-variant assessment.

2. Figures 4 and 5 show the time-variant reliability and time-variant redundancy, respectively, for both the brittle and ductile systems, considering different values of the initial central safety factor. The loading is increased by 1% per year and the resistance of bar 1 is degrading by 1% per year, while bar 2 remains intact. Statistical independence is considered between the resistances of the bars, i.e. $\rho(R_1, R_2) = 0$. The significant positive impact of increasing the initial central safety factor on the time-variant reliability and redundancy index of both systems is evident from the graphs. It can be noticed that the time-variant redundancy index sensitivity with respect to the initial safety factor reduces with time. This implies that the advantage of the higher initial central safety factor tends to diminish during the life of the structure.

3. Figures 6 and 7 present the time-variant reliability and time-variant redundancy of both the brittle and ductile systems, respectively, considering the following cases: (a) a load increase of 1% per year and degradation of 1% per year is imposed on the resistance of bar 1, (b) a load increase of 1% per year with no resistance degradation and (c) degradation of 1% per year is imposed on the resistance of bar 1 and no load increase is considered. In all three cases, an initial central safety factor of 2.5 is considered and bar 2 remains intact. Also, statistical independence is considered between the resistances of the bars (i.e. $\rho(R_1, R_2) = 0$). It can be concluded that the degradation in resistance has a higher impact on the time-variant reliability and the time-variant redundancy of both systems than the increase in load, except in the case

of the brittle system where the load increase has a higher impact on the time-variant redundancy.

4. Figure 8 shows the time-variant redundancy index for the brittle and ductile systems, where Bar 2 remains intact and the resistance of Bar 1 degrades considering different deterioration rates. The loading is increased by 1% per year and an initial central safety factor of 2.5 is used. Clearly, the time-variant redundancy index decreases in the ductile system case as the deterioration rate increases. However, the time-variant redundancy index of the brittle system is almost

not affected by the increase in the deterioration rate.

4.2. Example 2: series–parallel system

Most structures can be modelled as series–parallel systems. The time-variant redundancy of the simple series–parallel system shown in Figure 9 is investigated in this example. The yield stresses of the bars are assumed to be lognormal random variables, with the same mean and standard deviation of 10 and 2 kN/cm^2, respectively. The time-variant applied load is also assumed to be a lognormal random variable, with an initial mean of 10 kN and initial standard deviation of

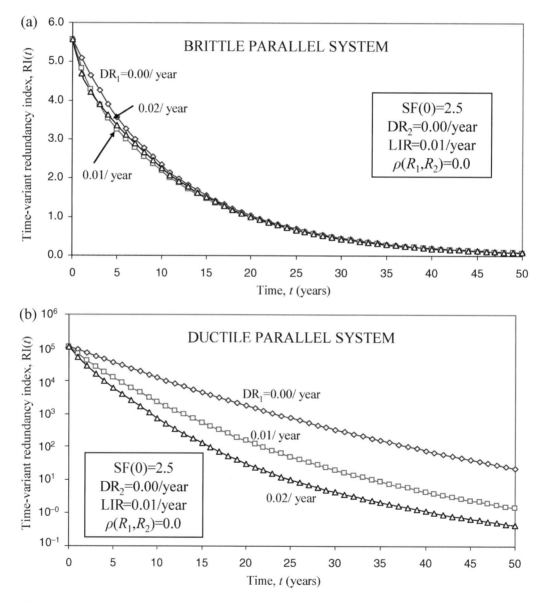

Figure 8. Effect of the deterioration rate of bar 1, DR$_1$, while bar 2 is kept intact, on the time-variant redundancy index, RI(t), of: (a) the brittle system and (b) the ductile system.

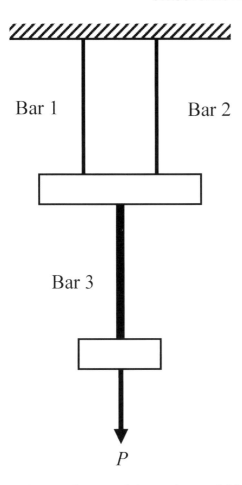

Figure 9. Structural system of three series–parallel bars.

1 kN. The two parallel bars have equal initial load sharing factors, i.e. $l_1(0) = l_2(0) = 0.5$, and the third bar has, of course, an initial load sharing factor $l_3(0) = 1$. Except where specified otherwise, the initial central safety factor for bars 1, 2, and 3 is the same $SF_1(0) = SF_2(0) = SF_3(0) = 2.5$. The initial cross-section of all three bars are calculated using Equation (19). Accordingly, the two parallel bars, bars 1 and 2, have equal initial means and equal initial standard deviations of their resistances and half that of bar 3, i.e. $\mu_{R1}(0) = \mu_{R2}(0) = 12.5$ kN, $\sigma_{R1}(0) = \sigma_{R2}(0) = 2.5$ kN, $\mu_{R3}(0) = 25$ kN and $\sigma_{R3}(0) = 5$ kN.

The mean and standard deviation of the resistance of the three bars at time t and their corresponding lognormal parameters are given by the degradation models in Equations (17–23). The time-variant load sharing factors of bars 1 and 2, $l_1(t)$ and $l_2(t)$, are computed using Equations (27) and (28), respectively. Clearly, the load sharing factor of bar 3 is $l_3(t) = 1$ at any time t. The mean of the load at time t is computed using Equation (24). The coefficient of variation of the applied load P, COV(P), is 10% and is assumed to be constant throughout the life of the structure. Therefore, the standard deviation of the applied load P

increases with time by the same rate of increase of the mean of the load P. Statistical independence is assumed between the applied load and the bar resistances. Two cases of correlation between the resistances of the bars are considered, namely statistical independence, i.e. $\rho(R_i, R_j) = 0$, and perfect correlation, i.e. $\rho(R_i, R_j) = 1$. The probabilities of failure are computed using the software RELSYS (Estes and Frangopol 1998) in which the average of the Ditlevsen bounds is the computed probability of failure. The following results are obtained:

1. Figure 10 shows the time-variant reliability and redundancy of the series–parallel system with ductile material properties for the following cases: (a) deterioration rate of 1% per year is imposed on one of the parallel bars, bar 1, and no load increase, (b) deterioration rate of 1% per year is imposed on the bar in series with the system, bar 3, and no load increase, and (c) a load increase rate of LIR = 1% per year with no deterioration. Statistical independence is assumed among the resistances of all three bars. This figure shows that all three cases considered reduce the time-variant reliability of the system. The highest impact, however, is due to the deterioration of bar 3. This is clearly due to the relative reliability importance of bar 3 in the system, where failure of bar 3 alone causes failure of the entire system. Deterioration of bar 3 also has the highest impact on the time-variant redundancy of the system. Figure 10 shows that the imposed deterioration on bar 1, however, results in increasing the time-variant redundancy index. This may seem counter-intuitive at first glance, but can be rationally justified as follows.

Figure 11 shows the initial probability of system failure and the initial redundancy index of the ductile series–parallel system versus the initial central safety factor of bar 3, $SF_3(0)$, while the initial central safety factors of both bars 1 and 2, $SF_1(0) = SF_2(0)$ are fixed at the values 1.5, 2.0 and 2.5. Accordingly, the two parallel bars, bars 1 and 2, have equal initial mean and equal initial standard deviation of their resistances. Their relation with respect to the mean and standard deviation of the resistance of bar 3, however, changes according to the change in $SF_3(0)$. For any central safety factor used for the parallel bars, both the initial reliability and initial redundancy index of the system improve by the increase in $SF_3(0)$. Evidently, improving the resistance of a component with high relative reliability importance, such as bar 3, improves the reliability and redundancy of the entire system. After a

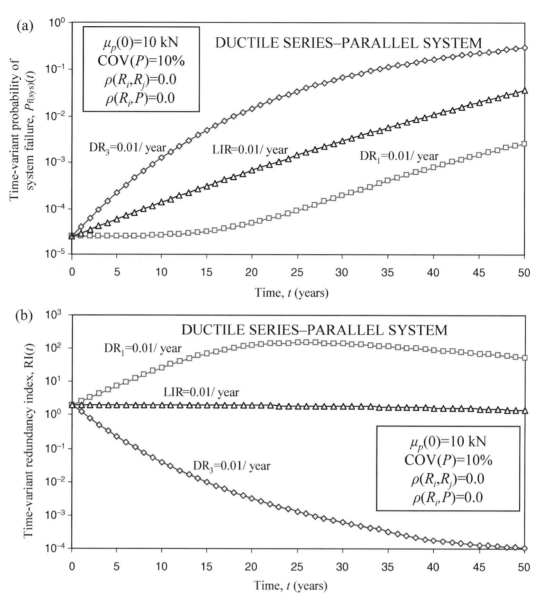

Figure 10. Effect of deterioration and load increase on the time-variant: (a) probability of system failure, $P_{f(sys)}(t)$, and (b) redundancy index, $RI(t)$, of the ductile series–parallel system with statistically independent bar resistances.

certain increase in $SF_3(0)$, however, for a given $SF_1(0) = SF_2(0)$, the increase in the initial reliability and/or the initial redundancy index stops at values equal to the initial reliability and/or initial redundancy index of the two parallel bar system alone, respectively. This can be explained by the fact that when bar 3 becomes relatively much stronger than bars 1 and 2, its reliability becomes too high to contribute to either the reliability or the redundancy of the system. It is interesting to note that the value of $SF_3(0)$ at which the reliability and redundancy reach their bounds is not too large compared to $SF_1(0) = SF_2(0)$. This is attributed in part to the fact that the area of bar 3 is twice that of bars 1 or 2 for the same central safety factor.

Figure 12 shows that increasing the initial central safety factors of bars 1 and 2, $SF_1(0) = SF_2(0)$, also improves the system reliability until they both become relatively too strong to affect the reliability of the system. For low values of $SF_1(0) = SF_2(0)$, and for a given value of $SF_3(0)$, the redundancy index improves as $SF_1(0) = SF_2(0)$ increases. Bar 3, in this case, is relatively too strong to affect the reliability of the system and the redundancy of the parallel subsystem improves with improving the resistance of its components. As $SF_1(0) = SF_2(0)$ approaches $SF_3(0)$ and increases beyond it, the redundancy index decreases because the probability that bar 3 will fail before bars 1 or 2 is growing. This decrease continues until some value of $SF_1(0) = SF_2(0)$ is reached, where the

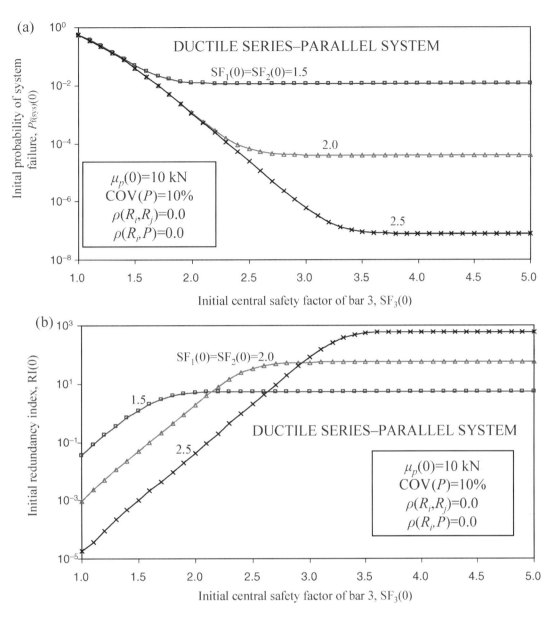

Figure 11. Effect of the initial safety factor of bar 3 on: (a) the initial probability of system failure, $P_{f(sys)}(0)$, and (b) the initial redundancy index, RI(0), of the ductile series–parallel system.

redundancy goes to zero because it becomes almost certain that bar 3 will fail before the other two bars, but the system can be redundant only if one of the parallel bars fails first. Conversely, decreasing $SF_1(0)$ and $SF_2(0)$ may increase or decrease the redundancy index, depending on the effects of this decrease on the relative resistance of bar 3. In terms of the time effects on the system, deterioration of the resistance of bars 1 and 2 results in the same consequences of decreasing $SF_1(0)$ and $SF_2(0)$. On the other hand, deterioration of the resistance of bar 3 results in the same consequences of decreasing $SF_3(0)$.

It is worth emphasising some of the findings from Figures 11 and 12. These figures stress on the fact that

the reliability may improve or remain the same, and that the redundancy may improve, remain the same or even degrade by increasing the resistance of a component, depending on the relative reliability importance of that component and the effect of that increase on the relative resistances of all the components in the system. This conclusion has strong economical impact on the design and repair practices of deteriorating civil infrastructure, since increasing the resistance of the components beyond a certain limit may provide no benefit to the reliability of the system and, therefore, is deemed a waste. Such increases may even harm the redundancy of the system. A trade off between central safety factors that optimise both the

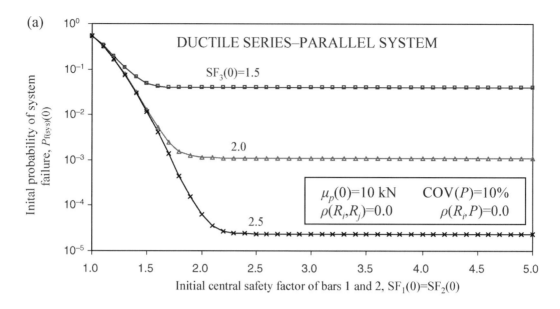

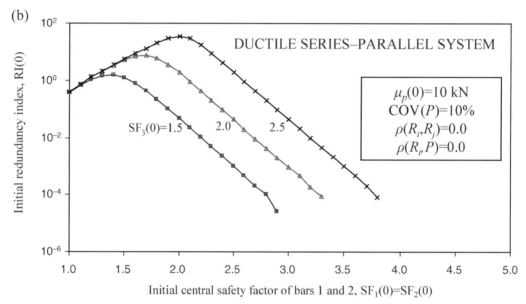

Figure 12. Effect of the initial safety factor of bars 1 and 2 on: (a) the initial probability of system failure, $P_{f(sys)}(0)$, and (b) the initial redundancy index, RI(0), of the ductile series–parallel system.

reliability and redundancy can be found. Clearly, a system-based design is more rational, efficient and economical than a component-based design. The fact that design codes are in, general, component-based designs urges the need to incorporate guidelines that take into account the reliability and redundancy of the entire system into these codes.

2. Figure 13 shows the time-variant reliability and redundancy of the series–parallel system with brittle material properties for the same cases considered in Figure 10. Statistical independence is also assumed between the resistances

of all three bars. Figure 14 shows these cases for ductile material properties and perfect correlation between all three bar resistances. Figure 15 shows them for the brittle material properties and perfect correlation between the three bar resistances. Figures 10, 13, 14 and 15 intend to display the effect of the extreme cases of correlation and both types of material properties on the time-variant reliability and redundancy index. The observations made on the time-variant reliability in Figure 10 can be also made in Figures 13, 14 and 15. In terms of the time-variant redundancy index, it can be

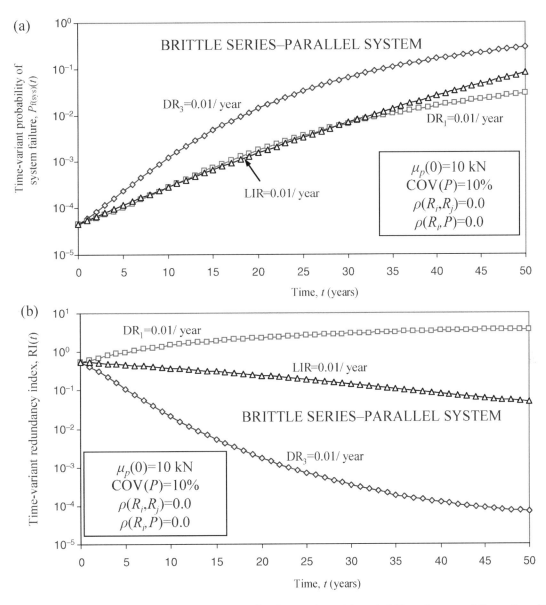

Figure 13. Effect of deterioration and load increase on the time-variant: (a) probability of system failure, $P_{f(sys)}(t)$, and (b) redundancy index, $RI(t)$, of the brittle series–parallel system with statistically independent bar resistances.

observed that perfect resistance correlation tends to reduce the time-variant redundancy index to zero at any time for both brittle and ductile systems with any deterioration or load increase case.

The reason that some observations may come as a surprise is that common sense dictates that 'the stronger the better'. In the logic of reliability and redundancy, this concept sometimes does not hold true. Other factors come in the picture when 'better' is defined in terms of reliability and redundancy. The main factor is the relative reliability importance of which component is becoming 'stronger', and how 'strong' it is

becoming. Providing more resistance to a component with high relative reliability importance is always advantageous, as long as it does not become too strong. Providing more resistance to a component with less relative reliability importance is advantageous, provided that the components with more reliability importance remain stronger. Perhaps it can be said in this case: 'the stronger the better, as long as the more important ones remain stronger, but not too strong'.

The implications of the observations and conclusions made from this example on the design of an efficient maintenance plan are significant. This knowledge can assist in choosing the most critical components in structural systems to be monitored, maintained and

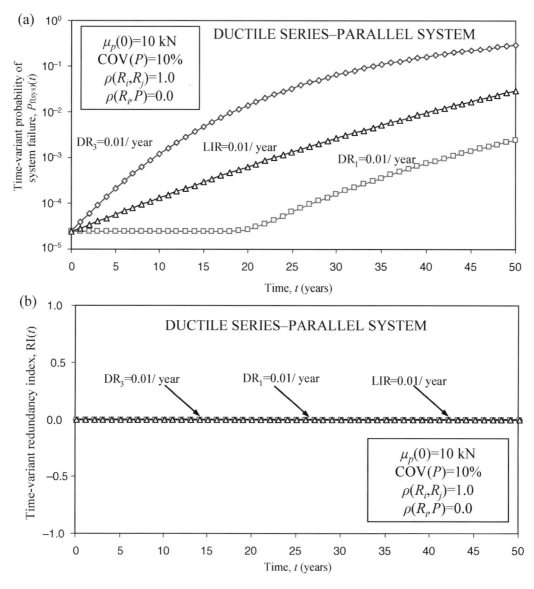

Figure 14. Effect of deterioration and load increase on the time-variant: (a) probability of system failure, $P_{f(sys)}(t)$, and (b) redundancy index, $RI(t)$, of the ductile series–parallel system with perfectly correlated bar resistances.

repaired. For example, in light of Figure 10, it is clear that maintaining bar 3 improves both the reliability and redundancy more than maintaining any other bar.

4.3. Example 3: truss system

The five-bar truss system shown in Figure 16 is investigated here to illustrate the application of time-variant system reliability and redundancy and their computations.

The random variables considered are the time-variant resistances of the bars, $R_1(t)$, $R_2(t)$, ..., $R_5(t)$ and the time-variant applied load $P(t)$. The bars are all assumed to have ductile lognormally distributed yield

stresses. Mean material yield stresses of 25 and 12.5 kN/cm^2 are assumed for tension and compression, respectively, for the five bars. An initial coefficient of variation of 10% is assumed for the initial resistances of all bars. An initial cross-sectional area of 3 cm^2 is assumed in all bars. The vertical, horizontal and diagonal bars are assigned the areas A_1, A_2 and A_3, respectively. The time-variant applied load is assumed to be a lognormal random variable, with an initial mean of 20 kN and initial standard deviation of 4 kN. The mean of the load at time t is computed using Equation (24) with a load rate increase LIR = 1% per year. The coefficient of variation of the applied load is COV(P) = 20% and is assumed to be constant

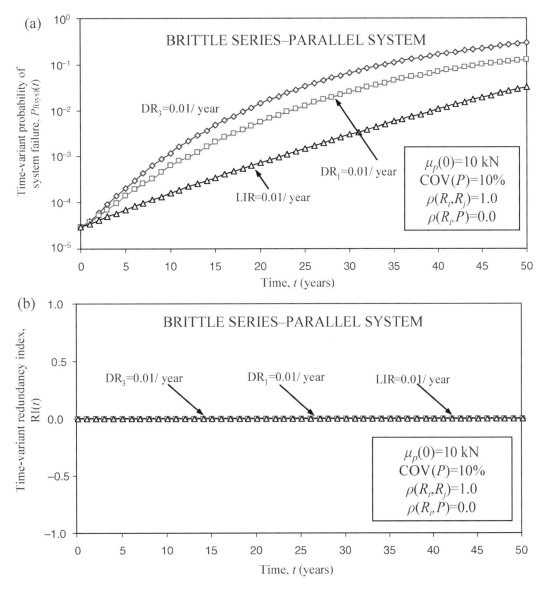

Figure 15. Effect of deterioration and load increase on the time-variant: (a) probability of system failure, $P_{f(sys)}(t)$, and (b) redundancy index, $RI(t)$, of the brittle series–parallel system with perfectly correlated bar resistances.

throughout the life of the structure. Therefore, the standard deviation increases with time at the same rate of increase of the mean of the load. Statistical independence is considered among the resistances of the bars, and between the resistances and the applied load. The probabilities of failure and first yield are found as the upper Ditlevsen bounds using the software **CALREL** (Liu *et al.* 1989). The following results are obtained:

1. Figure 17 plots the time-variant probability of system failure and time-variant redundancy index for the five-bar truss considering the following cases: (a) a deterioration rate of 1%

per year is imposed on A_1 while keeping A_2 and A_3 intact, (b) a deterioration rate of 1% per year is imposed on A_2 while keeping A_1 and A_3 intact and (c) a deterioration rate of 1% per year is imposed on A_3 while keeping A_1 and A_2 intact.

Clearly, the time-variant reliability and redundancy of the structure degrade rapidly with time. Interestingly, an imposed deterioration rate on A_3 compromises the time-variant reliability of the structure more than if the same deterioration rate is imposed on A_1 or A_2; meanwhile, the time-variant redundancy of the structure is less affected by the

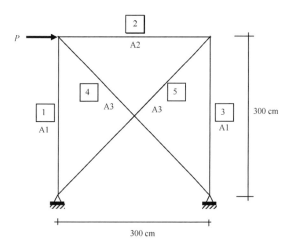

Figure 16.   Five-bar truss: geometry and loading.

deterioration of A_3. The redundancy of the structure at a point in time in the future, say at 20 years, may still be adequate, while its reliability may become significantly compromised. This clearly implies that the structural reliability and redundancy are different system performance indicators. The time-variant reliability of a structure may be more vulnerable to a type of deterioration than its time-variant redundancy. A critical threshold may be reached by either of them, i.e. reliability or redundancy, at different time instances during the life of the structure. This clearly urges the need of incorporating both the safety, manifested in the reliability, and the redundancy of the structure in any time-variant structural management framework to achieve an optimum inspection, repair and management plan. In addition,

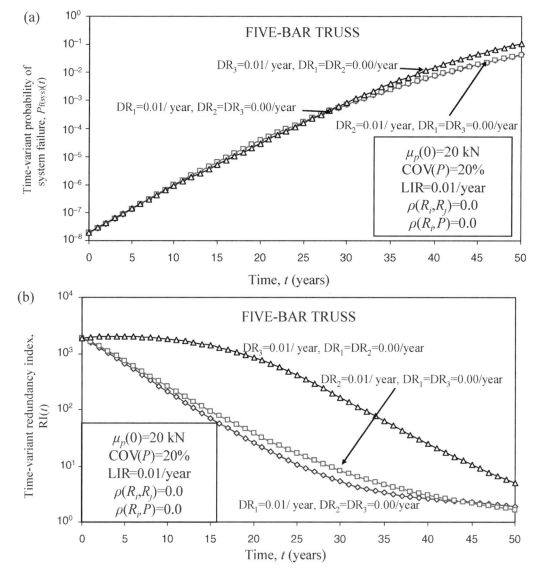

Figure 17.   Effect of the deterioration rate, DR_i, imposed on single area groups on the time-variant: (a) probability of system failure, $P_{f(sys)}(t)$, and (b) redundancy index, $RI(t)$.

investigating the sensitivity of the time-variant reliability and redundancy to single bar deterioration, as in Figure 17, helps identify the relative importance of the components to the system. The figure indicates that this importance is different with respect to the time-variant reliability than it is with respect to the time-variant redundancy.

2. Figure 18 plots the time-variant probability of system failure and time-variant redundancy index for the five-bar truss considering the following cases: (a) a deterioration rate of 1% per year is imposed on A_1 and A_2 while keeping

A_3 intact, (b) a deterioration rate of 1% per year is imposed on A_1 and A_3 while keeping A_2 intact, (c) a deterioration rate of 1% per year is imposed on A_2 and A_3 while keeping A_1 intact and (d) a deterioration rate of 1% per year is imposed on A_1, A_2 and A_3. The time-variant reliability of the structure is more vulnerable to deterioration of A_3 combined with either or both of A_1 and A_2. On the other hand, the time-variant redundancy is less vulnerable to deterioration of A_3 combined with either or both of A_1 and A_2. Therefore, if all bars are deteriorating and the bars with the area A_3

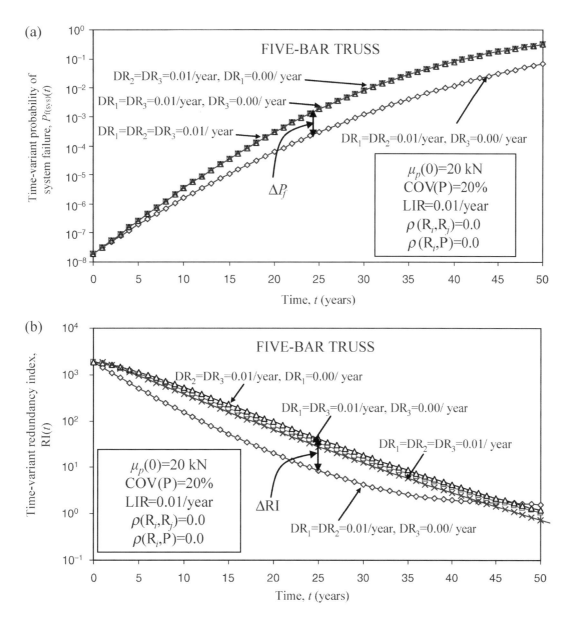

Figure 18. Effect of the deterioration rate, DR_i, imposed on combinations of area groups on the time-variant: (a) probability of system failure, $P_{f(sys)}(t)$, and (b) redundancy index, $RI(t)$.

are repaired at the year 25, for example, the probability of system failure is reduced by the amount ΔP_f, as shown in Figure 18a. Repairing any other bars will not significantly increase the reliability of the structure. However, repairing only the bars with area A_3 may reduce the redundancy index by ΔRI, as shown in Figure 18b. To overcome this dilemma, a better repair decision at year 25 may be to repair the bars with both areas A_2 and A_3. The relative cost benefit gained from repairing A_2, in addition to A_3, should also be considered. This is an issue better tackled by a multi-objective optimisation framework that incorporates the time-variant reliability, redundancy and cost in a complete mathematical platform (Frangopol and Liu 2007). This is an area suggested for further research.

5. Conclusions

This paper investigates the effects of structural deterioration and load increase on the time-variant reliability and redundancy of structures. A framework for generating structural reliability and redundancy profiles has been presented and implemented in numerical examples. The following conclusions are drawn:

(1) The time-variant reliability may degrade or remain the same, and the time-variant redundancy may degrade, remain the same or even improve with the deterioration of the resistance of a component, depending on the relative reliability importance of that component in the system and the effect of the deterioration on the relative resistance of all the components in the system.

(2) It is evident, in some cases, that the reliability and/or redundancy can reach intolerable limits at a certain time, sometime not too far from construction. Depending on these limits, the reliability may or may not become at risk before the redundancy of the structure does. This urges the need of incorporating both the reliability and redundancy in any structural time-variant assessment.

(3) Ductile systems provide higher time-variant reliability and redundancy than their brittle counterparts.

(4) Using components with lower resistance correlations and/or higher initial central safety factors provides greater reliability and redundancy to the ductile structural systems and greater redundancy to the brittle structural systems.

(5) The advantage of using higher initial central safety factor tends to diminish during the life of deteriorating structures.

(6) Time-variant reliability and redundancy usually do not have a one-to-one relation. This conclusion, among various others, emphasises the importance of incorporating both the time-variant reliability and redundancy in the assessment and design of structures.

(7) Investigating the sensitivity of the time-variant reliability and redundancy to single component deterioration helps identify the relative reliability importance of these components to the system.

(8) The time-variant reliability and redundancy assessment was shown to be helpful in making a more informed repair decision. A multi-objective optimisation framework, which incorporates the time-variant reliability, time-variant redundancy and cost in a complete mathematical platform, can improve the decisions made for structural maintenance planning. This is an area suggested for further research.

The above conclusions are derived from specific numerical examples. However, most of them can be generalised.

Acknowledgements

The support by grants from the Commonwealth of Pennsylvania, Department of Community and Economic Development, through the Pennsylvania Infrastructure Technology Alliance (PITA) is gratefully acknowledged. The support of the National Science Foundation through grant CMS-0639428 to Lehigh University is also gratefully acknowledged. Also, the support from the Federal Highway Administration Cooperative Agreement Award DTFH61-07-H-00040 and that of the Office of Naval Research, contract number N00014-08-0188 is gratefully acknowledged. The opinions and conclusions presented in this paper are those of the authors and do not necessarily reflect the views of the sponsoring organisations.

References

Ang, A.H.-S. and Tang, W.H., 1984. *Probability concepts in engineering planning and design: decision, risk and reliability*, vol. II. New York: John Wiley & Sons.

Ang, A.H.-S. and De Leon, D., 2005. Modeling and analysis of uncertainties for risk-informed decision in infrastructures engineering. *Structure and Infrastructure Engineering*, 1 (1), 19–31.

Bertero, R.D. and Bertero, V.V., 1999. Redundancy in earthquake-resistant design. *Journal of Structural Engineering*, 125 (1), 81–88.

Biondini, F., Frangopol, D.M., and Restelli, S., 2008. On structural robustness, redundancy and static indeterminacy. *In: Proceedings of the Structures Congress 2008*, 24–26 April, Vancouver, Canada [CD-ROM].

Cornell, C.A., 1967. Bounds on the reliability of structural systems. *Journal of Structural Division*, 93 (ST1), 171–200.

De, S.R., Karamchandani, A., and Cornell, C.A., 1990. Study of redundancy in near ideal parallel structural systems. *In*: A.H.-S. Ang, M. Shinozuka, and G.I. Schüeller, eds. *Structural Safety and Reliability*, vol. 2. New York: ASCE, 975–982.

Ditlevsen, O., 1979. Narrow reliability bounds for structural systems. *Journal of Structural Mechanics*, 7 (4), 453–472.

Ellingwood, B.R., 2005. Risk-informed condition assessment of civil infrastructure: state of practice and research issues. *Structure and Infrastructure Engineering*, 1 (1), 7–18.

Estes, A.C. and Frangopol, D.M., 1998. RELSYS: a computer program for structural systems reliability. *Structural Engineering and Mechanics*, 6 (8), 901–919.

Estes, A.C. and Frangopol, D.M., 2005. Life-cycle evaluation and condition assessment of structures. *Chapter 36. In*: W.-F. Chen and E.M. Lui, eds. *Structural engineering handbook*, second edition. CRC Press, 36-1–36-51.

Frangopol, D.M., ed., 1987. *Effects of damage and redundancy on structural performance*. New York: ASCE.

Frangopol, D.M., ed., 1989. *New directions in structural system reliability*. Boulder, CO: University of Colorado Press.

Frangopol, D.M., 1992. Bridge loading, reliability and redundancy: concepts and applications. *In*: R. Rackwitz and P. Thoft-Christensen, eds. *Reliability and optimization of structural systems 1991*. Berlin: Springer-Verlag, 1–18.

Frangopol, D.M. and Curley, J.P., 1987. Effects of damage and redundancy on structural reliability. *Journal of Structural Engineering*, 113 (7), 1533–1549.

Frangopol, D.M. and Iizuka, M., 1992. A survey of system redundancy measures. *In*: J. Morgan, ed. *ASCE structures congress 1992*. New York: ASCE, 153–156.

Frangopol, D.M. and Nakib, R., 1991. Redundancy in highway bridges. *Engineering Journal*, 28 (1), 45–50.

Frangopol, D.M., Izuka, M., and Yoshida, K., 1992. Redundancy measures for design and evaluation of structural systems. *Transactions of ASME, Journal of Offshore Mechanics and Arctic Engineering*, 114 (4), 285–290.

Frangopol, D.M. and Liu, M., 2007. Maintenance and management of civil infrastructure based on condition, safety, optimization, and life-cycle cost. *Structure and Infrastructure Engineering*, 3 (1), 29–41.

Frangopol, D.M., Strauss, A., and Kim, S., 2008. Bridge reliability assessment based on monitoring. *Journal of Bridge Engineering*, 13 (3), 258–270.

Frangopol, D.M., Gharaibeh, E.S., Hearn, G., and Shing, P.B., 1998. System reliability and redundancy in codified bridge evaluation and design. *In*: N.K. Srivastava, ed. *Structural Engineering World Wide 1998*. Amsterdam: Elsevier, paper reference T121-2 [CD-ROM].

Fu, G., 1987. *Lifetime structural system reliability*. Cleveland, OH: Department of Civil Engineering, University of Case Western Reserve University, Report no. 87-9.

Fu, G. and Frangopol, D.M., 1990. Balancing weight, system reliability and redundancy in a multi-objective optimisation framework. *Structural Safety*, 7 (2–4), 165–175.

Gharaibeh, E.S. and Frangopol, D.M., 2000. Safety assessment of highway bridges based on system reliability and redundancy. *In*: *16th congress of IABSE*, Lucerne, Switzerland, Congress report, 274–275 [CD-ROM].

Gharaibeh, E.S., Frangopol, D.M., and Enright, M.P., 2000a. Redundancy and member importance evaluation of highway bridges. *In*: R.E. Melchers and M.G. Stewart, eds. *Applications of statistics and probability, Proceedings of ICASP 8 Conference*, vol. 2, 12–15 December, Sydney, Rotterdam: Balkema, Rotterdam, 651–658.

Gharaibeh, E.S., Frangopol, D.M., Shing, P.B., and Hearn, G., 2000b. System function, redundancy, and component importance: feedback for optimal design. *In*: M. Elgaaly, ed. *Advanced technology in structural engineering*. Reston, VA: ASCE [CD-ROM].

Ghosn, M. and Frangopol, D.M., 2007a. Redundancy of structures: a retrospective. *In*: D.M. Frangopol, M. Kawatani, and C.-W. Kim, eds. *Reliability and optimization of structural systems: assessment, design, and life-cycle performance*, London, UK: Taylor & Francis, 91–100.

Ghosn, M. and Frangopol, D.M., 2007b. Structural redundancy and robustness measures and their use in assessment and design. *In*: J. Kanda, T. Takada, and H. Furuta, eds. *Applications of statistics and probability in civil engineering*, London, UK: Taylor & Francis, 181–182 [CD-ROM].

Ghosn, M. and Moses, F., 1998. *Redundancy in highway bridge superstructures*. Washington, DC: Transportation Research Board, NCHRP report 406.

Ghosn, M., Moses, F., and Frangopol, D.M., 2010. Redundancy and robustness of highway bridge superstructures and substructures. *Structure and Infrastructure Engineering*, (in press).

Hendawi, S. and Frangopol, D.M., 1994. System reliability and redundancy in structural design and evaluation. *Structural Safety*, 16 (1–2), 47–71.

Liu, D., Ghosn, M., and Moses, F., 2000. *Redundancy in highway bridge substructures*. Washington, DC: Transportation Research Board, NCHRP report 458.

Liu, P.L., Lin, H.-Z., and Der Kiureghian, A., 1989. *CALREL user manual*. Berkeley, CA: Department of Civil Engineering, University of California, Report no. UCB/SEMM-89/18.

Moan, T., 2005. Reliability-based management of inspection, maintenance and repair of offshore structures. *Structure and Infrastructure Engineering*, 1 (1), 33–62.

Moses, F., 1982. System reliability development in structural engineering. *Structural Safety*, 1 (1), 3–13.

Paliou, C., Shinozuka, M., and Chen, Y.-N., 1990. Reliability and redundancy of offshore structures. *Journal of Engineering Mechanics*, 116 (2), 359–378.

Rashedi, M.R. and Moses, F., 1983. *Studies on reliability of structural systems*. Cleveland, OH: Department of Civil Engineering, Case Western Reserve University, Report R83-3.

Rashedi, M.R. and Moses, F., 1988. Identification of failure modes in system reliability. *Journal of Structural Engineering*, 114 (2), 292–313.

Sorensen, J.D. and Frangopol, D.M., 2008. Advances in reliability and optimization of structural systems. *Structure and Infrastructure Engineering*, 4 (5), 325–412.

Yang, S.-I., Frangopol, D.M., and Neves, L.C., 2004. Service life prediction of structural systems using lifetime functions with emphasis on bridges. *Reliability Engineering and System Safety*, 86 (1), 39–51.

Yang, S.-I., Frangopol, D.M., and Neves, L.C., 2005. Optimum maintenance strategy for deteriorating structures based on lifetime functions. *Engineering Structures*, 28 (2), 196–206.

Redundancy and robustness of highway bridge superstructures and substructures

Michel Ghosn, Fred Moses and Dan M. Frangopol

Major advances have been recently achieved in developing methodologies for the structural analysis of cascading failures and in understanding the behaviour of different types of systems under suddenly applied extreme loads. Yet, a main issue related to defining objective measures of redundancy and quantifying the levels of redundancy that exist in structural systems remains vastly unresolved. This paper reviews the work done by the authors and their colleagues on the quantification of system redundancy of typical highway bridges and reassesses previously made proposals for including system redundancy and robustness during the structural design and safety evaluation of bridge superstructure and substructure systems. These proposals, which are based on system reliability principles, consider structural system safety, system redundancy and system robustness in comparison to member safety, and account for the uncertainties associated with determining member and system strengths as well as future loads in a consistent and rational manner.

1. Introduction

A 1985 *State of the art report on redundant bridge systems* concluded that, although analytical techniques to study the response of damaged and undamaged flexural systems to high loads are available, 'little work has been done on quantifying the degree of redundancy that is needed' (ASCE–AASHTO Task Committee 1985). More than two decades later, and following several tragic failures caused by blasts and natural hazards, issues related to structural safety, redundancy and robustness have gained increased importance. Recent and on-going studies have focused on determining the failure mechanisms of typical structural systems and are making recommendations to improve the overall system safety of civil structures, including buildings and bridges (NCST 2005). Yet, despite all the advances made in developing structural analysis methodologies and in understanding the behaviour of different types of systems under suddenly applied extreme load events, the issue raised in the 1985 state of the art report pertaining to the quantification of the required levels of redundancy remains vastly unresolved. In this paper, the authors review previously made recommendations on the quantification of system redundancy and describe a proposed approach for including system redundancy and robustness during the structural design and safety assessment of highway bridge superstructures and substructures. These proposals, which are based on system reliability principles, consider structural system safety, system redundancy and system robustness in comparison to member safety, and account for the uncertainties associated with determining member and system strengths, as well as future loads, in a consistent and rational manner. In this context, system redundancy is defined as the ability of the bridge system to redistribute the applied load after reaching the ultimate capacity of its main load-carrying members. Robustness is defined as the ability of the system to still carry some load after the brittle fracture of one or more critical components (De *et al.* 1990).

2. System reliability background

Current structural design specifications evaluate a structural system's safety based on the capacity of its weakest member. Because of the uncertainties associated with estimating member and system capacities, as well as the uncertainties associated with predicting future loads, the evaluation of structural safety should be based on probabilistic models. A common probabilistic measure of safety that is used in structural design and evaluation is the reliability index, β, which is related to the probability of failure, P_f, by:

$$P_f = \Phi(-\beta), \qquad (1)$$

where Φ is the cumulative Gaussian probability distribution function.

If the resistance, R, and the applied load, P, follow normal probability distributions, the reliability index is obtained as:

$$\beta = \frac{\bar{R} - \bar{P}}{\sqrt{\sigma_R^2 + \sigma_P^2}}, \qquad (2)$$

where $\bar{R}$ is the mean value of resistance; $\bar{P}$ is the mean value of the applied load effect; σ_R is the standard deviation of R; and σ_P is the standard deviation of P. On the other hand, if the resistance, R, and the applied load, P, follow lognormal distributions, the reliability index can be approximated using:

$$\beta = \frac{\ln\left(\frac{\bar{R}}{\bar{P}}\right)}{\sqrt{V_R^2 + V_P^2}}, \qquad (3)$$

where $V_R = \sigma_R / \bar{R}$ is the coefficient of variation (COV) of R and V_P is the COV of P.

The reliability index can be evaluated on a member-by-member basis or on a structural system basis. Traditionally, structural design codes are calibrated so that structural members uniformly achieve a target reliability index value, β_{target}, which is usually extracted from historically satisfactory designs. For example, a $\beta_{member\ target} = 3.5$ value was used as the target for the reliability index of structural members during the development of the AASHTO load and resistance factor design (LRFD) specifications (Nowak 1999).

The member-oriented approach for structural safety assessment can lead to the design of over-conservative systems by ignoring the ability of ductile and redundant systems to redistribute their loads when a single member reaches its maximum load-carrying capacity. On the other hand, the approach may lead to the design of non-robust systems, which may be susceptible to cascading failures and collapse after the accidental brittle failure of a single member (Frangopol and Curley 1987, Frangopol and Nakib 1991).

The relationship between system safety and member safety depends on the system's configuration and whether the members are in parallel or in series, the ductility of the members and the statistical correlation between the strengths of the members. To understand the relationship between member properties and the reliability of parallel multi-member systems, Hendawi and Frangopol (1994) analysed the two-member system described in Figure 1, considering the effect of post-failure behaviour and the ductility of the members, the correlation between member capacities and their ability to distribute the applied load. The member properties are represented in terms of the load versus deflection ($P - \delta$) relationships for members A and B, where K_i gives the stiffness in the linear elastic range.

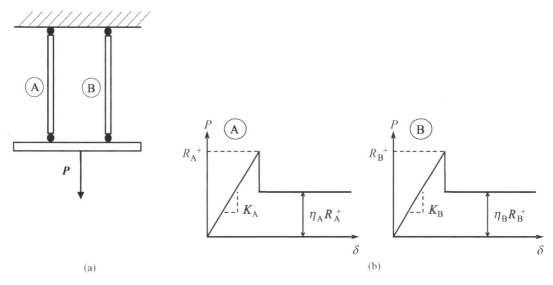

(a)   (b)

Figure 1.   Idealised two-member parallel system: (a) two-member parallel system and (b) force–deformation characteristics of members.

The residual strength $\eta_i R_i^+$ is the member capacity that remains after the ultimate capacity R_i^+ is reached. When $\eta_i = 1.0$, the member is elasto-plastic, when $\eta_i = 0$, the member is brittle.

In order to assess the system's capacity, a system safety factor, SSF, is defined as:

$$\text{SSF} = \frac{\sum_i \bar{R}_i}{\bar{P}}, \qquad (4)$$

where $\bar{R}_i$ is the mean resistance of member i and $\bar{P}$ is the mean of the total applied load. A deterministic member resistance-sharing factor, RSF_i, is defined in terms of each member's stiffness, K_i, as:

$$\text{RSF}_i = \frac{K_i}{\sum_i K_i}. \qquad (5)$$

A reliability analysis is performed for several hypothetical cases, assuming that the resistances, R, and the load, P, are normally distributed random variables. The results of the analyses are given in terms of the system's reliability index, β_{sys}, obtained from Equation (2). The results of the analyses are presented in Figures 2 and 3 for the two-member parallel system (see also Hendawi and Frangopol 1994).

Figure 2 shows the effect of the correlation in member strengths on the reliability index of the system, β_{sys}. A correlation coefficient $\rho(R_A, R_B)$ varying between 1.0, 0.5 and 0.0 is used to analyse the reliability of either a ductile or a brittle system, while the COV of the members and the load are set at $V_{RA} = V_{RB} = V_P = 15\%$. The load is assumed to be equally distributed to each of the two members (A and B) of the system so that $\text{RSF}_A = \text{RSF}_B = 0.5$. A brittle system is defined as a system whose members lose their load-carrying capacity when their ultimate strengths are reached, leading to $\eta_i = 0.0$ in the member force–deformation curve of Figure 1. A ductile system's members maintain their strengths when their ultimate capacity is reached so that $\eta_i = 1.0$. The results of the system reliability analysis show that the reliability of brittle systems is not significantly affected by the correlation between the strengths of the members. On the other hand, a higher correlation in member strengths leads to lower system reliability if the members are ductile. Thus, it would be generally conservative to assume that the members of a parallel system are fully correlated with $\rho(R_A, R_B) = 1.0$.

To study the effect of the load distribution represented by the RSF on the reliability of a parallel system, the RSF for one of the members is varied between 0 (indicating that this member does not carry

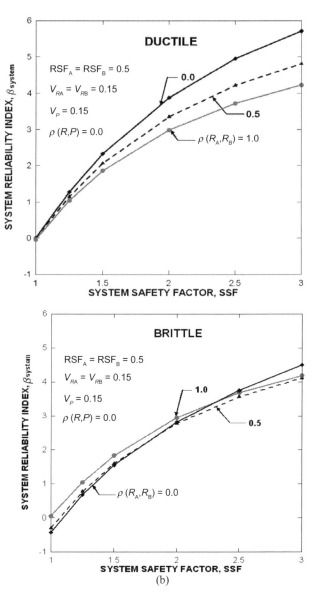

Figure 2. Effect of member resistance correlation and SSF on system reliability (adapted from Hendawi and Frangopol 1994).

any load while the second member carries the full load) and 0.5 (indicating that the two members equally share in carrying the applied load). The results of the reliability analysis shown in Figure 3 indicate that equal resistance sharing, as represented by $\text{RSF}_1 = \text{RSF}_2 = 0.5$ for a two-member system, produces maximum system reliability for ductile systems, but minimum system reliability for brittle systems. However, the reduction in the system reliability of the brittle system is relatively small. Thus, it would be conservative to apply uneven loading on the members of a ductile parallel system so that the applied load on one critical member is maximised. For a brittle system,

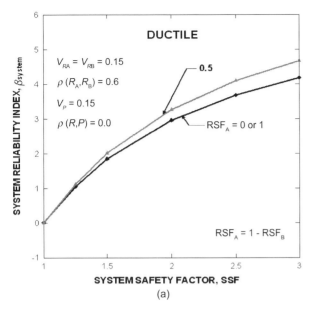

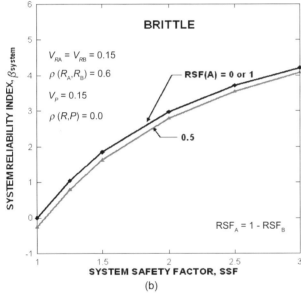

Figure 3. Effect of resistance sharing on system reliability (adapted from Hendawi and Frangopol 1994).

the effect of the load distribution is relatively small, and the system reliability is not significantly affected by the load distribution factor.

Figures 2 and 3 also show that maintaining a target system reliability index, β_{system}, can be easily achieved by properly controlling the system's safety factor, SSF. The SSF can be controlled through the scaling of the member's resistances R_i in the numerator of Equation (4) by applying a higher member safety factor. Thus, as shown in Figure 2 and assuming that the two members are ductile and uncorrelated with $\rho(R_A, R_B) = 0.0$ and assuming that a target system

reliability index $\beta_{system\ target} = 3.85$ is required, then the SSF should be set to equal to 2. If $\beta_{system\ target} = 4.92$ is required, then the SSF should be set to equal to 2.5. To increase SSF from 2.0 to 2.5, and consequently β_{system} from 3.85 to 4.92, it would be necessary to increase the safety factor of the members by 1.25.

Based on these and other similar analyses, Frangopol and Nakib (1991), Hendawi and Frangopol (1994), Ghosn and Moses (1992, 1998), Liu et al. (2001) recommended the application of a system factor, ϕ_s, during the design of structural systems to increase or reduce member resistances, R_i. The proposed system factor should be calibrated so that a given system would achieve a target system reliability value β_{system}. Thus, the codified design equations would take the form:

$$\phi_s \phi R_n \geq \sum_i \gamma_i P_i, \qquad (6)$$

where ϕ_s is the system factor; ϕ is the member resistance factor; R_n is the nominal member resistance; γ_i are the load factors; and P_i are the nominal design loads. The application of a system factor $\phi_s < 1.0$ would lead to increasing the system reliability of designs that traditionally have been shown to have low levels of system safety. Alternatively, $\phi_s > 1.0$ would lead to rewarding the design of systems with high safety levels by allowing their members to have lower capacities. The calibration of the appropriate system factors should be controlled by the required system reliability target. The approach is illustrated for the example previously worked out with two uncorrelated ductile members. In such a case, the application of a system factor $\phi_s = 1.0/1.25$ would lead to increasing the system reliability index from $\beta_{system} = 3.85$ to 4.92. The determination of the target system reliability index, $\beta_{system\ target}$, will depend on the limit state chosen to define the system's safety, as will be discussed in the next section.

3. General concepts of system safety, redundancy and robustness

Figure 4 gives a conceptual representation of the behaviour of a structure and the different levels of safety that should be considered. For example, the solid line labelled 'intact system' may represent the live load versus maximum vertical displacement of a ductile multi-girder bridge superstructure. Assuming that the live load applied has the configuration of the AASHTO HS-20 (AASHTO 2002) vehicle, then the first main member will fail when the HS-20 truck weight is multiplied by a factor LF$_1$. However, the ultimate capacity of the whole bridge is not reached

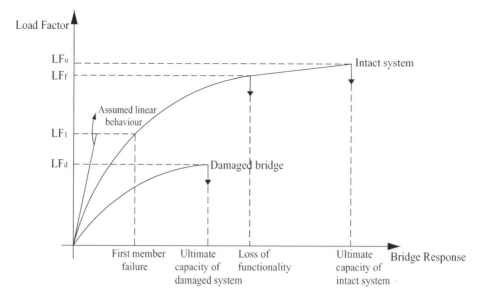

Figure 4. Representation of typical behaviour of bridge systems.

until the HS-20 truck weight is multiplied by a factor LF$_u$. Large vertical deformations rendering the bridge unfit for use are reached when the HS-20 truck weight is multiplied by a factor LF$_f$. At this point, the bridge is said to have lost its functionality. If the bridge has sustained major damage due to the brittle failure of one or more of its members, its behaviour after damage is represented by the curve labelled 'damaged bridge'. In this case, the ultimate capacity of the damaged bridge is reached when the weight of the HS-20 truck is multiplied by a factor LF$_d$. As an example, a bridge may be damaged when a fracture critical member has failed or after the brittle failure of a main load-carrying member due to collision or other impacts. Similar curves can be drawn to represent the behaviour of other systems subjected to other loading conditions, such as the loading of a bridge substructure under the effect of lateral loads.

The load multipliers, LF$_i$, provide deterministic estimates of critical limit states that describe the safety of a structural system. These load multipliers are usually obtained by performing an incremental non-linear finite-element analysis of the structure. Because of the presence of large uncertainties in estimating the parameters that control member properties, the bridge response and the applied loading more appropriate safety measures must be adopted based on probabilistic concepts. A common probabilistic measure of safety that is used in structural design and evaluations is the reliability index, β. For the lognormal model of Equation (3), the relationship between the reliability index and the load multipliers, LF, for a bridge

superstructure subjected to HS-20 truck loading can be established as:

$$\beta = \frac{\ln\left(\frac{R}{P}\right)}{\sqrt{V_R^2 + V_P^2}} = \frac{\ln\left(\frac{\overline{LF}\ HS20}{\overline{LL}\ HS20}\right)}{\sqrt{V_{LF}^2 + V_{LL}^2}} = \frac{\ln\left(\frac{\overline{LF}}{\overline{LL}}\right)}{\sqrt{V_{LF}^2 + V_{LL}^2}}, \quad (7)$$

where LF is the load multiplier obtained from the incremental analysis and LL HS20 is the expected maximum live load effect that will be applied on the superstructure within the appropriate return period. HS20 is the load effect of the nominal HS-20 design truck. V_{LF} is the COV of the bridge resistance, defined as the standard deviation divided by the mean value and V_{LL} is the COV of the applied live load. Both the resistance and the applied live load are expressed as a function of the HS-20 truck load effect, which can then be factored out. For example, during the calibration of the AASHTO LRFD, the maximum truck load effect expected to be applied on a single lane of a 18 m simple span bridge during a 75year design life is equal to LL$_{75}$ = 1.79 times the effect of an HS-20 vehicle.

Equation (7) lumps all the random variables that control the load-carrying capacity of a bridge structure into the load multipliers, LF. This approach, which assumes that the member strengths are fully correlated, gives conservative estimates of the reliability of structural systems, as demonstrated by the results obtained in the simplified two-parallel-member system of Figure 1. For more accurate results, the system's capacity, represented by the random variable, LF, can be expanded as a function of the basic random parameters that control member strength and stiffness, and the probability of failure can be generally represented by an equation of the form:

$$P_f = \Pr(LF_i < LL) = \Phi(-\beta). \qquad (8)$$

After expanding LF_i as a function of the basic random parameters, Equation (8) may be solved using first-order reliability methods (FORM) or similar techniques (Thoft-Christensen and Baker 1982, Melchers 1999). An example describing how this can be achieved is provided below.

Equations (7) or (8) can be used to determine the reliability index, β, for any member or system limit state defined in terms of the load multipliers LF_1, LF_f, LF_u or LF_d of Figure 4. In most typical applications, bridge specifications have focused on ensuring that bridge members meet minimum safety requirements and failure is defined as the point at which the first member reaches its ultimate load capacity. Thus, according to current practice, it is the most critical member's capacity represented by LF_1 that controls the design of a bridge structure. In reality, for many bridges, the failure of a main carrying member may not necessarily lead to bridge collapse, neither would it necessarily lead to sufficiently high deflections that would render the bridge unfit for use; such bridges are recognised as being redundant. Even after the brittle failure of a main load-carrying member, a bridge may still be able to carry significant levels of live load; such a bridge is known to be robust. Different types of bridges may have different degrees of redundancy or robustness, depending on their topology and material behaviour. Current specifications encourage the design of redundant and robust bridges. However, as mentioned earlier, there are currently no widely accepted objective measures of redundancy and robustness.

If redundancy is defined as the capability of a structure to continue to carry loads after the ductile failure of the most critical member, and robustness is defined as the capability of the system to carry some load after the brittle failure of a main load-carrying member, then comparisons between the load multipliers LF_u, LF_f, LF_d and LF_1 would provide objective measures of system redundancy and robustness. Thus, three deterministic measures of the system's capacity as compared to the most critical member's capacity, are defined as:

$$R_u = \frac{LF_u}{LF_1}, \quad R_f = \frac{LF_f}{LF_1} \quad \text{and} \quad R_d = \frac{LF_d}{LF_1}, \qquad (9)$$

where R_u is the system reserve ratio for the ultimate limit state; R_f is the system reserve ratio for the functionality limit state; and R_d is the system reserve ratio for the damage condition.

The system reserve ratios provide a nominal deterministic measure of bridge redundancy and robustness. For example, when the ratio $R_u = 1.0$ ($LF_u = LF_1$), the ultimate capacity of the system is equal to the capacity of the bridge to resist failure of its most critical member. Based on the definitions provided above, such a bridge is non-redundant. As R_u increases, the level of bridge redundancy increases. In addition, a redundant bridge should also be able to function without leading to high levels of deformations after the failure of one of its components. Thus, R_f provides another measure of redundancy. Similarly, a robust bridge structure should be able to carry some load after the brittle fracture of one of its members, and R_d would provide an objective measure of structural robustness. Note that system failure or collapse may be due to a variety of failure modes. For bridge superstructures or substructures, these failure modes include the formation of a collapse mechanism in the beam or column assemblies, the failure of the soil–foundation system, reaching large transverse or lateral deformations rendering the bridge non-functional, or the collapse of the structure following a brittle failure in a main load-carrying member or connection due to fracture, fatigue, shear, impact or blast. For a given bridge, each of the possible failure modes may be addressed separately and the corresponding LF_u, LF_f and LF_d values obtained from an incremental nonlinear load analyses. By comparing the system's LF values to LF_1, the system reserve ratios R_u, R_f and R_d are calculated for each mode.

Alternatively, and in order to take into consideration the uncertainties in estimating the system and member capacities, as well as the applied loads, probabilistic measures of redundancy can be defined. Hence, it might be useful to examine the differences between the reliability indices of the systems expressed in terms of β_{ultimate}, $\beta_{\text{functionality}}$, β_{damaged} and the reliability of the most critical member β_{member}. This could be done through, $\Delta\beta_u$, $\Delta\beta_f$ and $\Delta\beta_d$, which are respectively the relative reliability indices for the ultimate, functionality and damaged limit states and are defined as:

$$\Delta\beta_u = \beta_{\text{ultimate}} - \beta_{\text{member}},$$
$$\Delta\beta_f = \beta_{\text{functionality}} - \beta_{\text{member}}, \qquad (10)$$
$$\Delta\beta_d = \beta_{\text{damaged}} - \beta_{\text{member}}.$$

As an example, using the simplified lognormal reliability model of Equation (7) for a superstructure under the effect of vertical live loading and assuming that the COVs of LF_u, LF_f, LF_d and LF_1 are all equal to the same value, V_{LF}, the probabilistic and deterministic measures are found to be directly related to each other as:

$$\Delta\beta_u = \beta_{ultimate} - \beta_{member} = \frac{\ln\left(\frac{\overline{LF_u}}{LL_{75}}\right) - \ln\left(\frac{\overline{LF_1}}{LL_{75}}\right)}{\sqrt{V_{LF}^2 + V_{LL}^2}}$$

$$= \frac{\ln\left(\frac{\overline{LF_u}}{\overline{LF_1}}\right)}{\sqrt{V_{LF}^2 + V_{LL}^2}} = \frac{\ln(R_u)}{\sqrt{V_{LF}^2 + V_{LL}^2}},$$

$$\Delta\beta_f = \beta_{functionality} - \beta_{member} = \frac{\ln\left(\frac{\overline{LF_f}}{LL_{75}}\right) - \ln\left(\frac{\overline{LF_1}}{LL_{75}}\right)}{\sqrt{V_{LF}^2 + V_{LL}^2}}$$

$$= \frac{\ln\left(\frac{\overline{LF_f}}{\overline{LF_1}}\right)}{\sqrt{V_{LF}^2 + V_{LL}^2}} = \frac{\ln(R_f)}{\sqrt{V_{LF}^2 + V_{LL}^2}}$$

and

$$\Delta\beta_d = \beta_{damaged} - \beta_{member} = \frac{\ln\left(\frac{\overline{LF_d}}{LL_2}\right) - \ln\left(\frac{\overline{LF_1}}{LL_{75}}\right)}{\sqrt{V_{LF}^2 + V_{LL}^2}}$$

$$= \frac{\ln\left(\frac{\overline{LF_f}}{\overline{LF_1}}\frac{\overline{LL_{75}}}{\overline{LL_2}}\right)}{\sqrt{V_{LF}^2 + V_{LL}^2}} = \frac{\ln\left(R_d\frac{\overline{LL_{75}}}{\overline{LL_2}}\right)}{\sqrt{V_{LF}^2 + V_{LL}^2}}. \quad (11)$$

Note that, for damaged bridges under the effect of the live load LL, the calculation of the reliability index for the damaged system is executed using the 2 year maximum load represented by the load multiplier, LL_2, rather than the maximum load for the 75 year design life represented by the load multiplier, LL_{75}. This is based on the assumption that any major damage to a bridge should be detected during the mandatory biennial inspection cycle and thus no bridge is expected to remain damaged for more than 2 years.

The reliability indices determined using Equation (11) are conservative estimates obtained from an incremental analysis assuming that the bridge member's strengths are fully correlated and follow lognormal distributions. Alternatively, the reliability index values for the system and the member can be obtained in function of the basic random parameters using the response surface method (RSM).

The results of the reliability analysis can be used to calibrate the system factors, ϕ_s, of Equation (6) to ensure that highway bridges provide a minimum level of system safety under bridge overloading or after a main component's brittle failure. The first step in the process is to determine the target relative reliability levels that adequately redundant and robust systems should achieve. The next two sections illustrate how

this concept is applied for typical highway bridge superstructure and substructure configurations. The subsequent section describes how the approach can be applied for the redundancy analysis of non-typical bridge configurations.

4. Implementation in bridge multi-beam superstructure systems

To evaluate the safety of typical bridge superstructure configurations and understand the behaviour of typical highway bridge superstructures, Ghosn and Moses (1998) analysed a large number of prestressed concrete and steel I-beam bridges. Specifically, simple span prestressed concrete I-beam bridges ranging in span between 14 and 46 m having 4, 6, 8 or 10 parallel members spaced at 1.2, 1.8, 2.4, 3.0, 3.6 m centre on centre were selected for analysis using all possible combinations of the above lengths, number of beams and beam spacing. Two-span continuous prestressed concrete bridges having span lengths of 30, 36 and 46 m were analysed for all the numbers of beams and beam spacings listed above. Simple span and continuous steel I-beam bridges having the same span lengths, number of girders and girder spacings as those of the prestressed I-beam bridges were also analysed. In addition, simple span prestressed concrete spread box girder bridges 18, 30 and 46 m span in lengths with 2, 3 or 5 box girders were studied. Finally, nonlinear analysis was performed on prestressed concrete multi-box beam bridges of 12, 21 and 30 m lengths. Furthermore, an extensive sensitivity analysis was undertaken to study the effect of bridge skewness, deck bending and torsional stiffness, finite-element mesh refinement, variations in member strengths and member stiffness, the length of the plastic hinge and the presence of diaphragms.

To perform the analysis of all the bridge configurations, a nonlinear finite-element program was developed. For the analysis, all the bridges were modelled using a grillage mesh similar to that proposed by Zokaie et al. (1991). Material nonlinearity was handled by assuming that hinges may form at the ends of each beam element. These hinges were modelled as nonlinear rotational springs having moment versus rotation relationships extracted from experimental data for steel I-beam bridges and from the moment versus curvature relationships of prestressed concrete main members. The slab was modelled by a grid of reinforced concrete members whose nonlinear behaviours were also modelled using moment versus curvature relationships. To obtain the plastic rotation of concrete beams, the curvature is multiplied by a plastic hinge length L_p equal to the depth of the beam section. The use of experimental moment rotations curves for the steel beams allowed for the consideration of local buckling and the presence of residual stresses (Schilling 1989).

4.1. Incremental analysis example

As an example of the analysis performed, a 30 m six-beam prestressed concrete I-girder bridge is shown in Figure 5. The bridge was designed to satisfy the AASHTO LFD (AASHTO 2002) design specifications. The finite-element mesh used for the analysis is shown in Figure 6. The moment–curvature relationship for the composite main girders is shown in Figure 7 where M_i give the moments at critical points of the multi-linear curve and S_i give the slopes for each line segment. The initial loading consisted of the dead weight and a basic vehicular live load, represented by two side-by-side HS-20 trucks placed in the positions shown in Figures 5 and 6. During the analysis, the weights of the HS-20 trucks were incremented and the stiffness of each beam updated as the moments in each individual member reached critical values. In addition to the determination of the internal moments and forces at each load step, the analysis led to a load versus vertical displacement curve similar to that shown in Figure 4 with the following results. First member failure occurred when the HS-20 weights were multiplied by a factor $LF_1 = 3.67$. Ultimate capacity of the whole system was reached at $LF_u = 5.28$ when the external girder was crushed as the plastic rotation of the beam reached a value $\theta_{pmax} = 0.0247$ rad. To determine the appropriate value for the functionality limit, the literature was reviewed and different options were considered. It was determined that a deflection limit equal to span length/100 is appropriate to ensure traffic safety, while simultaneously avoiding visibly excessive permanent deformations. The 30 m bridge analysed in this example reached the deflection limit of 0.3 m at a load factor $LF_f = 3.65$.

To analyse the robustness of the superstructure, defined as the capacity of the damaged bridge to sustain live load after the brittle failure of one main girder, the external girder under the live load was completely removed from the model simulating situations, where the beam is shattered due to a collision or an impact. The incremental nonlinear analysis is then

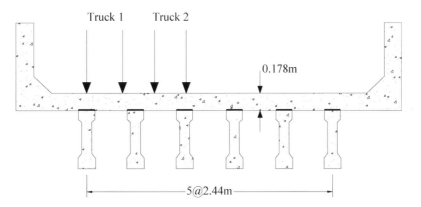

Figure 5. Cross-section of example 30 m prestressed concrete bridge.

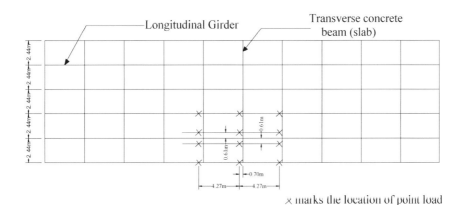

Figure 6. Finite-element mesh of example 30 m prestressed concrete bridge.

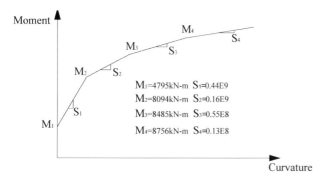

$M_1 = 4795$kN-m $S_1 = 0.44$E9
$M_2 = 8094$kN-m $S_2 = 0.16$E9
$M_3 = 8485$kN-m $S_3 = 0.55$E8
$M_4 = 8756$kN-m $S_4 = 0.13$E8

Figure 7. Typical post-yielding moment–curvature relationship for prestressed concrete members.

repeated for the damaged bridge, assuming that the slab and other secondary elements can still provide sufficient strength to redistribute the load to the remaining girders. This is deemed to be a realistic scenario based on observations made in the field about bridges whose exterior girders were badly damaged due to collisions or other types of fracture. In such a case, the ultimate capacity of the damaged bridge is reached when the HS-20 truck weights are multiplied by a factor $LF_d = 2.05$.

4.2. Simplified reliability analysis

To calculate the reliability index, β, for the four limit states represented by LF_1, LF_f, LF_u and LF_d, the results of the deterministic analysis are used in addition to the statistical data assembled during the calibration of the AASHTO LRFD specifications (Nowak 1999). For example, for the 30 m bridge analysed above, the mean value of the maximum live load expected in the 75 year design life of the bridge is represented by a load multiplier $LL_{75} = 1.89$ (or the expected maximum load is 1.89 times the HS-20 load effect), the mean value of LF_1 is obtained from the deterministic analysis, accounting for the fact that the bias is 1.05 times the nominal value obtained from the analysis. The value of $V_{LL} = 19\%$ is used, while a value of $V_{LF} = 14\%$ is extracted from the data given by Nowak (1999). This would lead to a reliability index for the failure of the first member to $\beta_{member} = 2.84$, which is of the same order of the value provided by Nowak (1999) for the 30 m LFD designed bridges having six beams at 2.4 m centre-to-centre. The reliability index for the ultimate capacity is calculated as $\beta_{ultimate} = 4.08$. For the functionality limit, the reliability index is calculated as $\beta_{functionality} = 2.82$. The damaged bridge resulted in a reliability index $\beta_{damaged} = 0.98$. Similar calculations are executed for all the bridge types and span lengths mentioned in the previous section. Using Equation (10), the relative reliability indices for this bridge example are obtained as $\Delta\beta_u = 1.24$, $\Delta\beta_f = -0.02$ and $\Delta\beta_d = -1.86$.

4.3. Advanced reliability analysis

The calculation of the reliability index executed, based on Equation (7), uses a simplified lognormal model that lumps all the random variables into two variables, which are the system capacity, expressed in terms of the load factor LF, and the applied live load, expressed in terms of the variable LL. A more exact analysis would require that the failure equation be explicitly expressed in terms of all the random variables that control the safety of the bridge. A closed-form expression for the failure function cannot be obtained directly and the finite-element method must be used to check whether particular values of the random variables would lead to the failure of the bridge system or not. Therefore, the Response Surface Method (RSM) is often used to obtain approximate expressions for the failure equation, based on the finite-element results. Ghosn *et al.* (1993) demonstrate how the method can be applied to bridge systems.

The RSM consists of executing finite-element analyses at pre-determined values of the random variables and uses a perturbation technique to determine how the response changes as each of the input variables is varied by a known amount. Given the results of the finite-element analysis, an approximate closed-form expression of the response is obtained using a first-order Taylor series expansion of the form:

$$f(x_i) = f(x_i^*) + \sum \frac{\partial f}{\partial x_i}\Big|_{x_i^*}(x_i - x_i^*), \qquad (12)$$

where x_i denotes the basic random variables that control the response of the system. In the calculations performed for this study, the basic random variables are the moment capacities of each member of the bridge, the dead load and the maximum plastic rotation that causes member crushing in longitudinal concrete members or the rupture of longitudinal steel girders. The parameter x_i^* in Equation (12) denotes the values around which the Taylor series is expanded. The operator $f(\ldots)$ represents the results of the finite-element analysis. In the first step of the analysis, $f(x_i^*)$ is the LF value obtained from the finite-element analysis when all the input variables are set at their nominal values. When analysing the ultimate capacity limit state of the 30 m bridge example discussed above, $f(x_i^*) = LF_u$ is equal to 5.28. The x_i^* values are subsequently updated from the nominal values to the values corresponding to the coordinates of the most likely failure point, which is identified by the First Order Reliability Method (FORM) algorithm (Thoft-Christensen and Baker 1982).

The partial derivatives used in Equation (12) are calculated using a finite difference approach such that:

$$\frac{\partial f}{\partial x_i}\Big|_{x_i^*} = \frac{f(x_i^* + 0.1x_i^*) - f(x_i^* - 0.1x_i^*)}{0.2x_i^*}. \quad (13)$$

The analyses of $f(x_i^* + 0.1x_i^*)$ and that of $f(x_i^* - 0.1x_i^*)$ are performed for all the six variables representing the strengths of the members (R_i), the variable representing the maximum plastic rotation at which concrete crushing occurs (θ_{max}) and the variable representing the dead load effect (D) produces. These random variables are used because they control the strength of the system. The input data for the bias, COV and probability distribution types for R and D, shown in Table 1, are adopted from the data used by Nowak (1999) during the calibration of the AASHTO LRFD code. The data for the rotation of the concrete at crushing was collected by the authors from the data provided by Mattock (1964). A Taylor series expansion can be given as:

$$LF_u = -0.826 + 227 \times 10^{-6} R_1 + 200 \times 10^{-6} R_2$$
$$+ 163 \times 10^{-6} R_3 + 112 \times 10^{-6} R_4$$
$$+ 85 \times 10^{-6} R_5 + 58 \times 10^{-6} R_6$$
$$+ 70\theta_{max} - 903 \times 10^{-6} D. \quad (14)$$

In Equation (14), the units used for the variables R_i and D are kN-m and θ_{max} is in radians.

The probability of failure and the reliability index, $\beta_{ultimate}$, are calculated using Equation (8) and a FORM algorithm based on the input data provided in Table 1. The reliability index obtained in this analysis is $\beta_{ultimate} = 4.34$. It is noted that this value is higher than the 4.08 value calculated using the simplified lognormal model. The difference is due to the fact that the reliability calculation performed using the RSM assumes that the finite-element analysis of the system is exact and there are no uncertainties associated with the modelling of the structure. In contrast, and more appropriately, the approximate model conservatively assumes that the ultimate load

capacity calculated has a COV at least equal to that associated with the determination of the individual member capacities. Thus, the results of the simplified lognormal model are deemed to be acceptable, although more conservative, than those of RSM. Furthermore, both analyses assume that all the member capacities are correlated, which adds to the conservatism of the results. If the bridge members are assumed to be uncorrelated, the reliability index will be slightly higher than calculated herein, as discussed by Ghosn et al. (1993) and seen in the simplified two-member system analysed above.

4.4. Analysis of results and determination of redundancy criteria

The results of the nonlinear structural analyses and the reliability index values obtained for the hundreds of bridge configurations performed by Ghosn and Moses (1998) in addition to the sensitivity analyses led to the following observations:

- The proposed measures of redundancy are robust in the sense that the redundancy ratios and the relative reliability indices of Equations (9) and (10) are not sensitive to variations in the stiffnesses of the individual members, member strengths, magnitude of the applied dead load, or for skew angles if they remain less than 45°. The individual values of the load factors LF_u, LF_f, LF_d and LF_1 may change, but, the ratios $R_u = LF_u/LF_1$, $R_f = LF_f/LF_1$ and $R_d = LF_d/LF_1$, as well as the relatively reliability indices $\Delta\beta_u$, $\Delta\beta_f$ and $\Delta\beta_d$, remain relatively constant. For example, the strength reserve ratios R_u, R_f and R_d, change by less than 10% when the slab strength and its stiffness are increased by 50% from their original values.
- The level of redundancy represented by R_u, R_f and R_d is not only a function of the number of parallel bridge girders, as accepted by current practice, but is a combination of the number of girders and girder spacing. For example, multi-girder bridges with wide spacings have low system reserve ratios independent of the number of beams. On the other hand, because the load

Table 1. Input data for reliability analysis.

Variable	Nominal value	Bias	COV	Distribution type
Main member resistances, R_1–R_6	9600 kN-m	1.05	7.5%	Lognormal
Dead load, D	3864 kN-m	1.05	10%	Normal
Maximum rotation, θ_{max}	0.0247	1.0	20%	Normal
Live load factor, LL	1.89	1.0	19%	Extreme type I

distribution factors of narrow bridges with very closely spaced beams is high, these bridges show little system reserve as the failure of one beam would cause a cascading failure of the adjacent beams.

- The presence of diaphragms helps in redistributing the load to the intact members when a bridge is damaged after the brittle failure of one of its members, thus leading to higher R_d ratios for bridges with diaphragms as compared to bridges without diaphragms. However, the presence of diaphragms does not significantly contribute to increasing R_u and R_f of undamaged bridges.
- Bridges whose members have low levels of ductility do not exhibit adequate levels of system reserve regardless of the number of supporting girders.
- As expected, bridges with a low number of girders show low levels of redundancy. In particular, the average redundancy ratios calculated for bridges with four girders are found to be $R_u = 1.30$, $R_f = 1.10$ and $R_d = 0.50$. Generally, bridges with more than four adequately spaced girders show higher system reserve ratios. Since, according to current practice, all four girders bridges are assumed to have adequate levels of redundancy, it is proposed to use the redundancy ratios corresponding to these bridges as the minimum redundancy ratios that any bridge configuration should satisfy to be considered adequately redundant. Such a calibration with currently accepted standards is typical for the development of new design specifications. The redundancy ratios $R_u = 1.30$, $R_f = 1.10$ and $R_d = 0.50$ correspond to relative reliability indices $\Delta\beta_u = 0.85$, $\Delta\beta_f = 0.25$ and $\Delta\beta_d = -2.70$.

Based on the results presented by Ghosn and Moses (1998) and summarised in this paper, the determination of whether a bridge provides an adequate level of redundancy can be done if a nonlinear analysis is performed, as described in the previous section, and the calculated redundancy ratios R_u, R_f and R_d are respectively compared to the threshold values 1.30, 1.10 and 0.50. Bridges that produce redundancy ratios higher than these minimum acceptable values may be classified as adequately redundant. Those that produce ratios lower than the threshold values should be classified as non-redundant. Accordingly, non-redundant bridges should be either redesigned with modified topological configurations to improve their levels of redundancy or alternatively should be penalised by requiring that their members and system strengths be higher than those of similar bridges that show adequate levels of redundancy.

Thus, a bridge that has a low level of redundancy can still be so 'over-designed' that collapse would have a low probability of occurring. The values of the system resistance factor, ϕ_s, can be calibrated for typical bridge configurations as discussed in the next section.

For bridge superstructures with non-typical configurations, the required system factor that should be applied on the member strengths to ensure adequate system safety should be proportional to the calculated ratio using:

$$\phi_s = \min\left(\frac{R_u}{1.30}, \frac{R_f}{1.10}, \frac{R_d}{0.50}\right). \qquad (15)$$

4.5. System factors for typical bridge superstructures

Although the steps involved in the calculation of the redundancy ratios and subsequently the determination of the applicable system factor are easy to follow when a nonlinear analysis program is available, the process may be too involved for routine applications on typical bridge configurations. Therefore, Ghosn and Moses (1998) have calibrated a set of system factors, ϕ_s, which can be directly applied during the design or the load capacity evaluation of existing bridge superstructures with typical configurations to ensure that the system capacities of these bridges are adequate. These system factors provide a measure of the system reserve strength as it relates to ductility, redundancy and operational importance.

The proposed system factors are calibrated such that the overall system safety of a bridge structure, as represented by its system's reliability index, meets a target value. Assuming that members of adequately redundant bridges are designed to satisfy the current AASHTO LRFD criteria, then the reliability index of their most critical members should be on the order of $\beta_{member} = 3.5$. The target system reliability index values that a system should meet are obtained based on the observation made earlier that adequately redundant bridges have been found to have relative reliability indices $\Delta\beta_u = 0.85$, $\Delta\beta_f = 0.25$ and $\Delta\beta_d = -2.70$. Specifically, given that adequately redundant bridge systems should have a $\Delta\beta_u = 0.85$, a set of system factors ϕ_s are calibrated so that the reliability index of an intact bridge system will be equal to $\beta_{ultimate} = 4.35$ (i.e. $3.50 + 0.85$). Similarly, the reliability index for a maximum deflection of span length/100 should be set at $\beta_{functionality} = 3.75$. Finally, the system reliability of a bridge that has lost a major member should meet a target reliability index value $\beta_{damaged} = 0.80$. As an example, the latter value implies that a superstructure that has sustained major damage or the complete loss in the load-carrying capacity of a main member should still have about 80% probability of survival under regular truck loading.

The introduction of the system factors ϕ_s into Equation (6) will serve to add member and system capacity to less redundant systems, such that the overall system reliability is increased. When adequate redundancy is present, a system factor, ϕ_s, greater than 1.0 may be used.

Based on the results of the analyses of the multi-girder bridge, Ghosn and Moses (1998) proposed that members of superstructures of bridges classified to be critical or that are susceptible to brittle damage, the system factor should be determined as $\phi_s = \min (\phi_{su}, \phi_{sd1})$ for multi-girder systems with diaphragms spaced at not more than 7600 mm and $\phi_s = \min (\phi_{su}, \phi_{sd2})$ for all other multi-girder systems.

For the superstructures of all other bridges, $\phi_s = \phi_{su}$, where ϕ_{su} is the system factor for superstructure ultimate capacity; ϕ_{sd1} is the system factor for damage of superstructures with regularly spaced diaphragms; and ϕ_{sd2} is the system factor for damage of super-structures with no diaphragms, as given in Table 2 for different number of beams and beam spacing. A minimum value of $\phi_s = 0.80$ is recommended, but in no instance should ϕ_s be taken as greater than 1.20. The upper and lower limits on ϕ_s have been proposed as temporary measures until the practising engineering community builds its confidence level in the proposed methodology.

Bridges susceptible to brittle damage include bridges with fatigue-prone details, those with members that are exposed to collisions from ships, vehicles and debris carried by swelling streams and rivers, or bridges that have a high security risk.

As was proposed in (AASHTO 2007) for the seismic design of bridges, Ghosn and Moses (1998) recommended that bridge owners, or those having jurisdiction, classify each bridge into one of three importance categories as follows: (a) critical bridges,

(b) essential bridges or (c) other bridges. Such classification should be based on social/survival and/or security/defence requirements following the guidelines for the design for earthquakes (AASHTO 2007). In addition, bridge owners should identify bridges that are susceptible to brittle damage, as mentioned earlier. The use of more stringent criteria for critical bridges is consistent with current trends to use performance-based design in bridge engineering practice.

Members in shear, as well as all joints and connections, are assigned a system factor $\phi_s = 0.80$. This assumes that the resistance factor ϕ was cali-brated to satisfy a target member reliability index $\beta_{member} = 3.5$. Since shear failures and connection failures are brittle, the application of a system factor $\phi_s = 0.80$ will increase the reliability index of the member and also that of the system to $\beta_{member} = \beta_{system} \approx 4.35$, which is the target value for system safety.

5. Implementation in bridge substructure systems

5.1. Redundancy in bridge substructures

While the safety of bridge superstructures is normally controlled by the effect of the live loads, the safety of bridge substructures is usually controlled by the effect of lateral loads, such as those of seismic loads. In most routine design situations, the engineer uses typical substructure configurations whose redundancy levels may not need to be evaluated on a case-by-case basis. In such situations, and following the approach proposed for superstructures, the designer may be able to apply a tabulated system factor ϕ_s during the check of the safety and redundancy of the system in order to ensure that members of substructure config-urations that are not sufficiently redundant have higher safety factors than members of substructure config-urations that are sufficiently redundant. In this case, the design-check equation for a substructure subjected

Table 2. System factors for girder superstructures.

Loading type	System/member type	ϕ_{su}	ϕ_{sd1}	ϕ_{sd2}
Bending of members of multi-girder systems	Two-girder bridges	0.85	0.80	0.80
	Three-girder bridges with spacing: ≤ 1.8 m	0.85	1.20	1.10
	Four-girder bridges with spacing: ≤ 1.2 rn	0.85	1.20	1.10
	Other girder bridges with spacing: ≤ 1.2 m	1.00	1.20	1.15
	All girder bridges with spacing: ≤ 1.8 m	1.00	1.20	1.15
	All girder bridges with spacing: ≤ 2.4 m	1.00	1.15	1.05
	All girder bridges with spacing: ≤ 3.0 m	0.95	1.00	0.90
	All girder bridges with spacing: ≤ 3.6 m	0.90	0.80	0.80
	All girder bridges with spacing: > 3.6 m	0.85	0.80	0.80
Slab bridges	For bending		1.00	
All bridge types	Members in shear		0.80	
All joints and connections			0.80	

to a dead load, D, live load, L, and lateral load, P, would take the format:

$$\phi_s \phi R_n = \gamma_d D_n + \gamma_l L_n + \gamma_p P_n, \qquad (16)$$

where ϕ_s is the system factor that is defined as a statistically based multiplier relating to the safety and redundancy of the substructure system; ϕ is the member resistance factor; R_n is the required nominal resistance capacity of the member; γ_d is the dead load factor; D_n is the nominal dead load effect; γ_l is the live load factor; L_n is the nominal live load effect; γ_p is the lateral load factor; and P_n is the nominal lateral load effect.

As was the case with superstructures, system factors for substructures can be calibrated using a reliability model, such that a system factor equal to 1.0 indicates that the bridge substructure under consideration will have a relative reliability index, $\Delta\beta$, equal to a target value which can be determined from the review of substructure configurations known to have acceptable levels of redundancy based on current practice and engineering judgement. The process is illustrated in the following subsections.

5.2. Example analysis of a substructure system

To illustrate the methodology proposed to calibrate substructure system factors, a representative example, consisting of a four-column bent, is presented in Figure 8. The substructure has columns that are 6.5 m high. The analysis is performed to study the redundancy of the system under the effect of lateral loads. These lateral loads may, for example, model the effects of seismic forces, wind, hydraulic or impact pressures. The analysis process increments the lateral load until system failure occurs. To perform the analysis, the vertical loads (dead lead + vehicular live loads) are set at their mean values. Table 3 lists the nominal (design) values for the input variables, as well as the mean (or expected) values for the material properties and the vertical loads. The bias gives the ratio of the mean to the nominal value. In addition, Table 3 gives the COVs (COV = standard deviation/ mean), providing a measure of the uncertainty associated with determining the random variables used during the analysis. In Table 3, the effects of the different material properties (concrete strength, yielding stress of steel) combine to produce the moment capacity of the column section. Two values for the ultimate section moment capacity are given. The first value assumes that the column is unconfined (labelled M_u). The second value is for the confined column (labelled M_c). The model used for the analysis assumes some strain hardening (strain hardening $H\varepsilon = 1\%$). The soil– foundation stiffnesses for the foundations below the piers are assumed to be fully correlated. The variables listed in the table without COVs are deterministic. Particularly, the geometric properties are all assumed to be well defined, such that any variability in their values are minimal and do not produce any noticeable effect on the reliability of the substructure. Notice that the ultimate concrete strain ε_u is assumed to be deterministic because the variability in this parameter has already been taken into consideration while accounting for the variability in M_u and M_c.

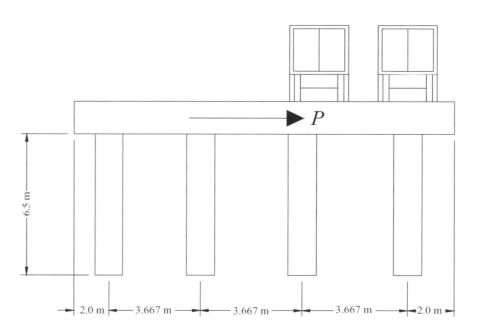

Figure 8. Configuration of four-column bent.

Table 3. Input data for example four-column substructure system (Liu *et al.* 2001).

	Variable	Symbol	Nominal Value	Mean	COV (%)	Bias
Reinforced concrete column moment capacity	Unconfined moment capacity	M_u	3841 kN · m	4345 kN · m	13	1.14
	Confined moment capacity	M_c	3508 kN · m	4000 kN · m	13	1.14
Foundation properties	Displacement stiffness	K_x	5.83E + 4 kN/m	5.83E + 4 kN/m	30	1.0
		K_y	7.78E + 4 kN/m	7.78E + 4 kN/m	30	1.0
		K_z	5.83E + 4 kN/m	5.83E + 4 kN/m	30	1.0
	Rotation stiffness	K_{xx}	1.00E + 10 kN/m/rad	1.00E + 10 kN/m/rad	30	1.0
		K_{yy}	1.00E + 10 kN/m/rad	1.00E + 10 kN/m/rad	30	1.0
		K_{zz}	1.87E + 6 kN/m/rad	1.87E + 6 kN/m/rad	30	1.0
	Ultimate moment	M_f	9.72E + 4 kN/m	9.72E + 4 kN/m	25	1.0
Loads	Dead load		5500 kN	5575 kN	10	1.05
	Dead load factor	γ_d	1.25			
	Wearing surface	WL	1485 kN	1485 kN	25	1.0
	Factor		1.50			
	Specific weight	γ_w	23.5 kN/m³	23.5 kN/m³	4	1.0
	Dead load factor	γ_d	1.25			
	Lane load	L_1	260 kN	260 kN	20	1.0
	Live load factor	γ_1	1.35			
	Truck load	L_t	325 kN	325 kN	20	1.0
	Live load factor	γ_1	1.35			
	Impact	I	1.33	1.33	20	1.0
Geometry	Steel area	A_s	420.45 mm²			
	Cover	C_s	50 mm			
	Column depth	B_c	1.0 m			
	Column width	W_c	1.0 m			
	Column height	H_c	6.5 m			
	Number of columns	#C	4			
	Column Spacing	Sp	3.667 m			
Control parameters	Maximum unconfined concrete strain	ε_u	0.003			
	Strain hardening	H_ε	1%			

In a first stage, the incremental pushover analysis is performed for a lateral load assuming the vertical loads (dead load and live load) remain fixed at their mean values. Figure 9 gives the plot for the lateral deflection (Δ) of the pier as a function of the applied lateral load. The responses obtained by setting the input variables at their mean values are given in Table 4. The calculations flag the lateral load at which different limit states are reached. These are:

(1) The load at which the first column reaches its ultimate bending strength, P_1^*;
(2) The load at which the shear capacity is reached in one of the columns, P_s^*;
(3) The load at which a hinge is formed in one of the columns assuming all the columns are unconfined, P_u^* (ductility exhausted);
(4) The load at which the lateral displacement reached a value equal to 2.5% of the column height, P_f^*;
(5) The loads at which a hinge is formed in one of the columns assuming all the columns are confined, P_c^*; and

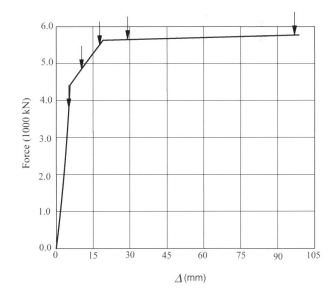

Figure 9. P−Δ for four-column bent.

(6) Vertical load which produces a moment equal to 30% of the ultimate moment capacity of one column, P_v^*.

Note that the calculations express the loads at which these limit states are reached. These P_i^* values represent the capacity of the system to resist failure in a given limit state, i. Failure occurs in a given mode when the applied load, P, is higher than the load capacity, P_i^*, for the limit state (failure mode) being considered.

5.3. Reliability analysis

The reliability-calibration of the system redundancy factors requires the knowledge of the mean and standard deviation (or COV) of the capacity of the structure to resist first-member failure and the ultimate capacity. In addition, the parameters of the expected loads are needed. Table 3 summarises the random variables that affect the determination of the capacity of the substructure. The bias, COV and probability distribution types for concrete member strengths under lateral forces and for live loads and dead weights are obtained from the values used by Nowak (1999) during the calibration of the AASHTO LRFD code. The statistical properties for the foundation stiffness are obtained from the work of Becker (1996). The geometric properties of the bridge columns, such as column height and spacing, are assumed to be well controlled and thus taken to be deterministic.

If an explicit closed-form expression describing the relationship between these variables and the substructure capacity were available, then the reliability calculations can be easily performed using a FORM algorithm. These calculations would provide the probability of failure, and the reliability index, β. Since such closed-form expressions are not available, an approximate expression can be obtained using the results of a perturbation analysis implemented into Equations (12) and (13) (Ghosn et al. 1993).

The results of Table 4 are obtained when the random variables representing the substructure properties and applied loads are set at their mean values. In addition to the column capacities, assuming confined columns, M_c, or unconfined columns, M_u, the random variables consist of the foundation stiffnesses, K_i, the

foundation moment capacity, M_f, as well as the vertical dead loads, DL, WL and γ_w, and the vertical live load, L_{max} (representing the summation of Lane load, L_L, Truck load, L_t and Impact I). The analysis produces a value of the horizontal load P_i^* that will generate a failure in mode i. Following the first analysis, which assumes that all the random variables are fixed at their mean values, the random variables are changed one at a time, to values equal to the mean minus one standard deviation, then to values equal to the mean plus one standard deviation. Hence, for the seven random variables, a total of 14 additional deterministic analyses are performed, where, for each analysis, one of the variables is perturbed from its original value. All the foundation stiffnesses are changed simultaneously because these variables are assumed to be fully correlated. In addition, the concrete strength and steel yielding stress are combined together to produce the member ultimate moment capacity. For each analysis, the value of the ultimate capacity of the substructure system, P^*, is obtained. For example, Table 5 presents the results obtained for the limit state corresponding to the formation of a hinge in the confined columns designated as P_c^*. The results of these deterministic sensitivity analyses are then used to fit a functional relationship between P^* and the seven random variables M_c, K_i, M_f, DL, WL, γ_w and L_{max}. The results in Table 5 show that changes in the random variable of foundation capacity M_f, did not influence the ultimate capacity, P_c^* at all.

Similar observations are made for the relationship between M_f and P^* for all the failure modes considered. This is because failure in the foundation did not occur for any of the cases analysed. This means that, based on current foundation design procedures,

Table 4.   Results of analysis of four-column bent.

Limit state	Symbol	Four-column bent (kN)
First member failure	P_1^*	3574
Shear failure	P_s^*	5843
Hinge in unconfined member	P_u^*	5499
Functionality, $\Delta = 2.5\%H$	P_f^*	5755
Hinge in confined member	P_c^*	5922
Vertical load which produces 30% of the column ultimate moment capacity	P_v^*	6792

Table 5.   Results of sensitivity analysis.

Variable changed (each by plus or minus one standard deviation)	P_c^* (kN) for four-column bent
No changes (variables are set at their mean values)	5922
$M_c + \sigma$	6463
$M_c - \sigma$	5274
$K_i + \sigma$	5864
$K_i - \sigma$	6136
$M_f + \sigma$	5922
$M_f - \sigma$	5922
DL $+ \sigma$	5950
DL $- \sigma$	5872
WL $+ \sigma$	5927
WL $- \sigma$	5925
$\gamma_w + \sigma$	5906
$\gamma_w - \sigma$	5896
$L_{max} + \sigma$	6026
$L_{max} - \sigma$	5899

the effect of the foundation capacity is not critical for the substructure configuration studied. Changes in foundation stiffness, K_i, produced a barely noticeable change in P^*. Similarly, the effects of the column dead weight represented by, γ_w, and the wearing surface, WL, are not important. From Table 5, the single most sensitive variable affecting system lateral capacity is the individual column capacity M_c. Based on the results shown in Table 5, one may find an approximate expression for P_c^* for the four-column bent as:

$$P_c^* = 971 + 1.14M_c - 0.00776K_i + 0.068\text{DL} + 0.00267\text{WL} + 5.340\gamma_w + 0.454L_{max}, \quad (17)$$

where the variables P_c^*, M_c, K_i, DL, WL, γ_w and L_{max} are expressed in kN and m. The mean of P_c^* can also be calculated by substituting the mean values of M_c, K_i, DL, WL, γ_w and L_{max} into Equation (17). For the example data, this will produce a mean of $P^* = 5925$ kN, which is very close to the value $P_c^* = 5922$ kN provided in Table 4. Note that a negative coefficient associated with the variable K_i indicates that the ultimate capacity decreases when the foundation stiffness increases. This is due to the fact that the change in the foundation stiffness changes the load distribution of the system leading to the over-loading of one member, which will cause it to fail at a lower load, thus producing a lower ultimate capacity. The ultimate capacity in this example is defined as the exhaustion of the ductility in one of the bridge columns. The positive coefficients in Equation (17) show that increasing the dead and live loads improves the strength of this system because of the increased moment capacity of each column when an additional axial load is added on the moment interaction diagram for points below the balanced point. The standard deviation of P_c^*, σ_{pc}, can be approximated from Equation (17) to give:

$$\sigma_{pc}^2 = (1.14\sigma_{Mc})^2 + (0.0076\sigma_{Ki})^2 + (0.068\sigma_{DL})^2 + (0.00267\sigma_{WL})^2 + (5.307\sigma_{\gamma w})^2 + (0.454\sigma_{Lmax})^2, \quad (18)$$

where σ_{Mc}, σ_{ki}, σ_{DL}, σ_{WL}, $\sigma_{\gamma w}$ and σ_{Lmax} are the standard deviations of M_c, K_i, DL, WL, γ_w and L_{max}, respectively. Using the data of Table 3, Equation (18) produces a standard deviation for P_c^* on the order of 614 kN or a COV on the order of 10%. This 10% COV should be increased to account for the modelling uncertainties when performing a nonlinear finite-element analysis. It is herein assumed that the effects

of the modelling uncertainties are associated with a COV = 8%. This 8% is an average value for the COVs associated with the professional analysis factor that accounts for the method of analysis (Ellingwood et al. 1980). The overall COV associated with estimating the strength of the bridge substructure becomes 13% $= \sqrt{(0.10)^2 + (0.08)^2}$. Assuming a COV of 50% for the applied lateral loads, which is approximately what has been traditionally used for seismic forces (Ellingwood et al. 1980), the reliability index for the substructure system for first member failure can then be obtained from:

$$\beta_{member} = \frac{\ln \frac{\overline{P_1^*}}{\overline{P}}}{\sqrt{V_P^2 + V_{P^*}^2}} = \frac{\ln \frac{3574}{\overline{P}}}{\sqrt{0.5^2 + 0.13^2}}. \quad (19)$$

Similar calculations can be performed for the reliability index for all the other limit states. For the ultimate bending capacity of substructures with confined columns, the reliability index becomes:

$$\beta_{ultimate} = \frac{\ln \frac{\overline{P_c^*}}{\overline{P}}}{\sqrt{V_P^2 + V_{P^*}^2}} = \frac{\ln \frac{5922}{\overline{P}}}{\sqrt{0.5^2 + 0.13^2}}. \quad (20)$$

The difference between the system and most critical member reliability indexes for the four-column bent is obtained from Equation (10) as $\Delta\beta_u = 0.97$. Note that the change in reliability indexes, $\Delta\beta_u$, is not a function of the mean value of the lateral load $\overline{P}$, as the subtraction of the logarithmic terms eliminates the denominator of both equations. Similar calculations are performed for the limit states for this four-column bent and other bents with different numbers of columns, column heights and foundation types. The results of these calculations are used to calibrate system factors, as will be explained in the next section.

5.4. Calibration of system factors for bridge substructures

The procedure followed to calibrate system factors, ϕ_s, for bridge substructures is similar to that used for the superstructures (Ghosn and Moses 1998, Liu et al. 2001). Specifically, following the review of many typical substructure configurations, with one-column, two-column and multi-column bents, supported by spread footings, drilled shafts or pile systems, in soft and stiff soils, it has been determined that substructures having $\Delta\beta_u = 0.50$ or higher would provide adequate substructure redundancy. Similarly, adequately redundant systems were those producing $\Delta\beta_f \geq 0.50$, whereby the functionality limit state is reached when the lateral deflection is equal to column height/50. For the damaged limit state, a damaged substructure that has lost a main load-carrying member should still produce a $\Delta\beta_d \geq -2.50$ to be considered sufficiently robust. For example, the four-column bridge bent analysed earlier has a relative reliability index for the ultimate limit state of confined columns $\Delta\beta_u = 0.97$, which is greater than 0.50, indicating that this bridge substructure's redundancy level is more than adequate. Hence, the safety level of this substructure's components may be decreased without affecting the overall system safety. This could be achieved by applying a system factor during the design process greater than 1.0 in Equation (16), so that bridges with four columns would be allowed to have a lower overall safety than bridges with fewer columns. The value of the system factor, ϕ_s, to be used in Equation (16) should be such that the reliability index of the member is decreased by $0.97 - 0.50 = 0.47$. This implies that the system factor, ϕ_s, associated with this configuration and for the limit state analysed should be equal to 1.27. By performing similar analyses for different bridge configurations and different limits states for all the configurations, a set of system factors for different substructure types and soil conditions is obtained, as shown in Table 6.

The system factors provided in Table 6 have been calibrated in NCHRP report 458 (Liu *et al.* 2001) to satisfy the system safety requirements for a set of typical bridge substructure configurations. This set includes bridges with concrete columns varying in height between 3.5 and 18 m and a vertical rebar reinforcement ratio of 1.85 to 2.3%. Soft soils are defined as soils that produce a standard penetration test (SPT) blow count (number of blows per 0.035 m penetration into the soil) of $N = 5$; normal soils are those with $N = 15$; and stiff soils are those with $N = 30$ or higher.

Members in shear, as well as all joints and connections, are assigned a system factor $\phi_s = 0.80$. This assumes that the member resistance factor, ϕ, was calibrated to satisfy a target member reliability index $\beta_{member} = 3.5$. Since a shear failure or a connection failure in a single member would cause the failure of the complete system, the application of a system factor $\phi_s = 0.80$ will increase the reliability index of the member and also that of the system so that $\beta_{member} = \beta_{system} > 4.0$, which is the value stipulated above for the target reliability of intact substructure systems.

According to Table 6, the design of abutments, piers and walls should be investigated for structural safety at the strength and extreme event limit states for each structural component and joint using Equation (16). For the substructures of bridges classified to be critical or susceptible to brittle damage, the system factor, ϕ_s, should be calculated using the direct analysis approach described below. For the substructures of bridges classified to be essential, the system factor is calculated as $\phi_s = \min(\phi_{sc}, \phi_{sf})$ for substructures with confined concrete members and $\phi_s = \min(\phi_{su}, \phi_{sf})$ for substructures with unconfined concrete members.

For the substructures of all other bridges, $\phi_s = \phi_{sc}$ for substructures with confined members and $\phi_s = \phi_{su}$ for substructures with unconfined members, where ϕ_{su} is the system factor for the ultimate capacity of substructures with unconfined concrete members; ϕ_{sc} is the system factor for the ultimate capacity of substructures with confined concrete members; and ϕ_{sf} is the system factor for the bridge substructure functionality.

Confined concrete columns are those that satisfy the provisions of Article 5.10 of the AASHTO LRFD (2007). Recommended values for ϕ_{su}, ϕ_{sc} and ϕ_{sf} for

Table 6. System factors for bridge substructures.

Substructure and member types	Foundation Soil type	Spread footing All soil	Drilled shaft			Piles		
			Soft	Normal	Stiff	Soft	Normal	Stiff
Walls, abutments and one-column bents in bending	ϕ_{su}	0.80	0.80	0.80	0.80	0.80	0.80	0.80
	ϕ_{sc}							
	ϕ_{sf}							
Two-column bents in bending	ϕ_{su}	0.95	0.85	0.90	0.95	0.90	0.95	1.00
	ϕ_{sc}	1.00	0.95	1.05	1.15	1.05	1.05	1.05
	ϕ_{sf}	1.00	0.80	0.85	1.00	0.95	0.95	0.95
Multi-column bents in bending	ϕ_{su}	1.00	0.90	0.95	1.00	0.95	1.00	1.05
	ϕ_{sc}	1.05	1.00	1.15	1.20	1.10	1.10	1.10
	ϕ_{sf}	1.05	0.80	1.05	1.20	1.05	1.05	1.05
All members in shear	ϕ_{su}	0.80	0.80	0.80	0.80	0.80	0.80	0.80
	ϕ_{sc}							
	ϕ_{sf}							
Bar pullout and other joint failures	ϕ_{su}	0.80	0.80	0.80	0.80	0.80	0.80	0.80
	ϕ_{sc}							
	ϕ_{sf}							

typical substructures with columns, piers, walls and abutments founded on spread footings, drilled shafts or piles are specified in Table 6 for soft, normal and stiff soils. For bridges with non-typical configurations, the direct analysis approach should be used. Bridges susceptible to brittle damage include substructures with members that are exposed to collisions from ships, vehicles and debris carried by swelling streams and rivers.

The use of more stringent criteria for critical and essential bridges is consistent with current trends to use performance-based design in bridge engineering practice. A minimum value of $\phi_s = 0.80$ is recommended, but ϕ_s should not be taken as greater than 1.20.

5.5. Push-over analysis of critical and non-typical substructure configurations

For bridges classified to be critical, and for bridges not covered in Table 6, the ϕ_s for structural components of a substructure can be calculated from the results of a nonlinear pushover analysis using Equation (21), where the system factor ϕ_s is obtained from:

$$\phi_s = \min\left(\frac{R_u}{1.20}, \frac{R_f}{1.20}, \frac{R_d}{0.50}\right). \qquad (21)$$

The values of R_u, R_f and R_d are obtained from Equation (9), where the values for the load multipliers are the results of a nonlinear pushover analysis, with LF_u as the lateral load factor that causes the failure of the substructure; LF_f as the lateral load factor that causes the total lateral deflection of the substructure to reach a value equal to average clear column height/50; LF_d as the lateral load factor that causes the failure of a damaged substructure; and LF_1 as the lateral load factor that causes the first member of the intact substructure to reach its limit capacity. The nonlinear pushover analysis of critical bridges should be performed by accounting for the nonlinear behaviour of the substructure and considering soil–foundation flexibility. The analysis requires the availability of a nonlinear analysis program that provides a lateral load versus lateral deflection curve and that adequately models the nonlinear behaviour of the substructure components up to crushing of concrete, rupture of steel or the formation of a collapse mechanism. The program should also be able to model the stiffness and the soil–foundation system by either the use of equivalent springs or actual modelling of the nonlinear foundation. The nonlinear pushover analysis should be executed by applying the factored loads (lateral and vertical live and dead loads) on a structural model of the substructure and incrementing the lateral loads until the failure of the first member. Member failure

criteria must be consistent with the bridge design specifications under effect. The factor by which the original lateral load is multiplied to cause the failure of the first member is defined as LF_1.

The nonlinear analysis is then continued beyond the failure of the first member until the lateral displacement at the top of the bent reaches a value equal to average clear column height/50. This displacement limit is defined as the functionality limit state. The factor by which the original load is multiplied to reach this functionality limit is defined as LF_f.

The analysis is further continued until one member reaches its maximum strain and concrete crushing ensues, or until a hinge collapse mechanism occurs. Unconfined concrete members crush at a strain of 0.003. Confined members crush at a strain of 0.015. These cases define the ultimate capacity limit state. The load factor by which the original lateral load is multiplied to reach the ultimate capacity is defined as LF_u.

The same process is repeated for a model of the damaged bridge substructure whenever consideration of the survival of a damaged bridge is required by the bridge owner (e.g. when the bridge is classified as critical or when a damage situation is considered likely). The damage scenario must be realistic and must be chosen in consultation with the bridge owner. Damage scenarios may include the complete loss of a column that may be subjected to risk of brittle failure from collisions by ships, vehicles, impact and blast or flooding debris. The analysis of the damaged bridge is executed in the same manner outlined above for the intact structure, but only the ultimate capacity limit states of the damaged bridge need to be checked. The load factor by which the original lateral load is multiplied to reach the ultimate capacity of the damaged structure is defined as LF_d.

Bridge substructures that produce redundancy ratios $R_u = 1.2$, $R_f = 1.2$ and $R_d = 0.5$, or higher, are classified as adequately redundant. Those that do not satisfy these criteria will have system factors ϕ_s less than 1.0 and require higher component safety levels. If bridge redundancy is sufficiently high, a system factor greater than 1.0 may be used. The ratios $R_u = 1.2$, $R_f = 1.2$ and $R_d = 0.5$ correspond respectively to $\Delta\beta_u = 0.5$, $\Delta\beta_f = 0.50$ and $\Delta\beta_d = -2.50$, which were used to calibrate the tabulated system factors for typical substructure configurations.

Satisfying the criteria for R_u, R_f and R_d is recommended for bridges classified to be critical. Satisfying the criteria for only R_u and R_f is recommended for essential bridges. Satisfying the criteria for only R_u is sufficient for all other bridges. The check of R_u verifies that a bridge's ultimate system capacity is at least 20% higher than the load level that will cause the failure of one member. The check of R_f verifies that a bridge's lateral deflection during the application of high loads is still acceptable, allowing the bridge to remain functional for emergency situations. The check of R_d verifies that a damaged bridge is still capable of carrying 50% of the load that a non-damaged bridge can carry before one member of the intact bridge fails.

6. Assessment of the redundancy and robustness of a truss bridge

6.1. Structural analysis procedure

The superstructure of a steel truss bridge is analysed to illustrate how a direct analysis can be applied to evaluate the redundancy and robustness of a non-typical structure for which no tabulated system factors are available. The through-truss bridge selected has two parallel trusses similar to the one shown in Figure 10. Table 7 gives a list of the truss members, along with their cross-sectional areas. The two parallel trusses of the bridge are connected by cross beams and diagonals supporting a concrete deck. The assessment of the truss's redundancy requires a three-dimensional nonlinear finite-element model of the truss. The nonlinear behaviour of the steel truss members is modelled by an elasto-plastic bilinear stress strain curve, assumed to be able to carry a maximum strain of 2×10^{-2} mm/mm. Although steel members may be able to sustain higher strain levels, these would create major local deformations causing failures at the connections. The same behaviour is assumed to hold for both tension and compression members, assuming that sufficient bracing is provided to prevent buckling failures. The cross beams and the deck are assumed to remain in the elastic range.

The incremental analysis is performed by initially placing two side-by-side HS-20 trucks, whose weights are gradually incremented until one member attains the limiting strain level of 0.02 mm/mm. For every load step, the maximum vertical deflection in the truss is obtained. A plot of load factor versus maximum vertical truss displacement is shown in Figure 11. The horizontal shift in the curve, which shows some vertical displacement when the load is zero, represents the displacement due to the dead loads. Curve a in Figure 11 shows that, at a load factor equal to 7.6, the first member (member 29 in Figure 10) reaches its yielding stress of 250 MPa. This load factor is defined as LF_1. At a load factor of 8.5, members 12 and 13 reach their yielding points. At a load factor of 9.0, the maximum vertical displacement in the truss reaches a value of span length/100. This load factor is defined as LF_f. Finally, at a load factor equal to 9.4, the maximum strain in member 29 reaches its specified ultimate value of 0.02 mm/mm. At that point, unloading in member 29 is assumed to occur and the analysis is stopped. This load factor is assumed to be the load capacity of the bridge before major damage occurs and the load factor of 9.4 is defined as LF_u.

To assess the robustness of the truss, the analysis is repeated after totally removing member 29. This

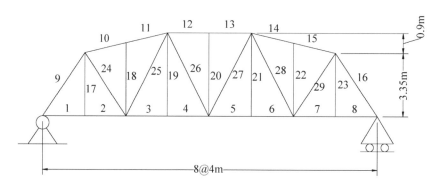

Figure 10. Profile of truss bridge.

Table 7. Cross-sectional areas of truss members.

No.	Area (cm^2)	No.	Area (cm^2)	No.	Area (cm^2)	No.	Area (cm^2)	No.	Area (cm^2)
1	160	7	160	13	280	19	125	25	85
2	160	8	160	14	260	20	85	26	85
3	250	9	245	15	260	21	125	27	85
4	250	10	260	16	245	22	85	28	85
5	250	11	260	17	100	23	100	29	110
6	250	12	280	18	85	24	110		

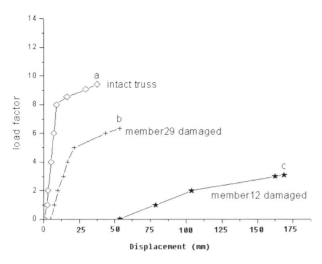

Figure 11. Load factor versus maximum displacement response of truss bridge.

member is the most critical loaded tension member. Removing it from the structural model simulates a damage scenario whereby this member is severely damaged due to external loading or due to fatigue or corrosion. The incremental analysis for the HS-20 AASHTO loading pattern shows that a load factor of 4.2 produces the yielding of member 28. At a load factor of 5.0, member 23 reaches its yielding point. Finally, at a load factor of 6.2, member 23 reaches a maximum strain of 0.02 mm/mm, indicating that unloading started rendering the damaged bridge unsafe for traffic. This load factor for the damaged bridge with member 29 removed is designated as LF_d. Curve b in Figure 11 shows the load factor versus displacement curve for this condition. Another possible damage scenario involves the damage of member 12. If member 12 is removed, the load factor versus displacement curve will be as shown in curve c of Figure 11. For this scenario, the load factor LF_d is equal to 3.23. The final value for LF_d that should be used for the analysis of the robustness of the truss is the lowest value from all possible damage scenarios, which for this bridge gives $LF_d = 3.23$.

6.2. Redundancy and robustness assessment

To assess the truss system's redundancy and robustness, the values of LF_u, LF_f and LF_d obtained from the incremental analysis, which are respectively given as 9.40, 9.0 and 3.23, are compared to the member load factor $LF_1 = 7.21$.

Using Equation (9), the system redundancy measures obtained are $R_u = LF_u/LF_1 = 1.30$, $R_f = LF_f/LF_1 = 1.25$ and $R_d = LF_d/LF_1 = 0.45$. These system redundancy measures are compared to the threshold values of Equation (15), which are equal to 1.30, 1.1 and 0.5. Thus, the redundancy reserve ratios are

obtained as $r_u = 1.30/1.30 = 1.0$, $r_f = 1.25/1.1 = 1.14$ and $r_d = 0.45/0.5 = 0.90$. Since the redundancy reserve ratio corresponding to the damaged limit state $r_d = 0.90$ is less than 1.0, this bridge's configuration is considered to be insufficiently robust. According to Equation (15), a new design of a bridge with a similar configuration should be associated with a system factor $\phi_s = 0.90$. However, since this existing bridge is already in operation, a check of its safety should be performed, as illustrated in the next section.

It should be emphasised that a bridge system, such as this truss bridge, which does not satisfy the redundancy and robustness criteria may still provide adequate levels of system safety for truck traffic if its members are conservatively designed when compared to the minimum code requirements. Therefore, the redundancy reserve ratios should be combined with a check of member capacity to verify whether the overall system safety is still adequate. Checking the member capacity can be performed according to the currently acceptable design criteria. For example, the next section illustrates how member safety is checked using the AASHTO WSD method (AASHTO 2002).

6.3. System safety analysis

The required member capacity according to AASHTO's WSD criteria is used as the basis of the calculations performed in this section to evaluate the overall safety of this bridge that has been shown not to provide sufficient levels of robustness. According to the WSD criteria, the required capacity of a steel member in tension should be:

$$0.55R_{req} = D_n + (1.0 + I)L_{HS20}, \qquad (22)$$

where R_{req} is the required member capacity; D_n is the nominal dead load; I is the impact factor; and L_{HS20} is the design live load effect. Using the results of the linear elastic analysis for member 29, which was found to be the most critical member of the truss, the dead load on this member is $D_n = 552$ kN, the impact factor is $I = 0.22$ and the AASHTO HS-20 live load produces an axial load $L_{HS20} = 300$ kN. These values are entered into Equation (22) to yield a required member capacity R_{req} equal to 1670 kN. Accordingly, the minimum code-stipulated load factor that will correspond to first member failure is:

$$LF_{1req} = \frac{R_{req} - D}{L_{HS\,20}} = 3.72. \qquad (23)$$

This value is compared to the actual load factor LF_1 obtained from the analysis of the actual bridge, which was earlier found to be equal to 7.21. The fact that the required LF_1 is 3.72, while the actual design provides a load factor of 7.21, shows that this bridge is over-designed and it provides a reserve ratio for member 29 of $r_1 = 7.21/3.72 = 1.94$, i.e. member 29 is overdesigned by 94%.

The fact that the critical member of this bridge system is overdesigned by a factor of 1.94, while the redundancy is below the threshold value by a factor of 0.90 indicates that overall system safety is still satisfied by a factor of 1.75 (i.e. 1.94×0.90). Hence, one can conclude that, even though this bridge geometry does not satisfy the redundancy and robustness criteria, the conservativeness of the bridge members imply that the system can still carry sufficient load, even when a critical member is totally damaged by an external loading event. Similar analyses must be performed for each critical member of this bridge.

7. Conclusions

This paper presents a framework for considering structural redundancy during the design and safety assessment of bridge superstructure and substructure systems. The framework consists of a method to penalise designs with insufficient levels of redundancy by requiring that their members be more conservatively designed than allowed by current standard specifications. This could be achieved by applying system factors in the codified member design-check equations, along with traditional member resistance factors. Tables of system factors are developed for typical configurations of multi-girder bridge superstructures and multi-column bridge bents. For typical superstructures, the values of the system factors depend on the number of girders, girder spacing, the presence of diaphragms and if the bridge is critical for security, emergency or national defence. The values of system factors for typical bridge substructures depend on the number of columns, the foundation type, the soil classification and the confinement of the columns, in addition to the classification of the bridge in terms of its criticality for social/survival and security/defence requirements. These tabulated system factors are calibrated using reliability techniques so that non-redundant bridges would provide a minimum level of system safety to meet a target system reliability index. System factors for bridges having non-typical configurations can be derived using a proposed incremental analysis procedure. The application of the incremental analysis approach is illustrated for a truss bridge example.

The development of the proposed methodology is based on a proposed set of objective measures of redundancy that compared system safety to member safety. These measures accounted for system reserve after the brittle or ductile failure of a main load-carrying member. The proposed approach is practical and can be implemented on a routine basis, as demonstrated in the truss bridge example and also by Hubbard et al. (2004). The proposed approach, originally developed for highway bridges, was also extended for applications to railway bridges by Wisniewsky et al. (2006). Further research is needed on the topic of redundancy and robustness of structural systems under uncertainty in connection with time-dependent performance and life-cycle optimization (Frangopol and Liu 2007, Okasha and Frangopol 2009).

References

American Association of State Highway and Transportation Officials (AASHTO), 2002. *Standard specifications for highway bridges.* 17th ed. Washington, DC: AASHTO.

American Association of State Highway and Transportation Officials (AASHTO), 2007. *Load and resistance factor design (LRFD) bridge design specifications.* 3rd ed. Washington, DC: AASHTO.

American Society of Civil Engineers-American Association of State Highway and Transportation Officials (ASCE-AASHTO) Task Committee, 1985. State of the art report on redundant bridge systems. *Journal of Structural Engineering, ASCE,* 111 (12), 2517–2531.

Becker, D.E., 1996. Development of the National Building Code of Canada. 18th Canadian geotechnical colloquium: limit stets design for foundations. *Canadian Geotechnical Journal,* 33, 984–1007.

De, S.R., Karamchandani, A., and Cornell, C.A., 1990. Study of redundancy of near-ideal parallel structural systems. *In*: A.H.-S. Ang, M. Shinozaka, and G.I. Schueller, eds. *Proceedings of ICOSSAR'89,* 7–11 August 1991. San Francisco, CA. *Structural Safety & Reliability.* ASCE, 2, New York, 975–982.

Ellingwood, B., Galambos, T., MacGregor, J.G., and Cornell, C.A., 1980. *Development of a probability-based load criterion for American National Standard.* Washington, DC: National Bureau of Standards, A58, NBS Special Publication 577.

Frangopol, D.M. and Curley, J.P., 1987. Effects of damage and redundancy on structural reliability. *Journal of Structural Engineering, ASCE,* 113 (7), 1533–1549.

Frangopol, D.M. and Nakib, R., 1991. Redundancy in highway bridges. *Engineering Journal,* 28 (1), 45–50.

Frangopol, D.M. and Liu, M., 2007. Maintenance and management of civil infrastructure based on condition, safety, optimization and life-cycle cost. *Structure and Infrastructure Engineering,* 3 (1), 29–41.

Ghosn, M. and Moses, F., 1992. Calibration of redundancy factors for highway bridges. *ASCE Speciality Conference on Probabilistic Mechanics and Structural & Geotechnical Reliability,* Denver, CO, July.

Ghosn, M. and Moses, F., 1998. *Redundancy in highway bridge superstructures.* Washington, DC: National Academy Press, National Cooperative Highway Research Program, NCHRP Report 406, Transportation Research Board.

Ghosn, M., Moses, F., and Khedekar, N., 1993. Response functions and system reliability of bridges. *In*: P.D. Spanos and Y.T. Wu, eds. *Proceedings of IUTAM Symposium,* 7–10 June. San Antonio. *Probabilistic structural mechanics: advances in structural reliability methods.* New York: Springer-Verlag.

Printed and bound by CPI Group (UK) Ltd, Croydon, CR0 4YY

28/10/2024

01780013-0001

Hendawi, S. and Frangopol, D.M., 1994. System reliability and redundancy in structural design and evaluation. *Structural Safety*, 16 (1–2), 47–71.

Hubbard, F., Shkurti, T., and Price, K.D., 2004. Marquette interchange reconstruction: HPS twin box girder ramps. *In*: *International Bridge Conference*, Pittsburgh, PA.

Liu, D., Ghosn, M., Moses, F., and Neuenhoffer, A., 2001. *Redundancy in highway bridge substructures*. Washington, DC: National Academy Press, National Cooperative Highway Research Program, NCHRP Report 458, Transportation Research Board.

Mattock, A.H., 1964. Rotational capacity of hinging region in reinforced concrete beams. *In*: *International symposium on flexural mechanics of reinforced concrete*, PCA, Miami, FL.

Melchers, R.E., 1999. *Structural reliability analysis and prediction*. 2nd ed. New York: Wiley.

National Construction Safety Team (NCST) on the Collapse of the World Trade Center Towers, 2005. *Federal building and fire safety investigation of the World trade Center disaster*. Washington, DC: US Department of Commerce, Draft Report, NIST NCSTAR 1.

Nowak, A.S., 1999. *Calibration of LRFD bridge design code*. Washington, DC: National Academy Press, NCHRP report 368, Transportation Research Board.

Okasha, N.M. and Frangopol, D.M., 2009. Time-variant redundancy of structural systems. *Structure and Infrastructure Engineering*, DOI: 10.1080/15732470802664514.

Schilling, C.G., 1989. *A unified autostress method*. Washington, DC: American Iron and Steel Institute, Report 51.

Thoft-Christensen, P. and Baker, M.J., 1982. *Structural reliability and its applications*. New York: Springer-Verlag.

Wisniewsky, D.F., Casas, J.R., and Ghosn, M., 2006. Load-capacity evaluation of existing railway bridges based on robustness quantification. *Journal of Structural Engineering International*, 16 (2), 161–166.

Zokaie, T., Osterkmap, T.A., and Imbsen, R.A., 1991. *Distribution of wheel loads on highway bridges*. Washington, DC: Transportation Research Board, NCHRP Project 12–26. Final report.

Effects of post-failure material behaviour on redundancy factor for design of structural components in nondeterministic systems

Benjin Zhu and Dan M. Frangopol

This paper investigates the effects of post-failure material behaviour on redundancy factor for the design of structural components in nondeterministic systems. The procedure for evaluating the redundancy factors of components of ductile and brittle nondeterministic systems is demonstrated using systems consisting of two to four components. The effects of the number of brittle components in a mixed (ductile–brittle) system and the post-failure behaviour factor on the redundancy factor are also studied using these systems. An efficient approach for simplifying the system model in the redundancy factor analysis of brittle systems with a large number of components is proposed. In order to generate standard tables to facilitate the design process, the redundancy factors and the associated component reliability indices of nondeterministic ductile and brittle systems with large number of components are calculated considering three correlation cases and two probability distribution types. Finally, a bridge example is used to demonstrate the application of the redundancy factor in the design of steel girders taking into account their post-failure behaviour.

Notations

A, B, D, F:	event
$E_c(R)$:	mean resistance of a single component
$E_{cs}(R)$:	mean resistance of a component in a system
g:	performance function
G:	performance function matrix
M:	bending moment
N:	number of components in a system
N_s:	number of series components in a system
N_p:	number of parallel components in a sub-parallel system
N:	normal distribution
LN:	lognormal distribution
P:	load
R:	resistance of a component
V:	coefficient of variation of a random variable
β_c:	reliability index of a single component
β_{cs}:	reliability index of a component in a system
β_{sys}:	reliability index of a system
η_R:	redundancy factor
ϕ_s:	system factor
ϕ_R:	system factor modifier
ρ:	correlation coefficient
δ:	post-failure behaviour factor

1. Introduction

Structural systems are designed to maintain adequate levels of serviceability and safety. In general, redundancy is introduced to some extent in the design of structures and infrastructures to ensure the required serviceability and safety levels. Research on structural redundancy has been extensively performed in the recent decades (Cavaco, Casas, & Neves, 2013; Frangopol & Curley, 1987; Frangopol & Klisinski, 1989a, 1989b; Frangopol & Nakib, 1991; Fu & Frangopol, 1990; Ghosn & Moses, 1998; Ghosn, Moses, & Frangopol, 2010; Liu, Ghosn, Moses, & Neuenhoffer, 2001; Rabi, Karamchandani, & Cornell, 1989; Tsopelas & Husain, 2004; Wen & Song, 2004). Several redundancy measures were proposed to quantify the ability of a structural system to redistribute the applied loads after the failure of its critical members. Frangopol and Curley (1987) defined the system redundancy as the ratio of the reliability index of the intact system, β_{intact}, to the difference between β_{intact} and the reliability index of the damaged system, $\beta_{damaged}$. In Rabi et al. (1989), the redundancy is defined by comparing the probability of system collapse to the probability of member failure. Ghosn and Moses (1998) proposed three reliability measures of redundancy that are defined in terms of relative reliability indices. Because redundancy is regarded as an important performance indicator, studies on the evaluation of system redundancy in real structures have been conducted. Tsopelas and Husain (2004) studied the system redundancy in reinforced concrete buildings. Wen and Song (2004) investigated the redundancy of special moment resisting frames and dual systems under seismic excitations. Kim (2010) evaluated the redundancy of steel box-girder bridge by using a nonlinear finite element model.

Although redundancy is a desired structural property, it is not common to find guidance in structural design codes on how to incorporate redundancy in the design process quantitatively. With better understanding of structural behaviour and system reliability-based design (Ang & Tang, 1984; Thoft-Christensen & Baker, 1982; Thoft-Christensen & Murotsu, 1986), Load and Resistance Factor Design (LRFD) replaced Allowable Stress Design within the last decades. LRFD approach (Babu & Singh, 2011; Ellingwood, Galambos, MacGregor, & Cornell, 1980; Hsiao, Yu, & Galambos, 1990; Lin, Yu, & Galambos, 1992; Paikowsky, 2004) provides an improved rational basis to incorporate system reliability and redundancy concepts as it considers the uncertainties associated with the resistance and load effects using separate factors. American Association of State Highway and Transportation Officials LRFD Bridge Design Specifications (AASHTO, 1994, 2012) are examples of a design code that incorporates redundancy in the design process quantitatively.

These specifications account for redundancy by means of a factor (η_R) from the load effects side in the strength limit state equation for a bridge component. Based on a rough classification of component redundancy level, generic values (0.95, 1.0 and 1.05) were recommended for this redundancy factor (AASHTO, 1994, 2012). Nevertheless, selecting the appropriate value for this factor is a very complex task. Furthermore, this redundancy factor does not account for several parameters including system modelling type (i.e. series, parallel), correlation among the resistances of components and post-failure behaviour of components (Ghosn & Moses, 1998; Hendawi & Frangopol, 1994). These parameters must be considered in establishing redundancy factors for design.

Zhu and Frangopol (2014b) investigated the reliability of systems considering the post-failure behaviour (i.e., ductile, brittle) of their components. In addition, Zhu and Frangopol (2014a) provided redundancy factors for systems of various types (i.e. series, parallel) associated with different correlation cases. Redundancy factor, as a multiplier of the resistance rather than the load effects, was defined by Zhu and Frangopol (2014a) as the ratio of the mean resistance of a component in a system when the system reliability is prescribed to the mean resistance of the same component when its reliability index is the same as that of the system. However, the post-failure behaviour of components was not considered by Zhu and Frangopol (2014a) in the evaluation of this redundancy factor. This is the main topic of this paper.

As indicated previously, Zhu and Frangopol (2014b) investigated the effects of post-failure material behaviour on system reliability without considering its effects on system redundancy. The post-failure behaviour of components should be considered in establishing the system redundancy factors for several reasons. First of all, failure of bridge systems is associated with inelastic deformations and the ability of components carrying loads beyond their elastic limits should be considered in evaluating redundancy factors. Second, systems that consist of ductile components are more redundant than those with brittle components. In reality, structural systems may assemble a combination of ductile and brittle components. The material behaviour of each component must be incorporated in redundancy evaluation of the overall system. Finally, the availability of alternative load paths within a system depends on the ductility capacity of its components. Identifying the redistribution of the loads in a system with excessively loaded components and the new load paths after partial failure requires the information on component post-failure behaviour. Therefore, it is essential to consider the post-failure behaviour of components in determining redundancy factors.

This paper investigates the redundancy factor of different systems considering the post-failure behaviour of components. Nondeterministic systems consisting of two to four components are used to illustrate the procedure for calculating the redundancy factors of ductile and brittle systems and to study the effects of the number of brittle components in a mixed (ductile-brittle) system and the post-failure behaviour factor on the redundancy factor. Most structures in practical cases consist of dozens or hundreds of components. The redundancy factors and the associated component reliability indices of nondeterministic ductile and brittle systems with large number of components are calculated with respect to three correlation cases and two probability distribution types for the purpose of generating standard tables to facilitate the design process. Finally, a bridge example is investigated to demonstrate the application of the obtained ductile redundancy factors in the design of steel girders.

It should be noted that although the redundancy of parallel systems considering post-failure behaviour have been studied in Hendawi and Frangopol (1994) and Okasha and Frangopol (2010), the maximum number of components of the ductile or brittle systems analysed was four while the parallel systems investigated in this paper consist of a large number of components (i.e. up to 500). Because the complexity of the redundancy analysis increases with the number of components (especially for the brittle parallel system), an approach for estimating the redundancy factor of the brittle parallel system with large number of components is proposed. In addition, the redundancy factor of mixed systems consisting of both ductile and brittle components that were not investigated in previous studies are analysed in this paper.

2. Brief review of the proposed redundancy factor

Before starting the evaluation of the redundancy factor for ductile and brittle systems, its definition and the procedure for calculation are briefly reviewed in this section. A redundancy factor that accounts for the redundancy from the resistance side in the strength limit state was proposed in Zhu and Frangopol (2014a). It was defined as the ratio of the mean resistance of a component in a system when the system reliability is prescribed to the mean resistance of the same component when its reliability index is the same as that of the system; the effects of several parameters on the redundancy factor were studied and conclusions were drawn from the sensitivity analysis.

The procedure for calculating the redundancy factor is summarised as follows:

(1) Given the probability distribution type of resistance R and load P, the mean value of load $E(P)$, the coefficients of variation of resistance and load $V(R)$ and $V(P)$ and the predefined component reliability index β_c, determine the mean resistance of the component $E_c(R)$;

(2) Given the correlations among the resistances of components and the prescribed system reliability index β_{sys}, evaluate the mean resistance of the components in the system $E_{cs}(R)$;

(3) Calculate the redundancy factor $\eta_R = E_{cs}(R)/E_c(R)$.

The redundancy factor η_R proposed in Zhu and Frangopol (2014a), the system factor modifier ϕ_R proposed by Hendawi and Frangopol (1994), the system factor ϕ_s proposed by Ghosn and Moses (1998) and the factor relating to redundancy in strength limit state of the AASHTO bridge design specifications are of the same nature because all of them serve as a reward or penalty factor. Bridges with sufficient redundancy are rewarded by allowing their components to have less conservative designs while bridges that are non-redundant are required to have higher component capacities.

However, the definitions of these factors are different. The system factor in Ghosn and Moses (1998) is calculated by comparing the strength reserve ratios associated with different limit states with the required redundancy ratios that are determined through a system reliability calibration to provide adequate redundancy. The system factors are calibrated for girder beam bridges with up to 12 beams and box beam bridges with up to 11 box girders. Three different values based on a general classification of the redundancy levels are recommended for the factor relating to redundancy in the AASHTO specifications (AASHTO, 1994, 2012). As mentioned previously, the redundancy factor proposed in Zhu and Frangopol (2014a) is defined as the ratio of the mean resistance of a component when the system reliability index is prescribed to the mean resistance of the component when its reliability index is

equal to the prescribed system reliability (e.g. 3.5). Compared with other factors, the redundancy factor proposed in Zhu and Frangopol (2014a) is more rational and specific because it takes into account the effects of several parameters, such as system type, correlation among the resistances of components, number of components in the system (up to 500) and probability distribution type of loads and resistances.

3. Redundancy factors of systems considering post-failure material behaviour

The systems investigated in Zhu and Frangopol (2014a) do not consider the post-failure behaviour of components. The limit state equations of components in the systems are similar to that of a single component. The failure modes of systems accounting for the post-failure material behaviour of their components are affected not only by the system type but also by the failure sequence of components. Therefore, the step for identifying the failure modes and the associated limit state equations is more complicated than that associated with the systems without considering the post-failure behaviour. Several systems consisting of two, three and four components are used herein as examples to illustrate the process of evaluating the redundancy factors of systems considering post-failure behaviour.

3.1. Redundancy factors of ductile systems

As mentioned previously, the first step in determining the redundancy factor in a system is to find the mean resistance of its component when the component reliability is prescribed as $\beta_c = 3.5$. Consider a single ductile component whose resistance R and load P are modelled as normally distributed random variables. The coefficients of variation of resistance and load $V(R)$ and $V(P)$, and the mean value of load $E(P)$ are assumed to be 0.05, 0.3 and 10, respectively. Therefore, the mean resistance of component $E_c(R)$ is found to be 21.132.

For a system consisting of two ductile components which are identical with the single component just mentioned, two different systems can be formed: series and parallel. Because failure of the series system can be caused by failure of any component, the redundancy factor of series system is not affected by the post-failure behaviour of the components. Therefore, the evaluation of redundancy factors in this paper is mainly focused on the parallel and series-parallel systems.

For a two-component parallel system subject to load $2P$, the resistances of its ductile components are denoted as R_1 and R_2, respectively. Three correlation cases among the resistances of components are considered herein: (a) $\rho(R_1, R_2) = 0$, no correlation; (b) $\rho(R_1, R_2) = 0.5$, partial

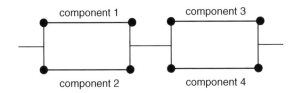

Figure 1. Four-component series-parallel system.

correlation and (c) $\rho(R_1, R_2) = 1.0$, perfect correlation. The statistical parameters associated with R and P are the same as those associated with the single component mentioned previously. By using the Monte Carlo Simulation (MCS) based method described in Zhu and Frangopol (2014a), the mean resistances $E_{cs}(R)$ of ductile components in the two-component parallel system associated with the three correlation cases are found to be 20.810, 20.950 and 21.132, respectively. The corresponding redundancy factors η_R are 0.985, 0.991 and 1.0, respectively. Consequently, the reliability indices of components in the system β_{cs} associated with three correlation cases are 3.40, 3.45 and 3.50, respectively.

Next, the three- and four-component parallel system and the 2×2 series-parallel system (Figure 1) are studied. The loads applied on these systems are $3P$, $4P$ and $2P$, respectively. By performing the same procedure used in the two-parallel system, the mean resistances, redundancy factors and reliability indices of components associated with three- and four-component parallel and series-parallel systems are presented in Table 1.

It is noticed from the results associated with two- to four-component systems that (a) as the number of components in the parallel system increases, the redundancy factor and component reliability index decrease slightly in the no correlation and partial correlation cases; (b) increasing the correlation among the resistances of components leads to higher redundancy factor and component reliability index in the parallel system and (c) compared with the four-component parallel system, the redundancy factor and component reliability index associated with the series-parallel system are higher in the no correlation and partial correlation cases.

3.2. Redundancy factors of brittle systems

Contrary to ductile components, brittle components do not take loads after their fracture failure; therefore, the applied loads will distribute to other remaining components in brittle systems. Due to this property, different failure sequences lead to different load distributions and thus different failure modes in brittle systems. All the possible failure modes must be accounted for and the associated limit state equations need to be identified to determine the redundancy factors. The two-component parallel system described in the previous section is used herein to demonstrate the procedure for calculating the redundancy factors of brittle systems.

Assuming both components are brittle, two different failure modes are anticipated and their respective limit state equations are given as follows:

$$g_1 = R_1 - P = 0, \quad g_3 = R_2 - 2P = 0 \quad (1)$$

and

$$g_2 = R_2 - P = 0, \quad g_4 = R_1 - 2P = 0. \quad (2)$$

Assuming the same statistical parameters of the normally distributed resistances and load as those described previously (e.g. $V(R_i) = 0.05$; $V(P) = 0.3$), the mean resistances of brittle components in the two-component parallel system associated with the three correlation cases are 21.585 if $\rho(R_i, R_j) = 0$; 21.481 if $\rho(R_i, R_j) = 0.5$ and 21.132 if $\rho(R_i, R_j) = 1.0$. The associated redundancy factors are 1.021, 1.017 and 1.0, respectively.

Similarly, the failure modes of three-component parallel systems can be identified, as shown in Figure 2. The limit state equations associated with all the failure modes are:

$$g_1 = R_1 - P = 0, \quad g_2 = R_2 - P = 0,$$
$$g_3 = R_3 - P = 0, \quad (3)$$

$$g_4 = R_2 - 1.5P = 0, \quad g_5 = R_3 - 1.5P = 0,$$
$$g_6 = R_1 - 1.5P = 0, \quad (4)$$

Table 1. $E_{cs}(R)$, η_R and β_{cs} of three- and four-component ductile systems associated with normal distribution.

	System		
Correlation	Three-component parallel system $E_{cs}(R)$; η_R; β_{cs}	Four-component parallel system $E_{cs}(R)$; η_R; β_{cs}	Four-component 2×2 series-parallel system $E_{cs}(R)$; η_R; β_{cs}
$\rho(R_i, R_j) = 0$	20.699; 0.980; 3.37	20.660; 0.978; 3.36	21.160; 1.001; 3.51
$\rho(R_i, R_j) = 0.5$	20.910; 0.989; 3.44	20.893; 0.989; 3.43	21.231; 1.005; 3.53
$\rho(R_i, R_j) = 1$	21.132; 1.000, 3.50	21.132; 1.000; 3.50	21.132; 1.000; 3.50

Note: $V(R) = 0.05$; $V(P) = 0.3$; $\beta_c = 3.5$; $\beta_{sys} = 3.5$; $E_c(R) = 21.132$.

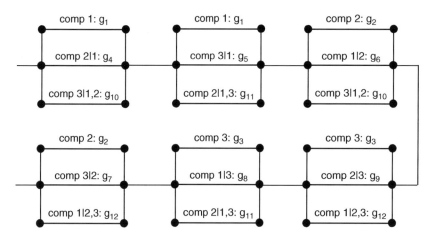

Figure 2. Failure modes of three-component brittle parallel system.

$$g_7 = R_3 - 1.5P = 0, \quad g_8 = R_1 - 1.5P = 0,$$
$$g_9 = R_2 - 1.5P = 0,$$
(5)

$$g_{10} = R_3 - 3P = 0, \quad g_{11} = R_2 - 3P = 0,$$
$$g_{12} = R_1 - 3P = 0.$$
(6)

The redundancy factors of the three-component parallel system associated with three correlation cases when R and P follow normal distribution are found to be 1.033 if $\rho(R_i, \ R_j) = 0$; 1.026 if $\rho(R_i, \ R_j) = 0.5$ and 1.0 if $\rho(R_i, \ R_j) = 1.0$.

It is observed from the results related to the two- and three-component brittle systems that (a) the redundancy factor of the parallel system becomes smaller as the correlation among the resistances of components increases and (b) in the no correlation and partial correlation cases, the redundancy factors associated with the two-component parallel system are less than those associated with the three-component parallel system.

3.3. Redundancy factors of mixed systems

The systems investigated previously consist of only ductile or brittle components. However, there are some cases where both types of material behaviour are included in the system. One of the examples is the steel truss railway bridge in Kama

River of Russia. Its superstructure consists of multi-span steel trusses while its substructure has many single column piers that are made of stones. Therefore, it is necessary to study the redundancy factors of systems having both ductile and brittle components (called 'mixed systems'). Mixed systems consisting of two, three and four components are used herein to investigate the redundancy factors.

For the two-component mixed parallel system, there is only one combination possible: one component is ductile and the other one is brittle (denoted as '1 ductile & 1 brittle'). As more components are included in the mixed system, the number of combinations increases. For the three-component parallel system, two mixed systems are considered: 1 ductile & 2 brittle and 2 ductile & 1 brittle. Similarly, three mixed systems can be formed for four-component parallel system: 1 ductile & 3 brittle, 2 ductile & 2 brittle and 3 ductile & 1 brittle. For the four-component 2×2 series-parallel system, there are two combinations associated with the 2 ductile & 2 brittle case: (a) 2 ductile & 2 brittle Case A, where two ductile components are located in the same sub-parallel system and (b) 2 ductile & 2 brittle Case B, where two ductile components are located in two sub-parallel systems, as shown in Figure 3. Therefore, four different mixed systems can be formed for the 2×2 series-parallel system: 1 ductile & 3 brittle, 2 ductile & 2 brittle Case A, 2 ductile & 2 brittle Case B and 3 ductile & 1 brittle.

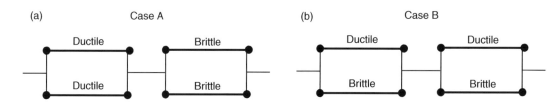

Figure 3. Four-component series-parallel systems: (a) 2 ductile & 2 brittle Case A and (b) 2 ductile & 2 brittle Case B.

Table 2. $E_{cs}(R)$, η_R and β_{cs} of mixed systems associated with the case $\rho(R_i, R_j) = 0$ when R and P are normal distributed.

System		$E_{cs}(R)$	η_R	β_{cs}
2-Component parallel system	1 ductile & 1 brittle	21.280	1.007	3.55
3-Component parallel system	1 ductile & 2 brittle	21.630	1.024	3.65
	2 ductile & 1 brittle	21.300	1.008	3.55
4-Component parallel system	1 ductile & 3 brittle	21.850	1.034	3.71
	2 ductile & 2 brittle	21.640	1.024	3.65
	3 ductile & 1 brittle	21.319	1.009	3.56
4-Component series-parallel system (2 × 2 SP system)	1 ductile & 3 brittle	21.850	1.034	3.71
	2 ductile & 2 brittle Case A	21.680	1.026	3.66
	2 ductile & 2 brittle Case B	21.680	1.026	3.66
	3 ductile & 1 brittle	21.440	1.015	3.59

Note: $V(P) = 0.3$; $V(R) = 0.05$; $\beta_c = 3.5$; $\beta_{sys} = 3.5$; $E_c(R) = 21.132$.

Assuming the resistances of components and the loads are normally distributed random variables with the coefficients of variation 0.05 and 0.3, respectively, the mean resistances, redundancy factors and reliability indices of components of the mixed systems associated with the no correlation and partial correlation cases are presented in Tables 2 and 3, respectively. In the perfect correlation case ($\rho(R_i, R_j) = 1.0$), $\eta_R = 1.0$ and $\beta_{cs} = 3.5$ for all the mixed systems. It is found from these tables that (a) the redundancy factors of the parallel systems are all at least 1.0 due to the existence of brittle component(s) in the systems and (b) for the 2×2 series-parallel system, the redundancy factors associated with the two cases in which the number of brittle components is two are the same; this means that the redundancy factor is not affected by the location of the brittle components in this series-parallel system.

Figure 4 shows the effects of the number of brittle components in the parallel system on the redundancy factor. It is noticed that (a) as the number of brittle components in the parallel system increases, the redundancy factor becomes larger in the no correlation and partial correlation cases and (b) as the correlation among the resistances of components increases, the

redundancy factor increases in the ductile case but decreases in the mixed and brittle cases.

3.4. Effects of post-failure behaviour factor on the redundancy factor

The post-failure behaviour factor δ of a material describes the percentage of remaining strength after failure. The value of δ varies from 0 (i.e. brittle) to 1 (i.e. ductile). The previous sections focus on the redundancy factors associated with only the two extreme post-failure behaviour cases. However, in addition to the ductile and brittle materials, there are some materials whose post-failure behaviour factors are between 0 and 1. Therefore, it is necessary to study the redundancy factors associated with these intermediate post-failure behaviour cases. In this section, parallel systems consisting of two to four components are used to investigate the effects of post-failure behaviour factor on the redundancy factor.

The post-failure behaviour factors of all components are assumed to be the same. The resistances and load associated with the components are considered as normally distributed variables with the coefficients of variation equal to 0.05 and 0.3, respectively. After

Table 3. $E_{cs}(R)$, η_R and β_{cs} of mixed systems associated with the case $\rho(R_i, R_j) = 0.5$ when R and P are normal distributed.

System		$E_{cs}(R)$	η_R	β_{cs}
2-component parallel system	1 ductile & 1 brittle	21.260	1.006	3.53
3-component parallel system	1 ductile & 2 brittle	21.530	1.019	3.62
	2 ductile & 1 brittle	21.290	1.007	3.55
4-component parallel system	1 ductile & 3 brittle	21.700	1.027	3.67
	2 ductile & 2 brittle	21.550	1.020	3.62
	3 ductile & 1 brittle	21.318	1.009	3.55
4-component series-parallel system (2 × 2 SP system)	1 ductile & 3 brittle	21.700	1.027	3.67
	2 ductile & 2 brittle Case A	21.585	1.021	3.63
	2 ductile & 2 brittle Case B	21.585	1.021	3.63
	3 ductile & 1 brittle	21.420	1.014	3.59

Note: $V(P) = 0.3$; $V(R) = 0.05$; $\beta_c = 3.5$; $\beta_{sys} = 3.5$; $E_c(R) = 21.132$.

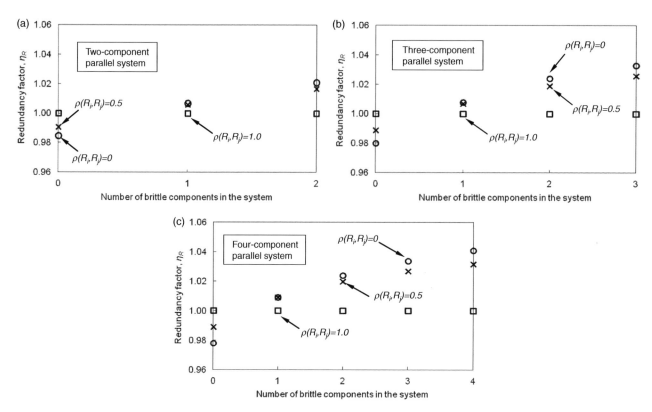

Figure 4.   Effects of number of brittle components on the redundancy factor in the parallel systems consisting of (a) two components, (b) three components and (c) four components.

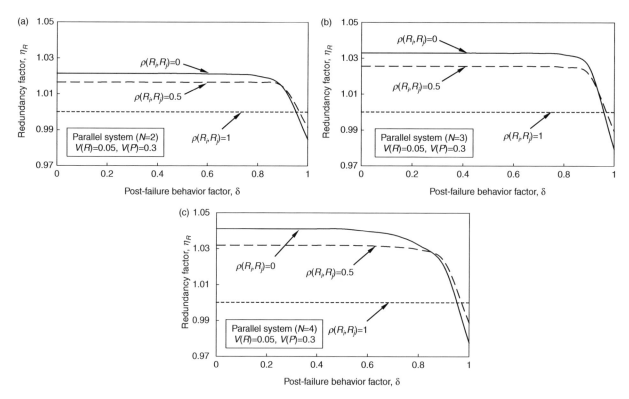

Figure 5.   Effects of post-failure behaviour factor δ on redundancy factor η_R in the parallel systems consisting of (a) two components, (b) three components and (c) four components.

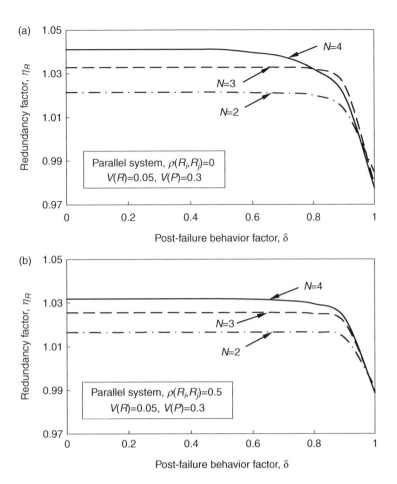

Figure 6.   Effects of post-failure behaviour factor δ on redundancy factor η_R in different parallel systems: (a) no correlation case and (b) partial correlation case.

identifying the failure modes of the parallel system and formulating the associated limit state equations, the redundancy factors of the two-, three-, and four-component parallel systems associated with different post-failure behaviour factors are calculated using the MCS-based method. The results are plotted in Figures 5 and 6.

It is noted that: (a) as δ increases from 0 to 1 in the no correlation and partial correlation cases, η_R in the three systems first remains the same and then decreases dramatically; (b) as the correlation among the resistances of components becomes stronger, the region of δ during which η_R remains the same increases; (c) η_R is not affected by δ in the perfect correlation case; (d) the differences in the redundancy factors associated with the three systems are almost the same for $\delta < 0.6$ and become less significant with increasing δ above 0.6 and (e) the redundancy factors reach almost the same value when δ is close to 1.0 (i.e. ductile).

During the calculation of the redundancy factor, the mean resistance of components ($E_{cs}(R)$) when the system reliability index is 3.5 is obtained. Substituting $E_{cs}(R)$ into the component reliability analysis yields the reliability

indices of components. Figures 7 and 8 show the effects of the post-failure behaviour factor on the component reliability index in the three parallel systems associated with three correlation cases. Most of the conclusions drawn from these two figures are similar to those regarding redundancy factors obtained from Figures 5 and 6. Moreover, it is seen that the reliability index of components when $\delta = 0$ (i.e. brittle) is greater than 3.5, while its value when $\delta = 1.0$ (i.e. ductile) is less than 3.5. This is because brittle systems are much less redundant than ductile systems and, therefore, a larger redundancy factor ($\eta_R > 1.0$) needs to be applied to penalise the brittle components by designing them conservatively ($\beta_{cs} > 3.5$); while in the ductile case, smaller redundancy factors ($\eta_R < 1.0$) can be used to achieve a more economical component design ($\beta_{cs} < 3.5$).

4.   Redundancy factors of ductile and brittle systems with many components

Redundancy factors associated with the ductile and brittle systems consisting of no more than four components were investigated previously. The results show that η_R is

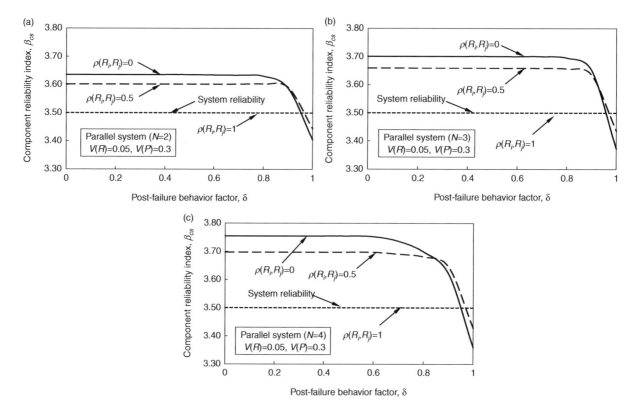

Figure 7. Effects of post-failure behaviour factor δ on component reliability index in the parallel systems consisting of (a) two components, (b) three components and (c) four components.

affected by the number of components in the system. Although the conclusions obtained from the previous sections provide information about the effects of the number of components on the redundancy factors of ductile and brittle systems with small number of components, it is difficult to determine the redundancy factors for systems with a large number of components based on this information. In this context, it is necessary to evaluate the redundancy factors of ductile and brittle systems that consist of many components so that standard tables of redundancy factors can be generated to facilitate the component design process.

4.1. Redundancy factors of ductile systems with many components

As stated previously, only the parallel and series-parallel systems are studied in this paper because the redundancy factors of series systems are independent of the material behaviour of its components. The redundancy factors associated with N-component series systems have been provided in Zhu and Frangopol (2014a). Based on the conclusion that the redundancy factor associated with a certain system is not affected by the mean value of the applied load, which was obtained in Zhu and Frangopol (2014a), the following assumption is made for the loads acting on the investigated parallel and series-

parallel systems: (a) for an N-component parallel system, the load it is subject to is $N \cdot P$, where P is the load applied to a single component which is used to calculate $E_c(R)$ and (b) for an N-component series-parallel system that has N_s series component and N_p parallel component ($N_p = 5$, 10 and 20 herein), the load acting on it is $N_p \cdot P$. In this way, the load effect of each component in the intact parallel and series-parallel system is P so that the obtained mean resistance $E_{cs}(R)$ can be compared with $E_c(R)$ to calculate the redundancy factor η_R.

The reliability of ductile systems with many components has been addressed in Zhu and Frangopol (2014b). A ductile component continues to carry its share of the load equal to its capacity after it fails. Therefore, for an N-component ductile parallel system, the load acting on an intact component j after m components in the system fail is $(N \cdot P - \sum_{i=1}^{m} R_i)/(N - m)$. This value is not affected by the failure sequence of the m components. Because the failure modes of ductile systems are independent of the failure sequence, the limit state equation of an N-component parallel system can be written as (Zhu and Frangopol 2014b):

$$g = \sum_{i=1}^{N} R_i - N \cdot P = 0. \qquad (7)$$

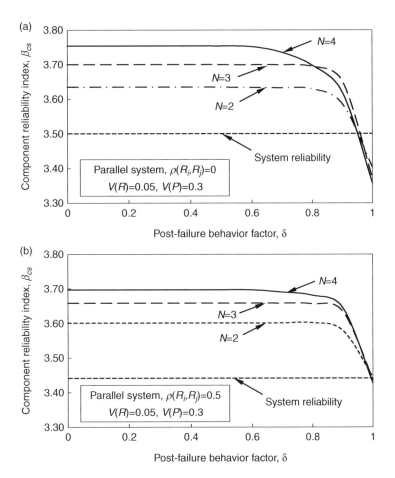

Figure 8. Effects of post-failure behaviour factor δ on component reliability index in the (a) no correlation case and (b) partial correlation case.

For an $N_p \times N_s$ series-parallel system which has N_s possible failure modes, the limit state equation associated with failure mode k is (Zhu and Frangopol 2014b)

$$g = \sum_{i=N_p \cdot (k-1)+1}^{N_p \cdot k} R_i - N_p \cdot P = 0, \qquad (8)$$

where $N_p \cdot (k-1) + 1$ and $N_p \cdot k$ denote the first and last component in the kth sub-parallel system, respectively. With the identified limit state equations of the N-component ($N = 100, 200, 300, 400$ and 500) ductile parallel and series-parallel systems and the assumed coefficients of variation of resistance and load equal to

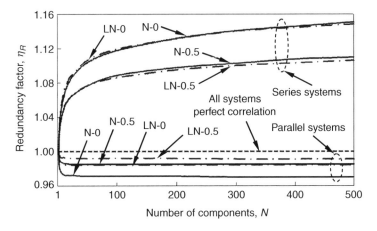

Figure 9. Effects of number of components on the redundancy factor in ductile systems.
Note: 'N' denotes normal distribution; 'LN' denotes lognormal distribution; '0' denotes $\rho(R_i, R_j) = 0$; '0.5' denotes $\rho(R_i, R_j) = 0.5$.

Table 4. $E_{cs}(R)$ and η_R of ductile systems associated with the case $\rho(R_i, R_j) = 0$.

System		Normal distribution		Lognormal distribution	
		$E_{cs}(R)$	η_R	$E_{cs}(R)$	η_R
100-Component system	Series system	23.626	1.118	30.457	1.120
	Parallel system	20.519	0.971	26.759	0.984
	5×20 SP system	21.428	1.014	27.928	1.027
	10×10 SP system	21.026	0.995	27.439	1.009
	20×5 SP system	20.794	0.984	27.085	0.996
200-Component system	Series system	23.921	1.132	30.784	1.132
	Parallel system	20.519	0.971	26.759	0.984
	5×40 SP system	21.576	1.021	28.119	1.034
	10×20 SP system	21.132	1.000	27.575	1.014
	20×10 SP system	20.878	0.988	27.248	1.002
300-Component system	Series system	24.112	1.141	31.028	1.141
	Parallel system	20.498	0.970	26.759	0.984
	5×60 SP system	21.639	1.024	28.227	1.038
	10×30 SP system	21.195	1.003	27.656	1.017
	20×15 SP system	20.921	0.990	27.303	1.004
400-Component system	Series system	24.217	1.146	31.137	1.145
	Parallel system	20.498	0.970	26.759	0.984
	5×80 SP system	21.703	1.027	28.255	1.039
	10×40 SP system	21.238	1.005	27.711	1.019
	20×20 SP system	20.942	0.991	27.303	1.004
500-Component system	Series system	24.323	1.151	31.246	1.149
	Parallel system	20.498	0.970	26.759	0.984
	5×100 SP system	21.745	1.029	28.309	1.041
	10×50 SP system	21.280	1.007	27.738	1.020
	20×25 SP system	20.963	0.992	27.357	1.006

Note: $E(P) = 10$; $V(P) = 0.3$; $V(R) = 0.05$; $\beta_c = 3.5$; $\beta_{sys} = 3.5$; $E_{c,N}(R) = 21.132$; $E_{c,LN}(R) = 27.194$.

0.05 and 0.3, respectively, the redundancy factors associated with two probability distribution types (i.e. normal and lognormal) and three correlation cases (i.e. $\rho(R_i, R_j) = 0$, 0.5 and 1.0) when the system reliability is 3.5 can be obtained using the MCS-based method.

In the perfect correlation case ($\rho(R_i, R_j) = 1.0$), $\eta_R = 1.0$ and $\beta_{cs} = 3.5$ for different types of systems with different number of components associated with both normal and lognormal distributions. This is because for systems whose components are identical and perfectly correlated, the system can be reduced to a single component; therefore, the redundancy factors in this correlation case do not change with the system type and the number of components.

Tables 4 and 5 present the redundancy factors associated with the correlation cases $\rho(R_i, R_j) = 0$ and 0.5, respectively, for the parallel and series-parallel ductile systems along with the results of series systems to facilitate the comparison analysis. In these tables, $E_{c,N}(R) = 21.132$ and $E_{c,LN}(R) = 27.194$ denote the mean resistance of a single component with 3.5 reliability index when its R and P follow normal and lognormal distributions, respectively. These results are also plotted in Figure 9 which shows the effects of number of components on the redundancy factors of series and parallel ductile systems.

It is observed that: (a) the effect of N on η_R in the parallel ductile system depends on the value of N: when N is small ($N \leq 5$), increasing N leads to lower η_R in the parallel system, and the change is less significant as the correlation among the resistances of component increases; however, when $N > 5$, η_R remains almost the same as N increases; (b) for the series-parallel ductile systems that have the same number of parallel components (N_p is the same in these series-parallel systems), η_R increases with N; (c) as the correlation among the resistances of components becomes stronger, η_R decreases and increases in the series and parallel system, respectively and (d) in the series system, the redundancy factors associated with normal and lognormal distributions are very close; however, in the parallel system, the differences in the redundancy factors associated with these two probability distribution cases are more significant.

The component reliability indices β_{cs} of the N-component ductile systems associated with the normal and lognormal cases are shown in Table 6. The results are also plotted in Figure 10 to directly display the effects of N on the component reliability index. It is found that: (a) the effects of N and $\rho(R_i, R_j)$ on the reliability index of components are similar to those on the redundancy factors just discussed and (b) in the series and parallel systems, the component reliability index associated with normal distribution is

Table 5. $E_{cs}(R)$ and η_R of ductile systems associated with the case $\rho(R_i, R_j) = 0.5$.

System		Normal distribution		Lognormal distribution	
		$E_{cs}(R)$	η_R	$E_{cs}(R)$	η_R
100-Component system	Series system	23.013	1.089	29.533	1.086
	Parallel system	20.815	0.985	26.976	0.992
	5 × 20 SP system	21.449	1.015	27.847	1.024
	10 × 10 SP system	21.195	1.003	27.439	1.009
	20 × 5 SP system	21.026	0.995	27.221	1.001
200-Component system	Series system	23.203	1.098	29.777	1.095
	Parallel system	20.815	0.985	26.976	0.992
	5 × 40 SP system	21.555	1.020	27.928	1.027
	10 × 20 SP system	21.259	1.006	27.575	1.014
	20 × 10 SP system	21.090	0.998	27.248	1.002
300-Component system	Series system	23.309	1.103	29.913	1.100
	Parallel system	20.815	0.985	26.949	0.991
	5 × 60 SP system	21.660	1.025	28.010	1.030
	10 × 30 SP system	21.322	1.009	27.602	1.015
	20 × 15 SP system	21.132	1.000	27.330	1.005
400-Component system	Series system	23.414	1.108	29.995	1.103
	Parallel system	20.815	0.985	26.949	0.991
	5 × 80 SP system	21.703	1.027	28.037	1.031
	10 × 40 SP system	21.343	1.010	27.629	1.016
	20 × 20 SP system	21.132	1.000	27.357	1.006
500-Component system	Series system	23.457	1.110	30.077	1.106
	Parallel system	20.815	0.985	26.949	0.991
	5 × 100 SP system	21.703	1.027	28.064	1.032
	10 × 50 SP system	21.364	1.011	27.656	1.017
	20 × 25 SP system	21.132	1.000	27.357	1.006

Note: $E(P) = 10$; $V(P) = 0.3$; $V(R) = 0.05$; $\beta_c = 3.5$; $\beta_{sys} = 3.5$; $E_{c,N}(R) = 21.132$; $E_{c,LN}(R) = 27.194$.

higher and lower than that associated with lognormal distribution, respectively.

4.2. Redundancy factors of brittle systems with many components

As indicated previously, for the evaluation of redundancy factors of brittle systems, all the possible failure modes and associated limit state equations need to be identified and accounted for to perform a correct reliability analysis. An approach to compute the reliability of brittle systems

with many components has been formulated in Zhu and Frangopol (2014b). The number of failure modes for an N-component brittle parallel system is N factorial ($N!$). When N is small ($N \leq 4$), the approach described in the previous section for determining the failure modes and limit state equations can be used; however, when $N > 4$, the number of failure modes will exceed 120 and it becomes difficult and computationally expensive to consider all the failure modes and associated limit states. Therefore, an alternative approach that can be combined with MATLAB (Mathworks, 2009) is introduced herein.

Table 6. Component reliability index β_{cs} of ductile systems.

System		Normal distribution		Lognormal distribution	
		$\rho(R_i, R_j) = 0$	$\rho(R_i, R_j) = 0.5$	$\rho(R_i, R_j) = 0$	$\rho(R_i, R_j) = 0.5$
100-Component system	Series system	4.23	4.05	3.88	3.77
	Parallel system	3.32	3.40	3.44	3.48
200-Component system	Series system	4.31	4.10	3.91	3.80
	Parallel system	3.32	3.40	3.44	3.48
300-Component system	Series system	4.36	4.14	3.94	3.81
	Parallel system	3.31	3.40	3.44	3.47
400-Component system	Series system	4.40	4.16	3.95	3.82
	Parallel system	3.31	3.40	3.44	3.47
500-Component system	Series system	4.42	4.18	3.96	3.83
	Parallel system	3.31	3.40	3.44	3.47

Note: $E(P) = 10$; $V(P) = 0.3$; $V(R) = 0.05$; $\beta_c = 3.5$; $\beta_{sys} = 3.5$; $E_{c,N}(R) = 21.132$; $E_{c,LN}(R) = 27.194$.

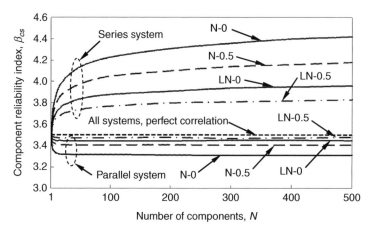

Figure 10. Effects of number of components on the reliability index of components in ductile systems.
Note: 'N' denotes normal distribution; 'LN' denotes lognormal distribution; '0' denotes $\rho(R_i, R_j) = 0$; '0.5' denotes $\rho(R_i, R_j) = 0.5$.

The number of limit state equations of the three-component parallel system is 12, as shown in Equations (3)–(6). It is noticed that some limit state equations can be merged because they are actually the same (i.e. g_4 and g_9, g_5 and g_7 and g_6 and g_8). After merging the identical ones, the limit state equations associated with the three-component parallel system are renumbered as follows:

$$g_1 = R_1 - P = 0, \quad g_2 = R_2 - P = 0,$$
$$g_3 = R_3 - P = 0, \tag{9}$$

$$g_4 = R_1 - 1.5P = 0, \quad g_5 = R_2 - 1.5P = 0,$$
$$g_6 = R_3 - 1.5P = 0, \tag{10}$$

$$g_7 = R_1 - 3P = 0, \quad g_8 = R_2 - 3P = 0,$$
$$g_9 = R_3 - 3P = 0. \tag{11}$$

It is seen that the number of the limit state equations after merging is nine. Similarly, the four-component parallel system has 16 limit state equations after merging.

Therefore, the number of the limit state equations associated with an N-component parallel system is N^2. The failure modes of the three-component parallel system with renumbered limit state equations are shown in Figure 11. It is observed that (a) g_1, g_2 and g_3 correspond to the cases where component 1, 2 and 3 fails first, respectively; (b) g_4, g_5 and g_6 correspond to the cases where component 1, 2 and 3 fails second, respectively and (c) g_7, g_8 and g_9 correspond to the cases where component 1, 2 and 3 fails last, respectively. Therefore, for an N-component brittle parallel system, its limit state equations can be formulated as a matrix (Zhu and Frangopol 2014b):

$$G = \begin{bmatrix} g_1 & g_2 & \cdots & g_N \\ g_{N+1} & g_{N+2} & \cdots & g_{2N} \\ \cdots & \cdots & \cdots & \cdots \\ g_{N(N-1)+1} & g_{N(N-1)+2} & \cdots & g_{N^2} \end{bmatrix}. \tag{12}$$

The element $G(i, j)$ in this matrix denotes that the failure sequence of component j is i. The limit state equation

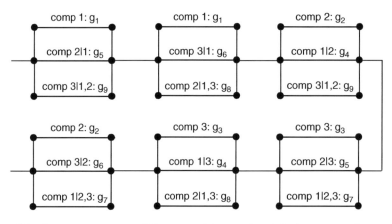

Figure 11. Failure modes of the three-component parallel system with renumbered limit state equations.

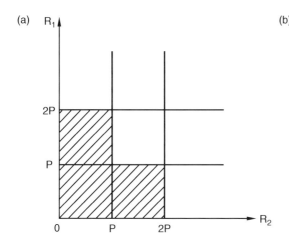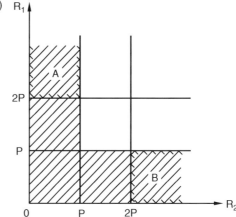

Figure 12. Sample space of (a) event F_1 and (b) event F_2.

associated with the element $G(i,j)$ in the matrix is given by (Zhu and Frangopol 2014b)

$$G(i,j) = R_j - \frac{N \cdot P}{N - i + 1} = 0, \qquad (13)$$

where $i, j = 1, 2, 3, \ldots, N$. For example, the coordinate of the limit state equation g_2 in the matrix is (1, 2) (i.e. first row and second column); therefore, g_2 represents the case where component 2 fails first, and the associated limit state equation is $R_2 - P = 0$ (see Equation (9)). Similarly, $g_{N(N-1)+1}$ stands for the case in which component 1 fails last; when $N = 3$ (three-component system), $g_7 = G(3, 1) = R_1 - 3P = 0$ (see Equation (11)).

After placing all the limit state equations into a matrix, the failure modes of the N-component parallel system can be easily obtained by selecting N elements that are in different rows and columns from the matrix and the set consisting of these selected N elements is one possible failure mode of the system. For example, the limit state equation matrix of the aforementioned three-component parallel system is

$$G = \begin{bmatrix} g_1 & g_2 & g_3 \\ g_4 & g_5 & g_6 \\ g_7 & g_8 & g_9 \end{bmatrix}, \qquad (14)$$

where g_i $(i = 1, 2, \ldots, 9)$ are defined in Equations (9)–(11). According to the selection process indicated previously, six possible failure modes can be found from the matrix: (a) $g_1 \rightarrow g_5 \rightarrow g_9$; (b) $g_1 \rightarrow g_6 \rightarrow g_8$; (c) $g_2 \rightarrow g_4 \rightarrow g_9$; (d) $g_2 \rightarrow g_6 \rightarrow g_7$; (e) $g_3 \rightarrow g_4 \rightarrow g_8$ and (f) $g_3 \rightarrow g_5 \rightarrow g_7$. These are the same as the failure modes shown in Figure 11.

By using the limit state equation matrix G, the process of generating limit state equations associated with different failure sequences and identifying the failure

modes can be achieved with MATLAB codes. The procedure for estimating the redundancy factor in brittle systems using this approach is summarised as follows:

(a) Determine the limit state equation of component j when it fails at the sequence i using Equation (13), where $i, j = 1, 2, \ldots, N$;

(b) Form the limit state equation matrix G by defining g_k $(k = 1, 2, \ldots, N^2)$ using the format shown in Equation (12);

(c) Identify all the combinations, each consisting of N elements located in different rows and columns of the matrix G; the obtained combinations are the failure modes of the system;

(d) Based on the obtained limit state equations and failure modes, and other statistical information associated with the resistances and load, the mean resistance of component when the system reliability index is prescribed (i.e. 3.5) can be determined;

(e) Calculate the redundancy factor.

This approach is used to compute the redundancy factor of the brittle parallel systems with up to eight components. However, the nine-component parallel system has 362,880 different failure modes and the reliability analysis becomes very time consuming and in most servers (such as a Dell Precision R5500 rack workstation equipped with two six cores X5675 Intel Xeon processors with 3.06 GHz clock speed and 24 GB DDR3 memory) the memory usage is exceeded. Therefore, in order to calculate the redundancy factors of brittle parallel systems consisting of more than eight components, another method has to be introduced.

Consider the two-component brittle parallel system described previously. It has two different failure modes and the associated limit state equations are shown in Equations (1) and (2). Its system failure can be expressed

Table 7. $E_{cs}(R)$ and η_R of brittle systems associated with the case $\rho(R_i, R_j) = 0$.

System		Normal distribution		Lognormal distribution	
		$E_{cs}(R)$	η_R	$E_{cs}(R)$	η_R
5-Component system	Series system	22.125	1.047	28.581	1.051
	Parallel system	22.125	1.047	28.581	1.051
10-Component system	Series system	22.484	1.064	29.098	1.070
	Parallel system	22.506	1.065	29.125	1.071
	5×2 SP system	22.506	1.065	29.125	1.071
15-Component system	Series system	22.738	1.076	29.342	1.079
	Parallel system	22.738	1.076	29.342	1.079
	5×3 SP system	22.738	1.076	29.342	1.079
20-Component system	Series system	22.865	1.082	29.560	1.087
	Parallel system	22.865	1.082	29.560	1.087
	5×4 SP system	22.865	1.082	29.560	1.087
	10×2 SP system	22.865	1.082	29.560	1.087
25-Component system	Series system	22.992	1.088	29.641	1.090
	Parallel system	22.992	1.088	29.641	1.090
	5×5 SP system	22.992	1.088	29.641	1.090
50-Component system	Series system	23.330	1.104	30.104	1.107
	Parallel system	23.330	1.104	30.104	1.107
	5×10 SP system	23.330	1.104	30.104	1.107
	10×5 SP system	23.330	1.104	30.104	1.107

Note: $E(P) = 10$; $V(P) = 0.3$; $V(R) = 0.05$; $\beta_c = 3.5$; $\beta_{sys} = 3.5$; $E_{c,N}(R) = 21.132$; $E_{c,LN}(R) = 27.194$.

in terms of components failure events as (Zhu and Frangopol 2014b)

$$F_1 = [(g_1 < 0) \cap (g_3 < 0)] \cup [(g_2 < 0) \cap (g_4 < 0)]. \tag{15}$$

Denoting the event $g_i < 0$ as D_i, the above equation can be rewritten as

$$F_1 = (D_1 \cap D_3) \cup (D_2 \cap D_4). \tag{16}$$

The probability of event F_1 is approximately equal to the probability of the following event F_2:

$$F_2 = (D_1 \cup D_2) \cap (D_3 \cup D_4)$$
$$= [(g_1 < 0) \cup (g_2 < 0)]$$
$$\cap [(g_3 < 0) \cup (g_4 < 0)]. \tag{17}$$

This is explained using Figure 12. The sample spaces generated by the events F_1 and F_2 are shown in Figure 12 (a) and (b), respectively. It is seen that

$$F_2 = F_1 \cup A \cup B, \tag{18}$$

where event A is $(R_1 > 2P) \cap (R_2 < P)$ and event B is $(R_2 > 2P) \cap (R_1 < P)$, as shown in the Figure 12(b).

Because R_1 and R_2 have the same mean value and standard deviation, the probabilities of occurrence of events A and B are very small and can be neglected. Therefore, the event F_2 in Equation (17) can be used to find the failure

probability of the two-component brittle parallel system. Extending this conclusion to the N-component brittle parallel system yields the system failure event as follows (Zhu and Frangopol 2014b):

$$F = [(g_1 < 0) \cup (g_2 < 0) \cup \ldots$$
$$\cup (g_N < 0)] \cap [(g_{N+1} < 0) \cup (g_{N+2} < 0) \cup \ldots$$
$$\cup (g_{2N} < 0)] \cap \ldots \cap [(g_{N(N-1)+1} < 0)$$
$$\cup (g_{N(N-1)+2} < 0) \cup \ldots \cup (g_{N^2} < 0)]. \tag{19}$$

where $g_1, g_2, \ldots, g_N^2$ are the performance functions listed in Equation (12). Therefore, by simplifying the system model from an $N! \times N$ series-parallel system to an $N \times N$ series-parallel system, the redundancy factors can be computed for brittle parallel systems having a large number of components.

It should be noted that this approach for estimating the failure probability of the brittle parallel system with many components is based on the assumption that the resistances of components in the system are the same. With this assumption, the limit state equations can be merged to form the limit state equation matrix for the failure modes identification and failure probability estimation. In most practical cases, the components in parallel positions are usually designed to have the same (or very similar) dimensions (e.g. pier columns, beam girders). Therefore, if the material of the components is brittle, the system failure probability can be approximately evaluated using this approach.

Table 8. $E_{cs}(R)$ and η_R of brittle systems associated with the case $\rho(R_i, R_j) = 0.5$.

System		Normal distribution		Lognormal distribution	
		$E_{cs}(R)$	η_R	$E_{cs}(R)$	η_R
5-Component system	Series system	21.914	1.037	28.200	1.037
	Parallel system	21.914	1.037	28.200	1.037
10-Component system	Series system	22.189	1.050	28.608	1.052
	Parallel system	22.189	1.050	28.608	1.052
	5×2 SP system	22.189	1.050	28.608	1.052
15-Component system	Series system	22.358	1.058	28.771	1.058
	Parallel system	22.337	1.057	28.771	1.058
	5×3 SP system	22.337	1.057	28.771	1.058
20-Component system	Series system	22.463	1.063	28.880	1.062
	Parallel system	22.442	1.062	28.880	1.062
	5×4 SP system	22.442	1.062	28.880	1.062
	10×2 SP system	22.442	1.062	28.880	1.062
25-Component system	Series system	22.527	1.066	28.962	1.065
	Parallel system	22.527	1.066	28.962	1.065
	5×5 SP system	22.527	1.066	28.962	1.065
50-Component system	Series system	22.759	1.077	29.288	1.077
	Parallel system	22.759	1.077	29.288	1.077
	5×10 SP system	22.759	1.077	29.288	1.077
	10×5 SP system	22.759	1.077	29.288	1.077

Note: $E(P) = 10$; $V(P) = 0.3$; $V(R) = 0.05$; $\beta_c = 3.5$; $\beta_{sys} = 3.5$; $E_{c,N}(R) = 21.132$; $E_{c,LN}(R) = 27.194$.

With the coefficients of variation of resistances and load being 0.05 and 0.3, respectively, the redundancy factors associated with two probability distribution types (i.e. normal and lognormal) and three correlation cases (i.e. $\rho(R_i, R_j) = 0$, 0.5 and 1.0) are calculated with respect to the N-component ($N \leq 50$) parallel and series-parallel brittle systems. Similar to the ductile systems, the redundancy factors associated with the perfect correlation case (i.e. $\rho(R_i, R_j) = 1.0$) in the brittle systems are also 1.0. The redundancy factors associated with the other two correlation cases are shown in Tables 7 and 8. Figure 13 plots the effects of number of components on the redundancy factors of brittle series and parallel systems.

It is noted that: (a) the redundancy factors η_R of the brittle parallel systems are all greater than 1.0, which implies that the brittle components have to be designed conservatively ($\beta_{cs} > 3.5$) even in the parallel systems; (b) when the number of brittle components are fixed, η_R associated with series, parallel and series-parallel systems are the same; this indicates that for an N-component brittle structure, η_R is independent of the system type; (c) as the number of components in the brittle system increases, η_R associated with all types of systems become larger; (d) η_R of all types of systems decreases when the correlation among the resistances of components becomes stronger; (e) in the no correlation case, η_R associated with the

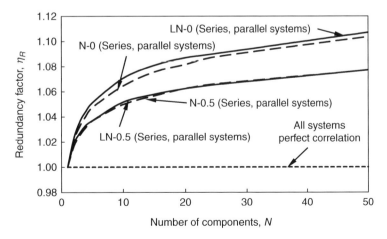

Figure 13. Effects of number of components on the redundancy factor in brittle systems.
Note: 'N' denotes normal distribution; 'LN' denotes lognormal distribution; '0' denotes $\rho(R_i, R_j) = 0$; '0.5' denotes $\rho(R_i, R_j) = 0.5$.

Table 9. Component reliability index β_{cs} of brittle systems.

System		Normal distribution		Lognormal distribution	
		$\rho(R_i, R_j) = 0$	$\rho(R_i, R_j) = 0.5$	$\rho(R_i, R_j) = 0$	$\rho(R_i, R_j) = 0.5$
5-component system	Series system	3.79	3.73	3.67	3.62
	Parallel system	3.79	3.73	3.67	3.62
10-component system	Series system	3.90	3.81	3.72	3.67
	Parallel system	3.90	3.81	3.72	3.67
15-component system	Series system	3.97	3.86	3.76	3.69
	Parallel system	3.97	3.86	3.76	3.69
20-component system	Series system	4.01	3.89	3.78	3.70
	Parallel system	4.01	3.89	3.78	3.70
25-component system	Series system	4.04	3.91	3.79	3.71
	Parallel system	4.04	3.91	3.79	3.71
50-component system	Series system	4.14	3.98	3.84	3.74
	Parallel system	4.14	3.98	3.84	3.74

Note: $E(P) = 10$; $V(P) = 0.3$; $V(R) = 0.05$; $\beta_c = 3.5$; $\beta_{sys} = 3.5$; $E_{c,N}(R) = 21.132$; $E_{c,LN}(R) = 27.194$.

lognormal distribution case is higher than that associated with the normal distribution case and (f) in the partial correlation case, η_R associated with normal and lognormal distributions are almost the same.

It should also be noted that although the redundancy factors of the N-component series and parallel brittle systems are identical, the designs of components (i.e. the mean resistances) in the two systems are not the same. This is because the loads applied on the series and parallel systems are different when calculating the redundancy factors. The mean resistances of the N-component series and parallel systems listed in the tables are computed with respect to the loads P and $N \cdot P$, respectively. Therefore, when the load is fixed, the mean resistance associated with the brittle parallel system is lower than that associated with the series system, which clearly indicates that the parallel system is more economical than the series system.

The component reliability indices of the N-component brittle systems when the system reliability indices are 3.5

are presented in Table 9 and Figure 14. It is seen that (a) for the brittle systems, increasing the number of components leads to higher reliability indices of components in both series and parallel systems; (b) in the no correlation and partial correlation cases, the reliability indices of components associated with the normal distribution are higher than those associated with the lognormal distribution and (c) as the correlation among the resistances of components increases, the component reliability indices in both series and parallel systems decrease.

5. A bridge example

A bridge example is presented herein to demonstrate the application of the proposed redundancy factor by taking into account the post-failure material behaviour in the design of steel girders. This bridge was used for demonstrating (a) the procedure for evaluating the reliability of systems consisting of equally reliable components (Zhu and Frangopol 2014b) and (b) the

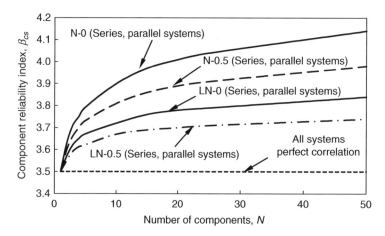

Figure 14. Effects of number of components on the reliability index of components in brittle systems. Note: 'N' denotes normal distribution; 'LN' denotes lognormal distribution; '0' denotes $\rho(R_i, R_j) = 0$; '0.5' denotes $\rho(R_i, R_j) = 0.5$.

application of the redundancy factor proposed by Zhu and Frangopol (2014a) without taking into consideration the effects of post-failure material behaviour. The bridge is simply supported with a span of 20 m. The deck consists of 18 cm of reinforcement concrete and an 8 cm surface layer of asphalt. The deck is supported by four I-beam steel girders whose dimensions are assumed the same. The goal of the design is to determine the bending resistance of the girders using the proposed redundancy factors taking into account the effects of post-failure material behaviour.

Because the bridge is simply supported, the maximum bending moment due to both dead loads and live loads occurs at the mid-span cross-section of the girders. The limit state equations associated with the flexure failure of girder i at the mid-span cross-section is (Zhu and Frangopol 2014a, 2014b)

$$g_i = M_{U,i} - M_{L,i} = 0, \qquad (20)$$

where $M_{U,i}$ and $M_{L,i}$ are the ultimate moment capacity and bending moments acting on girder i, respectively. Assuming the cross-section of the girder is uniform along the length, the ultimate moment capacity at mid-span cross-section, $M_{U,i}$, will govern the design. In order to find the mean value of the ultimate moment capacity of each girder, the maximum bending moments of girders due to dead and live loads need to be determined.

As shown in Zhu and Frangopol (2014a): (a) the total bending moments associated with the exterior and interior girders are $M_{L,ext} = 3407$ kN m and $M_{L,int} = 3509$ kN m, respectively; (b) in the limit state (see Equation (20)), both the moment capacity and load effect are assumed to be normally distributed random variables with the coefficients of variation of $M_{U,i}$ and $M_{L,i}$ equal to 0.05 and 0.3, respectively; (c) the total bending moments are used as the mean values of the $M_{L,i}$ in Equation (20); and (d) based on statistical information on the capacity and load effect, the mean resistances of exterior and interior girder when their reliability indices are 3.5 are: $E_c(M_{U,ext}) = 7200$ kN m (exterior girder) and $E_c(M_{U,int}) = 7415$ kN m (interior girder). The larger value between the mean resistances of exterior and interior girders is used as the final mean resistance of the girders: $E_c(M_U) = 7415$ kN m.

As mentioned previously, the redundancy factor associated with series system is independent of the post-failure behaviour of components. Therefore, only the parallel and series-parallel systems are considered herein. For the investigated four girders, the four-component parallel system and the 2×3 series-parallel system can be formed based on two different definitions of system failure: (a) the system fails only if all girders fail (parallel system) and (b) the system fails if any two adjacent girders fail (series-parallel system). Because the girders are made of steel which is a ductile material, the redundancy factors associated with the ductile case will be applied.

The redundancy factors η_R of the four-component ductile parallel system associated with three correlation cases are provided in Table 1. By performing the procedure for calculating the redundancy factors of ductile systems, η_R of the 2×3 series-parallel system can also be determined with respect to three correlation cases as: 1.009 if $\rho(R_i, R_j) = 0$, 1.012 if $\rho(R_i, R_j) = 0.5$ and 1.0 if $\rho(R_i, R_j) = 1.0$. By multiplying the mean resistances of girders obtained previously by these redundancy factors, the designed mean resistances of girders in parallel and series-parallel systems are obtained, as listed in Table 10.

It is noticed that (a) for both systems, the designed resistances of girders associated with the partial correlation case are higher than those associated with the no correlation case; (b) in the no correlation and partial correlation cases, the designed mean resistances of girders in the parallel system are less than those in the series-parallel system and (c) the designed mean resistances of girders associated with the perfect correlation case are the same for both systems considered.

With the designed resistances of girders, the total bending moments due to dead and live loads, and the associated statistical parameters, the reliability indices of the exterior β_{ext} and interior β_{int} girders in the parallel and series-parallel systems are evaluated, as listed in Table 10. It is found that: (a) in both systems, the reliability indices associated with interior girders are less than those associated with the exterior girders; (b) compared with the parallel system, the reliability indices of girders of series-parallel systems are higher and (c) as the correlation among resistances of girders increases, the reliability indices of girders increase in the parallel system.

6. Conclusions

This paper investigates the redundancy factors of systems considering the post-failure behaviour of components. Systems consisting of two to four components are used as examples to demonstrate the procedure for evaluating the redundancy factors of ductile, brittle and mixed systems. The effects of number of brittle components in a system and the post-failure behaviour factor on the redundancy factor are also studied using these systems. In order to generate standard tables to facilitate the design process, the redundancy factors of N-component nondeterministic ductile and brittle systems with a large number of components are calculated with respect to three correlation cases and two probability distribution types. Finally, a bridge example is studied to demonstrate the application of the ductile redundancy factor in the design of steel girders. For the systems analysed, the following conclusions are drawn:

1. An approach for simplifying the system model in the redundancy factor analysis of brittle systems is

Table 10. The designed mean resistances of girders and the reliability indices of the exterior and interior girders.

System type	Correlation case	Designed mean resistance of girders (kN m)	β_{ext}	β_{int}
Parallel system	$\rho(R_i, R_j) = 0$	7252	3.55	3.36
	$\rho(R_i, R_j) = 0.5$	7333	3.61	3.43
	$\rho(R_i, R_j) = 1.0$	7415	3.69	3.50
Series-parallel system	$\rho(R_i, R_j) = 0$	7482	3.74	3.56
	$\rho(R_i, R_j) = 0.5$	7504	3.77	3.57
	$\rho(R_i, R_j) = 1.0$	7415	3.69	3.50

proposed. By reducing the $N! \times N$ series-parallel system model to the $N \times N$ series-parallel system model, this approach makes it possible to calculate the redundancy factor of components in brittle parallel systems with a large number of components.

2. The redundancy factors associated with series, parallel and series-parallel systems in the brittle case are the same; this indicates that for an N-component brittle structure, the redundancy factor is independent of the system modelling type.

3. For the ductile parallel system consisting of only a few components, increasing N leads to a significant decrease of the redundancy factor. However, as N continues increasing, this decrease becomes insignificant.

4. Increasing the correlation among the resistances of components leads to higher redundancy factors in the ductile parallel system, respectively. The difference in the redundancy factors between the normal and lognormal distributions is more significant in the ductile parallel than in the series system.

5. The redundancy factors of the mixed parallel systems are at least 1.0 due to the existence of brittle component(s) in the systems. As the number of brittle components increases in an N-component mixed system, the redundancy factor becomes larger and closer to the redundancy factor associated with the brittle case. Increasing the correlation among the resistances of components leads to a lower redundancy factor in the mixed parallel systems.

6. As the post-failure behaviour factor increases in the no correlation and partial correlation cases, the redundancy factor of the parallel system initially remains the same and then decreases dramatically.

Acknowledgements

The support from the US Federal Highway Administration Cooperative Agreement 'Advancing Steel and Concrete Bridge Technology to Improve Infrastructure Performance' Project Award DTFH61-11-H-00027 to Lehigh University is gratefully acknowledged. The opinions and conclusions presented in this paper are those of the authors and do not necessarily reflect the views of the sponsoring organisation.

References

American Association of State Highway and Transportation Officials. (1994). *LRFD bridge design specifications* (1st ed.). Washington, DC: Author.

American Association of State Highway and Transportation Officials. (2012). *LRFD bridge design specifications* (6th ed.). Washington, DC: Author.

Ang, A.H-S., & Tang, W.H. (1984). *Probability concepts in engineering planning and design* (Vol. 2). New York, NY: Wiley.

Babu, S.G.L., & Singh, V.P. (2011). Reliability-based load and resistance factors for soil-nail walls. *Canadian Geotechnical Journal, 48*, 915–930.

Cavaco, E.S., Casas, J.R., & Neves, L.A.C. (2013). Quantifying redundancy and robustness of structures. *Proceedings of IABSE workshop on safety, failures and robustness of large structures*. International Association for Bridge and Structural Engineering (IABSE), Zurich, Switzerland.

Ellingwood, B., Galambos, T.V., MacGregor, J.G., & Cornell, C.A. (1980). *Development of a probability-based load criterion for American National Standard A58*. NBS Special Publication 577. Washington, DC: U.S. Dept of Commerce.

Frangopol, D.M., & Curley, J.P. (1987). Effects of damage and redundancy on structural reliability. *ASCE Journal of Structural Engineering, 113*, 1533–1549.

Frangopol, D.M., & Klisinski, M. (1989a). Material behavior and optimum design of structural systems. *ASCE Journal of Structural Engineering, 115*, 1054–1075.

Frangopol, D.M., & Klisinski, M. (1989b). Weight–strength–redundancy interaction in optimum design of three-dimensional brittle–ductile trusses. *Computers and Structures, 31*, 775–787.

Frangopol, D.M., & Nakib, R. (1991). Redundancy in highway bridges. *Engineering Journal, American Institute of Steel Construction, 28*, 45–50.

Fu, G., & Frangopol, D.M. (1990). Balancing weight, system reliability and redundancy in a multiobjective optimization framework. *Structural Safety, 7*, 165–175.

Ghosn, M., & Moses, F. (1998). *Redundancy in highway bridge superstructures* (NCHRP Report 406). Washington, DC: Transportation Research Board.

Ghosn, M., Moses, F., & Frangopol, D.M. (2010). Redundancy and robustness of highway bridge superstructures and substructures. *Structure and Infrastructure Engineering, 6*, 257–278.

Hendawi, S., & Frangopol, D.M. (1994). System reliability and redundancy in structural design and evaluation. *Structural Safety, 16*, 47–71.

Hsiao, L., Yu, W., & Galambos, T. (1990). AISI LRFD method for cold-Formed steel structural members. *Journal of Structural Engineering, 116*, 500–517.

Kim, J. (2010). *Finite element modeling of twin steel box-girder bridges for redundancy evaluation* (Dissertation). The University of Texas at Austin, Austin, TX.

Lin, S., Yu, W., & Galambos, T. (1992). ASCE LRFD method for stainless steel structures. *Journal of Structural Engineering, 118*, 1056–1070.

Liu, D., Ghosn, M., Moses, F., & Neuenhoffer, A. (2001). *Redundancy in highway bridge substructures* (National Cooperative Highway Research Program, NCHRP Report 458, Transportation Research Board). Washington, DC: National Academy Press.

MathWorks. (2009). *Statistical toolbox*. MATLAB Version 7.9. The MathWorks Inc., Natick, MA.

Okasha, N.M., & Frangopol, D.M. (2010). Time-variant redundancy of structural systems. *Structure and Infrastructure Engineering: Maintenance, Management, Life-Cycle Design and Performance, 6*, 279–301. doi:10.1080/15732470802664514

Paikowsky, S.G. (2004). *Load and resistance factor design (LRFD) for deep foundations* (NCHRP Report 507). Washington, DC: Transportation Research Board.

Rabi, S., Karamchandani, A., & Cornell, C.A. (1989). Study of redundancy of near-ideal parallel structural systems. *Proceedings of the 5th international conference on structural safety and reliability*. pp. 975–982. ASCE: New York, NY.

Thoft-Christensen, P., & Baker, M.J. (1982). *Structural reliability theory and its applications*. Berlin: Springer-Verlag.

Thoft-Christensen, P., & Murotsu, Y. (1986). *Application of structural systems reliability theory*. Berlin: Springer-Verlag.

Tsopelas, P., & Husain, M. (2004). Measures of structural redundancy in reinforced concrete buildings. II: Redundancy response modification factor RR. *Journal of Structural Engineering, 130*, 1659–1666.

Wen, Y.K., & Song, S.-H. (2004). Structural reliability/redundancy under earthquakes. *Journal of Structural Engineering, 129*, 56–67.

Zhu, B., & Frangopol, D.M. (2014a). Redundancy-based design in nondeterministic systems. In D.M. Frangopol & Y. Tsompanakis (Eds.), *Safety and maintenance of aging infrastructure*, Chapter 23. Boca Raton, FL: CRC Press. doi:10.1201/b17073-24 (in press).

Zhu, B., & Frangopol, D.M. (2014b). Effects of postfailure material behavior on the reliability of systems. *ASCE-ASME Journal of Risk and Uncertainty in Engineering Systems, Part A: Civil Engineering*, doi:10.1061/AJRUA6.0000808 (in press).

Time-variant redundancy and failure times of deteriorating concrete structures considering multiple limit states

Fabio Biondini [ORCID] and Dan M. Frangopol

ABSTRACT

Structural redundancy and load redistribution capacity are desirable features to ensure suitable system performance under accidental actions and extreme events. For deteriorating structures, these features must be evaluated over time to account for the modification of the redistribution mechanisms due to damage processes. In particular, the identification of the local failure modes and prediction of their occurrence in time is necessary in order to maintain a suitable level of system performance and to avoid collapse. In fact, repairable local failures can be considered as a warning of damage propagation and possible occurrence of more severe and not repairable failures. In this paper, failure loads and failure times of concrete structures exposed to corrosion are investigated and life-cycle performance indicators, related to redundancy and elapsed times between sequential failures, are proposed. The effects of the damage process on the structural performance are evaluated based on a methodology for life-cycle assessment of concrete structures exposed to diffusive attack from environmental aggressive agents. The uncertainties involved are taken into account. The proposed approach is illustrated using two applicative examples: a reinforced concrete frame building and a reinforced concrete bridge deck under corrosion. The results demonstrate that both failure loads and failure times can provide relevant information to plan maintenance actions and repair interventions on deteriorating structures in order to ensure suitable levels of structural performance and functionality during their entire life-cycle.

Introduction

Structural reliability and durability of civil engineering structures and infrastructure facilities are essential to the economic growth and sustainable development of countries. However, aging, fatigue and deterioration processes due to aggressive chemical attacks and other physical damage mechanisms can seriously affect structure and infrastructure systems and lead over time to unsatisfactory structural performance (Clifton & Knab, 1989; Ellingwood, 2005). The economic impact of these processes is extremely relevant (ASCE, 2013; NCHRP, 2006) and emphasises the need of a rational approach to life-cycle design, assessment and maintenance of deteriorating structures under uncertainty based on suitable reliability-based life-cycle performance indicators (Biondini & Frangopol, 2014; Frangopol & Ellingwood, 2010; Saydam & Frangopol, 2011; Zhu & Frangopol, 2012). This need involves a major challenge to the field of structural engineering, since the classical time-invariant structural design criteria and methodologies need to be revised to account for a proper modelling of the structural system over its entire life-cycle by taking into account the effects of deterioration processes, time-variant loadings, and maintenance and repair interventions under uncertainty (Biondini & Frangopol, 2008a, 2016; Frangopol, 2011; Frangopol & Soliman, 2016).

In structural design the level of performance is usually specified with reference to structural reliability. However, when aging and deterioration are considered, the evaluation of the system performance under uncertainty should account for additional probabilistic indicators aimed at providing a comprehensive description of the life-cycle structural resources (Barone & Frangopol, 2014a, 2014b; Biondini & Frangopol, 2014, 2016; Frangopol & Saydam, 2014). The availability of stress redistribution mechanisms and the ability to mitigate the disproportionate effects of sudden damage under accidental actions, abnormal loads and extreme events, are often investigated in terms of structural redundancy (Bertero & Bertero, 1999; Biondini, Frangopol, & Restelli, 2008; Frangopol & Curley, 1987; Frangopol, Iizuka, & Yoshida, 1992; Frangopol & Nakib, 1991; Fu & Frangopol, 1990; Ghosn, Moses, & Frangopol, 2010; Hendawi & Frangopol, 1994; Husain & Tsopelas, 2004; Paliou, Shinozuka, & Chen, 1990; Pandey & Barai, 1997; Schafer & Bajpai, 2005; Zhu & Frangopol, 2013, 2015), structural vulnerability and robustness (Agarwal, Blockley, & Woodman, 2003; Baker, Schubert, & Faber, 2008; Biondini, Frangopol, & Restelli, 2008; Biondini & Restelli, 2008; Ellingwood, 2006; Ellingwood & Dusenberry, 2005; Ghosn et al., 2010; Lind, 1995; Lu, Yu, Woodman, & Blockley, 1999; Starossek & Haberland, 2011), and seismic resilience (Bocchini

& Frangopol, 2012a, 2012b; Bruneau et al., 2003; Chang & Shinozuka, 2004; Cimellaro, Reinhorn, & Bruneau, 2010; Decò, Bocchini, & Frangopol, 2013).

However, depending on the damage propagation mechanism, continuous deterioration may also involve alternate load redistribution paths and disproportionate effects, which can affect over time structural reliability and other performance indicators, including redundancy, robustness, resilience, and sustainability (Biondini, 2009; Biondini & Frangopol, 2014; Biondini, Camnasio, & Titi, 2015; Biondini, Frangopol, & Restelli, 2008; Biondini & Restelli, 2008; Decò, Frangopol, & Okasha, 2011; Enright & Frangopol, 1999; Frangopol & Bocchini, 2011; Furuta, Kameda, Fukuda, & Frangopol, 2004; Okasha & Frangopol, 2009, 2010; Sabatino, Frangopol, & Dong, 2015; Zhu & Frangopol, 2012, 2013). The effects of continuous deterioration can be particularly relevant for concrete structures exposed to the diffusive attack from aggressive agents, such as chlorides and sulfates, which may involve corrosion of steel reinforcement and deterioration of concrete (Bertolini, Elsener, Pedeferri, & Polder, 2004; CEB, 1992).

For reinforced concrete (RC) structures under corrosion the identification of the local failure modes and of their occurrence in time provides useful information in order to maintain a suitable level of system performance and to avoid collapse over their lifetime. In fact, repairable local failures can be considered as a warning of damage propagation and possible occurrence of more severe and not repairable failures (Mori & Ellingwood, 1994). Structural redundancy is a key performance indicator to this purpose, since it measures the ability of the system to redistribute among its active members the load which can no longer be sustained by other damaged members after the occurrence of a local failure (Biondini, Frangopol, & Restelli, 2008; Frangopol & Curley, 1987; Frangopol et al., 1992). However, this indicator refers to a prescribed point in time and does not provide a direct measure of the failure rate, which depends on the damage scenario and damage propagation mechanism (Biondini & Frangopol, 2008b, 2014).

Failure times and time intervals between subsequent failures, or elapsed time between failures, should be computed to provide complete information about the available resources after occurrence of local failures (Biondini, 2012; Biondini & Frangopol, 2014). In fact, the elapsed time between subsequent failures can be considered as a measure of system redundancy in terms of rapidity of evacuation and/or ability of the system to be repaired right after a critical damage state is reached. More specifically, the identification of all the local failure modes up to collapse and their occurrence in time could be helpful to plan emergency procedures, as well as maintenance and repair interventions to ensure suitable levels of life-cycle system performance and functionality.

In the following, failure loads and failure times of concrete structures under corrosion are investigated. Criteria and methods for the definition of life-cycle performance indicators related to redundancy and elapsed times between subsequent failures are proposed. The effects of the damage process on the structural performance are evaluated by using a proper methodology for life-cycle assessment of concrete structures exposed to diffusive attack from environmental aggressive agents (Biondini,

Bontempi, Frangopol, & Malerba, 2004a, 2006). The uncertainties in the material and geometrical properties, in the physical models of deterioration processes, and in the mechanical and environmental stressors, are taken into account in probabilistic terms. The proposed approach is illustrated through the assessment of structural redundancy and elapsed time between failures of a RC frame building and a RC bridge deck under corrosion. The goal is to show that both failure loads and failure times may provide important information to protect, maintain, restore and/or improve the life-cycle structural resources of deteriorating concrete structures.

Time-variant failure loads and failure times

A failure of a system is generally associated with the violation of one or more limit states. Focusing on RC frame systems, limit states of interest may be the occurrence at the material level of local failures associated to cracking of concrete and/or yielding of steel reinforcement, which represent warnings for initiation of damage propagation, as well as attainment of failures associated with the ultimate capacity of critical cross-sections and/or system collapse (Malerba, 1998).

Time-variant failure loads and redundancy

Denoting $\lambda \geq 0$ a scalar load multiplier, the limit states associated to the occurrence of a series of sequential failures $k = 1,2,\dots$ can be identified by the corresponding failure load multiplier λ_k (Biondini, 2012). Since the structural performance of RC structures deteriorates over time, the functions $\lambda_k = \lambda_k(t)$ need to be evaluated by means of structural analyses taking into account the effects of the damage process (Biondini et al., 2004a).

The ability of the system to redistribute the load after the failure $k = i$ up to the failure $k = j$ depends on the reserve load carrying capacity associated to the failure load multipliers $\lambda_i = \lambda_i(t)$ and $\lambda_j = \lambda_j(t)$:

$$\Delta\lambda_{ij}(t) = \lambda_j(t) - \lambda_i(t) \geq 0 \tag{1}$$

Therefore, the following quantity can be assumed as time-variant measure of redundancy between subsequent failures:

$$\Lambda_{ij}(t) = \Lambda(\lambda_i, \lambda_j) = \frac{\lambda_j(t) - \lambda_i(t)}{\lambda_j(t)} \tag{2}$$

The redundancy factor $\Lambda_{ij} = \Lambda_{ij}(t)$ can assume values in the range $[0;1]$. It is zero when there is no reserve of load capacity between the failures i and j ($\lambda_i = \lambda_j$), and tends to unity when the failure load capacity λ_i is negligible with respect to λ_j ($\lambda_i \ll \lambda_j$).

It is worth noting that the classical measure of redundancy refers to the ability of the system to redistribute the load after the occurrence of the first local failure, reached for $\lambda_1 = \lambda_1(t)$, up to structural collapse, reached for a collapse load multiplier $\lambda_c = \lambda_c(t)$. For the sake of brevity, the time-variant redundancy factor between the first failure and collapse is denoted $\Lambda = \Lambda(t)$ (Biondini & Frangopol, 2014):

$$\Lambda(t) = \Lambda(\lambda_1, \lambda_c) = \frac{\lambda_c(t) - \lambda_1(t)}{\lambda_c(t)} \tag{3}$$

It is also noted that redundancy is often associated with the degree of static indeterminacy of the structural system. However, it has been demonstrated that the degree of static indeterminacy is not a consistent measure for structural redundancy (Biondini & Frangopol, 2014; Biondini, Frangopol, & Restelli, 2008; Frangopol & Curley, 1987). In fact, structural redundancy depends on many factors, such as structural topology, member sizes, material properties, applied loads and load sequence, among others (Frangopol & Curley, 1987; Frangopol & Klisinski, 1989; Frangopol & Nakib, 1991). Moreover, the role of these factors may change over time due to structural deterioration, both in deterministic and probabilistic terms (Biondini, 2009; Biondini & Frangopol, 2014; Okasha & Frangopol, 2009).

Failure times and elapsed time between failures

Structural redundancy refers to a prescribed point in time and does not provide information on the failure sequence and failure rate over the structural lifetime. Failure times should be computed to this purpose and the time interval between subsequent failures, or the elapsed time between failures, could represent an effective indicator of the damage tolerance of the system and its ability to be repaired after local failures.

The failure times T_k associated to the occurrence of sequential failures $k = 1,2,\ldots$ can be evaluated by comparing the time-variant failure load multipliers $\lambda_k = \lambda_k(t)$ to prescribed time-variant target functions $\lambda_k^* = \lambda_k^*(t)$ as follows (Biondini, 2012):

$$T_k = \min \left\{ \, t \mid \lambda_k(t) < \lambda_k^*(t) \, \right\} \qquad (4)$$

After a local failure $k = i$, the ability of the system to delay the failure $k = j$ depends on the elapsed time between these failures occurring at times T_i and T_j (Figure 1):

$$\Delta T_{ij} = T_j - T_i \geq 0 \qquad (5)$$

For the sake of brevity, the elapsed time between the first failure, occurring for $\lambda_1 = \lambda_1(t)$ at time T_1, and the structural collapse, reached for $\lambda_c = \lambda_c(t)$ at time T_c, is denoted ΔT (Biondini, 2012; Biondini & Frangopol, 2014):

$$\Delta T = T_c - T_1 \qquad (6)$$

This is an important performance indicator, since it provides the residual lifetime after the first damage warning, for example associated with the formation of a plastic hinge, and identifies the time to global failure due to the activation of a set of plastic hinges leading to structural collapse (Biondini & Frangopol, 2008b).

As mentioned previously, the elapsed time between failures can also be considered as a measure of system redundancy in terms of rapidity of evacuation and/or ability of the system to be repaired right after a critical damage. However, even though they are related concepts, elapsed time between failures and structural redundancy refer to different system resources.

Role of the uncertainties

The geometrical and material properties of the structural systems, the mechanical and environmental stressors, and the parameters of the deterioration processes are always uncertain. Consequently, life-cycle prediction models have to be formulated in probabilistic terms and all parameters of the model have to be considered as random variables or processes. Therefore, the time evolution of the failure loads $\lambda_k = \lambda_k(t)$ and the corresponding failure times T_k, with $k = 1,2,\ldots$, are also random variables or processes, as shown in Figure 2 where uncertainties are associated with initial load capacity, damage initiation, deterioration rate, load capacity after maintenance/repair interventions, and failure time without or with maintenance/repair (Frangopol, 2011; Biondini & Frangopol, 2016). Therefore, a lifetime probabilistic analysis is necessary to investigate the time-variant effects of uncertainty on both redundancy factors $\Lambda_{ij} = \Lambda_{ij}(t)$ and elapsed time intervals ΔT_{ij} between the failures i and j.

Deterioration modelling: a review

A life-cycle probabilistic-oriented approach to design and assessment of structural systems must be based on a reliable and effective modelling of structural deterioration mechanisms. Deterioration models could be developed on empirical bases, as it is generally necessary for rate-controlled damage processes, or founded on mathematical descriptions of the underlying physical

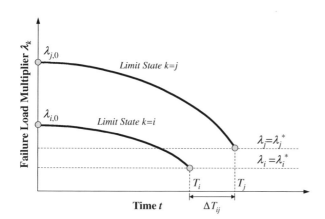

Figure 1. Time-variant failure load multipliers λ_i and λ_j, failure times T_i and T_j associated with the two limit states $\lambda_i = \lambda_i^*$ and $\lambda_j = \lambda_j^*$, and elapsed time ΔT_{ij} between these two sequential failures.

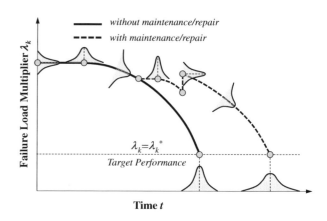

Figure 2. Time-variant failure load multiplier with uncertainties associated with initial load capacity, damage initiation, deterioration rate, load capacity after maintenance and repair interventions, and failure time without or with maintenance/repair.

mechanisms, as it is often feasible for diffusion-controlled damage processes (Ellingwood, 2005).

The latter is the case of interest for RC structures exposed to chloride ingress, where damage induced by the diffusive attack may involve corrosion of steel reinforcement and deterioration of concrete (Bertolini et al., 2004; CEB, 1992). The modelling of these processes should account for both the diffusion process and the related mechanical damage, as well as for the coupling effects of diffusion, damage and structural behaviour.

Diffusion process

The diffusion of chemical components in solids can be described by the Fick's laws which, in the case of a single component diffusion in isotropic, homogeneous and time-invariant media, can be reduced to the following second order linear partial differential equation (Glicksman, 2000):

$$D\nabla^2 C = \frac{\partial C}{\partial t} \qquad (7)$$

where D is the diffusivity coefficient of the medium, $C = C(\mathbf{x}, t)$ is the concentration of the chemical component at point $\mathbf{x} = (x, y, z)$ and time t, $\nabla C = \mathbf{grad}\, C(\mathbf{x}, t)$ and $\nabla^2 = \nabla \cdot \nabla$.

For one-dimensional diffusion, the Fick's equation is amenable to be solved analytically. This analytical solution is a convenient mathematical tool for practical applications (fib, 2006). However, the actual diffusion processes in concrete structures are generally characterised by two- or three-dimensional patterns of concentration gradients. For this reason, a numerical solution of the Fick's diffusion equation may be necessary for accurate life-cycle assessment of corroding RC structures (Titi & Biondini, 2016).

In this study, the diffusion equation is solved numerically by using cellular automata (Wolfram, 1994). With reference to a regular uniform grid of cells in two dimensions, the Fick's model can be reproduced at the cross-sectional level by the following evolutionary rule (Biondini et al., 2004a):

$$C_{ij}^{k+1} = \phi_0 C_{ij}^k + \frac{1 - \phi_0}{4}(C_{i,j-1}^k + C_{i,j+1}^k + C_{i-1,j}^k + C_{i+1,j}^k) \quad (8)$$

where the discrete variable $C_{ij}^k = C(\mathbf{x}_{ij}, t_k)$ represents the concentration of the component in the cell (i, j) at point $\mathbf{x}_{ij} = (y_i, z_j)$ and time t_k, and $\phi_0 \in [0;1]$ is a suitable evolutionary coefficient. In particular, to regulate the process according to a given diffusivity D, the grid dimension Δx and the time step Δt of the automaton must satisfy the following relationship:

$$D = \frac{1 - \phi_0}{4}\frac{\Delta x^2}{\Delta t} \qquad (9)$$

A proof is given in Biondini, Frangopol, and Malerba (2008).

The deterministic value $\phi_0 = 1/2$ usually leads to a good accuracy of the automaton. However, the local stochastic effects in the mass transfer can be taken into account by assuming ϕ_0 as random variable. The stochastic model also allows to simulate the interaction between diffusion process and mechanical behaviour of the damaged structure. Further details can be found in Biondini et al. (2004a).

Corrosion process

The most relevant effect of corrosion in concrete structures is the reduction of the cross-section of the reinforcing steel bars. The area $A_s = A_s(t)$ of a corroded bar can be represented as follows (Biondini et al., 2004a):

$$A_s(t) = [1 - \delta_s(t)] A_{s0} \qquad (10)$$

where A_{s0} is the area of the undamaged steel bar and $\delta_s = \delta_s(t)$ is a dimensionless damage index which provides a measure of cross-section reduction in the range [0; 1].

Effects of corrosion are not limited to damage of reinforcing steel bars. In fact, the formation of oxidation products may lead to propagation of longitudinal splitting cracks and concrete cover spalling (Al-Harthy, Stewart, & Mullard, 2011; Cabrera, 1996; Vidal, Castel, & François, 2004). In this study, the local deterioration of concrete is modelled by means of a degradation law of the effective resistant area of concrete matrix $A_c = A_c(t)$ (Biondini et al., 2004a):

$$A_c(t) = [1 - \delta_c(t)] A_{c0} \qquad (11)$$

where A_{c0} is the area of undamaged concrete and $\delta_c = \delta_c(t)$ is a dimensionless damage index which provides a measure of concrete deterioration in the range [0; 1]. However, it may be not straightforward to relate the damage function $\delta_c = \delta_c(t)$ to the amount of steel mass loss. For this reason, in some cases, it could be more convenient to model the concrete deterioration due to splitting cracks and cover spalling through a reduction of the concrete compression strength (see Biondini & Vergani, 2015).

Additional effects of corrosion may occur depending on the type of corrosion mechanisms, i.e. uniform corrosion, localised (pitting) corrosion, or mixed type of corrosion (Stewart, 2009; Zhang, Castel, & François, 2010). As an example, depending on the amount of steel mass loss, non uniform corrosion may involve a remarkable reduction of steel ductility (Almusallam, 2001; Apostolopoulos & Papadakis, 2008) and a limited reduction of steel strength (Du, Clark, & Chan, 2005). Further information for a proper modelling of these effects can be found in Biondini and Vergani (2015). In this study, the effects of uniform corrosion only are investigated.

Damage rates

For diffusion-controlled damage processes, the deterioration rate depends on the time-variant concentration of the diffusive chemical components. In such processes, damage induced by mechanical loading interacts with the environmental factors and accelerates both diffusion and deterioration. Therefore, the dependence of the deterioration rate on the concentration of the diffusive agent is generally complex, and the available information about environmental factors and material characteristics is usually not sufficient for a detailed modelling. Despite the complexity of the problem at the microscopic level, simple coupling models can often be successfully adopted at the macroscopic level in order to reliably predict the time evolution of structural performance (Biondini & Frangopol, 2008b; Biondini, Frangopol, & Malerba, 2008; Biondini et al., 2004a).

Based on available data for sulfate and chloride attacks (Pastore & Pedeferri, 1994) and correlation between chloride content and corrosion current density in concrete (Bertolini et al., 2004; Liu & Weyers, 1998; Thoft-Christensen, 1998), a linear relationship between rate of corrosion in the range 0–200 µm/ year and chloride content in the range 0–3% by weight of cement could be reasonable for structures exposed to severe environmental conditions. In this study, the time-variant damage indices $\delta_c = \delta_c(\mathbf{x}, t)$ and $\delta_s = \delta_s(\mathbf{x}, t)$ are related to the diffusion process by assuming a linear relationship between the rate of damage and the mass concentration $C = C(\mathbf{x}, t)$ of the aggressive agent:

$$\frac{\partial \delta_c(\mathbf{x}, t)}{\partial t} = \frac{C(\mathbf{x}, t)}{C_c \Delta t_c} = q_c C(\mathbf{x}, t) \qquad (12)$$

$$\frac{\partial \delta_s(\mathbf{x}, t)}{\partial t} = \frac{C(\mathbf{x}, t)}{C_s \Delta t_s} = q_s C(\mathbf{x}, t) \qquad (13)$$

where C_c and C_s are the values of constant concentration leading to a complete damage of the materials after the time periods Δt_c and Δt_s, respectively. The damage rate coefficients $q_c = (C_c \Delta t_c)^{-1}$ and $q_s = (C_s \Delta t_s)^{-1}$ depend on both the type of corrosion mechanism and corrosion penetration rate. The initial conditions $\delta_c(\mathbf{x}, t_{cr}) = \delta_s(\mathbf{x}, t_{cr}) = 0$ with $t_{cr} = \min\{t \mid C(\mathbf{x}, t) \geq C_{cr}\}$ are assumed, where C_{cr} is a critical threshold of concentration (Biondini et al., 2004a).

Structural analysis considering time effects

The lifetime structural performance is evaluated by means of structural analysis considering time-variant parameters (Biondini & Vergani, 2015; Biondini et al., 2004a). The formulation is based on the general criteria and methods for nonlinear analysis of concrete structures (Malerba, 1998). At cross-sectional level, the vector of the stress resultants (axial force N and bending moments M_z and M_y):

$$\mathbf{r} = \mathbf{r}(t) = \begin{bmatrix} N & M_z & M_y \end{bmatrix}^T \qquad (14)$$

and the vector of the global strains (axial elongation ε_0 and bending curvatures χ_z and χ_y):

$$\mathbf{e} = \mathbf{e}(t) = [\varepsilon_0 \ \chi_z \ \chi_y]^T \qquad (15)$$

are related, at each time instant t, as follows:

$$\mathbf{r}(t) = \mathbf{H}(t) \mathbf{e}(t) \qquad (16)$$

The time-variant stiffness matrix $\mathbf{H} = \mathbf{H}(t)$ of the RC cross-section under corrosion is derived by integration over the composite area of the materials, or by assembling the contributions of both concrete $\mathbf{H}_c = \mathbf{H}_c(t)$ and steel $\mathbf{H}_s = \mathbf{H}_s(t)$:

$$\mathbf{H}(t) = \mathbf{H}_c(t) + \mathbf{H}_s(t) \qquad (17)$$

$$\mathbf{H}_c(t) = \int_{A_c(x)} E_c(y, z, t) \, \mathbf{B}(y, z) \, [1 - \delta_c(y, z, t)] \, \mathrm{d}A \qquad (18)$$

$$\mathbf{H}_s(t) = \sum_m E_{sm}(t) \, \mathbf{B}_m [1 - \delta_{sm}(t)] \, A_{sm} \qquad (19)$$

where the symbol m refers to the mth reinforcing bar located at point (y_m, z_m) in the centroidal principal reference system (y, z) of the cross-section, $E_c = E_c(y, z, t)$ and $E_{sm} = E_{sm}(t)$ are the moduli of the materials, $\mathbf{B}(y, z) = \mathbf{b}(y, z)^T \mathbf{b}(y, z)$ is a linear operator matrix, and $\mathbf{b}(y, z) = [1 \ -y \ z]$.

It is worth noting that the vectors $\mathbf{r}$ and $\mathbf{e}$ have to be considered as total or incremental quantities based on the nature of the stiffness matrix $\mathbf{H}$, which depends on the type of formulation adopted (i.e. secant or tangent) for the generalised moduli of the materials associated with the stress–strain nonlinear constitutive laws.

The proposed cross-sectional formulation can be extended to formulate the characteristics of RC beam finite elements for time-variant nonlinear and limit analysis of concrete structures under corrosion. Details can be found in Biondini et al. (2004a), Biondini & Frangopol (2008b), Biondini and Vergani (2015).

Applications

The proposed approach is applied to the probabilistic assessment of structural redundancy and elapsed time between failures of a RC frame and a RC bridge deck under corrosion.

Constitutive laws of the materials

The constitutive behaviour of the materials is described in terms of stress–strain nonlinear relationships. For concrete, the Saenz's law in compression and a bilinear elastic-plastic model in tension are assumed, with: compression strength f_c; tension strength $f_{ct} = .25 f_c^{2/3}$; initial modulus $E_{c0} = 9500 f_c^{1/3}$; peak strain in compression $\varepsilon_{c0} = .20\%$; strain limit in compression $\varepsilon_{cu} = .35\%$; strain limit in tension $\varepsilon_{ctu} = 2 f_{ct}/E_{c0}$. For steel, a bilinear elastic-plastic model in both tension and compression is assumed, with yielding strength f_{sy}, elastic modulus $E_s = 210$ GPa, and strain limit $\varepsilon_{su} = 1.00\%$ associated with bond failure. In this way, the constitutive laws are completely defined by the material strengths f_c and f_{sy}.

Probabilistic modelling

The probabilistic analysis accounts for the uncertainty in both the geometrical and mechanical characteristics of the structural systems and in the parameters which define the deterioration processes. At cross-sectional level, the probabilistic model of the mechanical behaviour, diffusion process and damage propagation mechanism considers as random variables the material strengths f_c and f_{sy} of concrete and steel, respectively, the coordinates (y_p, z_p) of each nodal point $p = 1, 2, \ldots,$ of the cross-section, the coordinates (y_m, z_m) and diameter $\varnothing_m$ of each steel bar $m = 1, 2, \ldots,$ the diffusivity coefficient D, and the damage rate coefficients q_c and q_s. Nominal values are assumed as mean values. The probabilistic distributions and coefficients of variation are listed in Table 1 (Biondini et al., 2006; Sudret, 2008; Vismann & Zilch, 1995). The input random variables are uncorrelated to emphasise the effects of the lack of correlation on the investigated output random variables (Harr, 1996). Moreover, high values of the coefficient of

Table 1. Probability distributions and coefficients of variation (nominal values are assumed as mean values μ).

Random variable ($t = 0$)	Type	C.o.V.
Concrete strength, f_c	Lognormal	5 MPa/μ
Steel strength, f_{sy}	Lognormal	30 MPa/μ
Coordinates of nodal points, (y_p, z_p)	Normal	5 mm/μ
Coordinates of steel bars, (y_m, z_m)	Normal	5 mm/μ
Diameter of steel bars, $\varnothing_m$	Normal*	.10
Diffusivity, D	Normal*	.10
Concrete damage rate, $q_c = (C_c \Delta t_c)^{-1}$	Normal*	.30
Steel damage rate, $q_s = (C_s \Delta t_s)^{-1}$	Normal*	.30

*Truncated distributions with non negative outcomes.

variation are adopted for the random variables which mainly influence the time-variant uncertainty of the corrosion damage, such as the steel bar diameter and damage rates.

The lifetime probabilistic analysis is carried out by Monte Carlo simulation. The required accuracy of the simulation process is achieved through a posteriori estimation of the goodness of the sample size based on a monitoring of the time-variant statistical parameters of the random variables under investigation.

RC frame

The lifetime structural performance of the RC frame shown in Figure 3 is investigated in terms of redundancy and elapsed time between failures. The nominal material strengths are $f_c = 40$ MPa for concrete in compression and $f_{sy} = 500$ MPa for the yield of reinforcing steel. The frame is subjected to a dead load $q = 32$ kN/m applied on the beam and a live load λF acting at top of the columns, with $F = 100$ kN.

The frame system is designed with cross-sectional stiffness and bending strength capacities much larger in the beam than in the columns. Moreover, shear failures are avoided over the lifetime by a proper capacity design (Celarec, Vamvatsikos, &

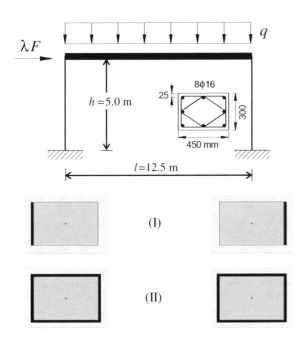

Figure 3. Reinforced concrete frame: geometry, structural scheme, cross-section of the columns, loading condition, grid of the diffusion model, and exposure scenarios with (I) columns exposed on one side and (II) columns exposed on four sides.

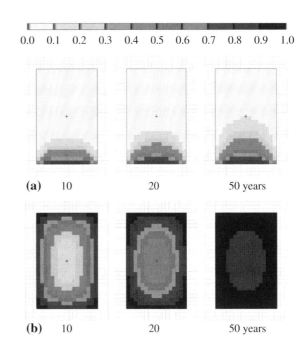

Figure 4. Maps of concentration $C(\mathbf{x}, t)/C_0$ of the aggressive agent after 10, 20, and 50 years from the initial time of diffusion penetration (nominal frame system): (a) scenario (I) with exposure on one side; (b) scenario (II) with exposure on four sides.

Dolšek, 2011; Titi & Biondini, 2014). In this way, a shear-type behaviour can be assumed, with the critical regions where plastic hinges are expected to occur located at the ends of the columns.

The structure is subjected to the diffusive attack from an aggressive agent located on the external surfaces of the columns with concentration C_0. The two exposure scenarios shown in Figure 3 are considered, with (I) columns exposed on the outermost side only, or (II) columns exposed on the four sides. A nominal diffusivity coefficient $D = 10^{-11}$ m²/sec is assumed. Figure 4 shows the deterministic maps of concentration $C(\mathbf{x}, t)/C_0$ for the two investigated exposure scenarios after 10, 20, and 50 years from the initial time of diffusion penetration.

The corrosion damage induced by diffusion is evaluated by taking the stochastic effects in the mass transfer into account (Biondini et al., 2004a). Corrosion of steel bars with no deterioration of concrete is assumed, with nominal damage parameters $C_s = C_0$, $\Delta t_s = 50$ years, and $C_{cr} = 0$. This model reproduces a deterioration process with severe corrosion of steel, as may occur for carbonated or heavily chloride-contaminated concrete and high relative humidity, conditions under which the corrosion rate can reach values above 100 µm/year (Bertolini et al., 2004).

Figure 5 shows the evolution over a 50-year lifetime of the failure load multipliers $\lambda_1 = \lambda_1(t)$ and $\lambda_c = \lambda_c(t)$ associated to the reaching of first local yielding of steel reinforcement and structural collapse of the frame system, respectively. The failure loads are computed at each time instant under the hypotheses of linear elastic behaviour up to first local yielding, and perfect plasticity at collapse.

The time evolution of the redundancy factor $\Lambda = \Lambda(t)$ of the frame system for the two investigated scenarios is shown in Figure 6. It is noted that for case (I) redundancy increases over time, even if the structural performance in terms of load capacity decreases. This is because the bending strength of the critical cross-sections corroded on the compression side deteriorates

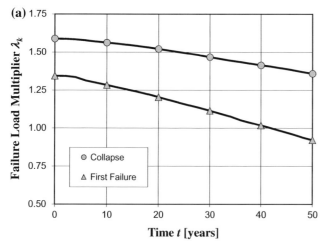

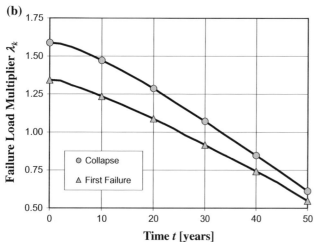

Table 2. Failure times T_1 and T_c and elapsed time ΔT associated to different target load values λ^*.

λ^*	T_1 [years]	T_c [years]	ΔT [years]
.70	41.4	46.4	5.0
.80	36.2	42.1	5.9
.90	30.8	37.7	6.9
1.00	25.1	33.2	8.1
1.10	19.2	28.6	9.4
1.20	12.4	24.0	11.6
1.30	4.9	19.3	14.5

Figure 5. Time evolution of the load multipliers λ_1 and λ_c associated with the reaching of first local yielding of steel reinforcement and collapse, respectively: (a) scenario (I) with exposure on one side; (b) scenario (II) with exposure on all sides.

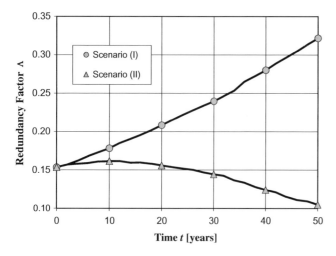

Figure 6. Time evolution of the redundancy factor Λ for scenario (I) with exposure on one side, and scenario (II) with exposure on all sides.

cross-section. On the contrary, redundancy mainly decreases over time for case (II). Therefore, case (II) is the worst damage scenario for structural redundancy.

The time evolution of the failure loads (Figure 5) indicates that the exposure scenario (II) may be critical with respect to structural collapse. Therefore, for this scenario it is of interest the assessment of the failure times T_1 and T_c associated to the occurrence of the first yielding and collapse, respectively, as well as the elapsed time $\Delta T = T_c - T_1$ between such failures. Failure times and elapsed times associated to different target values $\lambda^* = \lambda_1^* = \lambda_c^*$ are listed in Table 2 for the nominal system. These values indicate that after local failures a significant rapidity of repair may be required under severe exposures. Moreover, it can be noticed that the failure times decrease as the target load multiplier increases. The elapsed time between failures shows instead an opposite trend. Therefore, the availability of a larger reserve of load capacity with respect to the design target is beneficial to delay the occurrence of failures, but it may require prompter repair actions after local failures occur.

The effects of the uncertainty are investigated based on the probabilistic information given in Table 1. The two sets of random variables associated to each column are preliminarily assumed as uncorrelated to emphasise the effects of the uncertainty. Figure 7 shows the probability mass functions (PMFs) of the failure times T_1 and T_c (Figure 7(a)) and elapsed time ΔT (Figure 7(b)) for two deterministic values of the target load multiplier, $\lambda^* = 1.00$ and $\lambda^* = .75$, based on a sample of 2000 Monte Carlo realisations. For $\lambda^* = 1.00$ the failure times T_1 and T_c are characterised by mean and standard deviation values lower than the values obtained for $\lambda^* = .75$. On the contrary, the mean value of the elapsed time ΔT is higher for $\lambda^* = 1.00$ than for $\lambda^* = .75$, with a small difference in terms of dispersion. These results confirm that a suitable reserve of load capacity with respect to the design target allows to delay the possible occurrence of failure events, but it demands for higher promptness and rapidity in the recovery actions.

It is worth noting that the effects of randomness on the reserve load capacity $\Delta\lambda = \lambda_c - \lambda_1$ lead to mean values of the elapsed time ΔT higher than the nominal deterministic values. Moreover, the strong correlation between the failure load multipliers λ_1 and λ_c is beneficial to achieve a lower variance of the elapsed time ΔT than the variance of the failure times T_1 or T_c.

The influence of correlation is also studied by assuming the two sets of random variables associated to each column as fully correlated. The results lead to conclusions similar to the case of uncorrelated variables, with small changes in the probabilistic parameters of the investigated performance indicators. As an example, the mean and standard deviation values obtained for the elapsed time ΔT are $\mu = 11.1$ years and $\sigma = 4.2$ years for $\lambda^* = 1.00$, and $\mu = 7.3$ years and $\sigma = 2.6$ years for $\lambda^* = .75$.

more slowly compared to the cross-sections corroded on the tension side. Therefore, the collapse load multiplier λ_c, which depends on the bending strengths of all critical cross-sections, has a lower deterioration rate than the load multiplier at first yielding λ_1, which is associated with the failure of a single

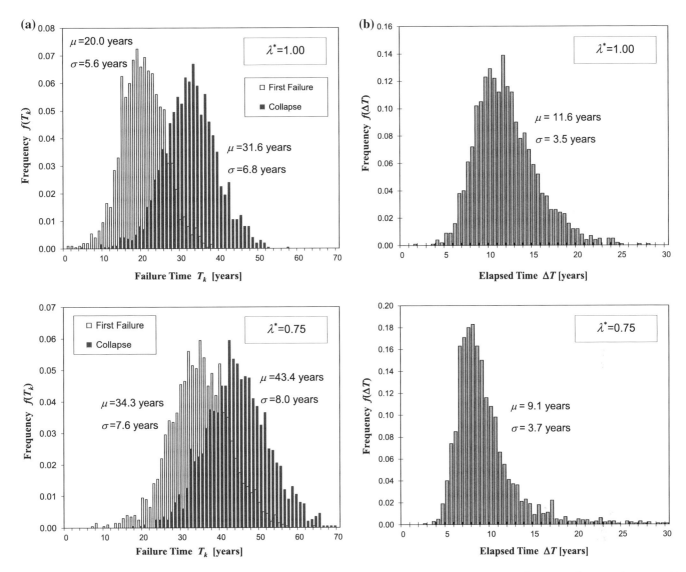

Figure 7. PMFs of (a) failure times T_1 and T_c and (b) elapsed time between failures ΔT for two values of the target load multiplier, $\lambda^* = 1.00$ and $\lambda^* = .75$, under scenario (II) with exposure on all sides.

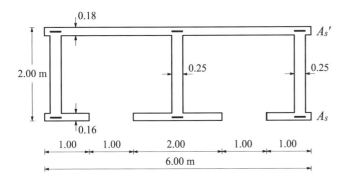

Figure 8. Reinforced concrete bridge deck: geometry of the cross-section and main steel reinforcement $A_s' = 48\varnothing 28$ mm and $A_s = 21\varnothing 28$ mm (additional reinforcing steel bars: $130\varnothing 8$ mm in the top slab; $60\varnothing 8$ mm in the bottom slab).

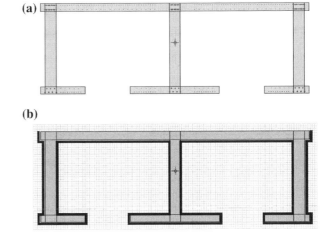

Figure 9. (a) Structural model of the bridge deck cross-section with indication of the steel reinforcement (259 steel bars: $48\varnothing 28$ mm + $130\varnothing 8$ mm in the top slab; $21\varnothing 28$ mm + $60\varnothing 8$ mm in the bottom slab). (b) Grid of the diffusion model and exposure scenario.

RC bridge deck

The lifetime structural performance of a RC bridge deck is investigated at cross-sectional level in terms of redundancy and elapsed time between failures. The geometry of the concrete cross-section

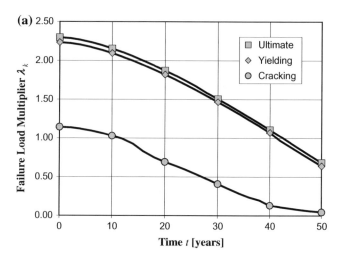

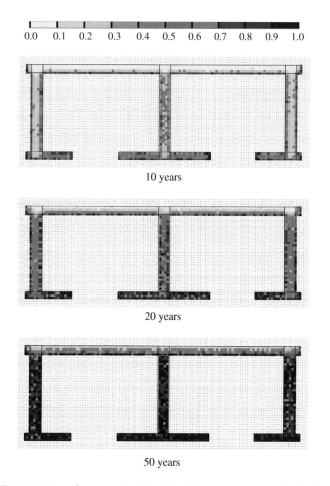

Figure 10. Maps of concentration $C(\mathbf{x}, t)/C_0$ of the aggressive agent after 10, 20, and 50 years from the initial time of diffusion penetration (nominal bridge deck under stochastic diffusion).

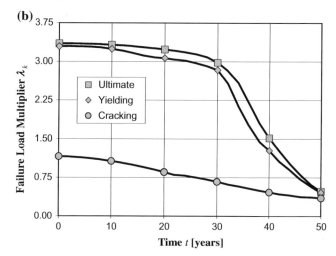

Figure 12. Time evolution of the failure load multipliers λ_k, with $k = 1,2,3$, at concrete first cracking ($k = 1$), steel first yielding ($k = 2$), and cross-section ultimate capacity ($k = 3$): (a) positive and (b) negative bending moment.

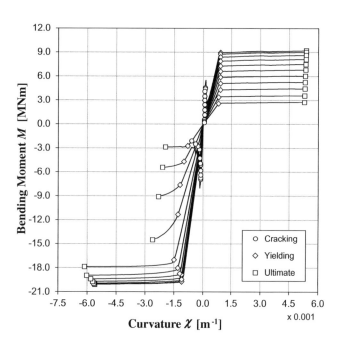

Figure 11. Time evolution of the nominal bending moment M vs. curvature χ diagrams over a 50-year lifetime ($\Delta t = 5$ years), with indication of the points associated to first cracking of concrete, first yielding of steel reinforcement, and ultimate flexural capacity of the cross-section.

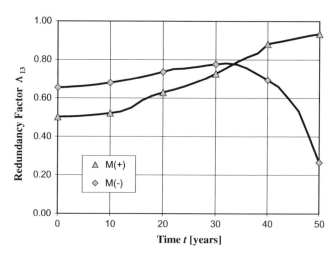

Figure 13. Time evolution of the nominal redundancy factor Λ_{13} between the states (1) and (3) associated to first concrete cracking and cross-section ultimate capacity, respectively, for positive and negative bending moment.

426

Table 3. Failure times T_1, T_2, T_3, and elapsed times between failures ΔT_{12}, ΔT_{23} [years]: (1) concrete first cracking; (2) steel first yielding and (3) cross-section ultimate capacity.

$\lambda^* = 1.0$	T_1	T_2	T_3	ΔT_{12}	ΔT_{23}
M^+	11.5	41.4	42.6	29.9	1.2
M^-	13.7	42.6	44.1	28.9	1.5

and the location of the main steel reinforcement in the top and bottom slabs are shown in Figure 8. The nominal dimensions are: width = 6.00 m; depth = 2.00 m; web thickness = .25 m; top slab thickness = .18 m; bottom slab thickness = .16 m. The steel reinforcement located in the top slab consists of 48 bars with nominal diameter $\varnothing$ = 28 mm, and 130 bars with $\varnothing$ = 8 mm. The steel reinforcement located in the bottom slab consists of 21 bars with $\varnothing$ = 28 mm and 60 bars with $\varnothing$ = 8 mm. Figure 9(a) shows the structural model of the cross-section, with detailed location of the steel bars. The nominal material strengths are f_c = 30 MPa for concrete in compression and f_{sy} = 300 MPa for the yield of reinforcing steel.

The bridge deck cross-section is subjected to the diffusive attack from an aggressive agent located with concentration C_0 on the external surface exposed to the atmosphere. The diffusion model and the exposure scenario are shown Figure 9(b). A nominal diffusivity coefficient $D = 10^{-11}$ m²/sec is assumed. Figure 10 shows the stochastic maps of concentration $C(\mathbf{x}, t)/C_0$ after 10, 20, and 50 years from the initial time of diffusion penetration.

A severe corrosion damage scenario is assumed, with nominal parameters $C_c = C_s = C_0$, Δt_c = 25 years, Δt_s = 50 years, and $C_{cr} = 0$. The mechanical damage induced by diffusion over a 50-year lifetime is shown in Figure 11 in terms of nominal bending moment M vs. curvature χ diagrams computed by assuming the bridge deck axially unloaded.

The results shown in Figure 11 indicate that damage significantly affect the flexural performance of the cross-section, both for positive and negative bending moments. Deterioration of structural performance is mainly due to the severe exposure of the bottom slab. For positive bending moment, the corrosion of the reinforcing steel bars in tension located in the bottom slab leads to a progressive bending strength deterioration over the lifetime, with no significant changes in the curvature ductility. For negative bending moment, the lower corrosion rate of the reinforcing steel bars in tension located in the top slab involves

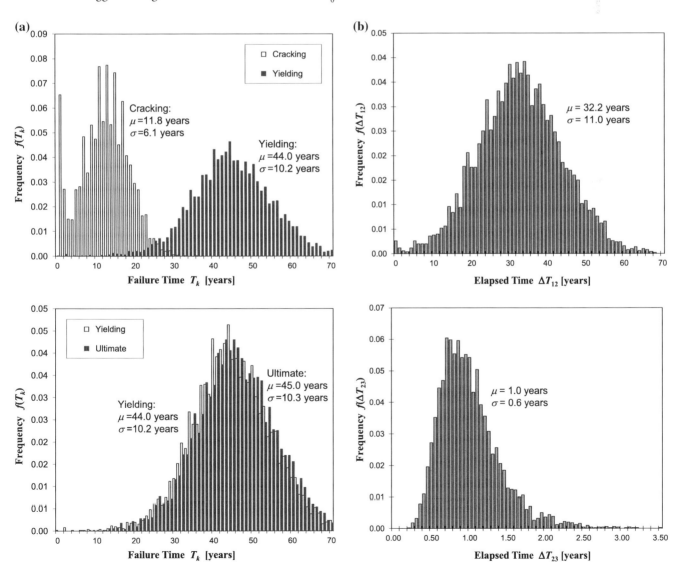

Figure 14. PMFs of (a) failure times T_1, T_2, T_3, and (b) elapsed times between failures ΔT_{12} and ΔT_{23} associated to the occurrence of the sequential limit states of (1) concrete first cracking, (2) steel first yielding and (3) cross-section ultimate capacity, for positive bending moment.

a limited reduction of bending strength over the first years of lifetime. However, after about 30 years of lifetime the severe deterioration of concrete in compression in the bottom slab causes a remarkable progressive decrease of structural performance in terms of both bending strength and curvature ductility.

At cross-sectional level, limit states of interest are the occurrence of local failures associated to cracking of concrete and yielding of steel reinforcement, which are warnings for initiation of damage propagation, as well as the attainment of the ultimate flexural capacity of the cross-section defined by the strain limits $\varepsilon_c = -\varepsilon_{cu}$ and/or $|\varepsilon_s| = \varepsilon_{su}$. The points associated to the limit states of (1) concrete first cracking, (2) steel first yielding, and (3) cross-section ultimate capacity, are indicated on the capacity curves shown in Figure 11. The corresponding time evolution of the $k = 1,2,3$, failure load multipliers $\lambda_k = \lambda_k(t)$, computed for the design values $M^+ = 4$ MNm and $M^- = -6$ MNm of the bending moments λM^+ and λM^-, is shown in Figure 12.

The reserve of load capacity after concrete cracking ensures a suitable level of overall structural redundancy for both positive and negative bending moment, as shown in Figure 13 in terms of redundancy factor $\Lambda_{13} = \Lambda_{13}(t)$ between (1) cracking and (3) ultimate capacity of the cross-section. For positive bending moment, redundancy increases continuously over time, even if the structural performance in terms of load capacity decreases. For negative bending moment, redundancy exhibits a moderate increase during the first period of exposure to damage, and rapidly decreases after about 30 years of lifetime. These results indicate that corrosion of steel reinforcement in tension, even though it involves a reduction of load capacity, may have beneficial effects in terms of structural redundancy. Contrary, the effects of deterioration of concrete in compression are generally detrimental to structural redundancy.

The reserve of load capacity after steel yielding is instead very limited and does allow for significant redundancy in between yielding and ultimate states. In this case the elapsed time between failures provides useful information about the available time to repair after a local yielding occurs.

With reference to a target load multiplier $\lambda^* = 1.0$, the failure times T_1, T_2, T_3, and the related elapsed times between failures ΔT_{12}, ΔT_{23}, associated to the occurrence of the sequential limit states of (1) concrete first cracking, (2) steel first yielding, and (3) cross-section ultimate capacity, are listed in Table 3 for the nominal system under positive and negative bending moments.

The uncertainty effects on these performance indicators are investigated also in probabilistic terms based on the probabilistic information given in Table 1. Based on 5000 Monte Carlo simulations, Figure 14 shows the PMFs of the failure times T_1, T_2, T_3 (Figure 14(a)), and elapsed time between failures ΔT_{12}, ΔT_{23} (Figure 14(b)) for the case of positive bending moment.

The deterministic and probabilistic results confirm that a remarkable rapidity of repair may be required after occurrence of a severe local failure, such as yielding of steel reinforcement. Moreover, failure loads and failure times associated to concrete cracking or other minor local failure events could provide warnings of more severe future damage states or critical threats and to support in this way the decision-making process for maintenance and repair planning.

Conclusions

Failure loads and failure times of deteriorating RC structures have been investigated. Life-cycle performance indicators, related to time-variant structural redundancy and elapsed times between sequential failures occurring under continuous deterioration processes, have been formulated. The effects of the damage process on the structural performance have been evaluated by considering uncertainties based a methodology for life-cycle assessment of concrete structures exposed to the diffusive attack from environmental aggressive agents. The proposed approach has been applied to the assessment of structural redundancy and elapsed time between failures of a RC frame and a RC bridge deck under corrosion.

The results show that the prediction of the local and global failure modes and of their occurrence in time provides useful information on the remaining life-cycle of deteriorating RC structures. In fact, after local failures occur, a very fast repair may be required under a severe exposure scenario to prevent structural collapse. Failure times and time intervals between subsequent failures must be computed for this purpose, since other damage-tolerance performance indicators, such as structural redundancy, do not provide a direct measure of the failure rate.

Therefore, failure times and elapsed time between failures are important performance indicators to be used jointly with other performance measures, such as reliability, redundancy, robustness, resilience, and sustainability for a rational approach to life-cycle design, assessment and maintenance of deteriorating structure and infrastructure systems. This approach is clearly more demanding than standard time-invariant design procedures, since it involves the modelling of complex deterioration processes and the evaluation of several performance indicators over the structural lifetime. For this reason, reliable deterioration modelling of materials and structural components and computationally efficient structural analysis procedures considering time effects, as those presented in this paper, are essential to a robust prediction of the time-variant structural performance and to support and advance the civil engineering profession in this field.

However, it is worth mentioning that deterioration models are generally very sensitive to change in the probabilistic parameters of the input random variables and their robust validation and accurate calibration are difficult tasks to be performed due to the limited availability of data. Further efforts aimed at gathering new data from both existing structures and experimental tests are crucial for a successful calibration and implementation in practice of the presented approach. Also efforts in the modelling of nonlinear structures using finite elements with time-variant properties (Biondini & Vergani, 2015; Biondini et al., 2004a), probabilistic finite element analysis (Biondini, Bontempi, Frangopol, & Malerba, 2004b; Biondini & Frangopol, 2008b; Teigen, Frangopol, Sture, & Felippa, 1991a, 1991b), reliability-based inspections (Onoufriou & Frangopol, 2002), probabilistic importance assessment of structural components (Gharaibeh, Frangopol, & Onoufriou, 2002), and developing improved models for cost and risk estimation of maintenance actions (Frangopol & Kong, 2001; Saydam & Frangopol, 2015) are necessary to ensure protection of civil infrastructure facilities over time at minimum life-cycle cost.

Disclosure statement

No potential conflict of interest was reported by the authors.

ORCID

Fabio Biondini (iD) http://orcid.org/0000-0003-1142-6261

References

Agarwal, J., Blockley, D. I., & Woodman, N. J. (2003). Vulnerability of structural systems. *Structural Safety, 25*, 263–286.

Al-Harthy, A. S., Stewart, M. G., & Mullard, J. (2011). Concrete cover cracking caused by steel reinforcement corrosion. *Magazine of Concrete Research, 63*, 655–667.

Almusallam, A. A. (2001). Effect of degree of corrosion on the properties of reinforcing steel bars. *Construction and Building Materials, 15*, 361–368.

Apostolopoulos, C. A., & Papadakis, V. G. (2008). Consequences of steel corrosion on the ductility properties of reinforcement bar. *Construction and Building Materials, 22*, 2316–2324.

ASCE. (2013, March). *Report card for America's infrastructure*. Reston, VA: American Society of Civil Engineers.

Baker, J. W., Schubert, M., & Faber, M. H. (2008). On the assessment of robustness. *Structural Safety, 30*, 253–267.

Barone, G., & Frangopol, D. M. (2014a). Life-cycle maintenance of deteriorating structures by multi-objective optimization involving reliability, risk, availability, hazard and cost. *Structural Safety, 48*, 40–50.

Barone, G., & Frangopol, D. M. (2014b). Reliability, risk and lifetime distributions as performance indicators for life-cycle maintenance of deteriorating structures. *Reliability Engineering & System Safety, 123*, 21–37.

Bertero, R. D., & Bertero, V. V. (1999). Redundancy in earthquake-resistant design. *Journal of Structural Engineering, 125*, 81–88.

Bertolini, L., Elsener, B., Pedeferri, P., & Polder, R. (2004). *Corrosion of steel in concrete*. Weinheim: Wiley-VCH.

Biondini, F. (2009). *A measure of lifetime structural robustness*. Proceedings of the SEI/ASCE Structures Congress, Austin, TX, USA, April 30–May 2. In L. Griffis, T. Helwig, M. Waggoner, & M. Hoit (Eds.), *Structures Congress 2009*. ASCE, CD-ROM.

Biondini, F. (2012). Discussion: Time-variant redundancy of ship structures, by Decò, A., Frangopol, D.M., & Okasha, N.M. *Society of Naval Architects and Marine Engineers Transactions, 119*, 40.

Biondini, F., Bontempi, F., Frangopol, D. M., & Malerba, P. G. (2004a). Cellular automata approach to durability analysis of concrete structures in aggressive environments. *Journal of Structural Engineering, 130*, 1724–1737.

Biondini, F., Bontempi, F., Frangopol, D. M., & Malerba, P. G. (2004b). Reliability of material and geometrically nonlinear reinforced and prestressed concrete structures. *Computers & Structures, 82*, 1021–1031.

Biondini, F., Bontempi, F., Frangopol, D. M., & Malerba, P. G. (2006). Probabilistic service life assessment and maintenance planning of concrete structures. *Journal of Structural Engineering, 132*, 810–825.

Biondini, F., Camnasio, E., & Titi, A. (2015). Seismic resilience of concrete structures under corrosion. *Earthquake Engineering and Structural Dynamics, 44*, 2445–2466.

Biondini, F., & Frangopol, D. M. (Eds.). (2008a). *Life-cycle civil engineering*. Boca Raton, FL, London, New York, NY, Leiden: CRC Press, Taylor & Francis Group, A.A. Balkema.

Biondini, F., & Frangopol, D. M. (2008b). Probabilistic limit analysis and lifetime prediction of concrete structures. *Structure and Infrastructure Engineering, 4*, 399–412.

Biondini, F., & Frangopol, D. M. (2014). Time-variant robustness of aging structures. Chapter 6. In D. M. Frangopol & Y. Tsompanakis (Eds.), *Maintenance and safety of aging infrastructure* (pp. 163–200). London:CRC Press, Taylor & Francis Group.

Biondini, F., & Frangopol, D. M. (2016). Life-cycle performance of deteriorating structural systems under uncertainty: Review. *Journal of Structural Engineering*. doi:10.1061/(ASCE)ST.1943-541X.0001544

Biondini, F., Frangopol, D. M., & Malerba, P. G. (2008). Uncertainty effects on lifetime structural performance of cable-stayed bridges. *Probabilistic Engineering Mechanics, 23*, 509–522.

Biondini, F., Frangopol, D. M., & Restelli, S. (2008). *On structural robustness, redundancy and static indeterminacy*. Proceedings of the SEI/ASCE 2008Structures Congress, Vancouver, BC, Canada, April 24–26, 2008. ASCE, CD-ROM.

Biondini, F., & Restelli, S. (2008). *Damage propagation and structural robustness*. First International Symposium on Life-Cycle Civil Engineering (IALCCE'08), Varenna, Italy, June 10–14. In F. Biondini & D. M. Frangopol (Eds.), *Life-cycle civil engineering* (pp. 565–570). CRC Press, Taylor & Francis Group.

Biondini, F., & Vergani, M. (2015). Deteriorating beam finite element for nonlinear analysis of concrete structures under corrosion. *Structure and Infrastructure Engineering, 11*, 519–532.

Bocchini, P., & Frangopol, D. M. (2012a). Optimal resilience- and cost-based postdisaster intervention prioritization for bridges along a highway segment. *Journal of Bridge Engineering, 17*, 117–129.

Bocchini, P., & Frangopol, D. M. (2012b). Restoration of bridge networks after an earthquake: Multicriteria intervention optimization. *Earthquake Spectra, 28*, 426–455.

Bruneau, M., Chang, S. E., Eguchi, R. T., Lee, G. C., O'Rourke, T. D., Reinhorn, A. M., Shinozuka, M., Tierney, K., Wallace, W. A., & Wintefeldt, D. V. (2003). A framework to quantitatively assess and enhance the seismic resilience of communities. *Earthquake Spectra, 19*, 733–752.

Cabrera, J. G. (1996). Deterioration of concrete due to reinforcement steel corrosion. *Cement and Concrete Composites, 18*, 47–59.

CEB. (1992). *Durable concrete structures – Design guide*. CEB Bulletin d'Information No. 183, Comité Euro-International du Béton. London: Thomas Telford.

Celarec, D., Vamvatsikos, D., & Dolšek, M. (2011). Simplified estimation of seismic risk for reinforced concrete buildings with consideration of corrosion over time. *Bulletin of Earthquake Engineering, 9*, 1137–1155.

Chang, S. E., & Shinozuka, M. (2004). Measuring improvements in the disaster resilience of communities. *Earthquake Spectra, 20*, 739–755.

Cimellaro, G. P., Reinhorn, A. M., & Bruneau, M. (2010). Framework for analytical quantification of disaster resilience. *Engineering Structures, 32*, 3639–3649.

Clifton, J. R., & Knab, L. I. (1989). *Service life of concrete*. NUREG/CR-5466. Washington, DC: U.S. Nuclear Regulatory Commission.

Decò, A., Frangopol, D. M., & Okasha, N. M. (2011). Time-variant redundancy of ship structures. *SNAME Journal of Ship Research, 55*, 208–219.

Decò, A., Bocchini, P., & Frangopol, D. M. (2013). A probabilistic approach for the prediction of seismic resilience of bridges. *Earthquake Engineering and Structural Dynamics, 42*, 1469–1487.

Du, Y. G., Clark, L. A., & Chan, A. H. C. (2005). Residual capacity of corroded reinforcing bars. *Magazine of Concrete Research, 57*, 135–147.

Ellingwood, B. R. (2005). Risk-informed condition assessment of civil infrastructure: State of practice and research issues. *Structure and Infrastructure Engineering, 1*, 7–18.

Ellingwood, B. R. (2006). Mitigating risk from abnormal loads and progressive collapse. *Journal of Performance of Constructed Facilities, 20*, 315–323.

Ellingwood, B. R., & Dusenberry, D. O. (2005). Building design for abnormal loads and progressive collapse. *Computer-Aided Civil and Infrastructure Engineering, 20*, 194–205.

Enright, M. P., & Frangopol, D. M. (1999). Reliability-based condition assessment of deteriorating concrete bridges considering load redistribution. *Structural Safety, 21*, 159–195.

fib (2006). *Model code for service life design*. Bulletin No. 34. Lausanne: Fédération internationale du béton/International Federation for Structural Concrete.

Frangopol, D. M. (2011). Life-cycle performance, management, and optimisation of structural systems under uncertainty: Accomplishments and challenges. *Structure and Infrastructure Engineering, 7*, 389–413.

Frangopol, D. M., & Bocchini, P. (2011). *Resilience as optimization criterion for the bridge rehabilitation of a transportation network subject to earthquake*. Proceedings of the SEI/ASCE Structures Congress, Las Vegas, NV, USA, April 14–16. In D. Ames, T. L. Droessler, & M. Hoit (Eds.), *Structures Congress 2011* (pp. 2044–2055). ASCE, CD-ROM.

Frangopol, D. M., & Curley, J. P. (1987). Effects of damage and redundancy on structural reliability. *Journal of Structural Engineering, 113*, 1533–1549.

Frangopol, D. M., & Ellingwood, B. R. (2010). Life-cycle performance, safety, reliability and risk of structural systems, Editorial. *Structure Magazine*. Chicago, IL: National Council of Structural Engineering Associations, NCSEA.

Frangopol, D. M., Iizuka, M., & Yoshida, K. (1992). Redundancy measures for design and evaluation of structural systems. *Journal of Offshore Mechanics and Arctic Engineering, 114*, 285–290.

Frangopol, D. M., & Klisinski, M. (1989). Weight-strength-redundancy interaction in optimum design of three-dimensional brittle-ductile trusses. *Computers and Structures, 31*, 775–787.

Frangopol, D. M., & Kong, J. S. (2001). Expected maintenance cost of deteriorating civil infrastructures. In D. M. Frangopol & H. Furuta (Eds.), *Life-cycle cost analysis and design of civil infrastructure systems* (pp. 22–47). Reston, VA: ASCE.

Frangopol, D. M., & Nakib, R. (1991). Redundancy in highway bridges. *AISC Engineering Journal, 28*, 45–50.

Frangopol, D. M., & Saydam, D. (2014). Structural performance indicators for bridges. Chapter 9. In W.-F. Chen & L. Duan (Eds.), *Bridge engineering handbook – Second Edition, Vol. 1 – Fundamentals* (pp. 185–205). Boca Raton, FL, London: CRC Press, Taylor & Francis Group.

Frangopol, D. M., & Soliman, M. (2016). Life-cycle of structural systems: recent achievements and future directions. *Structure and Infrastructure Engineering, 12*(1), 1–20.

Fu, G., & Frangopol, D. M. (1990). Balancing weight, system reliability and redundancy in a multiobjective optimization framework. *Structural Safety, 7*, 165–175.

Furuta, H., Kameda, T., Fukuda, Y., & Frangopol, D. M. (2004). Life-cycle cost analysis for infrastructure systems: Life-cycle cost vs. safety level vs. service life. Keynote Paper. In D. M. Frangopol, E. Brühwiler, M. H. Faber, & B. Adey (Eds.), *Life-cycle performance of deteriorating structures: Assessment, design and management* (pp. 19–25). Reston, VA: ASCE.

Gharaibeh, E. S., Frangopol, D. M., & Onoufriou, T. (2002). Reliability-based importance assessment of structural members with applications to complex structures. *Computers & Structures, 80*, 1111–1131.

Ghosn, M., Moses, F., & Frangopol, D. M. (2010). Redundancy and robustness of highway bridge superstructures and substructures. *Structure and Infrastructure Engineering, 6*, 257–278.

Glicksman, M. E. (2000). *Diffusion in solids*. New York, NY: Wiley.

Harr, M. E. (1996). *Reliability-based design in civil engineering*. Mineola, NY: Dover.

Hendawi, S., & Frangopol, D. M. (1994). System reliability and redundancy in structural design and evaluation. *Structural Safety, 16*, 47–71.

Husain, M., & Tsopelas, P. (2004). Measures of structural redundancy in reinforced concrete buildings. I: Redundancy indices. *Journal of Structural Engineering, 130*, 1651–1658.

Lind, N. C. (1995). A measure of vulnerability and damage tolerance. *Reliability Engineering & System Safety, 48*(1), 1–6.

Liu, Y., & Weyers, R. E. (1998). Modeling the dynamic corrosion process in chloride contaminated concrete structures. *Cement and Concrete Research, 28*, 365–379.

Lu, Z., Yu, Y., Woodman, N. J., & Blockley, D. I. (1999). A theory of structural vulnerability. *The structural engineer, The Institution of Structural Engineers, 77*, 17–24.

Malerba, P. G. (Ed.). (1998). *Analisi limite e non lineare di structure in calcestruzzo armato* [Limit and nonlinear analysis of reinforced concrete structures]. Udine: International Centre for Mechanical Sciences (CISM). (In Italian).

Mori, Y., & Ellingwood, B. R. (1994). Maintaining reliability of concrete structures. I: Role of inspection/repair. *Journal of Structural Engineering, 120*, 824–845.

NCHRP (2006). *Manual on service life of corrosion-damaged reinforced concrete bridge superstructure elements* (National Cooperative Highway Research Program, Report 558). Washington, DC: Transportation Research Board.

Okasha, N. M., & Frangopol, D. M. (2009). Lifetime-oriented multi-objective optimization of structural maintenance considering system reliability, redundancy and life-cycle cost using GA. *Structural Safety, 31*, 460–474.

Okasha, N. M., & Frangopol, D. M. (2010). Time-variant redundancy of structural systems. *Structure and Infrastructure Engineering, 6*, 279–301.

Onoufriou, T., & Frangopol, D. M. (2002). Reliability-based inspection optimization of complex structures: A brief retrospective. *Computers & Structures, 80*, 1133–1144.

Paliou, C., Shinozuka, M., & Chen, Y. (1990). Reliability and redundancy of offshore structures. *Journal of Engineering Mechanics, 116*, 359–378.

Pandey, P., & Barai, S. (1997). Structural sensitivity as a measure of redundancy. *Journal of Structural Engineering, 123*, 360–364.

Pastore, T., & Pedeferri, P. (1994). La corrosione e la protezione delle opere metalliche esposte all'atmosfera [Corrosion and protection of metallic structures exposed to the atmosphere]. *L'edilizia, 1994*, 75–92. (In Italian).

Sabatino, S., Frangopol, D. M., & Dong, Y. (2015). Sustainability-informed maintenance optimization of highway bridges considering multi-attribute utility and risk attitude. *Engineering Structures, 102*, 310–321.

Saydam, D., & Frangopol, D. M. (2011). Time-dependent performance indicators of damaged bridge superstructures. *Engineering Structures, 33*, 2458–2471.

Saydam, D., & Frangopol, D. M. (2015). Risk-based maintenance optimization of deteriorating bridges. *Journal of Structural Engineering, 141*(4), 04014120, 1–10.

Schafer, B. W., & Bajpai, P. (2005). Stability degradation and redundancy in damaged structures. *Engineering Structures, 27*, 1642–1651.

Starossek, U., & Haberland, M. (2011). Approaches to measures of structural robustness. *Structure and Infrastructure Engineering, 7*, 625–631.

Stewart, M. G. (2009). Mechanical behaviour of pitting corrosion of flexural and shear reinforcement and its effect on structural reliability of corroding RC beams. *Structural Safety, 31*, 19–30.

Sudret, B. (2008). Probabilistic models for the extent of damage in degrading reinforced concrete structures. *Reliability Engineering and System Safety, 93*, 410–422.

Teigen, J. G., Frangopol, D. M., Sture, S., & Felippa, C. A. (1991a). Probabilistic FEM for nonlinear concrete structures. I: Theory. *Journal of Structural Engineering, 117*, 2674–2689.

Teigen, J. G., Frangopol, D. M., Sture, S., & Felippa, C. A. (1991b). Probabilistic FEM for nonlinear concrete structures. II: Applications. *Journal of Structural Engineering, ASCE, 117*, 2690–2707.

Thoft-Christensen, P. (1998). Assessment of the reliability profiles for concrete bridges. *Engineering Structures, 20*, 1004–1009.

Titi, A., & Biondini, F. (2014). Probabilistic seismic assessment of multistory precast concrete frames exposed to corrosion. *Bulletin of Earthquake Engineering, 12*, 2665–2681.

Titi, A., & Biondini, F. (2016). On the accuracy of diffusion models for life-cycle assessment of concrete structures. *Structure and Infrastructure Engineering, 12 *, 1202–1215.

Vidal, T., Castel, A., & François, R. (2004). Analyzing crack width to predict corrosion in reinforced concrete. *Cement and Concrete Research, 34*, 165–174.

Vismann, U., & Zilch, K. (1995). Nonlinear analysis and safety evaluation by finite element reliability method. In: *New developments in non-linear analysis method*, CEB Bulletin d'Information No. 229 (pp. 49–73). Lausanne: Comité Euro-International du Béton.

Wolfram, S. (1994). *Cellular automata and complexity – Collected papers*. Reading, MA: Addison-Wesley.

Zhang, R., Castel, A., & François, R. (2010). Concrete cover cracking with reinforcement corrosion of RC beam during chloride-induced corrosion process. *Cement and Concrete Research, 40*, 415–425.

Zhu, B., & Frangopol, D. M. (2012). Reliability, redundancy and risk as performance indicators of structural systems during their life-cycle. *Engineering Structures, 41*, 34–49.

Zhu, B., & Frangopol, D. M. (2013). Risk-based approach for optimum maintenance of bridges under traffic and earthquake loads. *Journal of Structural Engineering, 139*, 422–434.

Zhu, B., & Frangopol, D. M. (2015). Effects of post-failure material behaviour on redundancy factor for design of structural components in nondeterministic systems. *Structure and Infrastructure Engineering, 11*, 466–485.

Index

Note: Page numbers in *italics* refer to figures
Page numbers in **bold** refer to tables

agent cost 104
airborne chlorides: attenuation of 158–159, *159*; hazard associated with 159–160, *159*, *160*, 217–218, *218*, 247–248
Akgül, F. 8, 32
Akiyama, M. 200, 217, 218, 242
aleatory uncertainty 5, 14, 98
algorithms 99–100
Alonso, C. 213
American Association of State Highway and Transportation Officials (AASHTO) 8, 335, 399
Ang, A.H.-S. 101, 288
Aoyama, M. 218
area method 169
asset management 12, 40
attenuation relationships 247, 275
autoregressive modelling 59
availability function 111, 112

Babler, R. 200
Barone, G. 110
bascule bridges *see* movable bridges
Bastidas-Arteaga, E. 48
Bayesian updating 57, 98, 100, 250, 319–320; conjugate priors 321; with large SHM samples 323–324, *324*; Metropolis-Hasting algorithm 321–322; non-conjugate priors 321; slice sampling algorithm 322; test example 322–323, *323*, **323**; updating random variable *X* 320–321
Beck, M. 200
Berto, L. 241
bicriterion optimisation 301, *305*, **305**, 306–307, *306*, 307, 313
bi-objective optimisation 134–135, *134*, **135**
Bocchini, P. 32, 34, 35, 37, 38, 56
bottleneck assumption 34, 35
Bressolette, P.H. 48
bridge damage index (BDI) 73, 276, 279
bridge maintenance management 101, 102
bridge maintenance planning 99–100; breakdown of maintenance cost *90*; concrete bridge model 88–89, *88*, **88**, *89*; effects of repair, restoring, and reconstruction **90**; maintenance strategies and life-cycle cost 89; material, labour, and scaffold costs **91**; multi-objective genetic algorithm 90–92, *92*; multi-objective problem 90; numerical example 92, *93*, 94–95, *94–95*; Pareto optimal solutions *92*
bridge management programs 13
bridge management systems (BMSs) 98, 101
bridge networks 30–32; and hospitals *see* healthcare–bridge network system; from individual structures to network 34–35; layout *36*; maintenance management 103–105, *105*, *106*; maintenance optimisation 38–39, *39*; network analysis and performance indicators 36–38; networks of networks 40; post-event recovery and resilience 39–40; reliability 14; risk assessment of 52–53; spatially distributed systems 32–34; time-dependent problems 35–36; time-dependent reliability 38, *38*

bridge piers *see* seismic performance of reinforced concrete structures
bridges, combined inspection/monitoring planning of: description *311*, **311**, 312–313; optimum balance of cost and expected damage detection delay 313–314, **313**
bridges, life-cycle performance/cost of 66–68; climate change 75–76, *75*; design 100–101; effects of maintenance *67*; extreme events 72–75, *73*, *74*, *75*; fatigue and fracture 72; integration of SHM and updating in bridge management 76, *77*; life-cycle cost 69, *70*; live load and corrosion 72; performance evaluation and prediction 68–71; probabilistic life-cycle optimisation 76–78, *77*, *78*; reliability 68–69; risk 69–70; sustainability 70–71, *71*; utility 71, *71*
BRIDGIT 68, 101
brittle systems, redundancy factors of 399, 401–402, *402*; with many components 409–414, *410*, *411*, **412**, *413*, **413**, *414*, **414**
Brockhoff, D. 131, 132
Bruneau, M. 35, 39
Burkert, A. 200

capacity reliability 104
Carlsson, C. 131
central safety factor 356–357, *359*, *360*, 362, 363, 366
change point detection method 59
Chateauneauf, A. 48
Chen, N.-Z. 10
Cheng, F.Y. 101
Christa McAuliffe Bridge 337–339, *338*, *339*
civil infrastructure 96–98; bridge maintenance management 101; design of bridges and buildings 100–101; deterioration mechanism and modelling 98–99; maintenance management at network 103–105, *105*, *106*; maintenance management at project level 102–103, *103*; maintenance management systems integrated with health monitoring 105–107; numerical optimization for maintenance management 99–100; performance indicators 98
civil infrastructure, life-cycle cost of 100–101, 229–230; genetic algorithms 230; multi-objective genetic algorithm 230–232, *231*, *233*; numerical example 232; road network 235–236; seismic risk 233–234, *234*, **234**; structural performance 230
clamping force, rivet 176, 179, 181, *181*, 182, *182*
climate change 49, 68; effect on bridges 75–76, *75*
component analysis 6, 110–112
component reliability 7, 68–69
concrete bridge model 88–89, *88*, **88**, *89*
concrete structures 143–144; arch bridge 148, *149–150*, **150**, *151–152*, 151–153, *153–155*, **154**; equilibrium and compatibility conditions 144–145; lifetime performance of deteriorating cross-sections 147–148; limit analysis of framed structures 144–146, *144*, *146*; modelling of structural damage 147; nonlinear structural analysis 148; probabilistic analysis and lifetime prediction 151–153, *154–155*; probabilistic prediction of structural lifetime 148; reference statistical quantities *144*; reference systems *144*; simulation of diffusion process 147; static and kinematic

approach (duality) 146; time-variant limit analysis 146, 151, *153, 154*; yield conditions and flow rule 145–146, *146*; *see also* reinforced concrete (RC) structures
condition assessment of movable bridges 335–336, *336*
condition index 98
condition-state models 13
confidence intervals 320
conjugate priors 321
connectivity reliability 104
Connor, R.J. 294–295
coordinate transformation matrix 145
Cornell, C.A. 69
corrective maintenance 33
correlation-based objective reduction approach 131, *132*
corrosion 47–49, 157, 216; bridges 72, 88, 89; cracking, of reinforced concrete structures *see* reinforced concrete (RC) structures, steel weight loss/corrosion cracking of; in deteriorating reinforced concrete structures 421; initiation and propagation under uncertainty 132; probability of crack occurrence due to steel corrosion 160–163, *160*, **161**, *162*, *163*; and reliability 9; and seismic performance *see* seismic performance of reinforced concrete structures
crack growth model 54
crack occurrence due to steel corrosion 161–162
crawl tests 58–59
critical distance 169, 172
cumulative long-term cost 104
cumulative probability of failure 50–51, *51*, 111, *113*, 115, *117*, 250, 325
cycle counting methods 187
Czarnecki, A.A. 8

damage-based safety margin 129, *129*
damage detection 59; role of SHM and inspection information in 57–59
damage detection delay 54–55, 300–301; expected 128–129, 302–303, *302, 304*, 307–312, 313–314
decision-making, life-cycle 56–57, 109; computations involved in decision support tool *111*
Decò, A. 51, 52
deformation-induced fatigue 168
Denton, S. 98
DesRoches, R. 255
deterioration modelling of reinforced concrete structures 420–421; corrosion process 421; damage rates 421–422; diffusion process 421
deterministic design 12
Devine, E.A. 193
diffusion process, in deteriorating concrete structures 147, 421
digital image analysis, steel weight loss estimation by *203*, 205–207
dominance relation-based objective reduction approach 131–132
Dong, Y. 52, 57, 70, 77, 110, 217
Dousti, A. 216
ductile systems, redundancy factors of 399, 400–401, *401*, **401**; with many components 406–409, *407*, **408**, **409**, *410*
dynamic amplification factor (DAF) 172

earthquakes 32, 52, 73, 74; healthcare–bridge network system 274–283; and life-cycle cost of civil infrastructure 229–237; performance analysis of Tohoku-Shinkansen viaducts 254–272; resilience 72; seismic performance of reinforced concrete structures 239–251
ε-constraint method 99–100, 311
Efstathiou, J. 37
element-level condition rating method 52, 68
Ellingwood, B.R. 8, 35
empirical cumulative distribution function (ECDF) 292, 296, *297*
empirical probability density function 223
Enright, M.P. 9, 69
environmental metric of sustainability 70–71

epistemic uncertainty 5, 14, 98, 220, 250
essential maintenance (EM) 19–20, 33, 34, 49, 55–56, 67, 112–113, 319
Estes, A.C. 8, 56
Evans, S.P. 36
event tree analysis 19–20, 54, 56, 101, 104, 302
exponential decay function 275, 279
exponential utility function 57, *112*, 116
extreme events 32, 52, 66; effect on bridges 72–75, *73*, *74*, *75*; space and time interconnection among *34*
extreme value distributions (EVDs) 288, 289, 290, 321; parameter estimation 295–298; single, transformation of multiple distributions into 291–294, **292**, *293*, *294*, *295*

failure load/failure time *see* reinforced concrete (RC) structures, deteriorating
failure path approach *see* member replacement method
Fang, C. 242
fatigue 49, 57, 300; assessment, of riveted railway bridge connections *see* riveted railway bridge connections, fatigue assessment of; bridges 72; limit states 189–190; probability of fatigue damage detection 301–302; ship structures *see under* high-speed ship structures; *see also* high-speed ship structures; inspection/monitoring planning, combined
fatigue damage ratio 186
fatigue strength 186
Fick's laws of diffusion 99, 147, 421
finite difference approach 385
finite element (FE) analysis 69, 168, 169–170, *170*, *171*, 317, 318, 325
finite element-incremental curvature method 9
first-failure based reliability 9–10
first-order reliability method (FORM) 8, 192, 354, 381, 384, 385, 390
Fisher, J.W. 313
Florida Department of Transportation (FDOT) 336
flow rule 145–146, *146*
fracture, bridges 72
fracture mechanics 49
fragility curves 73, 160, 162, *162*, 218–219, *219*, 245–246, *246*, 254–255, 262–264, **264**, 268–269, *270–271*, 275, 279, *279*
Frangopol, D.M. 7, 9, 14, 21–22, 32–35, 38, 54–56, 58, 70, 100–101, 157, 217, 229, 294, 300, 318, 399–400, 406
friction of moving bridges 343–345, *345*, *346*
Frieze, P.A. 187
Fullér, R. 131
Fully Connected Ratio (FCR) 37
functionality assessment, hospital 276–277, **280**, *281*
Furuta, H. 231

Gamma distribution 291
Gardoni, P. 241
Gaussian distribution 288, *289*, 291, 321, 324
generalised extreme value (GEV) probability density function 307
genetic algorithms (GAs) 19, 23, 38, 55, 56, 90, 97, 98, 100, 110, 229, 230, 310; *see also* multi-objective genetic algorithm (MOGA)
Ghosn, M. 382, 384, 385, 386, 387
Gindy, M. 17, 289, 291
global temperature 75
Goebbels, J. 200
Golroo, A. 38
graph theory 36
gravitational model 37
Great East Japan earthquake (2011): acceleration time history measured during *259*; damage investigation of Shinkansen viaducts 257–262; damage states of Shinkansen viaducts after 258, *261*; damages to Tohoku-Shinkansen viaducts before retrofit during *262*; response accelerations during *259*, *269*; *see also* Tohoku-Shinkansen viaducts
ground motion prediction equations *see* attenuation relationships

Gumbel distribution 288, 289, *289*, **290**, 292
Gylltoft, K. 242

Hamada, H. 223
Hattori, N. 265
hazard function 111–112
healthcare–bridge network system 73, *74*, 274–275; bridge seismic vulnerability 275–276; earthquake scenarios 275; hospital damage assessment 280–281, *281*, **281**; hospital functionality assessment 276–277, **280**, *281*; illustrative example 278–282; layout *278*; link seismic damage assessment 276; seismic performance of bridges and links 279–280, *280*, **280**; system level performance assessment 277–278, *278*, 281–282, *281–282*, **282**
heavy movable structures *see* movable bridges
Hendawi, S. 377
Hess, P. 9
heuristic algorithms 100
high-cycle fatigue failure 72
high-speed ship structures 185–186; fatigue reliability analysis 189–192, *190*, **190**, 194–196, *195*, **195**, *196*; fatigue resistance and loads 186–189, *192*, 193–194, *193*, *194*; probabilistic lifetime loads prediction for fatigue 188–189, *189*, 191; sea load estimation based on simulation/monitoring 186–187; segmented model *191*, 192–193; S–N approach and Miner's rule 186, *186*; stress range bin histogram and probability density functions 187–188, 191
Highway Bridge Replacement and Rehabilitation Program (HBRRP) 14
highway bridges, assessment and performance prediction of 287–288; characterising monitoring-based distributions and their uncertainty 289–294; Lehigh River Bridge 294–298, *295*, *296*, **296**, *297*, **298**; observation of transformation of single known distribution 289–291, **290**, *291*; observation on transformation of multiple distributions into a single EVD 291–294, *292*, *293*, **294**, *295*; theoretical background and its application 288–299
highway bridges, redundancy of 376; behaviour of bridge systems 379–380, *380*; multi-beam superstructure systems 382–387, *383*, *384*, **385**, *387*; substructures 387–394, *388*, *389*, **389**, **390**, **392**; system safety, redundancy and robustness 379–382; truss bridge 394–396, *394*, **394**, *395*
hospital damage index (HDI) 277, 280–281
hospitals *see* healthcare–bridge network system
hot-spot stress approach 168, 169
Hyogo-Ken Nanbu earthquake (1995) 255
hyper-parameters 320

I-39 Bridge 325; Bayesian updating 328–330, *328*, *329*, *330*; lifetime reliability, updating *330*, 331; monitoring program and data extraction 327–328, *327*; system performance analysis 325–327, *326*, **326**, *327*
Idealised Structural Unit Method 9
Iida, Y. 104
incremental curvature method 10
incremental method 8, 10, 354, 358
incremental nonlinear finite element analyses (INL-FEA) 8, 9, 383–385, *383*, *384*, **385**
infrastructure management 47; programs 13; role of optimisation in *19*
inspection 54–56, 127, 128–129; information, role in bridge management 76; information, role in performance updating and damage detection 57–59, *58*; movable bridges 336–337, *337*, *338*
inspection/monitoring planning, combined 300–301; application to existing bridge *311*, **311**, 312–314, **313**; application to naval ship **306**, 307–312, *307*, *308*, *309*, *310*, **310**; bicriterion optimisation *305*, **305**, 306–307, *306*; cost 303, 305–306; expected damage detection delay 302–303, *302*, *304*; prediction of crack growth 301; probability of fatigue damage detection 301–302
interdependent infrastructure systems *see* healthcare–bridge network system

Isecke, B. 200
Iwate-Miyagi Nairiku earthquake (2008) 268, *269*
Japan: damage to bridges due to recent earthquakes in 255–257; seismic design and retrofit of bridges in 255, 257, *258*; *see also* Tohoku-Shinkansen viaducts
jetty structures, reinforced concrete deck slab of 216–217, *217*; deterioration states **221**; hazard assessment associated with airborne chloride 217–218, *218*; illustrative example *221*, *222*, **222**, 223–224, *223*, **223**, *224*; life-cycle reliability estimation 220–224; Marko model 220–221, *221*; reliability assessment 218–219, *219*, *220*; reliability-based durability design 219, *221*; Sequential Monte Carlo Simulation 221–223
Joetsu-Shinkansen viaducts 256, *257*
joint high-speed sealift ship (JHSS) 186; model *191*, 192–193; primary vertical bending moment *192*

Kanakubo, T. 242
Kaplan, P. 185
Kawamura, C. 161
Khan, Fazlur R. 47
Kim, S. 21, 22, 54, 55, 300
Kim, S.-H. 268
kinetic architecture 334
Kobayashi, K. 210, 242
Kong, J.S. 101
Kumar, R. 241
Kwon, K. 54

LAMP 187
Latin hypercube sampling 8, 9, 52, 318
least squares method 320–321
Lee, Y.-J. 36
Lehigh River Bridge 294–298, *295*, *296*, **296**, *297*, **298**
life-cycle bridge management framework, integration of SHM in 316–317; Bayesian updating 319–324; building finite element models 317; end of service life 319; extreme value prediction 325; flowchart *317*, *318*; I-39 Bridge 325–330; life-cycle performance, computing 317–319; maintenance optimisation 319; modelling of live load 324–325; structural health monitoring 319; updating life-cycle performance 319
life-cycle cost (LCC) 16–17, 52, 53, 89, 91, *93*, 97; of bridges 69, *70*, 91, *93*, 94; of civil infrastructure *see* civil infrastructure, life-cycle cost of; minimisation, optimum solution based on *18*; service life management 130; *see also* bridges, life-cycle performance/cost of
life-cycle management (LCM) 5–6, 19, 46–47, 66, 76; inspection, maintenance and retrofit optimisation 54–56, *54*, *55*; integrated management, framework *6*, *77*; integration of structural health monitoring in 10–18; performance of structures and systems 47–53; post-hazard functionality restoration 56; procedure *48*; risk 51–53; role of SHM and inspection information in performance updating and damage detection 57–59, *58*; under uncertainty, computational framework *48*; utility-based life-cycle decision-making 56–57; *see also* bridges, life-cycle performance/cost of
life-cycle performance 5, 9, 10, 319; of bridges *see* bridges, life-cycle performance/cost of; index profile, with/without monitoring *10*; profile, under uncertainty *16*
life-cycle reliability estimation, of RC deck slab of jetty structure 220–224
life cycle utility-informed maintenance planning 109–110; attributes evaluation 113–116, *114–116*, **115**; benefit 116; case study 119–123, *121*, **121**, *122*, **122**, *123*, *124*; consequences 114–116; cost 114; illustrative example 119; lifetime functions 110–113, *111–114*; multi-attribute utility 116–118; single attribute utility assignment 116; tri-objective optimisation 118–119, *119*, **119**, 121; utility assessment 116–118, *117–118*, **118**; utility associated with performance benefit 118
lifelines 40
lifetime distributions 110; effects of essential maintenance 112–113

lifetime functions 9, 50, 56, 69, 109, 110, 318; component analysis 110–112; system analysis 112

lifetime prediction of concrete structures 151–153, *154–155*

lifetime sea loads *see* high-speed ship structures

likelihood function 58, 320, 321, 322, *324*

limit analysis 143; of framed structures 144–146, *144*, *146*

limit states 6–7, 8, 33, 50, 173, 354, 360, 389, 415; fatigue 189–190; *see also* reinforced concrete (RC) structures, deteriorating

linear normalisation technique 231, 232

line method 169

link damage index (LDI) 73

Liu, M. 14, 20, 37, 101, 325

live loads on bridges 72; modelling of 17–18, 324–325

load resistance factored design (LRFD) 12, 377, 380, 392, 399

load sharing factor 356–357, 359, 362, 365

lognormal distribution 188, 357, 363

low-cycle fatigue failure 72

lower bound theorem 146

'low-probability high-consequence' events 30, 31

Lua, J. 9

Lundgren, K. 242

maintenance, life-cycle 54–56, *54*; *see also* life cycle utility-informed maintenance planning

maintenance management 97; bridges 33; civil infrastructure 99–100, 102–107; systems, integrated with health monitoring 105–107

maintenance optimisation 19–20, 318, 319; bridge networks 38–39, *39*

maintenance plan *see* bridge maintenance planning

marine environment, reinforced concrete structures in 157–158; attenuation of airborne chlorides 158–159, *159*; durability design factors *163*, **164**, 165; evaluation of probability of crack occurrence due to steel corrosion 160–163; hazard associated with airborne chlorides 159–160, *159*, *160*; occurrence of steel corrosion and corrosion cracking *158*; performance function for crack occurrence 161–162, *162*; performance function for steel corrosion 160–161, *160*, **161**, *162*; reliability assessment 162–163, *162*, *163*; reliability-based design criterion 163–165; seismic capacity 241–246, *241–247*

Markov chain models/process 52, 56, 58, 68, 99, 217, 220–221

matrix-based system reliability method 34, 40

Matsuda, T. 218

McCarthy, J.R. 294–295

member replacement method 8, 9, 69

Messervey, T.B. 13, 14, 289

Metropolis-Hasting (MH) algorithm 321–322, 323

Mikata, Y. 210

Miner's rule 170, 186, 189

minimum-cut set method 326

Mirza, M. 199

mixed systems, redundancy factors of 402–403, *402*, **403**, *404*

Miyamoto, A. 20

Moan, T. 10, 57

Monte Carlo simulation 8–9, 15, 37, 98, 162, 172–173, 246, 250, *250*, 264, 268, 277, *278*, 279, 354

Moradian, M. 216

Mori, Y. 8, 35

Moses, F. 382, 385, 386, 387

movable bridges 334–335; analysis of inspection data 336–337, *337*, *338*; Christa McAuliffe Bridge 337–339, *338*, *339*; and condition assessment 335–336, *336*; Florida's inventory of *336*; Sunrise Boulevard Bridge 339–347, *340*, **341–343**, *344–348*; types of *336*

multi-attribute utility 57, 67, 71, *71*, 77, 109, 110, 116–118

multi-hazard approach 31

multi-objective genetic algorithm (MOGA) 14, 38, 87, 88, 90, *92*, 131, 135, 136, 229; application to maintenance planning 90–92, *92–94*; life-cycle cost of civil infrastructure 230–232, *231*, *233*

multi-objective optimisation 90, *92*, 97, 98, 232; in civil infrastructure maintenance management 97, 98, 99, 100, 101, 104; of optimum service life management 130–132

Nakagawa, T. 162

Nassif, H. 17, 289, 291

National Bridge Inventory (NBI), United States 14, 35; condition rating system 68

naval ship, combined inspection/monitoring planning of: optimum balance of cost and expected damage detection delay 307–312, *308*, *309*, *310*, **310**; ship hull structure subjected to fatigue **306**, 307, *307*

network disconnectedness 37

network reliability 104

network robustness index (NRI) 37

Neves, L.C. 20

Ng, A.K.S. 37

Nguyen, M.N. 75

Nielse, N. 265

Niigataken-Chuetsu earthquake (2004) 256–257, *257*

non-conjugate priors 321

non-destructive evaluation (NDE) 11, 13

non-dominated sorting method 232, 310

nonlinear finite element analysis 52

nonlinear pushover analysis 393

nonlinear structural analysis, of concrete cross-sections 148

Nowak, A.S. 8

objective functions 91, 133

objective reduction approach 127–128, 131–132

offshore structures, reliability of 9, 10

Okasha, N.M. 7, 9, 19, 20, 36, 56, 58, 317, 318, 325, 327

optimisation 47; bicriterion *305*, **305**, 306–307, *306*, 307, 313; life-cycle 54–56, *55*, 76–78, *77*, *78*; post-hazard functionality restoration 56; role in infrastructure management *19*; SHM planning under uncertainty *21*; solutions, decision space for *20*; of structural systems under uncertainty 18–22; trade-off solutions between two conflicting objectives *22*; *see also* multi-objective optimization

Oyado, M. 210, 242

Padgett, J.E. 52, 255

Paik, J.K. 187

Palsson, R. 199

Pareto front-based approach 128, 131

Pareto optimal solutions *20*, 38, *39*, 55, *55*, 78, *78*, 90, *92*, *120*, **120**, 121, **121**, *122*, 123, *123*, *124*, 131, 231–232, *231*, *305*, 306, 310–311, **310**

Particle Swarm technique 319

peak counting method 187, 188, 191, 194

peak floor acceleration (PFA) 276, 280

peak ground acceleration (PGA) 258, 268–269, 275, 276, 279, 280

peak picking method 289, 290

performance functions 158, 160–162, *160*, *161*, *162*, 354, 355, 358, 359–360

pitting corrosion model 132

plastic hinge analysis 242, *243*

Plos, M. 242

point-in-time method 357, 358

point method 169

Poisson point process 7, 8, 69, 249

Pontis 52, 68, 101

post-event recovery, bridge networks 39–40

post-failure material behaviour, effects on redundancy factor 398–399, 403–405, *404–405*, *406*, *407*; bridge example 414–415, **416**; brittle systems 401–402, *402*; brittle systems with many components 409–414, *410*, *411*, **412**, *413*, **413**, *414*, **414**; ductile systems 400–401, *401*, **401**; ductile systems with many components 406–409, *407*, **408**, **409**, *410*; mixed systems 402–403, *402*, **403**, *404*; redundancy factor 400

post-hazard functionality restoration 56
precipitation 75
preference-based approach 130–131
preventive maintenance (PM) 19, 20, 33, 49, 56, 67, 319
Priestley, M.J.N. 250
prior probability density function 320, 322
probabilistic design 12
probabilistic network analysis 32
probabilistic seismic hazard assessment (PSHA) 247, 248
probability density function (PDF) 67, 110, 222–223, 280, **281**;
    high-speed ship structures 187–188, 191; of time-to-failure
    50, *51*
probability mass functions (PMFs) 424, *425*, *427*, 428
probability of failure of system 6–7, 50, 69, 175, 354, 377,
    381; under hazard 51; parallel system *358*, *359*, *361*, 363;
    series–parallel system *366*, *367*, *368*, *369*, *370*, *371*;
    time-based 129; truss system 371, *372*, 373, *373*
probability of fatigue damage detection 301–302
probability of satisfactory performance 7
proof load testing 316

Qi, L. 162
quad-objective optimisation **135**, 136–137, *136*, **136**, *137*, **137**

R^2 statistic 321
railway bridges *see* riveted railway bridge connections, fatigue
    assessment of
rainflow counting method 176
Rashetnia, R. 216
redundancy 7, 8, 23, 50, 355–356; factor 400; *see also* post-failure
    material behaviour, effects on redundancy factor; of highway
    bridges *see* highway bridges, redundancy of; structural 353,
    355–356, 420; *see also* time-variant redundancy
reinforced concrete (RC) structures 47–49, 70, 97; arch bridge
    148, *149–150*, **150**, *151–152*, 151–153, *153–155*, **154**; jetty
    structures *see* jetty structures, reinforced concrete deck slab
    of; in marine environment *see* marine environment, reinforced
    concrete structures in; seismic performance of *see* seismic
    performance of reinforced concrete structures; *see also* concrete
    structures
reinforced concrete (RC) structures, deteriorating 418–419; bridge
    deck 425, *425*, *426*, 427–428, *427*, **427**; constitutive laws of
    materials 422; failure times and elapsed time between failures 420,
    420; frame system 423–424, *423*, *424*; modelling 420–422;
    probabilistic modelling 422–423, **423**; role of uncertainties 420,
    420; structural analysis considering time effects 422; time-variant
    failure loads and redundancy 219–220
reinforced concrete (RC) structures, steel weight loss/corrosion
    cracking of 199–200; accuracy of estimation method **203**,
    207–208; digital image analysis *203*, 205–207; effect of stirrups
    211, 213; effect of water-to-cement ratio *210*, 213; electrolytic
    experiment 201, *2101*; experimental plan 200, *200*, **200**; image
    enhancement before analysis *202*, 204–205; materials and concrete
    mix proportion 201, **201**; relationship between steel weight loss
    and crack widths *211–212*, 213; spatial variability of steel weight
    loss and crack widths *204–209*, 208–210; specimen fabrication
    procedure 201; surface crack width measurement 201–202, *201*;
    trend of steel weight loss and crack widths 210, *210*, 213; X-ray
    photogram acquisition procedure *201*, 202–204, *202*
Reinhorn, A. 35
reliability 7–10, 11, 23, 33, 50, 54; assessment, of reinforced
    concrete structures in marine environment 162–163, *162*, *163*;
    -based design 13, 157, 163–165, *163*, **164**; -based durability
    design, RC deck slab of jetty structures 219, *221*; -based life-cycle
    cost optimization 100, 101; -based life-cycle maintenance
    management 101; bridge networks 34, 35, 36, 38, *38*; bridges
    68–69; comparison of monitoring paths and critical monitoring
    points *16*; fatigue 189–192, *190*, **190**, 194–196, *195*, **195**, *196*; *see
    also* riveted railway bridge connections, fatigue assessment of;
    Lehigh River Bridge 297, **298**; life-cycle reliability estimation, of

RC deck slab of jetty structure 220–224; life-cycle seismic
    reliability 249–251, *249*, *250*; network 104; performance, of
    bridges 32–33; time-dependent monitoring paths *15*
reliability importance factor (RIF) 14
reliability index 7, 15, 33, 50, 69, 98
RELSYS program 186, 190, 192, 195, 365
RELTSYS program 9, 69
required maintenance (RM) 33
resilience 53, *53*, 56, 68; bridge networks 35, 37, 39–40; earthquake
    72; life-cycle *73*
resistance degradation 356–360, *357*
resistance-sharing factor (RSF) 378, *379*
response surface method (RSM) 8, 10, 69, 318, 381, 384
rigid plastic constitutive laws 145, *146*
risk 51–53, 67, 110, 114; bridges 69–70; resilience 53, *53*;
    sustainability 52–53
riveted railway bridge connections, fatigue assessment of 167–168;
    critical volumes for theory of critical distances 176, *177*; damage
    scenarios 176, *181–182*, 181–183; fatigue load spectra 176, *178*,
    180; fatigue reliability 172–173, 177–181, *179–180*; finite element
    analysis 169–170, *170*, *171*; loading random variables 172, *172*;
    reliability analysis framework 169–176; resistance random
    variables 170–172, **172**; system, modelling 173–175, *174*, *175*;
    theory of critical distances 168–169
road network, life-cycle cost optimisation for 235; calculation
    235–236, **236**; maximum acceleration 236, *236*, *237*; model 235,
    235, *236*; repair cost 235–236; three network models *235*, *236*,
    236; user cost 235
robustness 7, 23, 376, 381; of highway bridge superstructure systems
    383–384, 394–395

Sabatino, S. 57, 77, 110
Saetta, A. 241
safety index 98
safety margins 7, 129, *129*
Saito, Y. 242
Sánchez-Silva, M. 49, 241
San Fernando earthquake (1971) 274
Sano, K. 223
Sanriku-Minami earthquake (2003) 255–256, *256*; acceleration time
    history measured during *259*; damage states of Shinkansen
    viaducts after 258, *261*; response accelerations during *259*, *269*
Sato, T. 265, 266
Saydam, D. 51–52, 217
SCORES 185
Scott, D.M. 37
scour 74–75, *75*
sea-level rise 75–76
second-order reliability method (SORM) 354
seismic damage cost 101
seismic damage probability 234
seismic hazard curve 247, *247*
seismic performance analysis 72–73
seismic performance of reinforced concrete structures 239–240;
    framework for computing the life-cycle seismic reliability *240*;
    hazard assessment 246–248, *247*, *248*; life-cycle seismic reliability
    including corrosion damage 249–251, *249*, *250*; seismic capacity
    in marine environment 241–246, *241–247*
seismic probabilistic hazard assessment (SPHA) 157–158, 217
seismic reliability analysis 242
seismic risk analysis 229; life-cycle cost of civil infrastructure
    233–234, *234*, **234**
Seki, H. 162
semi-probabilistic design 12
sensitivity analysis 34, 153, 390, **390**
Sequential Monte Carlo Simulation (SMCS) 217, 221–223
series-parallel system model 8
serviceability 8, 14, 30; bridge networks 32, 35
service life *10*, 16, 19, 319; assessment 35; bridges 91, *93*, 94;
    prediction, ship structures *see under* high-speed ship structures

service life management, optimum 127–128; application to existing highway bridge 132–137; bi-objective optimisation 134–135, *134*, **135**; correlation among objective functions 133, **134**; efficient multi-objective optimisation 130–132; expected damage detection delay 128–129; extended service life under uncertainty 129–130, *130*; formulation of objective functions 132–133; initial service life estimation 132, **133**; life-cycle cost analysis 130; objectives, design variables, required estimations and given conditions in **128**; quad-objective optimisation **135**, 136–137, *136*, **136**, *137*, **137**; single-objective optimisation 133–134, *134*; time-based safety margin and probability of failure 129; tri-objective optimisation 135, *135*, **135**

shear strength 250
Shekarchi, M. 216
Shimomura, T. 210
Shinozuka, M. 6, 34, 268
ship structures: reliability of 9–10, 58; *see also* high-speed ship structures
Sikora, J.P. 185, 187
Simioni, P. 241
single-objective optimisation 133–134, *134*
slice sampling algorithm 322
small displacements hypothesis 144
Smith, C. 9
S–N method/curve 72, 168, 169, 170, 173, 179, 180, 186, *186*, 189, 190, 193
social metric of sustainability 70
Soliman, M. 58
Song, J. 34
Song, R. 57
Sozen, M. 265
spatially distributed systems 32–34, 35
SPECTRA 185, 187
steel corrosion, in reinforced concrete structures: in marine environment 160–163, *160*, **161**, *162*, *163*; visualization by X-ray 243, *244*
steel jacketing 254, 255, 258, *262*, 264–265, 282
steel weight loss *see* reinforced concrete (RC) structures, steel weight loss/corrosion cracking of
Stewart, M.G. 75, 199
stiffness matrix 148, 422
still water loads 187
stirrups, and steel corrosion *211*, 213
stochastic dynamic programming 14
Strauss, A. 8
stress range bin histograms 187–188, 191, *193*, 194, 195
structural health monitoring (SHM) 46, 300; bridge networks 40; high-speed ship structures 185, 187; information, role in performance updating and damage detection 57–59, *58*; integration in bridge management 76, *77*, 316–331
structural health monitoring (SHM), integration in life-cycle management 10–18; monitoring within life-cycle context 15–18; possible effect of SHM on load effect and resistance *17*; supporting paradigm 11–12; top-down approach 13–15; using monitoring as catalyst to improve existing design and management methodologies 12–13
structural redundancy 353, 355–356, 420
structural stress method 169
structure management system (SMS) 105–106
substructures, bridge: calibration of system factors 391–394, **392**; example analysis of 388–390, *388*, *389*, **389**; redundancy in 387–388; reliability analysis 390–391, **390**
Sunrise Boulevard Bridge 339–340; balance and friction 343–345, *345*, *346*; considerations for operational and environmental monitoring **343**; design of monitoring system 340, 341; SHM considerations for mechanical/electrical components **341–342**; SHM considerations for structural components **341**; wind temperature and effects on bridge balance 345, *346*, 347, *347*, *348*; wireless communication scheme and mechanical room *344*

superstructure systems, bridge multi-beam 382–383; analysis of results and determination of redundancy criteria 385–386; incremental analysis 383–385, *383*, *384*, **385**; system factors for 386, **387**
survivor function 51, *51*, 111, 112, 357
sustainability 52–53, 67, *68*; bridges 70–71, *71*; metrics *71*
system analysis 112, 167, *327*
system-based safety measures 6
system factors 386, **387**, 391–394, 400
system redundancy 7, 355–356, 376, 381; definition of 398; time-invariant 355–356; *see also* time-variant system redundancy
system reliability 7–9, 15, 23, 68–69, 167–168, 376–379; bridge substructures 390–391, **390**; of highway bridge superstructure systems 381–382, 384, 385, *385*, 386–387; matrix-based 34, 40; time-invariant 353–355, *355*; two-member parallel system *377*, *378*, *379*; *see also* time-variant system reliability
system safety 376, 377, 380, 395–396
system safety factor (SSF) 378, *378*, 379

Taheri, S. R. 216
Takeda, T. 265
Takiguchi, M. 265
Tamai, S. 265, 266
Tamura, T. 223
Tang, W.H. 288
Taniguchi, O. 223
Tapia, C. 52
Taylor series 384, 385
theory of critical distances (TCD) 168–169, 181; critical volumes for 176, *177*
time-based safety margin 129, *129*
time-dependent performance 52; deterioration, of civil infrastructure 99
time-dependent reliability 9, 17, 32, 250
time-invariant system redundancy 355–356
time-invariant system reliability 353–355, *355*
time-to-failure 50, 110–111, 112, 119; probability density function of 50, *51*
time-variant failure loads 419–420
time-variant limit analysis 146; deterministic 151
time-variant system redundancy 7, 353, 356, 360–361; of deteriorating RC structures *see* reinforced concrete (RC) structures, deteriorating; parallel system *358*, *359*, *360*, *361*, 362–364, *362*, *364*; and resistance degradation 356–360, *357*; series–parallel system 364–370, *365*, *366*, *367*, *368*, *369*, *370*, *371*; truss system 370–374, *372*, *373*
time-variant system reliability 356; parallel system *359*, *361*, 362–364; and resistance degradation 356–360, *357*; series–parallel system 364–370, *366*, *369*, *370*, *371*; truss system 370–374, *372*, *373*
Tohoku-Shinkansen viaducts 254–255; comparison of damage states before and after seismic retrofit 257–260, *259–262*; damage during Sanriku-Minami earthquake (2003) 255, *256*; dynamic response analysis 264–268, *265–269*; fragility curves for as-built and retrofitted viaducts 262–264, **264**, 268–269, *270–271*; seismic retrofit by means of steel jackets *262*; seismic specifications and impact of aftershock 261–262, *263*
Torii, K. 218
total probability theorem 277, 320
total travel distance (TTD) 37
total travel time (TTT) 37, 38
traditional algorithms 99–100
traffic loads 74–75
transition probability 217, 220, 221
transportation asset management 40
transportation networks 14, 30, 31, 33, 36–37, 73, 275–276; maintenance management 103–105, *105*, *106*; *see also* bridge networks
travel time (healthcare–bridge network system) 277, 281–282

travel time reliability 104

tri-objective optimisation 118–119, *119*, **119**, 121, 135, *135*, **135**

truss bridge: cross-sectional areas of truss members **394**; load factor *vs.* maximum displacement response of *395*; redundancy and robustness assessment 395; structural analysis 394–395; system safety analysis 395–396; time-variant redundancy 370–374, *372*, *373*

uncertainty: in deteriorating reinforced concrete structures 420, *420*; extended service life under 129–130, *130*; life-cycle management under 50; sources of 97–98

uncertainty, structural systems under 5–6; integration of structural health monitoring in life-cycle management 10–18; role of optimisation 18–22; system performance assessment and prediction 6–10

upper bound theorem 146

user cost 104, 236

user equilibrium 37

utility-informed life-cycle decision-making 56–57, 71, *71*, 77, *77*; *see also* life cycle utility-informed maintenance planning

vector autoregressive modelling 59

Verma, R. 250

vertical failure 74

Vidal, T. 213

visual inspections 10–11, 13, 68, 97, 98, 217, 316

Vitaliani, R. 241

volume method 169

vulnerability 50

waiting time (healthcare–bridge network system) 277, 281–282, *281*

Wakabayashi, H. 104

Wang, C.S. 168

Wang, X. 75

water-to-cement ratio *210*, 213

wave-induced loads 187, 188

Weibull distribution 110, 111, 119, 188, *193*, 194, 307, *307*, 325–326

weigh in motion (WIM) studies 17, 289, 291

weighted sum method 100, 130, 136

wind temperature, and moving bridges 345, *346*, 347, *347*, *348*

Xiao, Y. 250

X-ray 200, 202, 240, 243–245, *244*; photogram acquisition *201*, 202–204, *202*

Yamamoto, T. 210

Yamamoto, Y. 242

Yavari, S. 277

Yeo, G.L. 69

yielding criterion 145–146, *146*

Zhou, Y. 34

Zhu, B. 51, 399, 400, 406, 415

Zitzler, E. 131, 132